The McGraw-Hill Environmental Biotechnology Series

Microbial Ecology

Principles, Methods, and Applications

Morris A. Levin
Maryland Biotechnology Institute
University of Maryland
Baltimore, Maryland

Ramon J. Seidler
U.S. Environmental Protection Agency
Corvallis, Oregon

Marvin Rogul
Maryland Biotechnology Institute
University of Maryland
Baltimore, Maryland

McGraw-Hill, Inc.
New York St. Louis San Francisco Auckland Bogotá
Caracas Lisbon London Madrid Mexico Milan
Montreal New Delhi Paris San Juan São Paulo
Singapore Sydney Tokyo Toronto

Library of Congress Cataloging-in-Publication Data

Microbial ecology : principles, methods, and applications / [edited
  by] Morris A. Levin, Ramon J. Seidler, Marvin Rogul.
        p.   cm.—(The McGraw-Hill environmental biotechnology
  series)
    Includes index.
    ISBN 0-07-037506-2
    1. Microbial ecology.   I. Levin, Morris A.   II. Seidler, Ramon J.
  III. Rogul, Marvin.   IV. Series: Environmental biotechnology.
  QR100.M516   1992
  576'.15—dc20   91-25208

ISBN   0-07-037506-2

The editing supervisor for this book was Stephen M. Smith and the
production supervisor was Suzanne W. Babeuf. It was set in Century
Schoolbook by McGraw-Hill's Professional Book Group composition
unit.

Printed and bound by R. R. Donnelley & Sons Company.

Contents

Part 2   Genetic Transfer and Stability

Chapter 19. Practical Considerations of Nucleic Acid Hybridization and Reassociation Techniques in Environmental Analysis   393

Part 3   Fate and Transport

Chapter 20. Overview: Fate and Transport of Microbes   423

Chapter 27. Field Sampling Design and Experimental Methods for the Detection of Airborne Microorganisms

Part 4   Ecosystems Effects

Chapter 28. Overview: Identifying Ecological Effects from the Release of Genetically Engineered Microorganisms and Microbial Pest Control Agents

Chapter 29. Methods for Evaluating the Effects of Microorganisms on Biogeochemical Cycling

Part 5   Effects on Individual Organisms (Nontarget)

Part 6   Decontamination and Mitigation

Contributors

Richard L. Anderson *USEPA Environmental Research Laboratory, Duluth, Minnesota* (CHAP. *31*)

John L. Armstrong *Biotechnology/Microbial Ecology Program, USEPA Environmental Research Laboratory, Corvallis, Oregon* (CHAP. *24*)

Ronald M. Atlas *University of Louisville, Louisville, Kentucky* (CHAP. *2*)

Tamar Barkay *Microbial Ecology and Biotechnology Branch, USEPA Environmental Research Laboratory, Gulf Breeze, Florida* (CHAP. *33*)

Gerard F. Barry *Department of Biological Sciences, Monsanto Company, St. Louis, Missouri* (CHAP. *8*)

Shoshana Bascomb *Baxter Healthcare Corp., MicroScan Division, West Sacramento, California* (CHAP. *6*)

Harvey Bolton, Jr. *Pacific Northwest Laboratory, Richland, Washington* (CHAP. *29*)

Myron K. Brakke *Department of Plant Pathology, University of Nebraska, Lincoln, Nebraska* (CHAP. *46*)

A. Breen *Department of Microbiology, University of Tennessee, Knoxville, Tennessee; and Center for Environmental Biotechnology, Knoxville, Tennessee* (CHAP. *19*)

John D. Briggs *Department of Entomology, The Ohio State University, Columbus, Ohio* (CHAPS. *34, 39*)

Susan Brown *Microbial Genetics Division, Pioneer Hi-Bred International, Inc., Johnston, Iowa* (CHAP. *44*)

Jeffrey J. Byrd *Division of Natural Science and Mathematics, St. Mary's College of Maryland, St. Mary's City, Maryland* (CHAP. *5*)

C. Lee Campbell *Department of Plant Pathology, North Carolina State University, Raleigh, North Carolina* (CHAP. *35*)

Rita R. Colwell *Maryland Biotechnology Institute, University of Maryland, College Park, Maryland* (CHAPS. *1, 4, 5, 6*)

C. R. Cripe *USEPA Environmental Research Laboratory, Sabine Island, Gulf Breeze, Florida* (CHAP. *23*)

Stephen M. Cuskey *USEPA Environmental Research Laboratory, Sabine Island, Gulf Breeze, Florida (Deceased)* (CHAPS. *17, 50*)

Donald H. Dean *Department of Biochemistry, The Ohio State University, Columbus, Ohio* (CHAP. *44*)

Thomas C. Dockendorff *Department of Microbiology, University of Tennessee, Knoxville, Tennessee* (CHAP. *19*)

David J. Drahos *BP Technologies, Inc., Stone Mountain, Georgia* (CHAP. *8*)

Anne Fairbrother *USEPA Environmental Research Laboratory, Corvallis, Oregon* (CHAP. *45*)

Stephen K. Farrand *Departments of Plant Pathology and Microbiology, University of Illinois at Urbana/Champaign, Urbana, Illinois (CHAP. 16)*

Susan W. Fisher *Department of Entomology, The Ohio State University, Columbus, Ohio (CHAP. 39)*

James K. Fredrickson *Pacific Northwest Laboratory, Richland, Washington (CHAPS. 28, 29)*

Michael A. Gealt *Department of Bioscience and Biotechnology, Drexel University, Philadelphia, Pennsylvania (CHAP. 15)*

D. Haefele *Microbial Genetics Division, Pioneer Hi-Bred International, Inc., Johnston, Iowa (CHAP. 49)*

Charles Hagedorn *Department of Crop and Soil Environmental Sciences, Virginia Polytechnic Institute and State University, Blacksburg, Virginia (CHAPS. 26, 28)*

Carol A. Hendrick *Microbial Genetics Division, Pioneer Hi-Bred International, Inc., Johnston, Iowa (CHAP. 44)*

Stephen C. Hern *USEPA Environmental Monitoring Systems Laboratory, Las Vegas, Nevada (CHAPS. 22, 27)*

William E. Holben *Center for Microbial Ecology and Department of Crop and Soil Sciences, Michigan State University, East Lansing, Michigan (CHAPS. 4, 20)*

Mary A. Hood *Department of Biology, University of West Florida, Pensacola, Florida (CHAPS. 20, 25)*

Clarence I. Kado *University of California, Davis, California (CHAP. 18)*

James L. Kerwin *Botany Department, University of Washington, Seattle, Washington (CHAP. 37)*

Donald A. Klein *Colorado State University, Fort Collins, Colorado (CHAP. 30)*

Ivor T. Knight *Department of Biology, James Madison University, Harrisonburg, Virginia (CHAP. 4)*

Jonathan Lamptey *Microbial Genetics Division, Pioneer Hi-Bred International, Inc., Johnston, Iowa (CHAP. 44)*

Richard E. Lenski *Center for Microbial Ecology, Michigan State University, East Lansing, Michigan (CHAP. 9)*

Morris A. Levin *Maryland Biotechnology Institute, University of Maryland, Baltimore, Maryland (CHAP. 1)*

Cynthia Liebert *Technology Research, Inc., USEPA Environmental Research Laboratory, Gulf Breeze, Florida (CHAP. 33)*

Bruce Lighthart *USEPA Environmental Research Laboratory, Corvallis, Oregon (CHAP. 22)*

J. Lindemann *Lindemann Consulting, El Cerito, California (CHAP. 49)*

S. E. Lindow *Department of Plant Pathology, University of California, Berkeley, California (CHAP. 49)*

Sarah A. McIntire *Biology Department, Texas Woman's University, Denton, Texas (CHAP. 10)*

Russel H. Meints *Department of Botany and Plant Pathology, Oregon State University, Corvallis, Oregon (CHAP. 48)*

Robert V. Miller *Department of Microbiology, Oklahoma State University, Stillwater, Oklahoma (CHAPS. 7, 11)*

Timothy J. Miller *Department of Molecular Genetics, SmithKline Beckman Animal Health Products, King of Prussia, Pennsylvania (CHAP. 43)*

Richard Y. Morita *Department of Microbiology, College of Science and College of Oceanography, Oregon State University, Corvallis, Oregon (CHAP. 21)*

O. A. Ogunseitan *Department of Microbiology, University of Tennessee, Knoxville, Tennessee; and Center for Environmental Biotechnology, Knoxville, Tennessee (CHAP. 19)*

Ronald H. Olsen *Department of Microbiology and Immunology, University of Michigan Medical School, Ann Arbor, Michigan (CHAP. 17)*

Susan B. O'Morchoe *Department of Biochemistry and Biophysics and the Program in Molecular Biology, Stritch School of Medicine, Loyola University of Chicago, Maywood, Illinois (CHAP. 13)*

J. G. Packard *Graduate Program in Ecology, University of Tennessee, Knoxville, Tennessee; and Center for Environmental Biotechnology, Knoxville, Tennessee (CHAP. 19)*

Norberto Palleroni *New York University Medical School, New York, New York (CHAP. 1)*

P. H. Pritchard *USEPA Environmental Research Laboratory, Sabine Island, Gulf Breeze, Florida (CHAP. 23)*

David C. Sands *Department of Plant Pathology, Montana State University, Bozeman, Montana (CHAPS. 34, 35)*

Dennis J. Saye *Department of Biochemistry and Biophysics and the Program in Molecular Biology, Stritch School of Medicine, Loyola University of Chicago, Maywood, Illinois (CHAP. 13)*

G. S. Sayler *Department of Microbiology and Graduate Program in Ecology, University of Tennessee, Knoxville, Tennessee; and Center for Environmental Biotechnology, Knoxville, Tennessee (CHAP. 19)*

Ramon J. Seidler *USEPA Environmental Research Laboratory, Corvallis, Oregon (CHAPS. 14, 22, 27)*

John A. Shadduck *Office of the Dean, Texas Veterinary Medical Center, College of Veterinary Medicine, Texas A&M University, College Station, Texas (CHAP. 38)*

Lyle Shannon *Department of Biology, University of Minnesota, Duluth, Minnesota (CHAP. 31)*

Jessup M. Shively *Department of Biological Sciences, Clemson University, Clemson, South Carolina (CHAP. 42)*

Joel P. Siegel *Center for Economic Entomology, Medical Entomology Program, Illinois National History Survey, Champaign, Illinois (CHAP. 38)*

J. Skujiņš *Department of Biology, Utah State University, Logan, Utah (CHAP. 47)*

Anne Spacie *Department of Forestry and Natural Resources, Purdue University, West Lafayette, Indiana (CHAP. 36)*

Linda D. Stetzenbach *Environmental Research Center, University of Nevada, Las Vegas, Nevada (CHAPS. 22, 27)*

Gregory J. Stewart *Department of Biology, University of South Florida, Tampa, Florida (CHAP. 12)*

Guenther Stotzky *Department of Biology, New York University, New York, New York (CHAPS. 29, 40)*

Frieda B. Taub *School of Fisheries, College of Ocean and Fishery Sciences, University of Washington, Seattle, Washington (CHAP. 32)*

James M. Tiedje   *Center for Microbial Ecology and Department of Crop and Soil Sciences, Michigan State University, East Lansing, Michigan (*CHAPS. *4, 20)*

Nancy J. Tomes   *Microbial Genetics Division, Pioneer Hi-Bred International, Inc., Johnston, Iowa (*CHAP. *44)*

James L. Van Etten   *Department of Plant Pathology, University of Nebraska, Lincoln, Nebraska (*CHAP. *48)*

Peter A. Vandenbergh   *Microlife Technics, Sarasota, Florida (*CHAP. *41)*

Anne K. Vidaver   *Department of Plant Pathology, University of Nebraska, Lincoln, Nebraska (*CHAP. *40)*

Michael V. Walter   *Research and Development, Texaco Inc., Beacon, New York (*CHAP. *14)*

Sara F. Wright   *USDA–ARS Soil Microbial Systems Laboratory, Beltsville, Maryland (*CHAP. *3)*

Gerben J. Zylstra   *Center for Agricultural Molecular Biology, Rutgers University, New Brunswick, New Jersey (*CHAP. *17)*

Preface

Elevated expenditures for biotechnology research dealing with environmentally oriented products (e.g., pesticides and waste-treatment products) have resulted in large numbers of petitions for permits and licenses at federal and state regulatory agencies to conduct field tests involving the release of genetically engineered microorganisms. As data bases and other more traditional sources of information are reviewed, it is becoming increasingly apparent to researchers and regulators that ecological measurements and information are the most essential elements in assessing the risks of such releases.

Frustration and difficulties in finding this material have served to sensitize investigators and government officials to the need for a repository of factual information and current methodology in microbial ecology. This book represents a response by microbiologists and allied scientists to bring this knowledge together in a guide to researchers and regulators alike.

The text compiles, describes, and references procedures and concepts being used by environmental scientists in microbial ecology. The need for specific, reliable, and effective methods is essential to the development of protocols for evaluating releases of microbial pest control agents and other environmental applications of either naturally occurring or genetically altered microorganisms.

An advisory group consisting of representatives from the biotechnology scientific community, federal agencies involved in regulating biotechnology products, and public interest groups helped formulate the boundaries of this book, establish its organization, and select the experts who would be responsible for overseeing each of its six parts. The editors wish to acknowledge the many valuable contributions of the advisory group, which consisted of Dr. Mary Ann Danello (Food and Drug Administration), Dr. Robert Frederick and Dr. Elizabeth Milewski (EPA), Dr. Mary Gant (Executive Office of the President; OSTP), Dr. Doug McCormick (*Bio/Technology*), Dr. Margaret Mellon (Environmental Law Institute), and Dr. Richard Parry, Jr. (USDA). The editors, in addition, wish to gratefully acknowledge the financial support of the EPA's Office of Research and Development. However, this book does not represent the official position or opinion of the U.S. Environmental Protection Agency or any agency with which a contributing author may be affiliated.

The selection of part coordinators was especially difficult since there are many persons who have made major contributions to the field of microbial and molecular ecology. The efforts of these coordinators in selecting chapter authors and reviewing the chapters were instrumental in the successful completion of this project.

Finally, the editors thank Dr. Edwin L. Schmidt, Dr. M. J. Sadonsky, and Dr. B. K. Kinkle, who reviewed the entire manuscript and provided many constructive comments to individual authors and part coordinators. Their efforts significantly improved the quality of individual chapters and the overall content of the end product.

Morris A. Levin
Ramon J. Seidler
Marvin Rogul

Introduction

Morris A. Levin

Ramon J. Seidler

The earliest studies of microbial ecology can be traced to the work of
Warrington at the Rothamstead Experimental Station in England and
Winogradsky in France (14). They were the pioneer investigators who
first described the ways microbial action affected the nitrogen cycle.
Their efforts, followed by other early workers in the field, led Bernard
Dixon to point out that the most successful biotechnology teams de-
veloping environmental applications and measuring effects will be
those with a microbial ecologist on board (10).

An understanding of the principles, applications, and methods of
microbial ecology is essential to measure, predict, and understand the
consequences of environmental releases of microorganisms. However,
knowledge of genetics, physiology, and metabolism are also required
to complement the principles of microbial ecology in order to permit a
holistic approach to environmental risk assessment.

Principles of genetics, metabolism, and physiology are being examined at the molecular biology level in laboratories where variables can
be controlled. Research to establish principles of microbial ecology are
expected to be established in the field or perhaps, to an extent, in contained environments (microcosms) that simulate a portion of the natural environment. Safe and meaningful environmental applications of
microorganisms require knowledge for predicting or anticipating
qualitative trends that describe survival, dispersal, and interactions
between and among populations of microbes, plants, and animals.
However, microbial ecology has not enjoyed the funding levels or attention required to produce the variety of methods and data necessary
to quantitatively evaluate the range of microorganisms being considered for environmental applications (7, 8).

This book was prepared in response to the needs of federal and state regulatory agencies, biotechnology companies, and research institutions to provide guidelines and procedures for gathering data with which to evaluate efficacy of newly developed products and to conduct risk assessments prior to the environmental application of microorganisms. Important lessons can be learned by applying environmental risk assessment techniques to microbial ecology studies. The overall objective was to assemble current techniques and concepts to devise and conduct field releases to maximize the kinds and quality of data gathered.

Background

The need to assess the efficacy and conduct risk assessments in terms of ecological effects of released microbes has existed since the first microbial pesticide was registered in 1948 (*Bacillus popilliae,* to control Japanese beetle larvae). Technical advances and greater commercial interest in environmental aspects of molecular biology over the past decade have resulted in increased environmental application of microbes, resulting in increased needs for guidelines for environmental risk assessment. Only thirteen microbial pesticides were registered between 1948 and 1980 by the USEPA, while approximately 100 applications for field trials of engineered, indigenous, or nonindigenous microbes have been received by EPA within the past 7 years.

Increased interest in the use of microorganisms in the environment as waste degraders, mineral solubilizers, pesticides, fertilizers, and as crop protectors has created awareness of the need for small-scale field testing of a wide variety of microorganisms in many environments. As can be seen in Table 1, over 3000 patents in areas related to environmental application were granted between 1980 and 1984. Many industrial sectors are represented. The rate of patent application in biotechnology has been increasing since 1986. The potential number of microbes in commerce appears to be increasing. Thus an updated guide to appropriate methodology to conduct environmental risk assessments will be a continuing need.

It was clear from the outset that there was no central or codified source identifying and evaluating acceptable methods which were germane to all possible conditions under which field trials might be conducted. The plethora of habitats and potential industrial applications preclude easy prediction of the type of microorganism or habitat which might be involved. Advances in ability to engineer microorganisms have resulted in increased interest in environmental use of recombinant microorganisms, resulting in the need for more complex risk assessment analyses. These factors underscore the need for a general guide to principles, applications, and methods.

The advisory group recognized at the outset that it was not possible

TABLE 1   Organism/Use Matrix*—General Use Code

Genus	A	B	C	E	G	M	N	O	P	R	W
Bacillus	17	97	53	18	30	1	1	3	307	8	83
Clostridium	—	269	204	1	42	1	—	4	14	3	49
Pseudomonas	3	9	44	1	16	14	—	5	57	7	34
Escherichia	2	48	45	10	13	8	5	4	134	1	28
Saccharomyces	—	96	75	1	51	6	—	5	16	—	31
Candida	—	74	58	—	29	2	—	2	35	2	57
Aspergillus	—	13	20	3	2	2	—	—	132	1	72
Klebsiella	—	75	21	—	14	—	5	7	4	—	24
Alcaligenes	—	—	31	—	6	—	3	8	23	1	55
Aerobacter	—	49	30	—	9	—	—	—	1	—	3
Zymomonas	—	33	25	—	23	—	—	1	—	—	5
Streptomyces	—	15	14	—	1	2	—	—	20	—	33
Kluyveromyces	—	16	15	—	13	—	—	—	10	—	20
Methylosinus	—	1	49	—	—	—	—	1	4	—	8
Thiobacillus	—	—	—	—	—	42	1	10	5	1	4
Penicillium	—	—	—	2	—	2	—	6	23	1	26
Proteus	—	14	8	—	18	—	—	—	2	—	3

General use: A, agricultural chemicals; B, conversion of biomass; C, industrial chemical production; E, monitoring/measurement/biosensor; G, energy; M, mining/metal recovery; N, nitrogen fixation; O, other; P, polymer/macromolecule production; R, enhanced oil recovery; W, waste/pollutant degradation.
*Microorganisms most likely to be used.
SOURCE: Report of the Biotechnology Science Advisory Committee, 1987, p. 45.

or even advisable to attempt to produce an all-inclusive volume describing every method for all possible measurements. The body of literature was too vast and diffuse. It quickly became clear that some of the literature has been reviewed from a variety of perspectives for specific purposes over many years. These reviews contain descriptions of techniques for making specific microbial measurements (2, 11, 12, 18) in particular environments. In many cases it was not readily apparent that data or methods from one environment could be extrapolated to another. However, it was considered possible to collect general principles, methods, and applications and describe, where possible, application to assessing the risk associated with environmental uses of microbial biotechnology products.

This book is designed for use with all types of microbes. The principles and methods described will be useful to industrial scientists seeking to estimate efficacy or evaluate potential risks associated with potential products. Academic scientists will use the same information when seeking validation of a hypothesis in experiments conducted in the field. The book focuses on principles and methods developed for dealing with nonengineered microbes as surrogates for genetically engineered microorganisms (GEMs) because little data are available to describe the effectiveness of many of the methods in environmental situations with engineered microorganisms.

Data Quality

Importance of quality assurance and control

The need for assuring that data are of uniform and known high quality is crucial in all areas of experimental science and is especially important in fieldwork. Unfortunately, there is no equivalent in the discipline of microbial ecology that compares to the Clinical Laboratory Improvement Act (CLIA) of 1967 that resulted in mandatory quality control procedures for laboratories involved in interstate commerce (12). Many of the issues which surfaced as a result of implementation of CLIA are applicable to microbial ecology. For example, some of the quality assurance and control (QA/QC) requirements in clinical microbiology which are also relevant to laboratories specializing in microbial ecology include the development of stringent record-keeping procedures, quality assurance standards for individual laboratories, quality control standards for common laboratory procedures, and the monitoring and training of personnel. Passage of CLIA established regulations which resulted in the adoption of standards for laboratory accreditation. With the higher standards and better focus on QA/QC came more consistent and higher-quality data. The benefits of QA/QC practices were clearly recognized. Although this book addresses numerous technical issues, it does not discuss QA/QC methodology.

CLIA resulted in a number of regulations. However, the need for some of these has been questioned (15). Many of the questions seem to be related to the need for mandatory testing at frequent intervals of materials which were produced following standardized procedures (e.g., biochemical tests such as Voges-Proskauer, indole, etc., and media for antibiotic susceptibility testing). Several reports have indicated that a more flexible quality assurance procedure would result in the same high-quality data (13, 17). It has also been pointed out that reducing the test load (frequency of testing) reduced costs by over 50 percent with no measurable impact on data quality. It is clear that the percent of time a microbiologist is required to spend dealing with QA/QC issues increases as the size of the laboratory decreases. Small laboratories can spend up to 10 percent of their effort on QA/QC activities, while in a large laboratory only 4 percent of the available work time is allocated to meet these needs. Many handbooks describing appropriate QA/QC procedures and protocols have been produced. Because CLIA deals specifically with clinical laboratories, there has been no pressure to produce similar manuals for fieldwork.

Nevertheless, most observers recognize the value of well-defined procedures for QA/QC when gathering field data. The Federal Bureau of Investigation has recently moved into the area of forensic use of molecular biology techniques and published four articles in its *Crime Laboratory Digest,* on quality control issues associated with sample gathering and analyses, especially when molecular biology techniques are involved (1, 6, 9, 16). Many of the techniques (e.g., sample handling and protection, polymerase chain reaction, use of restriction endonucleases) are common to microbial ecology. Numerous texts and articles describe QA/QC in great detail (3–5).

The importance of maintaining the quality of field research data dictates that the most important features be discussed here, although an intensive treatment of QA/QC is beyond the scope of this book.

Implementation of quality assurance procedures

The frequency of conducting a particular test and what constitutes proper quality assurance are closely related. If a test is conducted by a researcher or technician frequently (dozens or hundreds of times daily), it may not be necessary to include a full range of control tests each time. If the test is conducted rarely (e.g., once a month), appropriate controls (both to assure that any medium has been properly prepared and stored as well as positive and negative controls for comparison purposes) must be included each time. Each laboratory must consider the costs involved and the need for appropriate measures to assure that high-quality data are obtained. The cost of contracting out less frequently conducted procedures should be compared to the cost of the appropriate quality assurance measures. The advisability of conducting quality assurance procedures internally or sending samples to a referral laboratory should also be considered in establishing a quality assurance program.

Collection and transport of samples, evaluation of sample quality, and the decision to assay or dispose of the samples are issues which must be resolved in each experimental procedure. Difficult issues such as whether to process a sample which arrives from the field in a damaged condition or after unexpected delay and possible exposure to extreme conditions (freezing or excess heat or moisture) should be decided in advance and not left to chance. Careful and complete records of sample handling, transport, and disposition must be maintained.

Recording activities

Protocols for recording time of arrival and sample condition on arrival must be in place. Handling, storage conditions, and time delay before assay must also be recorded and should be standardized at the laboratory level.

Development and maintenance of appropriate records is especially important if the possibility of utilization of the data in legal proceedings is contemplated. An entire data set can be invalidated if it cannot be demonstrated that proper procedures to preclude unauthorized persons from tampering with the samples were in place. Appropriate legal authorities should be consulted prior to developing protocols for accepting such samples for assay.

Quality control

In addition to the adoption of a quality assurance program, awareness of the importance of quality control procedures must be indoctrinated within the laboratory. This includes generating a list of procedures conducted by personnel and equipment that will be subjected to surveillance for quality control purposes. These activities are diverse and include, for example, monitoring media for expiration dates, incubators for changes in temperature and humidity, measuring devices for accuracy, and autoclaves for proper pressure and temperature settings and recordings. Limits of accuracy for any new test procedures must be established.

The intervals between monitoring are a function of the type of activity, and many authors have described the need for and implementation of such monitoring activities (3–5, 9). Development and maintenance of records is important for verification of monitoring procedures and troubleshooting in case of anomalous results.

Organization of This Book

This book comprises six parts; each contains an overview chapter prepared by the part coordinators. Each overview describes the topic area and presents the scope of the part. The overview chapter discusses the development of methods in each particular area, the current areas being emphasized in terms of active research, and speculates about future research direction. The overviews are followed by chapters describing concepts, methods, and limitations of the particular topic area. Selection of the specific topics and authors was a cooperative effort of the part coordinators, the advisory group, and the editors. Attempts were made to identify areas that had been or are being intensively investigated and, thus, where sufficient information was available for review. In each part the chapters were grouped by topic area and provide a common pattern wherever possible. Each part begins with general issues and then describes specific cases or applications. In some chapters, the topics are organized by medium, e.g., soil or water, or by applicability to plant, insect, or animal investigations.

Part 1 examines ways to identify and enumerate microbes and specifically detect recombinant DNA in environmental situations. Drs. R. R. Colwell, N. Palleroni, and W. E. Holben have collaborated to produce a part which describes the isolation and identification of microbes using methods ranging from selective media through the use of fluorescence microscopy to gene probes. This part describes general methods which may be termed "traditional" and proceeds to the analysis of gene pools. Current techniques for classification are described and the methods are compared in terms of applicability to types of microorganisms. The part concludes with an illustration of classification of streptomycetes and actinomycetes, using some of the described methodology.

Part 2 is a guide to methods for evaluating gene transfer and genetic stability in situ. Methods to measure the transfer of genes are described, with numerous examples, relative to a variety of experimental situations. In the overview chapter, Dr. R. Miller, the part coordinator, explains the need to understand the impact of recombinant DNA in natural situations and describes the type of data required to meet those needs. The chapter includes descriptions and examples of the means of natural exchange of genetic material, practical considerations for estimating exchange rates, and the discussion of the potential for gene transfer in water, wastewater, soil, and on plants.

Part 3 deals with the fate and transport of microbes in soil, water, and air. Drs. W. E. Holben and M. A. Hood, the part coordinators, describe the state of monitoring, measuring, and derivative procedures in the overview chapter. The importance of energy supply and source to bacterial survival is then described, followed by a series of chapters characterizing microbial persistence in terrestrial and aquatic systems and the use of microcosms as test systems. Chapters dealing with microcosms highlight the advantages and disadvantages of extrapolating from laboratory to field using microcosm data. The part closes with chapters describing dispersal and survival of aerosolized microorganisms and some of the methods used to detect and enumerate airborne microorganisms.

In the overview chapter of Part 4, Drs. C. Hagedorn and J. K. Fredrickson discuss methods used to estimate ecosystem effects. The part includes chapters dealing with population dynamics, effects on geochemical cycles, and the use of microcosms to measure effects. The occurrence and distribution of metal-resistant bacteria are presented as an example of an effect of pollution and of application of specific measurement methods.

In Part 5, Drs. D. C. Sands and J. D. Briggs deal with the measurement of effects on nontarget organisms in the overview chapter. They discuss the importance of estimating nontarget effects and relate this to estimating environmental impacts and regulatory decision making. Chapters include descriptions of measurement techniques for use in estimating effects of microbes on plants, fish, crustaceans, birds, mammals, and insects.

Concepts of risk management are coordinated in Part 6 by Drs.

A. K. Vidaver and G. Stotzky. This part deals with the containment and mitigation of released microbes. In the overview chapter, the co-ordinators describe the concepts needed to geographically contain and mitigate the effects of microbes in the environment. Control and mitigation of bacteria, algae, fungi, and viruses are discussed in the contributed chapters. Examples involved in the control of degradative as well as metal solubilizing bacteria are presented. This part includes a description of biological containment with the use of suicide genes.

References

1. Adams, D. 1988. Validation of the FBI procedure for DNA analysis. *Crime Laboratory Digest* 15:106–109.
2. Atlas, R., and B. Bartha. 1988. *Microbial Ecology*, McGraw-Hill, New York.
3. Bartlett, R. C. 1974. *Medical Microbiology: Quality, Cost and Clinical Relevance.* Wiley, New York.
4. Bartlett, R. C., V. D. Allen, D. J. Blazevic, C. T. Dolan, V. R. Dowell, T. L. Gaven, S. L. Inhorn, G. L. Lombard, J. M. Matsen, D. M. Melven, H. M. Sommers, M. T. Suggs, and B. S. West. 1978. Clinical microbiology. In S. L. Inhorn (ed.), *Clinical Microbiology in Quality Assurance Practices for Health Laboratories.* American Public Health Association, Washington, D.C., pp. 871–1005.
5. Blazevic, D. J., C. T. Hall, and M. E. Wilson. 1976. In A. Balows (ed.), *Comitech3.* American Society for Microbiology, Washington, D.C., pp. 1–21.
6. Budowle, B., H. A. Deadman, R. S. Murch, and F. S. Baechtel. 1988. Review articles: An introduction to the methods of DNA analysis under investigation in the FBI laboratory. *Crime Laboratory Digest* 15:8–22.
7. Colwell, R. R. 1985. The role of microbial ecology in biotechnology and risk assessment. In H. O. Halvorsen, D. Pramer, and M. Rogul (eds.), *Engineered Organisms in the Environment.* American Society for Microbiology, Washington, D.C., pp. 2–3.
8. Colwell, R. R., C. Somerville, I. Knight, and W. Straube. 1988. Detection and monitoring of genetically-engineered micro-organisms. In M. Sussman, C. H. Collins, F. A. Skinner, and D. E. Stewart-Tull (eds.), *The Release of Genetically Engineered Microorganisms.* Academic Press, London, pp. 47–61.
9. Brown, B. (ed.). 1988. Issue on DNA implementation. *Crime Laboratory Digest*, Vol. 15, Suppl. 1.
10. Dixon, B. 1983. The need for microbial ecologists. *Bio/Technology* 1:45.
11. Environmental Protection Agency. *Microbiological Methods for Monitoring the Environment: Wastes and Waste Water,* USEPA—600/8-78-017.
12. *Federal Register,* vol. 33, 15 October 1968.
13. Jones, R. N., D. C. Edson, and J. V. Marymount. 1982. Evaluation of antimicrobial susceptibility test proficiency by the College of American Pathologists Survey Program. *American Journal of Clinical Pathology* 78:168–172.
14. Lechevalier, H. A., and M. Solotrovosky. 1965. *Three Centuries of Microbiology.* McGraw-Hill, New York, pp. 260–280.
15. Bartlett, R. C. 1985. Quality control in clinical microbiology. In Lennette, E., A. Balows, W. J. Hauser, and H. J. Shadomy (eds.), *Clinical Microbiology.* American Society for Microbiology, Washington, D.C., pp. 14–24.
16. Mudd, J. L., and L. A. Presley, 1988. Quality control in DNA typing: A proposed protocol. *Crime Laboratory Digest* 15:109–114.
17. Nagel, J. G., and L. G. Kunz, 1973. Needless testing of quality assured commercially prepared media. *Applied Microbiology* 26:31–37.
18. United States Public Health Service. Shellfish Sanitation Methods. 1962 USPHS.
19. Wood, F. 1965. *Marine Microbial Ecology,* Chapman and Hall, London.

Detection,
Identification,
Classification,
and Enumeration

Overview: Historical Perspective, Present Status, and Future Directions

Rita R. Colwell

Morris A. Levin

Norberto Palleroni

Introduction

The commercial applications of recombinant DNA (rDNA) techniques are many and varied, ranging from the pharmaceutical to the food and manufacturing industries. Some applications, such as bioremediation or agriculture, may involve introduction of genetically modified organisms (GEMs) into the environment. For the latter, risks associated with environmental introductions must be assessed (28). Monitoring for GEMs in the environment as part of such risk assessment is a top priority to ensure public health and environmental safety. Extensive public debate has already taken place in anticipation of the need for monitoring and information on the survival, dispersal, and effects of GEMs in the environment (28, 45, 46). In Part 1 the significant and highly relevant aspects of monitoring GEMs, such as detection, identification, and enumeration in the environment after planned introductions, are reviewed.

Unfortunately, there are major gaps in the information available on microbial community structure and function in natural environmental systems. A resurgence of interest in general microbiology methods and microbial ecology has resulted in new methods and conceptual approaches. During the last five years, many of the older methods, e.g., plate counting and MPN analysis, have been modified, and new methods, such as gene probes and in situ DNA:DNA hybridization have been developed, requiring that the advantages and disadvantages of each method, old and new, must be carefully considered.

By virtue of the intended use, selection of a method in an environmental application is case-dependent. That is, the microorganism to be introduced, the environment into which it will be introduced, and related environmental factors must be evaluated, and a case-specific judgment made. It can be stated unequivocally that, at present, every single cell of any culture released to the environment, whether into the soil, water, or air, cannot be recovered after it has been introduced. However, methods are available which allow monitoring of population size, dispersion, and persistence. In general, information gathered by such monitoring is in many cases sufficient for estimating the fate and effect, if any, of introduction of modified microorganisms to the environment, when data from previous introductions of related nonengineered forms are already available in significant quantity. Methods are described in this part to provide detail appropriate for decision making in release experiments.

Available methods

The major drawback of most of the methods currently available for detecting GEMs, or any microorganism for that matter, in the environment is that they require culturing of the microorganisms to be monitored. Culture methods can be useful for detection and monitoring of microorganisms in the environment, but two significant disadvantages must be emphasized. The first is that culture methods do not necessarily track genes of interest, i.e., cells originally carrying genes. Although initially detectable by the available culturing methods, the genes may be lost or transferred with the result that they will go undetected. Secondly, isolation and culture of microorganisms from samples collected from air, water, or soil are difficult if the microorganism is easily overgrown by other bacteria in the sample, if it is fastidious in nutrient requirements, or if it has entered a viable but nonculturable state (13, 71).

Direct methods for detection, enumeration, and monitoring, on the other hand, although providing relatively greater efficiency of detection, generally do not permit differentiation of viable from dead microorganisms. Modification of direct microscopic methods to include immunological procedures, i.e., highly specific, fluorescent-labeled monoclonal antibodies, make it possible to target specific microorganisms precisely. Although such procedures do not track the genes, in many cases, when coupled with the use of nalidixic acid, assessment of viability is possible (37).

Gene probes that allow tracking of the genome have been developed, and the number of such probes has increased substantially in recent years. Species-specific probes and, more importantly, probes for genetically modified sequences of GEMs have been produced. Chemical methods of detection, such as spent-medium analysis and chemical analysis of characteristic structural and metabolic compounds or elements, have also been proposed. A recent review by Herbert (30) discusses methods for enumerating and detecting microorganisms in the environment.

Metabolic or structural and functional properties of a GEM of interest can serve as a "tag." However, it can be concluded, based on published data, that an accurate and reliable approach to the problem of the direct detection and monitoring of GEMs in the environment requires a combination of fluorescent-labeled monoclonal antibody and epifluorescent microscopy, together with a gene probe. However, the technology of detection and monitoring is moving swiftly, and improvements, both in sensitivity and selectivity in all methodologies, are expected within the next few years. Ultimately, the least expensive, easiest to use, and most reliable of methods will prevail.

Ideal methods

In seeking the ideal method to detect and monitor GEMs, certain requirements for information must be recognized and met. For example, survival rates of GEMs within the spectrum of conditions likely to occur in a specific release area and in the surrounding geographical environment must be determined. It is also important to know the following: (1) the reproduction rate of GEMs in the environment into which they have been introduced; (2) whether the GEM of interest survives and reproduces in or on target organisms or sites and/or nontarget organisms or sites in the environment; (3) whether the GEM is able to spread from the release area and, if so, by what mechanisms; (4) whether the GEM can establish itself in nontarget species or materials; and (5) the sensitivity of the method employed to monitor the populations of modified, target, and nontarget organisms. This range and depth of information is essential for assessment of risk, at least until a more complete data bank of environmental releases can be accumulated.

At present, the combination of fluorescent monoclonal antibody with epifluorescent microscopy procedures and a specific gene probe offers the most sensitive and specific detection method for GEMs, in most situations in natural environments. Methodology and equipment for a very large variety of microorganisms is available, including counting chambers for algae, protozoa, fungi, bacteria, and microalgae using direct counting and employing epifluorescent microscopy, counting by Coulter counter, fluorescent antibody methods, and micro-ELISA. Viable counting methods, and indirect methods, such as measurement of adenosine triphosphate, lipopolysaccharides, muramic acid, photosynthetic pigments, extractable liquid phosphate, and other, miscellaneous methods which are successfully employed in microbial ecology, have been described (30).

Comparison of Methods

Part 1 provides an overview of currently employed methods, with their application and limitations. To reiterate, culture methods employing selective media, colorimetric reactions, and spent-medium analysis require growth of the microorganism or GEM and identification and enumeration using culture methods based on phenotypic characteristics. Specific techniques and problems associated with the culturing approach are described in Chap. 2. It must be emphasized that culture methods do not necessarily track the genome. To monitor the genome, a specific gene or complex (cassette) of genes, gene probes, restriction enzyme analysis, nucleic acid sequencing, or insertion of marker elements are the methods of choice. Furthermore, culture methods may fail if the microorganism, once released into the environment, becomes dormant, i.e., remains viable but cannot be cultured by any of the currently available methods for enrichment or nonspecific culture (13, 70).

Detection can be accomplished by direct immunofluorescent-epifluorescent microscopy, with confirmation by gene probe or other molecular genetic methods. Thus, direct microscopy, coupled with use of a specific gene probe, offers the greatest sensitivity and specificity at the present time. As part of a USEPA program to develop monitoring methods, including identification, detection, and enumeration procedures, existing methods have been examined and evaluated (41). Table 1.1 presents an overview of the strengths and weaknesses of commonly used and newly developed techniques.

Cultural: *Escherichia coli* as an example

A useful example for discussion of relative merits of culture methods is that of the coliforms and, in particular, *Escherichia coli,* as reviewed by Colwell et al. (15). Examination of environmental samples for fecal indicators and directly for *E. coli* has been conducted since the turn of the century (64). Detection of *E. coli* is based mainly on culturability,

TABLE 1.1   Methods for Detection, Identification, Enumeration, and Monitoring of Microorganisms in Environmental Samples*

Methods of detection	Advantages	Disadvantages	Cost	Applications
Plating/culturing	Fairly sensitive Selective, colorimetric or enriched media can be employed Operationally feasible and easy Follows genome Amenable to statistics Spent media can be analyzed	Requires culture of organisms Detects phenotypic expression Inserted markers may be unstable Limited by media, temperature, etc. Long incubation period may be needed Will not detect dormant or nonculturable organisms	Relatively inexpensive	Gerhardt (23) Barry (5)
Direct microscopy/ monoclonal antibody tagging	High specificity High sensitivity where concentration is possible Direct detection of organisms that cannot be cultured Acridine orange direct staining, direct viable count†; fluorescent antibody can be employed	Low sensitivity if sample cannot be concentrated No differentiation between living and dead if DVC† method does not apply Time consuming unless automated Interference from organic slimes Antigen gene may be unstable in environment	Relatively expensive, if automated costs can be very high Preparation of monoclonal antibodies is time consuming and expensive	Bohlool and Schmidt (10) Irby et al. (33) Brayton and Colwell (11) Zelibor et al. (71)
Protein product fingerprinting	Highly specific	Low sensitivity	Relatively expensive	Summers and Hoops (61)
Gene probes	Permits detection of genotype	Low sensitivity	Relatively expensive	Barkay et al. (4) Holben et al. (31) Knight et al. (36)
rRNA sequencing and fingerprinting	High specificity	Low sensitivity	Expensive; requires extensive data base	Loh et al. (42) MacDonell and Colwell (43) Stahl et al. (59) Giovannoni et al. (25)
Flow cytometry	Highly promising in ability to automate individual cell analysis	Little work done to date to assess sensitivity	Expensive	Shapiro (56)

*Adapted from Colwell et al. (15).
†Kogure et al. (37).

i.e., the ability to grow at 44°C, produce acid and gas from lactose, and indole from tryptophane (1, 2). Detection, isolation, and enumeration of *E. coli* by traditional methods usually requires an enrichment step, using liquid media or membrane filtration and incubation on a solid medium, in each case followed by subculture and additional testing of positive cultures or colonies. Havelaar and During (29) found that direct plating yielded better recoveries of *E. coli* than membrane filter methods, using 0.1% sodium lauryl sulfate agar. McClure and Roberts (44) recently examined the effect of incubation time and temperature on growth of *E. coli*. They employed two-dimensional gradient plates and staining, laser densitometry, and computer graphics. The results of these studies, combined with the abundant historical literature on the subject, show that a familiar organism such as *E. coli* is still recovered only with variable success when culture methods are employed. The efficiency and accuracy of recovery depends on temperature, nutrient, method of concentration of cells, salt concentration, and a host of related environmental factors.

Heterotrophic bacteria

The effect of methodology on the recovery of heterotrophic bacteria has also been demonstrated. A comparison of media and of spread and pour plate methods for enumeration of heterotrophic bacteria in drinking water (24) showed that spread-plating followed by incubation for seven days yielded a greater count (520-fold) than the standard method employed in the United Kingdom (use of pour plates containing yeast extract agar and three-day incubation). Thus, standard procedures for nonengineered bacteria are highly inefficient, at best. This problem, and the well-known phenomenon of lack of phenotypic similarity of bacteria grown in the laboratory, compared with strains of the same species in the natural environment, namely atypical forms of *E. coli* in streams (53), casts serious doubt on the usefulness of culture methods for detection and enumeration of bacteria, especially when the problems of viable but nonculturable cells (13), as discussed, are included.

Compared with culturing methods, greater precision in detection can be achieved by direct microscopy and use of fluorescent monoclonal antibodies prepared against cell components. Immunological techniques for detection, identification, and enumeration of microbes in environmental samples are described in this section. This approach permits direct detection of specific organisms in environmental samples and has been used to identify *Rhizobium* spp. in soil (10) and to assess serological diversity of marine ammonia and nitrite-oxiding bacteria in seawater (66). Coupled with a procedure using nalidixic acid and yeast extract, which results in the enlargement of living cells and, thereby, allowing their identification (11, 37), direct viable counts can be achieved.

The degree of cross-reactivity between antisera prepared against one strain and other related strains of the same physiological type or species can be a problem in the immunological approach, both with environmental and laboratory isolates. In some cases, very little intergenus cross-reactivity was found between different strains (6), while other workers (34) were able to identify at least four serotypes among nitrite oxidizers of identical morphology. Studying a variety of discernibly different morphological types under similar culture conditions, Ward and Carlucci (66) found greater cross-reactivity between antisera prepared against one strain and other related strains of the same physiological type. It should be noted that the marine ammonium-oxidizing population is less diverse than soil populations of these organisms. Identification problems related to lack of similarity are discussed in Part 1 in the chapter on numerical taxonomy, in which the advantages and disadvantages of using numerical taxonomic procedures, the importance of using reference strains, and the need to develop a data base for environmental isolates are pointed out.

It is possible that the water-column environment in the sea may present less habitat diversity than the soil environment, resulting in a relatively less diverse microbial assemblage, but this is a hypothesis which the newer methods of genetic probes may confirm or disprove (26, 67). In any case, the immunofluorescent-epifluorescent microscopy approach is of value in the direct detection, enumeration, and monitoring of microorganisms in the environment. Advances in microscopic techniques are presented in detail in this section. Conclusions that can be drawn are that the problem of species cross-reaction and intraspecies antigenic variation will reduce both the specificity and sensitivity of the method, but these problems are not insurmountable.

Hybridization with gene probes relies on base pairing between homologous nucleic acid sequences, which allows identification of specific sequences homologous with the labeled DNA probe. This method has been used to detect particular organisms in environmental samples (4, 30, 36, 42) and is described below in this section.

Specific DNA probes for identification of *Campylobacter jejuni* (38), enterotoxigenic *E. coli* (51), and *Salmonella typhi* (52) have recently been developed and, no doubt, many more such reports will reach the literature. Those cited here are a few of the many examples available. A variety of applications is possible, and probes have been used to detect specific organisms in environmental samples and pathogenic organisms in water systems and foods. Recent studies have shown the value of DNA probes in elucidating the ecology of the rumen (3, 63) and in food (35). Schleifer et al. (54) employed cloned ribosomal RNA genes from *Pseudomonas aeruginosa* as probes for conserved DNA sequences.

In general, hybridization probes used for microbial identification are highly specific and, in general, comprise cloned genes from specific organisms. Unfortunately, specific hybridization probes may necessitate culture of the organism of interest to obtain sufficient cells for the probe to be effective.

Use of the polymerase chain reaction (PCR) obviates the necessity of culturing to amplify the genetic material being monitored. If recombinant DNA sequences in GEMs are to be detected by DNA probes, these sequences must be present in very high copy number in a sample, otherwise the target sequence concentration will be below the detection threshold of conventional probes. Somerville et al. (57) developed a method for enhancing sensitivity, based on the use of specific hybridization probes as primers for chain elongation. In this method, specificity is determined by probe hybridization, while sensitivity is enhanced by polymerization into adjacent genes. The results showed that sensitivity can be enhanced by at least three orders of magnitude over conventional probing methodology. The PCR technique, described in Chap. 4, offers PCR as a solution to this problem. Using PCR it is possible to replicate specific DNA sequences to the desired concentrations in short time periods.

Indirect extraction of 5S and 16S rRNA from environmental samples for sequencing has been carried out (14, 43, 47). At present, these methods are time consuming and require sophisticated laboratory procedures. However, sequencing methods are rapidly becoming streamlined. In addition, the data base for sequence comparison by computer is expanding rapidly, although it is not yet sufficient for every large-scale environmental application. Since access to computerized data base containing sequences is becoming a reality, e.g., Genebank (62), it is possible that direct extraction and sequence analysis may offer a powerful method for detection and monitoring of GEMs in the future (22).

A somewhat different approach was taken by Holben et al. (31), who developed a protocol to obtain purified bacterial DNA from a soil sample. This technique is described in Chap. 4. The presence of extracellular DNA in fresh water and seawater has been known for decades, and recent work by Paul et al. (48) in surveys of offshore and subsurface waters has shown that the lowest dissolved DNA values are found in offshore regions, with concentrations increasing shoreward and in estuarine plumes. In general, dissolved DNA from marine sources has been shown to possess a wide range of molecular sizes (0.12–35.2 kb; 19, 20). Somerville et al. (57) developed a procedure for direct extraction of particulate-bound nucleic acid from seawater which has been successful in detecting the presence of bacteria, otherwise nonculturable, in water samples, such as hydrothermal vent fluids (60). In addition to precise and accurate detection and monitoring of microorganisms released into the environment, a reliable, reproducible, and information-rich taxonomy is required and, from that taxonomy, an identification system. Substantial progress has been made in microbial systematics during the past three decades, with improvement achieved from the applications of numerical taxonomy, analytical chemistry, and molecular genetics to systematics. Probabilistic identification keys for clinically important bacteria have reached a relatively advanced stage of development and have also gained widespread acceptance. However, for soil and aquatic systems, microorganisms making up the naturally occurring communities remain very poorly characterized, and identification schemes for these are, at best, rudimentary.

Application of Methods

The research needs for improved microbial systematics in microbial ecology are most critical for species that make up the soil, water, and atmospheric populations of the natural environment. Although it could be argued that it is possible to monitor, with reasonable assurance, an introduced GEM during the initial stages of introduction, it is not clear that long-term monitoring will be possible without a good knowledge base for naturally occurring species. Thus, without an adequate data base, species interactions, including horizontal and vertical exchanges of genetic imitation, may not be measurable with sufficient precision and reproducibility to meet the objections of even the most reasonable critics of GEM introductions.

The kind and value of information needed for evaluating environmental introductions of introduced microorganisms can be understood by considering the actinomycetes as a useful example. In spite of the moldlike appearance of many actinomycetes (to which, in fact, they owe the *mycete* part of their name), these are truly prokaryotic organisms with typical bacterial cell structure of the gram-positive type. Actinomycetes are morphologically complex and are common in soil and polluted waters. In contrast to many other bacteria, they usually form filamentous cells, with various degrees of branching. In a large subgroup (sporoactinomycetes), the filaments may produce spores by fragmentation. The spores are either single (as in *Micromonospora*) or numerous, in chains which can be naked or covered with a sheath (*Streptomyces*), or enclosed in sporangia. Some of the members of the sporangiate actinomycetes (for instance, *Actinoplanes*) have spores capable of acquiring active swimming mobility after brief immersion in water. The morphological variety in this group of actinomycetes is more pronounced than in most other bacterial groups, and can be considered a very useful guide for the identification of taxa.

Actinomycetes as a prototype

A second subgroup (nocardioform-actinomycetes) is composed of organisms which do not sporulate and display the lowest degree of branching or no branching at all. This subgroup is morphologically less complex than the former, and it is not uniquely circumscribed by morphological criteria. Thus, coryneform bacteria are similar and have a tendency to give branched forms, but are not actinomycetes.

A third subgroup includes organisms able to produce a mycelium that divides in two planes to give coccoid cells, which may become motile. Examples in this subgroup are the genera *Dermatophilus* and *Geodermatophilus*.

During the last two decades, actinomycetes have been introduced into the stream of phylogenetic studies on which modern taxonomies of bacteria are being based.

Two sets of techniques are now available, one of which permits detection of affinities among closely related taxa, while the second defines distances among suprageneric taxonomic hierarchies. Thus, sporo-actinomycetes can be divided into three main ribosomal RNA similarity groups. The genera included in these groups are (1) *Actinoplanes, Amorphosporangium* (at present considered to be synonymous with *Actinoplanes*), *Ampullariella,* and *Micromonospora*; (2) *Planomonospora, Planobispora,* and *Streptosporangium*; and (3) *Chainia, Elytrosporangium, Kitasatoa, Microellobosporia, Streptomyces,* and *Streptoverticillium* (58).

The most common aerobic actinomycetes to be encountered in direct isolations from soil are members of the genus *Streptomyces*. A large number of species have been described and identification of new isolates is one of the most difficult tasks in determinative bacteriology. Fortunately, the work of Williams, Goodfellow, and their collaborators (69) has simplified the situation, with substantial reduction of the number of species of this most important genus. Even so, the road to a precise species concept for this group of organisms is paved with formidable obstacles. In their fundamental paper, bases for a scheme of classification of *Streptomyces* and related genera, which is useful for identifying streptomycetes in natural habitats, are presented. Criteria given by these authors comprise the basis of the treatment included in Bergey's *Manual of Systematic Bacteriology* (7).

The following is a list of properties used for the description of streptomycetes, taken from Williams et al. (69): presence of spores, spore chain morphology, spore ornament, color of the aerial spore mass, pigment of substrate mycelium, pigment diffusing into the medium, antimicrobial activity against gram-positive and gram-negative organisms, yeasts, *Streptomyces* strains and molds; enzymatic activities (proteolysis, lipolysis, hydrolysis of pectin, chitin, lecithin, hippurate, nitrate reduction, etc.), degradation of various compounds of large and small molecular weight, resistance to antibiotics, growth at different temperatures, growth in the presence of inhibitors (such as sodium chloride, phenol, tellurite, thallous acetate), and utilization of a number of compounds as sources of carbon or nitrogen.

The distribution of a total of 139 unit characters among 394 type cultures has been used by these authors for calculation of similarity coefficients which permit the circumscription of clusters of strains and cluster groups (see Chap. 6). The strains could be arranged into 10 cluster groups which, in turn, were subdivided into 73 composite and 28 single-member clusters. The largest cluster group (A) contained most (72 percent) of the *Streptomycea* strains examined, but at the same time it included other quite different genera such as *Actinosporangium, Microellobosporia, Saccharopolyspora,* and others. Even though these organisms were placed at the periphery of the cluster group, other *Streptomyces* species (*S. rimosus, S. lavendulae, S. viridoflavus,* etc.) were excluded and assigned to other groups.

Similarly, another large cluster group (F) was dominated by the genus *Streptoverticillium,* but in addition it included *Streptomyces* and *Kitasatoa* species. It is evident that in spite of the simplification that this classification has meant with respect to the number of species names, the scheme is not free from difficulties.

Since this work is based on the use of numerical taxonomic procedures, which appears, at this moment, as the most reasonable measure of classification, the authors later proposed (39, 68) probabilistic matrices as examples of methodologies for identification of taxa in this large and complex group of organisms. Clearly the actinomycetes demonstrate the difficulties which can be encountered in identifying soil isolates and differentiating GEMs.

As a guide to the identification of members of other actinomycete groups, Goodfellow and Cross (27) prepared a classification scheme and a valuable review on structure and taxonomy of *Streptomyces* has been written by Dietz (20).

Progress has been made in the development of methods for the specific or preferential isolation of members of various groups of actinomycetes, and for the reduction of the numbers of associated bacteria that interfere with the isolation of actinomycetes with particular emphasis on sample treatments preceding the isolation proper (17).

At present, strains of some *Streptomyces* species are being actively used in gene-cloning projects for which there are available many plasmids, satisfactory methodology for their manipulation (32), and extensive information on the technology of large-scale fermentation. Thirty-three patents were issued over a four-year period for *Streptomyces* with potential environmental application (see Introduction). For all the above reasons, members of the group can be seriously considered as subjects for experimental release of engineered strains into natural habitats, resulting in a need for risk assessment information. The following discussion is based on considerations raised and discussed by Rafii et al. (50), using many of the types of data discussed above, concerning the limitations and risks associated with their use and by Crawford (16) who discusses potential environmental applications and summarizes research demonstrating that genetic exchange occurs readily among *Streptomyces* in soil and the potential for exchange between recombinant *Streptomyces* and native soil bacteria.

Streptomyces strains have demonstrated a capacity for survival over long periods of time in experiments under normal soil conditions. Wang et al. (65) have demonstrated that engineered strains may survive for as long as indigenous strains of the same genus and that in some cases they may become very active participants in the turnover of organic matter, thus suggesting that laboratory organisms are not necessarily less hardy than wild strains of the autochthonous flora.

In recent years gene exchange in soil has been demonstrated for various groups of organisms. Although transduction occurs in *Bacillus* in sterilized soil, conjugation and transformation are genetic mechanisms observed to operate in untreated soil. Conjugation (in untreated soil) and transduction (in sterile soil) occur in *Streptomyces* and may be operative genetic exchange mechanisms under natural conditions. Of the two, in theory, the most important should be transduction, which does not require cell-to-cell contact and the genes are protected by the capsid structure of the viruses during their passage from cell to cell. However, gene transfer by transduction in soil has not yet been observed for *Streptomyces*.

As stated above, the *Streptomyces* can offer a good example of the difficulties frequently encountered in identification of *Streptomyces* strains isolated from nature or reisolated after release. It is appropriate to warn the worker about the marked and significant instability of some of the *Streptomyces*. Properties of sporulation, formation of aerial mycelium, pigmentation, production of antibiotics, and resistance to various deleterious agents are all properties highly variable during normal handling of cultures grown in the laboratory. The frequency of these variations often can be increased by relatively mild treatments which are not truly mutagenic.

Hypervariability, an interesting case of genetic instability, has been attributed to amplification by Leblond et al. (40). These authors have observed two levels of instability, an initial event of genetic instability, followed by the emergence of progeny without preponderant phenotype, which is the consequence of hypervariability. In *S. lividans,* two pathways for instability have been identified, the Cml → Arg (susceptibility to arginine auxotrophy among the chloramphericol-sensitive variants), and Tet → Ntr (in which strains defective in the regulation of production of glutamine synthetase appear frequently among the tetracycline sensitive clones) (8, 21).

In some instances, the changes are reversible, suggesting participation of transposable elements, and in others, total loss of properties indicates either plasmid-born genetic determinants, or chromosomal deletions. Synthesis of tyrosinase, argininosuccinate synthetase, streptomycin phosphotransferase, and A-factor, are all coded for by genes which delete at high frequency from the *Streptomyces* genome (55). In industrial practice, mass selection directed to high product yields sometimes results in hyperproducing strains that are atypical in phenotype when compared with the original isolates. Unfortunately, the studies of these phenomena have not advanced sufficiently to allow their control for practical purposes. High-producing strains or strains resistant to high levels of antibiotics may originate as a consequence of gene amplification.

Amplifications may appear without application of selective pressure, and may be correlated with the production of deletions in neighbor regions of the genome. As a consequence of these phenomena, 40 percent of the characteristics used in taxonomy or *Streptomyces* may be unstable, and variants may easily be classified in different taxonomic groups (55). This is why it is very important to warn that unless a comprehensive characterization of strains under study is carried out, the intrinsic instability of many characteristics of *Streptomyces* strains may convert the practice of taxonomy to an exercise in futility.

In general, *Streptomyces* strains may show high mutability of some genomic regions, a property suggestive of mechanisms of evasion from the stabilizing forces acting on the bacterial genome. Genetic experiments may not give a precise idea of the magnitude of the changes involved, since the geneticist for the most part focuses attention on one or a few genes at a given time.

It can also be reasonably inferred that conjugation must take place among *Streptomyces* in soil, since many strains isolated from nature carry conjugative plasmids. Working with sterile soil, plasmid-recipient transconjugants have been recovered by Rafii et al. (50). Intraspecies transfer gives the highest frequency, but at best is significantly lower than transfer under laboratory conditions, where cell-to-cell contact can be controlled more rigorously. Mobilization of nonconjugative plasmids also has been shown to occur in sterile soil. Plasmid exchange was most frequent in nutrient-amended, low-moisture soils, and the moisture level appeared to be an important factor. Two- to threefold differences in plasmid transfer were observed when comparing effects of low to high moisture content (9). The most favorable temperature was 30°C.

Since triparental crosses are effective in mobilizing plasmids under laboratory conditions, it is also possible that such process takes place in soil, in which case, nonconjugative plasmids in an engineered strain may be mobilized by other plasmids present in indigenous *Streptomyces* strains. Working with *S. lividans* and *S. parvulus*, Rafii and Crawford (49) evaluated the effect of the host strain on frequency of plasmid transfer, demonstrating that the frequency of transfer was determined by the recipient. One possible consequence of *Streptomyces* gene transfer in soil which has been considered a potential risk is the transmission of *new* antibiotic-resistant determinants to members of the natural microflora. It has been proposed that antibiotic resistance genes may have originated in antibiotic-producing microorganisms as a protective mechanism, which places *Streptomyces* in a prominent position, since more than half of the many antibiotics are produced by strains of this group. From *Streptomyces,* the resistance genes could have been transferred to other members of the soil microflora. *Pseudomonas* and *Streptomyces* share similar environmental preferences: well-aerated soils with a pH near neutrality and high concentrations of organic matter. Because of their powerful degradative activity toward high-molecular-weight compounds that are refractory to attack by other microorganisms, *Streptomyces* may be a source of low-molecular-weight compounds utilizable as carbon sources by versatile *Pseudomonas* strains, and coexistence in the same niches may be further ensured by the tolerance of *Pseudomonas* to *Streptomyces* secondary metabolites. In fact, *Pseudomonas* is a genus whose members are notoriously resistant to antibiotics, and part of this resistance can often be attributed to plasmids carrying resistance markers for many antibiotics, metals, and other toxic agents.

The problem of transfer of resistance markers from *Streptomyces* to other organisms has been investigated (12), but *Pseudomonas* has not been suggested as a possible recipient, in spite of its ecological requirements and the widespread presence of resistance factors in many strains of the group.

Conclusions

In summary, detection, enumeration, identification, and classification of GEMs introduced to the environment offer challenges because of the relative infancy of molecular microbial systematics and ecology. The tasks are formidable but not insurmountable. The example given here of the richness of information offered by *Streptomyces* spp. for use as GEMs in environmental applications is useful in that it is, in many ways, a model for soil and water microorganisms one can expect to be employed as GEMs in environmental applications in the future. The need, therefore, is for methods suitable for environmental biotechnology.

Those methods of best value at the present time are discussed in the following chapters. No doubt, as this book is being written, better methods that are more efficient, more precise, simpler, and more rapid are being devised. The future promises detection, enumeration, identification, and classification to be much simplified, with cell sorting, collection, identification by computer, and direct nucleic acid sequencing. The most sophisticated methods of today are sure to be antiquated very quickly as molecular systematics and molecular microbial ecology flourish in the years ahead.

References

1. American Public Health Association. 1985. *Standard Methods for the Examination of Water and Wastewater,* 16th ed. Washington, D.C.
2. Anon. 1983. The bacteriological examination of drinking water supplies. Reports on Public Health and Medical Subjects No. 71. HMSO, London.
3. Attwood, G. T., R. A. Lockington, G.-P. Xue, and J. D. Brooker. 1988. Use of a unique gene sequence as a probe to enumerate a strain of Bacteroides ruminicola introduced into the rumen. *Applied and Environmental Microbiology* 54:534–539.
4. Barkay, T., D. L. Fouts, and B. H. Olson. 1985. Preparation of a DNA gene probe for detection of mercury, resistance genes in gram-negative bacterial communities. *Applied and Environmental Microbiology* 49:686.
5. Barry, G. F. 1986. Permanent insertion of foreign genes into the chromosomes of soil bacteria. *Bio/Technology* 4:446–449.
6. Belser, L. W., and E. L. Schmidt. 1978. Diversity in the ammonia-oxidizing nitrifier population of a soil. *Applied and Environmental Microbiology* 36:584–588.
7. Buchanan, R. E., and N. E. Gibbons (eds.). 1989. *Bergey's Manual of Systematic Bacteriology,* vol 4. William and Wilkins, Baltimore.
8. Betzler, M., P. Dyson, and H. Schrempf. 1987. Relationship of an unstable *argG* gene to a 5.7 kilobase amplifiable DNA sequence in *Streptomyces lividans. Journal of Bacteriology* 169:4804–4810.
9. Bleakley, B. H., and D. L. Crawford. 1989. The effects of varying moisture and nutrient levels on the transfer of a conjugative plasmid between Streptomyces species in soil. *Canadian Journal of Microbiology* 35:544–549.
10. Bohlool, B. B., and E. L. Schmidt. 1980. The immunofluoresence approach in microbial ecology. *Advances in Microbial Ecology* 4:203.
11. Brayton, P. R., and R. R. Colwell. 1987. Fluorescent antibody staining method for enumeration of viable environmental *Vibrio cholerae 01. Journal of Microbiology Methods* 6:309–314.
12. Chater, K. F., D. J. Henderson, M. J. Bibb, and D. A. Hopwood. 1988. Genome flux in *Streptomyces coelicolor* and other streptomycetes and its possible relevance to the evolution of mobile antibiotic resistance determinants. In A. J. Kingsman, K. F. Chater, and S. M. Kingsman (eds.), *Transposition.* Cambridge University Press, London.
13. Colwell, R. R., P. R. Brayton, D. J. Grimes, D. B. Roszak, S. A. Huq, and L. M.

Palmer. 1985. Viable but nonculturable *Vibrio cholerae* and related pathogens in the environment: Implications for release of genetically engineered microorganisms. *Bio/Technology* 3:817–820.

14. Colwell, R. R., M. MacDonell, and D. Swartz. 1989. Identification of an antarctic endolithic microorganism by 5S RNA sequence analysis. *International Journal of Systemic Microbes* 11:182–186.

15. Colwell, R. R., C. Somerville, I. Knight, and W. Straube. 1988. Detection and monitoring of genetically-engineered micro-organisms. In M. Sussman, C. H. Collins, F. A. Skinner, and D. E. Stewart-Tull (eds.), *The Release of Genetically Engineered Microorganisms*. Academic Press, London.

16. Crawford, D. L. 1988. Development of recombinant Streptomyces for biotechnological and environmental use. *Biotechnology Advance* 6:183–206.

17. Cross, T. 1982. Actinomycetes: a continuing source of new metabolites. *Developments in Industrial Microbiology* 23:1–18.

18. de Flaun, M. F., and J. H. Paul. 1986. Hoechst 33258 staining of DNA in agarose gel electrophoresis. *Journal of Microbiological Methods* 5:265–270.

19. de Flaun, M. F., J. H. Paul, and W. H. Jeffrey. 1987. Distribution and molecular weight of dissolved DNA in subtropical estuarine and oceanic environments. *Marine Biology Progress Series* 38:65–73.

20. Dietz, A. 1986. Structure and taxonomy of *Streptomyces*. In S. W. Queener and L. E. Day (eds.), *The Bacteria*. Academic Press, Orlando, pp. 1–25.

21. Dyson, P., M. Betzler, T. Kumar, and H. Schrempf. 1987. Biochemical and genetic instability of spontaneous genetic instability and DNA amplification in *Streptomyces lividans*. In M. Alevic, D. Hranuelli, and Z. Toman (eds.), *Genetics of Industrial Microorganisms*, Part B. Pliva, Zagreb, pp. 57–65.

22. Fox, G. E., E. Stackebrandt, R. B. Hespell, J. Gibson, J. Mamiloff, T. A. Dyer, R. S. Wolfe, W. E. Balch, R. S. Tanner, L. J. Magrum, L. B. Zablen, R. Blakemore, R. Gupta, L. Bonen, B. J. Lewis, D. A. Stable, K. R. Lirehrsen, K. N. Chen, and C. R. Woese. 1980. The phylogeny of procaryotes. *Science* 109:457.

23. Gerhardt, P. 1981. *Manual of Methods for General Microbiology*. American Society for Microbiology, Washington, D.C.

24. Gibbs, R. A., and C. R. Hayes. 1988. The use of R2A medium and the spread plate method for the enumeration of heterotrophic bacteria in drinking water. *Letters in Applied Microbiology* 6:19–22.

25. Giovannoni, S. J., E. F. Delong, G. J. Olsen, and N. R. Pace. 1988. Phylogenetic group-specific oligodeoxynucleotide probes for identification of single microbial cells. *Journal of Bacteriology* 170:720–726.

26. Giovannoni, S. J., T. B. Britschgi, C. L. Moyer, and K. G. Sheild. 1990. Genetic diversity in Saragasso Sea bacteria. *Nature* 345:60–62.

27. Goodfellow, M., and T. Cross. 1984. Classification. In M. Goodfellow, M. Mordarski, and S. T. Williams (eds.), *The Biology of Actinomycetes*. Academic Press, London, pp. 7–164.

28. Halvorson, H. O., D. A. Pramer, and M. Rogul (eds.). 1985. *Release of Engineered Organisms*. American Society for Microbiology, Washington, D.C.

29. Havelaar, A. H., and M. During. 1988. Evaluation of the Anderson Baird-Parker direct plating method for enumerating *Escherichia coli* in water. *Journal of Applied Bacteriology* 64:89–98.

30. Herbert, R. A. 1990. Methods for enumerating microorganisms and determining biomass in natural environments. In R. Grigorova and J. R. Norris (eds.), *Methods in Microbiology*. Academic Press, New York, pp. 1–39.

31. Holben, W. E., J. K. Jansson, B. K. Chelm, and J. M. Tiedje. 1988. DNA probe method for the detection of specific microorganisms in the soil bacterial community. *Applied and Environmental Microbiology* 54:703–711.

32. Hopwood, D. A., M. J. Bibb, K. F. Chater, T. Kieser, C. J. Bruton, H. M. Keiser, D. J. Lydiate, C. P. Smith, J. M. Ward, and H. Schrempf. 1985. *Genetic Manipulation of Streptomyces: A Laboratory Manual*. The John Innes Foundation, Norwich.

33. Irby, W. S., Y. S. Huang, C. Y. Kawanishi, and W. M. Brooks. 1985. Immunoblot analysis of exposure polypeptides from some entomorphic microsporidia. *Virology* 143:370.

34. Josserand, A., and J. C. Cleyet-Marel. 1979. Isolation from soils of *Nitrobacter* and evidence for novel serotypes using immunofluorescence. *Microbial Ecology* 5:197–205.

35. Kasper, C. W., and C. Tartera. 1990. Methods for detecting bacteria in food and water. In R. Grigova and J. R. Norris (eds.), *Methods in Microbiology*. Academic Press, New York, pp. 437–541.

36. Knight I. T., C. W. Shultz, C. W. Kaspar, and R. R. Colwell. 1990. Direct detection of *Salmonella* spp. in estuaries using a DNA probe. *Applied and Environmental Microbiology* 56:1059–1066.

37. Kogure, K., U. Simidu, and N. Taga. 1979. A tentative direct microscopic method for counting living marine bacteria. *Canadian Journal of Microbiology* 25:415–419.

38. Korolik, V., P. J. Coloe, and V. Krishnapillai. 1988. A specific DNA probe for the identification of *Campylobacter jejuni*. *Journal of General Microbiology* 134:521–529.

39. Langham, C. D., S. T. Williams, P. H. A. Sneath, and A. M. Mortimer. 1989. New probability matrices for identification of *Streptomyces*. *Journal of General Microbiology* 135:121–135.

40. Leblond, P., P. Demuyter, L. Moutier, M. Laakel, B. Decaris, and J. M. Simonet. 1989. Hypervariability, a new phenomenon of genetic instability, related to DNA amplification in *Streptomyces ambofaciens*. *Journal of Bacteriology* 171:419–423.

41. Levin, M. A., R. Seidler, A. W. Bourquin, J. R. Fowle III, and T. Barkay. 1987. EPA developing methods to assess environmental release. *Bio/Technology* 5:38–45.

42. Loh, L. C., J. J. Hamm, C. Y. Kawanishi, and E. S. Hiiang. 1982. Analysis of the *Spodoptera frugiperda* nuclear polyhedrosis virus genome by restriction endonucleases and electron microscopy. *Journal of Virology* 44:747–752.

43. MacDonell, M. T., and R. R. Colwell. 1984. Identical 5S RRNA nucleotide sequence of *Vibrio cholerae* strains representing temporal, geographical, and ecological diversity. *Applied and Environmental Microbiology* 48:199.

44. McClure, P. J., and T. A. Roberts. 1987. The effect of incubation time and temperature on growth of *Escherichia coli* on gradient plates containing sodium chloride and sodium nitrite. *Journal of Applied Bacteriology* 63:401–407.

45. National Academy of Sciences. 1987. *Introduction of Recombinant DNA-Engineered Organisms into the Environment: Key Issues*. National Academy Press, Washington, D.C.

46. National Research Council. 1989. *Field Testing Genetically Engineered Organisms*. National Academy Press, Washington, D.C.

47. Olsen, G. J., D. J. Lane, S. J. Giovannoni, N. R. Pace, and D. A. Stahl. 1986. Microbial ecology and evolution: a ribosomal RNA approach. *Annual Reviews of Microbiology* 40:337–365.

48. Paul, J., M. F. Deflaun, W. H. Jeffrey, and A. W. David. 1988. Seasonal and diel variability in dissolved DNA and in microbial biomass and activity in a subtropical estuary. *Applied and Environmental Microbiology* 54:718–827.

49. Rafii, F., and D. L. Crawford. 1989. Donor/recipient interactions affecting plasmid transfer among streptomyces species: A conjugative plasmid will mobilize nontransferable plasmids in soil. *Current Microbiology* 19:115–121.

50. Rafii, F., D. L. Crawford, B. H. Bleakley, and Z. Wang. 1989. Assessing the risks of releasing recombinant *Streptomyces* in soil. *Microbiological Sciences* 5:358–362.

51. Romick, T. L., J. A. Lindsay, and F. F. Busta. 1987. A visual DNA probe for detection of enterotoxigenic *Escherichia coli* by colony hybridization. *Letters in Applied Microbiology* 5:87–90.

52. Rubin, F. A., D. J. Kopecko, K. F. Noon, and L. S. Baron. 1985. Development of a DNA probe to detect *Salmonella typhi*. *Journal of Clinical Microbiology* 22:600–605.

53. Rychert, R. C., and G. R. Stephenson. 1981. Atypical *Escherichia coli* in streams. *Applied and Environmental Microbiology* 41:1276–1278.

54. Schleifer, K. H., W. Ludwig, J. Kraus, and H. Festl. 1985. Cloned ribosomal ribonucleic acid genes from *Pseudomonas aeruginosa* as probes for conserved deoxyribonucleic acid sequences. *International Journal of Systematic Bacteriology* 35:231–236.

55. Schrempf, H., P. Dyson, M. Betzler, T. Kumar, and P. Groitl. 1987. Amplification and deletion of DNA sequences in *Streptomyces*. In M. Alevic, D. Hranuelli, and Z. Toman (eds.), *Genetics of Industrial Microorganisms*, Part B. Pliva, Zagreb, pp. 177–184.

56. Shapiro, H. M. Flow cytometry in laboratory microbiology: New directions. *ASM News* 56:584.

57. Somerville, C., I. T. Knight, W. L. Straube, and R. R. Colwell. 1988. Probe-directed, polymerization-enhanced detection of specific gene sequences in the environment. REGEM 1, Abstract 70.

58. Stackebrandt, E., B. Wunner-Fussl, V. J. Fowler, and K. H. Schleifer. 1981. Deoxyribonucleic acid homologies and ribosomal ribonucleic acid similarities among sporeforming members of the order *Actinomycetales*. *International Journal of Systemic Bacteriology* 31:420–431.

59. Stahl, D. A., D. J. Lane, G. J. Olsen, and N. D. Pace. 1985. Characterization of a Yellowstone hot spring microbial community by 5S RNA sequences. *Applied and Environmental Microbiology* 49:1379.

60. Straube, W. L., J. W. Deming, C. C. Somerville, R. R. Colwell, and J. A. Baross. 1990. Particulate DNA in smoker fluids: Evidence for the existence of populations in hot hydrothermal systems. *Applied and Environmental Microbiology* 56:1440–1445.

61. Summers, M. D., and P. Hoops. 1980. Radioimmunoassay analysis of Baculovirus granulins and polyhedrins. *Virology* 103:89.

62. Swartz, D. G., M. T. MacDonnell, and R. R. Colwell. 1989. *Biomolecular Data: A Resource in Transition*. Oxford University Press, Oxford.

63. Taiinock, G. W. 1988. Molecular genetics: A new tool for investigating the microbial ecology of the gastrointestinal tract? *Microbial Ecology* 15:239–256.

64. Theroux, M. R., E. F. Eldridge, and W. L. Mallmann. 1943. *Analysis of Water and Sewage*. McGraw-Hill, New York.

65. Wang, Z., D. L. Crawford, A. L. Pometto III, and F. Rafii. 1989. Survival and effects of wild-type, mutant, and recombinant *Streptomyces* in a soil ecosystem. *Canadian Journal of Microbiology* 35:535–543.

66. Ward, B. B., and A. F. Carlucci. 1985. Marine ammonia- and nitrite-oxidizing bacteria: Serological diversity determined by immunofluorescence in culture and in the environment. *Applied and Environmental Microbiology* 50:194–201.

67. Ward, D. M., R. Weller, and M. M. Bateson. 1990. 16S Rna sequences reveal numerous undetected microorganisms. *Nature* 345:63–65.

68. Williams, S. T., M. Goodfellow, E. M. H. Wellington, J. C. Vickers, G. Alderson, P. H. A. Sneath, M. J. Sackin, and R. Mortimer. 1983. A probability matrix for identification of some streptomycetes. *Journal of General Microbiology* 129:1815–1830.

69. Williams, S. T., M. Goodfellow, G. A. Alderson, E. M. H. Wellington, P. H. A. Sneath, and M. J. Sackin. 1983. Numerical classification of *Streptomyces* and related genera. *Journal of General Microbiology* 129:1743–1813.

70. Xu, H. S., N. Roberts, F. L. Singleton, R. W. Attwell, D. J. Grimes, and R. R. Colwell. 1982. Survival and viability of non-culturable *Escherichia coli* and *Vibrio cholerae* in the estuarine and marine environment. *Microbial Ecology* 8:313–323.

71. Zelibor, J., M. Tamplin, and R. R. Colwell. 1987. A method for measuring bacterial resistance to metals employing epifluorescent microscopy. *Journal of Microbiological Methods* 7:143–145.

Chapter 2

Detection and Enumeration of Microorganisms Based upon Phenotype

Ronald M. Atlas

The detection and enumeration of microorganisms based upon phenotypic characteristics almost always depend upon procedures that permit the growth of the targeted microorganisms in vitro. This means that those microorganisms to be specifically detected and/or enumerated must be recovered from environmental samples in ways that maintain their viability and allow them to express a distinguishable phenotype when grown in the laboratory. Recognizable and distinctive phenotypes must be expressed during in vitro culture for the detection of specific microorganisms. In some cases, the observation of a single unique characteristic may be sufficient to recognize the specifically targeted microbe. A genetically engineered microorganism (GEM) with a unique phenotype, such as the ability to metabolize DDT aerobically to CO_2 or the failure to produce a peptide that initiates ice crystal formation (ice-minus phenotype), could be selected based upon such a single distinguishing characteristic. In other cases, it may be necessary to determine a pattern of multiple characteristics in order to distinguish the specific targeted microbe from all others. For example, clinical identification of pathogenic microorganisms frequently relies upon a distinctive pattern of about 20 phenotypic characteristics. Regardless of whether one or more phenotypic characteristics must be examined, culture methods must be designed to provide both efficient recovery of the targeted microbe and its unambiguous recognition.

The unambiguous detection of a microbe based upon phenotype is a difficult task because even distantly related organisms can have similar phenotypes for some characteristics, making the selection of key identifying features critical. Distinguishing among closely related microbes is particularly difficult because of the limited number of characteristics that can be selected for separating target organisms from all others. Mutations can also alter the phenotype of a target microbe, obscuring its recognition.

In the case of GEMs that may be identical to indigenous microbes except for the recombinant addition or deletion of genes, specific detection of the GEM may be accomplished by observing the expression of the recombinant genes, so that the unique recombinant phenotype can be observed. However, because it is not always practical to observe the phenotypic characteristics specifically determined by the recombinant genes, additional marker genes can be included in the recombinant organism. In fact, readily detectable marker genes, such as genes coding for antibiotic resistance, often are included during the creation of GEMs to facilitate detection of the recombinants that are formed. Such marker genes also can be invaluable for the environmental tracking of deliberately released GEMs, but consideration must be given to the effects of adding specific marker genes on the risk of releasing particular GEMs and on the ecological survival of those GEMs.

Detection and Enumeration of Viable Microorganisms

The classical approaches for detecting microorganisms, which have been used for over a century, are to place viable microbial cells onto a solid medium (plating procedures) or into a liquid broth (enrichment procedures), containing all of the nutrients essential for the growth of the targeted microorganisms, and to incubate the inoculated cultures under conditions that favor the growth of those microbes so that their phenotypes can be observed.

Plating procedures rely upon the ability to separate microorganisms so that individual microbial cells are deposited at discrete locations on a solid medium; when the deposited microorganisms reproduce, they form discrete colonies that comprise clones of the original microorganisms. Plating procedures accomplish two things. First, they separate individual microbes from the multitude of microorganisms in a natural microbial community so that the phenotypes of individual microbes can be determined, which is the basis of the pure culture methods that are the mainstay of microbiology. Second, they amplify the signal (phenotype) to be observed through cell reproduction. The phenotypic characteristics of the millions of presumably identical cells in a colony can be observed much more readily than those of individual microbial cells.

Plating procedures are effective for detecting a particular microbe when that target microbe constitutes a significant portion of the microbial community, but microorganisms constituting extremely low proportions of a community can be overlooked by plating procedures. In cases where a microbe makes up only a small fraction of a community, growth in liquid cultures under conditions that specifically favor the growth of the target microbe over the growth of other nontarget microbes can be used to enrich for the target microbe. Such enrichment cultures are also useful for detecting GEMs.

Besides simply detecting microorganisms, plating and enrichment cultural procedures can be used to enumerate numbers of target microorganisms. The two basic approaches for the detection and enumeration of microorganisms based upon phenotypic characteristics are the viable plate count and the most-probable-number (MPN) techniques (3). Viable plating methods involve growing microorganisms on or in a solidified nutritive medium, most frequently an agar-based medium. MPN procedures usually use a liquid medium to support the growth of microorganisms (3). Often MPN procedures are enrichment cultures, designed to favor the growth of microorganisms with specific phenotypes.

Sampling and processing

Since the phenotypic detection of microbes, including GEMs, depends upon the recovery of viable microorganisms, sampling and processing procedures must be designed to ensure the maintenance of the targeted microorganism's viability (5, 34). The targeted numbers determined using an enumeration procedure should accurately reflect the numbers actually present in the environment at the time of sampling. In certain habitats, such as within the water column of some aquatic habitats, where microbial populations are free-living and relatively evenly dispersed, samples do not require extensive processing. In other habitats, where microbial populations are heterogeneously distributed and live in tight association with surfaces, such as in soils and sediments where many microorganisms form microcolonies that are adsorbed onto particles, extensive processing is needed to separate microbial cells from nonliving particles and/or the surfaces of other organisms.

Preparing soil and sediment samples for viable count procedures normally involves attempting to separate the microorganisms from the particles and establishing a homogeneous suspension by adding a diluent and agitating to distribute microorganisms evenly. Dispersants may also be added to enhance separation of the microorganisms. The efficiency of recovery of viable microorganisms is highly dependent upon the chemical composition and osmotic strength of the diluent, the mixing time, temperature, and degree of agitation used in the processing procedure (21, 29). Processes that most effectively desorb microbes from particles can cause cell lysis, and a balance between desorption and cell survival must be achieved in order to maximize the efficiency of viable cell recovery.

In many cases, serial dilutions are used to achieve appropriate concentrations of microorganisms, e.g., concentrations where individual microbial cells will be deposited at discrete locations on a solid medium. The temperature and osmotic strength of the diluent should not cause loss of cell viability. In cases where filters are used to concentrate microbial cells, consideration must be given to possible toxicity in selecting appropriate membrane filters (18, 27).

Of necessity, the recovery of viable microbes requires that the conditions used during processing be compatible with the physiological tolerance ranges of those microbes being recovered. The use of incompatible conditions during processing, such as exposure of strict anaerobes to air or psychrophiles to temperatures above 20°C, would lead to a loss of viability and an underestimation and potential failure to detect the targeted microorganisms. Clearly, sample collection and processing should be aimed at maintaining the numbers of viable targeted microorganisms actually present in the sample. There are, however, no universal sampling and processing conditions, and, therefore, sampling and processing procedures must be individually designed with regard to the characteristic properties of the habitat, the way the targeted microbial population is distributed, the relative concentration of the targeted microbes in the sample, and the physiological tolerance ranges of the targeted microorganisms.

Viable plating procedures

Plating methods are useful for the detection and enumeration of specific bacterial populations from environmental samples (9, 21, 26). Plate-count procedures are most frequently used for the enumeration of bacteria, but may also be used for other groups of microorganisms including fungi. However, the use of viable plate-count procedures for counting fungi is strongly biased (31). Spores and single-celled vegetative forms (e.g., yeast and yeastlike fungi) yield high numbers by plate-count procedures, while filamentous forms yield low numbers. Filamentous fungi also tend to overgrow agar plates, often making differentiation of single colonies impossible.

Two approaches used in plate-count procedures are the surface-spread-plate technique and the pour-plate technique (7). In the surface-spread-plate technique, the suspension of microorganisms is uniformly spread over the surface of the solidified medium. In the pour-plate technique the agar medium is melted and held at approximately 40°C and a suspension of microorganisms is added to the liquid agar medium, which then is poured into a Petri dish or other suitable container and allowed to solidify with the microorganisms dispersed throughout the agar. The pour-plate method limits exposure to oxygen and this protects some anaerobes and microaerophiles, but the targeted organisms must be able to survive exposure to 40°C for at least several minutes. The use of the surface-spread plate is simpler because the plates can be prepared in advance and also because it avoids possible death of cells due to exposure to 40°C.

The use of the agar plate count in microbial ecology has been severely criticized for its unreliability (7, 34), but this criticism is largely unjustified. The real problem with the viable-plate count lies in the misuse of the technique and/or misinterpretation of the data generated by viable-plate counts and not in the applicability of the technique for the detection and enumeration of specific target microorganisms. The viable-plate-count procedure, by its very nature, is selective and, therefore, cannot yield accurate data on total numbers of microorganisms in environmental samples. The conditions of the plate-count procedure must meet the nutritional and physiological growth requirements of the microorganisms being enumerated, but meeting the requirements of one physiological type of microorganism by necessity excludes other physiological types, e.g., using aerobic incubation conditions precludes the growth of strict anaerobes. However, once the selectivity of the plate-count procedure is recognized, this technique can be used effectively to yield valuable data on specific microorganisms (14). A wide variety of media and incubation conditions can be used for the enumeration of different microorganisms (24).

By adjusting the chemical composition of the growth medium and incubation conditions, plating procedures are designed to be differential and/or selective to detect specific microorganisms. Selective plating inhibits the growth of some microbes while permitting the growth of others, and thus allows the detection and counting of only a selected population, whereas differential plating procedures allow the growth of multiple populations, but permit differentiation of one population of interest from the others based upon the expression of a recognizable phenotype (24). Because the ability of a particular microorganism to grow on or in a particular medium is based on its ability to utilize specific organic and inorganic nutrients, a culture medium can be made selective on the basis of its constituents (e.g., the particular carbon sources, the salt concentration, etc.). Also the incubation conditions can be made selective by adjusting the physical growth conditions (e.g., temperature, oxygen concentration, etc.). The addition of inhibitory compounds to a medium suppresses the development of the majority of microorganisms, enabling the development of the desired species to occur. As examples, 0.5% (w/v) bile salts are used to inhibit nonintestinal bacteria, allowing the growth of *Escherichia coli,* and 0.03% (w/v) cetrimide is useful for the selective enrichment of *Pseudomonas aeruginosa.* Media containing antibiotics have been used extensively; for example, actidione agar is used for the isolation and enumeration of bacteria in samples containing large numbers of yeasts and fungi. Specific antibiotics are also used for detecting antibiotic-resistant microorganisms.

Because the phenotype of the microorganism depends upon the specific pathways for the metabolism of the nutrients in the medium, specific dyes, redox indicators, etc., can be included in the medium to reflect the metabolism of specific nutrients by different populations and hence, the differentiation of target microorganisms. In some cases, a medium is designed to be both selective and differential. For example, eosine methylene blue (EMB) agar is a differential and selective medium used to grow gram-negative enteric organisms, such as *E. coli* (42). The dyes in EMB (eosine and methylene blue) inhibit the growth of gram-positive bacteria, making it selective, and also act as indicators of lactose fermentation, making it differential.

The enumeration and monitoring of specific bacterial populations in environmental samples by selective plating and enrichment techniques has some limitations. Several problems with viable plating for detection of GEMs have been described by Sayler et al. (37). Aside from the problem of specificity for detecting recombinant DNA–containing organisms, the conventional plate-count procedure has serious limitations because it requires the growth of the organisms under the specific conditions used in the particular plate-count procedure and, thus, is subject to criticism for underestimating populations in samples (3). Conventional culture techniques are constrained by the lack of a universal medium that permits the growth of all potential host organisms within the sample, by the requirement for growth of target organisms, and by the fact that the fate of the recombinant genes may be independent of that of the original host organism (20). Problems encountered with plate counts include: poor or slow growth of the population(s), no growth due to the need for a cometabolic substrate or auxotrophic nutrient requirement(s), the need for another population to contribute to cross-feeding, lack of sensitivity to population(s) representing a small proportion of the total population, colony-colony inhibition, and the failure to detect poorly selectable phenotypes that are not expressed on the specific medium (37). Additionally, some microorganisms have been found to be viable, but nonculturable in environmental samples (11, 35). Plate counts underestimate or produce false-negative results if the organisms are sufficiently stressed at the immediate time of sampling and analysis.

In plating methods for the detection and/or enumeration of GEMs, cultivation media are usually formulated to take advantage of specific traits of the GEM, such as nutritional capabilities and/or resistances to specific antibiotics, so that growth of the GEM is favored over other non-recombinant strains. Generally, a visible phenotypic characteristic (e.g., pigmentation), the ability to use a specific substrate, or resistance to a certain antibiotic or heavy metal is used to differentiate the target GEM from other nontarget microorganisms. Viable plating techniques for monitoring GEMs are primarily designed for the selection of the host microorganisms containing the recombinant genes rather than the specific recombinant DNA (rDNA) sequence itself. Selection for those strains harboring the specific rDNA sequences usually is indirect, which means that selection is based upon determinants, such as antibiotic resistance associated with the host organism's chromosome, cloning vector, or subcloned genes, rather than the specific rDNA gene sequence of the GEM. As a consequence, conventional plating techniques generally are limited in terms of specificity for rDNA sequences and must be considered only to be presumptive tests, requiring confirmation of rDNA presence by other methods, such as gene probes. For these reasons, applications and limitations of conventional cultivation approaches must be examined in developing an rDNA monitoring strategy.

Despite the limitations inherent in viable plating methods, it is likely that viable culture procedures will have a necessary role in tracking GEMs because they allow for the selection of specific populations and amplify low-density populations that may exist in various environments. Both differential and selective plating techniques offer sensitivity for organisms capable of growth in the cultivation medium. Such selective plating methods were employed to track the first deliberate experimental release of a genetically engineered microbe, an ice-minus strain of *Pseudomonas syringae* (38), which could be distinguished based upon its resistance to antibiotics.

MPN method

Like the plate-count procedure, the MPN procedure has been widely used for the enumeration of specific microbial populations. The MPN technique for determining viable numbers of microorganisms employs a statistical approach in which successive dilutions are performed to reach an extinction point (1, 10, 12). In the MPN technique, replicates, usually 3 to 10, of each dilution are made and the pattern of positive and negative scores is recorded. A statistical table based on a Poisson distribution is used to determine the most probable number of viable

microorganisms present in the original sample. Different MPN procedures employ different criteria for establishing positive and negative
scores. In many procedures, a tube is scored positive when there is visible growth, i.e., an increase in turbidity. Other procedures employ
more quantitative criteria, such as an increase in protein concentration, or differential criteria, such as the production of acids from carbohydrates or the production of [^{14}C] carbon dioxide from radiolabeled
substrates (2, 4, 12).

MPN techniques have several advantages and disadvantages when
compared to viable-plate-count procedures (12). The use of a liquid
culture medium in most MPN techniques eliminates the need for a so-
lidifying agent, such as agar, and thus the problem of contamination
with organic compounds. This is of particular advantage when enumerating viable numbers of microorganisms capable of utilizing a particular organic compound as sole carbon source, e.g., in the enumeration of hydrocarbon-utilizing microorganisms. In one study, however,
where the enumeration of petroleum-degrading organisms by spread-
plate counts on oil agar was compared with an MPN method using a
liquid medium enriched with oil, it was found that the specific microorganisms detected by the different techniques were quite different
(8); *Pseudomonas* species were predominant in liquid cultures,
whereas a greater variety of genera were present on the solid media.

Despite some of the problems with the MPN method, this approach
for enumeration can be useful for the detection and enumeration of
GEMs released in the environment, particularly if a unique, selectable physiological trait is present. The relative simplicity of the MPN
method, combined with its ability to measure substrate utilization,
can make this method useful for monitoring rDNA hosts. By proper
dilutions, number of tubes inoculated, dilution series employed, composition of the medium, etc., different physiological groups and, in
some cases, species of recombinant organisms, can be detected and
enumerated.

The MPN technique is subject, however, to most of the same inherent problems as the plate-count procedure; for example, it is necessary
for the microorganisms to reproduce in order to establish viability and
be counted as positive scores, and it is impossible to establish a medium and incubation conditions which will allow for simultaneous
growth of all viable microorganisms in the sample. Microorganisms
must be separated initially into individual reproductive units in order
to satisfy the statistical assumptions used in this technique because
aggregates of microorganisms lead to an underestimation of numbers
of viable microorganisms.

In comparison to plate-count techniques, MPN techniques generally
are more laborious, employing many tubes of liquid media in order to

establish the necessary dilutions to reach a point of extinction. Large numbers of replicates are needed to increase accuracy. However, the use of microtiter plates, coupled with automated spectrophotometric readings to establish positive and negative scores, has greatly simplified the use of MPN techniques in some procedures (36). The MPN technique is less accurate than the plate-count procedure in the sense that it establishes a most probable number and confidence limit, rather than an actual number of reproductive units.

Genetically Engineered Markers

To improve the detection capabilities of viable-plating and MPN methods, it is possible to include specific genetic markers in the GEMs that can be easily detected by either viable-plating or MPN procedures in which the phenotype of the genetic marker is expressed. Modern genetic recombination techniques make the labeling of organisms in unique ways possible. As pointed out by Jain et al. (20), to be suitable for use as a marker (1) the gene should be expressed in the host; (2) the gene product should be stable in the host; (3) the gene product should be transported to its final location; and (4) the gene product should be readily detectable by a standard, reliable assay, or provide a selection technique, or both. With the development of marker traits directly linked to the rDNA sequences of interest, cultivation techniques will also have practical applications for direct rDNA monitoring. Potentially, several types of markers can be used, including chromogenic markers, markers of metabolic traits, antibiotic-resistance markers, and heavy-metal-resistance markers. Selective or enrichment media can easily be based on specific gene expression, such as growth on a medium containing ampicillin to detect organisms with an ampicillin-resistance genetic marker. The choice of one or more antibiotics or heavy metals as resistance markers upon which selection can be based should be such that the growth of susceptible organisms (non-GEMs) is suppressed, while the GEM is able to grow from a small initial population to a population size that can be readily detected. In many cases, the use of at least two markers as selective traits is necessary to ensure a high degree of specificity in the recovery process.

Pigmentation and bioluminescence

The pigmentation of bacterial colonies can be used as a means of detecting target bacteria, and it may be possible to transfer the genes for the production of characteristic pigments as markers for GEMs. The genes for the production of yellow pigments have been transferred

from *Erwinia herbicola* into *E. coli,* producing easily recognizable recombinant *E. coli* (33). The genes for the production of the red pigment
of *Serratia marcescens* and the brick-red fluorescent pigment of
Bacteroides melaninogenicus are also potential candidates for such
markers (20).

Another way to produce a visibly detectable phenotype is to use the *lux*
genes from *Vibrio fischeri,* which code for bioluminescence (25, 32).
Bioluminescence can readily be observed, making it an excellent phenotypic marker. Detection of the light produced by strains transformed
with *lux* is extremely sensitive, making this a good marker for tracking
GEMs that are deliberately released into the environment. To facilitate
cloning of *lux* genes into a target microbe, a gene fusion of *lux* genes with
bacteriophage Mu, called mini-Mu*lux,* has been made (16). Using this
mini-Mu*lux,* it is possible to insert *lux* genes at random in the host chromosome or plasmid by virtue of Mu phage transposition, which has been
demonstrated by inserting the *lux* genes into the *lac* and *area* genes of *E.
coli* (16). The strain of *E. coli* formed using mini-Mu*lux* is able to support
the functioning of the *lux* proteins and so becomes a bioluminescent
strain of *E. coli. Lux* genes have also been added to transposons that have
been used to insert these bioluminescence genes into target organisms to
create bioluminescent strains of *Agrobacterium, Pseudomonas,* and
Rhizobium (6).

The *lux* genes are particularly useful because they do not confer a
selective advantage on the host organism and do not contribute to
pathogenicity. The functioning of the *lux* system, however, requires a
large metabolic input from the host cell and may reduce the competitiveness of those GEMs containing the *lux* genes. Also, the *lux* genes
may not be expressed or the luciferase enzymes may not be functional
in many bacterial species, limiting its applicability.

Antibiotic resistance and heavy metal tolerance

The genes coding for resistance to antibiotics and heavy metal toler-
ances are especially attractive as marker genes because of the ease of
their detection. The resistance genes for mercury, for example, have
been well studied (41) and shown to be transferable (22). These genes
for heavy metal resistance are potentially useful markers because organisms containing them can be readily detected with a selective culture medium that contains the heavy metal at a concentration that
inhibits growth of other microorganisms.

Antibiotic-resistance markers, which are widely used in genetic engineering to select recombinants, also are potentially useful markers.
Many plasmid vectors that are used for delivering recombinant genes

contain antibiotic-resistance genes so that selective media containing antibiotics can be used for the selection of those organisms with the plasmids and the antibiotic-resistance genes they contain. The specificity of detection can be improved by using two antibiotics so that there is a double marker for the recombinant microorganism. Antibiotic-resistance markers can also be used to detect recombinant microorganisms in environmental samples. For example, the 2,4,5-T–degrading *Pseudomonas cepacia* AC1100 is resistant to several antibiotics and can be readily detected on a medium containing nalidixic acid; a nalidixic acid–containing medium has been used to monitor the fate of this organism in freshwater microcosms (40). Antibiotic resistance, in this case rifampicin resistance, was also used as the marker for tracking the aerial dispersal and epiphytic survival of ice-minus *Pseudomonas syringae,* the first genetically engineered bacterium deliberately released into the environment (28).

Care, however, must be given to confirm the identification of the target organism, because some naturally occurring microorganisms also contain antibiotic-resistance genes. There is also concern about increasing the environmental distribution of antibiotic-resistance genes as plasmids are frequently transferred among microorganisms, particularly Gram-negative bacteria, raising the possibility that antibiotic-resistance genes used as markers for GEMs will move into pathogens and cause public health problems. Furthermore, adding an antibiotic-resistance marker can, in some cases, alter the competitiveness of a GEM in the environment (13, 19, 39). Compeau et al. (13), for example, found that while many pseudomonads with an added rifampicin-resistance gene were equally competitive in soil systems with those lacking the rifampicin-resistance gene, in some cases, rifampicin-resistant pseudomonads had much shorter survival times in soil than comparable nonrifampicin-resistant pseudomonads; they concluded that antibiotic-resistance-marked strains should be compared with wild-type parental strains before being used as markers for environmental monitoring studies. In another study on the survival of *E. coli* with and without a transposon containing a kanamycin-resistance marker in aerosols and in a farm environment, Marshall et al. (30) found enhanced survival for the antibiotic-resistant strains in a few cases, again suggesting the need for caution in using markers that may alter the relative environmental survival times of GEMs.

Substrate utilization

A relatively easy method for detecting a GEM that has been created with a unique metabolic activity is to rely upon the observed phenotypic expression of that activity. For example, the genetically engineered bacterium *P. cepacia* AC1100 is able to metabolize the pesticide 2,4,5-T, a metabolic activity not commonly found in naturally occurring microorganisms. By using a medium with 2,4,5-T as the carbon source, it is possible to detect 2,4,5-T–degrading bacteria and thus to track *P. cepacia* AC1100 with a relatively high degree of specificity (40). The method is not, however, without its problems. In the case of *P. cepacia* AC1100, growth on a 2,4,5-T medium is slow and the organism sometimes fails to grow, making this method for enumeration unreliable. Also, any naturally occurring 2,4,5-T degraders would be mistaken for the target organism unless additional confirmatory tests were performed to verify the identity of organisms growing on the 2,4,5-T medium. Furthermore, contaminants in the medium could permit the growth of nontarget and non-2,4,5-T–degrading microorganisms. Studies on the detection of hydrocarbon-utilizing microorganisms have repeatedly shown that the simple observation of growth is inadequate to establish the ability of a microorganism to metabolize a substrate in a medium and that additional confirmatory tests, such as chemical analyses, are necessary to confirm a microorganism's metabolic capabilities.

In some cases, a distinctive metabolic product is formed that can readily be observed and used to confirm the metabolic capabilities of the observed organism. For example, Ensley et al. (17) were able to detect recombinant *E. coli* containing the genes from *Pseudomonas putida* that code for the conversion of naphthalene to salicylic acid because the recombinant *E. coli* was able to grow on naphthalene as the sole carbon source and to excrete indigo as a metabolic product; indigo has a distinctive blue color and the observation of blue colonies growing on a medium with naphthalene as the carbon source is a direct method of detecting the recombinant *E. coli*. This approach, however, cannot be generalized and is limited to certain engineered organisms whose metabolisms produce readily detectable products that are uniquely associated with recombinant genes.

A more general approach is to include a second marker gene that is not specific to the primary recombinant genes, but which can easily be included in the rDNA and the expression of which can easily be detected. An excellent example of such genetic engineering for the creation of a recognizable marker based upon substrate utilization was the introduction of the *lac*Z and *lac*Y genes of *E. coli* (encoding β-galactosidase and lactose permease, respectively) into a *Pseudomonas fluorescens* strain, thereby creating a phenotypically Lac$^+$ *Pseudomonas* (15). In this case, the genetically engineered pseudomonad strain can be distinguished from other pseudomonads by its ability to grow on lactose-minimal agar. Alternatively, primary selection can be done on a medium which contains the chromogenic substrate 5-bromo-

4-chloro-3-indolyl-β-D-galactoside (X-gal). The recombinant pseudo-monad hydrolyzes X-gal, producing blue colonies, whereas natural pseudomonads produce white colonies, providing a quick and easy means of detecting the recombinant pseudomonads. A small-scale field test was conducted to examine the sensitivity and reliability of using the *lacZY* genes to track GEMs in soil has been conducted, successfully demonstrating the applicability of this approach (23).

References

1. Alexander, M. 1965. Most-probable-number method for microbial populations. In C. A. Black (ed.), *Methods of Soil Analysis. Part 2, Chemical and Microbiological Properties.* American Society of Agronomy, Madison, Wis., pp. 1467–1472.
2. Alexander, M. 1965. Nitrifying bacteria. In C. A. Black (ed.), *Methods of Soil Analysis. Part 2, Chemical and Microbiological Properties.* American Society of Agronomy, Madison, Wis., pp. 1484–1486.
3. Atlas, R. M. 1982. Enumeration and estimation of biomass of microbial components in the biosphere. In R. G. Burns and J. H. Slater (eds.), *Experimental Microbial Ecology.* Blackwell, Oxford, pp. 84–102.
4. Atlas, R. M. 1979. Measurement of hydrocarbon biodegradation potentials and enumeration of hydrocarbon-utilizing microorganisms using carbon-14 hydrocarbon-spiked crude oil. In J. W. Costerton and R. R. Colwell (eds.), *Native Aquatic Bacteria: Enumeration, Activity, and Ecology.* ASTM Special Technical Publication No. 695, American Society for Testing and Materials, Philadelphia, pp. 196–204.
5. Board, R. G., and D. W. Lovelock. 1973. *Sampling—Microbiological Monitoring of Environments.* Academic Press, London.
6. Boivin, R., F. P. Chalifour, and P. Dion. 1988. Construction of a Tn5 derivative encoding bioluminescence and its introduction in *Pseudomonas, Agrobacterium,* and *Rhizobium. Molecular and General Genetics* 213:50–55.
7. Buck, J. D. 1979. The plate count in aquatic microbiology. In J. W. Costerton and R. R. Colwell (eds.), *Native Aquatic Bacteria: Enumeration, Activity and Ecology.* ASTM Special Technical Publication No. 695, American Society for Testing and Materials, Philadelphia, pp. 19–28.
8. Calomiris, J. J., B. Austin, J. D. Walker, and R. R. Colwell. 1977. Enrichment for estuarine petroleum-degrading bacteria using liquid and solid media. *Journal of Applied Bacteriology* 42:135.
9. Clark, F. E. 1965. Agar-plate method for total microbial count. In C. A. Black (ed.), *Methods of Soil Analysis. Part 2, Chemical and Microbiological Properties.* American Society of Agronomy, Madison, Wis., pp. 1460–1466.
10. Cochran, W. G. 1950. Estimation of bacterial densities by means of the "most probable number." *Biometrics* 2:105–116.
11. Colwell, R. R., P. R. Brayton, D. J. Grimes, D. B. Roszak, S. A. Huq, and L. M. Palmer. 1985. Viable but nonculturable *Vibrio cholerae* and related pathogens in the environment: Implications for release of genetically engineered microorganisms. *Bio/Technology* 3:817–820.
12. Colwell, R. R. 1979. Enumeration of specific populations by the most-probable-number (MPN) method. In J. W. Costerton and R. R. Colwell (eds.), *Native Aquatic Bacteria: Enumeration, Activity and Ecology.* ASTM Special Technical Publication No. 695, American Society for Testing and Materials, Philadelphia, pp. 56–64.
13. Compeau, G. B., J. Al-Achl, E. Platsouka, and S. B. Levy. 1988. Survival of rifampin-resistant mutants of *Pseudomonas fluorescens* and *Pseudomonas putida* in soil systems. *Applied Environmental Microbiology* 54:2432–2438.
14. Costerton, J. W., and G. G. Geesey. 1979. Which populations of aquatic bacteria should we enumerate? In J. W. Costerton and R. R. Colwell (eds.), *Native Aquatic Bacteria: Enumeration, Activity and Ecology.* ASTM Special Technical Publication No. 695, American Society for Testing and Materials, Philadelphia, pp. 7–18.

15. Drahos, D. J., B. C. Hemming, and S. McPherson. 1986. Tracking recombinant organisms in the environment: β-galactosidase as a selectable non-antibiotic marker for fluorescent pseudomonads. *Bio/Technology* 4:439–444.
16. Engebrecht, J., M. Simon, and M. Silverman. 1985. Measuring gene expression with light. *Science* 227:1345.
17. Ensley, B. D., B. J. Ratzkin, T. D. Offlund, M. J. Simon, L. P. Wackett, and D. T. Gibson. 1983. Expression of naphthalene oxidation genes in *Escherichia coli* results in the biosynthesis of indigo. *Science* 222:167–169.
18. Green, B. L., E. Clausen, and W. Litsky. 1975. Comparison of the new Millipore HC with conventional membrane filters for the enumeration of faecal coliform bacteria. *Applied Microbiology* 30:697–699.
19. Hagedorn, C. 1979. Relationship of antibiotic resistance to effectiveness in *Rhizobium trifolii* populations. *Soil Science Society of America Journal* 43:921–925.
20. Jain, R. K., R. S. Burlage, and G. S. Sayler. 1988. Methods for detecting recombinant DNA in the environment. *CRC Critical Reviews in Biotechnology* 8:33–84.
21. Jensen, V. 1968. The plate count technique. In T. R. G. Gray and D. Parkinson (eds.), *The Ecology of Soil Bacteria.* Liverpool University Press, Liverpool, pp. 158–170.
22. Kelly, W. J., and D. C. Reanney. 1984. Mercury resistance among soil bacteria: Ecology and transferability of genes encoding resistance. *Soil Biology and Biochemistry* 16:1–8.
23. Kluepel, D. A., E. L. Kline, J. Mueller, D. J. Drahos, G. Barry, B. C. Hemming. 1990. Evaluation of the risks associated with the release of soil-borne genetically engineered bacteria into the environment. *Abstract Agro 155.* April 22–27 National Meeting of the American Chemical Society, Boston, Mass.
24. Krieg, N. R. 1981. Enrichment and Isolation. In P. Gerhardt (ed.), *Manual of Methods for General Bacteriology.* American Society for Microbiology, Washington, D.C., pp. 112–142.
25. Legocki, R. P., M. Legocki, T. O. Baldwin, and A. A. Szalay. 1986. Bioluminescence in soybean root nodules demonstration of a general approach to assay gene expression in-vivo by using bacterial luciferase. *Proceedings of the National Academy of Sciences USA* 83:9080–9084.
26. Levin, M. A., R. Seidler, A. W. Bourquin, J. W. Fowle III, and T. Barkay. 1986. EPA developing methods to assess environmental release. *Bio/Technology* 5:38–45.
27. Lin, S. D. 1976. Evaluation of Millipore HA and HC membrane filters for the enumeration of indicator bacteria. *Applied Environmental Microbiology* 32:300–302.
28. Lindow, S. E., G. R. Knudsen, R. J. Seidler, M. V. Walter, V. W. Lambou, P. S. Amy, D. Schmedding, V. Prince, and S. Hern. 1988. Aerial dispersal and epiphytic survival of *Pseudomonas syringae* during a pretest for the release of genetically engineered strains into the environment. *Applied Environmental Microbiology* 54:1557–1563.
29. Litchfield, C. D., J. B. Raker, J. Zindulis, R. T. Watanabe, and D. J. Stein. 1975. Optimization of procedures for the recovery of heterotrophic bacteria from marine sediments. *Microbial Ecology* 1:219–233.
30. Marshall, B., P. Flynn, D. Kamely, and S. B. Levy. 1988. Survival of *Escherichia coli* with and without ColE1::Tn5 after aerosol dispersal in a laboratory and a farm environment. *Applied Environmental Microbiology* 54:1776–1783.
31. Menzies, J. D. 1965. Fungi. In C. A. Black (ed.), *Methods of Soil Analysis. Part 2, Chemical and Microbiological Properties.* American Society of Agronomy, Madison, Wis., pp. 1502–1505.
32. O'Kane, D. J., W. L. Lingle, J. E. Wampler, M. Legocki, R. P. Legocki, and A. A. Szalay. 1988. Visualization of bioluminescence as a marker of gene expression in *Rhizobium* infected soybean root nodules. *Plant Molecular Biology* 10:387–400.
33. Perry, K. L., T. A. Simonitch, K. J. Harrison-Lavoie, and S. Liu. 1986. Cloning and regulation of *Erwinia herbicola* pigment genes. *Journal of Bacteriology* 168;607.
34. Postgate, J. R. 1969. Viable counts and viability. In J. R. Norris and D. W. Robbons (eds.), *Methods in Microbiology,* vol. 1. Academic Press, London, pp. 611–628.

35. Roszak, D. B., and R. R. Colwell. 1987. Survival strategies of bacteria in the natural environment. *Microbiology Reviews* 51:365–379.
36. Rowe, R., R. Todd, and J. Waide. 1977. Microtechnique for most-probable-number analysis. *Applied Environmental Microbiology* 33:675–680.
37. Sayler, G. S., M. S. Shields, A. Breen, E. T. Tedford, S. Hooper, K. M. Sirotkin, and J. W. Davis. 1985. Application of DNA:DNA colony hybridization to the detection of catabolic genotypes in environmental samples. *Applied Environmental Microbiology* 49:1295–1303.
38. Seidler, R. J., and S. Hern. 1988. Special report: The release of ice minus recombinant bacteria at California test sites. EPA Environmental Research Laboratory, Corvallis, Oregon.
39. Smith, M. A., and J. M. Tiedje. 1980. Growth and survival of antibiotic-resistant denitrifier strains in soil. *Canadian Journal of Microbiology* 26:854–856.
40. Steffan, R. J., A. Breen, R. M. Atlas, and G. S. Sayler. 1989. Monitoring genetically engineered microorganisms in freshwater microcosms. *Journal of Industrial Microbiology* 4:441–446.
41. Summers, A. O. 1986. Organization, expression, and evolution of genes for mercury resistance. *Annual Reviews in Microbiology* 40:607–634.
42. Wolf, H. W. 1972. The coliform count. In R. Mitchell (ed.), *Water Pollution Microbiology*. Wiley, New York, pp. 333–345.

Chapter 3

Immunological Techniques for Detection, Identification, and Enumeration of Microorganisms in the Environment

Sara F. Wright

Introduction

Biological molecules which recognize microorganisms can be generated by vertebrate antibody-forming systems. Animals immunized
with a microorganism, a complex structure composed of many antigens, will produce antibody-secreting cell lines. Each antibody recognizes a small portion, approximately six amino acids or six mono-
saccharide units (31), of an antigen complex. Antiserum obtained from
an immunized animal contains antibodies against several to many different microbial antigens and is considered to be a polyclonal antibody
probe. Monoclonal antibody probes result from the fusion of an individual antibody-secreting cell and a malignant cell to produce an im-
mortal hybrid cell which secretes a single type of specific antibody.

Polyclonal antisera have been used for identification and tracing of
microorganisms for the past three decades. They are considered to be
stable probes because they consist of mixtures of antibodies which recognize different antigens. Loss of expression of a single antigen generally does not result in failure of the polyclonal antiserum to react

with a microorganism. However, such antisera may be cross-reactive with related organisms, and this presents a problem in the use of these probes to identify specific strains of an organism.

Recent advances in production technology have made monoclonal antibody probes options to consider. Antibodies which react with antigens specific for a microbial strain can be selected and used sequentially or mixed together to produce a defined polyclonal antiserum.

Perspectives on choosing monoclonal or polyclonal antibody probes and limitations of both types of antisera will be presented in this chapter. Production schemes for both types of antisera will be described briefly with references to published methods. Also, uses and limitations of selected assays to detect antigen-antibody reactions appropriate for tracing microorganisms will be presented.

Immunoglobulins

Antigenic stimulation of antibodies

The vertebrate lymphocyte population recognizes up to millions of amino acid or polysaccharide antigenic determinants. This inherent ability to recognize so many different antigens, by a mechanism first proposed by Burnet (5), is because a population of lymphocytes (B cells) with set recognition sites is always present. A specific B cell is stimulated by a complementary antigenic determinant to produce antibodies. Clonal expansion of an antibody-producing cell line results in large amounts of antibody being released in the peripheral blood.

Some regions of antigens are more immunogenic than others. Immunodominance of antigenic determinants is at least partly dependent on the animal species and the individual animal. However, as a general rule, antigenic determinants which occur as repeating identical units or are highly aggregated favor strong antibody responses (20).

Immunoglobulin molecules

Antibodies are proteins (immunoglobulins) which possess variable and constant amino acid sequence regions. The antigen-binding specificity of the molecule resides in the identical variable regions which are in the two prongs of a molecule generally represented in the shape of a "Y." The basic unit of an antibody is a dimer containing four polypeptides—two light and two heavy chains of amino acids (Fig. 3.1). Distinctly different kinds of constant regions are found in the heavy chains of immunoglobulins, thus differentiating the molecules into classes of antibodies. Immunoglobulin G (IgG) and immunoglobulin M (IgM) are the classes most commonly resulting from im-

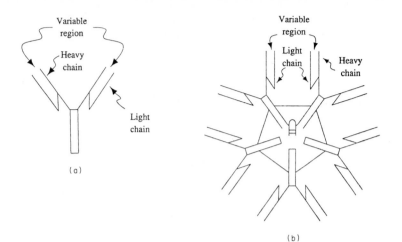

Figure 3.1   The basic unit of an antibody molecule showing heavy and light chains of amino acids and the variable region that is complementary to the antigenic determinant recognized by the antibody. IgG (*a*) and IgM (*b*) molecules are illustrated.

munization with bacteria. The first antibody to be produced in the primary response to an antigen is IgM. After immunization and booster injections there is a switch to antibodies which express the γ chain (IgG) constant region rather than the μ chain constant region (IgM).

Affinity of antibodies to antigens

The capacity of an antibody to bind to an antigen is called affinity. It is the summation of attractive and repulsive noncovalent intermolecular forces resulting from the interaction of the antigen and antibody binding sites. Early-response antibodies often have lower affinities for antigens than do antibodies which develop later. However, the early-response antibodies are more specific, i.e., they show less cross-reaction with related or unrelated molecules than do later and higher-affinity antibodies (39).

Microbial Antigens

Microbes in the environment

The general uses of antibodies against microbes from environmental samples have been for identification or enumeration of an organism which (1) is a member of a population which can be selectively cultured and then confirmed as a subset of the population, (2) is studied in situ and must be identified as cells in a mixture of organisms, or (3) cannot be selectively cultured from environmental samples and must be identified as individual cells.

Microbial antigens have different ranges of immunological specificity, i.e., type-specific antigens which occur among one, two, or three strains of a species and group or species-specific antigens which are shared by larger numbers of related microorganisms. The use of antibodies as probes to trace microorganisms in the environment is based upon the ability to detect an immunologically distinct marker. If a genetically modified member of the naturalized population is to be added back to an environment, it is unlikely that antibody probes will differentiate such an organism from the naturalized strain unless the modification also causes expression of a distinctly different and detectable antigen.

Electrophoresis

Gel electrophoresis profiles of whole-cell extracts or lipopolysaccharide (LPS) from cultured isolates of the naturalized flora may be useful as indicators of the potential immunological diversity of a population. Separation of proteins according to size is routinely accomplished by sodium dodecylsulfate polyacrylamide gel electrophoresis (SDS-PAGE). Though not as definitive as protein separation by size, LPS fragments will also separate in SDS-PAGE gels. Gradient or nongradient SDS-PAGE profiles of *Rhizobium* isolates have been studied in relation to immunological differences among populations or selected strains (13–15, 29, 30, 67).

Extracted LPS subjected to SDS-PAGE may be used to indicate the potential to obtain specific antibodies for identification of microbes. LPSs from *Rhizobium, Salmonella, Shigella, E. coli,* and *Pseudomonas* have been studied by SDS-PAGE (7, 10, 12, 21, 24, 47, 71). Carlin and Lindberg (7) produced monoclonal antibodies against *Shigella flexneri* O-antigenic polysaccharides which demonstrated superior specificity when compared with rabbit polyclonal antiserum.

Heterogeneity of rhizobial LPS has been demonstrated by SDS-PAGE (8), and LPS from wild and transposon mutants of rhizobia also have been compared (11). Other analyses of LPS (8, 9) may be helpful in determining potential antigenic differences among isolates.

Protein profiles are routinely stained by Coomassie stain. However, silver staining SDS-PAGE profiles of whole cell extracts may provide greater sensitivity than Coomassie stain (42). Also, a modified silver stain is used to reveal LPS in SDS-PAGE preparations (59).

Protein bands from electrophoresis profiles can be excised and used as the antigen to produce monoclonal or polyclonal antibodies (17, 20). LPS fractions separated by gel filtration (9) or electrophoresis might be used as a source of antigen.

Monoclonal versus Polyclonal Antibodies

After the decision to use antibodies as probes has been made, one must make a decision about the use of polyclonal or monoclonal antibodies. There are several areas to consider carefully before choosing one of the two different tools. Figure 3.2 contrasts the preparation of monoclonal and polyclonal antisera and the products obtained from each procedure. Differences between the two methods are presented in

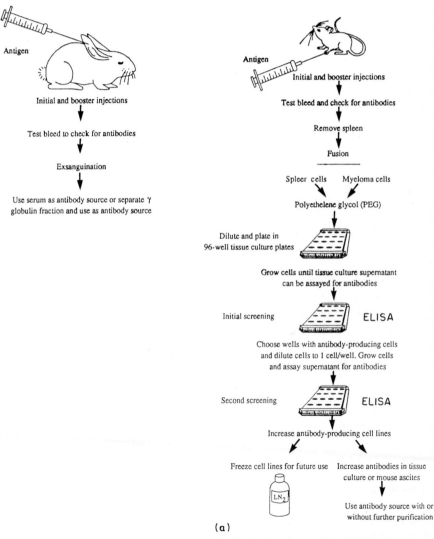

(a)

Figure 3.2   The (a) process and (b) products of polyclonal and monoclonal antibody production are compared.

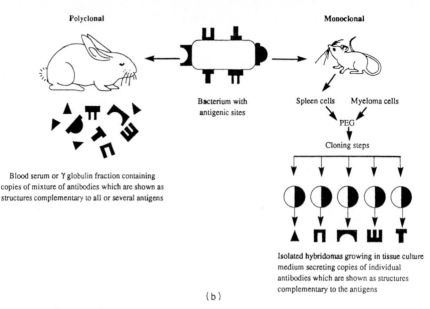

Figure 3.2   (*Continued*)

Table 3.1. Consultation with an immunologist or an antibody-production facility may be helpful in making a decision.

Animal responses

The response of animals to immunization varies and can affect the ability of the investigator to raise useful antisera. However, particular techniques of immunization may overcome this problem. Background information on aspects of antiserum production such as the form of the antigen to be injected, the animal to use, route of immunization, injection schedule, and manner of bleeding are covered in many references (6, 35, 55, 56, 63, 66). Preparation of antigens specifically for monoclonal antibody production against pili, toxin, O-antigen, and other polysaccharide antigens have been described (37, 38).

For routine polyclonal antibody production, most investigators choose New Zealand white rabbits. A suggested protocol for injection, test bleeding, and exsanguination of rabbits is described by Warr (63).

Monoclonal antibody production requires the use of a mouse or rat strain for which a myeloma cell (malignant or fusion cell) line is available. The BALB/c mouse strain is used most commonly. Detailed descriptions and illustrations of immunization procedures are given by Zola (72). As with polyclonal antibody production in rabbits, the best immunization protocol may be obtained by a trial-and-error process,

TABLE 3.1 A Comparison of Some Characteristics and Aspects of Production to Consider When Choosing Polyclonal or Monoclonal Antibodies

Characteristics of antibodies and aspects of production to consider	Polyclonal antibodies	Monoclonal antibodies
Specificity	Variety of antibodies directed against an antigen as well as antibodies which do not react with the antigen of interest. Multiple antibodies against the antigen of interest.	An antibody of single specificity and immunoglobulin class.
Reactivity	Generally can be detected by all assays.	May not precipitate the antigen unless the determinant is present in multiple copies.
Sensitivity	Varies.	Varies.
Affinity	Range of affinities.	Varies. May be very high. If it is low, this may present difficulties with assays.
Batch reaction variation	Reaction may vary from animal to animal.	No batch variation. Virtually immortal product available in large amounts. A standard reagent.
Time for production	May be as short as 4 to 6 weeks if the animal is highly responsive to the antigen.	May require 2 months to obtain a product after the fusion.
Cost	Low.	High.
Skill required for production	Moderate.	High.
Early or late response antibodies to antigen	Generally late.	Early or late.

although recommendations by Zola (72) and Goding (20) are helpful. The point of divergence in methodology for polyclonal and monoclonal antibody production is after immunization. Exposure to the antigen stimulates antibody-producing cells and antibodies are detectable in peripheral blood. For polyclonal antibodies, the investigator obtains blood serum and uses it with or without further modification to detect microbial antigens. For monoclonal antibody production, the investigator uses the antigen-stimulated lymphocytes which are producing antibodies to an immunogen (Fig. 3.2).

Monoclonal Antibody Production

Production of monoclonal antibodies was first demonstrated by Kohler and Milstein (33). Detailed discussions of the background and theory are given in several texts on methodology (20, 25, 72). A brief discussion will be presented here.

After immunization of a mouse with an antigen or antigen complex, the animal's spleen is removed as the source of antibody-producing cells. A suspension of splenic lymphocytes is fused with myeloma tumor cells by the action of polyethylene glycol (PEG) on cell membranes. Tissue culture medium supplemented with drugs ensures that only hybrids between the tumor cell and the normal cells grow. Fused cells are diluted and grown in 96-well tissue culture plates. Microscopic plaques of dividing cells can be seen on the bottom of the wells 4 to 5 days postfusion. After 10 to 15 days of growth, culture supernatants are tested for the presence of the desired antibody. Selected cell-line clones are isolated by limiting dilution. These clones may be injected into the peritoneum of a mouse to produce antibody-rich ascites or the cell lines may be grown in tissue culture medium and the supernatant used as a source of antibodies. Cell lines may be frozen in liquid nitrogen and kept indefinitely.

Fusion of an antibody-producing cell with a myeloma cell is a random event. The probability of obtaining an antibody to a specific antigen is related to the number of antigen-stimulated lymphocytes, the number of hybridomas resulting from the fusion procedure, and the genetic stability of the hybridoma.

Initiation of monoclonal antibody production presents the investigator with what seem to be interrelated compound problems. Decisions about the immunogen and preparation of the antigen to screen antibodies may seem to be somewhat perplexing, and at least a few trial-and-error attempts at production or screening may be required.

The first consideration in production of monoclonal antibodies is the assay to detect the recognition of antigen and antibody. The assay should accommodate rapid screening of several thousand samples, depending on the number of different antigens to be assayed and the number of microtiter wells containing growing hybrid cells. The indirect enzyme-linked immunosorbent assay (ELISA) (20, 60, 72) commonly is used to detect IgG or IgM antibodies against microbial antigens. It is important to have a dependable ELISA before the production process is initiated to ensure detection of appropriate antibodies.

ELISA

Some considerations to address in the use of ELISA during production of monoclonal antibodies are given below. A thorough presentation of the practice and theory of enzyme immunoassays is given by Tijssen (58).

Antigen. Sufficient antigen to carry through the entire production protocol in a timely manner should be on hand or readily generated. Careful consideration should be given to the choice of antigens for the initial screening to get the most information possible on selective or cross-reactive binding to a multiantigenic immunogen. There is usually sufficient supernatant for ELISA against three antigens at the initial screening (Fig. 3.2). To separate a choice antibody-producing cell line from nonproducing cells growing in the same well, cloning should be initiated as soon as possible.

ELISA reagents. Reagents are described in detail by Tijssen (58). However, if ELISA is being initiated in a laboratory, it is suggested that a reagent kit be purchased to standardize the test and to detect antibodies in the test-bleeds during the immunization procedure. Such kits are available from companies specializing in supplies for biotechnology and immunology.

Polyvalent enzyme-linked antimouse antiserum (containing anti-IgG and -IgM) is useful to screen for IgG or IgM antibodies if there is no preference for a specific class, only a specific reaction. Such products are available from biochemical supply companies.

Concentration of antigen and attachment to ELISA plates. It is difficult to give a recommendation on antigen concentration to use for initial screening. If a multiantigenic organism was used as the immunogen, the concentration of a single antigen may be only a small fraction of the total protein or polysaccharide content of the organism. In general, highly concentrated particulate antigens may wash off plates during the ELISA procedure and should be avoided. Bacteria suspended in phosphate-buffered saline (PBS) adjusted to a concentration $O.D._{450} = 0.010$ at 50 μl per well for rhizobia or the particulate and soluble contents of five vesicular-arbuscular mycorrhizal spores per well have been used (67, 68). Titration of antigen against test-bleed antiserum is helpful to indicate an antigen concentration for screening.

Indirect ELISA usually is performed using 96-well polyvinyl chloride or polystyrene plates. There are limitations to the use of plastic (58), but convenience, cost, and generally acceptable results are important factors to consider.

Recommendations for binding of antigen to plates are incubation overnight at 4°C, 1 hour at room temperature, or until dry at 37°C. The nature of the antigen should dictate the procedure for attachment, but often this is not known at the onset of antibody production when a multiantigenic cell is used to elicit antibodies. Covalent attachment of antigens which adsorb poorly to plastic is discussed by Tijssen (58) and Goding (20). LPS from rough strains of bacteria may require magnesium to facilitate binding to a plastic solid phase for ELISA (27).

Positive and negative antibody control wells should be included on each plate, but these become a part of the circular problem encountered if there is no supply of antibody against the injected antigen. Test-bleed antiserum which gives a positive reaction can be used until antibody-producing cell lines are available.

Colorimeter for reading ELISA plates.   Many different models and prices of ELISA plate readers are available. These are useful for quantification of the intensity of reactions but are not essential for production of monoclonal antibodies. Qualitative notation of positive reactions is often all that is necessary.

Fusion procedure

Protocols for fusion of antibody-producing cells and myeloma cells are abundant (20, 25, 43, 72). Several of these references (20, 24, 72) give comprehensive descriptions of all phases of monoclonal antibody production.

Myeloma cell lines for fusion are available from the American Type Culture Collection or may be obtained from other investigators who are producing monoclonal antibodies. Commonly used cell lines are P3-NS1/1-Ag4-1 (NS-1) or X63-Ag8.653 (NP-3).

Successful fusions are highly dependent on the use of proper media and chemicals. Many biochemical suppliers now carry lines of hybridoma-tested tissue culture products and chemicals, and it is recommended that those initiating monoclonal antibody production use these.

Intrasplenic immunization

A small amount of immunogen injected directly into the spleen of a mouse or a rat will elicit antibodies for monoclonal antibody production (54). As little as 20 μg of protein or 2.5×10^5 cells have been found to be sufficient to elicit a high proportion of hybridomas secreting antibodies against cell surface antigens and soluble proteins. Both IgM and IgG classes of immunoglobulins are obtained by using this procedure. This procedure is an option to consider when the amount of antigen is limited.

Commercial or university biotechnology centers as producers of antibodies

Many universities and several commercial companies will produce monoclonal or polyclonal antibodies for a user, or they will produce antibodies from hybridomas provided by the user. These laboratories will usually allow the purchaser to define the desired refinement of a product in order to provide a range of prices. The purchaser of such services and products should understand the production schemes for antibody probes, particularly the screening for monoclonal antibodies when a multiantigenic organism is used as the immunogen.

Antibody Standardization and Purification

Blood serum containing polyclonal antibodies from rabbits or large-scale tissue culture supernatant containing monoclonal antibodies (20) may be used without further purification to detect antigens by agglutination, immunodiffusion, or ELISA. Appropriate dilutions of the antiserum to be used for an assay are determined by titration using the homologous antigen as the target. For ELISA a box titration may be performed by diluting the antigen across the plate in each row and the antibody down the plate in each column. The lowest concentration of antigen giving a strongly positive reaction with the highest dilution of antiserum indicates the sensitivity of the test and the dilution of antibody to use as a working solution.

Adsorbed polyclonal antiserum

Adsorption with cross-reactive antigens to enhance the specificity of polyclonal antiserum is common (64). A polyclonal antiserum is incubated with antigens which are cross-reactive with the antiserum. Incubation is usually for 30 minutes on ice, and then the serum is separated from particulate adsorbing material by centrifugation. Soluble antigens must be coupled to an insoluble matrix and purification carried out by affinity chromatography (64). The ratio of serum to adsorbent must be determined for each case, and serum activity against the adsorbing antigen should be monitored until the undesired reactivity has been removed. Adsorbed serum should always be checked carefully for specificity using the same assay for which it will ultimately be employed. Adsorption sometimes yields highly specific antiserum which rivals that of a monoclonal antibody–producing cell line. However, adsorbed antiserum also can have a very low reactivity with a specific antigen.

Purification of antibodies

Purification of immunoglobulins preparatory to conjugation with dyes or enzymes can be accomplished by ammonium sulfate precipitation, gel filtration, ion exchange chromatography, or protein-A affinity chromatography. Choice of method may be based on quantity of the source of the antibody, cost, or immunoglobulin class. Texts on antibody production present purification protocols and precautions for use of specific procedures (20, 25, 64).

Testing whether monoclonal antibodies recognize the same antigenic site

It is highly desirable to use several monoclonal antibodies, all having different specificities for strain-specific immunodeterminants, to identify a microorganism. Western-blot analysis of antibodies may be used to probe electrophoretically separated cellular components with antibodies (20) to determine whether antibodies recognize different bands, and therefore presumably different antigenic sites. If the same band is recognized by different antibodies, it is necessary to determine whether these antibodies recognize the same or distinct regions of the macromolecular antigen. Analysis of antibody specificity is based on the principle that antibodies specific for one antigenic determinant will compete for the same binding site. Most assays to test for competition are based on the use of a labeled antibody competing with an unlabeled antibody (28, 61). Friguet et al. (18) present a convenient ELISA assay based on additivity of enzymatic activity observed when two monoclonal antibodies bind to different antigenic determinants.

Selected Assays

Immunofluorescence

Direct or indirect fluorescent antibody techniques are adaptable procedures for tracing and enumerating microbes in soil or the environment. Fluorescent dye–labeled immunoglobulin may be used to identify cultured organisms, organisms in situ, or populations of organisms in soil or water. In the indirect technique reaction with unlabeled antiserum is followed by labeled globulin specific for the animal species used to produce the first antiserum. The indirect method must be carefully controlled to assure that detection of the antigen is specific.

Labeling immunoglobulin with fluorescent dyes and use of such products by direct immunofluorescence to detect microorganisms on buried slides, on root surfaces, in nodules, or in soil are given by Bohlool and Schmidt (3) and Bohlool (2). This method has also been used to trace bacteria in water (22). Problems and limitations of the immunofluorescence approach to microbial ecology are presented by Bohlool and Schmidt (3). General discussions on immunofluorescence are given in several texts (20, 34).

Detection of viable but nonculturable *Escherichia coli* and *Vibrio cholerae* in microcosms by using immunofluorescence (70) demonstrates the utility of this technique in direct detection of microorganisms in aquatic and marine ecosystems. The nonculturable stage, which many microorganisms undergo when placed in stressful environments (49), can complicate studies in which microbial ecologists are attempting to enumerate specific populations in aquatic and marine environments. Immunofluorescence microscopy can be used to enumerate these populations without the need for culturing the samples. Those organisms which can be cultured from the environment, but only with difficulty, can be detected and enumerated more easily using immunofluorescence microscopy, as in the case of the *Bacteroides fragilis* group of anaerobic bacteria (16).

Distribution of *Vibrio cholerae* in the natural environment of a cholera-endemic region has been studied using immunofluorescence and employing a highly specific monoclonal antibody (4). On the basis of O antigens, *V. cholerae* can be subdivided into serovars, and 83 O serogroups have been recognized with the cholera vibrio being assigned to O1 (50). Two antigenic forms of *V. cholerae,* Ogawa and Inaba, with the antigenic formula AB(C) and AC, respectively, are recognized. Shamida and Sakazaki (53) reported a serogroup on non-O1 *V. cholerae,* Hakata, which possess the C (Inaba) factor but not B (Ogawa) and A factors of *V. cholerae* O1. Shamida et al. (52) reported a group of marine vibrios that possess the B and C factors of *V.*

cholerae O1 as well as their own major antigen factor, but the A factor has not, so far, been found in any vibrios other than *V. cholerae* O1. Thus the A-subunit-specific monoclonal antibody used by Brayton and Colwell (4) allows detection and enumeration of this organism in environmental samples when cultural methods are ineffective. The same monoclonal has been used to investigate the association of *V. cholerae* O1 with phytoplankton (57) and to understand the relationship between phytoplankton blooms and numbers of *V. cholerae* in estuarine waters of regions where cholera is endemic.

Indirect immunofluorescence microscopy has also been used to detect and enumerate *Campylobacter pylori* in water, including the viable but nonculturable stage, by employing a monoclonal antibody (M. Shahamat, R. R. Colwell, C. Paszko-Kolva, and A. Pearson. 1989. Production and characterization of monoclonal antibodies to *Campylobacter pylori*. In *Abstracts of the Annual Meeting of the American Society for Microbiology,* Session V24. American Society for Microbiology, Washington, D.C., p. 490.). *Legionella pneumophila* also enters a viable but nonculturable stage and can be detected by using direct immunofluorescence microscopy (26). The eukaryote *Giardia* has been detected in aquatic systems by using an immunofluorescence technique developed by Sauch (51).

ELISA

Enumeration of rhizobia by ELISA in peat and in soil has been reported (40, 48). Direct ELISA was useful for enumerating populations in peat, and the indirect ELISA, using the antibody to capture the antigen, was used to quantify *Rhizobium* populations in soil when the population exceeded 1×10^3 cells per gram of soil (40). Fluorescent ELISA results compare with those from the antibiotic-resistance technique when soil populations exceeded 10^4 cells per milliliter (48). ELISA also is used to identify rhizobial strains in root nodule material (19, 67).

ELISA has been used to detect and enumerate viral pathogens in water and wastewater. Nasser and Metcalf have developed an amplified ELISA for the detection of hepatitis A virus in estuarine samples (41), and Guttman-Bass et al. compared several ELISA techniques for detecting rotaviruses in water (23).

Precipitation and agglutination tests

These tests generally are used to identify organisms immunologically which can be obtained in pure culture. Principles of these reactions are discussed in several texts (35, 45, 61, 65). The ring precipitation test detects soluble antigens by a reaction which results in visible aggregation of antigen and antibody in a solution. This test has, for the most part, been replaced by immune diffusion.

Immune diffusion tests detect precipitation reactions between antigens and antibodies in a gel. One- or two-dimensional immune diffusion tests may be used. The classic double-diffusion technique of Ochterlony (46) has been used widely for qualitative analysis of antigen-antibody reactions. A typical procedure is given by Wang (62).

Dot-immunoblot tests

A dot blot is a qualitative assay which detects antigen bound to nitrocellulose membrane. The test is essentially an indirect ELISA which employs a precipitating dye and the substrate for the enzyme-linked antibody. The test has been used to identify strains of rhizobia (1). The dot-blot method is also used to identify strains of bacteria growing as colonies on a Petri dish in a mixture with other morphologically similar colonies (44). Antigen from colonies is blotted on an overlaying membrane, and the membrane is subjected to ELISA to reveal the number and location of specific colonies. Detection of rhizobia by this method depends upon the retention of cells on a membrane (nitrocellulose) which is subjected to the different solutions and multiple washes of an ELISA.

The dot-blot method works well with a soluble antigen from a vesicular-arbuscular mycorrhizal fungus (69). Antigen from spores attached to plant roots is released onto nitrocellulose by crushing the root with a rolling pin. After inactivating endogenous peroxidase, an indirect ELISA is performed using peroxidase-labeled second antibody. The final reaction is precipitation of a dye where the antigen was released and bound to the nitrocellulose.

Immunomagnetic particles to extract
bacteria from mixed cultures

Magnetic polymer particles to which immunoglobulins are attached have been used to separate an antigenically distinct *E. coli* from a mixed culture of five different O serogroups of *E. coli* (36). Monoclonal antibodies to the distinct antigen were used as the probe attached to the magnetic particles. Innovative procedures such as this may be useful for isolation of selected bacteria from mixed cultures.

Acknowledgments

The contributions of Ivor T. Knight and Rita R. Colwell to this chapter were greatly appreciated.

References

1. Ayanaba, A., K. D. Weiland, and R. M. Zablotowicz. 1986. Evaluation of diverse antisera, conjugates, and support media for detecting *Bradyrhizobium japonicum* by indirect enzyme-linked immunosorbent assay. *Applied and Environmental Microbiology* 52:1132–1138.
2. Bohlool, B. B. 1987. Fluorescence methods for study of *Rhizobium* in culture and in situ. In G. H. Elkan (ed.), *Symbiotic Nitrogen Fixation Technology.* Marcel Dekker, New York, pp. 127–155.
3. Bohlool, B. B., and E. L. Schmidt. 1980. The immunofluorescence approach in microbial ecology. In M. Alexander (ed.), *Advances in Microbial Ecology*, vol. 4. Plenum, New York, pp. 203–241.
4. Brayton, P. R., and R. R. Colwell. 1987. Fluorescent antibody staining method for enumeration of viable environmental *Vibrio cholerae* O1. *Journal of Microbiological Methods* 6:309–314.
5. Burnet, F. M. 1957. A modification of Jerne's theory of antibody production using the concept of clonal selection. *Australian Journal of Science* 20:67–69.
6. Campbell, D. H., J. S. Garvey, N. E. Cremer, and D. H. Sussdorf. 1974. *Methods in Immunology*. W. A. Benjamin, Reading, Mass.
7. Carlin, N. I. A., and A. A. Lindberg. 1983. Monoclonal antibodies specific for O-antigenic polysaccharides of *Shigella flexneri:* Clones binding to II, II:3,4, and 7,8 epitopes. *Journal of Clinical Microbiology* 18:1183–1189.
8. Carlson, R. W. 1984. Heterogeneity of *Rhizobium* lipopolysaccharides. *Journal of Bacteriology* 158:1012–1017.
9. Carlson, R. W., R. E. Sanders, C. Napoli, and P. Albersheim. 1978. Host-symbiont interactions III. Purification and partial characterization of *Rhizobium* lipopolysaccharides. *Plant Physiology* 62:912–917.
10. Carlson, R. W., R. Shatters, J.-L. Duh, E. Turnbull, B. Hanley, B. G. Rolfe, and M. A. Djordjevic. 1987. The isolation and partial characterization of the lipopolysaccharides from several *Rhizobium trifolii* mutants affected in root hair infection. *Plant Physiology* 84:421–427.
11. Carlson, R. W., and M. Yadav. 1985. Isolation and partial characterization of the extracellular polysaccharides and lipopolysaccharides from fast-growing *Rhizobium japonicum* USDA 205 and its nod⁻ mutant, HC205, which lacks the symbiotic plasmid. *Applied and Environmental Microbiology* 50:1219–1224.
12. Darveau, R. P., and R. E. W. Hancock. 1983. Procedure for isolation of bacterial lipopolysaccharides from both smooth and rough *Pseudomonas aeruginosa* and *Salmonella typhimurium* strains. *Journal of Bacteriology* 155:831–838.
13. Demezas, D. H., and P. J. Bottomley. 1984. Identification of two dominant serotypes of *Rhizobium trifolii* in root nodules of uninoculated field-grown subclover. *Soil Science Society of America Journal* 48:1067–1071.
14. Dughri, M. H., and P. J. Bottomley. 1983. Effect of acidity on the composition of an indigenous soil population of *Rhizobium trifolii* found in nodules of *Trifolium subterraneum* L. *Applied and Environmental Microbiology* 46:1207–1213.
15. Dughri, M. H., and P. J. Bottomley. 1984. Soil acidity and the composition of an indigenous population of *Rhizobium trifolii* in nodules of different cultivars of *Trifolium subterraneum* L. *Soil Biology and Biochemistry* 16:405–411.
16. Fiksdal, L., and J. D. Berg. 1987. Evaluation of a fluorescent antibody technique for the rapid enumeration of *Bacteroides fragilis* groups of organisms in water. *Journal of Applied Bacteriology* 62:377–383.
17. Flaster, M. S., C. Schley, and B. Zepster. 1983. Generating monoclonal antibodies against excised gel bands to correlate immunocytochemical and biochemical data. *Brain Research* 277:196–199.
18. Friguet, B., L. Djavadi-Ohaniance, J. Pages, A. Bussard, and M. Goldberg. 1983. A convenient enzyme-linked immunosorbent assay for testing whether monoclonal antibodies recognize the same antigenic site. Application to hybridomas specific for the β_2-subunit of *Escherichia coli* tryptophan synthase. *Journal of Immunological Methods* 60:351–358.
19. Fuhrmann, J., and A. G. Wollum II. 1985. Simplified enzyme-linked

immunosorbent assay for routine identification of *Rhizobium japonicum* antigens. *Applied and Environmental Microbiology* 49:1010–1013.

20. Goding, J. W. 1986. *Monoclonal Antibodies: Principles and Practice*, 2d ed. Academic Press, New York.
21. Goldman, R. C., and L. Leive. 1980. Heterogeneity of antigenic-side-chain length in lipopolysaccharide from *Escherichia coli* 0111 and *Salmonella typhimurium* LT2. *European Journal of Biochemistry* 107:145–153.
22. Grimes, D. J., and R. R. Colwell. 1983. Survival of pathogenic organisms in the Anacostia and Potomac Rivers and the Chesapeake Bay Estuary. *Journal of the Washington Academy of Science* 73:45–50.
23. Guttman-Bass, N., Y. Tchorsch, and E. Marva. 1987. Comparison of methods for rotavirus detection in water and results of a survey of Jerusalem wastewater. *Applied and Environmental Microbiology* 53:761–767.
24. Hitchcock, P. J., and T. M. Brown. 1983. Morphological heterogeneity among *Salmonella* lipopolysaccharide chemotypes in silver-stained polyacrylamide gels. *Journal of Bacteriology* 154:269–277.
25. Hurrell, J. G. R. 1983. *Monoclonal Hybridoma Antibodies: Techniques and Applications*. CRC Press, Boca Raton, Fla.
26. Hussong, D., R. R. Colwell, M. O'Brien, M. Weiss, A. D. Pearson, R. M. Weiner, and W. D. Burge. 1987. Viable *Legionella pneumophila* not detectable by culture on agar media. *Bio/Technology* 5:947–950.
27. Ito, J. I., A. C. Wunderlich, J. Lyons, C. E. Davis, D. G. Guiney, and A. I. Braude. 1980. Role of magnesium in the enzyme-linked immunosorbent assay for lipopolysaccharides of rough *Escherichia coli* strain J5 and *Neisseria gonorrhoeae*. *Journal of Infectious Disease* 142:532–537.
28. Ivanyi, J., J. A. Morris, and M. Keen. 1985. Studies with monoclonal antibodies to mycobacteria. In A. J. L. Macario and E. Conway de Macario (eds.), *Monoclonal Antibodies against Bacteria*, vol. 1. Academic Press, New York, pp. 59–90.
29. Jenkins, M. B., and P. J. Bottomley. 1985. Evidence for a strain of *Rhizobium meliloti* dominating the nodules of alfalfa. *Soil Science Society of America Journal* 49:326–328.
30. Jenkins, M. B., and P. J. Bottomley. 1985. Composition and field distribution of the population of *Rhizobium meliloti* in root nodules of uninoculated field-grown alfalfa. *Soil Biology and Biochemistry* 17:173–179.
31. Kabat, E. A. 1968. *Structural Concepts in Immunology and Immunochemistry*. Holt, Rinehart and Wilson, New York.
32. Kogure, K., U. Simidu, and N. Taga. 1979. A tentative direct microscopic method for counting living marine bacteria. *Canadian Journal of Microbiology* 25:415.
33. Kohler, G., and C. Milstein. 1975. Continuous cultures of fused cells secreting antibody of predefined specificity. *Nature* 256:495–497.
34. Knapp, W., K. Holubar, and G. Wicks. 1978. *Immunofluorescence and Related Staining Techniques*. Elsevier, New York.
35. Kwapinski, J. B. G. 1972. *Methodology of Immunochemical and Immunological Research*. Wiley, New York.
36. Lund, A., A. L. Hellemann, and F. Vartdal. 1988. Rapid isolation of K88+ *Escherichia coli* by using immunomagnetic particles. *Journal of Clinical Microbiology* 26:2572–2575.
37. Macario, A. J. L., and E. Conway de Macario. 1985. *Monoclonal Antibodies against Bacteria*, vol. 1. Academic Press, New York.
38. Macario, A. J. L., and E. Conway de Macario. 1985. *Monoclonal Antibodies against Bacteria*, vol. 2. Academic Press, New York.
39. Marchalonis, J. J. 1982. Structure of antibodies and their usefulness to nonimmunologists. In J. J. Marchalonis and G. W. Warr (eds.), *Antibody as a Tool*. Wiley, New York, pp. 3–20.
40. Nambair, P. T. C., and V. Anjaiah. 1985. Enumeration of rhizobia by enzyme-linked immunosorbent assay (ELISA). *Journal of Applied Bacteriology* 58:187–193.
41. Nasser, A. M., and T. G. Metcalf. 1987. An A-ELISA to detect hepatitis A virus in estuarine samples. *Applied and Environmental Microbiology* 53:1192–1195.
42. Oakley, B. R., D. R. Kirsch, and N. R. Morris. 1980. A simplified ultrasensitive sil-

ver stain for detecting proteins in polyacrylamide gels. *Analytical Biochemistry* 105: 361–363.

43. Oi, V. T., and L. A. Herzenberg. 1980. Immunoglobulin-producing hybrid cell lines. In B. B. Mishell and S. M. Shiigi (eds.), *Selected Methods in Cellular Immunology.* W. H. Freeman, San Francisco, p. 351.

44. Olsen, P. E., and W. A. Rice. 1989. *Rhizobium* strain identification and quantification in commercial inoculants by immunoblot analysis. *Applied and Environmental Microbiology* 55:520–522.

45. Otterness, I., and F. Karush. 1982. Principles of antibody reactions. In J. J. Marchalonis and G. W. Warr (eds.), *Antibody as a Tool.* Wiley, New York, pp. 97–137.

46. Ouchterlony, O., and L. A. Nilssen. 1978. Immunodiffusion and immunoelectrophoresis. In D. M. Weir (ed.), *Handbook of Experimental Immunology.* Vol. 1. *Immunochemistry,* 3d ed. Blackwell, Oxford, pp. 19.1–19.44.

47. Palva, E. T., and P. H. Makela. 1980. Lipopolysaccharide heterogeneity in *Salmonella typhimurium* analyzed by sodium dodecylsulfate/polyacrylamide gel electrophoresis. *European Journal of Biochemistry* 107:137–143.

48. Renwick, A., and D. G. Jones. 1985. A comparison of the fluorescent ELISA and antibiotic resistance identification techniques for use in ecological experiments with *Rhizobium trifolii. Journal of Applied Bacteriology* 58:199–206.

49. Roszak, D. B., and R. R. Colwell. 1987. Survival strategies of bacteria in the natural environment. *Microbiology Reviews* 51:365–379.

50. Sakazaki, R., and T. J. Donovan. 1984. Serology and epidemiology of *Vibrio mimicus.* In T. Bergan (ed.), *Methods in Microbiology,* vol. 16. Academic Press, London, pp. 271–289.

51. Sauch, J. F. 1985. Use of immunofluorescence and phase-contrast microscopy for detection and identification of *Giardia* cysts in water samples. *Applied and Environmental Microbiology* 50:1434–1438.

52. Shamida, T., R. Sakazaki, and M. Que. 1987. A bioserogroup of marine vibrios possessing somatic antigen factors in common with *Vibrio cholerae* O1. *Journal of Applied Bacteriology* 62:453–456.

53. Shamida, T., and R. Sakazaki. 1988. A serogroup of non-O1 *Vibrio cholerae* possessing the Inaba antigen of *Vibrio cholerae* O1. *Journal of Applied Bacteriology* 64:141–144.

54. Spitz, M. 1986. Single-shot intrasplenic immunization for the production of monoclonal antibodies. *Methods in Enzymology* 121:33–41.

55. Stewart-Tull, D. E. S., and M. Davies. 1985. *Immunology of the Bacterial Cell Envelope.* Wiley, New York.

56. Sutherland, I. W. 1977. Immunochemical aspects of polysaccharide antigens. In L. E. Glynn and M. W. Steward (eds.), *Immunochemistry: An Advanced Textbook.* Wiley, New York.

57. Tamplin, M. L., A. L. Gauzens, A. Huq, D. A. Sack, and R. R. Colwell. Attachment of *Vibrio cholerae* serogroup O1 to zooplankton and phytoplankton of Bangladesh waters. *Applied and Environmental Microbiology* 56:1977–1980.

58. Tijssen, P. 1987. Practice and theory of enzyme immunoassays. In R. H. Burdon and P. H. van Knippenberg (eds.), *Laboratory Techniques in Biochemistry and Molecular Biology.* Elsevier, New York.

59. Tsai, C.-M., and C. E. Frasch. 1982. A sensitive silver stain for detecting lipopolysaccharides in polyacrylamide gels. *Analytical Biochemistry* 119:115–119.

60. Voller, A., D. Bidwell, and A. Bartlett. 1980. Enzyme-linked immunosorbent assay. In N. R. Rose and H. Friedman (eds.), *Manual of Clinical Immunology,* 2d ed. American Society for Microbiology, Washington, D.C., pp. 359–371.

61. Wagener, C., U. Fenger, B. R. Clark, and J. E. Shively. 1984. Use of biotin-labeled monoclonal antibodies and avidin-peroxidase conjugates for the determination of epitope specificities in a solid-phase competitive enzyme immunoassay. *Journal of Immunological Methods* 68:269–274.

62. Wang, A.-C. 1982. Methods of immunodiffusion, immunoelectrophoresis, precipita-

tion, and agglutination. In J. J. Marchalonis and G. W. Warr (eds.), *Antibody as a Tool*. Wiley, New York, pp. 139–161.

63. Warr, G. W. 1982. Preparation of antigens and principles of immunization. In J. J. Marchalonis and G. W. Warr (eds.), *Antibody as a Tool*. Wiley, New York, pp. 21–58.

64. Warr, G. W. 1982. Purification of antibodies. In J. J. Marchalonis and G. W. Warr (eds.), *Antibody as a Tool*. Wiley, New York, pp. 59–96.

65. Weaver, R. W., and L. R. Frederick. 1982. *Rhizobium*. In A. L. Page, R. H. Miller, and D. R. Keeney (eds.), *Methods of Soil Analysis, Part 2*, 2d ed. American Society of Agronomy, Madison, Wis., pp. 1043–1070.

66. Williams, C. A., and M. W. Chase. 1967. *Methods in Immunology and Immunochemistry. Vol. 1: Preparation of Antigens and Antibodies*. Academic Press, New York.

67. Wright, S. F., J. G. Foster, and O. L. Bennett. 1986. Production and use of monoclonal antibodies for identification of strains of *Rhizobium trifolii*. *Applied and Environmental Microbiology* 52:119–123.

68. Wright, S. F., J. B. Morton, and J. E. Sworobuk. 1987. Identification of a vesicular-arbuscular mycorrhizal fungus by using monoclonal antibodies in an enzyme-linked immunosorbent assay. *Applied and Environmental Microbiology* 53:2222–2225.

69. Wright, S. F., and J. B. Morton. 1989. Detection of VAM fungal root colonization by using a dot-immunoblot assay. *Applied and Environmental Microbiology* 55:761–763.

70. Xu, H.-S., N. Roberts, F. L. Singleton, R. W. Attwell, D. J. Grimes, and R. R. Colwell. 1982. Survival and viability of nonculturable *Escherichia coli* and *Vibrio cholerae* in the estuarine and marine environment. *Microbial Ecology* 8:313–323.

71. Zevenhuizen, L. P. T. M., I. Scholten-Koerselman, and M. A. Posthumus. 1980. Lipopolysaccharides of *Rhizobium*. *Archives of Microbiology* 125:1–8.

72. Zola, H. 1987. *Monoclonal Antibodies: A Manual of Techniques*. CRC Press, Boca Raton, Fla.

Chapter 4

Nucleic Acid Hybridization Techniques for Detection, Identification, and Enumeration of Microorganisms in the Environment

Ivor T. Knight

William E. Holben

James M. Tiedje

Rita R. Colwell

Introduction

The unique base sequences of the nucleic acids of organisms and the exquisite specificity with which these sequences hybridize to complementary sequences have been exploited to produce a set of powerful techniques for detection and identification of microorganisms. Techniques such as DNA:DNA hybridization, DNA:RNA hybridization, DNA and RNA sequencing, DNA fingerprinting, and the use of gene probes for detection of specific sequences have been employed widely in the clinical setting and, more recently, in the field of microbial ecology.

Such techniques offer new ways to approach previously untenable problems in microbial ecology. For example, assessment of the diversity of microbial communities requires characterization of the predominant, yet nonculturable, populations in nature. The role of the environment in gene expression, the extent of gene exchange among microbial populations in the environment, and the taxonomic relationships among community members can also be assessed using these new methods. Molecular approaches to microbial ecology will, likely, also raise new questions which challenge our current understanding of the microbial world. The discovery of microorganisms which are adapted to "extreme" environments (10, 92), and the task of assessing the fate and risk of genetically engineered microorganisms released into the environment (8) represent new challenges which can be addressed using nucleic acid methodologies.

This chapter reviews hybridization methods, principally hybridization probes, and their application to the study of microbial ecology. The appropriate use of probes in hybridization assays requires a rudimentary understanding of the kinetics of nucleic acid renaturation and an appreciation for the conditions of hybridization affecting probe specificity. Various strategies for constructing and labeling probes are also reviewed and evaluated, as are methods for preparing target nucleic acids from environmental samples and other applications of nucleic acid hybridization in microbial ecology.

The current surge of interest in nucleic acid probes and their widespread application has generated several reviews of the subject. Of particular note is the review by Tenover (99) which, though the focus is on the use of probes in the clinical setting, is a comprehensive treatment useful to those who wish to use probes for detection of microorganisms in the environment. Reviews by Ogram and Sayler (64), Hazen and Jimenez (27), and Holben and Tiedje (33) are aimed at the use of probes in environmental microbiology. The objective of this chapter is to provide sufficient detail about the techniques presently available to permit selection of a strategy appropriate to specific environmental problems.

General Principles

The basis of the hybridization probe assay is the ability of single-stranded nucleic acid to form a stable, double-stranded structure via complementary base-pairing. The reaction, commonly termed "renaturation" or "annealing," occurs between two strands of DNA, two strands of RNA, or between strands of RNA and DNA. The specificity of the reaction is determined by hydrogen bonding between complementary bases in each strand, and those conditions which affect hydrogen bonding, such as temperature and ionic strength, influence the specificity of the reaction. For relatively short hybrids (less than 500 base pairs) stability of the hybrid is also dependent on its length. Nearly complementary single-stranded DNA molecules also form double-stranded structures of lesser stability. The rate of the reaction follows second-order kinetics and can be predicted under specific conditions. The publication by Wetmur and Davidson (104), now a classic, remains very useful, and principles outlined in their paper provide the rationale for procedures followed in many hybridization protocols.

Nucleic acid probes are molecules of single-stranded DNA or RNA which have been labeled, either chemically or radioactively, so that they can be detected in a mixture of renatured nucleic acids. The probe is incubated with denatured test DNA or RNA (often total cellular DNA) and, under appropriate conditions, hybridizes with regions of the test nucleic acid which are complementary. Excess, unhybridized probe is selectively removed and probe-target hybrids are detected using methods appropriate to the particular labeling strategy.

The most common format for the probe hybridization assay is the mixed-phase or membrane hybridization assay, pioneered by Southern (90). Test DNA or RNA is denatured and allowed to bind to a nitrosylated cellulose membrane (nitrocellulose), a nylon-based membrane, or a membrane composed of a nylon-nitrocellulose mixture. These membranes, available from molecular biological supply companies under a variety of trade names, bind single-stranded nucleic acids under elevated salt conditions (high ionic strength). The nucleic acid is bound by a combination of hydrophobic and electrostatic interactions and can be released by lowering the salt concentration. After the binding step the nucleic acids are fixed to the membrane by baking at 80°C or by ultraviolet irradiation. This binds the single-stranded nucleic acid irreversibly to the membrane, with the target sequence available for hybridization with probe nucleic acid. After incubation with the probe (hybridization), the unhybridized probe is washed from the filter and the remaining hybridized probe is measured. The kinetics of hybridization under conditions where one strand is bound to a solid matrix are similar to those where both strands are free in solution and the parameters which affect hybridization rate and hybrid stability are also similar. Discussion of the theory and practice of mixed-phase hybridization is provided by Meinkoth and Wahl (57).

The conditions which most affect hybrid formation rate and stability are temperature and ionic strength. These variables are manipulated in mixed-phase hybridization assays to maximize the rate of hybridization and modulate the specificity of the assay. Native DNA in solution will denature predictably with increasing temperature. The melting temperature (T_m) of a given probe-target hybrid is the temperature at which 50 percent of the nucleic acid is denatured (55) and is an important parameter used to determine the appropriate hybridization conditions. Hybridizations are usually performed at 20 to 30°C below T_m because the rate of hybridization reaches maximum in this range (104). The renaturation rate is also dependent on the ionic strength of the hybridization solution. For an electrolyte such as NaCl, which is used almost exclusively in hybridization assays, the rate of reaction increases dramatically with concentration, up to 0.4 M. Above 0.4 M the rate and ionic strength are essentially independent (57). For this reason, hybridization reactions in probe assays are carried out at NaCl concentrations between 0.5 and 1 M. Since the rate of hybridization increases with increasing concentration of probe, reducing the reaction volume to a minimum is desirable. The use of heat-sealed bags for large membranes allows small volumes to be used, while ensuring that all parts of the membrane come into contact with the hybridization solution. Addition of high-molecular-weight polymers, such as dextran sulfate, cause molecular crowding, thereby increasing reaction rates more than 10-fold (103). The hybridization time in the mixed-phase format can be estimated from probe length and concentration, using the formula for predicting $C_0t_{1/2}$, the time required for half of the single-stranded probe in solution to anneal to the target (53).

Since most probes are double-stranded DNA molecules which are first denatured before hybridization, a competing (and more favorable) reaction of probe-probe hybridization also occurs in the hybridization solution. For this reason it is recommended that hybridization be allowed to continue for two to three times the $C_0t_{1/2}$, which usually results in long hybridization times, e.g., from 12 to 72 hours, depending upon the probe. Solution hybridization, in which both target and probe are free in solution, occurs much more rapidly than mixed-phase hybridization. Newer methods allowing the probe-target hybridization to occur in solution, followed by hybrid capture using various affinity matrices show great promise (alternative hybridization formats are described below). Another method for abbreviating hybridization time is to use single-stranded probes, e.g., RNA transcript probes and M13-derived probes (32, 33).

The specificity of the mixed-phase probe assay is determined by the stability of the hybrid when the membrane is washed under proscribed conditions of ionic strength and temperature. Stringency of washing conditions is important. High stringency conditions require low salt concentration (0.1 to 0.3 M) and high temperature (25 to 30°C below the T_m of the probe in the case of probes greater than 500 nucleotides), while low stringency is achieved by raising the salt concentration and/or lowering the temperature. The higher the degree of complementarity between probe and target, the more stable the hybrid will be when washed under conditions of high stringency. To detect targets which hybridize to the probe with less fidelity, i.e., are less specific, wash stringency conditions are lowered. Conditions of washing, therefore, are a very important part of the mixed-phase hybridization assay. Probe length influences hybrid stability in the case of probes less than 500 nucleotides long, but the effect only becomes significant when the probe is less than 100 nucleotides (57). Short probes, such as chemically synthesized oligonucleotide probes, form hybrids of much lower stability than long probes and the wash temperatures used are generally much lower. The effect of wash conditions on hybrid stability can be predicted, using equations developed from empirical observations. The equations are useful for estimating percent mismatch between probe and target which can be tolerated under a given set of wash conditions. They are summarized by Meinkoth and Wahl (57). Figure 4.1 illustrates the effect of salt concentration on the T_m of duplex DNAs of varying degrees of homology. It must be emphasized that theoretical estimates, although based upon empirical observations, cannot provide precise predictions of the behavior of a probe. While much is known about nucleic acid renaturation, the process is not completely understood. Optimal conditions, therefore, must still be determined experimentally for each probe.

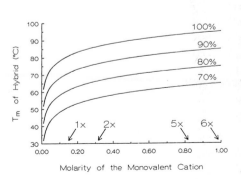

Figure 4.1 The theoretical effect of salt concentration on hybrid stability for DNA hybrids of varying sequence homology. The equations are plotted for DNA with a GC mol% of 50 and length greater than 500 base pairs. The equations are Delta T_m = 16.6(log M), where M = molarity of the monovalent cation, and Delta T_m = 100 − h, where h = % homology between strands. The curves for four different levels of homology are plotted and several concentrations of a commonly used wash buffer, SSC (53), are indicated. [*Equations from Meinkoth and Wahl (57).*]

Strategies for Probe Construction

Nucleic acid probes can be grouped into three major categories, based upon strategies used to construct them: total genomic DNA, cloned restriction fragments, and synthetic oligodeoxynucleotides.

Total genomic DNA

Total genomic DNA hybridization, the technique of using total genomic DNA of one organism as a target for hybridization with the total genomic DNA of another, has been used successfully in the field of bacterial systematics to determine relationships among closely related bacteria. Total genomic DNA hybridization has not been used routinely to identify and/or detect microorganisms because preparation of the DNA and quantification of percent homology are time consuming and do not lend well to large-scale screening.

Several investigators, however, have employed the principles of total genomic DNA hybridization and have streamlined the methods to develop assays using total genomic DNA probes (14, 31, 60). One advantage of this type of probe is that cloning is not involved in development of the probe. Morotomi et al. (60) and Hodgson and Roberts (31) applied whole cells to a membrane support, lysed the cells, and probed with radiolabeled chromosomal DNA to detect strains of *Bacteroides* and *Rhizobium*, respectively. Alternatively, the approach of Esaki et al. (14) was to apply purified chromosomal reference DNA to a membrane and label crude preparations of test organism DNA for hybridization with the DNA on the membrane. This approach was used successfully to identify members of the genuses *Campylobacter, Bacteroides, Vibrio, Flavobacterium, Sphingobacterium,* and *Legionella.* The advantage of this method is that the time-consuming step of producing purified reference DNA is performed only once, and reference DNA membranes can be produced and then stored for months until needed.

The obvious disadvantage of using total genomic DNA as a probe is the likelihood of cross-reactivity arising from homology between genes highly conserved between species. Grimont et al. (24) encountered this problem in the development of a probe for *Legionella pneumophila,* when they found that radiolabeled chromosomal DNA cross-hybridized with DNA from many other species belonging to several families. They overcame the problem by identifying cross-hybridizing fragments on a Southern blot of *L. pneumophila* DNA which had been digested with a restriction endonuclease. Regions on an agarose gel corresponding to the molecular weights of the cross-hybridizing fragments were excised and the remaining fragments, when pooled and tested, proved specific for *L. pneumophila.*

Cloned restriction fragments

The approach most widely used is the construction of probes from restriction fragments of DNA cloned into plasmid vectors. The various methods for constructing a probe using cloned DNA can be grouped into two categories: shotgun and targeted. An example of the shotgun strategy is the first commercially available probe for *Salmonella* spp., used to detect these microorganisms in food (17–19). The probe, marketed successfully by Gene Trak Systems, Framingham, Massachusetts, was produced by constructing a genomic DNA library from *Salmonella typhimurium* and testing the cloned restriction fragments for cross-hybridization with radiolabeled *E. coli* chromosomal DNA. Those clones which did not cross-hybridize were then tested for cross-hybridization with a battery of strains other than *Salmonella* spp. and those which did not cross-react were tested for specificity using hundreds of *Salmonella* strains. Those clones which hybridized with the largest number of strains were used to prepare the probe.

In the shotgun approach, little is known about the resultant probe DNA other than size and performance as a probe. The advantage of this approach is that it can be much less time consuming than the targeted approach, which involves identification and characterization of specific clones. A disadvantage is that one may never clone a fragment of desired specificity, since chance ultimately determines the size of the cloned fragments. A clone containing highly specific sequence may also contain less specific sequence and would thus be discarded. The shotgun approach can be an effective strategy, however, evidenced by those probes successfully developed using this approach (1, 48, 49, 82).

In contrast to the shotgun strategy for developing probes from cloned genomic libraries is the targeted strategy, which involves cloning a particular genetic determinant. That cloned fragment, or a portion thereof, is then used as a specific probe. In this case, the function of the gene is known and the function often determines the choice of clones for use as a probe. Most often, development of such probes is ancillary to the main goal of cloning a particular gene and determining sequence, role, and regulation of the gene. Examples are probes constructed from cloned pathogenic determinants (13, 30, 42, 61–63, 67, 80, 88), genes encoding resistance to antibiotics (98) or heavy metals (3), genes encoding degradative phenotypes (70, 84), and nitrogen-fixation determinants (86).

Probes that have been developed using a targeted strategy can be employed as function-specific probes to detect similar genes in different bacteria or they can be used as species-specific probes if the cloned gene sequence is conserved within a species but highly variable between species. Function-specific probes are useful tools in the study of microbial ecology and have been used both to assess the potential of a particular community for expressing a given phenotype and for detecting and culturing isolates possessing genes encoding given functions. The results of Pettigrew and Sayler (70) illustrate the advantages of using a function-specific probe to detect and enumerate pollutant-degrading organisms in freshwater sediment.

Cloned DNA determinants encoding ribosomal RNA (rRNA) have significant applications as species-specific or phylogenetic probes. These probes can be used to detect either rRNA determinants or the rRNA molecules themselves. Determinants encoding 5S, 16S, and 23S rRNA contain regions in which the sequence is highly conserved between species and regions in which it is highly variable. The variable regions comprise sequences which can be used as probes specific for a particular species, while the conserved regions offer probes with broader specificities. The application of cloned rRNA determinants in the field of microbial ecology is reviewed by Pace et al. (66) and the strategy of employing rRNA probes to detect and enumerate specific organisms in given environments has been successfully employed by many investigators (11, 44, 46, 47, 74, 85, 91). Stahl et al. (91) describe an experiment in which the authors monitored population changes of anaerobic bacteria in the rumen, using rRNA-targeted probes, illustrating the utility of phylogenetic probes as a means of studying specific populations in a given niche. Another advantage of rRNA-directed probes is that they are much more sensitive than DNA-directed probes. While a single cell may have only one or several copies of a chromosomal target, it can have up to 10,000 copies of an rRNA target depending upon the stage of growth of the cell and how many ribosomes it contains (11). Little is known, however, about rRNA expression in variable environmental conditions and, therefore, rRNA-directed probes, although highly sensitive, may not be useful for enumerating microorganisms in the environment.

Synthetic oligodeoxynucleotide probes

The development and simplification of DNA and RNA sequencing methods and the development of automated DNA synthesis machines have made possible the designing of probes of defined sequence and specificity. Synthetic oligodeoxynucleotide (oligomer) probes are generally from 15 to 50 bases long and can be more specific than restriction fragment probes. This is because instability introduced into a hybrid with a mismatched base pair is much higher for short hybrids than for long ones. It is possible to design an oligomer probe which will discriminate between two targets differing only by one base (102). Alternatively, one can synthesize mixtures of oligomers with different deoxynucleotides at one or more positions, creating a probe with a much broader specificity, as was done by Giovannoni et al. in constructing an rRNA-directed probe which detects members of the kingdom Eubacteria, but not members of the Archebacteria or Eukaryote kingdoms (23). A disadvantage of oligomer probes is that they are generally less sensitive than restriction fragment probes because less label can be incorporated per molecule.

Labeling Techniques and Hybridization Formats

While the specificity of hybridization probes is determined primarily by the choice of nucleic acid fragment or sequence used to make the probe and the conditions of the assay, the sensitivity of the probe assay is determined largely by the labeling and detection scheme employed in the assay. This includes the choice of label, the method used to incorporate the label into the probe, and the method used to detect the labeled probe. A comprehensive review by Matthews and Kricka covers DNA labeling techniques and the authors also compare sensitivities of labeling and detection methods (56).

The current confusion in the literature concerning definitions of sensitivity must be addressed. Reports of the development of a new probe or a new method for constructing probes appear with increasing frequency and authors usually report the limit of sensitivity for their probe(s). Confusion arises, however, when different authors report probe sensitivity using different units, making direct comparisons between probes difficult and, in some cases, impossible. Most often, probe sensitivity is reported as mass unit of DNA required for detection. This can be highly misleading because the nature of the target is not taken into account. If the target is a 500-base sequence on a 10,000-bp plasmid, then the target constitutes 5 percent of the mass of the DNA. On the other hand, if the same target is present once on a chromosome of 5×10^6 bp then it constitutes only 0.01 percent of the mass of the DNA. Since the probe detects only target, it is necessary to know whether the target is single copy or multicopy in order to evaluate the sensitivity of the probe. Confusion can be avoided by reporting mass units of target sequence required for detection rather than mass units of DNA. Using the target examples above, if sensitivity of the probe is reported as 1 pg of target sequence, then 20 pg of plasmid DNA or 10,000 pg of chromosomal DNA is needed to detect the target. Alternatively, the amount of DNA required for detection can be reported in molar units. This has been proposed as a standard method of reporting probe sensitivity (56) because it removes ambiguities associated with reporting sensitivity in mass units. One attomole of plasmid DNA would have the same number of targets as one attomole of chromosomal DNA, although the mass would be quite different.

The most practical way of reporting sensitivity of a probe is to state the number of organisms required for detection. This information should be derived experimentally. Predictions made by dividing the mass of nucleic acid required for detection by the average mass of nucleic acid per cell, while estimating a theoretical minimum number of cells required for detection, fails to consider variables affecting nucleic acid yield per cell, such as physiological condition of the cells and the environment from which they are recovered. Reporting sensitivity limits in terms of cell numbers required for detection can also be misleading if the target sequence exists in multiple copies per cell since the stated sensitivity applies only to that specific probe and organism and not the method in general.

The mixed-phase format

Several mixed-phase hybridization methods are useful in probe assays. One of the simplest is the dot blot or slot blot, in which denatured nucleic acid is applied to the membrane in discrete dot- or slot-shaped areas using a vacuum manifold apparatus, e.g., the "Minifold" apparatus (Schleicher and Schuel, Keene, New Hampshire). Up to 96 nucleic acid samples can be applied to a single membrane using this apparatus. Results of hybridization of probes with dot blots or slot blots are quantitative as well as qualitative. By applying a dilution series of known target DNA to the membrane and measuring the amount of hybridized probe to each dot, one can generate a standard curve for estimating the amount of hybridizable nucleic acid in an unknown sample from the amount of probe which is bound. Quantitative dot-blot hybridization is described by Kafatos et al. (40). Attwood et al. describe its use for enumerating specific microorganisms in a mixed community, in which an introduced strain of *Bacteroides ruminicola* is monitored in the rumen (1). When using the dot- or slot-blot format for quantitative purposes, it is important to determine how much DNA can be applied to the membrane without saturating the hybridization signal obtained. This is best determined empirically for each type of experiment as the proportion of target DNA to total DNA will affect this value. Background or nonspecific hybridization will be higher in the dot- or slot-blot format, as compared with the Southern transfer (described below), since all of the DNA (both target and nontarget DNA) is concentrated in a small area on the membrane. In the Southern transfer, target DNA is concentrated in a discrete area, while nonspecific signal is dispersed over a large area (32). Negative controls are, of course, required to determine background hybridization levels.

A technique known as colony-blot hybridization, developed by Grunstein and Hogness (25) for screening bacterial colonies for the presence of cloned genes, is used widely for rapid detection of target sequence in colonies growing on agar plates. Colonies are transferred from the plate to a membrane by placing the membrane directly on the plate and lifting off the bacterial cells attached to the membrane. The membranes are subjected to treatment with 0.5 N NaOH, which serves to lyse the cells and denature the DNA, which then binds to the membrane. The membranes are then neutralized and the DNA linked to the filter for subsequent hybridization analysis. Modifications of this procedure by Maas (52) allow for detection of single-copy genes via colony blots. Alternative to lifting colonies from the plate, the membrane can be placed on the plate and inoculated directly. Colonies grow on the membrane utilizing moisture and nutrients absorbed from the agar. Selective media have been employed, in conjunction with the colony-blot technique, to enumerate specific organisms in mixed environments and to probe them for selected target genes, using both most-probable-number (MPN) (21) and plate-count (38) enumerative schemes.

The Southern blot (for DNA) and the Northern blot (for RNA) techniques are, analytically, much more powerful. In Southern blot analysis double-stranded DNA fragments generated from restriction endonuclease digestion are first size-fractionated by agarose gel electrophoresis. These fragments are subsequently denatured and transferred to a solid support (usually a nylon or nitrocellulose membrane) which maintains the relative spatial relationship of each fragment to all others (90). Size fractionation is also an important feature of Northern blot analyses. RNA molecules (e.g., total RNA or messenger RNA) are size-fractionated by electrophoresis in agarose gels, in this case under denaturing conditions, then transferred to membranes for hybridization analyses (2, 53, 100). The analytical power of these procedures derives from the initial size-fractionation step which separates species of nucleic acid molecules based on their size. When two or more species contain the same or similar genes, that gene will have a slightly different sequence and/or be flanked by DNA having a different sequence. Therefore, complex mixtures of DNA obtained from microbial communities consisting of multiple populations can effectively be analyzed for the presence of specific populations or genes. This is possible because when the DNA is digested with an appropriate restriction endonuclease, the different-sized fragments, of different species, will be detected as different bands of hybridization. Similarly, this technique also allows the detection of genetic rearrangements and horizontal gene transfer since the transferred or rearranged gene will also be flanked by different sequence (39). Often, very closely related organisms can be differentiated on the basis of which size fragments in a particular digest hybridize to a specific probe. This is the concept behind DNA fingerprinting. An example is the different banding patterns achieved when the sequence specific for the cholera toxin structural gene is used to probe Southern blots of *Vibrio cholerae* O1 chromosomal DNA digested with the restriction enzyme Hind III. Isolates from cholera cases in the United States along the Gulf Coast have a pattern (fingerprint) which is distinctly different from isolates obtained from Bangladesh (42, 43).

Radioactive labels

The most common method of labeling nucleic acids is to incorporate radiolabeled nucleotides into the molecule using in vitro enzymatic reactions. ^{32}P and ^{35}S are the nuclides most often used. The classic method of Rigby et al. (77), known as nick translation, employs DNAse I to break the phosphodiester backbone of DNA and DNA polymerase I to remove nucleotides downstream of the nick while simultaneously polymerizing a nascent strand using exogenous nucleotide triphosphates. One or more of the nucleotide triphosphate species supplied contains ^{32}P at the alpha position and the result is DNA which is uniformly labeled with ^{32}P. The enzyme and substrate concentrations must be carefully defined to obtain maximum incorporation of label. Several molecular biological supply companies supply enzymes and reagents in kit form, e.g., Boehringer and Manheim Biochemicals, Indianapolis, Indiana; Pharmacia LKB Biotechnology, Inc., Piscataway, New Jersey; DuPont NEN Research Products, Wilmington, Delaware; and Amersham Corp., Arlington Heights, Illinois.

Another method of radiolabeling DNA is the method of Feinberg and Vogelstein (16), known as random priming or random hexamer priming. This method is rapidly gaining popularity because the procedure is significantly simpler than nick translation and, in general, incorporates more label. The method employs a random collection of short oligomers (hexamers) which hybridize to single-stranded DNA and serve as priming sites for the Klenow fragment of DNA polymerase I to polymerize a complementary strand using exogenous nucleotides, some of which are radiolabeled. The ratio of hexamer to DNA concentrations is critical and, again, kits are marketed by molecular biological supply companies.

Both methods can be used to radiolabel relatively long (more than 200 bp) restriction fragments, but are not suited to labeling oligomer probes. Oligomer probes are generally labeled only at either their 5' or 3' end. The 5' ends of synthesized oligomers are hydroxyl residues (as opposed to native DNA which has a phosphate group on the 5' end). This serves as a substrate for polynucleotide kinase which can transfer the gamma phosphate residue of a nucleotide triphosphate to the 5' end of the oligomer. If the nucleotide triphosphate supplied is labeled with ^{32}P at the gamma position then the result will be oligomer molecules with one $^{32}PO_4^{-2}$ at the 5' end. The 3' end can be labeled by addition of one or more radiolabeled nucleotides using the enzyme terminal transferase. When more than one nucleotide is added to the 3' end the reaction is called a tailing reaction, because a homogeneous polynucleotide tail has been added. This is one way to incorporate more than one label moiety per oligomer and, as long as the tail does not change the specificity of the resulting oligomer probe, the sensitivity of the probe is increased. Endlabeling protocols are described in literature distributed by molecular biology and biotechnology supply companies and in several recombinant DNA techniques textbooks, e.g., Sambrook et al. (83).

When preparing probes from cloned restriction fragments, one must be careful not to label the plasmid vector in which the fragment resides as well as the fragment itself. Otherwise, the probe will have a dual specificity. To avoid this error the fragment is usually cut from the vector, using a restriction endonuclease and is purified by gel electrophoresis, although this may not remove all traces of vector DNA from the preparation. There are some novel labeling procedures which make use of specialized vectors to prevent problems of contamination with vector DNA without the need to purify the fragment of interest. Two methods employ recombinant bacteriophage M13, a vector commonly used in DNA sequencing. Hu and Messing (36) describe a technique which uses the single-stranded form of M13 to produce a radiolabeled molecule which is single-stranded only in the region of the inserted restriction fragment. Since the vector sequence is double-stranded, it will not hybridize to a target and only the sequence of interest is free to hybridize to target. Holben et al. (32) have applied a method for generating single-stranded, vector-free probes in M13, with the added advantages of being highly sensitive owing to the high specific activity attainable by the labeling method and elimination of probe-probe hybridization by virtue of being single-stranded. Specialized vectors are used in a method currently marketed by Promega Corp. (Madison, Wisconsin) as the Riboprobe system. It employs RNA polymerase, rather than DNA polymerase, to synthesize labeled nucleic acid. A specific RNA polymerase promoter, such as the promoter of the bacteriophage SP6, is part of the vector sequence (58) and is located adjacent to the multiple cloning site (the site where the desired restriction fragment is inserted into the vector). After cutting the vector at the other end of the restriction fragment with appropriate restriction enzyme, SP6 RNA polymerase is used to make "runoff" RNA transcripts from the specific primer using the cloned DNA as the template. The preparation is then digested with DNAse I, which degrades all vector DNA, and leaves only the RNA probe intact. This system also produces single-stranded probes which cannot self-anneal in the hybridization solution.

Hybrids formed between radiolabeled probes and their target sequences can be detected using photographic films, by liquid scintillation counting, or using recently developed instruments for measuring beta emissions from hybridization membranes. Photographic films can be employed in various ways, depending upon the particular nuclide used as label. Autoradiography and fluorography are the two most commonly employed techniques for detecting radiolabeled probes. Autoradiography, in which direct exposures of the film are made by the emissions of the nuclide, is used widely to detect ^{32}P- and ^{35}S-labeled probes. Films used for both autoradiography and fluorography generally have an emulsion on both sides and emissions of ^{32}P have enough energy to penetrate the film plate and expose both emulsions, while the lower-energy emissions of ^{35}S exposes only one emulsion. Fluorography, the incorporation of scintillants into the sample, which then emit light when struck by radioactive emissions, is often used to enhance exposure of film by lower-energy emitters. These compounds can be sprayed onto a gel or membrane prior to exposure. Frequently one needs to detect smaller amounts of ^{32}P-labeled probe than can be detected using direct autoradiography, and there are options for increasing the sensitivity of the probe assay by intensifying the autoradiographic exposure. Laskey and Mills (50) evaluated the technique of preflashing the film with a short burst of light to hypersensitize it. They also evaluated the use of intensifying screens, plates coated with rare earth metals which absorb sample emissions and produce multiple photons which return to the film and expose the film photographically, in addition to the autoradiographic exposure. They found that when these techniques are employed together, autoradiography of ^{32}P is enhanced 10.5-fold and that as little as 50 cpm/cm^2 can be detected. The use of intensifying screens is termed indirect autoradiography to differentiate it from fluorography and direct autoradiography. A useful booklet by Laskey describing the theory and practice of autoradiography and fluorography has been published by Amersham Corp. (Laskey, R. A. 1984. Radioisotope detection by fluorography and intensifying screens. Amersham Corp., 2636 South Clearbrook Dr., Arlington Heights, Illinois.).

An alternative method for detecting radiolabeled probes is liquid scintillation counting. This approach is used to detect beta-particle emitters and can yield precise measurement, compared with autoradiography, which is qualitative. It must be mentioned, however, that laser densitometry can be used to analyze quantitatively an autoradiograph, since the darkness of the image is proportional to the amount of radioactivity in the sample. This relationship only holds true for film exposed with intensifying screens (indirect autoradiography) and not for film exposed by direct autoradiography (50). Liquid scintillation counting of dot-blot or slot-blot samples necessitates destruction of the membrane, however, since individual samples must be cut from the membrane to be counted.

Recently, several firms have begun marketing instruments (e.g., Betascope, Betagen Corp., Waltham, Massachusetts, and Ambis Radioanalytic Imaging System, Ambis Corp., San Diego, California) which directly detect and quantitate beta-particle emissions from two-dimensional samples such as hybridization membranes. Some of these are quite precise and sensitive, giving quantitative information 10 to 20 times faster than autoradiographic exposure. In contrast with liquid scintillation counting, the membrane remains intact during quantitation of hybridization signals and can, therefore, be used for subsequent hybridization by stripping off the probe and rehybridizing with another probe. In order to strip the probe, the membrane must not be allowed to dry after hybridization.

Radiolabeled nucleic acids maintain the same specificity as their unlabeled counterparts and can be detected both qualitatively and quantitatively. For these reasons they are highly suitable as probes. Problems arise, however, for two reasons. One is the hazard of using radioactive materials. ^{32}P emits relatively high-energy beta particles which can be a hazard for those individuals exposed to it through careless or extended usage. Lower-level emitters, while not dangerous via exposure, are certainly contamination hazards. Since they are hazards they require special training and facilities to be used safely. This precludes the general use of radiolabeled probes outside a laboratory setting. Another reason radiolabeled probes are problematic is that they decay, both radioactively and chemically, over time. Since the half-life of ^{32}P is 14.2 days and ^{32}P decays to an atom of sulfur, rendering the nucleic acid chemically unstable, probes labeled with this nuclide have the shortest shelf life of all. Once labeled with ^{32}P, a probe can be used with confidence only for approximately two weeks.

For these reasons nonradioactive labeling and detection systems are being developed. Since radiolabeled probes are so powerful in their precision and specificity, however, they remain the standard by which nonradioactive probes are measured.

Nonradioactive labels and alternative formats

The first practical alternative to radiolabeled probes was the biotin label and avidin-enzyme conjugate detection system of Leary et al. (51). The label is incorporated into DNA by the same in vitro enzymatic reactions used in radiolabeling, but the labeled nucleotide triphosphates are biotin analogues of either thymidine triphosphate (TTP) or uridine triphosphate (UTP). The system for detecting the labeled hybrids takes advantage of the strong affinity of biotin for avidin or streptavidin. Conjugates of streptavidin and alkaline phosphatase are used as reporter molecules and, when a chromogenic substrate such as 5-bromo,4-chloro-indoyl phosphate is added, a colored precipitate forms on the membrane, where the probe has hybridized. Sensitivities for this system are generally lower than can be obtained with radiolabeled probes labeled to a high specific activity and specificity has been a problem. Zwadyk et al. (106), in a comparison of two commercially available biotin-based detection systems with ^{32}P-labeled probes, observed that biotinylated probes are less specific and less sensitive than radiolabeled probes. However, Haas and Flemming (26) achieved improved specificity when colony blots were subjected to deproteinization prior to hybridization. DNA can be biotinylated directly, without the use of in vitro enzymatic reactions by using photobiotin, a photoactivatable moiety linked to biotin. When a solution of DNA and photobiotin are mixed and the mixture irradiated with visible light, one biotin molecule per 100 to 400 residues is incorporated. Commercial detection systems can then be used to detect the labeled probe (20). Oligomer probes can also be labeled chemically with biotin, with no effect on T_m, even for oligomers as short as 15 bp (7). Wilchek and Bayer have provided a comprehensive review of the chemistry and applications of avidin-biotin techniques, including nucleic acid probes (105).

Another method for preparing nonradioactively labeled probes is to conjugate an enzyme, such as alkaline phosphatase or horseradish peroxidase directly to the nucleic acid via a short linker arm and detect with a chromogenic substrate (75). Surprisingly, the presence of such a large molecule attached to the probe does not affect the specificity of the hybridization significantly, even when oligomers are similarly labeled (37). This strategy has been used successfully to develop a colormetric assay for detection of enterotoxigenic E. coli, using both a restriction fragment probe (78) and oligomer probes (87).

Immunological techniques have also been used in labeling and detection systems. One strategy is to label the nucleic acid with a hapten and detect it using a monoclonal antibody conjugated to alkaline phosphatase or horseradish peroxidase. Alternatively, instead of the enzyme being conjugated to the primary antibody, it is conjugated to a secondary antibody which binds to the monoclonal antibody. Such procedures are obviously analogous to enzyme-linked immunosorbant assays (ELISAs). Several base analogues have been used to serve as the immunoreactive moiety in the labeled nucleic acid. One of the most successful has been a purine with a dinitrophenyl group attached to the no. 8 carbon atom (101). DNA can also be modified chemically to produce a specifically immunoreactive probe (97). FMC Bioproducts, Rockland, Maine, markets a nonradioactive labeling kit based on the insertion of antigenic sulfone groups into the cytosine moieties of single-stranded DNA. The modified DNA is detected, after hybridization to the target sequence, by a sandwich, immunoenzymatic reaction. A monoclonal antibody, specific for sulfonated cytosine, binds to the probe and an enzyme-conjugated, antimouse immunoglobulin is used as the secondary antibody. A chromogenic substrate is then added to visualize the bound probe.

One of the most promising immunodetection strategies involves production of antibodies with specificity for double-stranded DNA or RNA:DNA hybrids. In this system, the probe nucleic acid need not be labeled or modified at all. Since the antibodies will bind only to double-stranded forms of nucleic acids, only probe-target hybrids will be detected. Boguslawski et al. (5) have produced an anti-DNA:RNA monoclonal antibody to detect hybridization of DNA probes to rRNA and Miller et al. have used this system to detect *E. coli* and *B. subtillis* with a sensitivity of 500 cells (59). Stollar and Rashtchian have also made an anti-DNA:RNA antibody and employed it in conjunction with a *Campylobacter*-specific probe (95).

Fluorogenic compounds, such as fluorescein and tetramethyl-rhodamine, have also been used successfully to label DNA (41) and RNA (76) for use as probes. A powerful method, using fluorogenically labeled, oligomer probes targeted to species-specific regions of 16S rRNA to differentiate a pathogenic bacterium from normal flora in the gut of wasps, has been demonstrated by Delong et al. (11). When this method was used in conjunction with eppifluorescence microscopy, the authors were able to detect single microbial cells.

Labeling and detection systems for nucleic acid probe assays represents an area of very active research, both in the public domain and in private industry. Much of this research is proprietary and does not reach the open literature or is slow to be published. It is possible, however, that nonradioactive labeling techniques may soon be as effective as radioactive probes. Linking of enzymes directly to probes or indirectly via intermediates such as antibodies or other affinity molecules will provide the same increased sensitivity that the ELISA demonstrates when compared with non-ELISA. Microbial ecologists who construct probes for detecting specific organisms or genotypes in the environment can look forward to a wide variety of choices for labeling and detection, with many systems being developed by companies eager to put simple, rapid, highly specific assays on the market.

Research into novel labeling and detection systems has given rise to new hybridization formats improving the overall performance of a probe assay. One new format, called sandwich hybridization, was developed by Ranki et al. to detect nucleic acids in crude samples (73). Instead of using one probe, two nonoverlapping portions of the probe sequence are cloned into two nonhomologous vectors. One vector (the capture DNA) is denatured and affixed to a solid support and the other (the probe DNA) is labeled and supplied in the hybridization mix with a crude nucleic acid sample. Only when target sequence is present to mediate the attachment of the labeled vector to the solid support is radioactivity detected on the filter. In this way, all available target DNA is concentrated by the capture DNA to a specific place on the solid matrix and the probe DNA has more targets with which to hybridize. Plastic supports, such as those used in immunoassays, are suitable for this methodology, making sandwich hybridization using nonradioactive probes a possibility for routine assay of crude samples, analogous to the ELISA (71).

As mentioned above, solution hybridization proceeds much faster than mixed-phase hybridization. Syvanen et al. (96) have devised a system which combines the advantages of sandwich hybridization with the high rate of a solution hybridization reaction. Briefly, antisulfonated DNA antibodies were coated in the wells of microtiter plates and used to capture complexes in which the capture DNA was modified with sulfone and the probe DNA was biotinylated. In this system, hybridization of target to capture DNA and probe DNA occurs in solution and the complexes are bound to the solid phase by affinity of the antibody for the capture DNA. Once the wells of the microtiter plate are washed of free DNA, an avidin-enzyme conjugate system is used to detect the biotinylated probe DNA. Obviously, many combinations of hybridization formats, labeling techniques, and detection schemes are possible given the current state of the art of probe assays.

Polymerase chain reaction

A technique, developed recently, for increasing the sensitivity of any DNA probe assay is the polymerase chain reaction (PCR) which uses a thermostable DNA polymerase (81) to selectively amplify the target sequence in a mixed DNA sample. Briefly stated, two oligomer primers specific for the 5′ flanking regions of a double-stranded DNA target sequence are combined with a sample of DNA; the mixture is then heated to 94°C to denature the DNA. The mixture is cooled to allow the primers to anneal and a DNA polymerase from the thermophilic bacterium *Thermus aquaticus* is added and the reaction is heated to 72°C, the optimal temperature for enzyme activity. New strands of target DNA are synthesized from the primers by the polymerase and the mixture is again heated to denature the DNA. The polymerase is stable at denaturing temperatures, thus when the temperature is again reduced to reanneal the primer and subsequently heated to 70°C, the enzyme again polymerizes new target DNA. As each cycle of heating and cooling is completed, the number of target DNA molecules increases geometrically. When the cycle is repeated 35 times, single-copy target sequences can be amplified by a factor of 10 million (81). Several companies (i.e., Perkin-Elmer, Inc., Norwalk, Connecticut) market programmable heat blocks which can automate the cycling, although the procedure can be carried out by moving samples between three heat blocks set to the appropriate temperatures. The successful use of this technique to amplify selected target sequences in environmental samples has been reported by Steffan et al. (93) using DNA extracted from soil. Precise quantitation of original target sequence in a sample using PCR is difficult since the efficiency of the amplification reaction varies with each sample. Nevertheless, this system shows great promise for detection of rare genes in environmental samples.

Recovery of Nucleic Acids from Environmental Samples

The application of hybridization techniques to studying the microbial ecology of a particular environment rests upon the ability to recover representative samples of the nucleic acids from that environment. Techniques to extract and purify nucleic acids from microorganisms grown in culture have been developed to the point where there are protocols for recovering purified nucleic acids from virtually any organism, provided it can be grown in culture. Most of these techniques are based on the classic extraction procedures of Marmur (chromosomal DNA and RNA; 54), Birnboim and Doly (plasmid DNA; 4), or Holmes and Quigley (plasmid DNA; 34) and can be found with various modifications in most molecular biology handbooks. DNA is generally more stable than RNA and high-quality RNA is much more difficult to obtain than DNA due to the presence of endogenous RNAses that are difficult to inactivate.

Nucleic acids can be extracted from organisms in the environment by culturing them and extracting their nucleic acids using conventional protocols. This is a prudent strategy as long as one is aware of the limitations of such an approach. However, it has long been evident to microbial ecologists that a large portion of the natural microbial population either cannot be cultured or is difficult to obtain in pure culture. In both the marine and soil environments, it has been estimated that only between 0.1 and 12.5 percent of the bacteria can be grown in culture (35). The situation is also complicated by organisms that normally grow readily on laboratory media which, when placed in a stressful environment, enter a viable but nonculturable stage (9, 79) and are thus not readily recoverable on laboratory media. It has also been shown that some genetic traits (potentially those to be probed for) can be lost when organisms are grown under certain conditions (29). This is not to say, however, that combining culture techniques and a hybridization strategy to detect certain organisms in the environment cannot work. On the contrary, it can be an extremely powerful technique. Two examples are the strategy employed by Drahos et al. for detecting recombinant *Pseudomonas fluorescens* released into an experimental field plot (12) and the work of Fredrickson et al. who enumerated mutant bacteria in soil using a combined MPN-hybridization approach (21).

The direct extraction of nucleic acids from environmental samples without the use of culture techniques provides a means of obtaining the total genetic complement of a given population, with representation of all organisms in the community. Methods for obtaining usefully intact and pure nucleic acids are still in their infancy, but some investigators have been able to extract successfully total microorganism nucleic acids from samples collected in selected environments.

Soil and sediment

In recent years several methods for the recovery of total community DNA from soils have been developed. These fall into two main methodologies. One is the direct extracting of DNA from the samples by lysis of the organisms in situ; the other is first to extract as many organisms as possible from the environmental matrix so that the DNA is protected until it is in a more defined environment.

Direct extraction of nucleic acids gives better yields of DNA than obtained when attempting to recover whole bacteria from soils and sediments. A disadvantage of the direct extraction methodologies has been contamination of recovered DNA by humic acids and other soil substances, which inhibit the efficiency of subsequent restriction endonuclease digestion and hybridization reactions. Also, the recovered DNA probably includes that of fungi, algae, protozoa, and other organisms in the sample, as well as free DNA not representing live organisms. G. Sayler and his coworkers (65) have developed a direct extraction protocol for aquatic sediments. The method obtained good yields of DNA but was rather labor-intensive, requiring large samples and volumes for extraction. Several investigators have developed modifications to this protocol which decrease processing time and sample sizes while increasing the number of samples which can be processed. Porteous and Armstrong (72) have developed a direct extraction protocol for 1-g samples of soil that are sufficiently pure to allow restriction digestion of the DNA. Hilger and Myrold (28) have further developed this protocol to allow PCR amplification of *Frankia* spp. genes from soil community DNA using gel purification and glassmilk column chromatography rather than equilibrium density centrifugation.

Holben et al. (32) have developed a method for recovering highly purified DNA from soils. The method involves recovery of the bacterial fraction prior to lysis. The methods for isolating the bacterial fraction from soils and sediments are derived from the differential centrifugation technique first described by Faegri et al. (15). The bacterial fraction is then lysed and the DNA purified for subsequent hybridization analyses. Steffan and Atlas have shown that DNA purified in this fashion is sufficiently pure to allow PCR amplification of specific genes (93). Some advantages of this approach are the removal of soil material and contaminants prior to lysis and that the DNA is less fragmented and more uniform in length, making restriction endonuclease digestion, agarose gel size fractionation, and Southern transfer and hybridization more efficient and easier to interpret. Also, the origin of the DNA is known. A disadvantage of this approach is that all of the bacteria in a sample are not recovered (32); however, the bacterial fraction isolated by differential centrifugation, and thus the isolated DNA, appears to be representative of the total bacterial population (6). The quantities of bacterial DNA isolated from soils and sediments represents larger numbers of bacteria than can be detected by viable count enumerations. Thus, nucleic acid methods are more likely to represent bacterial populations than are nonselective culturing methods (32, 65, 94).

Water

Strategies for extracting nucleic acids from microorganisms in aquatic and marine environments necessarily involve concentration of cells from the water sample before cell lysis. Unless cells are first concentrated in some way, the volumes required for adequate yield are, in general, unmanageable. The approach taken by most workers is to concentrate cells onto a micropore filter and lyse the cells on the filter in a small volume of liquid (22, 68, 69, 89). Paul and Myers extracted cellular DNA from aquatic samples without first concentrating the sample, but the quantities obtained are insufficient for molecular manipulations such as cloning or hybridization (68). Since the pore size of the filters used for concentrating the sample must be 0.22 to 0.45 μm in diameter to retain bacteria, filters become clogged with particulates and larger organisms before sufficient volume is filtered. This problem has been addressed in a variety of ways. Prefiltration using a high-capacity glass fiber pad permits several-fold higher volumes to be filtered (22, 89), however, from 5 to 20 percent of the bacterial population is also retained depending upon the particulate composition of the water and how often the prefilter is changed (89). Alternatively, a tangential-flow filtration system was used by Delong et al. to concentrate cells in 1640 liters of oligotrophic ocean water to a volume of several milliliters for subsequent extraction, hybridization and cloning of the DNA (E. F. Delong, S. J. Giovannoni, T. M. Schmidt, and N. R. Pace. 1988. The use of rRNA sequences to characterize picoplankton populations. In *Abstracts of the First International Symposium on Marine Molecular Biology*. Center of Marine Biotechnology, University of Maryland, Baltimore.). A third strategy is to use filters with large surface area (22) or filters designed specifically for high-capacity filtration (89).

The system of Somerville et al. (89) utilizes a high-capacity cylindrically mounted filter through which water is pumped using a peristaltic pump. The filter is housed in a chamber with a volume of 2 ml. Once the water is passed through the filter, a lysis buffer can be added into the chamber and lysis of the cells on the filter can be effected without further manipulation of the filter. The lysate is drawn off the housing and both DNA and RNA can be purified. The advantage is that significant volumes can be concentrated and the entire process is carried out in the disposable filter unit. Use of a portable peristaltic pump and the ability to stabilize quickly and freeze the filter units for later lysis and extraction make the system amenable to the field (89). The application of this system for detection of *Salmonella* in estuarine water using a hybridization probe has been demonstrated (45).

References

1. Attwood, G. T., R. A. Lockington, G. P. Xue, and J. D. Brooker. 1988. Use of a unique gene sequence as a probe to enumerate a strain of *Bacteroides ruminicola* introduced into the rumen. *Appl. Environ. Microbiol.* 54:534–539.
2. Ausubel, F. M., R. Bent, R. E. Kingston, D. D. Moore, J. A. Smith, J. G. Seidman, and K. Struhl. 1987. *Current Protocols in Molecular Biology*. Greene Publishing and Wiley, New York.
3. Barkay, T., D. L. Fouts, and B. H. Olson. 1985. Preparation of a DNA gene probe for detection of mercury resistance genes in gram-negative bacterial communities. *Appl. Environ. Microbiol.* 49:686–692.
4. Birnboim, H. C., and J. Doly. 1979. A rapid alkaline extraction procedure for screening recombinant plasmid DNA. *Nucleic Acids Res.* 7:1513–1523.

5. Boguslawski, S. J., D. E. Smith, M. A. Michalak, K. E. Mickelson, C. O. Yehle, W. L. Patterson, and R. J. Carrico. 1986. Characterization of a monoclonal antibody to DNA:RNA and its application to immunodetection of hybrids. *J. Immunol. Methods* 89:123–130.

6. Bone, T. L., and D. L. Balkwill. 1986. Improved flotation technique for microscopy of *in situ* soil and sediment microorganisms. *Appl. Environ. Microbiol.* 51:462–468.

7. Chollet, A., and E. H. Kawashima. 1985. Biotin-labelled synthetic oligodeoxyribonucleotides: Chemical synthesis and uses as hybridization probes. *Nucleic Acids Res.* 13:1529–1541.

8. Colwell, R. R., C. Somerville, I. Knight, and W. Straube. 1988. Detection and monitoring of genetically-engineered micro-organisms. In M. Sussman, C. H. Collins, F. A. Skinner, and D. E. Stewart-Tull (eds.), *The Release of Genetically-Engineered Micro-organisms*, Proceedings of the First International Conference. Academic Press, London, pp. 47–61.

9. Colwell, R. R., P. R. Brayton, D. J. Grimes, D. R. Roszak, S. A. Huq, and L. M. Palmer. 1985. Viable, but non-culturable *Vibrio cholerae* and related pathogens in the environment: Implications for release of genetically engineered microorganisms. *Bio/Technology* 3:817–820.

10. Colwell, R. R., M. T. MacDonell, and D. Swartz. 1989. Identification of an antarctic endolithic microorganism by 5S rRNA sequence analysis. *System. Appl. Microbiol.* 11:182–186.

11. Delong, E. F., G. S. Wickham, and N. R. Pace. 1989. Phylogenetic stains—ribosomal RNA-based probes for the identification of single cells. *Science* 243: 1360–1363.

12. Drahos, D. J., B. C. Hemming, and S. McPherson. 1986. Tracking recombinant organisms in the environment: Beta-galactosidase as a selectable non-antibiotic marker for fluorescent pseudomonads. *Bio/Technology* 4:439–444.

13. Echeverria, P., J. Seriwatana, O. Chityothin, W. Chaicumpa, and C. Tirapat. 1982. Detection of enterotoxigenic *Escherichia coli* in water by filter hybridization with three enterotoxin gene probes. *J. Clin. Microbiol.* 16:1086–1090.

14. Esaki, T., N. Takeuchi, S. Liu, A. Kai, H. Yamamoto, and E. Yabuuchi. 1988. Small-scale DNA preparation for rapid genetic identification of *Campylobacter* species without radioisotope. *Microbiol. Immunol.* 32:141–150.

15. Faegri, A., V. L. Torsvik, and J. Goksoyr. 1977. Bacterial and fungal activities in soil: separation of bacteria by rapid fractional centrifugation technique. *Soil Biol. Biochem.* 9:105–112.

16. Feinberg, A. P., and B. Vogelstein. 1983. A technique for radiolabeling DNA restriction endonuclease fragments to high specific activity. *Anal. Biochem.* 132:6–13.

17. Fitts, R. 1985. Development of a DNA hybridization test for the presence of *Salmonella* in foods. *Food Technol.* 39:95–102.

18. Fitts, R. 1986. Detection of foodborne microorganisms by DNA hybridization. In M. D. Pierson and N. J. Stern (eds.), *Foodborne Microorganisms and Their Toxins: Developing Methodology*. Marcel Dekker, New York, pp. 283–290.

19. Fitts, R., M. Diamond, C. Hamilton, and M. Neri. 1983. DNA:DNA hybridization assay for detection of *Salmonella* spp. in foods. *Appl. Environ. Microbiol.* 46:1146–1151.

20. Forster, A. C., J. L. McInnes, D. C. Skingle, and R. H. Symons. 1985. Nonradioactive hybridization probes prepared by the chemical labeling of DNA and RNA with a novel reagent, photobiotin. *Nucleic Acids Res.* 13:745–761.

21. Fredrickson, J. K., D. F. Bezdicek, F. J. Brockman, and S. W. Li. 1988. Enumeration of Tn5 mutant bacteria in soil by using a most-probable-number-DNA hybridization procedure and antibiotic resistance. *Appl. Environ. Microbiol.* 54:446–453.

22. Furhman, J. A., D. E. Comeau, A. Hagstrom, and A. M. Chan. 1988. Extraction from natural planktonic organisms of DNA suitable for molecular biological studies. *Appl. Environ. Microbiol.* 54:1426–1429.

23. Giovannoni, S. J., E. F. Delong, G. J. Olsen, and N. R. Pace. 1988. Phylogenetic group-specific oligodeoxynucleotide probes for identification of single microbial cells. *J. Bacteriol.* 170:720–726.

24. Grimont, P. A. D., F. Grimont, N. Desplaces, and P. Tchen. 1985. DNA probe specific for *Legionella pneumophila*. *J. Clin. Microbiol.* 21:431–437.
25. Grunstein, M., and D. S. Hogness. 1975. Colony hybridization: A method for the isolation of cloned DNAs that contain a specific gene. *Proc. Natl. Acad. Sci. USA* 72:3961–3965.
26. Haas, M. J., and D. H. Fleming. 1988. A simplified lysis method allowing the use of biotinylated probes in colony hybridization. *Anal. Biochem.* 168:239–246.
27. Hazen, T. C., and L. Jimenez. 1988. Enumeration and identification of bacteria from environmental samples using nucleic acid probes. *Microbiol. Sciences* 5:340–343.
28. Hilger, A. B., and D. D. Myrold. 1991. Method for the extraction of Frankia DNA from soil. *Agric. Ecosys. Environ.* (in press).
29. Hill, W. E., and C. L. Carlisle. 1981. Loss of plasmids during enrichment for *Escherichia coli*. *Appl. Environ. Microbiol.* 41:1046–1048.
30. Hill, W. E., W. L. Payne, and C. C. G. Auliso. 1983. Detection and enumeration of virulent *Yersinia enterocolitica* in food by DNA colony hybridization. *Appl. Environ. Microbiol.* 46:636–641.
31. Hodgson, A. L. M., and W. P. Roberts. 1983. DNA colony hybridization to identify *Rhizobium* strains. *J. Gen. Microbiol.* 129:207–212.
32. Holben, W. E., J. K. Jansson, B. K. Chelm, and J. M. Tiedje. 1988. DNA probe method for the detection of specific microorganisms in the soil bacterial community. *Appl. Environ. Microbiol.* 54:703–711.
33. Holben, W. E., and J. M. Tiedje. 1988. Applications of nucleic acid hybridization in microbial ecology. *Ecology* 69:561–568.
34. Holmes, D. S., and M. Quigley. 1981. A rapid boiling method for preparation of bacterial plasmid. *Anal. Biochem.* 114:193–197.
35. Hoppe, H. G. 1978. Relationships between active bacteria and heterotrophic potential in the sea. *Neth. J. Sea Res.* 12:78–98.
36. Hu, N. T., and J. Messing. 1982. The making of strand-specific M13 probes. *Gene* 17:271–277.
37. Jablonski, E., E. W. Moonmaw, R. H. Tullis, and J. L. Ruth. 1986. Preparation of oligodeoxynucleotide-alkaline phosphatase conjugates and their use as hybridization probes. *Nucleic Acids Res.* 14:6115–6128.
38. Jagow, J., and W. E. Hill. 1986. Enumeration by DNA colony hybridization of virulent *Yersinia enterocolitica* colonies in artificially contaminated food. *Appl. Environ. Microbiol.* 51:441–443.
39. Jansson, J. K., W. E. Holben, and J. M. Tiedje. 1989. Detection in soil of a deletion in an engineered DNA sequence by using DNA probes. *Appl. Environ. Microbiol.* 55:3022–3025.
40. Kafatos, F. C., C. W. Jones, and A. Efstradiadis. 1979. Determination of nucleic sequence homologies and relative concentrations by a dot hybridization procedure. *Nucleic Acids Res.* 7:1541–1552.
41. Kamur, A., P. Tchen, F. Roullet, and J. Cohen. 1988. Nonradioactive labeling of synthetic oligonucleotide probes with terminal deoxynucleotidyl transferase. *Anal. Biochem.* 169:376–382.
42. Kaper, J. B., H. B. Bradford, N. C. Roberts, and S. Falkow. 1982. Molecular epidemiology of *Vibrio cholerae* in the U.S. Gulf Coast. *J. Clin. Microbiol.* 16:129–134.
43. Kaper, J. B., and M. M. Levine. 1981. Cloned cholera enterotoxin genes and prevention of cholera. *Lancet* 2:1162–1163.
44. Kingsbury, D. T. 1985. Rapid detection of mycoplasmas with DNA probes. In D. T. Kingsbury and S. Falkow (eds.), *Rapid Detection and Identification of Infectious Agents*. Academic Press, New York, pp. 219–233.
45. Knight, I. T., S. Shults, C. W. Kaspar, and R. R. Colwell. 1990. Direct detection of *Salmonella* spp. in estuaries by using a DNA probe. *Appl. Environ. Microbiol.* 56:1059–1066.
46. Kohne, D., J. Hogan, V. Jonas, E. Dean, and T. H. Adams. 1986. Novel approach for rapid and sensitive detection of microorganisms: DNA probes to rRNA. In L.

Leive (ed.), *Microbiolog—1986*. American Society for Microbiology, Washington, D.C., pp. 110–112.

47. Krauss, J., W. Ludwig, and K. H. Schleifer. 1986. A cloned 23S rRNA gene fragment of *Bacillus subtilis* and its use as a hybridization probe of conserved character. *FEMS Microbiol. Lett.* 33:89–93.

48. Kuritza, A. P., C. E. Getty, P. Shaughnessy, R. Hess, and A. A. Salyers. 1986. DNA probes for identification of clinically important *Bacteroides* species. *J. Clin. Microbiol.* 23:343–349.

49. Kuritza, A. P., P. Shaughnessy, and A. A. Salyers. 1986. Enumeration of polysaccharide-degrading *Bacteroides* species in human feces by using species-specific DNA probes. *Appl. Environ. Microbiol.* 51:385–390.

50. Laskey, R. A., and A. D. Mills. 1977. Enhanced autoradiographic detection of ^{32}P and ^{125}I using intensifying screens and by hypersensitized film. *FEBS Lett.* 82: 314–316.

51. Leary, J. J., D. J. Brigati, and D. C. Ward. 1983. Rapid and sensitive colormetric method for visualizing biotin-labeled DNA probes hybridized to DNA or RNA immobilized on nitrocellulose: Bio-blots. *Proc. Natl. Acad. Sci USA* 80:4045–4049.

52. Maas, R. 1983. An improved colony hybridization method with significantly increased sensitivity for detection of single genes. *Plasmid* 10:296–298.

53. Maniatis, T., E. F. Firtsch, and J. Sambrook. 1982. *Molecular Cloning: A Laboratory Manual*. Cold Spring Harbor Laboratory, Cold Spring Harbor, New York.

54. Marmur, J. 1961. A procedure for the isolation of deoxyribonucleic acid from microorganisms. *J. Mol. Biol.* 3:208–218.

55. Marmur, J., and P. Doty. 1962. Determination of the base composition of deoxyribonucleic acid from its thermal denaturation temperature. *J. Mol. Biol.* 5: 109–118.

56. Matthews, J., and L. J. Kricka. 1988. Analytical strategies for the use of DNA probes. *Anal. Biochem.* 169:1–25.

57. Meinkoth, J., and G. Wahl. 1984. Hybridization of nucleic acids immobilized on solid supports. *Anal. Biochem.* 38:267–284.

58. Melton, D. A., P. A. Krieg, M. R. Rebagliati, T. Maniatis, K. Zinn, and M. R. Green. 1984. Efficient *in vitro* synthesis of biologically active RNA and RNA hybridization probes from plasmids containing a bacteriophage SP6 promoter. *Nucleic Acids Res.* 12:7035–7056.

59. Miller, C. A., W. L. Patterson, P. K. Johnson, C. T. Swartzell, F. Wogoman, J. P. Albarelli, and R. J. Carrico. 1988. Detection of bacteria by hybridization of rRNA with DNA-latex and immunodetection of hybrids. *J. Clin. Microbiol.* 26:1271–1276.

60. Morotomi, M., T. Ohno, and M. Masahiko. 1988. Rapid and correct identification of intestinal *Bacteroides* spp. with chromosomal DNA probes by whole-cell dot blot hybridization. *Appl. Environ. Microbiol.* 54:1158–1162.

61. Moseley, S. L., I. Huq, A. R. M. A. Alim, M. So, M. Samadpour-Motalebi, and S. Falkow. 1980. Detection of enterotoxigenic *Escherichia coli* by DNA colony hybridization. *J. Infect. Dis.* 142:892–898.

62. Moseley, S. L., P. Echeverria, J. Seriwatana, C. Tirapat, W. Chaicumpa, T. Sakuldaipeara, and S. Falkow. 1982. Identification of enterotoxigenic *Escherichia coli* by colony hybridization using three enterotoxin gene probes. *J. Infect. Dis.* 145:863–869.

63. Nataro, J. P., M. M. Baldini, J. B. Kaper, R. E. Black, N. Bravo, and M. M. Levine. 1985. Detection of an adherence factor of enteropathogenic *Escherichia coli* with a DNA probe. *J. Infect. Dis.* 152:560–565.

64. Ogram, A. V., and G. S. Sayler. 1988. The use of gene probes in the rapid analysis of natural microbial communities. *J. Industrial Microbiol.* 3:281–292.

65. Ogram, A., G. S. Sayler, and T. Barkay. 1987. The extraction and purification of microbial DNA from sediments. *J. Microbiol. Methods* 7:57–66.

66. Pace, N. R., D. A. Stahl, J. Lane, and G. L. Olsen. 1986. The use of rRNA sequences to characterize natural microbial populations. *Adv. Microb. Ecol.* 9:1–55.

67. Palmer, L., and S. Falkow. 1985. Selection of DNA probes for use in the diagnosis

of infectious disease. In D. T. Kingsbury and S. Falkow (eds.), *Rapid Detection and Identification of Infectious Agents*. Academic Press, New York, pp. 211–218.

68. Paul, J. H., and B. Myers. 1982. Fluorometric determination of DNA in aquatic microorganisms by use of Hoechst 33258. *Appl. Environ. Microbiol.* 43:1393–1399.

69. Paul, J. H., W. H. Jeffrey, and M. F. DeFlaun. 1987. Dynamics of extracellular DNA in the marine environment. *Appl. Environ. Microbiol.* 53:170–179.

70. Pettigrew, C. A., and G. S. Sayler. 1986. The use of DNA:DNA colony hybridization in the rapid isolation of 4-chlorobiphenyl degradative bacterial phenotypes. *J. Microbiol. Methods* 5:205–213.

71. Polsky-Cynkin, R., G. H. Parsons, L. Allerdt, G. Landes, G. Davis, and A. Rashtchian. 1985. Use of DNA immobilized on plastic and agarose supports to detect DNA by sandwich hybridization. *Clin. Chem.* 31:1438–1443.

72. Porteous, L. A., and J. L. Armstrong. 1991. Recovery of bulk DNA from soil using a rapid, small-scale extraction method. *Appl. Environ. Microbiol.* (in press).

73. Ranki, M., A. Palva, M. Virtanen, M. Laaksonen, and H. Soderlund. 1983. Sandwich hybridization as a convenient method for the detection of nucleic acids in crude samples. *Gene* 21:77–85.

74. Regensburger, A., W. Ludwig, and K. Schleifer. 1988. DNA probes with different specificities from a cloned 23S rRNA gene of *Micrococcus luteus*. *J. Gen. Microbiol.* 134:1197–1204.

75. Renz, M., and C. Kurz. 1985. A colormetric method for DNA hybridization. *Nucleic Acids Res.* 12:3435–3444.

76. Richardson, R. W., and R. I. Gumport. 1983. Biotin and fluorescent labeling of RNA using T4 RNA ligase. *Nucleic Acids Res.* 11:6167–6184.

77. Rigby, P. W. J., M. Dieckmann, C. Rhodes, and P. Berg. 1977. Labeling DNA to high specific activity *in vitro* by nick translation with DNA polymerase I. *J. Mol. Biol.* 113:237–251.

78. Romick. T. L., J. A. Lindsay, and F. F. Busta. 1987. A visual DNA probe for detection of enterotoxigenic *Escherichia coli* by colony hybridization. *Lett. Appl. Microbiol.* 5:87–90.

79. Roszak, D. B., and R. R. Colwell. 1987. Survival strategies of bacteria in the natural environment. *Microbiol. Rev.* 51:365–379.

80. Ruben, F. A., D. J. Kopecko, K. F. Noon, and L. S. Baron. 1985. Development of a DNA probe to detect *Salmonella typhi*. *J. Clin. Microbiol.* 22:600–605.

81. Saiki, R. K., D. H. Gelfand, S. Stoffel, S. J. Scharf, R. Higuchi, G. T. Horn, K. B. Mullis, and H. A. Erlich. 1988. Prime-directed enzymatic amplification of DNA with a thermostable DNA polymerase. *Science* 239:487–491.

82. Salyers, A. A., S. P. Lynn, and J. F. Gardner. 1983. Chromosomal DNA probes for identification of *Bacteroides* species. *J. Bacteriol.* 154:287–293.

83. Sambrook, J., E. F. Fritsch, and T. Maniatis. 1989. *Molecular Cloning: A Laboratory Manual*, 2d ed. Cold Spring Harbor Laboratory Press, Cold Spring Harbor, New York.

84. Sayler, G. S., M. S. Shields, E. T. Tedford, A. Breen, S. W. Hooper, K. M. Sirotkin, and J. W. Davis. 1985. Application of DNA-DNA colony hybridization to the detection of catabolic genotypes in environmental samples. *Appl. Environ. Microbiol.* 49:1295–1303.

85. Schleifer, K. H., W. Ludwig, J. Kraus, and H. Festl. 1985. Cloned ribosomal ribonucleic acid genes from *Pseudomonas aeruginosa* as probes for conserved deoxyribonucleic acid sequences. *Int. J. Syst. Bacteriol.* 35:231–236.

86. Schofield, P. R., M. A. Djordjevie, B. G. Rolfe, J. Shine, and J. M. Watson. 1983. A molecular linkage map of nitrogenase and nodulation genes in *Rhizobium trifilii*. *Mol. Gen. Genet.* 192:459–465.

87. Seriwatana, J., P. Echeverria, D. N. Taylor, T. Sakuldaipeara, S. Changchawalit, and O. Chivoratanoid. 1987. Identification of enterotoxigenic *Escherichia coli* with synthetic alkaline phosphatase-conjugated oligonucleotide probes. *J. Clin. Microbiol.* 25:1438–1441.

88. So, M., W. S. Dallas, and S. Falkow. 1978. Characterization of an *Escherichia coli* plasmid encoding for synthesis of heat-labile toxin: Molecular cloning of the toxin determinant. *Infect. Immun.* 21:405–411.

89. Somerville, C. C., I. T. Knight, W. L. Straube, and R. R. Colwell. 1989. Simple, rapid method for direct isolation of nucleic acids from aquatic environments. *Appl. Environ. Microbiol.* 55:548–554.

90. Southern, E. M. 1975. Detection of species specific sequences among DNA fragments separated by gel electrophoresis. *J. Mol. Biol.* 98:503–517.

91. Stahl, D. A., B. Flesher, H. R. Mansfield, and L. Montgomery. 1988. Use of phylogenetically based hybridization probes for studies of ruminal microbial ecology. *Appl. Environ. Microbiol.* 54:1079–1084.

92. Stahl, D. A., D. L. Lane, G. J. Olsen, and N. R. Pace. 1985. Characterization of a Yellowstone hot spring microbial community by 5S ribosomal RNA sequences. *Appl. Environ. Microbiol.* 49:1379–1384.

93. Steffan, R. J., and R. M. Atlas. 1988. DNA amplification to enhance detection of genetically engineered bacteria in environmental samples. *Appl. Environ. Microbiol.* 54:2185–2191.

94. Steffan, R. J., J. Goksøyr, A. K. Bej, and R. M. Atlas. 1988. Recovery of DNA from soils and sediments. *Appl. Environ. Microbiol.* 54:2908–2915.

95. Stollar, B. D., and A. Rashtchian. 1987. Immunochemical approaches to gene probe assays. *Anal. Biochem.* 161:387–394.

96. Syvanen, A. C., M. Laaksonen, and H. Soderlund. 1986. Fast quantification of nucleic acid hybrids by affinity-based collection. *Nucleic Acids Res.* 14:5037–5048.

97. Tchen, P., R. P. P. Fuchs, E. Sage, and M. Leng. 1984. Chemically modified nucleic acids as immunodetectable probes in hybridization experiments. *Proc. Natl. Acad. Sci. USA* 81:3466–3470.

98. Tenover, F. C. 1986. Studies of antimicrobial resistance genes using DNA probes. *Antimicrob. Agents Chemother.* 29:721–725.

99. Tenover, F. C. 1988. Diagnostic deoxyribonucleic acid probes for infectious diseases. *Clin. Microbiol. Rev.* 1:82–101.

100. Thomas, P. S. 1980. Hybridization of denatured RNA and small DNA fragments transferred to nitrocellulose. *Proc. Natl. Acad. Sci. USA* 77:5201–5205.

101. Vincent, C., P. Tchen, M. Cohen-Solal, and P. Kourilsky. 1982. Synthesis of 8-(2-4 dinitrophenyl 2-6 aminohexyl) amino-adenosine 5' triphosphate: Biological properties and potential uses. *Nucleic Acids Res.* 10:6787–6796.

102. Wallace, R. B., J. Shaffer, R. F. Murphy, J. Bonner, T. Hirose, and K. Itakura. 1979. Hybridization of synthetic oligodeoxynucleotides to PhiX 174 DNA: The effect of single base pair mismatch. *Nucleic Acids Res.* 6:3543–3557.

103. Wetmur, J. G. 1975. Acceleration of DNA renaturation rates. *Biopolymers* 14:2517–2524.

104. Wetmur, J., and N. Davidson. 1968. Kinetics of renaturation of DNA. *J. Mol. Biol.* 31:349–370.

105. Wilchek, M., and E. A. Bayer. 1988. The avidin-biotin complex in bioanalytical applications. *Anal. Biochem.* 171:1–32.

106. Zwadyk, P., Jr., R. C. Cooksey, and C. Thornsberry. 1986. Commercial detection methods for biotinylated gene probes: comparison with [^{32}P]-labeled DNA probes. *Curr. Microbiol.* 14:95–100.

Chapter 5

Microscopy Applications for Analysis of Environmental Samples

Jeffrey J. Byrd

Rita R. Colwell

Introduction

Historically, the microscope has been a standard tool of the microbiologist. It is used for examination of structures and morphology of microorganisms, as well as for enumeration. Despite the rush toward applying molecular genetic methods to microbial ecology, the microscope remains the primary tool for analysis of specimens collected from the environment. Detection and enumeration of microorganisms using bright-field, epifluorescent, and electron microscopes are described herein. Determination of the viability of microorganisms is also addressed, with respect to microscopy as a tool for such assessment. References are provided for the methods discussed, rather than providing an instructional manual for use of the various microscopes and specimen preparation. Thus, an overview of methods is presented so that the reader will have information for deciding on the methodology appropriate for the problem under study.

Morphology-Based Identification

The morphology and ultrastructure of an organism has been a basic method for identification of microorganisms historically. The Gram stain (46), which differentiates bacteria by their cell wall structure, is the most obvious example of the value of microscopy. Many methods are available for staining cells and determining their morphology, most of which are described by Doetsch (30) in detail. In general it is necessary to employ a staining procedure that will enhance specific features of the cell or population of cells.

In the mid-1950s, methods for direct observation of environmental samples utilized bright-field stains, e.g., methylene blue (57), Victoria pure blue (32) and 1% erythrosine in 5% phenol (59). The methods included filtering the sample and staining the filter with these bright-field stains. These early attempts at direct enumeration were not always successful because simple staining and bright-field microscopy do not allow differentiation of bacteria from detritus (39).

In addition to staining, other methods are used for detection and enumeration of microorganisms. Casida (21) utilized a modified metallurgical microscope to study unstained soil preparations and was able to describe soil organisms in situ. However, it is essential that one have good knowledge of soil systems and organisms making up the soil biota to use this microscope. Another method for direct enumeration of microorganisms in environmental samples makes use of the Petroff Hauser counting chamber. The chamber is a gridded slide that enables the observer to count the number of cells in each well. A drawback is that at least 10^7 cells per milliliter must be present in the sample for enumeration to be possible (39). To meet this criterion, some water samples may have to be concentrated before counting is possible.

Phase-contrast microscopy is an excellent method for examining environmental samples without staining. The phase-contrast microscope separates direct background light from the light passing through the object. This provides enough contrast to detect the morphology of a microorganism that would not otherwise be seen without staining. Richards and Krabek (93) utilized phase-contrast microscopy to examine bacteria that were retained on filters. At the time, this seemed like a promising method to count microorganisms directly in environmental samples. However, phase-contrast microscopy does not allow ready differentiation between cells and detritus (39). Therefore, phase-contrast microscopy is not a powerful enough method for enumerating organisms in environmental samples and is best used in pure culture analysis.

Although bright-field and phase-contrast microscopy provide information useful for enumeration of microorganisms, both are limited in application to identification and classification. Nevertheless, bright-field and phase-contrast microscopy each have a role in microbial ecology and represent an essential tool for the microbial ecologist.

Epifluorescent Microscopy

Total count

Detection and direct quantitation of microorganisms in the environment has been enhanced by the use of the fluorescent microscope. Traditional microscopic methods, such as phase-contrast microscopy, enable counting of up to 90 percent of the bacteria in polluted and eutrophic waters, mainly because in such environments these organisms tend to occur as large rods, i.e., responding to the elevated concentrations of nutrients present in the water (29). Methods other than fluorescence microscopy, for waters of low productivity, may yield underestimates of total counts by an order of magnitude or more because microorganisms in low-nutrient waters tend to occur as small, rounded-up rods or cocci. Ferguson and Rublee (34) reported that 80 percent of the bacteria in coastal water samples were small, coccoidal forms of 0.5 μm in diameter. These small forms were shown, in fact, to be small rods and vibrios when transmission electron microscopy was employed (120). Because of the significant contrast achieved with fluorescent microscopy, such morphological entities can be detected, even though they appear as very small cocci under the microscope. Fluorescent microscopy methods usually require specific staining of the microorganisms with a fluorochrome, followed by examination by epifluorescent microscopy.

Use of fluorochromes requires different procedures from those employed in bright-field microscopy. A significant difference, for example, is the necessity to prevent the fluorochrome from drying, so that an accurate count of the microorganisms can be achieved (53). The method described by Hobbie et al. (52) involves direct staining of the microorganisms with a fluorochrome, i.e., acridine orange (AO), and observation of the preparation by wet mount. Examination of slides was accomplished by both epifluorescent and phase-contrast microscopy because many particulates in the seawater naturally fluoresced green. Thus, confirmation of which fluorescent dots were cell-like had to be done by phase-contrast microscopy. Since quantitation is more difficult in wet-mount preparations, a solid support is required, e.g., membrane filters, for improved observation and enumeration.

In 1928, Cholodny devised a method for the concentration of bacteria by filtration. The residue from the Cholodny filter was transferred to a slide and examined by microscopy. Rasumov (90, 91) utilized Cholodny's method of filtering but developed a procedure for direct examination of the filter. By placing a drop of immersion oil on the filter, the oil rendered the filter transparent and able to be scanned directly by microscopy. Ehrlich (32) and Jannasch (57, 58) improved on this technique by washing the filter with water and drying the filter prior to examination. In both cases the filters were examined by bright-field microscopy.

The appropriate membrane filter to use for direct counting has been the subject of a number of studies. Cellulose filters were traditionally employed in the early studies. However, cellulose filters, although suitable for sterilization by filtration, create problems in viewing directly those microorganisms trapped on the filter. The cellulose matrix proved too rough and irregular for effective microscopy (14) and the matrix is interwoven, so much so that detection of organisms is obscured (29, 53, 115). With cellulose filters, the bacteria tend to become trapped in the filter matrix, and are thereby sheltered from detection. Fleischer et al. (37) first proposed the use of plastic filters possessing uniformly etched holes that provide a smooth surface for filtration and collection of microorganisms. Polycarbonate membrane filters were compared with cellulose membranes under the scanning electron microscope by Todd and Kerr (115). *Pseudomonas, Staphylococcus,* and *Bacillus* spp. were employed in the tests. The polycarbonate filter was found to provide better detection because of its smooth surface. A problem with polycarbonate membranes, however, is that portions of the membrane can be hydrophobic (53). To overcome this impediment, Hobbie et al. (53) suggested the use of 0.5% solution of surfactant, but only for exceptional cases, since the surfactant tends to increase overall fluorescence, thereby interfering with detection of the bacteria. Filters currently in use tend to be less hydrophobic, with the result that there is usually no need for alteration of the filter surface. Polycarbonate filters also autofluoresce; however, this decreases the contrast between filter and specimen, which is overcome by staining the membrane. Thus, a stain commonly employed to coat filters for epifluorescent microscopy is irgalan black, which greatly enhances contrast (53). Prestained filters are now available commercially and are popular despite the fact that staining involves very little effort or cost. Because of the advantages of the polycarbonate membrane, it has become the membrane of choice for enumeration of microorganisms in samples collected from the environment.

Many fluorochromes have been utilized for the enumeration of microorganisms but acridine orange (3,6-tetramethyl diaminoacridine) is the most common. Acridine orange (AO) has been used for direct staining since 1948, when Strugger (108) used it to differentiate between cells and detritus. AO is a nucleic acid–specific stain that fluoresces upon intercalation of the nucleic acid by the stain. At relatively low concentrations of AO, nucleic acids from both procaryotic and eukaryotic cells interact with AO in the same way (123). AO interacts differently with DNA and RNA because of their different secondary structure. RNA, a random coil matrix, allows interaction with many molecules of AO, forming dimers which fluoresce red-orange. AO will also fluoresce red-orange after reaction with single-stranded DNA or degraded DNA. The interaction of AO with double-stranded DNA, because of the rigid structure of the double helix, forms monomers, which fluoresce green. Background material which fluoresces as a result of interaction with AO tends to fluoresce orange-red.

Viable microorganisms were hypothesized to produce a green fluorescence and, thereby, be directly detectable. However, AO as a useful agent for determining the viability of microorganisms has not been universally accepted (60) because it was shown that degree and type of fluorescence are related to concentration and contact time with the fluorochrome, as much as viability at the time of staining. Also, loss of viability and shift in color of fluorescence do not correlate within the same time frames, resulting in inaccurate assumption of viability (64).

To utilize AO more effectively in enumeration of aquatic microorganisms, Hobbie et al. (53) developed a method that employs polycarbonate membranes. AO is added to the sample and the preparation is incubated for a standardized time. The membranes are stained with irgalan black and the stained sample is filtered through it. A drop of low-fluorescing immersion oil is added below and on top of the filter, which is placed on a slide. A coverslip is placed over the filter and examination is by fluorescent microscopy.

Preservation of the sample prior to staining is achieved by fixation with formalin. Formalin treatment will retain cells at the original concentrations for up to two weeks or longer (29). After filtration of the sample, the filter should be kept moist to obtain the most reliable results (29). To guard against inaccurately high counts, all reagents should be clarified by filtration. For increased shelf life of reagents, addition of 2% formalin is recommended (29) so that filtration is not needed before every staining.

AO staining has predominantly been utilized for enumeration of aquatic microorganisms. Attempts have been made to adapt the AO technique for use in soil systems (22, 70). Casida (22) found that AO could be used to detect *Sarcina* spp. immediately after addition to soil, but failed to induce fluorescence in a portion of the resident soil microbial population. Either the nonstaining microorganisms were attached to soil particles or possessed extracellular matrices. To determine if release of the microorganisms from the soil is efficient, McDaniel and Capone (70) compared four methods of dispersion of soil microorganisms, with sonication found to be the best for the soil system under study. They recommended comparing methods with each new soil system to be analyzed.

AO staining procedures have also become routine for enumeration of microorganisms in food (104) and milk (27). Epifluorescent counting methods for milk and food samples have been reviewed by Pettipher (82).

The AO procedure has been employed as a differential stain by Rodrigues et al. (94), who utilized the Gram staining procedure with AO as the counterstain. A combination of bright-field and epifluorescent microscopy revealed that bacteria fluorescing bright orange were gram-negative. The procedure has been applied to the study of raw milk and appears to have utility in detecting gram-negative organisms in environmental samples. However, extension to other systems has not been done in any extensive manner.

Dyes other than AO have been suggested for application in enumerating bacteria by fluorescence. Euchrysine 2GNX (E-2GNX), a fluorochrome that is an acridine derivative, was utilized by Jones (60). When compared with AO, E-2GNX yielded favorable results, e.g., E-2GNX gave consistently higher results than AO. Jones (60) pointed out that the method employed will influence fluorochrome staining significantly, with a result that counts may differ according to method used. Daley and Hobbie (29) did a comparative study between the two dyes and found each to yield comparable results. However, euchrysine-stained samples faded too rapidly to be counted in some cases.

A fluorochrome used previously to detect eukaryotic DNA is DAPI (4'6-diamidino-2-phenylindole; 26, 121). DAPI, a DNA-specific stain which fluoresces bright blue when bound to DNA, was found to be useful for detection of *Mycoplasma* spp. contaminating tissue culture cell lines (98). Eukaryotic cells readily take up the stain, with all of the stain concentrated in the nuclear region without cytoplasmic fluorescence. *Mycoplasma* contamination is readily detected in the cytoplasm of tissue culture cells by fluorescence of the *Mycoplasma*. DAPI was adapted for counting aquatic microorganisms by Porter and Feig (84), who found DAPI to work better than AO for eutrophic and seston-rich water samples. Clays, colloids, and detritus of such samples produce a reddish-orange background, which impairs AO counting. DAPI and AO were comparable in efficiency of enumeration of total organisms in aquatic samples. Porter and Feig (84) also reported that DAPI-stained slide counts remained constant for up to 24 weeks when the stained slides were stored at 4°C in the dark, compared with AO-stained slides which yielded decreased counts after storage for a week under the same conditions. Thus, DAPI is a preferred fluorochrome for processing samples to be examined other than directly after staining.

Another useful DNA-specific fluorochrome is Hoechst 33258 (bisbenzimide). Bisbenzimide binds to adenine and thymine-rich regions of the DNA, increasing in fluorescence after binding (66). Hoechst 33258 has been recommended for enumeration of bacteria on surfaces, especially if the surface tends to bind AO (77). An advantage of the bisbenzimide stain, Hoechst 33258, is that it is unaffected by detergent solutions, laboratory salts, or biological materials associated with DNA (23). Bergstrom et al. (8) compared AO, bisbenzimide, and acriflavine for enumerating bacteria in clear and humic water samples. Acriflavine is a DNA-staining dye used mainly for staining the nuclear material of eukaryotic cells in suspension and in conjunction with the Feulgen reaction (28). Bergstrom et al. (8) found the bisbenzimide to produce a nonfluorescent precipitate which, unfortunately, obscured many cells. Acriflavine and AO produced comparable results, with AO providing more effective detection of small cells.

Direct counts of microorganisms in soil are associated with a variety of problems. Babuik and Paul (5) utilized fluorescein isothiocyanate (FITC) as the fluorochrome for detection of bacteria in soil samples. FITC conjugates with proteins on the exterior surfaces of the cell and fluoresces green. According to Babuik and Paul (5), nonspecific fluorescence can be quenched by addition of sodium pyrophosphate, making the reagent highly specific. Thus, improved "accuracy" in quantitation of large and small organisms in soils of clay content ranging from 20 to 80 percent is achieved using FITC. Even though FITC appears to be promising, use of this stain to count organisms in environmental samples has not been widespread. However, FITC has been used very successfully in conjunction with antibodies for detection of specific antigen-carrying cells.

In summary, fluorescent staining to detect microorganisms in environmental samples by direct microscopy has achieved wide acceptance in microbial ecology. It is inexpensive, reliable, reproducible, and relatively unsophisticated in application. However, an epifluorescent microscope is required, which is an expensive initial investment, especially with the optical accessories that are required for direct counting methods. With a microscope available, the cost of assaying samples is minimal and the time spent in sample analysis can be as short as 10 minutes for laboratory workers familiar with, and experienced in, direct counting methodology.

Viable cell detection and enumeration

Microscopy has been used since 1887 to determine the viability of microorganisms, when Metchnikoff added vesuvin solution to leukocyte exudate containing bacteria (62). In the early and mid-fifties, extensive work was done to investigate the phenomenon of cell death versus viability (85). The concept of life cycles in bacteria was promulgated in the thirties through the fifties (10), e.g., round body formation (33, 51) and/or ultramicrovibrio (67) as a stage in the life cycle in the environment. Starvation survival, another concept of viability of cells in the environment, has been extensively reviewed by Morita (74). A recent review by Roszak and Colwell (97) provides a historical and practical point of view of methods for assaying viability. A few of the techniques are prominent in importance for detection and enumeration of viable microorganisms and are reviewed here.

Enumeration of viable microorganisms has traditionally been accomplished by plating samples on a nutrient medium and enumerating colonies (colony-forming units, CFU) appearing on the agar medium. The method was adapted for microscopy in a procedure by which an agar film is placed on a microscope slide and formation of microcolonies observed under the microscope (40, 86). Agar, in the original method, was poured into a ring on the slide (86), but this approach was modified so that a thin film of agar was coated across the length of the slide (40). Buchanan-Mappin et al. (19) utilized slide culture to enumerate bacteria in groundwater, but with little or no success, presumably because of the small number of organisms in groundwater samples and the possibility that outgrowth of organisms in groundwater required extended incubation. Although the slide culture method is more effective than plate counts, it is tedious and becomes unwieldy if sample numbers are large.

A method was developed by Kogure et al. (63) to determine viability, which is both useful and technically simple, and is referred to as the direct viable count (DVC) method. This method exploits the finding of Goss et al. (45) that nalidixic acid is an inhibitor of DNA synthesis. The nalidixic acid prevents cell division in gram-negative bacteria by suppressing DNA replication. Subsequently, cross-wall formation is obstructed because of lack of replication, so the cells elongate instead of divide. Kogure and coworkers (63) used yeast extract as a nutrient source and incubation was for 6 hours. This allowed detection of viable bacteria in seawater samples. Incubation of seawater samples beyond 12 hours often allows either growth of gram-negative, nalidixic acid–resistant bacteria (88) or possible inactivation of nalidixic acid by high salt concentrations. Therefore, it is impossible to quantitate, by extended incubation, slow-metabolizing bacteria in samples and these may go unnoticed in the 6 hours of incubation usually employed in the DVC procedure.

Peele and Colwell (80) tested the response of seawater samples to various substrates. Even though simple carbon and nitrogen sources were utilized, yeast extract and tryptone provided the most numerous substrate-responsive cells. The conclusion was drawn that requirements of environmental organisms for growth are wide-ranging and may not necessarily be satisfied by a single carbon or nitrogen source as is often used in heterotrophic uptake measurements of environmental samples.

DVC is technically a straightforward procedure by which one can detect dormant cells, i.e., cells physiologically responsive but not dividing. The disadvantage of the procedure is the subjective nature of what constitutes an elongated cell. Thus, good controls are critical, i.e., the examination of both an AO direct count (53) and a DVC of the same sample, to estimate the size and number of those organisms in the direct count that may be construed falsely as DVC-positive. The problem can be overcome by use of an image analyzer or a particle counter (124) and setting a minimum elongation size.

A microscopic method that allows detection of uptake of selected substrates is microautoradiography. The method employs a microscope slide coated with a photographic emulsion. The sample is incubated with a ^{3}H-substrate and, after addition of substrate, is filtered through a Nucleopore filter, as above. The filter is placed organism-side down on the emulsion. Upon removal of the filter, the bacteria adhere to the emulsion. The slide is developed in the dark so that the ^{3}H creates an image on the emulsion. The slide is stained with methylene blue (or suitable stain) and examined by light microscopy. Each field is analyzed in two ways: (1) by counting the number of stained organisms and (2) by counting the number of organisms associated with silver grains. The silver grains indicate substrate uptake by that particular organism. Since this procedure employs a photographic emulsion, most of the critical steps must be done in the dark. The method allows for direct viewing of substrate-responsive cells.

Microautoradiography has been a useful tool in relating viability to individual cells in ecological studies (18). The technique has been used to study water (54, 79) and soil samples (38), including plant-microbial interactions (89). However, it can be difficult to relate spots on the emulsion to actual cells (54, 81). Fliermans and Schmidt (38) employed immunofluorescence to detect specific cells in samples and were able to relate viability not only to specific cells, but to a specific population in a mixture. To determine total population of viable cells, Meyer-Reil (71) used AO to stain the cells. The use of a fluorescent stain provides enhancement of the total count of viable microorganisms. Increased sensitivity and improved detection of small cocci in environmental samples can be achieved by the inclusion of AO in this technique. Tabor and Neihof (109) were able to see no statistical difference in counts of a sample examined by AO direct counting (53) and AO microautoradiography counts. Therefore, this technique is able to generate accurate total counts of bacteria. AO adaptation was utilized by Tabor and Neihof (110) to demonstrate seasonal distribution of viable organisms in the Chesapeake Bay. Roszak and Colwell (96) took this method one step further by incorporating the DVC procedure and found that about 90 percent of those cells responsive by the DVC were metabolically active by microautoradiography. Microautoradiography has been criticized as only being useful as a qualitative technique (122), but with advances in photographic emulsions and increases in sensitivity and standardization of the technique, it can be utilized quantitatively (109). Microautoradiography is, overall, a straightforward procedure but can be tedious if numerous samples must be analyzed.

In addition to microautoradiography, which can be used to detect metabolically active cells, there is a procedure employing 2-(p-iodophenyl)-3-(p-nitrophenyl)-5-phenyl tetrazolium chloride (INT) to detect respiring organisms (125). INT is converted intracellularly by respiring organisms to INT-formazan. INT-formazan appears as a dark red spot within the cell when it is viewed by light microscopy. Total counts can be done simultaneously by using AO and switching back and forth from bright-field microscopy to epifluorescence. Dutton et al. (31) added malachite green to the INT protocol to enhance staining of the background. They were able to use the method to determine the number of respiring organisms in samples collected from a sewage plant during steps of sewage treatment. In a comparison of various viable cell enumeration methods, Quinn (88) found the INT and DVC procedures to yield comparable numbers of viable organisms, if no substrate was added in carrying out the INT procedure, as is the case in the Zimmerman et al. (125) procedure. If nicotinamide adenine dinucleotide or nicotinamide adenine dinucleotide phosphate is added as substrate in the INT procedure, the INT yields larger numbers of viable organisms than DVC. There is a question whether addition of these substrates presents assessment of in situ activity. One of the problems with the method is the difficulty in detecting spots within small coccoid cells when aquatic environmental samples are examined.

Overall, there are various methods to detect viability of bacteria in the environment. Which method is the most accurate has been the subject of debate for many years. Each method measures a different property of the cell and this should be considered in their value. If possible, more than one method should be used to determine viability in environmental samples. In environmental experiments, what should be done and what actually can be done are usually two different things. In this case, the researcher must decide for him- or herself what is the most accurate method for determination of viability in the system being studied.

Image analysis

Epifluorescent microscopy can be converted to automatic data analysis to relieve the shear tedium involved in examining a large number of samples. As early as 1952, image analysis was utilized to count microscopic coal particles (118). One application of the use of image analysis in epifluorescent microscopy was that of Pettipher and Rodrigues (83), who used image analysis to count AO-stained bacteria in milk. DAPI stain was used by Sieracki et al. (105) to detect and enumerate planktonic bacteria. They found that counts done by standard visual analysis and image analysis gave statistically equal results.

Image analysis can be a very powerful technique, but some constraints apply. The results of image analysis will be only as good as the image provided. Staining of each organism must be strong, with little or no background or extraneous fluorescence. This method has limited application to soil or sediment samples without extreme care being taken to removal of particulates (117). Another constraint is the necessity for a highly sensitive camera to detect low levels of light emitted from epifluorescent-stained samples. There is also some question as to the ability of the camera to differentiate between debris and cells when fluorescent counts are recorded. Overall, the expense of the system is made up by the welcome reduction in automation and time needed to analyze individual samples, as well as increased productivity in analysis offered by these systems.

Fluorescent antibody microscopy

Detection and enumeration of specific organisms has been advanced by the increased knowledge of immunology. The role of antibodies and specific recognition of antigens can be very useful in the detection of organisms. Present methods in the field of microscopy utilize antibodies after they have been fluorescent-labeled. Cells that are covered with the specific antigens stand out with dramatic clarity upon examination with a fluorescent microscope. For ease in examining environmental samples, the samples are filtered through a polycarbonate membrane previously stained with irgalan black.

Fluorescent-antibody labeling has been employed to examine environmental samples directly to detect various organisms. Table 5.1 provides a brief list of organisms presently being detected in the environment by this procedure. The main advantage of the method is its sensitivity, but specificity and ease with which the procedure can be accomplished are also important. Specificity is the key question when dealing with antibodies. The possibility of cross-reaction with other organisms must always be considered and controlled by use of highly specific monoclonal or well-absorbed polyclonal antibodies. This point

TABLE 5.1   Bacteria Detected by Immunofluorescence in Environmental and Clinical Samples

Organism	Environment	Reference
Ammonia-oxidizing bacteria	Marine	Ward and Carlucci (119)
Azotobacter	Soil	Tchan and DeVille (112)
Bacteriodes fragilis	Water	Fiksdal and Berg (36)
Bdellovibrio	Culture	Schelling and Conti (100)
Campylobacter jejuni	Clinical	Chan et al. (24)
Chromobacterium lividum	Plant tissue	Bettelheim et al. (9)
Clavibacter michiganense	Plant	Stead (106)
Clostridium septicum	Soil	Garcia and McKay (42)
Erwinia amylovora	Plant	Miller (72)
Erwinia carotovora	Plant	Allen and Kelman (3)
Escherichia coli	Clinical	Ferrier (35)
Escherichia coli	Wastewater	Pugsley and Evison (87); Abshire (1)
Gardnella vaginalis	Clinical	Cano et al. (20)
Giardia	Water	Sauch (99)
Haemophilus pleuropneumonia	Clinical	Rosendal et al. (95)
Legionella spp.	Water	Hussong et al. (55)
Methanobacterium	Sediment	Strayer and Tiedje (107)
Methylomonas methanica	Water	Reed and Dugan (92)
Neisseria gonorrheae	Clinical	Ogawa (75)
Nitrate-oxidizing bacteria	Marine	Ward and Carlucci (119)
Nocardia erythropolis	Wastewater	Kurane (65)
Pseudomonas solanacearum	Plant	Stead (106)
Pseudomonas syringae	Plant tissue	Patton (76)
Rhizobium	Soil	Schmidt et al. (102)
Rickettsia	Clinical	Iida et al. (56)
Salmonella	Water	Thomason et al. (113)
Shigella	Clinical	Tsunematsu (116)
Sphaerophorus necrophorus	Clinical	Garcia et al. (43)
Spirochetes	Clinical	Magnarelli et al. (68)
Staphylococcus	Clinical	Tadokoro (111)
Streptococcus faecalis	Water	Abshire and Guthrie (2)
Streptococcus (Group B)	Clinical	Boyer et al. (15)
Sulfolobus acidocaldarius	Hot-spring	Bohlool and Brock (12)
Thermoplasma acidophilum	Coal refuse	Bohlool and Brock (11)
Thiobacillus ferrooxidans	Acid mine	Apel et al. (4)
Treponema hyodysenteriae	Clinical	Golikov et al. (44)
Vibrio cholerae	Water	Brayton et al. (17)
Xanthomonas campestris	Plant	Malin et al. (69)

was emphasized by H. L. Good (1972, master's thesis, University of Minnesota, Minneapolis) who attempted to stain specifically for vegetative *Azotobacter chroococcum* and found the cysts of the organism stained nonspecifically. The cysts of *Azotobacter* fit into the classification of "universal acceptor," as do some fungal spores (101) because they cross-react with any conjugate. Some of this nonspecific stain could be controlled by a preliminary staining with rhodamine-gelatin (13). Very little intergenus cross-reactivity was observed by Belser and Schmidt (7) between different strains of the same genus. Cross-reactivity between genera of the ammonia-oxidizing nitrifying bacteria did not occur with the organisms and samples used in their studies.

To reduce the problem of nonspecificity, monoclonal antibodies can be utilized effectively in immunofluorescence. A monoclonal antibody is an antibody produced by a single cell line that recognizes one, and only one, antigen. These antibodies are produced under the assumption that the organism somewhat possesses an antigen specific to only that organism. The significant disadvantage with the method is that viability of the organism cannot be determined, without using other methods, as cited above.

Brayton and Colwell (16) incorporated the direct viable count procedure of Kogure et al. (63) with fluorescent antibody labeling to detect viability (FA-DVC). The combination of the two procedures produces a valuable method for direct enumeration of viable bacteria in environmental samples. This method was successfully employed by Brayton et al. (17) in a field trial to detect viable *Vibrio cholerae* O1 in Bangladesh waters. Zelibor et al. (124) used an ELZONE particle counter (Particle Data, Inc., Elmhurst, Illinois) to remove the subjective nature of the direct viable count procedure. By incorporating an automated approach in the FA-DVC the overall ability to detect viability of a specific organism in the environment is enhanced.

Electron Microscopy

Electron microscopy is a powerful method for determining structures, internal and external, of microorganisms and in identifying viruses. Since culturing is not required, samples can be viewed directly for presence of viruses which, otherwise, may not have been detected at all. Direct detection of virus particles on grids circumvents the need for culturing (49, 73). Another advantage is the speed with which analysis can be accomplished. Depending on the methods used to prepare and view samples, analysis can be completed in 20 minutes to 2 hours. Third, if immunoelectron microscopy is not employed, virus-specific reagents are not required. A major drawback is the fact that virus particles must be present at concentrations of 10^5 to 10^6 per milliliter for visualization under the electron microscope. Thus, unique methods for concentration of viral particles are needed (49, 50). Both transmission electron microscopy (TEM) (49) and scanning electron microscopy (SEM) (50) can be used successfully for identification of viruses. A review of methods for staining and identification are presented by Miller (73). Detection and identification of specific viral particles can be accomplished by immune electron microscopy (61), a technique that allows direct viewing of antibody-virus interaction. Even though the technique of electron microscopy is useful in microbial ecology, laboratories may not have an electron microscope and a technician trained in virus morphology.

Electron microscopy is not as widely used to detect and identify bacteria as are viruses. Because bacteria are much larger than viruses, a standard light microscope will suffice. Nevertheless, electron microscopy is used for direct examination of bacteria in specialized environments, e.g., air-water interfaces, the rumen of cows, and soil systems. Fuerst et al. (41) described bacterioneuston, i.e., the bacteria found in the surface microlayer of seawater, by transmission electron microscopy (TEM). They were able to elucidate microcolonial associations and appendages of bacteria in surface microlayers.

The use of electron microscopy in soil microbiology is well documented. Gray (47) employed the stereoscan electron microscope to examine sand grains and humus particles collected from soil of a sand-dune planted with *Pinus nigra* var. *laricie,* as well as bacteria on the root surface of *Trifolium repens.* Hagen et al. (48) utilized SEM to study *Bacillus cereus* and *Staphylococcus aureus* inoculated into soil. Hagen et al. (48) found the technique to be useful for detecting the organisms by their shape. However, as noted above, bacteria must be present in a concentration of 10^7 to 10^{10} cells per gram of soil, or greater, to be viewed under the microscope. Therefore, organisms must be present in high ($>10^7$) numbers, or the sample must be concentrated for detection. Bae et al. (6) used TEM to study the microflora of soil, overcoming the concentration problem by using transmission microscopy instead.

Scanning electron microscopy has also been used widely to study aquatic samples. Aquatic bacteria, phytoplankton, and detritus have been examined directly by electron microscopy with good success (78, 103, 114). Bowden (14) utilized SEM as a quantitative tool for enumerating aquatic bacteria. Despite the power of the electron microscope, enumeration by epifluorescent microscopy is preferable since it provides reasonably good accuracy. With the advent of immuno-electron microscopy, it is now possible to identify organisms directly by electron microscopy.

In conclusion, detection and enumeration of bacteria by electron microscopy is both feasible and useful, but because of expense and time required for sample preparation, it is mainly utilized as a special-purpose method.

Summary

The microscope has been the main tool of the microbiologist since the advent of microbiology. Even today, as biotechnology has taken over a good portion of the focus of microbiology, the microscope still plays an important role. Direct enumeration and viability testing of microorganisms are two ways in which microscopy has enhanced the processing of environmental samples. This is beyond the usual morphology, cellular structure, and motility testing that traditionally the microscope has been used for. Methods for the microscope will continue to be developed for analysis of environmental samples, because of the ease, speed, and accuracy with which the samples can be processed.

References

1. Abshire, R. L. 1976. Detection of enteropathogenic *Escherichia coli* strains in wastewater by fluorescent antibody. *Can. J. Microbiol.* 22:364–378.
2. Abshire, R. L., and R. K. Guthrie. 1971. The use of fluorescent antibody technique for detection of *Streptococcus faecalis* as an indicator of faecal pollution in water. *Water Res.* 5:1089–1097.
3. Allen, E., and A. Kelman. 1977. Immunofluorescent stain procedures for detection and identification of *Erwinia carotovora* var. *atroseptica. Phytopathology* 67:1305–1312.
4. Apel, W. A., P. R. Dugan, J. A. Filppi, and M. S. Rheins. 1976. Detection of *Thiobacillus ferroxidans* in acid mine by indirect fluorescent antibody staining. *Appl. Environ. Microbiol.* 32:159–165.
5. Babuik, L. A., and E. A. Paul. 1970. The use of fluorescein isothiocyanate in the determination of the bacterial biomass of grassland soil. *Can. J. Microbiol.* 16:57–62.
6. Bae, H. C., E. H. Cota-Robles, and L. E. Casida, Jr. 1972. Microflora of soil as viewed by transmission electron microscopy. *Appl. Microbiol.* 23:637–648.
7. Belser, L. W., and E. L. Schmidt. 1978. Serological diversity within a terrestrial ammonia-oxidizing population. *Appl. Environ. Microbiol.* 36:589–593.
8. Bergstrom, I., A. Heinanen, and K. Salonen. 1986. Comparison of acridine orange, acriflavin, and bisbenzamide stains for enumeration of bacteria in clear and humic waters. *Appl. Environ. Microbiol.* 51:664–667.
9. Bettelheim, K. A., J. F. Gordon, and J. Taylor. 1968. The detection of a strain of *Chromobacterium lividum* in the tissue of certain leaf-nodulated plants by immunofluorescence technique. *J. Gen. Microbiol.* 54:177–184.
10. Bisset, K. A. 1955. *The Cytology and Life History of Bacteria.* E. and S. Livingstone, London.
11. Bohlool, B. B., and T. D. Brock. 1974. Immunofluorescence approach to the study of the ecology of *Thermoplasma acidophilum* in coal refuse material. *Appl. Microbiol.* 28:11–16.
12. Bohlool, B. B., and T. D. Brock. 1974. Population ecology of *Sulfolobus acidocaldarius.* II. Immunological studies. *Arch. Microbiol.* 97:181–194.
13. Bohlool, B. B., and E. L. Schmidt. 1968. Nonspecific staining: Its control in immunofluorescence examination of soil. *Science* 162:1012–1014.
14. Bowden, W. B. 1977. Comparison of two direct-count techniques for enumerating aquatic bacteria. *Appl. Environ. Microbiol.* 33:1229–1232.
15. Boyer, K. M., C. A. Gadzala, P. C. Kelly, L. C. Burd, and S. P. Gotoff. 1981. Rapid identification of maternal colonization with Group B Streptococci by use of fluorescent antibody. *J. Clin. Microbiol.* 14:550–556.
16. Brayton, P. R., and R. R. Colwell. 1987. Fluorescent antibody staining methods for enumeration of viable environmental *Vibrio cholerae* O1. *J. Microbiol. Methods* 6: 309–314.
17. Brayton, P. R., M. L. Tamplin, A. Huq, and R. R. Colwell. 1987. Enumeration of *Vibrio cholerae* O1 in Bangladesh waters by fluorescent-antibody direct viable count. *Appl. Environ. Microbiol.* 53:2862–2865.
18. Brock, M. L., and T. D. Brock. 1968. The application of micro-autoradiographic techniques to ecological studies. *Mitt. Int. Ver. Theor. Agnew. Limnol.* 15:1–29.
19. Buchanan-Mappin, J. M., P. M. Wallis, and A. G. Buchanan. 1986. Enumeration

and identification of heterotrophic bacteria in groundwater and in a mountain stream. *Can. J. Microbiol.* 32:93–98.

20. Cano, R. J., M. A. Beck, and D. V. Grady. 1983. Detection of *Gardnerella vaginalis* on vaginal smears by immunofluorescence. *Can. J. Microbiol.* 29:27–38.

21. Casida, L. E., Jr. 1969. Observations of microorganisms in soil and other natural habitats. *Appl. Microbiol.* 18:1065–1071.

22. Casida, L. E., Jr. 1971. Microorganisms in unamended soil as observed by various forms of microscopy and staining. *Appl. Microbiol.* 21:1040–1045.

23. Cesarone, C. F., C. Bolognesi, and L. Santi. 1979. Improved microfluorometric DNA determinations in biological material using 33258 Hoechst. *Anal. Biochem.* 100:188–197.

24. Chan, F. T. H., G. Stringel, and A. M. R. Mackenzie. 1983. Isolation of *Campylobacter jejuni* from an appendix. *J. Clin. Microbiol.* 18:422–424.

25. Cholodny, W. 1928. Contributions to the quantitative analysis of bacterial plankton. *Trav. Sta. Biol. Dniepre.* 3:157.

26. Coleman, A. W. 1979. Use of the fluorochrome 4'-6-diamidino-2-phenylindole in genetic and developmental studies of chloroplast DNA. *J. Cell. Biol.* 82:299–305.

27. Cousins, C. M., G. L. Pettipher, C. H. McKinnon, and R. Mansell. 1979. A rapid method for counting bacteria in milk. *Dairy Ind. Int.* 44:27–39.

28. Crissman, H. A., P. E. Mullaney, and J. A. Steinkamp. 1975. Methods and applications of flow systems for analysis and sorting of mammalian cells. *Methods Cell Biol.* 9:180–246.

29. Daley, R. J., and J. E. Hobbie. 1975. Direct counts of aquatic bacteria by a modified epifluorescence technique. *Limnol. Oceanogr.* 20:875–882.

30. Doetsch, R. N. 1981. Determinative methods of light microscopy. In P. Gerhardt, P. G. E. Murray, R. N. Costilow, E. W. Nester, W. A. Woods, N. R. Krieg, and G. B. Phillips (eds.), *Manual of Methods for General Bacteriology.* American Society for Microbiology, Washington, D.C., pp. 21–33.

31. Dutton, R. J., G. Bitton, and B. Koopman. 1983. Malachite Green-INT (MINT) method for determining active bacteria in sewage. *Appl. Environ. Microbiol.* 46:1263–1267.

32. Ehrlich, R. 1955. Technique for microscopic count of microorganisms directly on membrane filters. *J. Bacteriol.* 70:265.

33. Felter, R. A., R. R. Colwell, and G. B. Chapman. 1969. Morphology and round body formation in *Vibrio marinus. J. Bacteriol.* 99:326–335.

34. Ferguson, R. L., and P. Rublee. 1976. Contribution of bacteria to the standing crop of coastal plankton. *Limnol. Oceanogr.* 21:141–145.

35. Ferrier, G. R. 1986. The use of the fluorescent antibody test to detect *Escherichia coli* with K88 and 987P pilus antigens. *Vet. Microbiol.* 11:197–201.

36. Fiksdal, L., and J. D. Berg. 1987. Evaluation of a fluorescent antibody technique for the rapid enumeration of *Bacteroides fragilis* groups of organisms in water. *J. Appl. Bacteriol.* 62:377–383.

37. Fleischer, R. L., P. B. Price, and E. M. Symes. 1964. Novel filter for biological material. *Science* 143:249–250.

38. Fliermans, C. B., and E. L. Schmidt. 1975. Autoradiography and immunofluorescence combined for autoecology study of single cell activity with *Nitrobacter* as a model system. *Appl. Microbiol.* 30:676–684.

39. Francisco, D. E., R. A. Mah, and A. C. Rabin. 1973. Acridine Orange epifluorescent technique for counting bacteria in natural waters. *Trans. Am. Microsc. Soc.* 92:416–421.

40. Fry, J. L., and T. Zia. 1982. A method for estimating viability of aquatic bacteria by slide culture. *J. Appl. Bacteriol.* 53:189–198.

41. Fuerst, J. A., A. McGregor, and M. R. Dickson. 1987. Negative staining of freshwater bacterioneuston sampled directly with electron microscope specimen support grids. *Microb. Ecol.* 13:219–228.

42. Garcia, M. M., and K. A. McKay. 1969. On the growth and survival of *Clostridium septicum* in soil. *J. Appl. Bacteriol.* 32:362–370.

43. Garcia, M. M., D. H. Neil, and K. A. McKay. 1971. Application of immuno-

fluorescence to studies on the ecology of *Sphaerophorus necrophorus*. *Appl. Microbiol.* 21:809–814.

44. Golikov, A. V., G. V. Zenin, V. V. Bondik, and V. D. Bukhanov. 1984. Diagnosis of swine dysentery by fluorescent antibody method. *Veterinariya* 9:64–66.

45. Goss, W. A., W. H. Deitz, and T. M. Cook. 1965. Mechanism of action of nalidixic acid on *Escherichia coli*. II. Inhibitor of DNA synthesis. *J. Bacteriol.* 89:1068–1074.

46. Gram, C. 1884. Uber die isolierte Farbung der Schizomyceten in Schnitt and Trockenpraparaten. *Fortschr. Med.* 2:185–189.

47. Gray, T. R. G. 1967. Stereoscan electron microscopy of soil microorganisms. *Science* 155:1668–1670.

48. Hagen, C. A., E. J. Hawrylewicz, B. T. Anderson, V. K. Tolkacz, and M. L. Cephus. 1968. Use of the scanning electron microscope for viewing bacteria in soil. *Appl. Microbiol.* 16:932–934.

49. Hammond, G. W., P. R. Hazelton, I. Chuang, and B. Klisko. 1981. Improved detection of viruses by electron microscopy after direct ultracentrifuge preparation of specimens. *J. Clin. Microbiol.* 14:210–221.

50. Heinz, B. A., D. O. Cliver, and G. L. Hehl. 1986. Enumeration of enterovirus particles by scanning electron microscopy. *J. Virol. Meth.* 14:71–83.

51. Henrici, A. T. 1925. A statistical study of the form and growth of the cholera vibrio. *J. Infect. Dis.* 37:75–81.

52. Hobbie, J. E., O. Holm-Hansen, T. T. Packard, L. R. Pomeroy, R. W. Sheldon, J. P. Thomas, and W. J. Wiebe. 1972. A study of the distribution and activity of microorganisms in ocean water. *Limnol. Oceanogr.* 17:544–555.

53. Hobbie, J. E., R. J. Daley, and S. Jasper. 1977. Use of Nucleopore filters for counting bacteria by fluorescence microscopy. *Appl. Environ. Microbiol.* 33:1225–1228.

54. Hoppe, H.-G. 1976. Determination and properties of actively metabolizing heterotrophic bacteria in the sea investigated by means of autoradiography. *Mar. Biol.* 36:291–302.

55. Hussong, D., R. R. Colwell, M. O'Brien, E. Weiss, A. D. Pearson, R. M. Weiner, and W. D. Burge. 1987. Viable *Legionella pneumophila* not detectable by culture on agar media. *Bio/Technology* 5:947–952.

56. Iida, T., H. Kawashima, and A. Kawamura, Jr. 1966. Direct immunofluorescence for typing of Tsutsugamushi disease rickettsia. *J. Immunol.* 95:1129–1133.

57. Jannasch, H. W. 1953. Zur methode der quantitativen. Untersuchungen con Bakterienkuturen in flussigen Medien. *Arch. Mikrobiol.* 18:425.

58. Jannasch, H. W. 1958. Studies of planktonic bacteria by means of a direct membrane filter method. *J. Gen. Microbiol.* 18:609–620.

59. Jannasch, H. W., and G. E. Jones. 1959. Bacterial populations in seawater as determined by different methods of enumeration. *Limnol. Oceanogr.* 4:128–139.

60. Jones, J. G. 1974. Some observations on direct counts of freshwater bacteria obtained with a fluorescent microscope. *Limnol. Oceanogr.* 19:540–543.

61. Kapikian, A. Z., J. L. Dienstag, and R. H. Purcell. 1980. Immune electron microscopy as a method for the detection, identification, and characterization of agents not cultivable in an *in vitro* system. In N. R. Rose and H. Friedman (eds.), *Manual of Clinical Immunology*, 2d ed. American Society for Microbiology, Washington, D.C., pp. 70–83.

62. Knaysi, G. 1935. A microscopic method of distinguishing dead from living bacterial cells. *J. Bacteriol.* 30:193–206.

63. Kogure, K., U. Simidu, and N. Taga. 1979. A tentative direct microscopic method for counting living marine bacteria. *Can. J. Microbiol.* 25:415–420.

64. Korgaonkar, K. S., and S. S. Ranade. 1966. Evaluation of Acridine Orange fluorescence test in viability studies on *Escherichia coli*. *Can. J. Microbiol.* 12:185–190.

65. Kurane, R. 1986. Microbial degradation of phthalate esters. *Microbiol. Sci.* 3:92–97.

66. Latt, S. A., and G. Statten. 1976. Spectral studies on 33258 Hoechst and related bisbenzimadazole dyes useful for fluorescent detection of DNA synthesis. *J. Histochem. Cytochem.* 24:24–32.

67. MacDonnell, M. T., and M. A. Hood. 1984. Ultramicrovibrios in Gulf Coast estua-

rine waters: Isolation, characterization, and incidence. In R. R. Colwell (ed.), *Vibrios in the Environment*. Wiley, New York, pp. 551–562.

68. Magnarelli, L. A., J. F. Anderson, and W. A. Chappell. 1984. Antibodies to spirochetes in white-tailed deer and prevalence of infected ticks from foci of Lyme disease in Connecticut. *J. Wildl. Dis.* 20:21–26.

69. Malin, E. M., D. A. Roth, and E. L. Belden. 1983. Indirect immunofluorescent staining for detection and identification of *Xanthomonas campestris* pv. *phaseoli* in naturally infected bean seed. *Plant Dis.* 67:645–647.

70. McDaniel, J. A., and D. G. Capone. 1985. A comparison of procedures for the separation of aquatic bacteria from sediments for subsequent direct enumeration. *J. Microbiol. Methods* 3:291–302.

71. Meyer-Reil, L.-A. 1978. Autoradiography and epifluorescence microscopy combined for the determination of number and spectrum of actively metabolizing bacteria in natural waters. *Appl. Environ. Microbiol.* 36:506–512.

72. Miller, H. J. 1983. Some factors influencing immunofluorescence microscopy as applied in diagnostic phytobacteriology with regard to *Erwinia amylovora*. *Phytopathologisches Zeitung.* 108:235–241.

73. Miller, S. E. 1986. Detection and identification of viruses by electron microscopy. *J. Electron Microsc. Tech.* 4:265–301.

74. Morita, R. Y. 1982. Starvation-survival of heterotrophs in the marine environment. *Adv. Microb. Ecol.* 6:171–197.

75. Ogawa, H. 1977. Shigella. In A. Kawamura, Jr. (ed.), *Fluorescent Antibody Techniques and Their Application*. University Park Press, Baltimore, pp. 233–235.

76. Patton, A. M. 1964. The adaptation of the immunofluorescence technique for use in bacteriological investigations of plant tissue. *J. Appl. Bacteriol.* 27:237–242.

77. Paul, J. H. 1982. Use of Hoechst dyes 33258 and 33342 for enumeration of attached and planktonic bacteria. *Appl. Environ. Microbiol.* 43:939–944.

78. Pearl, H. W. 1973. Detritus in Lake Tahoe: Structural modification by attached microflora. *Science* 180:496–498.

79. Pearl, H. W. 1974. Bacterial uptake of dissolved organic matter in relation to detrimental aggregation in marine and freshwater systems. *Limnol. Oceanogr.* 19:966–972.

80. Peele, E. R., and R. R. Colwell. 1981. Application of a direct microscopic method for enumeration of substrate-responsive marine bacteria. *Can. J. Microbiol.* 27:1071–1075.

81. Peroni, C., and O. Lavarello. 1975. Microbial activities as a function of water depth in the Ligurian Sea: An autoradiographic study. *Mar. Biol.* 30:37–50.

82. Pettipher, G. L. 1986. Review: The direct epifluorescent filter technique. *J. Food Tech.* 21:535–546.

83. Pettipher, G. L., and U. M. Rodrigues. 1982. Semi-automated counting of bacteria and somatic cells in milk using epifluorescence microscopy and television image analysis. *J. Appl. Bacteriol.* 53:323–329.

84. Porter, K. G., and Y. S. Feig. 1980. The use of DAPI for identifying and counting aquatic microflora. *Limnol. Oceanogr.* 25:943–948.

85. Postgate, J. R. 1967. Viability measurements and the survival of microbes under minimum stress. In A. H. Rose and J. F. Wilkinson (eds.), *Advances in Microbial Physiology*, Vol. 1. Academic Press, London, pp. 1–23.

86. Postgate, J. R., J. E. Crumpton, and J. R. Hunter. 1961. The measurement of bacterial viabilities by slide culture. *J. Gen. Microbiol.* 24:15–24.

87. Pugsley, A. P., and L. M. Evison. 1974. Immunofluorescence as a method for the detection of *Escherichia coli* in water. *Can. J. Microbiol.* 20:1457–1463.

88. Quinn, J. P. 1984. The modification and evaluation of some cytochemical techniques for the enumeration of metabolically active heterotrophic bacteria in the aquatic environment. *J. Appl. Bacteriol.* 57:51–57.

89. Ramsay, A. J. 1974. The use of autoradiography to determine proportion of bacteria metabolizing in an aquatic habitat. *J. Gen. Microbiol.* 80:363–373.

90. Rasumov, A. S. 1932. Die direkte methode der Zahlung der bakterien in wasser und ihre vergleichung mit der kochschen plattenkultur methode. *Mikrobiologia* 1:131.

91. Rasumov, A. 1933. New methods for the quantitative and qualitative studies of water microflora. *Mikrobiologia* 2:346–352.
92. Reed, W. M., and P. R. Dugan. 1978. Distribution of *Methylomonas methanica* and *Methylosinus trichosporium* in Cleveland harbor as determined by an indirect fluorescent antibody-membrane filter technique. *Appl. Environ. Microbiol.* 35:422–430.
93. Richards, O. W., and W. B. Krabek. 1954. Visibilizing microorganisms on membrane filter surfaces. *J. Bacteriol.* 67:613.
94. Rodrigues, U. M., and R. G. Kroll. 1985. The direct epifluorescent filter technique (DEFT): Increased selectivity, sensitivity and rapidity. *J. Appl. Bacteriol.* 59:493–499.
95. Rosendal, S., L. Lombin, and J. DeMoor. 1981. Serotyping of detection of *Haemophilus pleuropneumoniae* by indirect fluorescent antibody technique. *Can. J. Comp. Med.* 45:271–274.
96. Roszak, D. B., and R. R. Colwell. 1987. Metabolic activity of bacterial cells enumerated by direct viable count. *Appl. Environ. Microbiol.* 53:2889–2983.
97. Roszak, D. B., and R. R. Colwell. 1987. Survival strategies of bacteria in the natural environment. *Microbiol. Rev.* 51:365–379.
98. Russell, W. C., C. Newman, and D. H. Williamson. 1975. A simple cytochemical technique for demonstration of DNA in cells infected with mycoplasmas and viruses. *Nature* 253:461–462.
99. Sauch, J. F. 1985. Use of immunofluorescence and phase-contrast microscopy for detection and identification of *Giardia* cysts in water samples. *Appl. Environ. Microbiol.* 50:1434–1438.
100. Schelling, M. E., and S. F. Conti. 1983. Serotyping of bdellovibrios by agglutination and indirect immunofluorescence. *Int. J. Syst. Bacteriol.* 33:816–821.
101. Schmidt, E. L. 1973. Fluorescent antibody technique for the study of microbial ecology. *Bull. Ecol. Res. Commun.* (Stockholm) 17:67–76.
102. Schmidt, E. L., R. O. Bankole, and B. B. Bohlool. 1968. Fluorescent antibody approach to the study of rhizobia in soil. *J. Bacteriol.* 95:1987–1992.
103. Schrader, H. 1971. Fecal pellets: Role of sedimentation of pelagic diatoms. *Science* 174:55–57.
104. Shaw, B. G., C. D. Harding, W. H. Hudson, and L. Farr. 1987. Rapid estimation of microbial numbers on meat and poultry by the direct epifluorescent filter technique. *J. Food Prot.* 50:652–657.
105. Sieracki, M. E., P. W. Johnson, and J. McN. Sieburth. 1985. Detection, enumeration, and sizing of planktonic bacteria by image-analyzed epifluorescence microscopy. *Appl. Environ. Microbiol.* 49:799–810.
106. Stead, D. E. 1987. Immunofluorescence techniques in plant pathology. In J. M. Grange, A. Fox, and N. L. Morgan (eds.), *Immunological Techniques in Microbiology*. Blackwell, Boston, pp. 129–136.
107. Strayer, R. F., and J. M. Tiedje. 1978. Application of the fluorescent-antibody technique to the study of a methanogenic bacterium in lake sediments. *Appl. Environ. Microbiol.* 35:192–198.
108. Strugger, S. 1948. Fluorescent microscope examinations of bacteria in soil. *Can. J. Res.* 26:188–193.
109. Tabor, P. S., and R. A. Neihof. 1982. Improved microautoradiographic method to determine individual microorganisms active in substance uptake in natural waters. *Appl. Environ. Microbiol.* 44:945–953.
110. Tabor, P. S., and R. A. Neihof. 1984. Direct determination of activities for microorganisms of Chesapeake Bay populations. *Appl. Environ. Microbiol.* 48:1012–1019.
111. Tadokoro, I. 1977. Staphylococci: An approach to analysis of microbial structure, invasion of alpha toxin into Ehrlich ascites tumor cells and formation of kidney

abscesses in mice. In A. Kawamura, Jr. (ed.), *Fluorescent Antibody Techniques and Their Applications.* University Park Press, Baltimore, pp. 223–225.

112. Tchan, Y. T., and R. R. DeVille. 1970. Application de l'immunofluorescence a l'etude des *Azotobacter* du sol. *Ann. Inst. Pasteur (Paris)* 118:665–673.

113. Thomason, B. M., J. W. Biddle, and W. B. Cherry. 1975. Detection of salmonellae in the environment. *Appl. Environ. Microbiol.* 30:764–767.

114. Todd, R. L., W. J. Humphreys, and E. P. Odum. 1973. The application of scanning electron microscopy in estuarine microbial research. In L. H. Stevenson and R. R. Colwell (eds.), *Estuarine Microbial Ecology.* Belle W. Baruch Library in Marine Science, no. 1. University of South Carolina Press, Columbia, pp. 115–125.

115. Todd, R. L., and T. J. Kerr. 1972. Scanning electron microscopy of microbial cells on membrane filters. *Appl. Microbiol.* 23:1160–1162.

116. Tsunematsu, Y. 1977. Gonorrhea. In A. Kawamura, Jr. (ed.), *Fluorescent Antibody Techniques and Their Application.* University Park Press, Baltimore, pp. 226–230.

117. Van Wambeke, F. 1988. Numeration et taille des bacteries planctoniques au moyen de l'analyse d'images couplee a l'epiflourescence. *Ann. Inst. Pasteur/ Microbiol.* 139:261–272.

118. Walton, W. H. 1952. Automatic counting of microscopic particles. *Nature (London)* 169:518–520.

119. Ward, B. B., and A. F. Carlucci. 1985. Marine ammonia- and nitrite-oxidizing bacteria: Serological diversity determined by immunofluorescence in culture and in the environment. *Appl. Environ. Microbiol.* 50:194–201.

120. Watson, S. W., T. J. Novitsky, H. L. Quinby, and F. W. Valois. 1977. Determination of bacterial number and biomass in the marine environment. *Appl. Environ. Microbiol.* 33:940–954.

121. Williamson, D. K., and D. J. Fennell. 1975. The use of fluorescent DNA-binding agent for detecting and separating yeast mitochondrial DNA. In D. M. Prescott (ed.), *Methods in Cell Biology,* Vol. 12. Academic Press, New York, pp. 335–354.

122. Wright, R. T. 1978. Measurement and significance of specific activity in the heterotrophic bacteria of natural waters. *Appl. Environ. Microbiol.* 36:297–305.

123. Yamabe, S. 1973. Binding of Acridine Orange with DNA. *Biochem. Biophys.* 154: 19–27.

124. Zelibor, J. L., Jr., M. Tamplin, and R. R. Colwell. 1987. A method for measuring bacterial resistance to metals employing epifluorescent microscopy. *J. Microbiol. Methods* 7:143–155.

125. Zimmermann, R., R. Iturriaga, and J. Becker-Birck. 1978. Simultaneous determination of the total number of aquatic bacteria and the number thereof involved in respiration. *Appl. Environ. Microbiol.* 36:926–935.

6

Application of Numerical Taxonomy in Microbial Ecology

Shoshana Bascomb

Rita R. Colwell

Introduction

Historically, identification and classification procedures employed in microbial ecology have included only classical methods for characterizing microorganisms and establishing species composition of environmental samples. During the last two decades, however, numerical taxonomy and, more importantly, molecular genetic methods have prevailed. Concomitantly, microbial ecology studies have involved analysis of community structure and species composition. Although numerical taxonomy was originally developed to deal with classically derived taxonomic information, the basic approach of using numerical or mathematical tools for handling voluminous and complex data is valid for both ecology and modern taxonomy. Computers are routinely applied in numerical taxonomy studies and are considered necessary in most ecological studies.

Theoretical aspects of numerical taxonomy have been recently reviewed (17, 18), as have practical aspects and specific problems of classification and identification (34, 42).

In general, taxonomy and systematics involve two basic processes: classification, i.e., the division of a *population* of individuals (described in terms of a given number of attributes) into a number of clusters, and identification, the assignment of an *individual* (described in terms of all, or some of the above attributes) into one of the previously defined clusters.

The number of clusters and the criteria used to create them depend on the purpose of a given analysis. In ecological studies, numerical taxonomic techniques may be used to classify populations found in a particular environment or to compare populations found in the same or different environments on the same or different sampling occasions. Genotypic and phenotypic parameters which may be used to characterize populations include morphological, biochemical, immunological, molecular, pathogenic, and growth requirement characteristics of the bacteria, as well as changes arising in or due to their environment. These attributes can be recorded as binary, multistate, or continuous data.

Genotypic information, obtained by nucleic acid hybridization techniques, provides a measure of relatedness between the genetic code of the isolates studied as well as with known taxa. However, hybridization techniques provide information relevant only to the part of the genome used for the DNA probe. Moreover, the phenotypic expression of some of these codes may be modulated by areas of the genome not revealed by the particular hybridization procedure, whereas phenotypic information, on the other hand, reveals only the part of the genome which is expressed under the particular artificial condition of the test. Thus, only a combination of phenotypic and genotypic information can provide comprehensive information on relatedness between bacteria.

Computer methods have been widely applied to phenotypic (5, 7, 9, 12, 40, 48) and genotypic (30, 32) attributes. Computers have also been employed on a more limited scale to serological or other typing techniques (10, 14, 15, 27, 43). A few applications of artificial intelligence software to bacterial identification have been reported (M. Chyle (ed.), *Abstracts of the 2nd Conference on Taxonomy and Automatic Identification of Bacteria,* Czechoslavakia Society for Microbiology of the Czechoslovak Academy of Sciences, Prague, 1987: B. Lefebvre and F. Gavini, Abstract 56; J. Valdes et al., Abstract 109). Computers may be employed for analysis of manually entered data as well as for acquisition and handling of instrument-generated data.

This chapter examines classification and identification techniques as applied to environmental studies as well as the use of computers for data acquisition and preparation of data for analysis.

Data Acquisition and Preparation

A large number of attributes can be considered for inclusion in taxonomic studies useful to microbial ecology. Whether an attribute is included in a taxonomic analysis is determined by the applicability and suitability of that attribute. While attributes which show a great degree of variability within the same individual tested on different occasions are, generally speaking, not useful in taxonomic studies, such attributes may be of vital importance for identification of an organism in a specific field study. Thus, resistance to a given antibiotic may be of transient nature and not valuable for establishing taxonomic relationships between different species of *Acinetobacter* but may be essential for identification of a strain employed in a field study. For a given field study, only a small set of attributes, e.g., utilization of specific substrates or degrative capabilities, may be of interest even though such a set may not be sufficient for a comprehensive classification of the isolates. Data gathered in environmental studies may be further complicated by a reaction to atmospheric, osmolarity, or temperature conditions, particularly when trying to relate freshly isolated strains to existing taxa.

Taxonomic data of populations takes the form of tables of various dimensions, with attributes (characters, tests, features, variables) occupying the columns and OTUs (operational taxonomic units: strains or taxa) occupying the rows. The number of attributes recorded per strain varies from 1 to over 200. Designing such tables for a given ecological analysis requires time and thought.

When creating a data file, it is useful to keep the records for an individual strain in lines of no more than 80 characters because some data analysis packages restrict the information to 80 alphanumeric characters per line, with no limit to the number of lines per record. An open-ended general system for computer coding of microbiological data was described by Rogosa, Krichevsky, and Colwell (36), describing the coding process, with examples of general forms for coding, and listings of coding allocations for bacterial attributes.

Manual input of data

Verbal information, such as species name of a strain, may be given in full but this type of information requires much record space and completely accurate spelling. An alternative is the use of numeric or alphabetic codes for genera and species names and other attributes which require verbal description. Checking manually entered information is essential and is preferably done by a second person.

A number of packages are available for entering information on individual strains using a predesigned format (DBase or spreadsheet-type packages). Some of these programs impose limitations on record length and on the total number of records. If the number of attributes is larger than the package limit, it is possible to have more than one data base file pertaining to the same groups. Alternatively, data can be entered using the editor or word-processing programs (in non-document mode) available on most personal computers (14).

Numerical methods are best suited to deal with complete and comprehensive sets of data, i.e., with results for every attribute available for every microorganism tested. This is not always possible, and missing values should be indicated at the data-entering stage.

Automated acquisition of data

Most modern instruments are now capable of sending discrete or continuous information directly, or via an interface, to a computer. An example of a discrete type of instrument is the Dynatech Autoreader (Dynatech Laboratories, Inc., 14340 Sullyfield Circle, Fairfax, Virginia 22021) used for the sequential measurement of fluorescence of 96 wells arranged in fixed positions on a panel. The information sent by the instrument consists of 96 separate numbers corresponding to the fluorescence of each well in the order in which they were measured.

An example of the continuous type of information is a chromatogram produced by HPLC (high-pressure liquid chromatography) instruments where the absorbance or fluorescence of the eluting phase is monitored continuously. The measurements taken by such a system are at very short time intervals, with the results appearing as a continuous trace plotted against time. Such information requires special programs for finding peaks and assigning them to the relevant attributes. A program dealing with data acquisition from gel electrophoresis and with classification of attributes has been published (24). Monitoring systems of this type are of value in ecological studies where catabolic tests are frequently "negative" (nonreactive) for isolates.

Output format

Results of taxonomic studies are subjected to both visual screening of raw data and computer analysis, each requiring different presentation of the data. Visual observation requires that the data are printed in clearly separated columns with headings indicating the identity of attributes and isolates. For computer handling of data a compact design saves memory. When data are gathered on different occasions, it is essential to identify the isolate and the date or occasion on every record. Furthermore, before attempting any manipulation of the data, the data must be examined for errors.

Data preparation

Binary data require no further manipulation once the information has been entered. Data from instruments may require some normalization procedure. The purpose of these manipulations is to remove variability which is unconnected with between-strain variability (4). In addition, because of memory limitations, it may be desirable to remove from the analysis isolates or attributes that give identical results for all attributes (42).

Clustering Models

Cluster analysis can be used to test a hypothesis for the homogeneity of the population (e.g., soil rhizosphere bacteria) or to generate a hypothesis on the number of clusters present in the population (e.g., noncontinuous communities in lake microflora). Determining species clusters in such analyses is very useful, particularly if the isolates do not fit easily into known species; results of cluster analysis can provide the investigator with a means of predicting attributes of an individual, if its class membership has been established. Cluster analysis is also frequently employed in environmental analyses to summarize information from samples for which the investigator has previous cluster analysis data.

Choice of OTUs and attributes

Before starting cluster analysis procedures, a selection of individuals and attributes must take place. In ecological studies, the greatest task is placing "boundaries" on the data set. The size of the study which will provide statistically valid information is the key to decision making. OTUs should be randomly chosen and, generally speaking, the bigger the sample, the more reliable the results. However, generating information on a large number of ecological isolates may not be possible and additional limitations may be imposed by the available computer memory and software.

Choice of attributes is even more complicated. When looking for "natural" clusters, selection of phenotypic attributes that are as different from each other as possible, is encouraged. Cluster analysis assumes that attributes will be independent of each other, and this is not easy to establish. For example, while the presence of β-galactosidase is necessary for the formation of acid from lactose, formation of acid requires the presence of a number of other enzymes. Therefore, although the orthonitrophenyl-beta-galactoside (ONPG) test value may influence the value of acid production from lactose, the relationship is, strictly speaking, undefined. Selection of divergent attributes will result in less homogeneous groups. Ecologically meaningful information would most likely result from selection of attributes related to ability to inhabit a particular environment or to ability to grow during different sampling occasions.

A taxonomic study is scientifically more informative if the taxa created can be related to already defined and known taxa. For this purpose reference strains (i.e., well-characterized strains deposited in culture collections) are often tested side by side with fresh isolates. However, culture collection strains do not always exhibit characteristics observed on original isolation, possibly because of changes occurring during preservation and the numerous subculturings necessary for maintenance, or they may differ because the strain is a clinical isolate selected for reasons related to pathogenicity. Some reference strains may not be able to grow under the specific test conditions required for environmental isolates and field isolates may not be reactive in tests designed for clinical specimens. In such circumstances, genotypic studies may be more helpful.

Measure of resemblance between pairs of OTUs

The first step in classification is the calculation of the resemblance between each pair of individuals. Resemblance can be expressed in terms of similarity, dissimilarity, or distance. Similarity is frequently used with qualitative (binary) data bases, while for quantitative (continuous) data bases the distance measure is used. Similarity and dissimilarity values are on a definite scale of 0 to 100 percent; distance values are unrestricted and can take any positive value.

A number of coefficients have been formulated for estimating resemblance between OTUs using binary attributes, differing in the way negative results are treated and whether special weight is attached to any test (42).

The simple matching coefficient, where the similarity is obtained by dividing the sum of common positive and common negative attributes by the sum of all attributes tested, is often used. Generally speaking, the validity of inclusion of attributes absent in both individuals has been questioned. This can be a serious problem in environmental studies where isolates may not be reactive in tests designed for clinical specimens. Thus, the similarity ratio (Jaccard coefficient, S_j) which excludes negative matches ultimately may prove more useful in ecological studies. Other coefficients exist which accord different weights to positive and negative matches. Studies comparing coefficients indicate that the S_j is fairly robust. A list of coefficients and formulations for calculating their values are available (CLUSTAN and TAXAN; see Table 6.1).

A number of options are available for calculating the distance between OTUs on the basis of quantitative attributes. Basically, the position of each individual is visualized in a multidimensional space where each attribute is represented by one axis (dimension), usually

TABLE 6.1   Some Statistical Packages for Classification

ARTHUR	Version 4.1 for VAX/VMS. InfoMetrix, Inc., Seattle, Wash. 98121.
BMDP	Dixon, W., and M. B. Brown (eds.) (1979). "BMDP-79 Biomedical Computer Programs P-series," 2d printing. University of California Press, Berkeley.
CLUSTAN	Wishart, D. (1982). *CLUSTAN User Manual,* 3d ed. Edinburgh University, Edinburgh.
EinSight	Pattern Recognition System, ver. 2.5 for MS-DOS computers. InfoMetrix, Inc., Seattle, Wash. 98121.
GENSTAT	Alvey, N. G., et al. (1977). GENSTAT: A General Statistical Program. Statistics Dept., Rothamsted Experimental Station, Harpenden.
MASLOC	Kaufman, L., and D. L. Massart (1981). *MASLOC Users' Guide.* Vrije Univ., Brussels.
NTSYS-PC	Rohlf, F. J. (1987). Numerical Taxonomy and Multivariate Analysis System for the IBM PC microcomputer (and compatibles) ver. 1.30. Applied Biostatistics Inc., 3 Heritage Lane, Setauket, N.Y. 11733.
Pirouette	Comprehensive chemometric package for MS-DOS system. InfoMetrix, Inc., Seattle, Wash. 98121.
SAS	*SAS Users' Guide,* Version 5. SAS Inst. Inc., Cary, N.C. 27512.
SPSS/PC+	Norusis, M. J. (1986). SPSS/PC+ for the IBM PC/XT/AT. SPSS/PC+ Inc., Chicago, Ill. 60611.
TAXAN	Colwell, R. R., and J. B. Kaper. University of Maryland Computer Program: Manual for Use, Dept. of Microbiology, College Park, Md. 20742.
TAXPAK	Sackin, M. J. (unpublished).

orthogonal to existing axes. Sometimes the information is condensed into a smaller number of axes. This can be achieved by principal component analysis with emphasis on the taxa (6), or by factor analysis (10, 23) emphasizing the similarities of the attributes. Options for calculation of some of these resemblance values are available in all packages which provide clustering algorithms. The CLUSTAN manual specifies whether coefficients are suitable for binary and/or continuous attributes. The number of coefficients available for mixed data sets is limited.

Resemblance between OTUs based on 5S rRNA data can also be obtained by employing a pairwise comparison of the base sequence of each pair and from electrophoretic data (30). Ordination methods can also be used for analyzing the data.

Methods of cluster formation

A taxonomic study starts with gathering results on individual strains, followed by grouping the strains into clusters. Cluster formation uses hierarchical, optimization, or density-seeking techniques. Two basic questions are asked, namely: How many groups exist in the population and what is the relatedness between these groups?

Hierarchical cluster formation can be divisive or agglomerative. A population can be divided into K clusters, K being no bigger than the number of individuals, or start with n clusters, each containing one individual, and join the clusters on the basis of a criterion. The choice of final number of clusters is subjective. These processes are repeated hierarchically, possibly by steps of predetermined range of resemblance measure (4). The most frequently used criteria for hierarchical formation of clusters are single, complete or average linkage, Ward's method, and the centroid method. Listings, and some explanation of the way these are calculated, can be found in CLUSTAN, SAS, SPSS, and TAXAN (Table 6.1). Results based on different linkage criteria may produce different numbers of clusters at different levels of similarity; moreover, the class membership of individual OTUs may also differ.

Optimization techniques are designed to deal with large populations. The population is divided into K groups, the value of K being determined either by the investigator or reached by optimizing a criterion related to the within—and between—cluster variability. Both K and the assignation of individuals to clusters may be iterated until an optimal solution is achieved (1). The groups thus formed are tested for within- and between-group variability. OTUs are transferred from one cluster to another until the between-group variability considerably exceeds that within groups. This may lead to error if the data have been previously sorted by group membership. However, the most widely spaced OTUs or those selected by the investigator may be used as nuclei.

Density-seeking protocols examine the distribution of individuals in multidimensional space and divide the space into densely or sparsely populated zones. The boundary of a cluster may be defined by a change in density.

Principal component analysis and principal coordinate analysis are also used for classification purposes. These procedures are usually applied to continuous data. In these procedures the aim is to express the information available on all attributes in terms of fewer dimensions. A large proportion of the variability can be accounted for by the first three principal components. Here again the choice of the number of clusters in the population is subjective.

Portraying results of clustering techniques

Clustering procedure starts with individual OTUs and produces clusters with a different number of OTUs in each, and with a different degree of resemblance between the clusters. Different ways of portraying the relationship between OTUs and clusters are available. Some experience is required for interpretation of the results obtained.

A shaded similarity matrix can be produced (Fig. 6.1) (11, 24, 45),

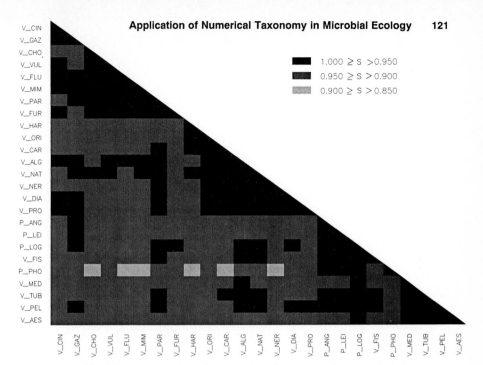

Figure 6.1 Graphic representation of taxonomic relatedness using a shaded similarity matrix, drawn by SAS graphics using data from West et al. (44). [*Reproduced with permission from Jacobs (25).*]

ideally with less than 100 OTUs. The OTUs are rearranged so that closely related individuals are placed near each other. The degree of resemblance is indicated by shading. The darker the shading the more similar are the OTUs. Presence of clusters can be observed, although clusters are not formally produced.

Graphical representation of the results can take the form of a dendrogram (tree) (Fig. 6.2a) where each individual is represented by a line. When a number of individuals merge, the cluster formed is represented by a single line (the branches of the tree). The lines representing individual or groups of OTUs are drawn parallel, the length of the line representing the degree of resemblance.

If the number of OTUs is large, portraying nonoverlapping clusters as black triangles whose size is proportional to the number of OTUs, is preferable (Fig. 6.2b). Portraying the relatedness between OTUs as a minimum spanning tree has been described (19). Phylogenetic information is frequently portrayed by similar types of trees (32).

Principal component analysis output may reveal overlap between clusters, providing further insight into the structure of the data. The relationship is commonly portrayed by plotting the data using two or more principal components (Fig. 6.3).

When optimization techniques are used the following information is available: cluster membership of each OTU, the distance of OTUs

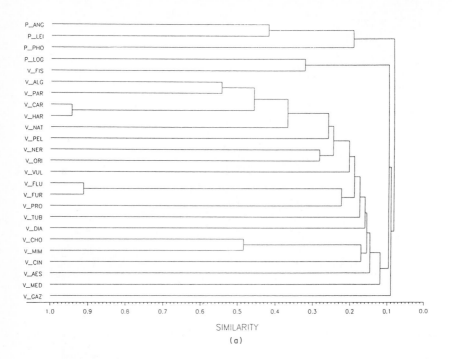

SIMILARITY

(a)

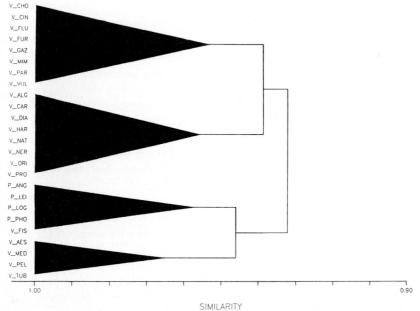

SIMILARITY

(b)

Figure 6.2 Graphic representation of taxonomic relatedness using a dendrogram, drawn by SAS graphics using data from West et al. (44). (a) Each operational taxonomic unit is represented by a single line; (b) groups of OTUs are clumped together in black triangles, the size of which is related to the number of OTUs. [*Reproduced with permission from Jacobs (25).*]

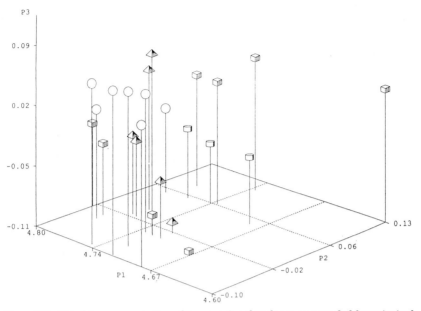

Figure 6.3 Graphic representation of taxonomic relatedness as revealed by principal component analysis using a three-dimensional plot and different symbols, drawn by SAS graphics using data from West et al. (44). [*Reproduced with permission from Jacobs (25).*]

from the cluster's centroid, as well as from its *K*-nearest neighbors (*K* being determined by the investigator).

Comparisons between clusters produced using hierarchical and optimization techniques of SPSS/PC (CLUSTER and QUICK CLUSTER procedures) are shown in Figs. 6.4 and 6.5, and Table 6.2. Optimization solutions for three to six clusters (Fig. 6.4), based on binary results from 44 two-hour enzyme tests on a total of 90 isolates belonging to *Vibrio* species were compared with a dendrogram produced using average linkage criterion (Fig. 6.5). The overall separation is fairly good, with the three species forming homogeneous clusters. However, the separation is not perfect, as some clusters contain representatives of all three taxa, regardless of the number of clusters chosen. Selecting a five-cluster solution for both methods, a similar number of OTUs are assigned to homogeneous and heterogeneous clusters (Table 6.2). The advantage of optimization techniques is that less memory is required. Thus, in this example, clustering of the 90 OTUs, using hierarchical and optimization techniques, required 48,856 and 5,224 bytes, respectively.

Distances between Final Cluster centers.

Cluster	1	2	3
1	.0000		
2	1.3781	.0000	
3	1.6879	1.9242	.0000

Distances between Final Cluster centers.

Cluster	1	2	3	4
1	.0000			
2	1.1651	.0000		
3	2.3777	2.4581	.0000	
4	2.3189	2.6581	1.9547	.0000

Distances between Final Cluster centers.

Cluster	1	2	3	4	5
1	.0000				
2	2.6555	.0000			
3	1.9458	2.0349	.0000		
4	1.5146	2.6150	2.5210	.0000	
5	2.0418	2.1307	2.0750	2.0509	.0000

Distances between Final Cluster centers.

Cluster	1	2	3	4	5	6
1	.0000					
2	2.5619	.0000				
3	1.9026	2.0471	.0000			
4	1.3155	2.7388	2.1098	.0000		
5	2.0636	1.4866	2.0518	2.3431	.0000	
6	2.5621	3.0342	3.2318	2.3962	2.5362	.0000

Figure 6.4 Distance between centroids of clusters formed using optimization procedures (SPSS/PC QUICK CLUSTER) for three, four, five, and six clusters, respectively. The distances between the clusters are larger in the five-cluster solution. This is indicated by the number of between-cluster distances that are less than 2.00. This number is least in the five-cluster solution.

Statistical packages available

A number of statistical packages are available for both classification and identification. A decision regarding which to use depends to a certain extent on the memory available on one's own PC. A list of some packages is given in Table 6.1, although it must be emphasized that such lists become outdated fairly quickly. BMDP, GENSTAT, SAS, and SPSS/PC+ are general statistical packages that can be used for manipulation of data, as well as for classification and identification purposes. ARTHUR, CLUSTAN, MASLOC, and TAXAN are dedicated to classification and/or identification and require other software for data preparation.

The SPSS package is "user-friendly," and requires very little programming experience. BMDP, GENSTAT, and SAS are favored by programmers and statisticians. CLUSTAN appears the most comprehensive, with the widest range of options, but it requires some programming experience and an advanced graphic output device.

The commonest output of classification studies is the dendrogram. To be of value it should be easily comprehensible and contain the resemblance scale and original label for each OTU, including previous cluster information. The dendrogram produced by SAS is very difficult to comprehend. However, Jacobs (25) has devised procedures for comprehensive high-quality graphic output within the SAS environment. Details of other available programs are given by Sakin (39).

Effect of choice of clustering technique on classification

The effect of different procedures for data transformation, choice of resemblance measure and method of cluster formation have been discussed (2, 4). Each affects the resultant classification. A number of methods should be used to validate the structure of the data set. Effect of inclusion of different types of attributes on classification is described by Mallory and Saylor (33). Comparisons between classifications obtained using genotypic and phenotypic data (31) show encouragingly good agreement.

Validation of classifications

Choice of the final number of clusters in the population and the appropriate level on the resemblance scale for choice of classification are subjective. For binary data, the similarity value of 85 percent is often chosen as equivalent to the species cutoff on the similarity scale. Such cutoff levels are not applicable to clustering techniques based on continuous attributes. Where the distance scales have a variable range,

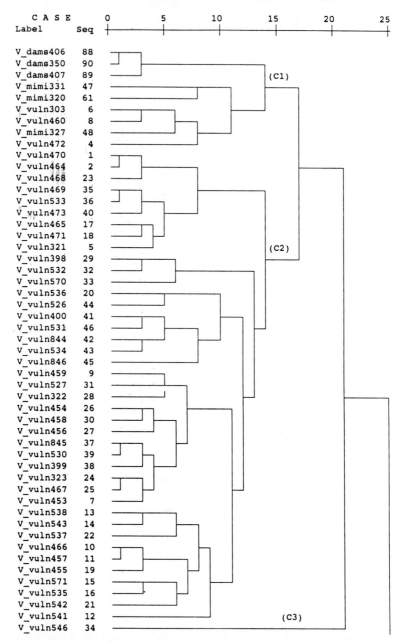

Figure 6.5 Dendrogram of OTUs belonging to 3 *Vibrio* species using 44 rapid enzyme tests as attributes. Produced using SPSS/PC CLUSTER routine. A five-cluster solution maintains the three species as separate groups (C2, C4, and C5) with minimum overlap. C1 contains OTUs of all three species, C3 contains only one OTU. Increasing the number of clusters causes splitting of the species into subgroups without removal of mixed clusters.

Dendrogram using Average Linkage (Between Groups)
(continued)
Rescaled Distance Cluster Combine

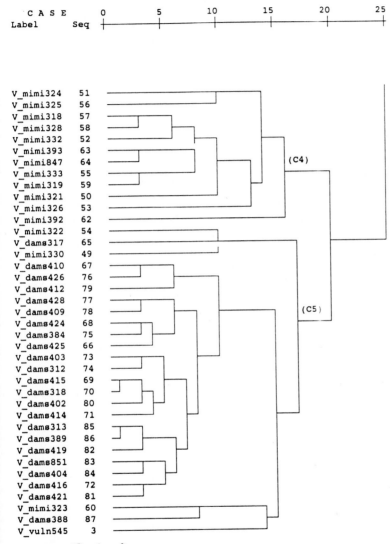

Figure 6.5 *(Continued)*

TABLE 6.2   Comparison between Classifications Obtained Using Hierarchical and Optimization Clustering Methods*

Bacterial taxon	Number of OTUs† tested	Clustering method	Number of clusters	Number of OTUs in homogeneous cluster	Number of OTUs in heterogeneous cluster
Vibrio vulnificus	46	Hierarchical	1	41	5
		Optimization	1	39	6
V. damsela	26	Hierarchical	1	23	3
		Optimization	1	22	4
V. mimicus	18	Hierarchical	1	12	6
		Optimization	1	11	7

*Using SPSS/PC CLUSTER and QUICK CLUSTER procedures.
†Operational taxonomic units.

the acuteness coefficient determined by the number of OTUs in the cluster, the homogeneity of the cluster, and the degree of separation from other clusters (44) should be considered.

A number of tests are used to determine the validity of the formed clusters. The RULE procedure of CLUSTAN can be used to choose the optimal number of nonoverlapping clusters or the distance of individuals from the centroid.

Ordination techniques provide a different perspective of the relationship between clusters by showing information on cluster structure using the value of each OTU on the first and second principal component axes. Circles of the clusters around the centroid, as well as cluster outlines, can be developed. An alternate approach would be to compare the classes formed by cluster analysis with cluster membership produced by other means, such as genotypic attributes (31).

Each of these methods may emphasize a different aspect of the population; therefore, a number of methods should be used to ensure comprehensive evaluation of the data.

With environmental studies, where groups are defined *de novo* and relationships to existing taxa are slightly nebulous, studies may need multiple-stage strategies, as suggested by Gyllenberg and Niemellä (22). They recommended testing of a limited number of isolates, on a large battery of tests, to determine those that provide differentiating information and to produce an initial temporary classification, followed by application of a limited number of tests to a larger number of isolates to verify the validity of the original clusters and define their homogeneity. A further step would be the use of phenotypic and genotypic studies to define the relationship between representative isolates from the ecological clusters and existing taxa. The diversity of microorganisms growing under variable conditions should be examined using both phenotypic and genotypic data. Classifications obtained using a mixture of these are more likely to be valid in the "natural" classification sense.

Clustering techniques rarely produce summarized information about the values of the different attributes of each cluster. Bryant (7) described three programs that were designed to (i) evaluate reproducibility of each attribute in replicate clusters, (ii) summarize the attributes of each cluster, and (iii) calculate overlap statistics between major clusters. Alternatively, the SPSS/PC + package has procedures to choose the most suitable attributes for identification purposes and provide the mean and standard deviation for each attribute in each cluster. These procedures will help in choosing the best set of attributes.

Identification Models

Once classification has been established, an identification (ID) system for assigning new OTUs to a cluster can be designed, building on information as it is gathered. The more frequently computerized identification is performed, the more important it is to make the process as cost effective as possible. ID matrices, which contain fewer attributes and/or require less manual operations, are more convenient and economical; this emphasizes the skill and care necessary for selection of taxa and choice of attributes.

Identification requires three decisions: the choice of the taxon which the unknown most resembles, determination of how much the unknown is nearer to that taxon in comparison with other taxa, and whether the unknown resembles the chosen taxon sufficiently to be considered a member.

Effectiveness of an ID depends on its ability to identify correctly unknowns belonging to one of the taxa included in the matrix, as well as its ability to recognize that an unknown is an outlier to one of the taxa, an intermediate between two of the taxa, or does not belong to any of the taxa of that matrix.

Identification models for binary data

Many classification studies have been performed with groups defined by natural classification (e.g., Enterobacteriaceae, Vibrionaceae). By contrast, ID schemes are often required for newly isolated environmental strains and the taxa found in most environments encompass more than one group. Schemes which require one matrix for identification of all taxa are preferable. Choice of taxa and attributes for ID matrix may be influenced by the environment from which the unknown OTUs originate (e.g., water, soil, etc.). Once the taxa have been chosen, the selection of attributes can proceed. The problem with an ID matrix containing widely divergent taxa is that attributes may not be applicable to all the taxa included.

All ID matrices are constructed so that each pair of groups differs by at least one attribute; however, this does not allow for aberrant behavior in strains. Most ID matrices therefore have a difference of at least two attributes between each pair.

An ideal attribute will be such that all members of each taxon will give the same result. It will also divide the taxa into two equal-sized groups (positive and negative for the attribute). If two attributes produce the same separation, one can be deleted. When attributes are considered collectively, the ability of each attribute to partition the taxa into two groups, different from those formed by previously chosen attributes, is the criterion for inclusion in an identification matrix.

If ideal attributes were available, the number required for an ID matrix, with one attribute difference between each pair, would be equal to the square root of the number of groups (20). However, attributes are rarely ideal. They do not give a uniform within-taxon result, only 70 to 99 percent of the OTUs give the same result, and the numbers of positive and negative taxa for each attribute are rarely equal. Consequently the number required is often several times greater than the minimal (square root of n) number.

Identification schemes for binary data can be hierarchical, using a dichotomous key, or simultaneous, according equal weight to all attributes. Dichotomous keys still appear in the literature because they are adequate when a small number of attributes are considered. Attribute selection and methods for computer construction of such keys have been described (28, 29, 35, 38, 46).

The order in which tests are applied is important. An error (aberrant strain) in an attribute positioned in the first stages of a dichotomous key will have a greater deleterious effect on identification than if it occurs toward the end. On the other hand, a hierarchical approach, using different attributes at different branches, may be more suitable for schemes encompassing more than one family or kingdom.

The probabilities of positive results of each taxon for each attribute are stored in the identification matrix. These may vary between zero and one although, in practice, the range is between 0.01 and 0.99. Addition of a third decimal digit may increase the separation but requires testing of a much larger number of isolates per taxon to be accurate.

Although Gyllenberg (21) has suggested principal component analysis for computer identification, the Bayes model (16, 29) is the one most frequently applied. One of its important elements is the inclusion of prior probability of occurrence of any taxon in the population studied. Inclusion of prior probabilities has rarely been practiced in bacterial identification. The arguments against its use include the difficulty in providing accurate estimates of occurrence, or the change of occurrence patterns with change in environmental conditions, and the risk of missing a very rare but highly significant taxon. It is nevertheless worth realizing that inclusion of prior probabilities may help in identification of abnormal isolates of common taxa.

The likelihood of the unknown belonging to each taxon is calculated by multiplication of the probabilities for all attributes. Criteria for acceptability are discussed below.

Identification models for continuous data

The ARTHUR package is probably the most comprehensive ID package providing a number of models. Identification routines are also available in BMDP, CLUSTAN, GENSTAT, SAS, and SPSS/PC+ packages (Table 6.1).

Although the Bayesian classification rule can also be used for quantitative data, it requires an enormous data bank to provide statistically significant values. In other methods summarized by Bascomb (4), the OTU-by-attribute matrix is visualized in a multidimensional space where each attribute occupies one dimension.

Criteria for acceptability of identification

Criteria for acceptability of identification of an OTU relating to either the chosen taxon or the resemblance to the ideal organism of that taxon have been suggested. The normalized likelihoods are calculated to determine the group to which the unknown most likely belongs (29).

When the values for an unknown are compared with a number of taxa, the unknown generally will be closer to one taxon than to all others. However, relative closeness is not a proof that the unknown is a member of that taxon. Criteria for acceptability are required. In the first successful computer identification study (29), two criteria were imposed: (i) if the unknown shows a similar distance to more than one group, it is unlikely to belong to either; and (ii) an upper limit established to the deviation from the ideal organism of that group. Conditional probability (13), taxonomic distance, standard error of this distance, and pattern distance (41) are also used to produce accurate estimates of the resemblance of an OTU to a taxon.

With quantitative data, acceptability criteria have been formulated only in the SIMCA (49) model, where an unknown is assigned to one group if it is within the 95 percent confidence limits of only one group. It is termed an intermediate if it is within the confidence limits of more than one group, or it is termed as unknown if it is outside the boundaries of all the groups.

In cases outside the acceptability levels, an indication of whether the organism is an intermediate, an outlier, or belongs to a taxon not included in the matrix is found by examining the closest group with relevant attributes and the resemblance to its ideal organism.

Additional testing

Identification models for binary data can operate using only some of the attributes in the matrix and using the remaining attributes only if identification is inconclusive. If acceptable probability levels for identification have not been reached, the attributes which will separate between the most likely groups are selected. Test selection based on ability to separate between the most likely groups, giving two test differences between each possible pair, have been described (29).

Similar models are not available for quantitative identification systems, where acquisition of data for all attributes is performed simultaneously and the results for all tests are available. However, this may result in the inclusion of tests which are meaningful to only part of the taxa included in the matrix. A generalized computer program for identification of bacteria using the Bayesian probability model, including test selection, was described (26).

Software available for identification

Options for identification based on the nearest-neighbor model are available in ARTHUR, TAXAN, and GENSTAT and could be applied to both binary and continuous data. MATIDN (41) can be applied using binary data and a probability matrix. The principle of application of Bayes' probability model to identification has been explained elsewhere (4, 5, 8, 16, 29, 40, 47).

Statistical packages (e.g., SPSS, BMDP, and ARTHUR) contain routines for calculating the discriminant function analysis on the basis of results for OTUs with known group membership and can store the resulting matrices for future use. These packages provide facilities for applying different criteria for inclusion of attributes in the analysis and for selection of a different number of attributes. It is also possible to divide the population into a learning set and a test set. The advantage of using these routines within general statistical packages is in the ability to apply other statistical routines to the data, before and after classification or identification, with a minimum of effort.

Assessment of performance of identification systems

Identification systems should be tested for ability to identify typical and atypical members of all taxa in the matrix and a few members of taxa not included but likely to be misidentified. However, this may sometimes be difficult to achieve with taxa for which only a few strains are known, as is common in ecological studies.

Data from freshly isolated strains rarely match those from ideal strains. Therefore, ID should be tested with both culture collection and freshly isolated strains, as the successful identification rate for the two groups may be different. Reproducibility of results for a selected number of typical and atypical strains within and between sites should also be established. Identification rates for each taxon and for the whole trial should be reported, as well as percentage of nonidentifiables. Reporting of agreement should indicate cases where identification was achieved only after using attributes not included in the matrix, for example, from serological tests.

Identification systems which perform well with all commonly occurring taxa, but are unable to identify rare ones, may not be suitable for a reference laboratory but may still be of use in ecological studies, as long as there is some indication of the unusual characteristics of the isolate.

Identification needs in many ecological studies are different from those in the clinical laboratory. Once the taxa inhabiting particular ecological niches have been determined, as well as their occurrence at different spatial, climatological, and chronological occasions, interest in repeated identification of freshly isolated strains from that environment is limited. The special needs for ecological studies are for comparisons within taxa, for cluster analysis, and for comparison of different classifications of populations. Populations may be compared on the basis of both the ranges of taxa found and the relative abundance of each taxon. Using analysis of variance techniques (37), statistical packages make it possible to explore ecological data in depth and to extract maximum information from data that may be costly to acquire. Factor analysis methods have been used to correlate environmental factors with bacterial attributes (23).

Commercial identification systems

Most commercial identification systems have been designed for use with clinical isolates. Therefore the attributes tested, and/or the probability matrix tables, may be unsuitable for environmental isolates. The exception is the MIDI Microbial Identification System (Hewlett Packard, Suite 115, Barksdale Prof. Center, Newark, Delaware) which determines the fatty acid contents of isolates and allows the investigator to create a data base specific to the population. Additionally, kits that test for activity of individual enzymes may be used with cell suspensions taken from cultures grown under conditions suitable for the environmental isolate. Studies with anaerobic taxa have shown that enzyme activity of these isolates can be determined under "normal" laboratory temperature, pressure, atmospheric, and cultural conditions. Moreover, using an experimental kit for tests of such enzymes, it was shown that the same tests could be used successfully for identification of fermentative and nonfermentative, gram-negative bacteria as well as bacilli, staphylococci, and streptococci without extensive baseline testing (3).

Conclusions

Numerical methods have been introduced into systematics to produce objective classifications and to cope with large numbers of OTUs and attributes. These objectives are also common to microbial ecology studies. Use of computers in identification enables the nonexpert to employ expertly designed identification systems. The current trend is toward mechanized execution of tests, automatic acquisition of data and on-line identification, all of which are additional aids to ecologists attempting to classify isolates and identify populations. Fresh isolates and testing systems that are not based on clinical criteria will require new data bases for ecological populations.

In spite of availability of mathematical models utilizing continuous data for both classification and identification, binary data have been used predominantly. Artificial intelligence software may help to apply expert knowledge in systems where information on differences between taxa does not lend itself to probability matrix format, for instance, when available for only a limited number of the taxa included in the matrix. Use of artificial intelligence models may be of particular value in ecological studies.

Numerical taxonomy has been helpful in ecological and environmental studies in the past. The tedium and burden of large-scale studies can be lifted with the newer techniques described herein, especially with automation and computer software. Populations can more readily be compared on the basis of both the ranges of taxa found and the relative abundance of each taxon. Using these techniques makes it possible to extract maximum information from data that may be costly to acquire in large-scale field studies.

References

1. Barnett, J. A., S. Bascomb, and J. C. Gower. 1975. A maximal predictive classification of Klebsielleae and of the yeasts. *J. Gen. Microbiol.* 86:93–102.
2. Bascomb, S. 1985. Comparison of transformation and classification techniques on quantitative data. In M. Goodfellow, D. Jones, and F. G. Priest (eds.), *Computer-Assisted Bacterial Systematics.* Academic Press, London, pp. 37–60.
3. Bascomb, S. 1987. Enzyme tests in bacterial identification. In R. R. Colwell and R. Grigorova (eds.), *Methods in Microbiology,* Vol. 19. Academic Press, London, pp. 105–160.
4. Bascomb, S. 1989. Computers in taxonomy and systematics. In T. N. Bryant and J. W. T. Wimpenny (eds.), *Computers in Microbiology, a Practical Approach.* IRL Press, Oxford, pp. 65–102.
5. Bascomb, S., S. P. Lapage, M. A. Curtis, and W. R. Willcox. 1973. Identification of bacteria by computer: Identification of reference strains. *J. Gen. Microbiol.* 77:291–315.
6. Brondz, I., and I. Olsen. 1990. Multivariate analyses of cellular carbohydrates and fatty acids of *Candida* and *Saccharomyces cerevisiae. J. Clin Microbiol.* 28:1854–1857.
7. Bryant, T. N. 1987. Programs for evaluating and characterizing bacterial taxonomic data. *Comput. Appl. Biosci.* 3:45–48.
8. Bryant, T. N., A. G. Capey, and R. C. W. Berkeley. 1985. Microcomputer assisted identification of *Bacillus* species. *Comput. Appl. Biosci.* 1:23–28.
9. Bryant, T. N., J. V. Lee, P. A. West, and R. R. Colwell. 1986. A probability matrix for the identification of species of vibrio and related genera. *J. Appl. Bacteriol.* 61:469–480.
10. Cinco, M., R. Dougan, and J. Stefanelli. 1977. Factor analysis of saprophytic serogroup Doberdo of *Leptospira biflex* and characterization of serovars *zoo* and *drahovce. Int. J. Syst. Bacteriol.* 27:63–65.
11. Colwell, R. R., D. Maneval, E. S. Remmers, E. L. Elliot, and N. E. Carlson. 1984. Ecology of the vibrios in Chesapeake Bay. In R. R. Colwell (ed.), *Vibrios in the Environment.* Wiley, New York, pp. 367–387.
12. Dawson, C. A., and P. H. A. Sneath. 1985. A probability matrix for the identification of vibrios. *J. Appl. Bacteriol.* 58:407–423.
13. Dybowski, W., and D. A. Franklin. 1968. Conditional probability and the identification of bacteria. A pilot study. *J. Gen. Microbiol.* 54:215–229.
14. Elliott, J. A., G. S. Bosley, G. M. Carlone, B. D. Plikaytis, and R. R. Facklam. 1987. Separation of *Haemophilus influenzae* type b subtypes by numerical analysis. *J. Clin. Microbiol.* 25:1476–1480.
15. Ferguson, M. W., K. L. Wycoff, and A. R. Ayers. 1988. Use of cluster analysis with monoclonal antibodies for taxonomic differentiation of phytopathogenic fungi and for screening and clustering antibodies. *Curr. Microbiol.* 17:127–132.
16. Friedman, R. B., D. Bruce, J. MacLowry, and V. Brenner. 1973. Computer-assisted identification of bacteria. *Am. J. Clin. Pathol.* 60:395–403.
17. Goodfellow, M., and R. G. Board (eds.). 1980. *Microbiological Classification and Identification.* Academic Press, London.
18. Goodfellow, M., D. Jones, and P. G. Priest. 1985. *Computer-Assisted Bacterial Systematics.* Academic Press, London.
19. Gower, J. C., and G. J. S. Ross. 1969. Minimum spanning trees and single linkage cluster analysis. *Appl. Statist.* 18:54–64.
20. Gyllenberg, H. G. 1963. A general method for deriving determination schemes for random collections of microbial isolates. *Ann. Acad. Scient. Fenn. A, IV, Biol.* 69:5–23.
21. Gyllenberg, H. G. 1965. A model for computer identification of micro-organisms. *J. Gen. Microbiol.* 39:401–405.
22. Gyllenberg, H. G., and T. K. Niemelä. 1975. New approaches to automatic identification of microorganisms. In R. J. Parkhurst (ed.), *Biological Identification with Computers.* Academic Press, London, pp. 121–136.
23. Holder-Franklin, M. A., and L. J. Wuest. 1983. Population dynamics of aquatic bac-

teria in relation to environmental change as measured by factor analysis. *J. Microbiol. Method.* 23:87–98.

24. Jackman, P. J. H., R. K. A. Feltham, and P. H. A. Sneath. 1983. A program in BASIC for numerical taxonomy of micro-organisms based on electrophoretic protein patterns. *Microbiol. Lett.* 1:209–227.
25. Jacobs, D. 1990. SAS/GRAPH software and numerical taxonomy. In *Proc. Fifteenth Annual SAS Users Group Int. Conf.* SAS Institute, Cary, N.C., pp. 1413–1418.
26. Jilly, B. J. 1988. Microcomputer application of Bayesean probability testing for the identification of bacteria. *Int. J. Biomed. Comput.* 22:107–119.
27. Kramer, S. M., N. P. Jewell, and N. E. Cremer. 1983. Discriminant analysis of data in enzyme immunoassay. *J. Immunol. Method.* 60:243–255.
28. Lapage, S. P., and S. Bascomb. 1968. Use of selenite reduction in bacterial classification. *J. Appl. Bacteriol.* 31:568–580.
29. Lapage, S. P., S. Bascomb, W. R. Willcox, and M. A. Curtis. 1970. Computer identification of bacteria. In A. Baillie and R. J. Gilbert (eds.), *Automation, Mechanization and Data Handling in Microbiology*. Academic Press, London, pp. 1–22.
30. MacDonell, M. T., and R. R. Colwell. 1985. Phylogeny of the Vibrionaceae, and recommendation for two new genera, *Listonella* and *Shewanella. Syst. Appl. Microbiol.* 6:171–182.
31. MacDonell, M. T., and R. R. Colwell. 1985. The contribution of numerical taxonomy to the systematics of Gram-negative bacteria. In M. Goodfellow, D. Jones, and F. G. Priest (eds.), *Computer-Assisted Bacterial Systematics*. Academic Press, London, pp. 107–135.
32. MacDonell, M. T., D. G. Swartz, B. A. Ortiz-Conde, G. A. Last, and R. R. Colwell. 1986. Ribosomal RNA phylogenies for the vibrio-enteric group of eubacteria. *Microbiol. Sci.* 3:172–178.
33. Mallory, L. M., and G. S. Saylor. 1984. Application of FAME (fatty acid methyl ester) analysis in the numerical taxonomic determination of bacterial guild structure. *Microbiol. Ecol.* 10:283–296.
34. Pankhurst, R. J. (ed.). 1975. *Biological Identification with Computers*. Academic Press, London.
35. Payne, R. W., and D. A. Preece. 1980. Identification keys and diagnostic tables: A review. *J. R. Statist. Soc.* A143:253–292.
36. Rogosa, M., M. I. Krichevsky, and R. R. Colwell. 1986. *Coding Microbiological Data for Computers,* Springer Series in Microbiology. New York.
37. Russek-Cohen, E., and R. R. Colwell. 1986. Application of numerical taxonomy procedures in microbial ecology. In R. L. Tate III (ed.), *Microbial Autoecology: A Method for Environmental Studies.* Wiley, New York, pp. 133–146.
38. Rypka, E. W., W. E. Clapper, I. G. Bowen, and R. Babb. 1967. A model for the identification of bacteria. *J. Gen. Microbiol.* 46:407–424.
39. Sakin, M. J. 1987. Computer programs for classification and identification. In R. R. Colwell and R. Grigorova (eds.), *Methods in Microbiology,* Vol. 19. Academic Press, London, pp. 459–494.
40. Schindler, Z., and J. Schindler. 1983. Numerical identification of micro-organisms using the HP-41C calculator. *Int. J. Bio-Med. Comput.* 14:17–22.
41. Sneath, P. H. A. 1979. Basic program for identification of an unknown with presence-absence data against an identification matrix of percent positive characters. *Comput. Geosci.* 5:195–213.
42. Sneath, P. H. A., and R. R. Sokal. 1973. *Numerical Taxonomy.* W. H. Freeman, San Francisco.
43. Thompson, C. J. 1987. New method of serotyping *Escherichia coli:* Mathematical development. *J. Clin. Microbiol.* 25:774–780.
44. Véron, M. 1974. Sur un critère de calcul du meilleur niveau de compare d'un dendrogramme de classification hiérarchique. *Ann. Microbiol.* (Inst. Pasteur) 125B: 29–44.

45. West, P. A., P. R. Brayton, T. N. Bryant, and R. R. Colwell. 1986. Numerical Taxonomy of Vibrios isolated from aquatic environments. *Int. J. Syst. Bacteriol.* 36: 531–543.

46. Wilcox, W. R., and S. P. Lapage. 1972. Automatic construction of diagnostic tables 6. *Comput. J.* 15:263–267.

47. Wilcox, W. R., S. P. Lapage, S. Bascomb, and M. A. Curtis. 1973. Identification of bacteria by computer: Theory and programming. *J. Gen. Microbiol.* 77:317–330.

48. Williams, S. T., M. Goodfellow, E. M. H. Wellington, J. C. Vickers, G. Alderson, P. H. A. Sneath, M. J. Sackin, and A. M. Mortimer. 1983. A probability matrix for identification of some streptomycetes. *J. Gen. Microbiol.* 129:1815–1830.

49. Wold, S. 1976. Pattern recognition by means of disjoint principal component models. *Pattern Recogn.* 8:127–139.

Part 2

Genetic Transfer and Stability

Overview: Methods for the Evaluation of Genetic Transport and Stability in the Environment

Robert V. Miller

Introduction

A major force for the change and evolution of a natural population is the introduction of new genetic elements into the gene pool of that population. The evaluation of the impact on the environment of the release of genetically engineered microorganisms must include an analysis of the potential of their genetically engineered sequences to be stably maintained in the natural community of microorganisms. The potential for intraspecific and interspecific transfer of these sequences as well as the relative fitness of the phenotype they produce will contribute to their probability of impacting the gene pool of the natural microbiota. As the genetic transfer systems of more and more bacterial species are investigated, it becomes clear that the potential for genetic transfer between cells of the same and other species and genera is much greater than has been previously suspected. Genetic transmission of modified genomes to natural populations of bacteria would appear to be a significant possibility. Few data are currently available on gene transfer in the environment (35, 42). Techniques for the evaluation of gene transmission in situ are only now being developed. A summary of the techniques that are now available and suggestions on their applications to environmental studies are presented in the following chapters.

Many of the systems described are models for environmental study and may not exactly reproduce the in situ situation. They often represent compromises between the controlled atmosphere of the laboratory and the uncontrolled conditions encountered in nature. These systems are presented simply as guides for the development of experimental protocols to address questions of environmental importance in specific ecosystems.

Background Development, Significance, History: Mechanisms of Gene Transfer in Bacteria

Three major systems of genetic transfer are recognized in bacteria. *Conjugation* is a parasexual process that requires direct cell-to-cell contact of donor and recipient cells (18, 80). *Transduction* (26) is the genetic transfer of DNA mediated by bacteriophages (viruses). *Transformation* leads to the acquisition of new genetic material by the absorption of free DNA from the external milieu (70). These genetic transfer systems are reflected in the methods routinely used in the laboratory as tools of genetic analysis, yet few data are available concerning the environmental impact of these processes. Few studies have been done to estimate the frequency of genetic transfer by these mechanisms in nature. Available data are summarized in several excellent reviews (35, 36, 42, 53, 62, 71, 72).

Since the Asilomar Conference of 1975 on the dangers of accidental release of genetically engineered organisms into the environment, it has been generally assumed that conjugation is the most probable and perhaps the only fertile method of genetic transfer in the environment. Not surprisingly, therefore, the evidence for environmental gene exchange is most convincing for conjugation (36, 47, 53, 62, 71, 72). Recent data (42, 46, 59, 60, 83) have demonstrated that transduction is also a potentially significant form of gene transfer in the environment. Transformation has not as yet been significantly investigated in natural ecosystems (42), but available data indicate that this form of gene transfer must also be considered in any evaluation of the potential for genetic transmission in situ.

The majority of environmental data collected to date has been epidemiological in nature. This means that the spread of a specific phenotype (usually associated with a plasmid) has been examined (9, 10, 20). While these studies have demonstrated the spread of a genetic element in a specific environment, most reports have not addressed the mechanism of gene transfer.

Many studies have been carried out that demonstrate the presence of plasmids capable of conjugation in natural environments (72). Hospital wastes, raw sewage, sewage effluents, fresh and marine water, animal feedlots, plants, and soils have all been shown to contain bacteria that can transfer these plasmids by conjugation. Many of these plasmids can be transferred to a wide variety of bacterial species in many genera (11, 51, 52). Other studies have demonstrated the presence of suitable recipients for conjugation in environmental samples (12, 40).

Several studies that clearly demonstrate the potential for transfer of genetic material by conjugation in the environment have been carried out. In situ transfer of plasmids by conjugation has been demonstrated in sewage treatment facilities (40), in freshwater environments (47), on plant surfaces (21), and in sterilized soils (76) by introduction of genetically marked plasmid donor and recipient strains. In vivo transfer of plasmids has been demonstrated in the intestines of various animal species including humans (72). In many cases, indigenous microbiota were eliminated to allow the establishment of the donor and recipient strains in the environmental system. But several studies in animals and in freshwater environments have demonstrated conjugation in the presence of the natural microbial community (47, 48, 68). Nontransmissible (Tra$^-$) plasmids have been shown to be mobilized by Tra$^+$ plasmids in several environmental studies (12, 40), and conjugal transfer of chromosomal DNA has been demonstrated in at least one in vivo study of gene transfer between strains of *Escherichia coli* inoculated into soil samples (78).

Because of the relative small host range of many bacteriophages and the fact that transduction is mediated by an external factor (the transducing phage), it has been generally assumed that transduction would be a less prevalent form of gene transfer in the environment. While fewer studies have addressed the frequency of transduction in the environment than have addressed conjugation, transductional transfer of both chromosomal and plasmid DNA has been shown to take place in the environment. Morrison et al. (46) demonstrated the transduction of chromosomally encoded streptomycin resistance (Sm^r) to an introduced recipient strain in environmental test chambers incubated in a freshwater lake. Saye et al. (59, 60) reported transduction of both plasmid and chromosomal DNA in similar freshwater field tests. They were able to demonstrate transduction in the presence, as well as in the absence, of the natural microbial community and found, as have other workers using conjugation as the mechanism of transfer, that the number of recoverable transductants was lower in the presence of the indigenous microbiota. In a model system containing the nonenvironmental bacterium *E. coli* and the coliphage P1, Zeph et al. (83) have demonstrated transduction in soil microcosms.

Transformation was the first mechanism of genetic exchange identified in the procaryotes. Griffith's original report in 1928 (16) described the transformation of avirulent pneumococcus to virulent forms in the lungs of infected rodents. Even though transformation has become an important tool in genetic engineering, little interest has been paid to the environmental aspects of the subject. In 1978, Graham and Istock (14, 15) reported that DNA introduced into soil samples could genetically transform *Bacillus subtilis* present in the soil samples. In laboratory simulations, Lorenz and Wackernagel (38) demonstrated the transformation of sand-attached cells by sand-absorbed DNA.

General Characteristics of Methods to Be Covered

Methods for the identification of gene transfer and stability in the environment have been adapted from microbial genetics, mocrobial ecology, and microbial population genetics and more recently from molecular biology and molecular genetics. In the chapters that follow, these methods will be reviewed and their adaptation to various ecosystems described. While the specific methods and protocols utilized will depend upon the gene transfer systems and ecosystems under study, a three-phased approach for the general design of gene transfer studies is recommended.

Phase 1: Idealized laboratory studies for the determination of maximal potentials

The rates of transfer of various plasmids (18, 80) as well as the frequency of transduction mediated by various phages (26, 59) differ even under ideal conditions. Hence, it is necessary to determine the idealized potential for gene transfer in the specific system under study by carrying out a series of experiments in the laboratory. These experiments should be done utilizing recognized methods for conjugation, transduction, or transformation so that the results obtained can be evaluated against the background of data available in the literature. In addition, various factors that may become important in the environmental studies can be evaluated. These may include the effects of temperature, cell concentration, metallic ions, selection criteria, and other variables.

Phase 2: Laboratory simulations— microcosm studies

The semicontrolled conditions afforded by the microcosm make it useful for the identification of factors that may affect results obtained in situ. Such laboratory simulations should be designed to mimic the natural environment under study as closely as possible. They afford a chance to evaluate known variables affecting the system under conditions that more closely reflect the environment than do the idealized experimental conditions employed in phase 1. They provide an opportunity to estimate both the sensitivity and specificity of the detection methods to be used in situ in phase 3. These studies will also allow for an estimation of the possible effects on enumeration and detection of the presence of the natural microbial community found at the field site (59). Microcosm studies should not, however, replace environmental studies. Several studies (47, 58–60) have shown that many factors operating in the environment cannot be modeled effectively in the laboratory using methods that are currently available.

Phase 3: Environmental studies

The environmental significance of data obtained in phases 1 and 2 should be validated by experimentation in situ whenever possible. The data obtained in the preliminary phases should be considered in designing protocols for this all-important portion of the study. In this phase, particular consideration should be paid to containment. Containment should ensure that general release of the test organisms will not take place, yet it should allow the conditions of the experiment to be as close as possible to those encountered in the uncontained environment. Methods for containment are considered in another section of this guide. Physical and biological background data on the test site should be collected and evaluated before field trial. During the trial, these data should be collected both from within the test (containment) chamber and from the surrounding milieu. They should be utilized to ensure that the conditions during the field trials are representative of the environment and field site under study.

Environmental Implications and Considerations

Compared to laboratory experimentation, genetic transfer in the environmental setting is uncontrolled, and a significant number of variables that are either unique or uncontrollable must be considered in designing experimental protocols and in evaluating the validity of the data obtained. While different variables may be more or less significant depending on the genetic system and environment under study, several are likely to be important in all studies. A partial list of these variables is identified here and should be considered in the design of all environmental studies.

Cell concentration and probability of association

Perhaps the most basic factor determining whether or not gene transfer will take place in a particular environment is the probability of association of the various components of the gene transfer system in that environment. Using a chemostat model, Levin et al. (34) demonstrated that conjugal transfer obeys a simple mass action model that is dependent on cell concentration. Transduction depends on the contact of a transducing particle with an appropriate recipient cell and hence would also be expected to obey a mass action model. Therefore, the concentration of the various biological components of the gene transfer system in an ecosystem is potentially the most important factor in determining whether or not genetic transfer will take place. However, both donors and recipients of conjugative plasmids can be isolated from numerous environmental sources (36, 42, 53, 62, 71, 72), and phages capable of transduction are present in the environment in numbers sufficient to serve as sources of transducing particles (2, 3, 26, 50, 69).

Studies conducted in situ have shown that the frequency of transfer by either conjugation (47) or transduction (59) is significantly influenced by both the absolute numbers of cells capable of donating genetic information and their relative concentration with respect to potential recipients of this genetic transfer. While laboratory experiments are done at high cell concentrations [10^8 to 10^9 colony-forming units (CFU) per milliliter], the concentrations of bacteria and bacteriophages found in nature are often lower (19, 26, 47, 50, 59, 69). The concentration of natural communities at the test site should be determined and experiments conducted at realistic cell densities.

Competence for transformation is also dependent on cell concentration in a number of bacteria (70). Transformational competence is induced in *Streptococcus* and *Bacillus* only after a threshold concentration of a small extracellular protein (competence factor) is reached in the cellular milieu (13).

Temperature

Any mass action phenomenon is dependent upon temperature. While this may theoretically eliminate some environments from serious consideration as venues for gene transmission, a vast majority of ecosystems are at temperatures that will theoretically allow gene transfer. Conjugal transfer is quite efficient between 15 and 37°C (53) and has been observed at temperatures as low as 7°C (1). Interestingly, observed rates of transfer of plasmids from soil bacteria to *E. coli* may be 1000 times higher at 28°C than at 37°C (53). Transposition in some systems has been shown to be more efficient at temperatures below 30°C than at 37°C (30). Transduction has been shown to occur at ambient lakewater temperatures of 20°C (59, 60).

Particulate matter

Particulate matter in the environment may have a significant effect on gene transfer. Conjugal transfer of many plasmids has been shown to be more efficient on solid surfaces than in liquids devoid of particulate matter (17). Particulate matter in aquatic environments may form surfaces on which gene transfer takes place (47, 58). In these environments, the majority of the microbial biomass as well as the majority of metabolic activity occurs on the surfaces of suspended particulate matter (19). Adsorption of DNA to sediment has been shown to protect it from nuclease degradation (37). DNA adsorbed to sand has been shown to be active in transformation of sand-associated cells (38). On the other hand, Singleton (67) demonstrated that colloidal clay strongly inhibits conjugal transfer in *E. coli*. Roper and Marshall (56) demonstrated that sorption of *E. coli* to small diameter (<0.6-μm) clay particles inhibits bacterium-bacteriophage interaction, while particles of larger spherical diameter (>2.6 μm) appear to stimulate this interaction.

Whether solid surfaces act to stimulate or deter genetic transfer in the environment remains an open question that should be more fully investigated. The effects of particulates must be considered in any environmental study.

Barriers to the entry of DNA into the recipient cell

The nature of the various mechanisms of gene exchange in the bacteria lead to unique restraints on the entry of exogenous genetic material into the potential recipient cell. These entry barriers may limit the free flow of DNA between bacterial species in the environment. Consideration of the unique barriers associated with each mechanism of gene transfer should be made in the development of any protocols for monitoring gene transfer.

Conjugation.   Conjugation requires the intimate interaction of the donor cell with the recipient. Initial contact is made between the pilus of the donor cell and the cellular envelope of the recipient (80). It appears that specific features must be present on the surface of the recipient cell to allow a successful interaction to take place. This requirement limits the host range of many plasmids to a relatively few closely related organisms. However, many plasmids are not restricted in their host range by barriers of entry. Plasmids of incompatibility groups N, P, and W (81) have broad host ranges and can transfer to a wide variety of gram-negative organisms. The plasmid RP1 has been shown to be transferred and maintained in organisms as distantly related as *E. coli* and *Rhizobium* sp. (52).

Transduction. Transduction is limited by the host range of the transducing phage. As with conjugation, initial adsorption of the transducing particle to the host requires the specific interaction of the virion with unique receptors on the surface of the potential host cell (26). While this often limits phages to infecting one or a small number of closely related species, many phages exist that have a broad host range (51). For example, the generalized transducing bacteriophage Mu, which replicates by transposition, is capable of infecting several genera of the family Enterobacteriaceae (7). Other examples of broad-host-range phages have been identified from environmental sources. Kelln and Warren (23) identified phages that formed plaques on a wide variety of pseudomonads. Thorne (73, 74) observed interspecies transduction by phage CP-54 between a variety of *Bacillus* sp., and Yu and Baldwin (82) demonstrated interspecies transduction in the staphylococci.

Transformation. Transformation is limited by the requirement that the recipient cell be competent for absorption of DNA. Transformation in the environment is limited to those species with the capacity to express natural competence. Competence development is a characteristic of the normal physiology of the species and is encoded by chromosomal genes (70). Natural competence is a property of a significant number of environmentally important species, both gram-positive and gram-negative (70).

In addition to competence, absorption of DNA in many species requires the presence of specific recognition sequences in the exogenous DNA (4). Only DNA that contains one or more of these sequences will be absorbed. This requirement acts to limit absorption to DNA of closely related organisms that contain the recognition sequence. The characteristics of competence development in the environmental species under study must be identified and considered in evaluating protocols for gene transfer studies.

Restriction—modification

Restriction-modification system(s) active in the potential host cell may act to limit heritable establishment of foreign DNA. These systems consist of two enzymes (54): (1) an endonuclease that upon recognition of a unique sequence of bases in a DNA molecule, cleaves the DNA and renders it biologically and genetically inactive; (2) a methylase that modifies any DNA recognition sequences in the host DNA, thus protecting them from cleavage by the restriction endonuclease.

While restriction-modification systems have the potential for imposing a significant barrier to gene exchange on the environment, several examples of the elimination of restriction have been documented. Environmental stresses have been shown to inhibit the restriction of DNA (55, 77). Temperate phages (22, 31) escape restriction by various mechanisms including the inactivation of the restriction endonuclease. In addition, restriction is seldom absolute. Often a few DNA molecules escape restriction by replicating too fast to be restricted (53) or simply by chance.

Incompatibility

A major factor affecting the stable establishment of plasmid DNA within a potential host is its possible incompatibility with any resident replicons (6). If two plasmids of the same incompatibility group infect the same cell, only one will be stably established. Interestingly, it is not always the superinfecting plasmid that is eliminated. Often the resident plasmid is lost instead. This phenomenon is likely to be important in regulating the free transfer of plasmid DNA throughout the environment. The incompatibility characteristics of the various DNA components utilized in model gene transfer systems must be considered in the model's development.

Superinfection immunity

The phenomenon of superinfection immunity among temperate bacteriophages is analogous to plasmid incompatibility. Often the presence of a prophage in a lysogen will inhibit the establishment of a similar or closely related prophage in the same cell. Superinfection immunity results from the repression of the expression of phage lytic function immediated following infection of the lysogen (26, 43). In many cases, absorption of DNA contained in the phage particle is not inhibited (59) and the potential for transduction of plasmid or bacterial chromosome DNA is unaltered in the lysogen. Hence, Saye et al. (59, 60) demonstrated that a lysogen is the best environmental recipient in transduction. This is probably observed because the transduced lysogen is protected from subsequent killing by infection with a viable phage particle. Superinfection immunity may actually serve to increase the observed frequency of successful transduction events in the environment and must be considered in the development of environmental models for transduction.

Fertility inhibition

Many plasmids inhibit their own transfer as well as that of other coresident plasmids by repressing expression of plasmid encoded transfer functions (53, 80). This phenomenon, known as fertility inhibition, has a potential for altering the frequency of plasmid transfer in the environment. However, fertility inhibition can be overcome in certain environments. Transitory derepression of fertility functions (i.e., *tra* gene expression) occurs in newly formed transconjugants and their immediate descendants (39).

DNA replication compatibility

Independent replicons such as plasmids must rely to a lesser or greater extent on host cell machinery for DNA replication. Many broad-host-range plasmids are efficient in replicating in a wide variety of bacterial species because they encode the auxiliary proteins necessary for their own replication (79).

Recombination

The transfer of chromosomal DNA by any of the various methods of genetic transfer in the bacteria requires the integration of the exogenous DNA into the host chromosome. The mechanism of this integration (homologous, generalized recombination) requires sequence homology between the endogenotes and exogenotes in the region of the crossover event. Hence, it has been argued that the insertion of engineered DNA into the chromosome will eliminate or significantly reduce the possibility of gene dissemination. While such a procedure may reduce gene transfer, it is unlikely to eliminate it. For recombination to take place, it is only necessary that DNA sequence homology be present in the regions flanking the inserted gene. The engineered gene sequence itself does not have to be homologous or even similar in nucleotide sequence to the potential endogenotic recipient. It has been estimated that in *E. coli* as few as 23 to 27 base pairs of homologous DNA sequence are required to allow a successful recombination event to take place (64). Many naturally occurring gene sequences appear to have been conserved during evolution by vastly divergent species of bacteria and can provide enough sequence homology to allow homologous recombination to proceed. For example, the DNA sequence of the *recA* genes of *E. coli* and *Pseudomonas aeruginosa* is greater than 85 percent similar (27, 28, 57). Several examples of the transfer and stable establishment of chromosomal DNA in environmental model systems have been reported (15, 46, 60, 70, 78).

The central protein needed for homologous recombination is the RecA protein, which acts as a synaptase to bring the homologous sequences into register. Expression of the *recA* genes of both *E. coli* (77) and *P. aeruginosa* (29, 44) is inducible by various stresses to the cell which may be operating in the environment. Hence, it is possible that rates of recombination measured in the laboratory under nonstressful conditions are underestimates of the rates that may be found in the environment. The recombinational potential of the recipient cell(s) must be considered in the development of model systems for genetic transfer in the environment, especially when the transfer of chromosomal DNA sequences is under investigation.

Consideration of potentials for gene expression

Most systems for the estimation of gene transmission and stability depend upon the detection of an altered phenotype in the recipient cell. After a gene is transferred to and established in a recipient strain or species, its product must be expressed in order to change the phenotype of the new host. Several factors may affect this expression. While failure of a gene to be expressed may inhibit detection of its transmission to a new species, it does not eliminate the gene from the gene pool nor does it eliminate the possibility of subsequent transfer of the gene to another strain or species in which the genetic information can be expressed. The expression of any gene to be used to monitor gene transfer or stability should be evaluated in possible recipients during the development of experimental protocols.

Transcription.   Transcription of the transferred gene is required for its expression. Thus the recipient cell must contain a sigma factor that will recognize and promote RNA synthesis from the promoter of the foreign gene. This is not always the case. For instance, many *P. aeruginosa* genes are not expressed in *E. coli* because their promoters are not read efficiently (27).

Posttranslational modification.   Often the gene product of a transferred gene must be posttranslationally modified before it is enzymatically or structurally active. Enzymes responsible for this modification may not be present in the heterologous species, and the protein product of the gene, although produced, may not be capable of eliciting the appropriate phenotype in the heterologous host. While the primary transcription product of many eukaryotic proteins can be produced when their genes are engineered into prokaryotic hosts, the proteins are not modified and their activity often cannot be detected in prokaryotic clones.

Other Factors That May Affect Gene Stability

Several factors may influence gene stability in the environment. These cannot always be modeled in the laboratory using present methods but must be carefully monitored as they may significantly modify measurement of apparent rates of gene dissemination.

Mutation

Mutation, the alteration of information contained in the genetic material, is a major source of genetic variation in the environment (51). Its consequence may be a change in phenotype. When alteration takes place in a genetic marker being used to evaluate gene transmission in the environment (46, 47, 59, 60), mutation may mask gene transfer events by altering the expected phenotype (45, 60). Therefore, rates of mutation must be evaluated by including the appropriate controls in all environmental studies.

Several processes of DNA repair that are clearly mutagenic are inducible by various forms of stress that may be encountered in the environment (77). While the effects of these systems on gene stability have not yet been evaluated in environmental studies, evidence is beginning to appear that they may act to increase rates of mutation in the environment above those that would be expected from laboratory simulations. Saye and Miller (58) found that the reversion frequency of several genetic markers they were using to evaluate chromosomal transduction in freshwater environments was so high in situ that it masked any transductional events that occurred. This was true even though these genes were stable in normal laboratory transductions and in laboratory simulations using lakewater and conditions designed to mimic the test site. This is a clear indication that stress-induced processes which lead to gene instability may be operating in the environment which cannot at present be modeled in the laboratory.

Clearly mutational events can alter not only the structure of the gene's product but may also affect gene regulation. Laboratory studies using *E. coli* and *Salmonella typhimurium* have demonstrated that mutations can alter patterns of gene expression leading to a dramatic alteration in the phenotype of the organism. Mutation has been demonstrated to activate cryptic genes in the genome leading to new and often unexpected phenotypes in the affected bacterium (8, 22, 49). The possible consequences of mutational modification of genetic potential and expression must be investigated in order to determine their contribution to potential risk from the release of genetically modified microorganisms into the environment.

Another factor that can affect the rate of genetic mutability is the existence of a number of so-called UV^r plasmids, which increase both the level of resistance of the host cell to DNA-damaging stress and the rate of mutagenesis stimulated by these stresses (32, 33, 66, 76). UV^r plasmids are often Tra^+ (33), and many of them are also fertility factors capable of mobilizing the host chromosome in conjugal transfer (32). These plasmids are frequently found in natural environments, and their acquisition by a genetically modified organism introduced into the environment from a naturally occurring strain may change both the ability of the introduced organism to survive in situ and its level of genetic instability.

Transposition

Reassortment of genetic material associated with movement of transposable (Tn) elements and insertion (Is) sequences may be the one most significant source of genetic variability in the environment. Tn elements have the potential for transmitting genetic loci to new replicons and for affecting expression of neighboring sequences as a consequence of their insertion (41). Transposition does not require homologous recombination for rearrangement of genetic material (63). Transposons are capable of initiating their own transfer from one replicon to another. In addition, they contain genes that confer a selectable phenotype on the host cell, and these genes are transferred as part of the transposon. If two replicons, one of which contains a transposable element, are present in the same cell, the element may be transferred to the second replicon without the necessity for sequence homology between the two genetic elements. This can lead not only to reassortment of linked genetic elements but also to insertional mutagenesis of the gene into which transposition takes place (25). Recently, transposons have been identified that are capable of mediating conjugal transfer (5). Transposition of these elements can confer a Tra^+ phenotype on a replicon that would otherwise be incapable of conjugation.

The movement of various transposable elements that is repressed under conditions of optimal growth has been shown to be increased by exposure of the host cell to stress. Transposition of the mercury-resistance encoding transposon Tn501 is induced by exposure to mercury (65). Tn3 transposition is stimulated when the growth temperature is lowered (30).

Transposition of Tn*917,* which codes for erythromycin resistance, is induced by exposure of the host cell of erythromycin (51, 75). The rate of transposition of one Tn element may be increased by the presence of a second Tn element operating in *trans* in the same cell (61). Therefore, rates of transposition can be significantly higher in the environment than the rates observed in the laboratory. Transposition potential must be considered in designing experimental protocols that utilize any genetic elements known to contain Tn elements, even when these elements have been altered to eliminate transposition in laboratory setting.

Conclusions

Evidence is accumulating that horizontal gene transmission occurs in the environment (35). The methods for detecting and measuring the occurrence of gene transfer and genome stability are in a stage of rapid development and testing. The methods presented in this book are examples of the current state of technology in this area. Much more work is needed to determine the environmental factors that are important in assessing these processes in situ. At present, one must assess each situation as an independent case and develop specific experimental protocols to deal with the unique features of the genes, organisms, and environments under investigation.

Overview of Part 2

Part 2 is divided into four general subdivisions. Each addresses a different aspect of the methodology and the problems that can occur in adapting it to environmental studies. The various aspects cannot be considered independently of the others. In designing a specific study, it may be necessary and certainly advisable to refer to a variety of the chapters to obtain information on classical laboratory methods, specific vectors and/or reporter systems, methods for analysis of data, and considerations specific for a certain ecosystem.

Part 2 is organized as follows:

Chapters 8 and 9: Methods for the chemical and physical evaluation of the genetic sequences under study. These chapters outline methods for assessing the purity of engineered genetic material. Methods for the identification and enumeration of genes in situ are outlined. Techniques that can be employed to measure the complexity of natural gene pools are also included.

Chapters 10 to 12: General and classical methods for the study of the mechanisms of gene transfer in bacteria. These chapters address classical methods for the study of gene transfer in bacteria. They outline procedures that should be employed in phase 1 of environmental studies. Environmental considerations and limitations are pointed out.

Chapters 13 to 16: Application to specific environments. These chapters outline data on gene transfer and stability in selected environments. Considerations for the development and analysis of protocols and data that are specific to these environmental settings are discussed. The limitations of the methods are described and enumerated.

Chapters 17 to 19: Reagents for use in environmental studies. These chapters discuss the use of standardized vectors for modified genetic sequences. Reporter genes and their application are discussed and their usefulness in environmental studies evaluated.

Acknowledgments

This endeavor was supported in part by cooperative agreements Nos. CR-815234 and CR-815282 with the Gulf Breeze Environmental Research Laboratory of the U.S. Environmental Protection Agency.

References

1. Bale, M. J., J. C. Fry, and M. J. Day. 1987. Plasmid transfer between strains of *Pseudomonas aeruginosa* on membrane filters attached to river stones. *J. Gen. Microbiol.* 133:3099–3107.
2. Bergh, O., K. Y. Borsheim, G. Bratbak, and M. Heidal. 1989. High abundance of viruses found in aquatic environments. *Nature* 340:467–468.
3. Borsheim, K. Y., G. Bratbak, and M. Heidal. 1990. Enumeration and biomass estimation of planktonic bacteria and viruses by transmission electron microscopy. *Appl. Environ. Microbiol.* 56:352–356.
4. Carlson, C. A., L. S. Pierson, J. J. Rosen, and J. L. Ingraham. 1983. *Pseudomonas stutzeri* and related species undergo natural transformation. *J. Bacteriol.* 153:93–99.
5. Clewell, D. B., and C. Gawron-Burke. 1986. Conjugative transposons and the dissemination of antibiotic resistance in streptococci. *Ann. Rev. Microbiol.* 40:635–659.
6. Datta, N. 1979. Plasmid classification: Incompatibility grouping. In K. A. Timmis and A. Pühler (eds.), *Plasmids of Medical, Environmental, and Commercial Importance.* Elsevier/North Holland, Amsterdam, pp. 3–12.
7. Dénarié, J., C. Rosenberg, B. Bergeron, C. Boucher, M. Michel, and M. Barate de Bertalmio. 1977. Potential of RP4::mu plasmids for in vivo genetic engineering of gram-negative bacteria. In Bukhari, A. I., J. A. Shapiro, and S. Adhya (eds.), *DNA Insertion Elements, Plasmids, and Episomes.* Cold Spring Harbor Laboratory, Cold Spring Harbor, N.Y., pp. 507–520.
8. Downs, D. M., and J. R. Roth. 1987. A novel P22 prophage in *Salmonella typhimurium. Genetics* 117:367–380.
9. Fisher, G. M., M. C. Kelsey, and E. M. Cooke. 1986. An investigation of the spread of gentamicine resistance in a district general hospital. *J. Med. Microbiol.* 22:69–77.
10. Fontaine, T. D., and A. W. Hoadley. 1976. Transferable drug resistance associated

with coliforms isolated from hospital and domestic sewage. *Health Lab. Sci.* 13:238–245.

11. Gauthier, M. J., F. Cauvin, and J. P. Breittmayer. 1985. Influence of salts and temperature on the transfer of mercury resistance from a marine pseudomonad to *Escherichia coli. Appl. Environ. Microbiol.* 50:38–40.

12. Gealt, M. A., M. D. Chai, K. B. Alpert, and J. C. Boyer. 1985. Transfer of plasmids pBR322 and pBR325 in wastewater from laboratory strains of *Escherichia coli* to bacteria indigenous to the waste disposal system. *Appl. Environ. Microbiol.* 49:836–841.

13. Goddgal, S. H. 1982. DNA uptake in *Haemophilus* transformation. *Ann. Rev. Genet.* 16:169–192.

14. Graham, J. B., and C. A. Istock. 1978. Genetic exchange in *Bacillus subtilis* in soil. *Mol. Gen. Genet.* 166:287–290.

15. Graham, J. P., and C. A. Istock. 1979. Gene exchange and natural selection cause *Bacillus subtilis* to evolve in soil culture. *Science* 204:637–639.

16. Griffith, F. 1928. The significance of pneumococcal types. *J. Hyg.* 27:113–159.

17. Haas, D., and B. W. Holloway. 1976. R factor variants with enhanced sex factor activity in *Pseudomonas aeruginosa. Mol. Gen. Genet.* 144:243–251.

18. Holloway, B. W. 1979. Plasmids that mobilize bacterial chromosome. *Plasmid* 2:1–19.

19. Iriberri, J., M. Unanue, I. Barcina, and L. Egea. 1987. Seasonal variation in population density and heterotrophic activity of attached and free-living bacteria in coastal waters. *Appl. Environ. Microbiol.* 53:2308–2314.

20. Joly, B., R. Cluzel, P. Henri, and J. Barjot. 1976. The resistance of "Pseudomonas" to antibiotics and heavy metal: Minimal inhibitory concentrations and genetic transfers. *Ann. Microbiol. (Paris)* 127:57–68.

21. Johnston, A. W., and J. E. Beringer. 1975. Identification of the rhizobium strains in pea root nodules using genetic markers. *J. Gen. Microbiol.* 87:343–350.

22. Kaiser, K., and N. E. Murray. 1980. On the nature of *sbcA* mutations in *E. coli* K12. *Mol. Gen. Genet.* 179:555–563.

23. Kelln, R. A., and R. A. Warren. 1971. Isolation and properties of a bacteriophage lytic for a wide range of pseudomonads. *Can. J. Microbiol.* 17:677–682.

24. Kilbane, J. J., and R. V. Miller. 1988. Molecular characterization of *Pseudomonas aeruginosa* bacteriophages: Identification and characterization of the novel virus B86. *Virology* 164:193–200.

25. Kleckner, N., J. Roth, and D. Botstein. 1977. Genetic engineering *in vivo* using translocatable drug-resistance elements. *J. Mol. Biol.* 116:125–159.

26. Kokjohn, T. A. 1989. Transduction: Mechanism and potential for gene transfer in the environment. In S. B. Levy and R. V. Miller (eds.), *Gene Transfer in the Environment.* McGraw-Hill, New York, pp. 73–97.

27. Kokjohn, T. A., and R. V. Miller. 1985. Cloning and characterization of the *recA* gene of *Pseudomonas aeruginosa. J. Bacteriol.* 163:568–572.

28. Kokjohn, T. A., and R. V. Miller. 1987. Characterization of the *Pseudomonas aeruginosa recA* analog and its protein product: *rec-102* is a mutant allele of the *P. aeruginosa* PAO *recA* gene. *J. Bacteriol.* 169:1499–1509.

29. Kokjohn, T. A., and R. V. Miller. 1988. Characterization of the *Pseudomonas aeruginosa recA* gene: The Les⁻ phenotype. *J. Bacteriol.* 170:578–582.

30. Kretschmer, P. J., and S. N. Cohen. 1979. Effects of temperature on translocation frequency of the Tn3 element. *J. Bacteriol.* 139:515–519.

31. Krishnapillai, V. 1971. A novel transducing phage: Its role in recognition of a possible new host-controlled modification system in *Pseudomonas aeruginosa. Mol. Gen. Genet.* 114:134–143.

32. Krishnapillai, V. 1975. Resistance to ultraviolet light and enhanced mutagenesis conferred by *Pseudomonas aeruginosa* plasmids. *Mutat. Res.* 29:363–372.

33. Lehrbach, P., A. H. C. Kung, B. T. O. Lee, and G. A. Jacoby. 1977. Plasmid modification of radiation and chemical-mutagen sensitivity in *Pseudomonas aeruginosa. J. Gen. Microbiol.* 98:167–176.

34. Levin, B. R., F. M. Stewart, and V. A. Rice. 1979. The kinetics of conjugative plasmid transmission: Fit of a simple mass action model. *Plasmid* 2:247–260.
35. Levy, S. B., and R. V. Miller. 1989. *Gene Transfer in the Environment.* McGraw-Hill, New York.
36. Linton, A. H. 1986. Flow of resistance genes in the environment and from animals to man. *J. Antimicrob. Chemother.* 18 (Suppl. C):189–197.
37. Lorenz, M. G., B. W. Aardema, and W. E. Krumbein. 1981. Interaction of marine sediment with DNA and DNA availability to nucleases. *Mar. Biol.* 64:225–230.
38. Lorenz, M. G., and W. Wackernagel. 1987. Adsorption of DNA to sand and variable degradation rates of adsorbed DNA. *Appl. Environ. Microbiol.* 53:2948–2952.
39. Lundquist, P. D., and B. R. Levin. 1986. Transitory derepression and the maintenance of conjugative plasmids. *Genetics* 113:483–497.
40. Mancini, P., S. Ferteis, D. Nave, and M. A. Gealt. 1987. Mobilization of plasmid pHSV106 from *Escherichia coli* HB101 in a laboratory-scale waste treatment facility. *Appl. Environ. Microbiol.* 53:665–671.
41. McClintock, B. 1984. The significance of responses of the genome to challenge. *Science* 226:792–801.
42. Miller, R. V. 1988. Potential for transfer and establishment of engineered genetic sequences in the environment. In J. Hodgson and A. M. Sugden (eds.), *Planned Release of Genetically Engineered Organisms* (Trends in Biotechnology and Trends in Ecology and Evolution Special Publication). Elsevier, Cambridge, U.K., pp. S23–S27.
43. Miller, R. V., and T. A. Kokjohn. 1987. Cloning and characterization of the *c1* repressor of *Pseudomonas aeruginosa* bacteriophage D3: A functional analog of phage lambda cI protein. *J. Bacteriol.* 169:1847–1852.
44. Miller, R. V., and T. A. Kokjohn. 1988. Expression of the *Pseudomonas aeruginosa recA* gene is inducible by DNA damaging agents. *J. Bacteriol.* 170:2385–2387.
45. Miller, R. V., T. A. Kokjohn, and G. S. Sayler. 1990. Environmental and molecular characterization of systems which affect genome alteration in *Pseudomonas aeruginosa.* In S. Silver, A. M. Chakrabarty, B. Iglewski, and S. Kaplan (eds.), *Pseudomonas: Biotransformations, Pathogenesis, and Evolving Biotechnology.* American Society for Microbiology, Washington, D.C., pp. 252–268.
46. Morrison, W. D., R. V. Miller, and G. S. Sayler. 1978. Frequency of F116 mediated transduction of *Pseudomonas aeruginosa* in a freshwater environment. *Appl. Environ. Microbiol.* 36:724–730.
47. O'Morchoe, S., O. Ogunseitan, G. S. Sayler, and R. V. Miller. 1988. Conjugal transfer of R68.45 and FP5 between *Pseudomonas aeruginosa* in a natural freshwater environment. *Appl. Environ. Microbiol.* 54:1923–1929.
48. Petrocherlou, V., J. Grinsted, and M. H. Richmond. 1976. R-plasmid transfer in vivo in the absence of antibiotic selection pressure. *Antimicrob. Agents Chemother.* 10:753–761.
49. Parker. L. L., P. W. Betts, and B. G. Hall. 1988. Activation of cryptic gene by excision of a DNA fragment. *J. Bacteriol.* 170:218–222.
50. Primrose, S. B., and M. Day. 1977. Rapid concentration of bacteriophage from aquatic habitats. *J. Appl. Bacteriol.* 42:417–421.
51. Reanney, D. 1976. Extrachromosomal elements as possible agents of adaptation and development. *Bacteriol. Rev.* 40:552–590.
52. Reanney, D. C. 1977. Genetic interaction and gene transfer. *Brookhaven Symposia in Biology,* No. 29.
53. Reanney, D. C., W. P. Roberts, and W. J. Kelly. 1982. Genetic interactions among microbial communities. In A. T. Bull and J. H. Slater (eds.), *Microbial Interactions and Communities,* vol. 1. Academic, New York, pp. 287–322.
54. Roberts, R. 1982. Restriction and modification enzymes and their recognition sequences. *Nucleic Acids Res.* 10:r117–r144.
55. Rolfe, B., and B. W. Holloway. 1966. Alterations in host specificity of bacterial deoxyribonucleic acid after an increase in growth temperature of *Pseudomonas aeruginosa. J. Bacteriol.* 92:43–48.

56. Roper, M. M., and K. C. Marshall. 1978. Effect of clay particle size on clay–*Escherichia coli*–bacteriophage interactions. *J. Gen. Microbiol.* 106:187–189.

57. Sano, Y., and M. Kageyama. 1987. The sequence and function of the *recA* gene and its protein in *Pseudomonas aeruginosa* PAO. *Mol. Gen. Genet.* 208:412–419.

58. Saye, D. J., and R. V. Miller. 1989. The aquatic environment: Consideration of horizontal gene transmission in a diversified habitat. In S. B. Levy and R. V. Miller (eds.), *Gene Transfer in the Environment.* McGraw-Hill, New York, pp. 223–259.

59. Saye, D. J., O. Ogunseitan, G. S. Sayler, and R. V. Miller. 1987. Potential for transduction of plasmids in a natural freshwater environment: Effect of plasmid donor concentration and a natural microbial community of transduction in *Pseudomonas aeruginosa. Appl. Environ. Microbiol.* 53:987–995.

60. Saye, D. J., O. A. Ogunseitan, G. S. Sayler, and R. V. Miller. 1990. Transduction of linked chromosomal genes between *Pseudomonas aeruginosa* strains during incubation in situ in a freshwater habitat. *Appl. Environ. Microbiol.* 56:140–145.

61. Schurter, W., and B. W. Holloway. 1987. Interactions between the transposable element IS21 on R68.45 and TN7 in *Pseudomonas aeruginosa* PAO. *Plasmid* 17:61–64.

62. Shaw, P. D. 1987. Plasmid ecology. In E. Nester and T. Kosuge (eds.), *Plant-Microbe Interactions: Molecular and Genetic Perspective.* Macmillan, New York, pp. 3–39.

63. Shapiro, J. A., S. L. Adhya, and A. I. Bukhari. 1977. Introduction: New pathways in the evolution of chromosome structure. In A. I. Bukhari, J. A. Shapiro, and S. L. Adhya (eds.), *DNA Insertion Elements, Plasmids, and Episomes.* Cold Spring Harbor Laboratory, Cold Spring Harbor, N.Y., pp. 3–11.

64. Shen, P., and H. V. Huang. 1986. Homologous recombination in *Escherichia coli:* Dependence on substrate length and homology. *Genetics* 112:441–457.

65. Sherratt, D. J. 1982. The maintenance and propagation of plasmid genes in bacterial populations. *J. Gen. Microbiol.* 128:655–661.

66. Simonson, C. S., T. A. Kokjohn, and R. V. Miller. 1990. Inducible UV repair potential of *Pseudomonas aeruginosa* PAO. *J. Gen. Microbiol.* 136:1241–1249.

67. Singleton, P. 1983. Colloidal clay inhibits conjugal transfer of R-plasmid R1drd-19 in *Escherichia coli. Appl. Environ. Microbiol.* 46:756–757.

68. Smith, M. G. 1975. *In vivo* transfer of R factors between *Escherichia coli* strains inoculated into the rumen of sheep. *J. Hyg.* 75:363–370.

69. Soyal, S. M., C. P. Gerba, and G. Bitton. 1987. *Phage Ecology.* Wiley, New York.

70. Stewart, G. J., and C. A. Carlson. 1986. The biology of natural transformation. *Ann. Rev. Microbiol.* 40:211–235.

71. Stotzky, G., and H. Babich. 1986. Survival of, and genetic transfer by, genetically engineered bacteria in natural environments. *Adv. Appl. Microbiol.* 31:93–138.

72. Stotzky, G., and H. Babich. 1984. Fate of genetically engineered microbes in natural environments. *Recomb. DNA Tech. Bull.* 7:163–188.

73. Thorne, C. B. 1968. Transducing bacteriophage of *Bacillus cereus. J. Virol.* 2:657–662.

74. Thorne, C. B. 1978. Transduction in *Bacillus thuringiensis. Appl. Environ. Microbiol.* 35:1109–1115.

75. Tomich, P. J., F. Y. An, and D. B. Clewell. 1980. Properties of erythromycin-inducible transposon Tn*917* in *Streptococcus talcalis. J. Bacteriol.* 141:1366–1374.

76. Trevors, J. T., and K. M. Oddie. 1986. R-plasmid transfer in soil and water. *Can. J. Microbiol.* 32:610–613.

77. Walker, G. C. 1984. Mutagenesis and inducible responses to deoxyribonucleic acid damage in *Escherichia coli. Microbiol. Rev.* 48:60–93.

78. Weinberg, S. R., and G. Stotzky. 1972. Conjugation and genetic recombination of *Escherichia coli* in soil. *Soil Biol. Biochem.* 4:171–180.

79. Wilkins, B. M., L. K. Chatfield, C. C. Wymbs, and A. Merryweather. 1985. Plasmid DNA primases and their role in bacterial conjugation. In D. R. Helinski, S. N.

Cohen, D. B. Clewell, D. A. Jackson, A. Hollaender, L. Hager, S. Kaplan, and J. Konisky (eds.), *Plasmids in Bacteria*. Plenum, New York, pp. 585–603.
80. Willetts, N., and R. Skurray. 1980. The conjugation system of F-like plasmids. *Ann. Rev. Genet.* 14:41–76.
81. Wong, F. H., and L. E. Bryan. Characteristics of PR5, a lipid-containing plasmid-dependent phage. *Can. J. Microbiol.* 24:875–882.
82. Yu, L., and J. N. Baldwin. 1971. Intraspecific transduction in *Staphylococcus epidermidis* and interspecific transduction between *Staphylococcus aureus* and *Staphylococcus epidermidis*. *Can. J. Microbiol.* 17:767–773.
83. Zeph, L. R., M. A. Onaga, and G. Stotzky. 1988. Transduction of *Escherichia coli* by bacteriophage P1 in soil. *Appl. Environ. Microbiol.* 54:1731–1737.

8

Assessment of Genetic Stability

David J. Drahos

Gerard F. Barry

Introduction

The successful application of a microorganism in a specific environment will be the product of a number of factors. Among the most important of these is the maintenance and expression of "critical" genes, and the stability and efficacy of the corresponding gene products. Generally these critical genes are considered to be those which impart a unique or key ability, or bring about a desired environmental effect. For example, the capacity of certain plant-colonizing bacteria to protect against disease infection is dependent upon the production of specific factors particularly effective against an incoming pathogen. Alternatively, a similar microbe engineered to produce an insect toxin will only be effective when the optimal degree and timing of the expression of the inserted toxin gene are achieved. Conceptually, the highest and most sustained levels of critical gene expression would appear the most desirable. However, the effect of sustained production on cell growth, physiology, or metabolic efficiency of the host strain may be significantly detrimental to longer-term survival in a highly competitive environmental niche. Unless the engineered trait confers some selective advantage for the host, which is often not the case, the complete or partial loss of the critical gene, or decrease in its expression level, must be considered.

The focus of this chapter is to review procedures that may be used in assessing the maintenance and functional expression in bacteria of a specific genetic element, whether indigenous or newly inserted. Certain methods useful in determining such stability are also applicable in monitoring for potential gene transfer between organisms, or in measuring the "permanence" of an introduced gene in the original host. Methods to evaluate the fitness and survival of the whole organism itself are described in Chap. 9.

Principles

Most genetic elements are introduced into bacteria by one of four primary pathways: plasmids, transposon insertion elements, transformation, or lysogenic phage. Of these, plasmid vectors typically afford the most rapid, practical, and versatile means, and as such are often the first vehicle of choice. However, decades of experience and observation have taught us that plasmids can behave quite dynamically under "real world" conditions (13). While certain specialized plasmid functions, such as colicinogeny (24), will occasionally be conserved through many generations, many others are rapidly lost or new ones may be acquired. Antibiotic resistance genes are most notorious for being rapidly disseminated among a bacterial population, even on a worldwide basis, particularly under selective pressure (8, 29). However, some studies indicate that plasmids containing heterologous DNA may be retained in *Escherichia coli* even under nonsterile, nonselective conditions (14, 15). However, the overall expectation is low that nonessential functions will be well maintained on a plasmid under native, nonselected growth, particularly in the face of any significant competition for niche establishment in the environment (26). Furthermore, the expression of a foreign gene on a plasmid can have a marked inhibition on the relative growth rate of the host cell (30, 36, 41).

For these reasons, certain specialized vectors have been constructed, some based on the use of transposable elements (28), which facilitate the insertion of foreign genes directly into the chromosome of the host bacterium (16, 40). However, due to their self-transmissible capacity, such elements (transposons) may not remain fixed in their insertion points and may be lost at a certain frequency, or further copies may accumulate (28). These problems may be largely overcome by using defective but complementable elements that effect the permanent, or one-way, delivery of desired genes into the host chromosome, with the concomitant loss of the original delivery vehicle (2, 3, 23).

Effective monitoring for the presence of the introduced gene can be accomplished either by assessing the presence of the gene itself or by relying on the expression of closely linked marker elements. Such marker elements, coinserted with the gene of interest, would have a relatively high probability of being either retained or lost along with the adjacent genes. Typically such monitoring procedures are the same or similar to those used in monitoring the presence of the host cell itself, which are reviewed in detail in Chap. 18 and are only briefly summarized here.

Current Practices and Trends

Monitoring gene presence

Antibiotic resistance factors.   A widely practiced method to monitor the presence of a target gene is to link it with an antibiotic resistance element. In practice, a gene expressing a transacting product capable of inactivating or inhibiting the cross-membrane transport of a specific antibiotic is placed in close genetic association with the gene of interest. Should the target gene be deleted from the genome, the likelihood is high that the antibiotic resistance marker will be lost as well. Unfortunately, this method has several significant drawbacks. Some antibiotic resistance enzymes themselves may have adverse effects on the host (10, 34). Further, a significant background resistance in soil and water samples already exists for many of the most useful antibiotics such as kanamycin, streptomycin, and tetracycline (11, 27, 49). Therefore, care must be exercised in determining in the test sample, or environment under study, the background level of organisms already resistant to a given antibiotic before the resistance gene marker or strategy is chosen.

Direct detection methods.   Use of antibiotic resistance markers requires the recovery of the organism carrying the target gene on defined laboratory media. In some instances, such as aquatic habitats, not all host cells are in a physiological state where they would readily readapt and grow on such media without some preconditioning or enrichment (9). For these cases, methods are under development that allow for the verification that the target gene is still being carried (or expressed) without the necessity for regrowth in the laboratory from a natural sample. These procedures include the use of nucleic acid hybridization probes, antibody probes against target gene products, and bioluminescence genes (see Chap. 1).

If the genetic element itself must be followed, or if the desired gene products are difficult to recognize or assess, then nucleic acid hybridization may be required. Generally, these procedures suffer from being relatively labor intensive and often less sensitive than selection for antibiotic resistance or nutritional factor utilization. However, significant information may be gained concerning the final integrity and genomic location of the introduced genetic element. For this procedure, described in more detail in Chap. 15, a specific DNA or RNA probe is prepared that is complementary to the genetic sequence of the target gene. This probe is labeled either radioactively or enzymatically during or after preparation. DNA from samples being analyzed must then be obtained. Depending on the relative concentration of the target gene and the type of material being sampled, this can be a significant task. Aquatic samples may require extensive concentration, while soil samples usually require a separation of the target organisms from extensive inorganic and nontarget material. A review of methods for the extraction of DNA from soil is presented in more detail in Chap. 4.

Once obtained, target cells are lysed either by sonication, detergent treatment, or temperature shock, and the DNA isolated by various phenolic, ethanol, or resin binding extraction methods. Purified DNA may then be cut to specific lengths with restriction endonucleases and separated by size by agarose gel electrophoresis. Whether this cutting and separation is done or not, the DNA is then mixed with the labeled probe so as to allow annealing of complementary nucleic acid elements. This mixing can be performed either in a buffered solution (solution hybridization), or it can be done after the target DNA is first fixed to a solid surface, such as a nitrocellulose membrane (44). Following annealing, nonassociated and loosely associated nucleic acid elements are removed either by washing the solid surface or by enzymatic digestion that specifically attacks unannealed material. The radioactive or enzymatic "signal" is then quantitated to determine the presence of the target gene.

Other direct detection methods include the use of antibody probes prepared specifically against a gene product of the target or linked marker gene (9) and the visualization of target gene expression by luminescent reporter genes (53, 55). The immunological detection methods (reviewed in Chap. 4) provide a high degree of specificity for the target gene product and do not require laboratory culturing. However, these methods can be time consuming and expensive and may be limited by background problems or insufficient target gene expression under various environmental conditions. Luminescent reporter genes have become more attractive with the recent development of practical vector systems based on the expression of certain bacterial luciferase enzymes (45, 46). In the past a major drawback of this technique has been the low sensitivity, as well as background problems due to the nonuniform quenching of photons by surrounding material. However, recent advances in instrument efficiency and cost effectiveness are rapidly diminishing these problems (45).

Nutritional and enzymatic markers.   An alternative to the use of antibiotic resistance factors and direct detection methods is the application of certain nutritional and enzymatic marker systems. The *xyl*E gene, which originated from the TOL plasmid pWWO of *Pseudomonas putida,* has found use in both gram-positive and gram-negative bacteria as an efficient and inexpensive indicator of gene expression (37, 60). The product of the *xyl*E gene, catechol 2,3-dioxygenase, catalyzes the conversion of catechol to the yellow-pigmented 2-hydroxymuconic semialdehyde. Bacterial colonies expressing *xyl*E turn yellow following an overspray with an aqueous solution of 0.5 *M* catechol. Since this is not a selective method, however, some preselection on antibiotic medium is necessary when natural samples are analyzed.

Among the better characterized and more widely used selective enzymatic marker systems is that employing the *E. coli* lactose operon genes *lac*Z and *lac*Y. The product of the *lac*Z gene is β-galactosidase, which cleaves the lactose molecule into glucose and galactose. β-galactosidase also cleaves the chromogenic dye "X-Gal" (5-bromo-4-chloro-3-indolyl-β-D-galactopyranoside), providing a brilliant blue color indication of the presence of the active *lac*Z gene. The *lac*Y gene, which codes for the enzyme lactose permease, allows bacteria to take up the lactose sugar efficiently. This *lac*ZY marker was originally developed for fluorescent pseudomonad bacteria, common soil organisms and efficient root colonizers. Fluorescent pseudomonads do not make β-galactosidase and subsequently cannot efficiently use lactose as a sole energy source (25). In certain environmental samples, such as soil, this method has proved rapid and effective in monitoring the presence of both the *lac*ZY-marked strain as well as genes closely linked to the inserted *lac*ZY element (2, 3, 18, 19). The selection of the *lac*ZY-marked fluorescent pseudomonad host strain itself is based on its new ability to grow on a minimal lactose medium while simultaneously producing its characteristic fluorescent pigment. By this method, fewer than 10 colony-forming units (CFUs) can be detected from a gram of typical field soil. Genes linked to the inserted *lac*ZY element can be traced indirectly using lactose selection followed by nucleic acid hybridization, based on probes specific for the target gene. Alternatively, this system can be applied to nonfluorescent bacteria that are still deficient for lactose utilization, provided that an antibiotic-resistant isolate (i.e., rifampicin or ampicillin) of such a strain is first selected. The combination of rifampicin resistance and lactose growth is an effective selection tool for most environmental samples, including soil, groundwater, plant extracts, and milk. It should be noted in caution, however, that some lac⁻ bacterial strains, such as *Rhizobium* species, are not amenable to selection based only on antibiotic resistance patterns. The advantages of the *lac*ZY system are its demonstrated effectiveness in bacterial monitoring in actual field tests (19; Brandt, E., EPA Premanufacture Notification No. P-87-1292 Report, 1988), its relative practicality, and the potential for applying this method to other similar bacterial systems. A series of versatile and efficient suicide plasmid delivery systems has been developed that greatly facilitates a permanent insertion of the *lac*ZY genes, as well as other chosen genes linked to the *lac*ZY element (2, 3). These are described in more detail below.

The disadvantages to using nutritional or enzymatic marker systems to follow gene maintenance or potential gene transfer include the requirement for marker gene expression in a recipient host and the necessity to distinguish potential recipients in the environment from the introduced, marked parent strain. Gene expression questions might be at least partially overcome by using alternate, or tandem, promoter systems. In addition, high expression promoters or the omega translational enhancer might be employed. These methods, as well as two other reporter gene systems, β-glucuronidase and levan sucrase, are described in Chap. 18.

Introduction of monitoring element

Traditionally, new genetic material has been introduced into bacteria using vectors of plasmids, transposons, or phage, or directly through recombination. Due to the lack of detailed understanding of the phage genetics of most bacteria, the use of phage as vectors is probably limited to only the common laboratory bacteria such as *E. coli* or *B. subtilis*. (However, see Chap. 11 for more precise consideration of phage.) This section will concentrate on the options available for plasmid vehicles and on those systems for chromosomal insertion. The choice of a plasmid versus chromosomal location will depend on the nature of the introduced gene(s) and vector and on the eventual use of the engineered strain.

Plasmid vehicles. Small, narrow-host-range plasmids such as the pUC (57) and the pBR (6) series serve as useful cloning vectors for the isolation of specific sequences but apart from the mobilizable vectors, such as pBR322 (56) (see "Chromosomal Insertion Procedures" in next section) and except as sources for generating large amounts of DNA for transformation, these vectors are generally of little use in the introduction of these new sequences into the target bacteria. Broad-host-range cloning and expression vectors have been developed for most gram-negative bacteria, and these have been based predominantly on the incompatibility groups IncP (IncP-1), IncQ (IncP-4), and

IncW. An excellent review of these vectors (and of those based on IncN and IncC) and a survey of their host ranges has been published recently (48). The reader is also directed to Chap. 17 for a detailed discussion of specific plasmids. In general, plasmids based on IncP and IncQ have been shown to have the broadest host ranges. Most IncP-derived vectors, but not all (23), have been rendered Tra⁻ as a result of reducing the size of the original plasmid (58 to 60 kb) and some small, stable IncP derivatives have been developed (47). These derivatives (and the IncQ and IncW plasmids) may be mobilized in triparental matings using the ColE1 Tra⁺ (IncP) plasmid pRK2013 (17). The most successful derivatives of the IncQ plasmids seem to be those where most of the manipulations have been carried out within the region encoding the Su-Sm operon (e.g., 59).

Methods

Chromosomal insertion procedures

In the chromosomal insertion systems described here, the new material is introduced along with complete or portions of transposons or flanked by homologous DNA derived from the target bacterium. In some of the approaches, combinations of transposition and homology are employed.

The new sequences may be inserted randomly using transposons such as Tn5, which has been shown to transpose in a very large number of bacteria (see Ref. 53, for exception) and which is believed to have a low specificity for insertion. In some cases the new genetic material is the transposon itself, introduced as a tracking element (21). In other cases, the new material is cloned within Tn5 on suicide delivery plasmids such as the pSUP series (50). Internal portions of the transposon may also be removed without affecting transposition ability (40). The exconjugants containing the new material are selected by resistance to the antibiotic whose resistance gene is carried on the transposon. These approaches result in a strain containing an active transposon, and although the transposition frequency of Tn5 from the genome may be quite low, the possibility remains that the element may be lost or transpose to a plasmid and thus spread more widely within the native population. Innovative approaches have been employed to inactivate the Tn5 element resident on the chromosome or to exploit the advantages of Tn5 while eliminating the more negative aspects (40).

Homologous recombination between cloned genomic fragments flanking a marker gene has also been employed. In one approach, "null" regions of the genome are located by identification of transposon insert sites with no discernible phenotype. Following cloning of this portion of the genome, a selectable marker gene is placed within the fragment (31) and the flanking homology is then used to deliver the new genetic material to the genome. While guaranteeing a high degree of stability and containment, this approach is applicable on a case-by-case basis, requiring the identification of such "null" regions for each target bacterium. In a related approach, the homologous sequences employed are those of an insertion sequence identified in the target bacterium (1).

Disarmed Tn7 systems

Principle. The transposon Tn7 differs from other transposons in having a very high specificity for insertion into discrete sites in bacterial chromosomes (the attTn7 site in *E. coli*, for example). Transposition of Tn7 is dependent on the products of 3 to 5 *tns* genes, which function efficiently in *trans* (58). These aspects of Tn7 have been used in the development of a series of vectors that allow the directed insertion of new genetic material to bacterial genomes and also provide a high level of insert stability and containment (2, 3, 23).

Procedures. If the aim of the experiment is to simply construct a marked strain, a number of vectors and procedures may be used. Tn7 transposition derivatives of IncQ plasmids were not isolated originally (4). However, following the cloning of the *E. coli* attTn7 into these plasmids, such derivatives could be isolated. In our hands, the most useful vector for the routine marking of soil isolates is an IncQ delivery plasmid containing a Tn7-*lac* element located in a cloned attTn7 fragment and also containing four of the five Tn7 *tns* genes (the Tn7-*lac* element encodes the fifth *tns* gene) (Fig. 8.1). This vector pMON7197 is mobilized using the IncP *tra* genes on pRK2013. Following a triparental mating, Lac$^+$ derivatives are selected directly on minimal media containing X-Gal. Incubation at 30°C for 4 to 6 days is usually sufficient, and Lac$^+$ isolates are isolated at a frequency of 10^{-7} to 10^{-8}, depending on the strain. If the initial selection applied is for Gm^r Lac$^+$, then the frequency of colony formation is usually 20 to 100 times higher and these will sometimes appear 1 to 2 days earlier. The vast majority of these colonies are usually Lac$^+$ transposition derivatives and retain the Gm^r plasmid portion of pMON7197. IncQ plasmids containing the *tns* genes are relatively unstable in the absence of antibiotic selection in all soil isolates examined. Usually one to two subcultures of the Lac$^+$ Gm^r isolates in lactose-minimal media are sufficient to recover Gm^s plasmidless isolates. A second useful vector, pMON7190 (Fig. 8.2), contains a Tn7-*lac* element in a cloned attTn7 and the Tn7 *tns* genes (all five genes in this case) in a pBR322

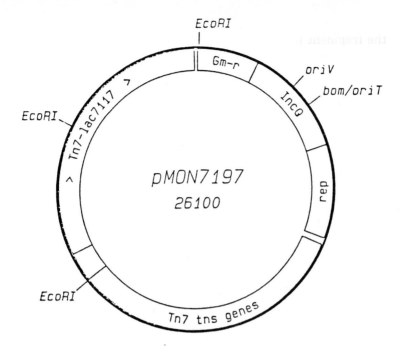

Figure 8.1 pMON7197 (= pMON7181::Tn7 − *lac*7117): Broad host-range Tn7-*lac* element (located in a cloned *att*Tn7 site) and the Tn7 *tns* genes.

derivative. This latter arrangement provides a "suicide" delivery of the marking element. Frequency of isolation of Lac$^+$ derivatives is usually lower with this vector, presumably because the plasmid is incapable of replication in the target and thus its "residence" time is much less than that of the IncQ plasmids.

Applications and variations

Preparation of target and *E. coli* strains.   It is often convenient to first isolate rifampicin-resistant derivatives of the target bacterium to facilitate the selection of exconjugants and counterselection of the *E. coli* hosts. Such derivatives may not be advised in some cases (12), and it may be more convenient to rely on the natural antibiotic resistances of the target bacteria or differences in nutritional requirements between the partners in the conjugation.

Many soil bacteria produce factors that are bacteriocidal, and as a practical matter it may be necessary to isolate or to identify resistant *E. coli* strains in order to ensure good survival during the matings. Since the nature of the toxic effect may not be fully understood, rather simple approaches are taken to isolate *E. coli* with the required resistance. Picking colonies from the cleared zone around a colony of the

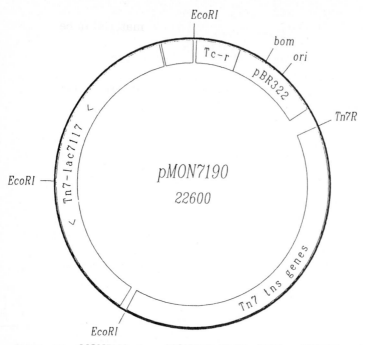

Figure 8.2 pMON7190 (= pMON7184::Tn7-*lac*7117): pBR322-based Tn7-*lac* "suicide" delivery plasmid. The plasmid may be transmobilized using pRK2013. This apporach is independent of the antibiotic sensitivities of the target bacteria.

inhibitory bacterium spotted on a lawn of *E. coli* is usually sufficient. Another approach is to select for growth of *E. coli* in the spent culture medium of the target bacterium.

Optimizing expression and evaluation of promoters. In a series of Tn7-*lac* elements (3), a number of different promoters were tested for expression in many bacteria. The criterion for usefulness of a particular construct was the relative growth and efficiency of plating of Tn7-*lac* derivatives on lactose and glucose minimal media. Constructs expressed from the *iucA* operon promoter (5) were generally the most widely successful, followed by those containing the *E. coli* trp promoter. Although employed in the original Tn7-*lac* element and apparently capable of very broad host-range expression, the *bla* promoter has not been used extensively due to difficulties of working with plasmids expressing the *lac* genes at high levels. The expression of the *iuc-lac* and *trp-lac* fusions may be down-regulated to some extent by the addition of $FeCl_3$ (75 µg/mL) or tryptophan (50 µg/mL) to the growth medium. In addition, ease of handling may be improved by choice of *E. coli* strain.

Linkage of cryptic and marker genes. Early versions of the Tn7 vectors required the direct cloning of any new genetic material to be carried into the cell into the carrier plasmids (2, 23). This cloning was far from trivial due to the large size of the plasmids and their low copy number and relatively poor choice of potential cloning sites. These difficulties have been circumvented by the use of a shuttle system. In the first stage, the Tn7-*lac* elements are located in the cloned *att*Tn7 on a high-copy *E. coli* minireplicon (e.g., pMON7117; Fig. 8.3). The number of potential cloning sites has also been increased. Following cloning of the new DNA sequences within the Tn7-*lac* element, the new derivative may then be tranposed to one of a number of *att*Tn7-containing plasmids that are then used to introduce the new sequences to the target bacterium. These delivery plasmids consist of a Bom⁺ narrow- or broad-host-range replicon containing both the *att*Tn7 site and the Tn7 *tns* genes. Transposition derivatives of these and the recombinant Tn7-*lac* derivatives are isolated as follows: a suitable *E. coli* strain (usually *E. coli* HB101; 7) is transformed in single step with all the plasmids involved; then, either plasmid DNA is prepared, *E. coli*

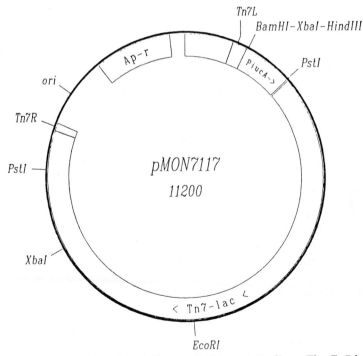

Figure 8.3 pMON7117: A high-copy Tn7-*lac* minireplicon. The Tn7-*lac* element contains a number of restriction sites suitable for the cloning of additional genes. The *lac*ZY genes are promoted from the *iuc*A promoter.

retransformed, and the new derivative identified by the relevant phenotype, or for the broad-host-range plasmids, the multiply transformed *E. coli* may be conjugated (using pRK2013) with a *Pseudomonas* sp. However, the isolation of the intermediate plasmid is usually not necessary. In practice, the multiply transformed *E. coli* is mated with the target bacterium, and the marked recombinant strain is then isolated simply by selection for Lac$^+$. The frequency of isolation of marked isolate is much lower (10 to 1000 times) but is much simpler and, except in specialized cases (an example is where conjugation frequencies might be expected to be extremely low), is usually the approach of choice.

Verification of recipient strain identity-integrity

Chromosomal fingerprint analysis

Principles. It is vital that accurate verification of the identity and integrity of a recipient strain be performed following insertion of a novel genetic element. It is equally important to determine whether a suspected recipient of genetic material, reisolated from an environmental sample, is indeed a different strain from the introduced organism. For bacteria, such analysis can be a formidable task for any appreciable number of test candidates using standard procedures. The advent of newer strain identification methods, such as computer-assisted gas chromatographic analysis of free fatty acids and immunological techniques (see Part 1), are beginning to address the requirement for accuracy, speed, and versatility.

This section describes a rapid, reliable method for the comparative identification of a wide variety of gram-negative and gram-positive bacteria, based on the analysis of chromosomal restriction patterns following gel electrophoresis. The sensitivity of the procedure allows differentiation of bacteria with nearly identical phenotypes, usually to the subspecies level. Should an inserted genetic element not be stably maintained, this tool can play an important role in the unambiguous verification that isolated strains no longer expressing the added gene are still, in fact, the original host bacterium. Furthermore, if genetic transfer is suspected between an introduced bacterium and an indigenous recipient, this method provides a rapid confirmation of the novel identity of the latter strain. Chromosomal fingerprinting also bypasses identification problems often caused by differential colony morphology, which can arise from altered gene expression caused by changes in culture or environmental conditions.

Prepared materials.

1. TEG buffer is: 10 mM Tris-HCl (pH 8.0), 50 mM EDTA (pH 8.0), and 1% glucose.

2. TE buffer is: 10 mM Tris-HCl (pH 8.0) and 1 mM EDTA (pH 8.0).

3. TAE running buffer (20-fold concentrated stock) is per liter: 387.2 g Tris base; 91.4 mL glacial acetic acid; 160 mL 0.5 M EDTA (pH 8.0); 1 μg/mL ethidium bromide.

4. Phenol-chloroform solution is equal portions of TE-saturated phenol with chloroform.

5. EcoRI buffer (tenfold concentrate) is 500 mM NaCl, 1 M Tris-HCl (pH 7.5), 50 mM MgCl$_2$.

6. RNase A is pancreatic RNase dissolved at 10 mg/mL in 10 mM Tris-HCl (pH 7.5) and 15 mM NaCl, heated at 100°C for 15 min and allowed to cool slowly to room temperature before dispensing into aliquots, which are stored at -20°C.

Procedure. Culture bacterial cells to be analyzed from an isolated colony to 2 to 4 mL LB broth (or equivalent medium) at 28°C (or appropriate temperature), overnight with shaking. Pellet 800 μL of overnight culture in an Eppendorf tube for 2 min in the microfuge, decant and discard supernatant. Resuspend pellet in 250 μL TEG buffer. Lyse cells by adding 30 μL 10% (w/v) sodium dodecyl sulfate, inverting to mix, and heating at 65°C for 10 min.

DNA is extracted by first adding 300 μL phenol-chloroform to the lysed bacterial suspension, vortexing briefly, and spinning 2 min in the microfuge. A glass pasteur pipette is used to slowly withdraw the upper DNA-containing phase. Here the viscous DNA near the interface is withdrawn by moving the pipette tip slightly up the side of the tube while drawing in the DNA mass. Any nonviscous clear upper-phase solution is also withdrawn and retained with the DNA in a new tube. The extraction is repeated by adding 300 μL phenol-chloroform to the decanted DNA from the first extraction. Only 150 to 200 μL of the upper-phase solution is withdrawn from the second extraction, with care taken to avoid the interface material (should it be difficult to obtain this volume of the upper phase, a smaller amount of starting cell culture should be used).

DNA is then precipitated by the addition of 1.0 mL ethanol (95 to 100%), and mixing *slowly* by inversion 3 to 4 times. Here the DNA is contained in a clearer zone (phase) which forms at the bottom of the tube. After 3 to 4 slow inversions, the upper phase will become cloudy and the lower phase (DNA) will be reduced in size to approximately 150 to 200 μL. Do not allow the DNA phase to form a pellet. The upper, cloudy phase is removed with a pasteur pipette and discarded. The DNA is washed by adding 1 mL 70% (v/v) ethanol and mixing several times by inversion. The DNA will form a small, loose precipitate, which is pelleted by spinning briefly (20 s) in the microfuge. Decant the supernatant with a pasteur pipette and discard. Add 1 mL 95% ethanol, spin 2 min, decant and discard supernatant, and dry lightly under vacuum (4 min).

The DNA is digested with EcoRI by resuspending in 44 μL TE buffer, then adding 4 μL 10XEcoRI buffer followed by 1 to 2 μL EcoRI (usually 20 units/μL). This is incubated at 37°C for 60 min. After restriction, 1 μL of RNase A is added, with incubation for 5 to 10 min, 37°C.

Restricted DNA is analyzed by submarine gel electrophoresis using 0.8% (w/v) agarose in TAE buffer, with TAE also as a running buffer. The gel dimensions should be at least 10 cm in length and 3 to 5 mm thick. Usually 10 to 15 μl of the DNA sample, combined with 2 μL bromophenol blue dye, is sufficient for clearly visualizing the banding pattern, although this volume may be varied in subsequent analyses depending on the relative concentration of the DNA. Samples are loaded into preformed wells approximately 0.5 to 0.8 cm in length and 0.8 to 1.5 mm thick, giving a 20 to 25 μL capacity. The gel is run at 100 to 110 V, until the blue dye front has progressed 8 to 10 cm from the loading point. Standard DNA fragments of known size (i.e., λ/ HindIII DNA) are included for gel to gel comparisons. Although ethidium bromide is present in the gel and running buffer, occasionally restaining the gel after electrophoresis in 1 μg/mL ethidium bromide in water, then destaining for at least 15 min in water gives improved band visualization.

Verification of host performance following gene insertion

Principle. The reevaluation of host performance following gene insertion is a practical requirement, and in the case of soil bacteria the ultimate test is the retention of colonization ability and environmental effects. The theoretical considerations are discussed in detail in Chap. 9.

Procedure. Our routine evaluations before proceeding to the final tests are chosen primarily for their ease. The growth rates of the parent and marked derivative should be unaltered in minimal medium, to identify any debilitation as a result of the marking. In the case of strains marked with the *lac*ZY system (see "Nutritional and Enzymatic Markers" in preceding section) the efficiency of plating should be the same on minimal lactose media and on the more traditional richer plating media, to verify the usefulness of the marking system for that particular host strain.

Gene insertion verification

Principle.   Once the phenotypic characteristics of the recipient strain are confirmed, various hybridization and sequence analysis procedures can be used to verify the presence of the inserted gene. If single-site chromosomal insertion has been used to place the target gene into the recipient cell (i.e., the Tn7 insertion system described above), then hybridization analysis can verify the point on the chromosome where insertion has occurred. This method requires the preparation of a nucleotide probe that is specific either to the gene being followed or to a closely linked marker element. A detailed description of probe construction and hybridization protocols, as well as a discussion on the reliability and limitations of this methodology, is given in Chap. 15. Hybridization methods may also be coupled with the expression of certain selectable marker genes, such as *lac*ZY, to provide an estimate of potential transfer of genetic material in the field.

Probably the ultimate and most accurate confirmation of the presence and integrity of the inserted gene is sequence analysis. For most applications this can be an arduous task, and usually unnecessary if overall confirmation of gene presence by hybridization methods can be achieved. Almost always sequencing will involve the recloning of the target back from the transformed host cell onto various plasmid vehicles. For chromosomally inserted genes, the use of linked selective marker elements will greatly aid this retrieval step. Generally, two basic methods are followed for the actual sequencing, the Maxam-Gilbert technique (35) and the Sanger procedure (43). Numerous variations and improvements in these methods have been derived and described in detail (34, 39, 52, 61).

Insert purity

Analysis of gene expression

Principle.   In the ideal situation, the inserted DNA fragment has been characterized completely. In practical terms, this means that the complete nucleotide sequence is known or that at least the origin of the fragment is completely understood and can be verified with routine restriction mapping. A fully sequenced fragment may be analyzed by computer programs that search for open reading frames and expression signals. However, an insert from an organism whose molecular genetic rules are not fully understood may not be accurately analyzed by these approaches.

Procedure.   Complete characterization of the insert could also involve analysis for the presence of potential translation products. Earlier schemes for this purpose include the examination of proteins produced in *E. coli* maxicells, and the advantages and caveats of this approach have already been discussed. More recent approaches overcome the problems of failure of expression of heterologous genes in *E. coli* (42, 51). In these, the insert fragment would be cloned in both orientations behind the T7 promoter and then introduced into an *E. coli* strain capable of expressing the T7 polymerase. Expression from the cloned insert is then examined in the presence of rifampicin and ^{35}S-methionine. Since the T7 polymerase recognizes few sequences as terminators and since translation is not usually a block to expression in *E. coli,* the coding ability of the insert may be evaluated with great confidence. A related approach is to examine expression from the T7 promoter *in vitro* using the commercially available kits. Good expression *in vitro* may be better with small insert fragments and may not work very well with polycistrons or very large inserts. Unstable or poorly labeled proteins may be detected in this way. Altering the labeling times for the *in vivo* approach may also answer the same question.

Insert stability

Measurement under laboratory conditions.   For routine evaluation of insert stability, a very simple and practical approach is taken. For a complete evaluation, the reader is referred to Chap. 9. The permanence of the Tn7-*lac* in the genome is tested by the introduction of plasmids containing the *att*Tn7 site but differing in containing or lacking the Tn7 *tns* genes. Recovery of Tn7-*lac* derivatives only in the presence of the *tns* genes indicates that the insertion-excision event is dependent on the *tns* genes and argues for a high level of containment and permanence.

Stability of the Lac$^+$ phenotype is evaluated by plating aliquots of the marked strain on nonselective media containing XGal, following a large number of generations in nonselective media. Ideally, a chemostat might be used, but for practical purposes serial culture is used (see 47).

Measurement in native environment.   Recently it has become possible to assess that stability of certain nonnative inserted genes under actual field conditions in soil bacteria. Two field studies have been initiated that involve the application of recombinant bacteria carrying heterologous genes.

The first of these tests was initiated in late 1987 and was designed specifically to determine the fate and stability of an engineered *P. aureofaciens* strain, as well as the inserted genes themselves (22). In this study, the Tn7-*lac*ZY genetic element (see "Monitoring Gene Presence" in preceding section) was inserted onto the chromosome of a wheat root–colonizing bacterium. An extensive monitoring program over an 18-month period was set up to monitor the survival, spread, carry-over on rotated crops, and potential genetic transfer of the *lac*ZY element to a limited number of similar endogenous organisms in the field. Following seed inoculation, wheat plants were sampled first at weekly (1 to 4 weeks postplanting), then at biweekly intervals, to determine the presence and numbers of inoculated *lac*ZY-marked bacteria. As part of this analysis, samples collected from the plant roots were analyzed on a medium (containing rifampicin and nalidixic acid) selective for the inoculated host bacterium, and also indicative of the expression of the *lac*ZY genes, by inclusion of the chromogenic dye X-Gal. Thus any instability of the inserted genetic element is assessed by scoring for white colonies among the blue colonies of the expressing, parental strain. No white bacterial colonies were observed among greater than 10^3 blue colonies analyzed in samples taken 22 weeks after field inoculation, or during analysis of similar numbers at any of the 16 other previous sampling times. This indicates that the *lac*ZY marker element remains stably integrated and expressed in the original host bacterium.

A second study, initiated in April 1988, involves the application of *Rhizobium meliloti,* modified to contain a plasmid carrying an amplified copy of the host strain's own *nif* genes to enhance nitrogen fixation capacity. The monitoring strategy in this test is designed to determine the persistence (stability) of the inserted plasmid in the field, as well as microbial dispersal and plasmid transfer (20, 22).

Summary and Conclusions

There have been significant recent advances in the stable introduction and monitoring of novel genetic elements into bacteria intended for environmental applications. Several of these methods have relatively broad applications to a number of typical field bacteria and avoid complications arising from the use of more traditional insertion and monitoring procedures. This chapter has summarized certain specific methods that have proved relatively successful not only under laboratory conditions, but in actual field studies as well.

Certainly a key factor in the accurate assessment of gene stability under environmental conditions will be the development of monitoring strategies that couple the positive aspects of the direct detection techniques with those of the selective recovery methods. For example, the population level of a marked bacteria in soil samples might be first assessed as a measure of recoverable colony-forming units on nonantibiotic selective media using the *lac*ZY gene system, which is relatively rapid and sensitive. These same samples could then be analyzed using fluorescent monoclonal antibodies directed against the product of the *lac*Y gene, lactose permease, which has been shown integrated in the outer membrane of expressing *Pseudomonas* and *E. coli* hosts (18, 38). Currently this latter procedure is more time-consuming and exacting, although the potential may exist to augment this technique with much more sensitive detection capabilities and rapid automated cell-sorting procedures (45). The coupling of such procedures would presumably permit a determination of the fraction of the total (*lac*Y-expressing) bacterial population present which is in a "recoverable" state (on minimal lactose medium). Should this fraction prove relatively constant over defined growth periods and environmental conditions, the more practical, selective recovery procedure could be used with greater confidence on a "first measure" basis for other similar tests.

By analogy, similar results might be obtained through the combination of hybridization probe technology with selective recovery methods. In addition, linking bioluminescence genes, such as the luciferases from *Vibrio* (46), with the selectable *lac*ZY system could provide the same dual verification potential. Extending this analogy further, the development of in situ direct detection methods might be possible, if based on the active presence of unique gene products or compounds. For example, a laser-induced excitation of molecules directly on or in the plant root could provide a very rapid and accurate indication of the spatial density and precise physical location of introduced microorganisms. This method could be coupled with the cleavage of a specific substrate by the *lac*Z gene product. "Laser-detectable" genes might afford an alternate and powerful means of measuring stability and detecting genetic exchange in environmental samples.

As discussed, the discovery that fluorescent pseudomonads were unable to utilize lactose as a sole carbon source (18) led to the development of the selective *lac*ZY tracking method. The observation was also made that few, if any, soil bacteria will grow readily on minimal lactose medium containing rifampicin. Therefore, this broadens the potential application of the *lac*ZY system to numerous other soil bacteria that are inherently deficient in lactose utilization and for which rifampicin-resistant derivatives can be selected.

References

1. Acuna, G., A. Alvarez-Morales, M. Hahn, and H. Hennecke. 1987. A vector for site-directed, genomic integration of foreign DNA into soybean root-nodule bacteria. *Plant Mol. Biol.* 9:41–50.
2. Barry, G. F. 1986. Permanent insertion of foreign genes into the chromosomes of soil bacteria. *Bio/Technology* 4:446–449.
3. Barry, G. F. 1988. A broad host-range shuttle system for gene insertion into the chromosome of Gram-negative bacteria. *Gene* 71:75–84.
4. Barth, P. T., N. Datta, R. W. Hedges, and N. J. Grinter. 1976. Transposition of a deoxyribonucleic acid sequence encoding trimethoprim and streptomycin resistance from R483 to other replicons. *J. Bacteriol.* 125:800–810.
5. Bindereif, A., and J. B. Neilands. 1985. Promoter mapping and transcriptional regulation of the iron assimilation system of plasmid ColV-K30 in *Escherichia coli* K-12. *J. Bacteriol.* 162:1039–1046.
6. Bolivar, F., R. L. Rodriguez, P. J. Greene, M. C. Betlach, H. L. Heynecker, H. W. Boyer, J. H. Crosa, and S. Falkow. 1977. Construction and characterization of new cloning vehicles. II. A multipurpose cloning system. *Gene* 2:95–113.
7. Boyer, H. W., and D. Rolland-Dussoix. 1969. A complementation analysis of the restriction and modification of DNA in *Escherichia coli. J. Mol. Biol.* 41:459–472.
8. Brock, T. D. 1974. *Biology of Microorganisms.* Prentice-Hall, Englewood Cliffs, N.J., pp. 326–355.
9. Colwell, R. R., P. R. Brayton, D. J. Grimes, D. B. Roxzak, S. A. Huq, and L. M. Palmer. 1985. Viable but non-culturable *Vibrio cholerae* and related pathogens in the environment: Implications for release of genetically engineered microorganisms. *Bio/Technology* 3:817–820.
10. Chopra, I., S. W. Shales, M. W. Ward, and L. J. Wallace. 1981. Reduced expression of Tn10-mediated tetracycline resistance in *Escherichia coli* containing more than one copy of the transposon. *J. Gen. Microbiol.* 126:45–54.
11. Cole, M. A., and G. H. Elkan. 1979. Multiple antibiotic resistance in *Rhizobium japonicum. Appl. Environ. Microbiol.* 37:867–870.
12. Compeau, G., B. J. Al-Achi, E. Platsouka, and S. B. Levy. 1988. Survival of rifampicin-resistant mutants of *Pseudomonas fluorescens* and *Pseudomonas putida* in soil systems. *Appl. Environ. Microbiol.* 54:2432–2438.
13. Datta, N. 1985. Plasmids as organisms. In D. R. Helinski, S. N. Cohen, D. B. Clewell, D. A. Jackson, and A. Hollaender (eds.), *Plasmids in Bacteria.* Plenum Press, New York, pp. 3–16.
14. Devanas, M. A., D. Rafaeli-Eshkol, and G. Stotzky. 1986. Survival of plasmid-containing strains of *Escherichia coli* in soil: Effect of plasmid size and nutrients on survival of hosts and maintenance of plasmids. *Curr. Microbiol.* 13:243–246.
15. Devanas, M. A., and G. Stotzky. 1986. Fate in soil of a recombinant plasmid carrying a *Drosophila* gene. *Curr. Microbiol.* 13:279–284.
16. DeVos, G. F., G. C. Walker, and E. R. Signer. 1986. Genetic manipulations in *Rhizobium meliloti* utilizing two new transposon Tn5 derivatives. *Mol. Gen. Genet.* 204:485–491.
17. Ditta, G., S. Stanfield, D. Corbin, and D. R. Helinski. 1980. Broad host range DNA cloning system for Gram-negative bacteria: Construction of a gene bank of *Rhizobium meliloti. Proc. Natl. Acad. Sci. USA* 77:7347–7351.
18. Drahos, D. J., B. C. Hemming, and S. McPherson. 1986. Tracking recombinant organisms in the environment: Beta-galactosidase as a selectable, non-antibiotic marker for fluorescent pseudomonads. *Bio/Technology* 4:439–443.
19. Drahos, D. J., G. F. Barry, B. C. Hemming, E. J. Brandt, H. D. Skipper, E. L. Kline, D. A. Kluepfel, T. A. Hughes, and D. T. Gooden. 1988. Pre-release testing procedures: US field test of a *lac*ZY-engineered soil bacterium. In M. Sussman, C. H. Collins, F. A. Skinner, and D. E. Stewart-Tull (eds.), *The Release of Genetically Engineered Micro-organisms.* Academic Press, London, pp. 181–191.

20. Ezzell, C. 1987. EPA clears the way for release of nitrogen-fixing microbe. *Nature* 327:90.
21. Fredrickson, J. K., D. F. Bezdicek, F. J. Brockman, and S. W. Li. 1988. Enumeration of Tn5 mutant bacteria in soil using a most-probable-number-DNA hybridization procedure and antibiotic resistance. *Appl. Environ. Microbiol.* 54:446–453.
22. Gaertner, F., and L. Kim. 1988. Current applied recombinant DNA projects. In J. Hodgson, and A. M. Sugden (eds.), *Planned Release of Genetically Engineered Organisms* (Trends in Biotechnology/Trends in Ecology and Evolution Special Publication). Elsevier, Cambridge, pp. S4–S7.
23. Grinter, N. J. 1983. A broad host-range cloning vector transposable to various replicons. *Gene* 21:133–143.
24. Hardy, K. G., G. G. Meynell, and J. E. Dowman. 1973. Two major groups of colicin factors: Their evolutionary significance. *Mol. Gen. Genet.* 125:217–250.
25. Hemming, B. C., and D. J. Drahos. 1984. β-galactosidase, a selectable non-antibiotic marker for fluorescent pseudomonads. *J. Cell. Biochem.* (Suppl.) 8B:252.
26. Itakura, K., T. Hirose, R. Crea, A. Riggs, H. Heyneker, F. Bolivar, and H. W. Boyer. 1977. Expression in *Escherichia coli* of a chemically synthesized gene for the hormone somatostatin. *Science* 198:1056–1063.
27. Kelch, W. J., and J. S. Lee. 1978. Antibiotic resistance patterns of gram-negative bacteria isolated from environmental sources. *Appl. Environ. Microbiol.* 36:450–456.
28. Kleckner, N. 1981. Transposable genetic elements in prokaryotes. *Ann. Rev. Genet.* 15:341–404.
29. Labigne-Roussel, A., J. Witchitz, and P. Courvalin. 1892. Evolution of disseminated Inc7-M plasmids encoding gentamicin resistance. *Plasmid* 8:215–231.
30. Lee, C. A., and M. H. Saier. 1983. Use of cloned *mtl* genes of *Escherichia coli* to introduce *mtl* deletion mutants into the chromosome. *J. Bacteriol.* 153:685–692.
31. Legocki, R. P., A. C. Yun, and A. A. Szalay. 1984. Expression of beta-galactosidase controlled by a nitrogenase promoter in stem nodules of *Aeschynomene scabra. Proc. Natl. Acad. Sci. USA* 81:5806–5810.
32. Lenski, R. 1992. Relative fitness: Its estimation and its significance for environmental applications of microorganisms. In M. A. Levin et al. (eds.), *Microbial Ecology.* McGraw-Hill, New York, pp. 183–198.
33. Levy, S. B. 1984. Resistance to the tetracyclines. In L. E. Bryan (ed.), *Antimicrobial Drug Resistance.* Academic, Orlando, Fla., pp. 191–240.
34. Maniatis, T., E. F. Fritsch, and J. Sambrook. 1982. *Molecular Cloning, a Laboratory Manual.* Cold Spring Harbor Laboratory, Cold Spring Harbor, N.Y.
35. Maxam, A. M., and W. Gilbert. 1977. A new method for sequencing DNA. *Proc. Natl. Acad. Sci.* 74:560–572.
36. Mizusawa, S., D. Court, and S. Gottesman. 1983. Transcription of the *sul*A gene and repression by *lex*A. *J. Mol. Biol.* 171:337–343.
37. Nakai, C., H. Kagamiyama, M. Nozaki, T. Nakazawa, S. Inouye, Y. Ebina, and A. Nakazawa. 1983. Complete nucleotide sequence of the metapyrocatechase gene on the TOL plasmid of *Pseudomonas putida* mt-2. *J. Biol. Chem.* 258:2923–2928.
38. Newman, M. J., D. L. Foster, T. H. Wilson, and H. R. Kaback. 1981. Purification and reconstitution of functional lactose carrier from *E. coli J. Biol. Chem.* 256:11804–11808.
39. Norrander, J., T. Kempe, and J. Messing. 1983. Construction of improved M13 vectors using oligodeoxynucleotide-directed mutagenesis. *Gene* 26:101–106.
40. Obukowicz, M. G., F. J. Perlak, K. Kusano-Kretzmer, E. J. Mayer, S. L. Bolten, and L. S. Watrud. 1986a. Tn5-mediated integration of the delta-endotoxin gene from *Bacillus thuringiensis* into the chromosome of root-colonizing pseudomonads. *J. Bacteriol.* 168:982–989.
41. Raibaud, O., and M. Schwartz. 1980. Restriction map of the *Escherichia coli mal* A region and identification of the *mal* T product. *J. Bacteriol.* 143:761–771.
42. Rosenberg, A. H., B. N. Lade, D. Chui, S. Lin, J. J. Dunn, and F. W. Studier. 1987.

Vectors for selective expression of cloned DNAs by T7 RNA polymerase. *Gene* 56: 125–135.

43. Sanger, F., S. Nicklen, and A. R. Coulson. 1977. DNA sequencing with chain-terminating inhibitors. *Proc. Natl. Acad. Sci.* 74:5463–5472.

44. Sayler, G. S., M. S. Shields, E. T. Tedford, A. Breen, S. W. Hooper, K. M. Sirotkin, and J. W. Davis. 1985. Application of DNA-DNA colony hybridization to the detection of catabolic genotypes in environmental samples. *Appl. Environ. Microbiol.* 49: 1295–1303.

45. Schauer, A. T. 1988. Visualizing gene expression with luciferase fusions. *Trends Biotech.* 6:23–27.

46. Schmetterer, G., C. P. Wolk, and J. Elhai. 1986. Expression of luciferases from *Vibrio harveyi* and *Vibrio fischeri* in filamentous cyanobacteria. *J. Bacteriol.* 167: 411–414.

47. Schmidhauser, T. J., and D. R. Helinski. 1985. Regions of broad-host-range plasmid RK2 involved in replication and stable maintenance in nine species of gram-negative bacteria. *J. Bacteriol.* 164:446–455.

48. Schmidhauser, T. J., G. Ditta, and D. R. Helinski. 1988. Broad-host-range plasmid cloning vectors for Gram-negative bacteria. In R. L. Rodriguez and D. T. Denhardt (eds.), *Vectors: A Survey of Molecular Cloning Vectors and Their Uses.* Butterworths, Boston, pp. 287–332.

49. Schroth, M. N., S. V. Thomson, and W. J. Moller. 1979. Streptomycin resistance in *Erwinia amylovora. Phytopathology* 69:565–568.

50. Simon R., U. Priefer, and A. Puhler. 1983. A broad host range mobilization system for in vivo genetic engineering: Transposon mutagenesis in Gram-negative bacteria. *Bio/Technology* 1:784–791.

51. Tabor, S., and C. C. Richardson. 1985. A bacteriophage T7 RNA polymerase/promoter system for controlled exclusive expression of specific genes. *Proc. Natl. Acad. Sci. USA* 82:1074–1078.

52. Tabor, S., and C. C. Richardson. 1987. DNA sequence analysis with a modified bacteriophage T7 DNA polymerase. *Proc. Natl. Acad. Sci. USA* 84:4767–4771.

53. Turner, P., C. Barber, and M. Daniels. 1984. Behaviour of the transposons Tn5 and Tn7 in *Xanthomonas campestris* pv. *campestris. Mol. Gen. Genet.* 195:101–107.

54. Turner, G. K. 1985. In K. Van Dyke (ed.), *Bioluminescence and Chemiluminescence: Instruments and applications.* CRC Press, Boca Raton, Fla., pp. 43–78.

55. Van Dyke, K. 1985. In K. Van Dyke (ed.), *Bioluminescence and Chemiluminescence: Instruments and Applications.* CRC Press, Boca Raton, Fla., pp. 79–82.

56. Van Haute, E., H. Joos, M. Maes, G. Warren, M. Van Montagu, and J. Schell. 1983. Intergeneric transfer and exchange recombination of restriction fragments cloned in pBR322: A novel strategy for the reversed genetics of the Ti-plasmids of *Agrobacterium tumefaciens. EMBO Journal* 2:411–417.

57. Vieira, J., and J. Messing. 1982. The pUC plasmids, an M13mp7-derived system for insertion mutagenesis and sequencing with synthetic universal primers. *Gene* 19: 259–268.

58. Waddell, C. S., and N. L. Craig. 1988. Tn7 transposition: Two transposition pathways directed by five Tn7-encoded genes. *Genes and Development* 2:137–149.

59. Werneke, J. M., S. G. Sligar, and M. A. Shuler. 1985. Development of broad host-range vectors for expression of cloned genes in *Pseudomonas. Gene* 38:73–84.

60. Zukowski, M. M., D. F. Gaffney, D. Speck, M. Kauffmann, A. Findeli, A. Wisecup, and J. Lecocq. 1983. Chromogenic identification of genetic regulatory signals in *Bacillus subtilis* based on expression of a cloned *Pseudomonas* gene. *Proc. Natl. Acad. Sci. USA* 80:1101–1105.

61. Zugursky, R., N. Baumeister, N. Lomax, and M Bergman. 1985. Rapid and easy sequencing of large linear double-standed DNA and supercoiled plasmid DNA. *Gene Anal. Techn.* 2:89–94.

Chapter 9

Relative Fitness: Its Estimation and Its Significance for Environmental Applications of Microorganisms

Richard E. Lenski

Introduction

When considering the environmental application of some microorganism, one of the most important series of questions to ask concerns the opportunity for *persistence* of the population after it has been introduced into the target environment (Committee on Scientific Evaluation of the Introduction of Genetically Modified Microorganisms and Plants into the Environment, 1989; Tiedje et al., 1989). Is it desirable for the introduced population to be self-sustaining? Or is it better if the introduced population performs its intended function and then dies out, being reintroduced only as need arises? The answer will depend, of course, on a comparison of the magnitudes of the additional benefits that may derive from prolonged persistence with the possible costs, if any, that might arise from potential adverse effects caused by persistence.

Once this comparison has been made, it is appropriate to ask: What efforts, if any, have been made to enhance or limit the persistence of the introduced population, as so desired? Such efforts may involve deliberately disabling the microorganism, for example, by incorporating some restrictive nutritional requirement into its genome, in order to limit its persistence; or they may involve selecting a strain that is particularly well suited to conditions in the target environment in order to extend its persistence.

And finally: What empirical data are there concerning the likelihood of indefinite persistence of the introduced microbial population in the target environment? For many environmental applications of microorganisms, there may already exist a sufficient body of information to reasonably exclude the possibility of adverse effects or to predict with some certainty the persistence of the introduced population (Committee on Scientific Evaluation of the Introduction of Genetically Modified Microorganisms and Plants into the Environment, 1989). However, in cases where the environmental application is less familiar, it may be necessary to evaluate empirically the likelihood of persistence of the introduced population.

A critical factor in determining the likelihood of persistence of any introduced organism is its *fitness* in the new environment. Fitness is a broadly inclusive term that encompasses the combined effects of all biotic and abiotic interactions on an organism's capacity to survive and reproduce in a particular environment. If some organism's fitness in a particular environment is such that each individual, on average, leaves less than one progeny, then we can anticipate that a population of such organisms will not persist indefinitely in that environment. By contrast, if an organism's fitness in some environment is such that each individual, on average, leaves more than one progeny, then we can reasonably expect that a population of these organisms, upon introduction into the environment, will become established and may persist indefinitely. Of course, density-dependent factors, such as resource limitation, must eventually come into play, so that no population can continue to increase indefinitely.

In many cases, a target environment will contain an ecologically self-sustaining population of an indigenous organism that is closely related to the organism proposed for introduction. Ecological interactions between closely related introduced and indigenous populations are likely to be particularly significant for the fate of the introduction; even slight differences in the ability to exploit resources or to escape adversity may affect the opportunity for persistence of the introduced population. Thus, the fitness of an introduced organism *relative* to a closely related indigenous population is likely to be especially useful in predicting the fate of an introduced population.

In this chapter, I show how *relative fitness* can be estimated experimentally and used operationally to predict whether or not an introduced microbial population will persist in a target environment that contains a related indigenous population. I also discuss several assumptions of this approach that may limit its usefulness in certain circumstances.

I do *not* attempt to summarize evidence from laboratory and field studies that may bear on the question of whether there are empirical trends in relative fitness that may be useful in evaluating the likelihood of persistence of an introduced microorganism. It has often been suggested, for example, that genetically engineered microorganisms will usually be less fit than their wild-type counterparts, owing to the energetic and physiological costs associated with carriage and expression of recombinant genes. Various points of view on this subject can be found elsewhere (Brill, 1985; Colwell et al., 1985; Regal, 1986, 1988; Davis, 1987; Sharples, 1987; Lenski and Nguyen, 1988; Committee on Scientific Evaluation of the Introduction of Genetically Modified Microorganisms and Plants into the Environment, 1989; Tiedje et al., 1989).

Definitions and Principles

Fitness is a term that is most widely used in the fields of population genetics and evolutionary biology. A textbook definition of fitness is "The average contribution of one allele or genotype to the next generation or to succeeding generations, compared with that of other alleles or genotypes" (Futuyma, 1986, p. 552). Differences in fitness between alleles or genotypes may cause the frequency of some allele or genotype within the population to change systematically with time. Differences in fitness between alleles or genotypes reflect systematic differences in either mortality or reproduction, which in turn reflect systematic differences in ecological properties such as the ability to compete for limiting resources, susceptibility to predation, and so on. Therefore, *fitness must be viewed as a property of an allele or a genotype that depends upon the environmental circumstances.*

The process of systematic change in the frequency of alleles or genotypes due to differences in fitness is referred to as *selection*. Other processes may also cause changes in the frequency of alleles or genotypes, including *mutation* (which, broadly speaking, includes transposition and segregation of extrachromosomal elements), *recombination* (here, taken to mean intergenomic exchange by processes including gametic fusion, conjugation, transformation, and transduction) and *genetic drift*. Genetic drift can be defined as "Random changes in the frequencies of two or more alleles or genotypes within a population" (Futuyma, 1986, p. 552). Thus, drift differs from selection in that changes in the frequencies of alleles or genotypes are due to chance events, rather than to systematic differences in ecological properties such as competitive ability.

As implied by its earlier definition, fitness is best regarded as a relative property, not an absolute one. Therefore, it usually makes more sense to state that "genotype j is more fit than genotype k under environmental conditions x, y, and z" than to say that "genotype j is fit" or that "genotype k is unfit." A *selection coefficient* is used to provide a quantitative measure of the difference in relative fitness between two genotypes or alleles. A selection coefficient has units of inverse time and indicates the rate at which one genotype or allele replaces another.

In this chapter, we are primarily interested in organisms that reproduce clonally, often with little or no intergenomic recombination. I shall use the term *strain* to refer to different clonally reproducing genotypes. If one strain is more fit than another in some particular environment, then it would seem reasonable to expect that the less fit strain would eventually be lost from that environment. Indeed, this may often be the case, even when the less fit strain could be sustained indefinitely in the absence of the more fit strain. However, various circumstances have been shown to permit the coexistence of two (or more) strains (or alleles). Several of the most important mechanisms that promote such *genetic polymorphism* are described below.

Selective neutrality: Two strains may coexist almost indefinitely if they are equally fit in a particular environment. In such cases, the strains are said to be selectively neutral. Genetic drift may nonetheless cause some change in the relative frequency of two selectively neutral strains, including even the extinction of one or the other, especially when the population size of one or both of the strains is very small.

Balance between selection and migration, mutation, or gene transfer: A strain that is less fit than another can be maintained by recurring mutation or by migration from another source population (including repeated releases into a target environment). Gene transfer can also maintain an allele or extrachromosomal element in a population in the face of opposing selection (Stewart and Levin, 1977; Levin and Rice, 1980).

Frequency-dependent selection: There are ecological circumstances in which the relative fitness of two strains depends upon their relative frequencies. If strain j is more fit than strain k when strain j is rare, and if strain k is more fit when it is rare, then this frequency-dependent selection actively promotes stable coexistence of the two strains. Consider, for example, the following situation. One strain degrades a substance in the environment that is toxic to a second strain, but there is some "cost" associated with the degradative function (Lenski and Hattingh, 1986). The degradative activities of the first strain reduce the concentration of toxin and thereby promote the growth of the second strain at the expense of the first.

When the frequency of the first strain becomes too low, however, toxins can accumulate to a level where the first strain is again more fit than the second. Such opposing ecological feedbacks thus promote a stable genetic polymorphism.

Spatial heterogeneity: A related situation may occur when two strains differ in their ability to exploit an environment that is spatially heterogeneous. If each strain is fitter than the other in some part of the environment, then they may coexist even though one strain may appear to be fitter "on average" than the other. Coexistence is not an inevitable outcome in these cases and may depend upon other factors such as rates of dispersal.

"Hitchhiking": An allele that is less fit than another may nevertheless be maintained in a population by virtue of its association with a favorable allele elsewhere in the genome. Such associations are termed *linkage disequilibrium* and are especially prevalent in organisms, such as bacteria, that reproduce asexually and where other forms of intergenomic recombination (such as conjugation, transformation, and transduction) usually occur at fairly low rates. Linkage disequilibrium may also be exploited deliberately to prevent a strain intended for release in the environment from persisting. For example, a gene that permits a bacterium to perform some new and useful function in the environment might be placed into a genetic background that contains a nonfunctional allele at another locus, which prevents the bacterium from synthesizing some product essential for growth in the target environment. Linkage patterns can change with subsequent mutation and selection, however, as might happen if a mutation occurred that restored function to the other locus.

Specific Methods

We are interested in determining whether an introduced strain is likely to persist in a certain environment, which already contains a closely related indigenous population. We will assume the availability of a suitable test system, such as a microcosm (Pritchard and Bourquin, 1984), into which the introduced strain can be inoculated and from which samples can be obtained that allow one to enumerate and distinguish the introduced and indigenous populations. (The choice of the test system and the methods for enumerating and distinguishing the two populations need not concern us here, as they are discussed in other chapters.) The basic logic of such an experiment is straightforward. If the *relative* abundances of the two populations change in a systematic fashion with time, then one can infer that one or the other type is fitter in that particular environment. But if the relative abundances remain essentially constant, then one must conclude that the two types are equally fit, at least within the statistical resolution of the experiment.

A hypothetical data set is shown in Table 9.1; it comprises 11 samples, each containing a pair of densities corresponding to the introduced and indigenous populations. The time 0 sample corresponds to the time of inoculation of the introduced strain into the test environment. Table 9.1 also includes the ratio of the density of the introduced population to that of the indigenous population, computed for each sample.

In Fig. 9.1, the population densities have been transformed to a natural logarithmic scale and plotted against the time of the sample. The two lines indicate the least-squares linear regressions for the population densities; the slopes of these lines provide estimates of the rates of population increase or decrease. The estimated rate of change for the introduced population is -0.052 per unit time, while the estimated rate of change for the indigenous population is $+0.244$ per unit time.

Summary statistics for the regressions are given in Table 9.2; computational methods for regression statistics can be found in many texts (e.g., Kleinbaum and Kupper, 1978) and are widely available in computer software packages. A correlation coefficient provides one measure of the scatter of points about a regression line; it ranges from 0 when the fit of the data to the regression line is extremely poor, to $+1$ or -1 when the fit is extremely good. There is evidently considerable scatter in the data at hand. Standard errors provide a related measure of the statistical error in estimating a slope and permit hypothesis testing by means of a t-test. Neither of the slopes for the introduced and indigenous populations is significantly different from 0.

One can also compute the difference between two slopes, which in-

TABLE 9.1 Hypothetical Data Used to Compare Trends in Densities of Introduced and Indigenous Populations

Time	Density of introduced population	Density of indigenous population	Ratio of introduced to indigenous populations
0	1.55×10^5	1.91×10^6	8.12×10^{-2}
1	2.45×10^2	5.89×10^3	4.16×10^{-2}
2	4.47×10^5	1.55×10^7	2.88×10^{-2}
3	1.35×10^5	1.02×10^7	1.32×10^{-2}
4	8.71×10^3	1.35×10^6	6.45×10^{-3}
5	2.88×10^3	6.76×10^4	4.26×10^{-2}
6	3.39×10^4	1.23×10^6	2.76×10^{-2}
7	1.51×10^6	9.33×10^7	1.62×10^{-2}
8	8.32×10^3	1.78×10^6	4.67×10^{-3}
9	3.31×10^4	4.68×10^6	7.07×10^{-3}
10	3.09×10^3	3.02×10^6	1.02×10^{-3}

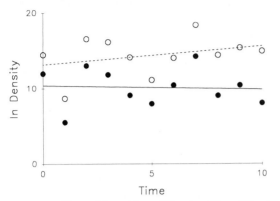

Figure 9.1   Natural logarithm of the densities of the introduced (solid circles) and the indigenous (open circles) populations versus time, using the data from Table 9.1. The lines indicate the least-squares linear regressions for the introduced (solid line) and indigenous (dashed line) populations, as given in Table 9.2. Neither slope is statistically significant from zero, nor are the slopes significantly different one from the other.

TABLE 9.2   **Summary of Statistical Analyses of Trends in Densities (Fig. 9.1) and Ratio (Fig. 9.2) of Introduced and Indigenous Populations**

	ln Density of introduced population	ln Density of indigenous population	ln Ratio of introduced to indigenous populations
Sample points	11	11	11
Correlation coefficient	0.068	0.310	0.785
Estimated y-intercept	10.391	13.154	-2.763
Estimated slope	-0.052	0.244	-0.296
Standard error of slope	0.254	0.249	0.078
t-statistic	-0.204	0.978	-3.803
Degrees of freedom	9	9	9
Significance level	$0.7 < p$	$0.3 < p$	$p < 0.01$
Estimated difference in slope	-0.296		
Standard error of difference	0.356		
t-statistic	-0.831		
Degrees of freedom	18		
Significance level	$0.4 < p$		

dicates the difference between the rates of change for the two populations. This difference in slopes provides an estimate of the selection coefficient, or difference in relative fitness, between the introduced strain and the indigenous population. From these data, the estimated selection coefficient is -0.296. Using the standard error of the difference in the two slopes (Kleinbaum and Kupper, 1978), one can then perform a *t*-test to determine whether the two slopes are significantly different from one another; this is equivalent to testing whether the selection coefficient is significantly different from 0. Owing to the rather larger standard error of the difference in slopes, one cannot reject the null hypothesis that the selection coefficient is zero (see Table 9.2).

There is nothing overtly incorrect in the preceding analysis, but it lacks statistical power (i.e., ability to discriminate) for reasons that we shall now see. In Fig. 9.2, the same set of data has been plotted, except that now the two densities have been converted to a single ratio prior to the natural logarithmic transformation. The line indicates the least-squares linear regression fit to the 11 sample points; summary statistics are provided in Table 9.2. The slope of this line provides an estimate of the selection coefficient (Dykhuizen and Hartl, 1983); using these data, we obtain a slope of -0.296, which is equal to the difference between the two slopes obtained from the separate regressions for each population. Note, however, that the scatter of the data points around the regression line calculated from the ratios (see Fig. 9.2) is much less than the scatter around the regression lines calculated from the densities (see Fig. 9.1). This is reflected by a much higher correlation coefficient and by a much smaller standard error of the slope (see Table 9.2). Indeed, the standard error is such that, based upon a *t*-test, one can claim with a high degree of statistical confidence

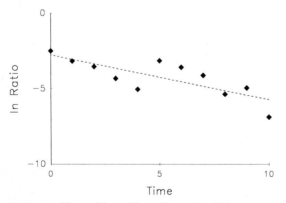

Figure 9.2   Natural logarithm of the ratio of the density of the introduced population to the density of the indigenous population versus time (Table 9.1). The line indicates the least-squares linear regression (Table 9.2). The slope is significantly less than zero ($p < 0.01$), which indicates that the introduced strain is less fit than the indigenous population under the prevailing environmental conditions.

($p < 0.01$) that the selection coefficient is less than zero. Thus, one can conclude that the introduced population is declining relative to the indigenous population at a significant rate.

Why is the selection coefficient based upon the difference in the two separately calculated rates of population change not significant, whereas the selection coefficient based upon the change in the ratio of the two population densities highly significant? The magnitudes of the two selection coefficients are comparable, but the standard error of the latter (0.078) is much smaller than the standard error of the former (0.356).

The smaller standard error arises because the fluctuations in the densities of the two populations are positively correlated in the hypothetical data set, and therefore the fluctuations in the two population densities tend to cancel one another. [By contrast, if the fluctuations in the two populations were independent (i.e., uncorrelated), then the standard error of the selection coefficient would not be reduced by calculating it from the rate of change in the ratio of the two densities.] There are two reasons to expect that, in real data sets, the fluctuations in the densities of two such populations would tend to covary in this manner. (1) Sampling variation: The densities of the two populations will often be estimated from the same physical sample, such as a plug of soil. Any uncontrolled variation in the volume of that sample, or in the efficiency of extraction of the microorganisms from it, will cause the estimated densities of the two populations to covary positively. (2) Environmental variation: Any temporal or spatial variation in environmental qualities, such as temperature or resource concentration, will cause the densities of the two populations to covary positively, provided that the two populations respond similarly to such variables.

These sources of variation are likely to be ubiquitous and will be especially important in natural or seminatural conditions, such as may exist in a microcosm. In effect, one has an internal control for these sources of variation when one analyzes the ratio of the densities of the introduced and indigenous populations, which is lacking when one analyzes separately the densities of the two populations. Therefore, in order to achieve greater statistical power, it is recommended that selection coefficients be estimated from the rate of change in the ratio of the two population densities.

The primary function of statistical inference is to formalize the degree of one's confidence in some conclusion. In the preceding analysis, a single estimate of the selection coefficient was computed by linear regression of the natural logarithm of the ratio of two population densities against time. The standard error of the slope was then used to determine whether that single estimate of the selection coefficient was significantly different from zero.

TABLE 9.3   Hypothetical Set of Five Replicate
Estimates of the Selection Coefficient and
a Summary of Their Statistical Analysis

Replicate estimates	-0.296
	-0.495
	-0.607
	-0.271
	-0.403
Sample mean	-0.414
Sample standard deviation	0.140
Standard error of the mean	0.063
t-statistic	-6.616
Degrees of freedom	4
Significance level	$p < 0.01$

An alternative approach is to obtain several estimates of the selection coefficient, each based upon an independent experimental replicate. Selection coefficients would be calculated by linear regression as before, but no significance level would be attached to any single estimate. Instead, the standard error of the selection coefficient would be calculated from the sample standard deviation based upon the several independent estimates. One could then use a t-test to determine whether the *mean* of the several estimates of the selection coefficient was significantly different from zero. Table 9.3 provides a hypothetical set of five replicate estimates of the selection coefficient and the corresponding statistical analysis. Statistical inferences based on somewhat less intensive sampling of many independent experimental units are generally more reliable than those based on exhaustive sampling of a single experimental unit, and so this second approach is usually preferable. Hurlbert (1984) provides a useful discussion of the importance of proper replication in ecological experiments.

Assumptions

The preceding methods of analysis, like any others, make several assumptions. It is important to be aware of the major assumptions and the circumstances under which the assumptions might be seriously violated. In some instances, it may be possible to modify the methods so that the results and conclusions are less sensitive to certain assumptions.

Assumption 1: The physiological states and ecological circumstances of the introduced and indigenous populations are comparable.   If this assumption is met, then any significant difference in fitness must be due to genetic differences. For example, we inferred from our hypothetical data that the introduced strain was less fit than the indigenous population under the prevailing set of environmental conditions. Implicit in this conclusion is the assumption that there were no differences between the introduced and indigenous populations with respect to their physiological states and ecological circumstances. In fact, however, the indigenous population may be physiologically acclimated to the environmental conditions in a particular test system, whereas a recently introduced population may experience some death or delay in growth as it acclimates to those conditions. Also, some fraction of the introduced population may be inoculated into microenvironments that are unsuitable for survival or growth of that organism (e.g., soil interstices for a bacterium that lives in the film associated with soil particles), thereby producing a high rate of decline at the start of an experiment. Either of these artifacts may cause one to systematically underestimate the introduced strain's true fitness in that environment.

One might get around this problematic assumption in a couple of different ways. First, one could exclude from the calculation of the selection coefficient one or more sample ratios obtained at the beginning of an experiment, if visual inspection (or, better yet, an appropriate statistical test) indicates nonlinearity in a plot of the logarithmically transformed ratio against time. Alternatively, one might redesign the experiment so as to avoid this assumption. This could be accomplished by admixing and simultaneously introducing two populations into the test system: one the strain of interest, and the other a strain isolated from the indigenous population and then genetically marked such that it can be distinguished from its counterparts in situ. One would then calculate the selection coefficient from the rate of change in the ratio of the densities of the strain of interest and the marked indigenous strain, with greater confidence that the two populations being compared have experienced similar trauma during their inoculation into the test system. In essence, the marked indigenous strain provides a control for the effects of physiological acclimation and ecological circumstance, which is otherwise lacking. (Of course, one should also perform an additional control to determine whether the genetic marker used to identify the indigenous strain affects its relative fitness.)

Assumption 2: The change in the ratio of the two population densities is due to selection. The ratio of the two population densities may be affected by processes other than selection. Genetic drift (see previous section) is another process that may affect the ratio of two strains, although it is usually an important influence only when one or both populations are extremely small. Also, genetic drift should not generally bias the estimation of the selection coefficient; rather, it introduces another source of statistical error but, by definition, does not cause systematic deviations in favor of one strain or another.

Potentially more serious violations of this assumption may arise when genetic processes that convert one genotype into another, such as plasmid segregation or conjugation, occur at a high rate. In these instances, the rate of change in the relative abundances of the two strains may not be adequately described by a single parameter, and more complicated analyses may be required. Lenski (1991) presents methods that can be used to distinguish the effects on population dynamics of selection from the effects of segregation. Stewart and Levin (1977) and Levin and Rice (1980) present mathematical models of bacterial conjugation that may be adapted to disentangle selection and gene transfer. In theory, either segregation or gene transfer can permit two strains to coexist, even if one is more fit, provided that the less fit type is regenerated de novo by the relevant genetic process. However, if selection coefficients are large relative to the rates governing these genetic processes, as may often be the case, then these genetic processes should have little effect on an estimate of the selection coefficient obtained from a relatively short-term experiment.

Assumption 3: The selection coefficient is constant. The utility of relative fitness as a criterion for predicting the fate of an introduced strain is affected only very little by certain kinds of variation in the selection coefficient, but it may be much more sensitive to other sources of variation. Of particular concern are those instances in which the sign of the selection coefficient, rather than simply its magnitude, may vary.

Figure 9.3 illustrates the effects of three different types of variation in the selection coefficient on the short-term persistence of an introduced population. In each case, the introduced strain is usually less fit than the indigenous population (i.e., the selection coefficient is less than zero), but occasionally the introduced strain is more fit (i.e., the selection coefficient is greater than zero). And in all three cases, the observed dynamics over the short term (e.g., 20 or so time units) would seem to indicate that the introduced strain will not persist.

In the first and simplest case, the variation in the selection coefficient occurs temporally over a relatively short scale. In particular, imagine that the introduced strain has a -0.1 selection coefficient relative to the indigenous population for 99 percent of the time. For the other 1 percent of the time, however, the introduced strain has a $+0.05$ selection coefficient relative to the indigenous population. Over the long term, the introduced population is expected to decline relative to the indigenous population at a rate that is simply the average of the variable rates in time, i.e., $(0.99)(-0.10) + (0.01)(+0.05) =$

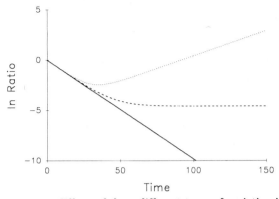

Figure 9.3 Effects of three different types of variation in
the selection coefficient on the persistence of an introduced
population. The natural logarithm of the ratio of the density of an introduced population to the density of an indigenous population is plotted against time. *Solid line:* Fine-
scale temporal variation in the selection coefficient.
Dashed line: Fixed spatial variation in the selection coefficient. *Dotted line:* Genetic variation in the selection coefficient. In each of the three cases that is illustrated, the
introduced strain is usually less fit than the indigenous
population (selection coefficient = −0.1), but occasionally
the introduced strain is more fit (selection coefficient =
+0.05). See text for further details.

−0.0985. Therefore, over both the short term and the long term, the
introduced population declines toward extinction.

In the second case, the variability in the selection coefficient exists
spatially and remains fixed in time. This variation may arise from differences in the qualities of certain habitats or microenvironments with
respect to their suitability for one population or the other. For example,
imagine that the introduced strain has a −0.1 selection coefficient relative to the indigenous population in 99 percent of the habitats. In the
other 1 percent of the habitats, the introduced strain has a +0.05 selection coefficient relative to the indigenous population. (The average densities supported per unit volume in the habitats are assumed to be equal
in this example, and the two populations are initially distributed randomly across the habitats.) Over the short term, the introduced population declines relative to the indigenous population. Over the long term,
however, the introduced and indigenous populations stably coexist by
virtue of their differential utilization of the two types of habitats.
Frequency-dependent selection (see previous section) may similarly give
rise to stable coexistence after a period of initial change.

In the third case, there is genetic variation in the selection coefficient such that some subset of the introduced population is more fit than the indigenous population, even though the majority is much less fit. In particular, imagine that 99 percent of the individuals in the introduced population have a −0.1 selection coefficient relative to the indigenous population, while the other 1 percent of the individuals in the introduced population have a +0.05 selection coefficient relative to the indigenous population. These differences are assumed to be heritable (i.e., genetically determined). Over the short term, the introduced population once again declines relative to the indigenous population. Over the long term, however, the introduced population increases relative to the indigenous population and may even cause its extinction, owing to the fitness advantage that accrues to an initial minority of the introduced population.

A hypothetical example of how this situation might arise is as follows. The introduced strain has been modified genetically to provide it with the enzymatic functions necessary to utilize as a resource some compound in the environment not available to the indigenous population. At the same time, the introduced strain has been deliberately handicapped ecologically by the incorporation of some restrictive nutritional requirement. If a spontaneous mutant that lost the ecological handicap occurred in the introduced population, then the mutant subpopulation would have the advantage of the strain intended for introduction, without the disadvantage, and hence it could increase unexpectedly. Kim et al. (1991) have developed quantitative models that may be useful in predicting the likelihood of a reversal in the outcome of selection due to secondary genetic changes in an introduced population.

Summary and Conclusions

The ability to predict whether a microbial population that may be released into the environment will persist or disappear is an important consideration when evaluating the possible benefits and risks of an application, particularly one that is unfamiliar. An important factor in determining the likelihood of persistence of the introduced microorganism is its fitness in the target environment. Fitness is a broadly inclusive term that encompasses the combined effects of all biotic and abiotic interactions on an organism's capacity to survive and reproduce in a particular environment.

In many cases, a target environment may already contain an ecologically self-sustaining population of an indigenous microorganism that is closely related to the microorganism proposed for introduction. In such cases, the difference in fitness between the introduced and indigenous microorganisms will be especially important in determining the fate of the introduced population. The difference in fitness between two clonally reproducing genotypes, or strains, is termed a selection coefficient. Selection coefficients have units of inverse time and indicate the rate at which one strain replaces another. This chapter has illustrated the basic methods used to estimate selection coefficients.

The basic design of an experiment to estimate the selection coefficient, or difference in fitness, for two strains is quite simple. Populations of the two strains are mixed together in some initial ratio in a test environment. At various time points, samples are obtained from the test system and the ratio of the two population densities in each sample is determined. If one strain is more fit than the other in the test environment, then this ratio should increase or decrease with time in a systematic fashion; if the ratio of the population densities remains essentially constant, then one must conclude that the two strains are equally fit in the test environment, at least within the statistical limit of resolution.

It should be emphasized that these methods of estimating selection coefficients are widely applicable to a variety of different test environments, ranging from simple laboratory culture systems, such as shake flasks and chemostats, to more complex seminatural systems, such as microcosms. These methods may also be useful in monitoring field trials. In principle, all that is required to measure a selection coefficient is the ability to monitor the ratio of two strains sharing a common environment.

Selection coefficients that are estimated in this manner invariably have an internal control; whether a particular sample or experimental unit is nutrient rich or poor, for example, the same is true for both strains. Because of this internal control, differences in relative fitness that are calculated directly from rates of change in the ratio of two population densities will often be more accurate than differences in fitness calculated from separate measurements of the growth properties for each strain.

Of course, these methods are not without their assumptions. The most serious violations may occur when the introduced and indigenous populations are not in comparable physiological states or ecological circumstances; when segregation or transfer of genes, and not selection, is primarily responsible for the change in the relative densities of the strains; and when the sign of the selection coefficient varies as a consequence of spatial heterogeneity, frequency-dependent selection, or secondary genetic changes in the less fit population. In such cases, successful prediction of the fate of an introduced microbial population may require more complex analyses or changes in the experimental design.

Acknowledgments

This chapter is respectfully dedicated to the memory of Steve Cuskey. Preparation of this chapter was supported by a cooperative research grant from the Gulf Breeze Laboratory of the U.S. Environmental Protection Agency (CR-815380) and by an award from the National Science Foundation (BSR-8858820).

References

Brill, W. J. 1985. Safety concerns and genetic engineering in agriculture. *Science* 227:381–384.

Colwell, R. K., E. A. Norse, D. Pimentel, F. E. Sharples, and D. Simberloff. 1985. Genetic engineering in agriculture. *Science* 229:111–112.

Committee on Scientific Evaluation of the Introduction of Genetically Modified Microorganisms and Plants into the Environment. 1989. *Field Testing Genetically Modified Organisms: Framework for Decisions.* National Academy Press, Washington, D.C.

Davis, B. D. 1987. Bacterial domestication: Underlying assumptions. *Science* 235:1329–1335.

Dykhuizen, D. E., and D. L. Hartl. 1983. Selection in chemostats. *Microbiol. Rev.* 47:150–168.

Futuyma, D. J. 1986. *Evolutionary Biology*, 2d ed. Sinauer Associates, Sunderland, Mass.

Hurlbert, S. H. 1984. Pseudoreplication and the design of ecological field experiments. *Ecol. Monogr.* 54:187–211.

Kim, J., L. R. Ginzburg, and D. E. Dykhuizen. 1991. Quantifying the risks of invasion by genetically engineered organisms. In L. R. Ginzburg (ed.), *Assessing Ecological Risks of Biotechnology.* Butterworth-Heineman, Boston, pp. 193–214.

Kleinbaum, D. G., and L. L. Kupper. 1978. *Applied Regression Analysis and Other Multivariable Methods.* Duxbury Press, North Scituate, Mass.

Lenski, R. E. 1991. Quantifying fitness and gene stability in microorganisms. In L. R. Ginzburg (ed.), *Assessing Ecological Risks of Biotechnology.* Butterworth-Heinemann, Boston, pp. 173–192.

Lenski, R. E., and S. E. Hattingh. 1986. Coexistence of two competitors on one resource and one inhibitor: A chemostat model based on bacteria and antibiotics. *J. Theor. Biol.* 122:83–93.

Lenski, R. E., and T. T. Nguyen. 1988. Stability of recombinant DNA and its effects on fitness. In J. Hodgson and A. M. Sugden (eds.), *Planned Release of Genetically Engineered Organisms (Trends in Biotechnology / Trends in Ecology and Evolution Special Issue).* Elsevier, Cambridge, England, pp. 18–20.

Levin, B. R., and V. A. Rice. 1980. The kinetics of transfer of nonconjugative plasmids by mobilizing conjugative factors. *Genet. Res., Camb.* 35:241–259.

Pritchard, P. H., and A. W. Bourquin. 1984. The use of microcosms for evaluation of interactions between pollutants and microorganisms. *Adv. Microb. Ecol.* 7:133–215.

Regal, P. J. 1986. Models of genetically engineered organisms and their ecological impact. In H. A. Mooney and J. A. Drake (eds.), *Ecology of Biological Invasions of North America and Hawaii.* Springer-Verlag, New York, pp. 149–162.

Regal, P. J. 1988. The adaptive potential of genetically engineered organisms in nature. In J. Hodgson and A. M. Sugden (eds.), *Planned Release of Genetically Engineered Organisms (Trends in Biotechnology / Trends in Ecology and Evolution Special Issue).* Elsevier, Cambridge, England, pp. 36–38.

Sharples, F. E. 1987. Regulation of products from biotechnology. *Science* 235:1329–1332.

Stewart, F. M., and B. R. Levin. 1977. The population biology of bacterial plasmids: A priori conditions for the existence of conjugationally transmitted factors. *Genetics* 87:209–228.

Tiedje, J. M., R. K. Colwell, Y. L. Grossman, R. Hodson, R. E. Lenski, R. N. Mack, and P. J. Regal. 1989. The planned introduction of genetically engineered organisms: Ecological considerations and recommendations. *Ecology* 70:298–315.

Chapter 10

Analysis of Conjugation in Bacteria

Sarah A. McIntire

Introduction and History

Introduction

Conjugation is defined as the one-way transfer of DNA between donor and recipient bacteria by a mechanism that requires cell-to-cell contact and is resistant to deoxyribonuclease. This process is usually encoded by a plasmid (but see Conjugative Transposons in this section), a nonessential extrachromosomal DNA molecule that can replicate autonomously in the bacterial cell. The fertility factor F of *Escherichia coli* K-12 was the first plasmid identified and has the best characterized conjugation system, although many other conjugative plasmids are known in both gram-positive and gram-negative bacteria.

History

The discovery of the F factor in the early 1950s was a direct result of the earlier studies of Lederberg and Tatum (100, 101) that demonstrated recombination of genetic markers between strains of *E. coli* K-12. The recombinants were thought to result from cell fusion until Hayes (71) and Lederberg et al. (99) demonstrated that conjugative recombination is a one-way transfer of genetic information from donor to recipient cells. Their further studies determined that the donor state is conferred by the fertility agent F, a DNA molecule that is transferred to recipient cells independently of the bacterial chromosome (29, 72). These investigations were possible because (1) the *E. coli* K-12 strain happened to contain the F factor; (2) the F factor transfers at a high frequency; (3) the F factor contains four copies of insertion sequences that allow it to recombine into the bacterial chromosome and mobilize chromosomal genes.

Some F^+ strains were isolated that transferred chromosomal genes 1000 times more frequently than the parental F^+ strain. These Hfr (*h*igh *f*requency of *r*ecombination) strains were used by Wollman and Jacob (166) to demonstrate that each Hfr strain transfers its chromosome in the same specifically oriented and linear way. Each Hfr, therefore, represents the insertion of the F factor into a particular site on the chromosome such that chromosomal genes on one side of the insertion site are transferred preferentially to a suitable recipient. An array of Hfr strains, each carrying F at a different site, was an invaluable tool in mapping the *E. coli* chromosome.

Some derivatives of Hfr strains carry a cytoplasmic F factor that, in the recombination event removing it from the chromosome, removed some of the adjacent chromosomal genes as well as the F factor. These F' (F prime) factors are usually named with the chromosomal genes carried, for example, F'*lac,* F'*his,* F'*pro.* They have been useful tools in chromosome mapping, gene complementation, and gene cloning studies.

Plasmids that carry a variety of genetic information have been described. These plasmid-determined characteristics generally fall into three broad categories: (1) resistance properties, such as antibiotic, heavy metal, ultraviolet, and phage or bacteriocin resistance; (2) metabolic properties, such as carbohydrate utilization, protein degradation, nitrogen fixation, and degradation of complex compounds; and (3) pathogenic or symbiotic properties, such as antibiotic and bacteriocin resistance, toxin production, colonization antigen synthesis, and tumor-inducing properties. A list of known plasmid-determined characteristics is given in Stanisich (138).

Many organisms with conjugative plasmids have been described also. Table 10.1 is a list of bacteria other than the commonly described *E. coli* and *Salmonella,* that have been shown (as of December 1988) to carry conjugative plasmids. In choosing the references, I attempted to include a review paper (where available) and a report that contains a workable conjugation method. The list of references is not meant to be complete for each organism but should give an inexperienced investigator a starting place for conjugation experiments.

As more plasmids were described, attempts were made to correlate such characteristics as transferability, host range, plasmid copy number, and phenotypes encoded. Especially among plasmids from gram-negative organisms, but also those from *Staphylococcus aureus* and *Streptococcus faecalis* (see Ref. 29), the most common classification is by incompatibility (Inc) grouping. If two plasmids are unable to stably

TABLE 10.1   Bacteria with Conjugative Plasmids

Bacteria	References
Gram-negative organisms:	
Aeromonas	3, 20
Acetobacter	154
Acinetobacter	76, 95
Agrobacterium	77, 78, 91, 137
Alcaligenes	54, 62, 63, 85, 153
Bacteroides	107, 117, 140, 142
Campylobacter	93, 144, 145
Citrobacter	146
Enterobacter	150
Erwinia	31, 32
Haemophilus	89
Klebsiella	155
Neisseria	136
Proteus	46, 147
Pseudomonas	2, 7, 17, 167
Rhizobium	9, 87
Serratia	88
Vibrio	12, 66
Yersinia	51
Gram-positive organisms:	
Arthrobacter	143
Bacillus	13, 30, 68
Clostridium	24, 117, 125
Corynebacterium	73
Rhodococcus	53
Staphylococcus	4, 5, 127
Streptococcus	28, 37, 81, 127, 134
Actinomycetes:	
Nocardia	113, 128
Streptomyces	79, 80, 120, 139

coexist in the same cell, they are termed incompatible and assigned to the same Inc group. Incompatibility occurs because the two plasmids have similar replication systems (see Incompatibility in the next section). It is generally accepted that incompatibility between two separately isolated plasmids is an indication of their evolutionary similarity. Conjugation systems seem to have evolved at the same time since most of the plasmids within a given Inc group encode similar conjugation systems. Many Inc groups have been described (48, 49), and some of their conjugation-related properties have been analyzed (16 and see Table 10.2).

Even though other systems exist, the F plasmid remains as the prototype of conjugative transfer. Initial analyses of conjugation in environmentally important organisms will rely on the methodology developed for the F plasmid. Therefore, a brief review of F conjugation is presented so that investigators wishing to study conjugation in these organisms can understand the difficulties involved in analyzing such a complex, multigenic system.

F factor as classic conjugative plasmid. More than 60 genes have been mapped on the F factor, which is a covalently closed, double-stranded DNA molecule that is 100 kilobase (kb) pairs in length (86). F is stably maintained in *E. coli* at a copy number of 1 to 2 per chromosome. Several plasmid genes are involved in DNA maintenance, replication, and partitioning (92, 118). Willetts and Skurray (163) reviewed the known properties of F and presented a complete genetic and physical map of the plasmid. Since recent reviews provide many details of the historic development of our current understanding of conjugation (84, 132, 163), I present here only a brief summary of the F factor conjugation process and the genes involved. The reader should refer to the above reviews for original references and detailed descriptions of the experiments that led to these conclusions.

The genes governing conjugation (*tra* genes) occupy a contiguous segment of 33.3 kb of the F DNA molecule. These genes can be divided into four functional groups: (1) *traA, L, E, K, B, V, C, W, U, F, Q, H, G* are involved in the elaboration of the pilus, an extracellular appendage that makes contact with a recipient cell; (2) *traN* and *traG* are required to stabilize that contact between mating cells; (3) *traM, Y, D, I* are involved with processing the DNA that is transferred; (4) *traJ* controls expression of the other transfer genes. The transfer region map shown in Fig. 10.1 is similar to previous maps (84, 163) except that *traZ* is no longer included (see below). Note that *traG* is included in functional groups 1 and 2; mutations in the amino terminus of the *traG* protein affect pilus synthesis, and mutations in the carboxyl terminus affect stable pairing. I* represents a protein, discussed below, that is made from a translational restart within the *traI* gene. The *tra* operon contains two additional genes, *traS* and *traT*, that are not re-

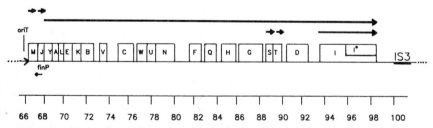

Figure 10.1   Map of F transfer genes. The boxes represent the sizes and locations of genes with known functions. Other genes of unknown function are given in the more detailed map of Ippen-Ihler and Minkley (84). The bottom line gives the kilobase coordinates of the *tra* region of F, with *oriT* at 66.7 and IS3 at 100/0. The arrow at *oriT* indicates the direction of DNA transfer. Heavy arrows above the boxes represent *tra* transcripts, and the arrow below represents the *finP* transcript.

quired for transfer, but instead express products that inhibit mating between two donor cells ("surface exclusion"). The *finP* gene product is involved in fertility inhibition and is discussed below. Other gene products encoded in this region have been identified and mapped, but no mutations exist that would define their functions. Since the *tra* alphabet is depleted, newly identified genes are called *trb* (84, 169).

Isolation of polar insertion mutations indicated that most *tra* genes are part of a single transcriptional unit, the *traYI* transcript. Other analyses detected weak internal promoters before *traS, traT,* and *traI*. In addition, the *traM, traJ,* and *finP* genes each have their own promoter, with *finP* being transcribed from the opposite strand. Wu and Ippen-Ihler (168) detected additional open reading frames, called *art*, on the strand opposite the *traYI* operon, but their significance is not known.

F donor cells possess one or two flexible pili that are 1 to 2 μm in length. These pilus structures are responsible for forming the contact between donor and recipient cells. The flexibility of the pilus probably accounts for efficient transfer in liquid medium, unlike other conjugative plasmids that elaborate rigid pili and transfer well only on solid surfaces. The working model is that the pilus tip binds unstably to a site on the recipient cell. In mating mixtures the donor and recipient cells form mating aggregates that are soon stabilized (requiring the products of *traN* and *traG*), with the cells brought into wall-to-wall contact. Only with the formation of stable aggregates of donor and recipient cells can transfer of DNA occur.

The structural gene for pilin is *traA,* which encodes a 121 amino acid polypeptide of 13,000 daltons. This propilin protein contains a 51 amino acid leader region that is removed to produce a 7000-dalton pilin subunit, which is then acetylated to make the antigenically recognized F pilin. Although the propilin molecule contains a typical signal peptidase recognition site that should be cleaved by the chromosomally encoded amino peptidase, cells that lack the plasmid-encoded *traQ* cannot process the 13,000-dalton *traA* product. Since a 51 amino acid leader is exceptionally long, one possibility may be that *traQ* crops the leader to a size recognized by the cellular amino peptidase (169). This could explain the observed small amount of an 8000-dalton polypeptide as a processing intermediate. An equally likely possibility is that the 94 amino acid hydrophobic *traQ* protein acts as a membrane receptor for the *traA* product, such that the 13,000-dalton propilin molecule inserts into the membrane in a conformation recognizable by the host amino peptidase (94). Further studies will determine the exact role of *traQ* in processing propilin.

Analysis of *tra* was greatly facilitated by the use of bacteriophages that attach only to the F pilus. These are called either pilus-specific, male-specific, or F-specific phages. For example, the F pilus-specific single-stranded RNA phages f2, R17, and QB adsorb to the sides of the F pilus, whereas the filamentous single-stranded DNA phages f1, M13, and fd adsorb to the pilus tip. Isolation of tra^- derivatives is assisted by including a selection for phage resistance in the protocol (83, 109). Other conjugation systems among gram-negative bacteria encode pili that are sensitive to other pilus-specific phages (see Table 10.2). Screening of a plasmid-containing strain with a panel of phages could provide a relatively quick means of determining the plasmid conjugation system. Methods for growing, titering, and using pilus-specific phages are given by Willetts (160).

Some of the early molecular studies of F transfer revealed that a strand- and site-specific nick is made at a site called the origin of transfer (oriT) and a unique single strand of DNA is transferred linearly into the recipient with the 5' end leading. As the DNA is unwound in the donor, the strand remaining in the donor is replicated, and the transferred strand is replicated in the recipient. The mechanism for circularizing the new double-stranded molecules is unknown.

There are four tra gene products known to be involved in the events that nick oriT and then unwind and move the linear single strand of DNA into the recipient. These proteins are encoded by traM, traY, traD, and traI. Using a λoriT clone, Everett and Willetts (58) developed an in vivo oriT-nicking assay and showed that nicking results from the combined activities of products from traY and traZ. traZ was mapped next to traI as the last gene in the operon. Revised mapping data, however, relocate the traI gene more promoter-distal and eliminate the traZ gene altogether (151). In addition, Traxler and Minkley (152) recently determined that traI encodes a bifunctional 180,000-dalton protein. The carboxyl-terminus of the molecule contains the unwinding activity (DNA helicase I), and the amino terminal half provides (with traY product) the site-specific endonuclease activity, the function formerly assigned to traZ. Thus, a complex of the products from traY and traI contains both nicking and unwinding activities. The I* product (see Fig. 10.1) results from a translational restart within the traI gene (151); the role of this 94,000-dalton protein is unknown at present.

The exact role of traM is unknown, although it is required for replication of the DNA strand remaining in the donor. It has been reported (Musgrave, Achtman, and Thompson, cited in 163) that traM protein binds specifically to oriT and may thus form the core of an oriT-nucleoprotein complex, perhaps with other DNA binding proteins (52), and act to trigger the nuclease and/or helicase activities. Once initiated, conjugative DNA synthesis is the responsibility of chromosomally encoded enzymes in both the donor and the recipient cells. Although not shown directly, the *traD* product, which is a DNA binding protein located in the inner membrane, may act in the export of DNA or may anchor the 3′ end of the strand remaining in the donor.

Models for the complex interactions of the various proteins with DNA in the *oriT* region have been proposed (58, 84, 152). When an in vitro assay for nicking is developed, the exact role of each component can be characterized.

F transfer genes are regulated at two levels:

1. By the product of *traJ*, which is required to activate transcription of the *traYI* operon. Mutations in *traJ* are thus pleiotropic, producing cells with no pili and that show no aspect of DNA metabolism or surface exclusion. The mechanism of *traJ* activation is not clear, although Willetts (159) showed reduced *traYI* mRNA levels in *traJ⁻* strains, indicating that control occurs at the transcriptional level.

2. By the fertility inhibition system FinOP, which represses the synthesis of *traJ* product. The absence of *traJ* product results in no expression of the other *tra* genes and, consequently, no transfer, i.e., the plasmid exhibits "fertility inhibition."

Fertility inhibition was observed in the early studies of antibiotic resistance plasmids such as R100. These plasmids did not transfer at the same high level as F and, in addition, inhibited the transfer of a coresident F molecule. The observation was that cells with a newly acquired R100 could transfer at a high level for a while, but soon become repressed. Dissection of this process showed that two diffusible products are required from the plasmid-encoded genes *finO* and *finP*. The F plasmid has *finP* but lacks a functional *finO* gene so that *traJ* protein is made constitutively. This explains the F plasmid's normally high transfer level. However, in cells carrying F and R100, the *finO* product from R100 interacts with the *finP* product from F to repress *tra* operon expression and thus inhibit F transfer. Among these F-like plasmids, the *finP* product is plasmid-specific, i.e., the *finP* product of R100 cannot substitute for the *finP* product of F. The *finP* product, a nontranslated RNA molecule, is encoded by the strand opposite *traJ*, beginning in the region corresponding to the amino terminus of the *traJ* protein.

Cheah and Skurray (35) and Yoshioka et al. (171) determined that F does carry a *finO* gene, but the gene has an IS3 insertion and is therefore nonfunctional. This is the IS3 that maps at the promoter-distal end of the F *tra* operon (see Fig. 10.1). F-like plasmids such as R100 carry the functional *finO* in this location. The R100 *finO* gene has been cloned and sequenced (108, 171), as has the *finO* gene of a nearly identical plasmid, R6-5 (33). Yoshioka et al. (171) and McIntire and Dempsey (108) visualized a 21,000-dalton protein from an open reading frame in this sequence, but McIntire and Dempsey (108) determined that internal out-of-frame deletions in the open reading frame retained full *finO* activity. Therefore, the exact nature of *finO* is still questionable, and the role of both *finO* and *finP* in fertility inhibition of F-like plasmids is unknown. Other unrelated plasmids also encode products that reduce F transfer but probably by a different mechanism (see 162).

The expression of transfer is influenced by chromosomal genes also (reviewed in 131). These genes were identified by several different groups and given different names. Mutations in chromosomal genes *sfrA* (first identified as *dye* and also called *cpxC* or *fexA*) or in *cpxA* and *cpxB* lead to a decrease in the amount of *traJ* protein, while mutations in *sfrB* (also called *rfaH*) result in a decreased transcription of the distal end of the *tra* operon. In addition, Dempsey (52) recently showed that a DNA binding protein called IHF (Integration Host Factor) is required for transfer of the F-like plasmid R100. It seems likely that a similar requirement will hold for F since the transfer systems of these two plasmids are so similar, but this has not been tested directly.

Analysis of transfer at the genetic level is complicated by the involvement of such a large number of genes and the lack of assays for the activities of each gene product. The only "assay" available (i.e., transfer of the plasmid) is several steps removed from the gene-protein level. It will be some time before all of the steps involved in this complicated process of conjugation are completely understood.

The *tra* organization of other plasmids is beginning to be examined. For example, both the 60-kb IncP plasmid RP4 (11) and a derivative of the 49-kb IncN plasmid R46 (165) have *tra* genes encoded in three separate regions, while the IncN plasmid pCU1 (149) has *tra* functions within a single 15-kb region. Thus, while transfer systems may be phenotypically similar, they may be very different at the genetic level.

Plasmid mobilization. Many plasmids are not conjugative, but some can be transferred by a coresident conjugative plasmid, and these are called mobilizable plasmids. This mobilization usually depends on the presence of an *oriT* site that is recognized by the coresident plasmid's transfer system. This aspect of conjugation was reviewed by Willetts and Wilkins (164). Commonly, these mobilizable plasmids are small and may also encode some but not all of the genes required for transfer. Because of the plasmid specificity of the interactions at *oriT*, not every conjugative plasmid will mobilize every nonconjugative plasmid.

Conjugative transposons. These are a relatively new class of conjugative elements that transfer DNA, mostly antibiotic resistance genes, even though the bacterial strains show no physical evidence of a plasmid. Soon after the emergence of clinical *Streptococcus pneumoniae* strains encoding multiple antibiotic resistances, Shoemaker et al. (130), and Buu-Hoi and Horodniceanu (28) reported the transfer of resistance determinants in these strains by a nonplasmid, DNase-resistant process. Since that time, similar observations have been made in other streptococcal species, including *S. faecalis* (61), *S. pyogenes* (98), *S. agalactiae* (25, 135), and *S. sanguis* (70).

Besides the streptococcal isolates, there have been reports of transferable resistance determinants with no detectable plasmid DNA in *Staphylococcus. aureus* (57), *Bacteroides* (103, 141), and *Clostridium* (104). A tetracycline resistance determinant in *Clostridium difficile* is homologous to a similar element in *Strep. faecalis* (69) and is probably identical. Although the *Staphylococcus* and *Bacteroides* elements share many similarities with the streptococcal conjugative transposons, currently they are called "nonplasmid conjugative systems" until the full extent of their similarity is ascertained.

Our current understanding of conjugative transposons was reviewed recently (39). Conjugative transposons apparently lack replication genes and therefore must depend on the chromosome or a plasmid for maintenance in a cell. Besides the antibiotic resistance, these elements encode both a conjugative function and the ability to transpose, and neither of these functions requires an intact recombination (Rec) system in either the donor or recipient cell. Recently the beginning of a genetic analysis was reported (170) with the isolation of Tn5 insertions into the conjugative transposon Tn916. The study showed that one end of Tn916 is required for conjugation.

Basically, however, the mechanisms of transposition and transfer are not understood. The current hypothesis contends that the element is spontaneously excised and then other functions, including conjugation, are expressed. The excision event is thought to be under negative control with excision functions repressed by a transposon-encoded product(s). The frequency of excision, however, does depend on the insertion site, i.e., excision is influenced by the flanking chromosome region (65). The form of the excised element (i.e., circular, linear, single-stranded, or double-stranded) is unknown, but the current model is of a double-stranded circular intermediate that is transferred in a plasmidlike process and incorporated into a new host genome.

The various conjugative transposons presently described range in size from 15 to 70 kb and encode an array of resistances, both singly and in combination. These include tetracycline, chloramphenicol, macrolide plus lincosamide plus streptogramin (MLS), kanamycin, cefoxitin, clindamycin, and erythromycin, with tetracycline resistance the most commonly described.

The best characterized element is the 16-kb tetracycline resistance transposon Tn916, which is capable of transferring at frequencies of 10^{-5} to 10^{-8} per donor and was first seen in S. faecalis (61). The resistance gene of Tn916 belongs to the TetM class of tetracycline resistance determinants (26, 27). This TetM class has now spread to many different organisms, including Neisseria gonorrhoeae (114), Gardnerella vaginalis (123), Mycoplasma hominis (124), Ureaplasma urealyticum (121), and Clostridium difficile (69). Comparison of the DNA sequences of the TetM determinants isolated from a pneumococcal strain (106) and from Ureaplasma (126) indicates 95 percent homology in the amino acids encoded, even though the strains were isolated far apart both in time (7 years) and geographically (Paris and Seattle).

Direct evidence that conjugation was the mechanism of TetM acquisition by these strains is lacking, although Roberts and Kenny (122) did demonstrate the conjugative transfer of Tn916 from S. faecalis to three strains of M. hominis in the laboratory. Also, several other conjugative transposons have been transferred experimentally among streptococcal species, as well as into and out of Lactobacillus, Bacillus, and various Staphylococcus species (36, 38, 56, 75). Such a broad host range is probably the major reason for the spread of multiple antibiotic resistance among these gram-positive organisms. It seems likely that other genera will acquire resistance determinants from conjugative transposons, and this will become a very significant mechanism of DNA transfer. Whether it is the same as plasmid-encoded conjugation systems remains to be determined.

General Considerations

The discovery of an unusual phenotype in a well-characterized strain is usually the first indication that a plasmid-determined trait might be involved. Attempts are usually made to correlate the unusual characteristic with the presence of a particular plasmid. Initial methodology may include plasmid curing, where loss of the plasmid leads to loss of the phenotype, and plasmid transfer, where the phenotypic characteristic is acquired by a recipient strain either by conjugation, transduction, or transformation. Many naturally occurring strains contain several plasmids, so care must be taken when attempting to correlate a particular plasmid with a given trait.

Quantitative conjugation experiments are usually done in one of two ways, either on a solid surface or in liquid. Among pili-producing conjugative plasmids, the optimum mating method is mostly determined by the type of pilus produced, either flexible or rigid. A rigid pilus, for example, is more easily broken, and bacteria with this type of pilus exhibit better conjugation frequencies when immobilized on a solid surface. Thus, Bradley (16) divides plasmids into three classes based on their optimum mating condition: surface obligatory, in which surface transfer is more than 2000 times better than in liquid; surface preferred, in which surface transfer is 45 to 450 times greater than in liquid; and universal, with equal transfer in both methods. The surface matings can be done either on agar plates or on filters as described below.

Table 10.2, expanded from Bradley (16), lists the pilus morphology, the pilus-specific phages, and preferred mating system of a representative plasmid from all presently described Inc groups. The IncP complex is almost exclusively *Pseudomonas* plasmids and contains several groups (IncP-1 to IncP-13) whose pili have been described by Bradley (17). He also points out that many *Pseudomonas* strains produce a filamentous metabolic material on the cell surface (not plasmid related) that seems to allow cells to stick together and enhance mating frequency, thus lowering their plate-broth ratios. Transfer of these plasmids from *E. coli* cells produced very different plate-broth ratios (D. E. Bradley, personal communication), since the surface stickiness did not occur in *E. coli*. Most IncP plasmids, however, do transfer significantly better in a surface system in either organism.

Note that cells with $IncI_1$ or $IncI_2$ plasmids were previously thought to utilize only thin, flexible pili for conjugation (16). Recent efforts (18) showed that these cells elaborate both thin, flexible and thick, rigid pili, and both types are plasmid-encoded. The thick, rigid pili are the conjugative organelles and mediate a typical surface obligatory system; however, the thin, flexible pili apparently stabilize mating aggregates such that transfer can occur also in liquid. Thus, the system is described as universal even though the actual conjugative organelle is surface obligatory.

When examining the conjugation properties of a newly identified plasmid, the studies should as far as possible be carried out in a host with well-defined characteristics and which lacks any other plasmids. Walter et al. (156) developed a combined mating technique that utilizes both surface and liquid methods to assess transferability in genetically engineered organisms. Application of such combined methodology to the investigation of newly recognized plasmids could facilitate their analysis.

Gram-negative organisms

In general, a conjugation experiment is done by mixing broth cultures of donor and recipient strains (see next section for surface or liquid conditions) and selecting recipient strains that now contain the

TABLE 10.2 Optimum Conjugation Method, Pilus Types, and Pilus-Specified Bacteriophages[a]

| | | | | | Bacteriophages | | |
| | | | | | | Filamentous | |
Mating system	Pilus morphology	Inc group	Representative plasmid	Ratio of transfer frequencies (plate/broth)	Isometric RNA	single-stranded DNA	Double-stranded DNA
Universal	Thin, flexible+rigid	I₁ complex[b,c]	R144	1	Iα	If1	—
		I₂[c]	R721	14	—	Ike	PR4
	Thick, flexible	F complex[d]	R100	0.7	f2	f1	—
		HI1	R27	5.5	pilHα	—	—
		HI2	R478	0.3	pilHα	—	—
		J	R391	0.9	—	—	J
		S	F₀lac	0.8	F₀lac	fd	—
		T	pIN25[e]	7.0	t	tf-1	—
		V	R753	0.4	—	—	—
	Rigid	HI3	MIP233[f]	0.4	—	—	—
Surface preferred	Thick, flexible	C	RA1	45	C-1	C-2	J
		D	R711G	180	D[g]	fd	J
		HII	pHH1457	66	pilHα	—	—
		X	R6K	250	—	X,X-2	—
Surface obligatory	Rigid	M	R831G	6×10^3	M	Ike,X	PR4
		N	N3	1×10^4	—	—	PR4
		P	RP1	2×10^3	PRR1	Pf3	PR4
		U	pAr-32	3×10^5	—	X	—
		W	Sa	4×10^4	—	X	PR4

[a]Data from references 16–23, 40–45, and Bradley (personal communication).

[b]Includes Inc groups I₁, I₁ + B, I₅, B, K, and Z. R144 is IncI₁ + B since it determines both I₁ and B phenotypes.

[c]All I₁ and I₂ plasmids determine both thin flexible and rigid pili, but both pilus types of Inc I₂ plasmids are different from any of the I₁ complex. Phages Iα and If1 adsorb to thin I₁ pili, whereas phages Ike and PR4 use only thick I₂ pili as receptors.

[d]Includes Inc groups F₁, F_II, F_III, and F_IV. R100 is Inc F_II.

[e]pIN25 has been chosen as the representative plasmid since the prototype, Rts1, is atypical in that it transfers at relatively low frequencies and determines few T pili.

[f]Although rigid in appearance. MIP233 pili are very short and might not be distinguishable from thick flexible pili. This would account for their ability to mediate conjugation in liquid.

[g]RNA phage D was unstable and has been lost.

plasmid. These plasmid-containing recipients are called transconjugants. Conditions must be such that the donor cells in the mixture cannot grow on the plating medium. This is often accomplished by using a recipient strain with a chromosomally determined antibiotic resistance to which the donor is sensitive. The most frequently used are streptomycin, nalidixic acid, or rifampicin resistances. Spontaneous mutations conferring these resistances can be obtained by plating on a medium containing the antibiotic. Specific methodology is given by Miller (110). In mating experiments, inclusion of the antibiotic in the plate medium allows growth of the transconjugants and prevents growth of the plasmid-containing donor strain. If suitably marked strains are available, use of donor and recipient strains with different auxotrophic markers allows selection on the appropriate minimal medium. For example, a mating mixture of a *trp his* Str^S donor carrying $F'lac^+$ and a *lac pro* Str^R recipient would be plated on a minimal salts medium containing proline, streptomycin, and lactose as a carbon source. This will select the $(F'lac^+)$ *pro* Str^R transconjugants.

Some precautions should be taken with various antibiotics. For example, ampicillin is unstable, and plates containing it should be used within 1 to 2 weeks. Magnesium ions are antagonists of tetracycline, and trimethoprim and the sulfonamides are inactivated in some nutrient media. Recipes for commonly used antibiotics are included in Maniatis et al. (105).

Quantitative matings are usually done with exponentially growing cells. If the organism is not well characterized, a few preliminary growth curves should determine the conditions for achieving this stage. For most enterobacteria, separately dilute overnight broth cultures of donor and recipient cells 1:20 into fresh broth and grow aerobically at 37°C to 2×10^8/mL. Growth is followed easily by using side-arm flasks and a Klett-Summerson colorimeter (660-nm red filter) or, alternatively, by monitoring the absorption at 620 nm. Once grown, the cells can be maintained on ice for about 1 h prior to mating.

Following mating (described in next section), plate 0.1-mL samples of appropriate dilutions of the mating mixture on selective medium, either with a sterile glass rod spreader or by adding the 0.1-mL sample to a tube containing 2.5 mL of molten (45°C) 0.7% agar, and pouring onto the plate. Dilutions are usually made in a phosphate-buffered saline containing 0.01% gelatin (8.5 g NaCl, 0.3 g anhydrous KH_2PO_4, 0.6 g anhydrous NA_2HPO_4, 0.1 g gelatin per liter of distilled water). Preliminary experiments with a new plasmid should establish the appropriate dilutions for plating.

If plating the diluted mixture on minimal medium, add 0.1 mL broth to each plated dilution to prevent metabolic shift-down conditions. When the donor and recipient cells have been grown in nutrient-rich medium and then are plated on the selective minimal medium, the result is a metabolic shift-down in growth conditions. Failure to account for these shift-down conditions can result in detecting as much as fivefold fewer transconjugants than are actually formed. This problem is prevented by the addition of broth to the plated mixture. Most laboratories find it easier to keep two sterile preparations of 0.7% agar, one with and one without the appropriate amount of broth.

Plate the donor culture alone on nutrient agar (usually a 10^{-5} dilution) to determine the viable count; use this number to calculate the efficiency of transfer as the number of transconjugants per donor cell. Confirm the continued presence of the plasmid in the donor by patching 25 to 50 colonies from the viable count plate onto medium selective for a plasmid trait. Some investigators determine the viable count of the donor strain by plating the mating mixture on a medium that is selective for the donor strain. This is usually done only when the mating time is exceptionally long and gives a more accurate count of the number of donor cells.

Any conjugation experiment must include the appropriate controls. Two sham matings should be done, one with the donor cells plus broth and the other with the recipient cells plus broth (with the cells in the same proportions as in the mating mixture). These sham matings are plated undiluted on the same selective medium used for the mating mixture. Neither the donor strain alone nor the recipient strain alone should grow on this medium that is selective for transconjugants.

Gram-positive organisms

Conjugative plasmids occur in several gram-positive organisms, including species of *Streptococcus, Staphylococcus, Bacillus,* and *Clostridium* (see Table 10.1 for references). The conjugative mechanism(s), however, are not clearly defined. The best described systems are among the streptococci, in which plasmid-determined characteristics have been described in most species found in animals or humans (reviewed in 37, 96, 97), and many of the plasmids are conjugative. The mechanism of transfer is not understood, but a pilus-type organelle has never been detected, and with a few exceptions, liquid matings are very unsuccessful. Quantitative analysis of conjugation is usually performed on membrane filters (74, 81) as described below.

In *S. faecalis*, however, two general categories of conjugative plasmids have been distinguished. One group contains plasmids identical to the other streptococcal plasmids mentioned above that transfer efficiently (10^{-2} to 10^{-4} transconjugants per donor) in filter matings and very poorly in broth (10^{-6} per donor). The other group transfers very efficiently in broth matings (10^{-3} to 10^{-1} per donor), and extensive analysis over the past 10 years shows that this broth transfer is regulated by sex pheromones.

Briefly, potential recipient strains of *S. faecalis* excrete low-molecular-weight peptide pheromones that induce the plasmid-containing donor cells to synthesize a surface adhesion that coats the cell surface and facilitates the formation of mating aggregates. These pheromone-induced aggregates appear as clumps in broth cultures, so the pheromone is referred to as a clumping-inducing agent (CIA). After aggregation, the plasmid DNA is transferred from donor to recipient by an as yet unknown mechanism, although it has been shown that CIA induces some functions directly related to the DNA aspects of transfer. Ike and Clewell (82) and Ehrenfeld and Clewell (56) have begun a genetic analysis of the transfer aspects of this system.

Mobilization

If testing a plasmid for mobilization, the simplest way is to use one of the methods described below and transfer several different conjugative plasmids into the strain carrying the nonconjugative plasmid. Then, further matings will determine whether each heterozygous strain can donate the characteristic encoded by the nonconjugative plasmid. If available, use a RecA⁻ host to prevent possible recombination between the two plasmids. If incompatibility prevents the construction of stable heterozygous strains, the experiment can be carried out using transient heterozygous cells as described in Transiently Heterozygous Donor Strains in the next section.

Biological-environmental parameters

Since most of the available information on conjugation concerns plasmids found in gram-negative organisms, especially F-like plasmids, the following comments relate to them, although similar problems may be encountered with plasmids from other organisms.

Incompatibility. Plasmid incompatibility is defined as the failure of two coresident plasmids to be stably maintained in the absence of external selection (116). Thus, if the introduction of a second plasmid destabilizes the inheritance of the first, the two plasmids are termed incompatible. In a recent review of plasmid incompatibility, Novick (115) concludes that incompatibility is due to the sharing of one or more elements of the plasmid replication or partitioning systems, and the consequent plasmid loss is due to interference with the ability of the plasmid to correct stochastic fluctuations in its copy number.

Plasmids in the same incompatibility (Inc) group usually have similar conjugation systems in that they encode pili that are morphologically, serologically, and genetically related (15). Incompatibility grouping is, therefore, a convenient means of classifying plasmids. Bradley et al. (22) suggest that examination of pilus morphology, along with characterization of the preferred mating method (broth or surface) and pilus-specific phage sensitivities, could limit the number of incompatibility groups to which an unknown plasmid may belong and thus provide a quick means of partial identification.

Incompatibility is a problem in conjugation studies when trying to determine whether two conjugation-deficient mutations are complementable. In this case, the analysis can be done using a transient heterozygous intermediate (see Transiently Heterozygous Donor Strains in the next section). This method is the basis of the original complementation data that defined the F *tra* genes (1, 83).

Surface exclusion.   As the term implies, this is a "surface"-related phenomenon in that mating between donor cells is inhibited, and no DNA is transferred. With F plasmid-containing cells, surface exclusion produces approximately a 400-fold decrease in recipient ability. The membrane proteins involved are the products of *traS* and *traT*, with the *traS* protein proposed to prevent the triggering of conjugal DNA synthesis in the donor (90). The *traT* protein is proposed to competitively bind to the pilus tip and thus interfere with the binding of the pilus to its normal receptor (111). The *traT* protein also has a role in the resistance of the cell to the complement-mediated bactericidal activity of serum (112).

The surface exclusion barrier may be bypassed, if necessary, by growing the plasmid-containing recipient culture to a late stationary phase. These cells, called F^- phenocopies, show a recipient ability almost equivalent to F^- cells (161).

Repression.   Most conjugative plasmids isolated from bacteria in nature, with the exception of those determining rigid pili, have repressed fertility functions, yielding conjugation frequencies of $< 10^{-3}$ per donor even under optimum conditions (see above discussion of fertility inhibition with plasmid F). This observation is often cited as a reason that plasmids cannot be maintained by infectious spread alone, and thus, for persistence in a population, there must be positive selection for a plasmid-determined characteristic. Lundquist and Levin (102) analyzed the population dynamics of both laboratory and wild plasmids in *E. coli* in a chemostat. The frequency of two of the wild plasmids increased rapidly even though they were repressed for pilus synthesis. These authors suggest that even though repressed plasmids have a survival advantage (i.e., resistance to male-specific bacteriophages), the retransfer by transitorily derepressed cells allows for the transmission of plasmids to naïve populations. Therefore, repression does not prevent the maintenance of these plasmids in a population by infectious transfer alone.

Willetts (158) showed that strains with a newly acquired repressible plasmid such as R100 synthesize a functioning conjugation system within 30 to 60 min and then transfer the plasmid at a derepressed level. The Fin products are also synthesized, but it requires about 5 to 6 h to dilute (by cell growth) the *tra* products and establish repression.

Plasmids with repressed transfer functions should be examined for transient derepression by mating for several hours with periodic dilutions to maintain optimum cell densities. This can be done with filter matings by resuspending mating cells from the filter, diluting the mixture, and collecting on a new filter for one more cycle of growth. For plasmids that transfer in liquid, dilute 10^4 overnight cells of the donor culture into a 1.0-mL broth along with 4×10^5 cells of a suitable recipient and incubate overnight. Dilute 1:20 with fresh broth, grow the mixture to exponential phase, and plate for transconjugants (160).

A method for isolating derepressed mutants following ethyl methane sulfonate treatment has been published (60). An alternative method (18) for plasmids that synthesize their own DNA primase utilizes a host strain carrying a temperature-sensitive DNA primase (*dnaG*) mutation. At the restrictive temperature these cells lack the chromosomal DNA primase that is required for DNA replication. The plasmid-encoded primase could substitute for the chromosomal enzyme, but this *tra*-encoded protein is lacking in a transfer-repressed cell. Spontaneously arising mutations in the repression system of the plasmid will allow synthesis of the plasmid primase, which can be utilized by the cell for growth at 42°C. Thus, growth at high temperature selects spontaneous transfer-derepressed mutants. Note that not all plasmids encode a DNA primase (see 157, 164).

Restriction.   Restriction is the mechanism by which a bacterial cell recognizes and degrades incoming foreign DNA. The basis for this recognition system is chemical modification of particular DNA bases by a cellularly encoded enzyme (14, 67). For example, in *E. coli* K-12, certain cytosine residues are methylated ("modification"), and any incoming DNA lacking these modified bases is degraded ("restricted"). Therefore, the possibility of a restriction barrier must be considered when attempting conjugation between species or genera. Restrictionless mutants are available in some well-characterized species (i.e., *E. coli* and *S. typhimurium*), but it may not be possible to examine a new plasmid in one of these species. If a restriction barrier is suspected, Curtiss (47) suggests that heating the recipient strain prior to mating for 5 to 10 min at 50°C or 15 to 20 min at 45°C will frequently eliminate the problem.

Temperature.   While most plasmids transfer optimally at 37°C, there are some exceptions. Most IncT plasmids have temperature-sensitive transfer systems (148), and conjugation experiments have been conducted at 30°C. Bradley and Whelan (23) showed that the type of pilus produced by IncT plasmids is determined by the growth temperature; at 30°C, long conjugative pili are made, and transfer occurs readily with both liquid or surface protocols. At 37°C, however, only short pili are synthesized, and transfer occurs only with the surface method.

In analyzing transfer of mercury resistance from marine pseudomonads, Gauthier et al. (64), found that temperature (30°C) and salinity (37 percent) were critical for optimum conjugation frequencies. In addition, some of the IncH plasmids encode temperature-sensitive conjugation systems (19, 145) with an optimum temperature of 26 to 30°C. Thus, before concluding that a new plasmid system is nonconjugative, mating experiments should be done at 25 to 30°C as well as at the standard 37°C.

Lethal zygosis.   Mixing *E. coli* F$^-$ cells with an excess (20:1) of Hfr donor cells leads to a decrease in the number of viable F$^-$ cells with an associated loss of recombinant transconjugants. Skurray and Reeves (133) showed that this phenomenon, termed lethal zygosis, is associated with metabolic changes in the recipient cell that could result from membrane alterations accompanying conjugation. F$^+$ and Hfr cells are insensitive to lethal zygosis, and the F gene encoding this phenotype (called Ilz for *i*mmunity to *l*ethal *z*ygosis) is *traT*. A second *ilz* gene mapping away from the *tra* region was proposed because *traJ*$^-$ strains are still phenotypically Ilz$^+$. Since small amounts of *traT* protein are still made from the *traT* promoter (34), there is probably no second *ilz* gene (119), and the mechanism of resistance is related to the presence of the *traT* protein in the cell membrane.

Specific Procedures

Liquid matings

Gram-negative organisms.   Mix 0.2 mL of the donor cells with 1.8 mL of the recipient cells (both grown as described in Gram-Negative Organisms in the preceding section) in a sterile 18 × 150 mm test tube. The large tube ensures adequate aeration of the mating mixture. Incubate the tube statically at the optimum temperature (usually 37°C, but see Temperature in the preceding section) for 45 min. The optimum mating time may vary and should be determined in preliminary experiments. To obtain quantitative data, however, the time element is critical as a longer mating period may measure retransfer of the plasmid from the recipients.

Streptococcal CIA methods

1. For the broth mating method (taken from 38), mix 0.05 mL of an overnight donor culture with 0.5 mL of the overnight recipient culture in 4.5 mL of broth. Incubate at 37°C with gentle shaking for 4 h, vortex, and plate on selective media. This incubation allows time for expression of CIA by the recipient cells and the aggregation response by the donor cells. Plate aliquots also on media selective for the donor strain to obtain a viable count for determining the transfer frequency per donor. Incubate plates for 48 h to detect transconjugants.

2. An alternative CIA mating method (55) utilizes preinduction of the donor cells with CIA and a shorter mating period. To prepare CIA, grow the CIA-producing recipient strain to the late exponential phase, centrifuge, filter the supernatant (0.22-μm pore size filter), and autoclave. For the induction, mix 0.9 mL of the CIA preparation with 0.9 mL broth and add 0.2 mL of a fresh late exponential culture of the donor cells. Make a similar preparation of recipient cells except use 1.8 mL of broth alone instead of the CIA-broth mixture. Incubate both cultures aerobically for 1 h at 37°C. For the mating, mix 0.2 mL of the donor cells with 1.8 mL of the recipient cells, incubate aerobically at 37°C for 10 min, votex, and plate dilutions as above.

Surface matings

Mix donor and recipient cells as described for liquid matings and then collect on a sterile membrane filter (pore size, 0.45 μm). A Swinnex-13 Millipore disk filter holder attached to a plastic syringe is a convenient means for filtering. With sterile forceps, carefully place the filter, cell side up, on a prewarmed nutrient (nonselective) plate and incubate the plate for 1 to 4 h (or longer as required). Suspend the cells in 2-mL broth with vigorous vortex mixing, dilute, and plate on selective medium as described above.

Plate matings

An alternative surface mating method, on agar plates without filtering the mixture, is somewhat less efficient than the filter mating method, which tends to force cells into contact. Spread 0.1-mL aliquots of donor cells (diluted appropriately) and recipient cells (undiluted) together on a plate selective for transconjugants and incubate as above. A variation of this technique is suitable for quantitative analysis (22). Nalidixic acid cannot be used as a counterselection in plate matings since it inhibits conjugation (10).

These are the standard methods used to quantitate conjugation in most organisms, although variations have been reported in which investigators attempted to mimic presumptive in vivo mating conditions. Bale et al. (7), for example, examined transfer of Mer^R by a *P. aeruginosa* plasmid by incorporating river stones as a base for the filter containing mating strains. They recently reported a method for studying in situ mating in river epilithon (6, 8).

Qualitative matings

When quantitative data are not necessary, one of several options can be used.

1. In strain construction, for example, a known conjugative plasmid can be transferred by mixing donor and recipient cells (either grown in broth overnight or suspended in broth from overnight plates) in approximately 1:1 ratio, incubating for 1 to 2 h and then streaking small aliquots for single colonies on selective media.

2. A replica-plate mating technique is useful for screening a large number of colonies for conjugation proficiency. Patch colonies with sterile flat-sided toothpicks in a grid pattern onto a nutrient plate and incubate for 6 h. Typical patch patterns for 25, 36, or 50 colonies are shown in Davis et al. (50). Concentrate an overnight culture of the recipient 10-fold and spread 0.1 mL on a selective plate. Transfer the patched colonies onto the recipient-spread plate using a sterile velveteen pad stretched over a block engineered to the diameter of the Petri dish. Transfer-proficient patches will appear following one (on nutrient medium) or two (on minimal medium) days' incubation.

3. To screen a few strains on one plate, streak a loopful of a 10-fold concentrated recipient strain down the center of the selective plate and allow to dry. Streak a loopful of each presumptive donor at right angles across the recipient. Following incubation, transconjugants appear in the streaked area after passing through the recipient cells.

Transiently heterozygous donor strains

This method provides a less quantitative means of testing mobilization (see Mobilization in preceding section) than if using stable heterozygous donors but is useful for screening a large number of strains. In this triparental cross, mate the strain containing a conjugative plasmid with the strain carrying the nonconjugative plasmid that is being tested for mobilization. Mate this mixture with the final recipient strain and plate on medium that is counterselective for both the donor and intermediate strains and selective for the final recipient and a marker carried by the nonconjugative plasmid.

Spot matings.   Spread 0.1 mL of an overnight culture of the final recipient onto the selective medium. Mix 0.05-mL aliquots of overnight cultures of each donor and intermediate strain on the surface of the recipient-spread plate. Incubate for 1 to 2 days at the appropriate temperature. This method allows several matings on one plate.

Filter matings.   Mix 0.1-mL aliquots of overnight cultures of the donor and intermediate strains. Add 0.8 mL of an overnight culture of the final recipient strain, filter the mixture, incubate the filter, and plate as described in Surface Matings in this section.

Summary and Conclusions

These basic methods for analyzing plasmid conjugation in the laboratory are suitable for many bacterial genera and can be adapted to the requirements of a particular organism (for example, salinity of the medium for marine bacteria). In general, the requirements for demonstrating conjugation include (1) a basic knowledge of the organism's optimum growth conditions; (2) donor and recipient strains with suitable genetic markers for selecting transconjugants; and (3) physical demonstration of a plasmid molecule. In addition, if transfer of genetic information occurs, it should be demonstrated that transformation or transduction are not involved. With careful attention to the possible biological or environmental variables mentioned above, the laboratory demonstration of conjugation should be accomplished readily.

Acknowledgments

I am grateful to D. Bradley, D. Clewell, W. Dempsey, K. Ippen-Ihler, M. Malamy, and E. G. Minkley for helpful discussions. I thank

D. Bradley for providing the information in Table 10.2 and D. Bradley and W. Dempsey for constructive comments on the manuscript. In addition, K. Ippen-Ihler, K. McArthur, and K. Uyeda provided timely computer assistance. I thank M. M. M. and C. D. M. for continuing encouragement and P. Kerby and Stella Fry for preparing the manuscript. Through 1989, work in my laboratory was funded by the Research Service of the U.S. Department of Veterans Affairs. Current support is provided by an Organized Research Award from Texas Woman's University.

References

1. Achtman, M., N. Willetts, and A. J. Clark. 1972. Conjugational complementation analysis of transfer-deficient mutants of F*lac* in *Escherichia coli*. *J. Bacteriol.* 110: 831–842.
2. Andreoni, V., and G. Bestetti. 1986. Comparative analysis of different *Pseudomonas* strains that degrade cinnamic acid. *Appl. Environ. Microbiol.* 52: 930–934.
3. Aoki, T., Yasutami, M., and J. H. Crosa. 1986. The characterization of a conjugative R-plasmid isolated from *Aeromonas salmonicida*. *Plasmid* 16:213–218.
4. Archer, G. L., J. P. Coughter, and J. L. Johnston. 1986. Plasmid-encoded trimethoprim resistance in staphylococci. *Antimicrob. Agents Chemother.* 29:733–740.
5. Archer, G. L., and J. L. Johnston. 1983. Self-transmissible plasmids in staphylococci that encode resistance to aminoglycosides. *Antimicrob. Agents Chemother.* 24:70–77.
6. Bale, M., M. Day, and J. Fry. 1988. Novel method for studying plasmid transfer in undisturbed river epilithon. *Appl. Environ. Microbiol.* 54:2756–2758.
7. Bale, M. J., J. C. Fry, and M. J. Day. 1987. Plasmid transfer between strains of *Pseudomonas aeruginosa* on membrane filters attached to river stones. *J. Gen. Microbiol.* 133:3099–3107.
8. Bale, M., J. Fry, and M. Day. 1988. Transfer and occurrence of large mercury resistance plasmids in river epilithon. *Appl. Environ. Microbiol.* 54:972–978.
9. Banfalvi, Z., E. Kondorosi, and A. Kondorosi. 1985. *Rhizobium meliloti* carries two megaplasmids. *Plasmid* 13:129–138.
10. Barbour, S. D. 1967. Effect of nalidixic acid on conjugational transfer and expression of episomal Lac genes in *Escherichia coli* K-12. *J. Mol. Biol.* 28:373–376.
11. Barth, P., N. J. Grinter, and D. Bradley. 1978. Conjugal transfer system of plasmid RP4: Analysis by transposon Tn7 insertions. *J. Bacteriol.* 133:43–52.
12. Bartowsky, E. J., G. Morelli, M. Kamke, and P. A. Manning. 1987. Characterization and restriction analysis of the P sex factor and the cryptic plasmid of *Vibrio cholerae* strain V58. *Plasmid* 18:1–7.
13. Battisti, L., B. D. Green, and C. B. Thorne. 1985. Mating system for transfer of plasmids among *Bacillus anthracis, Bacillus cereus,* and *Bacillus thuringiensis. J. Bacteriol.* 162:543–550.
14. Boyer, H. W., L. T. Chow, A. Dugaiczyk, J. Hedgepath, and H. M. Goodman. 1973. DNA substrate site for the *Eco*RII restriction endonuclease and modification methylase. *Nature (London)* 244:40–43.
15. Bradley, D. E. 1980. Morphological and serological relationships of conjugative pili. *Plasmid* 4:155–169.
16. Bradley, D. E. 1981. Conjugative pili of plasmids in *Escherichia coli* K-12 and *Pseudomonas* species. In S. B. Levy, R. C. Clowes, and E. L. Koenig (eds.), *Molecular Biology, Pathogenicity, and Ecology of Bacterial Plasmids.* Plenum Press, New York, pp. 217–226.

17. Bradley, D. E. 1983. Specification of the conjugative pili and surface mating systems of *Pseudomonas* plasmids. *J. Gen. Microbiol.* 129:2545–2556.

18. Bradley, D. E. 1984. Characteristics and function of thick and thin conjugative pili determined by transfer-derepressed plasmids of incompatibility group I_1, I_2, I_5, B, K, and Z. *J. Gen. Microbiol.* 130:1489–1502.

19. Bradley, D. E. 1986. The unique conjugation system of IncHI3 plasmid MIP233. *Plasmid* 16:63–71.

20. Bradley, D. E., T. Aoki, T. Kitao, T. Arai, and H. Tschape. 1982. Specification of characteristics for the classification of plasmids in incompatibility group U. *Plasmid* 8:89–93.

21. Bradley, D. E., F. A. Sirgel, J. N. Coetzee, R. W. Hedges, and W. F. Coetzee. 1982. Phages C-2 and J: IncC and IncJ plasmid-dependent phages, respectively. *J. Gen. Microbiol.* 128:2485–2498,

22. Bradley, D. E., D. E. Taylor, and D. R. Cohen. 1980. Specification of surface mating systems among conjugative drug resistance plasmids in *Escherichia coli* K-12. *J. Bacteriol.* 143:1466–1470.

23. Bradley, D. E., and J. Whelan. 1985. Conjugation systems of IncT plasmids. *J. Gen. Microbiol.* 131:2665–2671.

24. Brefort, G., M. Magot, H. Ionesco, and M. Sebald. 1977. Characterization and transferability of *Clostridium perfringens* plasmids. *Plasmid* 1:52–66.

25. Burdett, V. 1980. Identification of tetracycline-resistant R-plasmids in *Streptococcus agalactiae* (group B). *Antimicrob. Agents Chemother.* 18:753–760.

26. Burdett, V. 1986. Streptococcal tetracycline resistance mediated at the level of protein synthesis. *J. Bacteriol.* 165:564–569.

27. Burdett, V., J. Inamine, and S. Rajagopalan. 1982. Heterogeneity of tetracycline resistance determinants in *Streptococcus. J. Bacteriol.* 149:995–1004.

28. Buu-Hoi, A., and T. Horodniceanu. 1980. Conjugative transfer of multiple antibiotic resistance markers in *Streptococcus pneumoniae. J. Bacteriol.* 143:313–320.

29. Cavalli, L. L., J. Lederberg, and E. M. Lederberg. 1953. An infective factor controlling sex compatibility in *Bacterium coli. J. Gen. Microbiol.* 8:89–103.

30. Chapman, J. S., and B. C. Carlton. 1985. Conjugal plasmid transfer in *Bacillus thuringiensis.* In D. R. Helinski, S. N. Cohen, D. B. Clewell, D. A. Jackson, and A. Hollaender (eds.), *Plasmids in Bacteria.* Plenum Press, New York, pp. 453–467.

31. Chatterjee, A. K., and M. P. Starr. 1973. Gene transmission among strains of *Erwinia amylovora. J. Bacteriol.* 116:1100–1106.

32. Chatterjee, A. K., and M. P. Starr. 1977. Donor strains of the soft-rot bacterium *Erwinia chrysanthemi* and conjugational transfer of the pectolytic capacity. *J. Bacteriol.* 132:862–869.

33. Cheah, K.-C., A. Ray, and R. Skurray. 1984. Cloning and molecular analysis of the *finO* region from the antibiotic-resistance plasmid R6-5. *Plasmid* 12:222–226.

34. Cheah, K.-C., A. Ray, and R. Skurray. 1986. Expression of F plasmid *traT:* Independence of *traY-Z* promoter and *traJ* control. *Plasmid* 16:101–107.

35. Cheah, K.-C., and R. Skurray. 1986. The F plasmid carries an IS3 insertion within *finO. J. Gen. Microbiol.* 132:3269–3275.

36. Christie, P. J., R. Z. Korman, S. A. Zahler, J. C. Adsit, and G. M. Dunny. 1987. Two conjugation systems associated with *Streptococcus faecalis* plasmid pCF10: Identification of a conjugative transposon that transfers between *S. faecalis* and *Bacillus subtilis. J. Bacteriol.* 169:2529–2536.

37. Clewell, D. B. 1981. Plasmids, drug resistance, and gene transfer in the genus *Streptococcus. Microbiol. Rev.* 45:409–436.

38. Clewell, D. B., F. Y. An, B. A. White, and C. Gawron-Burke. 1985. *Streptococcus faecalis* sex pheromone (cAM373) also produced by *Staphylococcus aureus* and identification of a conjugative transposon (Tn*918*). *J. Bacteriol.* 162:1212–1220.

39. Clewell, D. B., and C. Gawron-Burke. 1986. Conjugative transposons and the dissemination of antibiotic resistance in streptococci. *Ann. Rev. Microbiol.* 40:635–659.

40. Coetzee, J. N., D. E. Bradley, J. Fleming, L. duToit, V. M. Highes, and R. W. Hedges. 1985. Phage pilHa: A phage which adsorbs to IncHI and IncHII plasmid-coded pili. *J. Gen. Microbiol.* 131:1115–1121.

41. Coetzee, J. N., D. E. Bradley, and R. W. Hedges. 1982. Phages Ia and I2-2: IncI plasmid-dependent bacteriophages. *J. Gen. Microbiol.* 128:2797–2804.

42. Coetzee, J. N., D. E. Bradley, R. W. Hedges, J. Fleming, and G. Lecatsas. 1983. Bacteriophage M: An incompatibility group M plasmid-specific phage. *J. Gen. Microbiol.* 129:2271–2276.

43. Coetzee, J. N., D. E. Bradley, R. W. Hedges, V. M. Hughes, M. M. McConnel, L. duToit, and M. Tweehuysen. 1986. Bacteriophages F$_o$lac h, SR,SF: Phages which adsorb to pili encoded by plasmids of the S-complex. *J. Gen. Microbiol.* 132:2907–2917.

44. Coetzee, J. N., D. E. Bradley, R. W. Hedges, M. Tweehuizen, and L. duToit. 1987. Phage tf-1: A filamentous bacteriophage specific for bacteria harbouring the IncT plasmid pIN25. *J. Gen. Microbiol.* 133:953–960.

45. Coetzee, J. N., D. E. Bradley, G. Lecatsas, L. duToit, and R. W. Hedges. 1985. Bacteriophage D: An IncD group plasmid-specific phage. *J. Gen. Microbiol.* 131:3375–3383.

46. Coetzee, J. N., N. Datta, and R. W. Hedges. 1972. R factors from *Proteus rettgeri*. *J. Gen. Microbiol.* 72:543–552.

47. Curtiss, R., III. 1981. Gene transfer. In P. Gerhardt, R. G. E. Murray, R. N. Costilow, E. W. Nester, W. A. Wood, N. E. Krieg, and G. B. Phillips (eds.), *Manual of Methods for General Bacteriology.* Am. Soc. Microbiol., Washington, D.C., pp. 243–265.

48. Datta, N. 1975. Epidemiology and classification of plasmids. In Schlessinger, D. (ed.), *Microbiology—1974.* Am. Soc. Microbiol., Washington, D.C., pp. 9–15.

49. Datta, N. 1979. Plasmid classification: Incompatibility grouping. In K. N. Timmis and A. Puhler (eds.), *Plasmids of Medical, Environmental, and Commercial Importance.* Elsevier, Amsterdam, pp. 3–12.

50. Davis, R. W., D. Botstein, and J. R. Roth. 1980. *Advanced Bacterial Genetics: A Manual for Genetic Engineering.* Cold Spring Harbor Laboratory, Cold Spring Harbor, N.Y.

51. deGrandis, S. A., and R. M. Stevenson. 1985. Antimicrobial susceptibility patterns and R plasmid-mediated resistance of the fish pathogen *Yersinia ruckeri*. *Antimicrob. Agents Chemother.* 27:938–942.

52. Dempsey, W. B. 1987. Integration host factor and conjugative transfer of the antibiotic resistance plasmid R100. *J. Bacteriol.* 169:4391–4392.

53. Desomer, J., P. Dhaese, and M. Van Montagu. 1988. Conjugative transfer of cadmium resistance plasmids in *Rhodococcus fascians* strains. *J. Bacteriol.* 170:2401–2405.

54. Don, R. H., and J. M. Pemberton. 1981. Properties of six pesticide degradation plasmids isolated from *Alcaligenes paradoxus* and *Alcaligenes eutrophus*. *J. Bacteriol.* 145:681–686.

55. Dunny, G., R. Craig, R. Carron, and D. Clewell. 1979. Plasmid transfer in *Streptococcus faecalis*. Production of multiple sex pheromones by recipients. *Plasmid* 2:454–465.

56. Ehrenfield, E. E., and D. B. Clewell. 1987. Transfer functions of the *Streptococcus faecalis* plasmid pAD1: Organization of plasmid DNA encoding response to sex pheromone. *J. Bacteriol.* 169:3473–3481.

57. El Solh, N., J. Allignet, R. Bismuth, B. Buret, and J. Fouace. 1986. Conjugative transfer of staphylococcal antibiotic resistance markers in the absence of detectable plasmid DNA. *Antimicrob. Agents Chemother.* 30:161–169.

58. Everett, R., and N. Willetts. 1980. Characterisation of an *in vivo* system for nicking at the origin of conjugal DNA transfer of the sex factor F. *J. Mol. Biol.* 136:129–150.

59. Finlay, B. B., L. S. Frost, and W. Paranchych. 1986. Origin of transfer of IncF plasmids and nucleotide sequences of the type II *oriT*, *traM*, and *traY* alleles from ColB4-K98 and the type IV *traY* allele from R100-1. *J. Bacteriol.* 168:132–139.

60. Finnegan, D. J., and N. S. Willetts. 1971. Two classes of F*lac* mutants insensitive to transfer inhibition by an F-like R factor. *Mol. Gen. Genet.* 111:256–264.

61. Franke, A. E., and D. B. Clewell. 1981. Evidence for a chromosome-borne resistance transposon (Tn*916*) in *Streptococcus faecalis* that is capable of "conjugal" transfer in the absence of a conjugative plasmid. *J. Bacteriol.* 145:494–502.

62. Friedrich, B., C. Hogrefe, and H. G. Schlegel. 1981. Naturally occurring genetic transfer of hydrogen-oxidizing ability between strains of *Alcaligenes eutrophus. J. Bacteriol.* 147:198–205.

63. Friedrich, B., C. Kortluke, C. Hogrefe, G. Eberz, B. Silber, and J. Warrelmann. 1986. Genetics of hydrogenase from aerobic lithoautotrophic bacteria. *Biochimie* 68:133–145.

64. Gauthier, M. J., F. Cauvin, and J. P. Breittmayer. 1985. Influence of salts and temperature on the transfer of mercury resistance from a marine pseudomonas to *Escherichia coli. Appl. Environ. Microbiol.* 50:38–40.

65. Gawron-Burke, C., and D. B. Clewell. 1982. A transposon in *Streptococcus faecalis* with fertility properties. *Nature* 300:281–284.

66. Gerbaud, G., A. Dodin, F. Goldstein, and P. Courvalin. 1985. Genetic basis of trimethoprom and 0/129 resistance in *Vibrio cholerae. Ann. Inst. Pasteur* 136B:265–273.

67. Gold, M., and J. Hurwitz. 1964. The enzymic methylation of ribonucleic acid and deoxyribonucleic acid. V. Purification and properties of the deoxyribonucleic acid-methylating activity of *Escherichia coli. J. Biol. Chem.* 239:3858–3865.

68. Gonzalez, J. M. Jr., B. J. Brown, and B. C. Carlton. 1982. Transfer of *Bacillus thuringiensis* plasmids coding for d-endotoxin among strains of *B. thuringiensis* and *B. cereus. Proc. Natl. Acad. Sci. USA* 79:6951–6955.

69. Hachler, H., F. H. Kayser, and B. Berger-Bachi. 1987. Homology of a transferable tetracycline resistance determinant of *Clostridium difficile* with *Streptococcus (Enterococcus) faecalis* transposon Tn*916*. *Antimicrob. Agents Chemother.* 31:1033–1038.

70. Hartley, D. L., K. R. Jones, J. A. Tobian, D. J. LeBlanc, and F. L. Macrina. 1984. Disseminated tetracycline resistance in oral streptococci: Implication of a conjugative transposon. *Infect. Immun.* 45:13–17.

71. Hayes, W. 1952. Genetic recombination in *Bact. coli* K-12: Analysis of the stimulating effect of ultraviolet light. *Nature* 169:1017.

72. Hayes, W. 1953. Observations on a transmissible agent determining sexual differentiation in *Bacterium coli. J. Gen. Microbiol.* 8:72–88.

73. Hendrick, C., W. Haskins, and A. Vidaver. 1984. Conjugative plasmid in *Corynebacterium flaccumfaciens* subsp. *oortii* that confers resistance to arsenite, arsenate, and antimony (III). *Appl. Environ. Microbiol.* 48:56–60.

74. Hershfield, V. 1979. Plasmids mediating multiple drug resistance in group B *Streptococcus:* Transferability and molecular properties. *Plasmid* 2:137–149.

75. Hill, C., C. Daly, and G. F. Fitzgerald. 1985. Conjugative transfer of the transposon Tn*916* to lactic acid bacteria. *FEMS Microbiol. Lett.* 30:115–119.

76. Hinchcliffe, E., and A. Vivian. 1980. Naturally occurring plasmids in *Acinetobacter calcoaceticus:* A P class R factor of restricted host range. *J. Gen. Microb.* 116:75–80.

77. Holsters, M., B. Silva, F. VanVliet, C. Genetello, M. deBlock, P. Dhaese, A. Depicker, D. Inze, G. Engler, and R. Villarroel. 1980. The functional organization of the nopaline *A. tumefaciens* plasmid pTiC58. *Plasmid* 3:212–230.

78. Hooykaas, P. J. J., C. Roobol, and R. A. Schilperoort. 1979. Regulation of the transfer of Ti plasmids of *Agrobacterium tumefaciens. J. Gen. Microbiol.* 110:99–109.

79. Hopwood, D. A., K. F. Chater, J. E. Dowding, and A. Vivian. 1973. Advances in *Streptomyces coelicolor* genetics. *Bacteriol. Rev.* 37:371–405.

80. Hopwood, D. A., D. J. Lydiate, F. Malpartida, and H. M. Wright. 1985. Conjugative sex plasmids of *Streptomyces*. In D. R. Helinski, S. N. Cohen, D. B. Clewell, D.

A. Jackson, and A. Hollaender (eds.), *Plasmids in Bacteria*. Plenum Press, New York, pp. 615–634.

81. Horaud, T., C. LeBouguenec, and K. Pepper. 1985. Molecular genetics of resistance to macrolides, lincosamides, and streptogramin B. (MLS) in streptococci. *J. Antimicrob. Chemother.* 16 (Suppl. A):111–135.

82. Ike, Y., and D. B. Clewell. 1984. Genetic analysis of the pAD1 pheromone response in *Streptococcus faecalis* using transposon Tn*917* as an insertional mutagen. *J. Bacteriol.* 158:777–783.

83. Ippen-Ihler, K., M. Achtman, and N. Willetts. 1972. Deletion map of the *Escherichia coli* K-12 sex factor F: The order of eleven transfer cistrons. *J. Bacteriol.* 110:857–863.

84. Ippen-Ihler, K. A., and E. G. Minkley, Jr. 1986. The conjugation system of F, the fertility factor of *Escherichia coli*. *Annu. Rev. Genet.* 20:593–624.

85. Jobling, M. G., and D. A. Ritchie. 1987. Genetic and physical analysis of plasmid genes expressing inducible resistance to tellurite in *Escherichia coli*. *Mol. Gen. Genet.* 208:288–293.

86. Johnson, D. A., and N. S. Willetts. 1983. Lambda transducing phages carrying transfer genes isolated from an abnormal prophage insertion into the *traY* gene of F. *Plasmid* 9:71–85.

87. Johnston, A. W. B., J. L. Beynon, A. V. Buchanan-Wollaston, S. M. Setchell, P. R. Hirsch, and J. E. Beringer. 1978. High frequency transfer of nodulating ability between strains and species of *Rhizobium*. *Nature* 276:635–636.

88. Katoh-Kanno, R., M. Kimura, T. Ikeda, and S. Kimura. 1986. Survey of modifying enzymes and plasmids in amikacin-resistant *Serratia marcescens*. *Microbiol. Immunol.* 30:509–519.

89. Kaulfers, P.-M., R. Laufs, and G. Jahn. 1978. Molecular properties of transmissible R factors of *Haemophilus influenzae* determining tetracycline resistance. *J. Gen. Microbiol.* 105:243–252.

90. Kingsman, A., and N. Willetts. 1978. The requirements for conjugal DNA synthesis in the donor strain during F*lac* transfer. *J. Mol. Biol.* 122:287–300.

91. Klapwijk, P. M., T. Scheulderman, and R. A. Schilperoort. 1978. Coordinated regulation of octopine degradation and conjugative transfer of Ti plasmids in *Agrobacterium tumefaciens:* Evidence for a common regulatory gene and separate operons. *J. Bacteriol.* 136:775–785.

92. Kline, B. C. 1985. A review of mini-F plasmid maintenance. *Plasmid* 14:1–16.

93. Kotarski, S. F., T. L. Merriwether, G. T. Tkalcevic, and P. Gemski. 1986. Genetic studies of kanamycin resistance in *Campylobacter jejuni*. *Antimicrob. Agents Chemother.* 30:225–230.

94. Laine, S., D. Moore, P. Kathir, and K. Ippen-Ihler. 1985. Genes and gene products involved in the synthesis of F-pili. In D. R. Helsinki, S. N. Cohen, D. B. Clewell, D. A. Jackson, and A. Hollaender (eds.), *Plasmids in Bacteria*. Plenum Press, New York, pp. 535–553.

95. Lambert, T., G. Gerbaud, and P. Courvalin. 1988. Transferable amikacin resistance in *Acinetobacter* spp. due to a new type of 3'-aminoglycoside phosphotransferase. *Antimicrob. Agents Chemother.* 32:15–19.

96. LeBlanc, D. J. 1981. Plasmids in streptococci: A review. In S. B. Levy, R. C. Clowes, and E. L. Koenig (eds.), *Molecular Biology, Pathogenicity, and Ecology of Bacterial Plasmids*. Plenum Press, New York, pp. 81–90.

97. LeBlanc, D. J., L. N. Lee, J. A. Donkersloot, and R. J. Harr. 1982. Plasmid transfer in streptococci (an overview). In D. Schlessinger (ed.), *Microbiology—1982*. Am. Soc. Microbiol., Washington, D.C., pp. 82–87.

98. Le Bouguenec, C., T. Horoud, G. Bieth, R. Colimon, and C. Dauguet. 1984. Translocation of antibiotic resistance markers of a plasmid-free *Streptococcus pyogenes* (group A) strain into different streptococcal hemolysin plasmids. *Mol. Gen. Genet.* 194:377–387.

99. Lederberg, J., L. L. Cavalli, and E. M. Lederberg. 1952. Sex compatibility in *E. coli*. *Genetics* 37:720–730.

100. Lederberg, J., and E. Tatum. 1946. Gene recombination in *E. coli*. *Nature* 158:558.

101. Lederberg, J., and E. L. Tatum. 1946. Novel genotypes in mixed cultures of biochemical mutants of bacteria. *Cold Spring Harbor Symp.* 11:113–114.

102. Lundquist, P. D., and B. R. Levin. 1986. Transitory derepression and the maintenance of conjugative plasmids. *Genetics* 113:483–497.

103. Macrina, F., T. Mays, C. Smith, and R. Welch. 1981. Non-plasmid associated transfer of antibiotic resistance in *Bacteroides*. *J. Antimicrob. Chemother.* 8:77–86.

104. Magot, M. 1983. Transfer of antibiotic resistances from *Clostridium innocuum* to *Clostridium perfringens* in the absence of detectable plasmid DNA. *FEMS Microbiol. Lett.* 18:149–151.

105. Maniatis, T., E. F. Fritsch, and J. Sambrook. 1982. *Molecular Cloning: A Laboratory Manual.* Cold Spring Harbor Laboratory, Cold Spring Harbor, N.Y.

106. Martin, P., P. Trieu-Cuot, and P. Courvalin. 1986. Nucleotide sequence of the tetM tetracycline resistance determinant of the streptococcal conjugative shuttle transposon Tn*1545*. *Nucleic Acids Res.* 14:7047–7058.

107. Martinez-Suarez, J. V., F. Baquero, M. Reig, and J. C. Perez-Diaz. 1985. Transferable plasmid-linked chloramphenicol acetyltransferase conferring high-level resistance in *Bacterioides uniformis*. *Antimicrob. Agents Chemother.* 28:113–117.

108. McIntire, S. A., and W. B. Dempsey. 1987. Fertility inhibition gene of plasmid R100. *Nucl. Acids Res.* 15:2029–2042.

109. McIntire, S. A., and N. Willetts. 1980. Transfer-deficient cointegrates of F*lac* and lambda prophage. *Mol. Gen. Genet.* 178:165–172.

110. Miller, J. H. 1972. *Experiments in Molecular Genetics.* Cold Spring Harbor Laboratory, Cold Spring Harbor, N.Y.

111. Minkley, E. G., Jr., and N. S. Willetts. 1984. Overproduction, purification, and characterization of the F *traT* protein. *Mol. Gen. Genet.* 196:225–235.

112. Moll, A., P. A. Manning, and K. N. Timmis. 1980. Plasmid-determined resistance to serum bactericidal activity: A major outer membrane protein, the *traT* gene product is responsible for plasmid-specified serum resistance in *Escherichia coli*. *Infect. Immun.* 28:359–367.

113. Moretti, P., G. Hintermann, and R. Hutter. 1985. Isolation and characterization of an extrachromosomal element from *Nocardia mediterranei*. *Plasmid* 14:126–133.

114. Morse, S. A., S. R. Johnson, J. W. Biddle, and M. C. Roberts. 1986. High-level tetracycline resistance in *Neisseria gonorrhoeae* is result of acquisition of streptococcal *tetM* determinant. *Antimicrob. Agents Chemother.* 30:664–670.

115. Novick, R. P. 1987. Plasmid incompatibility. *Microbiol. Rev.* 51:381–395.

116. Novick, R. P., R. C. Clowes, S. N. Cohen, R. Curtis III, N. Datta, and S. Falkow. 1976. Uniform nomenclature for bacterial plasmids: A proposal. *Bacteriol. Rev.* 40: 168–189.

117. Odelson, D. A., J. L. Rasmussen, J. Smith, and F. L. Macrina. 1987. Extrachromosomal systems and gene transmission in anaerobic bacteria. *Plasmid* 17:87–109.

118. Ogura, T., and S. Hiraga. 1983. Partition mechanism of F plasmid: Two plasmid gene-encoded products and a cis-acting region are involved in partition. *Cell* 32: 351–360.

119. Ou, J. T. 1980. Role of surface exclusion genes in lethal zygosis of *Escherichia coli* K-12 mating. *Mol. Gen. Genet.* 178:573–581.

120. Rafii, F., and D. Crawford. 1988. Transfer of conjugative plasmids and mobilization of a nonconjugative plasmid between *Streptomyces* strains on agar and in soil. *Appl. Environ. Microbiol.* 54:1334–1340.

121. Roberts, M. C., and G. E. Kenny. 1986. Dissemination of the *tetM* tetracycline resistance determinant to *Ureaplasma urealyticum*. *Antimicrob. Agents Chemother.* 29:350–352.

122. Roberts, M. C., and G. E. Kenny. 1987. Conjugal transfer of transposon Tn*916* from *Streptococcus faecalis* to *Mycoplasma hominis*. *J. Bacteriol.* 169:3836–3839.

123. Roberts, M. C., S. L. Hillier, J. Hale, K. K. Holmes, and G. E. Kenny. 1986. Tetracycline resistance and *tetM* in pathogenic urogenital bacteria. *Antimicrob. Agents Chemother.* 30:810–812.

124. Roberts, M. C., L. A. Koutsky, K. K. Holmes, D. J. LeBlanc, and G. E. Kenny.

1985. Tetracycline-resistant *Mycoplasma hominis* strains contain streptococcal *tetM* sequences. *Antimicrob. Agents Chemother.* 28:141–143.

125. Rood, J. I., V. N. Scott, and C. L. Duncan. 1978. Identification of a transferable tetracycline resistance plasmid (pCW3) from *Clostridium perfringens.* *Plasmid* 1: 563–570.

126. Sanchez-Pescador, R., J. T. Brown, M. Roberts, and M. S. Urdea. 1988. The nucleotide sequence of the tetracycline resistance determinant *tetM* from *Ureaplasma urealyticum.* *Nucl. Acids Res.* 16:1216–1217.

127. Schaberg, D. R., and M. J. Zervos. 1986. Intergeneric and interspecies gene exchange in gram-positive cocci. *Antimicrob. Agents Chemother.* 30:817–822.

128. Sensfuss, C., M. Reh, and H. G. Schlegel. 1986. No correlation exists between the conjugative transfer of the autotrophic character and that of plasmids in *Nocardia opaca* strains. *J. Gen. Microbiol.* 132:997–1007.

129. Shapiro, J. A. 1977. Appendix B: Bacterial plasmids. In A. I. Bukhari, J. A. Shapiro, and S. L. Adhya (eds.), *DNA Insertion Elements, Plasmids, and Episomes.* Cold Spring Harbor Laboratory, Cold Spring Harbor, N.Y., pp. 601–704.

130. Shoemaker, N. B., M. D. Smith, and W. R. Guild. 1980. DNase-resistant transfer of chromosomal *cat* and *tet* insertions by filter mating in pneumococcus. *Plasmid* 3: 80–87.

131. Silverman, P. M. 1985. Host cell-plasmid interactions in the expression of DNA donor activity by F⁺ strains of *Escherichia coli* K-12. *BioEssays* 2:254–259.

132. Silverman, P. M. 1986. The structural basis of prokaryotic DNA transfer. In M. Inouye (ed.), *Bacterial Outer Membranes as Model Systems.* John Wiley and Sons, Inc., New York.

133. Skurray, R. A., and P. Reeves. 1973. Physiology of *Escherichia coli* K-12 during conjugation: Altered recipient cell functions associated with lethal zygosis. *J. Bacteriol.* 114:11–17.

134. Smith, M. D., and W. R. Guild. 1980. Improved method for conjugative transfer by filter mating of *Streptococcus pneumoniae. J. Bacteriol.* 144:457–459.

135. Smith, M. D., and W. R. Guild. 1982. Evidence for transposition of the conjugative R-determinants of *Streptococcus agalactiae* B109. In D. Schlessinger (ed.), *Microbiology—1982.* Am. Soc. Microbiol., Washington, D.C., pp. 109–111.

136. Sox, T. E., W. Mohammed, E. Blackman, G. Biswas, and P. F. Sparling. 1978. Conjugative plasmids in *Neisseria gonorrhoeae.* Infect. Immun. 29:181–185.

137. Stachel, S. E., and P. C. Zambryski. 1986. *Agrobacterium tumefaciens* and the susceptible plant cell: A novel adaptation of extracellular recognition and DNA conjugation. *Cell* 47:155–157.

138. Stanisich, V. A. 1984. Identification and analysis of plasmids at the genetic level. In P. M. Bennett, and J. Grinsted (eds.), *Methods in Microbiology,* vol. 17. Academic Press, Orlando, Fla., pp. 5–32.

139. Stonesifer, J., P. Matsushima, and R. H. Baltz. 1986. High frequency conjugal transfer of tylosin genes and amplifiable DNA in *Streptomyces fradiae. Mol. Gen. Genet.* 202:348–355.

140. Tally, F. P., and M. H. Malamy. 1986. Resistance factors in anaerobic bacteria. *Scand. J. Inf. Dis.* (Suppl.)49:56–63.

141. Tally, F. P., M. J. Shimell, G. R. Carson, and M. H. Malamy. 1981. Chromosomal and plasmid-mediated transfer of clindamycin resistance in *Bacteroides fragilis.* In S. B. Levy, R. C. Clowes, and E. Koenig (eds.), *Molecular Biology, Pathogenicity, and Ecology of Bacterial Plasmids.* Plenum Press, New York, p. 51.

142. Tally, F. P., D. R. Snydman, S. L. Gorbach, and M. H. Malamy. 1979. Plasmid-mediated transferable resistance to clindamycin and erythromycin in *Bacteroides fragilis. J. Infect. Dis.* 139:83–88.

143. Tam, A. C., R. M. Behki, and S. U. Khan. 1987. Isolation and characterization of an s-ethyl-N,N-dipropylthiocarbamate-degrading *Arthrobacter* strain and evidence for plasmid-associated s-ethyl-N,N-dipropylthiocarbamate degradation. *Appl. Environ. Microbiol.* 53:1088–1093.

144. Taylor, D. E., S. A. DeGrandis, M. A. Karmali, and P. C. Fleming. 1981. Trans-

missible plasmids from *Campylobacter jejuni. Antimicrob. Agents Chemother.* 19: 831–835.

145. Taylor, D. E., and J. G. Levine. 1980. Studies of temperature-sensitive transfer and maintenance of the H incompatibility group plasmids. *J. Gen. Microbiol.* 116: 475–484.

146. Taylor, D. E., and A. O. Summers. 1979. Association of tellurium resistance and bacteriophage inhibition conferred by R plasmids. *J. Bacteriol.* 137:1430–1433.

147. Terawaki, Y., and R. Rownd. 1972. Replication of the R factor Rts1 in *Proteus mirabilis. J. Bacteriol.* 109:492–498.

148. Terawaki, Y., H. Takayasu, and T. Akiba. 1967. Thermosensitive replication of a kanamycin resistance factor. *J. Bacteriol.* 94:687–690.

149. Thatte, V., D. E. Bradley, and V. N. Iyer. 1985. N conjugative transfer system of plasmid pCU1. *J. Bacteriol.* 163:1229–1236.

150. Tran Van Nhieu, G., F. W. Goldstein, M. E. Pinto, J. F. Acar, and E. Collatz. 1986. Transfer of amikacin resistance by closely related plasmids in members of the family *Enterobacteriaceae* isolated in Chile. *Antimicrob. Agents Chemother.* 29:833–837.

151. Traxler, B. A., and E. G. Minkley, Jr. 1987. Revised genetic map of the distal end of the F transfer operon: Implications for DNA helicase I, nicking at *oriT*, and conjugal DNA transport. *J. Bacteriol.* 169:3251–3259.

152. Traxler, B. A., and E. G. Minkley, Jr. 1988. Evidence that DNA helicase I and *oriT* site-specific nicking are both functions of the F *traI* protein. *J. Mol. Biol.* 204:205–209.

153. Umeda, F., H. Min, M. Urushihara, M. Okazaki, and Y. Mirua. 1986. Conjugal transfer of hydrogen-oxidizing ability of *Alcaligenes hydrogenophilus* to *Pseudomonas oxalaticus. Biochem. Biophys. Res. Commun.* 137:108–113.

154. Valla, S., D. H. Coucheron, and J. Kjosbakken. 1987. The plasmids of *Acetobacter xylinum* and their interaction with the host chromosome. *Mol. Gen. Genet.* 208:76–83.

155. Walia, S. K., T. Madhavan, T. D. Chugh, and K. B. Sharma. 1987. Characterization of self-transmissible plasmids determining lactose fermentation and multiple antibiotic resistance in clinical strains of *Klebsiella pneumoniae. Plasmid* 17:3–12.

156. Walter, M. W., A. Porteous, and R. J. Seidler. 1987. Measuring genetic stability in bacteria of potential use in genetic engineering. *Appl. Environ. Microbiol.* 53:105–109.

157. Wilkins, B. M., L. K. Chatfield, C. C. Wymbs, and A. Merryweather. 1985. Plasmid DNA primases and their role in bacterial conjugation. In D. R. Helinski, S. N. Cohen, D. B. Clewell, D. A. Jackson, and A. Hollaender (eds.), *Plasmids in Bacteria.* Plenum Press, New York, pp. 585–603.

158. Willetts, N. S. 1974. The kinetics of inhibition of F*lac* transfer by R100 in *E. coli. Mol. Gen. Genet.* 129:123–130.

159. Willetts, N. 1977. The transcriptional control of fertility in F-like plasmids. *J. Mol. Biol.* 112:141–148.

160. Willetts, N. 1984. Conjugation. In P. M. Bennett and J. Grinsted (eds.), *Methods in Microbiology,* vol. 17. Academic Press, Orlando, Fla, pp. 33–59.

161. Willetts, N. S., and J. Maule. 1974. Interactions between the surface exclusion systems of some F-like plasmids. *Genet. Res.* 24:81–89.

162. Willetts, N., and R. Skurray. 1980. The conjugation system of F-like plasmids. *Annu. Rev. Genet.* 14:41–76.

163. Willetts, N., and R. Skurray. 1987. Structure and function of the F factor and mechanism of conjugation. In J. L. Ingraham, K. B. Low, B. Magasanik, F. C. Neidhart, M. Schaechter, and H. E. Umbarger (eds.), *Escherichia coli and Salmonella typhimurium: Cellular and Molecular Biology.* American Society of Microbiology, Washington, D.C., pp. 110–113.

164. Willetts, N., and B. Wilkins. 1984. Processing of plasmid DNA during bacterial conjugation. *Microbiol. Rev.* 48:24–41.

165. Winans, S. C., and G. C. Walker. 1985. Conjugal transfer system of the IncN plasmid pKM101. *J. Bacteriol.* 161:402–410.

166. Wollman, E. L., and F. Jacob. 1955. Sur le mechanisme du transfer de materiel

genetique au cours de la recombination chez *E. coli* K-12. *C. R. Acad. Sci. Paris* 240:2449–2452.

167. Wong, C. L., and N. W. Dunn. 1974. Transmissible plasmid coding for the degradation of benzoate and *m*-toluate in *Pseudomonas arvilla mt-2*. *Genet. Res.* 23:227–232.

168. Wu, J., and K. Ippen-Ihler. ᵣ ᵤ9. Nucleotide sequence of *traO* and adjacent loci in the *Escherichia coli* K-12 F-plasmid transfer operon. *J. Bacteriol.* 171:213–221.

169. Wu, J. H., D. Moore, T. Lee, and K. Ippen-Ihler. 1987. Analysis of *Escherichia coli* K-12 F factor transfer genes: *traQ, trbA,* and *trbB. Plasmid* 18:54–69.

170. Yamamoto, M., J. M. Jones, E. Senghas, C. Gawron-Burke, and D. B. Clewell. 1987. Generation of Tn5 insertions in streptococcal conjugative transposon Tn*916. Appl. Environ. Microbiol.* 53:1069–1072.

171. Yoshioka, Y., H. Ohtsubo, and E. Ohtsubo. 1987. Repressor gene *finO* in plasmids R100 and F: Constitutive transfer of plasmid F is caused by insertion of IS3 into F *finO. J. Bacteriol.* 169:619–623.

11

Methods for Evaluating Transduction: An Overview with Environmental Considerations

Robert V. Miller

Introduction: A Historical View of Transduction

After the discovery of conjugation in *Escherichia coli* (68, 69), Joshua Lederberg and coworkers turned their attention to *Salmonella typhimurium* in the hope of finding that a similar process occurred in that bacterium. They discovered that genetic transfer did take place between auxotrophic strains of this species, not only when cells of the two mutants were placed in contact with each other, but also when one strain was exposed to cell-free extracts of the other (67). Further investigation revealed that the transfer of genetic elements from one cell to the other by this process, termed *transduction,* was mediated by the packaging of the DNA in bacteriophage capsids. In this case, transfer was mediated by the bacteriophage P22 (124).

There are two major types of transduction. In *generalized* (or *unrestricted*) *transduction,* all regions of the chromosome or other genetic element present in the donor cell can be transferred with approximately the same frequency (124). *Specialized* (or *restricted*) *transduction* (87) allows the transfer of only a restricted number of genes within the host chromosome. In addition, the mechanisms for the production of the transducing particles are fundamentally different in the two types of systems.

Principles

Lysogeny

Transduction is most often mediated by *temperate* bacteriophages. While *virulent* bacteriophages are capable of only the lytic life cycle, a *temperate* phage can alternatively *lysogenize* its host. The temperate and lytic life cycles of bacteriophages are illustrated in Fig. 11.1. Particularly good detailed descriptions of bacterial viral life cycles are presented by Stent and Calendar (106) and by Ptashne (94).

When a bacterial virus infects a susceptible host, the lytic life cycle is usually initiated. Viral genes are expressed, leading to the replication of the viral nucleic acid and synthesis of viral proteins. Progeny phage genomes are subsequently packaged into shells (*capsids*) composed of viral proteins producing infective particles (*virions*). After a *latent period* during which the production of virions takes place, the host cell bursts releasing a number of progeny phage virions.

In the *temperate* or *lysogenic response,* the lytic functions of the viral genome are suppressed by a repressor (94), and the viral genome is carried quiescently as a *prophage* in the host cell. Because of the presence of the repressor in the host cell, the lysogenized cell is *immune* to infection by the same (and other closely related) phages (106). In most cases, immunity does not inhibit absorption of the phage DNA but simply inhibits its expression.

Bacterial cells carrying the prophage are referred to as *lysogens* because the infected host has the potential for producing phage particles at a later time through induction of the lytic cycle. Prophage induction occurs spontaneously at a low level in each cell generation. Many prophages can also be induced to the lytic response at high frequency af-

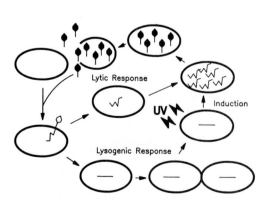

Figure 11.1 Life cycles of bacteriophages. The lytic response leads to the replication of virus nucleic acid and proteins and results in the lysis of the host cell with the release of progeny phase particles. Virulent bacteriophage can only elicit the lytic response in infected hosts. Temperate bacteriophages can elicit either the lytic or lysogenic response upon infection. In the lysogenic response, the viral genome establishes a symbiotic relationship with the host (lysogen) in which the viral genome (prophage) is replicated and segregated into progeny of the host cell. Various stresses to the lysogen may induce a lytic response.

ter exposure of the host cell to various stresses which damage DNA or inhibit DNA replication. Agents such as ultraviolet (UV) irradiation, mitomycin C, and nalidixic acid trigger the destruction of the phage repressor allowing expression of the lytic functions of the phage (94).

The prophages of various viruses exist in one of two states within the host cell. For example, the prophage of the *E. coli* virus λ becomes integrated into the host chromosome at a specific site (106). The prophage of *Pseudomonas aeruginosa* phage D3 appears to have at least two alternate but specific integration sites within the chromosome (21). The prophages of other viruses are not integrated into the chromosome at all but are carried in the lysogenized cell as extrachromosomal elements (82, 106).

Generalized transduction

Generalized transducing particles are produced by the inappropriate packaging of host DNA into a phage capsid in the place of the phage genome. Either chromosomal or plasmid DNA can be packaged and transferred (Fig. 11.2), and generalized transducing particles contain

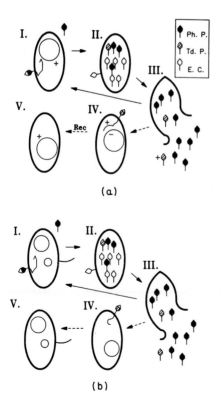

Figure 11.2 Generalized transduction. (*a*) Chromosomal transduction. I. A phage infects the *donor* producing a lytic response. II. Phage genomes are packaged into capsids (E. C.) to produce infective phage particles (Ph. P.). In some, portions of the host genome are packaged in place of the phage genome (see text), producing transducing particles (Td. P.). III. The donor cell lysis releasing phage particles and transducing particles. IV. A transducing particle is adsorbed onto a *recipient* bacterium and the DNA is absorbed. The transduced DNA recombines with the recipient's chromosome producing a transductant with altered phenotype. (*b*) Plasmid transduction. The steps are the same as for chromosomal transduction except that plasmid DNA (small circle) is packaged and transferred to the recipient.

only bacterial or plasmid DNA (55). They do not contain any phage DNA, and the transductants formed are not immune to superinfection by the mediating phage (97). Transducing particles can be produced either during *primary infection* of the bacterial host or by *induction* of a host-associated prophage. Virulent as well as temperate bacteriophages have been reported to carry out generalized transduction under appropriate conditions (60, 109). Environmentally, it is most likely that the majority of transduction will be mediated by temperate phages since environmental lysogens appear to form a reservoir for these viruses in situ (97).

Cotransduction

If two genes can be inherited by a recipient bacterium as the result of a single transductional event, they are said to be *cotransducible*. Because the geometry and size of the capsid restrict the amount of DNA which can be packaged, the pieces of bacterial DNA that are contained in the transducing particles are uniformly sized fragments of the host chromosome. For example, each transducing particle of the *P. aeruginosa* bacteria phage F116 contains approximately 2 percent of the *P. aeruginosa* chromosome (83, 90). If two genetic loci are positioned in close proximity on the host chromosome, they may be packaged into the same capsid and transferred as a single unit. Widely spaced markers cannot be transferred together in this manner. Cotransduction permits the estimation of relative linkage of two loci from the frequency with which they are inherited together. The greater the distance between the two genes, the lower the probability of their coinheritance (106).

Specialized transduction

The production of *specialized* transducing particles requires replacement of one or more phage genes with bacterial DNA. Transducing particles are produced only during the induction to lytic growth of a chromosomally integrated prophage. Their origin is in an illegitimate recombination event associated with the excision of the prophage from the host genome (106). Hence, only the bacterial DNA in close proximity to the phage integration site is transferred (Fig. 11.3). Specialized transducing particles contain a mixture of viral and host DNA, and the transductants produced contain a defective prophage that often renders them immune to superinfection (106).

Diversity of transduction as a mechanism of gene transfer in bacteria

Bacteriophages capable of mediating the exchange of genetic material have been identified for many environmentally important bacterial

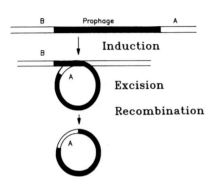

Figure 11.3 Production of a specialized transducing phage. A prophage that integrates into the host genome (see text) is induced by stress to the host cell. The first steps in induction require the excision of the prophage from the host cell genome via an illegitimate recombinational event. During this process, a portion of the phage genome is left in the bacterial chromosome and replaced by adjacent bacterial sequences. All progeny phage contain this hybrid genome. The produced phage may be *defective* due to the loss of the viral genes left behind in the bacterial chromosome.

species (60). A partial list of transduction systems that have been identified to date is presented in Table 11.1. This table outlines the most significant characteristics of the various systems and cites representative references.

Methods

Enumeration of phage

The most easily identified symptom of phage infection of a susceptible bacterium is *lysis*. Lysis is caused by the disruption of the bacterial cell envelope and results in the release of newly produced phage particles. If a virion is mixed with enough bacteria to produce a continuous layer of growth on an agar plate, a *plaque* results due to localized lysis of the bacterial lawn by the progeny viruses (Fig. 11.4). Bacterial viruses are normally enumerated by determining the concentration (*titer*) of *plaque-forming units* (PFU) in a particular sample.

Isolation of bacteriophage from nature

Numerous studies have reported significant titers of both temperate and lytic bacteriophages from various environmental sources including both marine (85) and freshwater (42, 91, 92) ecosystems, soils (18, 116), waste treatment facilities (12, 53, 58), and clinical isolates (84, 115). Both temperate and lytic phages can be isolated from these sources.

The methods for the isolation of phage particles from nature have been reviewed in *Phage Ecology* (103), and the reader is referred to this recent monograph for specific methods. In general, an environ-

TABLE 11.1 Transduction Systems

Host	Phage	Temperate life cycle	Generalized or specialized	Plasmid[a]	Inducing treatment	Interspecies[b]	Selected references
Achromobacter sp.[c]	α	Yes	G		UV		117
Acinetobacter sp.	P78	Yes	G		UV		46
Bacillus							
anthracis	CP-51	Yes	G	Yes		Yes	96
alesti	CP-54	Yes	G			Yes	111
cereus	CP-51	Yes	G	Yes		Yes	96, 110, 119
	CP-53	Yes	G				119
	CP-54	Yes	G				111
entomocidus-limassol	CP-54	Yes	G			Yes	111
galleriae	CP-54	Yes	G			Yes	111
megaterium	MP13	Yes	G				114
pumilus	φ75	Yes	?[d]	Yes			14
	PBP1	Yes	G	Yes	mitoC[e]		13, 71, 72
	PBS1	Yes	G	Yes			13, 73
	PMB1	Yes	G	Yes			13
subtilis	φ105	Yes	S	Yes	mitoC		7, 61, 99
	AR9	?	G	Yes			93
	SP	Yes	S				44, 70, 95, 122
	SPO2	Yes	?	Yes			61, 76
	SPP1	No	G	Yes			3, 19, 35, 36, 37, 38, 43, 11
sotto	CP-54	Yes	G			Yes	111

[a]Transduction of plasmid DNA has been established in this system.
[b]Interspecies transduction has been identified with this system.
[c]Species unspecified or unidentified.
[d]Chromosomal transduction has not been established in this system.
[e]mitoC: mitomycin C.
[f]Nal: nalidixic acid.
[g]This phage replicates by transposition.
[h]Nor: norfloxacin, a quinolone related to nalidixic acid.
[i]φDS1 is probably an independent isolate of F116.

TABLE 11.1 Transduction Systems (Continued)

Host	Phage	Temperate life cycle	Generalized or specialized	Plasmid[a]	Inducing treatment	Interspecies[b]	Selected references
Bacillus							
thompsoni	CP-51	Yes	G	Yes		Yes	111
	CP-54	Yes	G			Yes	111
thuringiensis	CP-51	Yes	G	Yes		Yes	96, 111
	CP-54	Yes	G			Yes	111
	CP-54Ber	Yes	G				66
	TP-13	Yes	G				8
	TP-18	Yes	G				8
Caulobacter							
crescentus	φCr30	No	G				9, 41
Corynebacter							
diphtheriae	γ	Yes	S				16
renale	RP28	Yes	G				48
Erwinia							
chrysanthemi	Erch-12	Yes	G		mitoC		23
Escherichia							
coli	P1	Yes	G	Yes	UV, mitoC, Nal[f]	Yes	10, 74, 120
	λ	Yes	S				5, 6, 87
	μ	Yes	G	Yes[g]		Yes	17, 39
Klebsiella							
aerogenes	P1	Yes	G			Yes	10
	PW52	Yes	G				75

[a]Transduction of plasmid DNA has been established in this system.
[b]Interspecies transduction has been identified with this system.
[c]Species unspecified or unidentified.
[d]Chromosomal transduction has not been established in this system.
[e]mitoC: mitomycin C.
[f]Nal: nalidixic acid.
[g]This phage replicates by transposition.
[h]Nor: norfloxacin, a quinolone related to nalidixic acid.
[i]φDS1 is probably an independent isolate of F116.

TABLE 11.1 Transduction Systems (Continued)

Host	Phage	Temperate life cycle	Generalized or specialized	Plasmid[a]	Inducing treatment	Interspecies[b]	Selected references
Myxococcus xanthus	Mx4	No	G				18, 45, 77
	Mx8	Yes	G				77
	Mx9	Yes	G				77
Proteus mirabilis	φ34	Yes	G				30
	5006M	Yes	S	Yes	UV	Yes	26, 27, 28, 64
	π1	Yes	G				104, 105
morganii	φM	Yes	G		UV		24
rettgeri	7/R49	Yes	G				63
vulgaris	107/69	Yes	G				29
	5006M	Yes	S	Yes	UV		26
Providencia sp.	PL25	Yes	S	Yes	UV	Yes	25, 31
Pseudomonas aeruginosa	B86	Yes	G		UV		59
	D3	Yes	S		UV, Nor[h]		21, 80
	D3112	Yes	G	Yes[g]	UV		1, 65
	F116	Yes	G	Yes	UV		50, 62, 82, 97
	φDS1[i]	Yes	G	Yes	UV		97
	G101	Yes	G				53
	P110	Yes	G				51

[a]Transduction of plasmid DNA has been established in this system.
[b]Interspecies transduction has been identified with this system.
[c]Species unspecified or unidentified.
[d]Chromosomal transduction has not been established in this system.
[e]mitoC: mitomycin C.
[f]Nal: nalidixic acid.
[g]This phage replicates by transposition.
[h]Nor: norfloxacin, a quinolone related to nalidixic acid.
[i]φDS1 is probably an independent isolate of F116.

TABLE 11.1 Transduction Systems (Continued)

Host	Phage	Temperate life cycle	Generalized or specialized	Plasmid[a]	Inducing treatment	Interspecies[b]	Selected references
Pseudomonas							
cepacia	CP75	Yes	G				78
putida	pf16h2	Yes	G		UV		22
	PP1	Yes	G				52
Rhizobium							
meliloti	11	Yes	G				100
	DF2	Yes	G				20
Salmonella							
typhimurium	μ	Yes	G	Yes[g]	UV	Yes	17, 39
	P22	Yes	Both		UV		109, 120
Serratia							
marcescens	HY	Yes	G				57
Shigella							
dysenteriae	P1	Yes	G	Yes		Yes	74
Staphylococcus							
aureus	φ11	Yes	G	Yes			40, 88
Streptococcus							
group A	A25	No	G	Yes		Yes	101, 113
	GT-234	Yes	G		UV	Yes	33
	Unnamed	Yes	G		mitoC	Yes	29, 32, 34, 54

[a]Transduction of plasmid DNA has been established in this system.
[b]Interspecies transduction has been identified with this system.
[c]Species unspecified or unidentified.
[d]Chromosomal transduction has not been established in this system.
[e]mitoC: mitomycin C.
[f]Nal: nalidixic acid.
[g]This phage replicates by transposition.
[h]Nor: norfloxacin, a quinolone related to nalidixic acid.
[i]φDS1 is probably an independent isolate of F116.

TABLE 11.1 Transduction Systems (Continued)

Host	Phage	Temperate life cycle	Generalized or specialized	Plasmid[a]	Inducing treatment	Interspecies[b]	Selected references
Streptococcus							
group C	GT-234	Yes	G		UV	Yes	33
	Unnamed	Yes	G		mitoC	Yes	32, 34, 115
group G	GT-234	Yes	G		UV	Yes	33
	Unnamed	Yes	G			Yes	34
Streptomyces							
hydroscopicus	SH10	Yes	G				108
olivaceus	1	Yes	G				2
venezuelae	φSV1	No	G				107
Vibrio							
cholerae	CP-T1	Yes	G		UV		89

[a]Transduction of plasmid DNA has been established in this system.
[b]Interspecies transduction has been identified with this system.
[c]Species unspecified or unidentified.
[d]Chromosomal transduction has not been established in this system.
[e]mitoC: mitomycin C.
[f]Nal: nalidixic acid.
[g]This phage replicates by transposition.
[h]Nor: norfloxacin, a quinolone related to nalidixic acid.
[i]φDS1 is probably an independent isolate of F116.

238

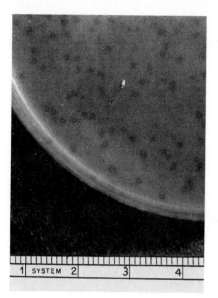

Figure 11.4 Plaques of temperate *P. aeruginosa* bacteriophage φ1407 (84). The *turbid* centers are produced by growth of lysogenized bacteria. Virulent bacteriophage produce *clear* plaques.

mental sample is taken and concentrated (91). Because the half-life of phage particles is often short (103), the concentrated sample should be titered as soon as possible using an appropriate indicator bacterium.

An interesting method for the isolation of broad-host-range phages has been described by Primrose et al. (92). This method utilizes a mixture of indicator bacteria of the species for which an interspecific phage is desired. Only phages that can productively infect all members of the mixed population will form plaques.

It should be noted that many phages have a temperature optimum for maximal infection which is often close to the temperature of their natural environmental niches (48, 92). Therefore, consideration should be given to selecting an appropriate incubation temperature for the identification of phages in environmental samples.

It is of particular importance to select appropriate indicator strains of the host bacterium. Various strains may be resistant to infection because of modification of the bacteriophage receptor, or they may be immune due to the presence of a related prophage. Many bacteria contain active or cryptic prophages within their genomes. Holloway (49) has estimated that almost all isolates of *P. aeruginosa* are lysogenic for at least one prophage and most are multiple lysogens. In addition, restriction-modification systems encoded by the host chromosome or by extrachromosomal elements present in the indicator strain (47, 56) may inhibit the growth of phages (see Chap. 7). Perhaps the easiest and best way to solve the problems of phage resistance, immunity, and restriction-modification is to utilize a number of independent isolates of the bacterial species in question as indicator strains in these studies.

Induction of lysogens

Many of the transducing phages described in Table 11.1 were originally isolated after the induction of naturally occurring lysogens by exposing the cell to a DNA-damaging stress (14, 23, 24, 33, 46, 117). General methods for the induction of potential lysogens utilizing either UV radiation, mitomycin C, or nalidixic acid are presented here. They must be adapted to the specific system in question. Important factors to consider in this adaption are the temperature of incubation, the dose of DNA-damaging stress, and the time allowed for the expression of the stress-induced response. Other modifications may also be necessary for specific bacterium-bacteriophage systems.

UV irradiation.   Cells to be induced should be inoculated into a 12-mL culture of rich medium such as L-broth (16) or nutrient broth (Difco) and incubated at the appropriate temperature for bacterial growth until the early exponential phase. It is important that the culture not be too dense if UV radiation is to be used as the inducing agent because high densities of cells will lead to shadowing. The cells should then be centrifuged in a table-top centrifuge and washed twice with 12 mL of saline or other minimal salts solution. It is important to remove the rich medium as aromatic compounds present in the medium will absorb UV light and protect the cells from induction.

One milliliter of the washed culture is removed and serially diluted for an estimation of total viable cell count. Ten milliliters are then transferred to a large sterile Petri dish (15 cm diameter) for irradiation. Remove 1.0 mL and transfer it to 4.0 mL of rich medium *prior* to UV exposure. This will serve as an unirradiated control to allow the estimation of the level of spontaneous phage release (if any) from the lysogen.

The remainder of the washed cells are then exposed to various UV doses (usually in the range of 5 to 100 J/m^2, depending on the sensitivity of the host species to UV killing). For each dose, transfer 1.0 mL to a foil-covered tube (to prevent photoreactivation) containing 4.0 mL of rich medium. Incubate all samples (with shaking) at an appropriate growth temperature for 2 to 3 h (actual times may vary depending on the species used). Be sure that they remain in the dark. After incubation, centrifuge the cultures to pellet unlysed cells and collect the supernatant fluids. These lysates should then be filtered through 0.45-μm membranes and refrigerated for storage. They should be titered as described below to determine the number of PFU/mL. They can also be used to estimate transducing potential (see below).

Induction by antimicrobial agents.   Inoculate a culture of appropriate growth medium with the strain to be induced and grow to early exponential phase. Divide the culture into 1.0-mL aliquots in foil-covered tubes. Introduce mitomycin C or nalidixic acid at various concentrations (usually in concentrations between 0.001 and 10 μg/mL for mitomycin C and 1 to 1000 μg/mL for nalidixic acid, depending on the sensitivity of the host cell for the antibiotic). One tube should receive *no drug*. This will be used as an untreated control to measure spontaneous levels of phage release. Allow all cultures to incubate for 2 to 3 h *in the dark* (to eliminate photoreactivation) and treat as described above for UV induction of prophage.

A general method for transductional analysis

The method presented here has been adopted from numerous descriptions of transduction. It is general and must be adapted to the specific system to be used. A number of specific considerations in the adaption of this method will be discussed in the following sections.

Preparation of transducing lysates.   The propagation of most phages is best achieved by the soft agar overlay method (6). In this method, bacteria and phages are mixed and added to a melted 0.7% agar medium that has been cooled to approximately 45°C.* This mixture is then poured onto a 1.5% agar plate containing an appropriate medium and allowed to solidify. After overnight growth, the phages are harvested by scraping the soft-agar overlay from the plate and placing it in a screw-capped tube. The phages are suspended in a few milliliters of rich medium such as L-broth. The agar is removed by low-speed centrifugation and the supernatant fluid (containing the phages) transferred to a new tube. Chloroform is often added to kill residual bacteria. If this is not done (and it should be noted that many phages are sensitive to chloroform), the *lysate* should be filtered through a 0.45-μm nitrocellulose filter to remove bacteria.

To prepare the transducing lysate, the donor strain of bacterium is grown to midlog in an appropriately rich medium such as L-broth or nutrient broth. The bacteria are then mixed with the phage at a multiplicity of infection (MOI, phage-to-bacterium ratio) of approximately

---

*It is important to be sure that the top agar is completely melted before use because small particles of unmelted agar may appear to be plaques.

0.1. Three milliliters of melted 0.7% agar medium (cooled to approximate 45°C) are added, and the mixture poured onto a thick 1.5% agar plate containing an appropriate growth medium, allowed to solidify, and incubated overnight.

The next day, the top-agar layer is scraped off with a spatula and transferred to a capped centrifuge tube containing 0.5 mL of chloroform, and the plate washed with 2 to 3 mL of medium, which is added to the centrifuge tube. The tube is closed, shaken vigorously, and spun on a bench-top centrifuge for 15 min. The supernatant fluid is transferred to another tube containing 0.5 mL $CHCl_3$ (if the phage is resistant to chloroform). This is the transducing lysate. To ensure that it contains only markers derived from the desired host, it should be recycled once using the same strain.

Assay (titering) of the phage (transducing) lysate.   A culture of an appropriate indicator strain should be prepared in rich medium such as L-broth or nutrient broth and grown to approximately 10^9 to 10^{10} CFU/mL. It should be spun down and resuspended in one-half volume of absorption buffer (see below).

Dilutions of the phage lysate should be made in an appropriate buffer and used immediately as many phages are unstable in minimal salts. Dilutions high enough to count individual plaques (often as high as 10^{-10}) must be made. One-tenth milliliter of each dilution should be mixed with 0.1 mL of indicator bacteria and incubated for 10 to 20 min at a temperature appropriate to the bacterium-bacteriophage system under study (37°C for *E. coli*). Top agar (1 to 2 mL) is then added and poured and spread quickly on an agar plate of appropriately rich medium. The plates are incubated overnight and the plaques counted the next day to determine the titer or PFU/mL.*

Transduction.   Grow a 10-mL culture of the recipient bacterium in rich medium to mid-exponential phase. Spin down the culture and resuspend it in the same medium.

Mix together bacteria and phage (diluted in rich medium) to obtain a MOI of 0.1. Incubate for 10 to 15 min and spin down the cells in a table-top centrifuge. Wash the cells in minimal salts to remove unadsorbed phage, centrifuge again, suspend in 0.1-mL minimal medium, and plate on various media selective for the genetic determinant(s) serving as markers for transduction. Incubate at least 2 days

---

*It is important to note that a plate showing confluent lysis of an indicator strain by a temperate bacteriophage may appear uninfected due to growth of lysogenized bacteria. It is therefore imperative to make and examine dilutions high enough to observe individual plaques.

(or longer for slow-growing environmental isolates) before counting colonies of transductants. Extended incubation is often necessary to allow for the phenotypic expression of the transduced gene.

Recipient cells unexposed to phage should be plated to determine the number of revertants present due to spontaneous mutation. Both the phage lysate and the media used should be plated to test for sterility.

Transduction frequency is usually reported as transductants per PFU. Alternatively, transductants per recipient can be used.

Variables to be considered in designing a specific transduction assay

Multiplicity of infection (MOI).   The frequency of transduction may vary with MOI. In general, the MOI should be kept low to reduce the probability of a recipient cell encountering both a transducing particle and an infectious phage particle. However, different phages have been shown to have different optimal MOIs for this process (18, 21, 81, 97). Because of these differences in efficiency of transduction, initial experiments should be done over a range of MOIs.

Temperature.   Bacteriophages often show a temperature optimum for both adsorption and replication (98). This temperature often reflects their ecological origin. It may not, however, correspond to the temperature optimum of their host (98). It is advisable to consider various temperatures of incubation for both the isolation of environmental bacteriophages and for the analysis of their potential to mediate transduction.

Ions.   Bacteriophages often require cations for adsorption to their hosts. Many phages will not adsorb in distilled water or at low concentrations of monovalent ions ($< 10^{-4}$ M), and many require specific divalent cations for maximal adsorption. The optimal concentrations permitting maximal adsorption efficiency are characteristic for each phage and each ion. Concentrations above the optimum may also reduce adsorption (112). The adsorption of transducing particles parallels the parental phage in their requirements for ions. The development of an appropriate adsorption buffer is crucial in maximizing transduction frequencies and the effects of various ions on this process should be investigated when a new transduction system is encountered.

Plasmid transduction—molecular size.   Phages generally encapsulate only a single DNA segment of the same size as their own genome (88).

There is certainly a maximal size to the DNA segment that can be packaged into a specific phage capsid, and there may be a minimal size as well. Saye et al. (97) found that transduction of plasmids by a variant of the *P. aeruginosa* phage F116 was more efficient if the molecular weight of the plasmid was similar to that of the phage genome. Smaller plasmids were transduced at lower frequencies. Novick et al. (88), on the other hand, found that small *Staphylococcus aureus* plasmids were efficiently transduced by $\phi11$ because they formed linear concatemers of larger molecular weight.

In analyzing plasmid transduction, it is essential to consider the molecular size of the plasmid to be used. If the molecular size of the phage genome is known, plasmids of similar size should be utilized initially. In an unknown system, plasmids of diverse molecular sizes should be tested.

UV treatment of transducing lysates. Many authors have found that treatment of transducing lysates with UV radiation prior to their use will increase the frequency of transduction sometimes as much as 10- to 50-fold (4, 11, 23, 30, 31, 63). Such increases are most likely due to the inactivation of the killing effects of the infectious bacteriophage particles present in the lysate (15, 41). Alternatively, it has been suggested that this treatment stimulates recombination within the host cell leading to increased incorporation of the transduced DNA into the recipient genome (11). UV treatment may increase the frequency of transduction and allow easier detection in initial studies.

Environmental considerations

Concentration of samples. Transduction frequencies of specific genes by generalized transduction are usually low ranging from 10^{-5} to 10^{-7} transductants per PFU (18, 88, 97). As titers of bacteriophages in natural environments may be lower than 10^7 PFU/mL, it is often desirable to concentrate samples when possible (e.g., from aquatic environments) in order to identify environmentally produced transductants. Concentration can be achieved by filtration through 0.45-μm filters, which can then be placed directly on a selective medium to allow the growth of colonies of environmentally produced transductants. This method has been used successfully by Saye et al. (97) in their demonstration of transduction in situ in a freshwater lake.

Marker selection. Because of the low frequency of transduction, it is advisable to select genetic markers in such a way as to take advantage of cotransduction. Cotransduction of closely linked loci will allow a more definitive identification of a unique transduced phenotype and reduce the background of revertants produced by spontaneous mutation.

With plasmids, this problem can be easily solved by using an extrachromosomal element encoding resistance to several antibiotics. In addition, plasmid transduction can be confirmed by restriction endonuclease digestion pattern analysis (97) and, if necessary, by Southern hybridization analysis (102). Selection of cotransducible chromosomal loci may be more problematic because of the necessity of mapping closely linked loci (see Cotransduction in preceding section). The potential benefits of contransductional analysis makes the effort expended to identify suitable genetic markers worthwhile.

Whenever possible, markers should also be chosen to allow for positive selection of the transductant. For example, the transduction of an amino acid auxotrophic recipient to prototrophy is preferable to scoring for the acquisition of an auxotrophic allele in a prototrophic recipient. Methods for the production and characterization of chromosomal mutations appropriate for this type of study have been described by Miller (79).

Inhibition of postsampling transduction during analysis of ecological samples. Because of the necessity of filtering environmental samples, there may be a considerable concentration of the components of the transduction system during analysis. This may lead to a situation that could promote the production of postsampling transductants during the analysis period.

Saye et al. (97) used a nalidixic acid-resistant recipient in a procedure designed to eliminate postsampling transduction. They found that the combination of treatment of the environmental sample with nalidixic acid before filtering to eliminate DNA synthesis in the donor, washing the filtrate several times to remove any free phage and transducing particles, and plating the filtered sample on selective medium containing nalidixic acid effectively eliminated any transduction on the sampling filter.

Conclusions

To date, few studies have assessed the frequency of transduction in the environment. Those reports that have examined transduction in situ present compelling evidence that there is a significant potential for gene exchange by this mechanism in the environment. Zeph et al. (123) have demonstrated P1 transduction between strains of *E. coli* inoculated into soil, even though this organism is not a true soil bacterium. Morrison et al. (86) demonstrated the transduction of chromosomal DNA in *P. aeruginosa* introduced into a freshwater lake in an environmental test chamber. Transduction was observed when either a transducing lysate or a lysogenic strain of *P. aeruginosa* was the source of the transducing particles.

Saye et al. (97) reported transduction of plasmid DNA in similar freshwater field tests. They tested three paradigms for the source of transducing particles and found that transduction frequencies were greatest when a lysogenic strain served as both the source of transducing phages and the recipient of transduced DNA. In these experiments, the plasmid was donated by an introduced nonlysogenic strain. Hence, transduction of the plasmid required (1) induction of temperate prophage from the recipient, (2) infection of the plasmid donor leading to the production of plasmid-containing transducing particles, and (3) reinfection of the original lysogen and subsequent establishment of the plasmid.

Transduction has been identified as a mechanism for gene transfer in a large number of environmentally significant bacterial species (Table 11.1). The presence of bacteriophages in natural environments is ubiquitous (103). Therefore, the conclusion is unavoidable that a great potential for gene transmission mediated by bacterial viruses exists in nature. It has led Novick (88) to speculate that, at least in some bacterial species, it may be the major mechanism for the natural dissemination of both chromosomal and plasmid DNA. It is clear that transduction cannot be ignored in studies designed to measure the environmental transmission of genetic elements.

Acknowledgments

This endeavor was supported in part by cooperative agreements Nos. CR-815234 and CR-815282 with the Gulf Breeze Environmental Research Laboratory of the U.S. Environmental Protection Agency.

References

1. Akhverdian, V. Z., E. A. Khrenova, M. A. Revlets, T. V. Gerasimova, and V. N. Krylov. 1985. The properties of transposable phages of *Pseudomonas aeruginosa* belonging to two groups distinguished by DNA-DNA homology. *Genetika* 21:735–747.
2. Alikhanian, S. I., T. S. Iljina, and N. D. Lomovskaya. 1960. Transduction in Actinomycetes. *Nature* 188:245–246.
3. Alonso, J. C., G. Lüder, and T. A. Trautner. 1986. Requirements for the formation of plasmid-transducing particles of *Bacillus subtilis* bacteriophage SPP1. *EMBO J.* 5:3723–3728.
4. Arber, W. 1960. Transduction of chromosomal genes and episomes in *Escherichia coli*. *Virology* 11:273–288.
5. Arber, W. 1983. A beginner's guide to lambda biology. In R. W. Hendrix, J. W. Roberts, F. W. Stahl, and R. A. Weisberg (eds.), *Lambda II*. Cold Spring Harbor Laboratory, Cold Spring Harbor, N.Y., pp. 381–394.
6. Arber, W., L. Enquist, B. Hohn, N. E. Murray, and K. Murrar. 1983. Experimental

methods for use with lambda. In R. W. Hendrix, J. W. Roberts, F. W. Stahl, and R. A. Weisberg (eds.), *Lambda II*. Cold Spring Harbor Laboratory, Cold Spring Harbor, N.Y., pp. 433–458.

7. Armentrout, R. W., and L. Rutberg. 1971. Heat induction of prophage φ105 in *Bacillus subtilis:* Replication of the bacterial and bacteriophage genomes. *J. Virol.* 8:455–468.

8. Barsomina, G. D., N. J. Robillard, and C. B. Thorne. 1984. Chromosomal mapping of *Bacillus thuringiensis* by transduction. *J. Bacteriol.* 157:746–750.

9. Bender, R. A. 1981. Improved generalized transducing bacteriophage for *Caulobacter crescentus. J. Bacteriol.* 148:734–735.

10. Bender, R. A., and L. C. Sambucitti. Recombination-induced suppression of cell division following P1-mediated generalized transduction in *Klebsiella aerogenes. Mol. Gen. Genet.* 189:263–268.

11. Benzinger, R., and P. E. Hartman. 1962. Effect of ultraviolet light on transducing phage P22. *Virology* 18:614–626.

12. Bitton, G. 1987. Fate of bacteriophages in water and wastewater treatment plants. In S. M. Soyal, C. P. Gerba, and G. Bitton (eds.), *Phage ecology*. Wiley, New York, pp. 181–195.

13. Bramucci, M. G., and P. S. Lovett. 1976. Low-frequency, PBS1-mediated plasmid transduction in *Bacillus pumilus. J. Bacteriol.* 127:829–831.

14. Bramucci, M. G., and P. S. Lovett. 1977. Selective plasmid transduction in *Bacillus pumilus. J. Bacteriol.* 131:1029–1032.

15. Buchanan-Wollaston, V. 1979. Generalized transduction in *Rhizobium leguminosarum. J. Gen. Microbiol.* 112:135–142.

16. Buck, G. A., and N. B. Groman. 1981. Genetic elements novel for *Corynebacterium diphtheriae:* Specialized transducing elements and transposons. *J. Bacteriol.* 148: 143–152.

17. Bukhari, A. I. 1976. Bacteriophage mu as a transposition element. *Annu. Rev. Genet.* 10:389–412.

18. Campos, J. M., J. Geisselsoder, and D. R. Zusman. 1978. Isolation of bacteriophage MX4, a generalized transducing phage of *Myxococcus xanthus. J. Mol. Biol.* 119: 167–178.

19. Canosi, U., G. Lüder, and T. A. Trautner. 1982. SPP1-mediated plasmid transduction. *J. Virol.* 44:431–436.

20. Casadesús, J., and J. Olivares. 1979. General transduction in *Rhizobium meliloti* by a thermosensitive mutant of bacteriophage DF2. *J. Bacteriol.* 139: 316–317.

21. Cavenagh, M. M., and R. V. Miller. Specialized transduction of *Pseudomonas aeruginosa* PAO by bacteriophage D3. *J. Bacteriol.* 165:448–452.

22. Chakrabarty, A. M., and I. C. Gunsalus. 1969. Autonomous replication of a defective transducing phage in *Pseudomonas putida. Virology* 38:92–104.

23. Chatterjee, A. K., and M. A. Brown. 1980. Generalized transduction in the enterobacterial phytopathogen *Erwinia chrysanthemi. J. Bacteriol.* 143:1444–1449.

24. Coetzee, J. N. 1966. Transduction in *Proteus morganii. Nature* 210:220.

25. Coetzee, J. N. 1975. Specialized transduction of kanamycin resistance in a providence strain. *J. Gen. Microbiol.* 88:307–316.

26. Coetzee, J. N. 1975. Transduction of a *Proteus vulgaris* strain by a *Proteus mirabilis* bacteriophage. *J. Gen. Microbiol.* 89:299–309.

27. Coetzee, J. N. 1976. Intra-species transduction with *Proteus mirabilis* high frequency transducing phages. *J. Gen. Microbiol.* 93:153–165.

28. Coetzee, J. N. 1977. High frequency cotransduction of a morganocinogenic plasmid and markers of R plasmids. *J. Gen. Microbiol.* 103:165–180.

29. Coetzee, J. N., H. C. de Klerk, and J. A. Smit. 1967. A transducing bacteriophage for *Proteus vulgaris. J. Gen. Virol.* 1:561–564.

30. Coetzee, J. N., and J. A. Smit. 1969. Restriction of a transducing bacteriophage in a strain of *Proteus mirabilis. J. Gen. Virol.* 4:593–607.

31. Coetzee, J. N., J. A. Smit, and O. W. Prozesky. 1966. Properties of providence and *Proteus morganii* transducing phages. *J. Gen. Microbiol.* 44:167–176.

32. Colón, A. E., R. M. Cole, and C. G. Leonard. 1970. Transduction in group A

streptococci by ultraviolet-irradiated bacteriophages. *Can. J. Microbiol.* 16:201–202.

33. Colón, A. E., R. M. Cole, and C. G. Leonard. 1971. Lysis and lysogenization of groups A, C, and G streptococci by a transducing bacteriophage induced from group G *Streptococcus. J. Virol.* 8:103–110.

34. Colón, A. E., R. M. Cole, and C. G. Leonard. 1972. Intergroup lysis and transduction by streptococcal bacteriophages. *J. Virol.* 9:551–553.

35. de Lencastre, H., and L. J. Archer. 1979. Transducing activity of bacteriophage SPP1. *Biochem. Biophys. Res. Commun.* 86:915–919.

36. de Lencastre, H., and L. J. Archer. 1980. Characterization of bacteriophage SPP1 transducing particles. *J. Gen. Microbiol.* 117:347–355.

37. de Lencastre, H., and L. J. Archer. 1981. Molecular origin of transducing DNA in bacteriophage SPP1. *J. Gen. Microbiol.* 122:345–349.

38. Deichelbohrer, I., J. C. Alonso, G. Lüder, and T. A. Trautner. 1985. Plasmid transduction by *Bacillus subtilis* bacteriophage SPP1: Effects of DNA homology between plasmid and bacteriophage. *J. Bacteriol.* 162:1238–1243.

39. Dénarié, J., C. Rosenberg, B. Bergeron, C. Boucher, M. Michel, and M. B. de Bertalmio. 1977. Potential for RP4::mu plasmids for in vivo genetic engineering of gram-negative bacteria. In A. I. Bukhari, J. A. Shapiro, and S. L. Adhya (eds.), *DNA Insertion Elements, Plasmids, and Episomes*, Cold Spring Harbor Laboratory, Cold Spring Harbor, N.Y., pp. 507–520.

40. Dyer, D. W., M. I. Rock, C. Y. Lee, and J. J. Iandolo. 1985. Generation of transducing particles in *Staphylococcus aureus. J. Bacteriol.* 161:91–95.

41. Ely, B., and R. C. Johnson. 1977. Generalized transduction in *Caulobacter crescentus. Genetics* 87:391–399.

42. Farrah, S. R. 1987. Ecology of phage in freshwater environments. In S. M. Soyal, C. P. Gerba, and G. Bitton (eds.), *Phage Ecology*. Wiley, New York, pp. 125–136.

43. Ferrari, E., U. Canosi, A. Galizzi, and G. Mazza. 1978. Studies on transduction process by SPP1 phage. *J. Gen. Virol.* 41:563–572.

44. Fink, P. S., and S. A. Zahler. 1982. Specialized transduction of the *ilvD-thyB-ilvA* region mediated by *Bacillus subtilis* bacteriophage SP. *J. Bacteriol.* 150:1274–1279.

45. Geisselsoder, J., J. M. Campos, and D. R. Zusman. 1978. Physical characterization of bacteriophage MX4, a generalized transducing phage for *Myxococcus xanthus. J. Mol. Biol.* 119:179–189.

46. Herman, N. J., and E. Juni. 1974. Isolation and characterization of generalized transducing bacteriophage for *Acinetobacter. J. Virol.* 13:46–52.

47. Hinkle, N. F., and R. V. Miller. 1979. pMG7-mediated restriction of *Pseudomonas aeruginosa* phage DNAs is determined by a class II restriction endonuclease. *Plasmid* 2:387–393.

48. Hirai, K., and R. Yanagawa. 1970. Generalized transduction of *Corynebacterium renale. J. Bacteriol.* 101:1086–1087.

49. Holloway, B. W. 1969. Genetics of *Pseudomonas. Bacteriol. Rev.* 33:419–443.

50. Holloway, B. W., J. B. Egan, and M. Monk. 1961. Lysogeny in *Pseudomonas aeruginosa. Aust. J. Exp. Biol. Med. Sci.* 38:321–329.

51. Holloway, B. W., and M. Monk. 1959. Transduction in *Pseudomonas aeruginosa. Nature* 184:1426–1427.

52. Holloway, B. W., and P. Van de Putte. 1968. Transducing phage for *Pseudomonas putida. Nature* 217:459–460.

53. Holloway, B. W., and P. Van de Putte. 1968. Lysogeny and bacterial recombination. In W. J. Peacock and R. D. Brock (eds.), *Replication and Recombination of Genetic Material*. Australian Academy of Science, Camberra, pp. 175–183.

54. Hyder, S. L., and M. M. Streitfeld. 1978. Transfer of erythromycin resistance from clinically isolated lysogenic strains of *Streptococcus pyogenes* via their endogenous phage. *J. Infect. Dis.* 138:281–286.

55. Ikeda, H., and J. Tomizawa. 1968. Prophage P1, an extrachromosomal replication unit. *Cold Spring Harbor Symp. Quant. Biol.* 33:791–798.

56. Jacoby, G. A., and L. Sutton. 1977. Restriction and modification determined by a *Pseudomonas* R plasmid. *Plasmid* 1:115–116.

57. Kaplan, R. W., and M. Brendel. 1969. Formation of prototrophs in mixtures of two auxotrophic mutants of *Serratia marcescens* HY by a transducing bacteriophage produced by some auxotrophs. *Mol. Gen. Genet.* 104:27–39.

58. Kelln, R. A., and R. A. J. Warren. 1971. Isolation and properties of a bacteriophage lytic for a wide range of pseudomonads. *Can. J. Microbiol.* 17:677–682.

59. Kilbane, J. J., and R. V. Miller. 1988. Molecular characterization of *Pseudomonas aeruginosa* bacteriophages: Identification and characterization of the novel virus B86. *Virology* 164:193–200.

60. Kokjohn, T. A. 1989. Transduction: Mechanism and potential for gene transfer in the environment. In S. B. Levy and R. V. Miller (eds.), *Gene Transfer in the Environment*. McGraw-Hill, New York, pp. 73–97.

61. Kopec, L. K., R. E. Yasbin, and R. Marrero. 1985. Bacteriophage SPO2-mediated plasmid transduction in transpositional mutagenesis within the genus *Bacillus*. *J. Bacteriol.* 164:1283–1287.

62. Krishnapillai, V. 1971. A novel transducing phage: Its role in recognition of a possible new host-controlled modification system in *Pseudomonas aeruginosa*. *Mol. Gen. Genet.* 114:134–143.

63. Krizsanovich, K., H. C. de Klerk, and J. A. Smit. 1969. A transducing bacteriophage for *Proteus rettgeri*. *J. Gen. Virol.* 4:437–439.

64. Krizsanovich-Williams, K. 1975. Specialized transduction of a leucine marker by *Proteus mirabilis* phage 5006M. *J. Gen. Microbiol.* 91:213–216.

65. Krylov, V. N., V. A. Akhverdian, V. G. Bogush, E. A. Khrenova, and M. Revlets. 1985. The modular structure of transposable phages genomes of *Pseudomonas aeruginosa*. *Genetika* 21:724–734.

66. Lecadet, M.-M., M.-O. Blondel, and J. Ribier. 1980. Generalized transduction in *Bacillus thuringiensis* var. *berliner* 1715 using bacteriophage CP-54Ber. *J. Gen. Microbiol.* 121:203–212.

67. Lederberg, J., E. M. Lederberg, N. D. Zinder, and E. R. Lively. 1951. Recombination analysis of bacterial heredity. *Cold Spring Harbor Symp. Quant. Biol.* 16:413–443.

68. Lederberg, J., and E. L. Tatum. 1946. Novel genotypes in mixed cultures of biochemical mutants of bacteria. *Cold Spring Harbor Symp. Quant. Biol.* 11:113–114.

69. Lederberg, J., and E. L. Tatum. 1946. Gene recombination in *E. coli. Nature* 158: 558.

70. Lipsky, R. H., R. Rosenthal, and S. A. Zanler. 1981. Defective specialized SP transducing bacteriophages of *Bacillus subtilis* that carry the *sup-3* or *sup-44* gene. *J. Bacteriol.* 148:1012–1015.

71. Lovett, P. S. 1972. PBP1: A flagella specific bacteriophage mediating transduction in *Bacillus pumilus*. *Virology* 47:743–752.

72. Lovett, P. S., D. Bramucci, M. G. Bramucci, and B. D. Burdick. 1974. Some properties of the PBP1 transduction system in *Bacillus pumilus*. *J. Virol.* 13:81–84.

73. Lovett, P. S., and F. E. Young. 1971. Linkage groups in *Bacillus pumilus* determined by bacteriophage PBS1-mediated transduction. *J. Bacteriol.* 106:697–699.

74. Luria, S. E., J. N. Adams, and R. C. Ting. 1960. Transduction of lactose-utilizing ability among strains of *E. coli* and *S. dysenteriae* and the properties of the transducing phage particles. *Virology* 12:348–390.

75. MacPhee, D. G., I. W. Sutherland, J. F. Wilkinson. 1969. Transduction in *Klebsiella*. *Nature* 221:475–476.

76. Marrero, R., F. A. Chiafari, and P. S. Lovett. 1981. SPO2 particles mediating transduction of a plasmid containing spo2 cohesive ends. *J. Bacteriol.* 147:1–8.

77. Martin, S., E. Sodergren, T. Masuda, and D. Kaiser. 1978. Systematic isolation of transducing phages for *Myxococcus xanthus*. *Virology* 88:44–53.

78. Matsumoto, H., Y. Itoh, S. Ohta, and Y. Terawaki. 1986. A generalized transducing phage of *Pseudomonas ceracia*. *J. Gen. Microbiol.* 132:2583–2586.

79. Miller, J. H. 1972. *Experiments in Molecular Genetics*. Cold Spring Harbor Laboratory, Cold Spring Harbor, N.Y.

80. Miller, R. V., and T. A. Kokjohn. 1988. Expression of the *recA* gene of *Pseudomonas aeruginosa* PAO is inducible by DNA damaging agents. *J. Bacteriol.* 170: 2385–2387.

81. Miller, R. V., and C.-M. C. Ku. 1978. Characterization of *Pseudomonas aeruginosa* mutants deficient in the establishment of lysogeny. *J. Bacteriol.* 134:875–883.

82. Miller, R. V., J. M. Pemberton, and A. J. Clark. 1977. Prophage F116: Evidence for extrachromosomal location in *Pseudomonas aeruginosa* PAO. *J. Virol.* 22:844–847.

83. Miller, R. V., J. M. Pemberton, and K. E. Richards. 1974. F116, D3 and G101: Temperate bacteriophages of *Pseudomonas aeruginosa*. *Virology* 59:566–569.

84. Miller, R. V., and V. J. Renta Rubero. 1984. Mucoid conversion by phages of *Pseudomonas aeruginosa* strains from patients with cystic fibrosis. *J. Clin. Microbiol.* 19:717–719.

85. Moebus, K. 1987. Ecology of marine bacteriophages. In S. M. Soyal, C. P. Gerba, and G. Bitton (eds.), *Phage Ecology*. Wiley, New York, pp. 137–156.

86. Morrison, W. D., R. V. Miller, and G. S. Sayler. 1978. Frequency of F116-mediated transduction of *Pseudomonas aeruginosa* in a freshwater environment. *Appl. Environ. Microbiol.* 36:724–730.

87. Morse, M. L. 1954. Transduction of certain loci in *Escherichia coli* K-12. *Genetics* 39:984–985.

88. Novick, R. P., I. Edelman, and S. Lofdahl. 1986. Small *Staphylococcus aureus* plasmids are transduced as linear multimers that are formed and resolved by replicative processes. *J. Mol. Biol.* 192:209–220.

89. Ogg, J. E., T. L. Timme, and M. M. Alemohammad. 1981. General transduction in *Vibrio cholerae*. *Infect. Immunol.* 31:737–741.

90. Pemberton, J. M. 1974. Size of the chromosome of *Pseudomonas aeruginosa* PAO. *J. Bacteriol.* 119:748–752.

91. Primrose, S. B., and M. Day. 1977. Rapid concentration of bacteriophages from aquatic habitats. *J. Appl. Bacteriol.* 42:417–421.

92. Primrose, S. B., N. D. Seeley, K. B. Logan, and J. W. Nicolson. 1982. Methods for studying aquatic bacteriophage ecology. *Appl. Environ. Microbiol.* 43:694–701.

93. Prozorov, A. A., T. S. Belova, and N. N. Surikov. 1980. Transformation and transduction of *Bacillus subtilis* strains with the Bsu R restriction-modification system by means of modified and unmodified DNA of pUB110 plasmid. *Mol. Gen. Genet.* 180:135–138.

94. Ptashne, M. 1986. *A Genetic Switch: Gene Control and Phage λ*. Cell Press and Blackwell Scientific Publications, Palo Alto, Calif.

95. Rosenthal, R., P. A. Toye, R. Z. Korman, and S. A. Zahler. 1979. The prophage of SP *c2dcitK₁*, a defective specialized transducing phage of *Bacillus subtilis*. *Genetics* 92:721–739.

96. Ruhfel, R. E., N. J. Robillard, and C. B. Thorne. 1984. Interspecies transduction of plasmids among *Bacillus anthracis, B. cereus,* and *B. thuringiensis*. *J. Bacteriol.* 157:708–711.

97. Saye, D. J., O. Ogunseitan, G. S. Sayler, and R. V. Miller. 1987. Potential for transduction of plasmids in a natural freshwater environment: Effect of plasmid donor concentration and a natural microbial community on transduction in *Pseudomonas aeruginosa*. *Appl. Environ. Microbiol.* 53:987–995.

98. Seeley, N. D., and S. B. Primrose. 1980. The effect of temperature on the ecology of aquatic bacteriophages. *J. Gen. Virol.* 46:87–95.

99. Shapiro, J. A., D. H. Dean, and H. O. Halvorson. 1974. Low-frequency specialized transduction with *Bacillus subtilis* bacteriophage φ105. *Virology* 62:393–403.

100. Sik, T., J. Horváth, and S. Chatterjee. 1980. Generalized transduction in *Rhizobium meliloti*. *Mol. Gen. Genet.* 178:511–516.

101. Skjold, S. A., W. R. Maxted, and L. W. Wannamaker. 1982. Transduction of the genetic determinant for streptolysin S in group A streptococci. *Infect. Immun.* 38: 183–188.

102. Southern, E. 1979. Gel electrophoresis of restriction fragments. *Methods Enzymol.* 68:152–176.

103. Soyal, S. M., C. P. Gerba, and G. Bitton. 1987. *Phage Ecology.* Wiley, New York
104. Stäber, H., and H. Böhme. 1971. Transduktion mit dem temperenten *Proteus mirabilis*-phagen π_1. I. Charakterisierung des transduktionssystems. *Zeit. Allg. Mikrobiol.* 11:221–230.
105. Stäber, H., and H. Böhme. 1971. Transduktion mit dem temperenten *Proteus mirabilis*-phagen π_1. II. Einfluss der vermehrungsmultipliizität auf die transduktionsfahigkeit des phagen. *Zeit. Allg. Mikrobiol.* 11:231–236.
106. Stent, G. S., and R. Calendar. 1978. *Molecular Genetics: An Introductory Narrative,* 2d ed. W. H. Freeman, San Francisco.
107. Stuttard, C. 1979. Transduction of auxotrophic markers in a chloramphenicol-producing strain of *Streptomyces. J. Ben. Microbiol.* 110:479–482.
108. Süss, F., and S. Klaus. Transduction in *Streptomyces hydroscopicus* mediated by the temperate bacteriophage SH10. *Mol. Gen. Genet.* 181:552–555.
109. Susskind, M. M., and D. Botstine. 1978. Molecular genetics of bacteriophage P22. *Microbiol. Rev.* 42:385–413.
110. Thorne, C. B. 1968. Transducing bacteriophage for *Bacillus cereus. J. Virol.* 2:657–662.
111. Thorne, C. B. 1978. Transduction in *Bacillus thuringiensis. Appl. Environ. Microbiol.* 35:1109–1115.
112. Tolmach, L. J. 1957. Attachment and penetration of cells by viruses. *Adv. Virus Res.* 4:63–110.
113. Totolyan, A. A., A. S. Boitsov, K. L. Kol, and V. I. Golubkov. 1981. Comparative characteristics of the transducing virulent streptococcal phages A25 and CA1. *Mol. Biol. (Moscow)* 15:894–900.
114. Vary, P. S., J. C. Garbe, M. Franzen, and E. W. Frampton. 1982. MP13, a generalized transducing bacteriophage for *Bacillus megaterium. J. Bacteriol.* 149:1112–1119.
115. Wannamaker, L. W., S. Almquist, and S. Skjold. 1973. Intergroup phage reactions and transduction between group C and group A streptococci. *J. Exp. Med.* 137:1338–1353.
116. Williams, S. T., A. M. Mortimer, and L. Manchester. 1987. Ecology of soil bacteriophages. In S. M. Soyal, C. P. Gerba, and G. Bitton (eds.), *Phage Ecology.* Wiley, New York, pp. 157–179.
117. Woods, D. R., and J. A. Thompson. 1975. Unstable generalized transduction in *Achromobacter. J. Gen. Microbiol.* 88:86–92.
118. Yasbin, R. E., and R. E. Young. 1974. Transduction in *Bacillus subtilis* by bacteriophage SPP1. *J. Virol.* 14:1343–1348.
119. Yelton, D. B., and C. B. Thorne. 1970. Transduction in *Bacillus cereus* by each of two bacteriophages. *J. Bacteriol.* 102:573–579.
120. Yoshikawa, M., and Y. Hirota. 1971. Impaired transduction of R213 and its recovery by a homologous resident R factor. *J. Bacteriol.* 106:523–528.
121. Yu, L., and J. N. Baldwin. 1971. Intraspecific transduction in *Staphylococcus epidermidis* and interspecific transduction between *Straphylococcus aureus* and *Straphylococcus epidermidis. Can. J. Microbiol.* 17:767–773.
122. Zahler, S. A., R. Z. Korman, R. Rosenthal, and H. E. Hemphill. 1977. *Bacillus subtilis* bacteriophage SP: Localization of the prophage attachment site, and specialized transduction. *J. Bacteriol.* 129:556–558.
123. Zeph, L. R., M. A. Onaga, and G. Stotzky. 1988. Transduction of *Escherichia coli* by bacteriophage P1 in soil. *Appl. Environ. Microbiol.* 54:1731–1737.
124. Zinder, N. D., and J. Lederberg. 1952. Genetic exchange in *Salmonella. J. Bacteriol.* 64:679–699.

12

Natural Transformation and Its Potential for Gene Transfer in the Environment

Gregory J. Stewart

Of the three classic mechanisms for genetic exchange in bacteria, natural transformation was the method first reported (1). Interestingly enough, even though this process was detected more than 20 years prior to both transduction (2) and conjugation (3), it remains the least understood, most likely due to its absence in *Escherichia coli* and the other enterics. Fortunately, the process of natural transformation has been of interest to a small number of research laboratories, and our understanding of this process has received the attention of these investigators for some time. Our understanding of the mechanism of natural transformation has increased dramatically, especially since the 1960s (see Ref. 5). In light of current interests in the release of genetically engineered microorganisms (GEMs) to the environment, an understanding of how bacteria exchange genes has taken on new importance. This chapter will review what is known about the processes of natural transformation, the limitations of our current knowledge about transformation with regard to the release of GEMs, and the physicochemical properties of DNA that may be important in predicting the potential for gene transfer by natural transformation in the environment. It will discuss current methods that have proved successful or that show promise for measuring transformation potential in the environment. Finally, it will summarize current information on environmental transformation and will attempt to identify important questions that should be addressed about the environmental importance of gene transfer in the environment.

Historical Perspectives and Introduction

The earliest report of a mechanism for exchange of genetic information in bacteria is also the earliest report of natural transformation. F. W. Griffith reported in 1928 that rough nonvirulent strains of pneumococcus (*Streptococcus pneumoniae*), when mixed with a heat-killed preparation of smooth, virulent strains, were capable of eliciting infection when injected into mice. Neither the rough-strain nor the smooth-strain preparation alone resulted in disease. Strains isolated from mice infected with the mixture were invariably smooth strains. Griffith proposed that a "transforming factor" in the heat-killed preparation converted the rough strain to smooth, although the nature of this factor was not determined (1). It is interesting to note that this earliest report of transformation is still one of the best published examples of genetic conversion of a strain in a natural environment. The nature of the transforming principle was not elucidated until sometime later when Avery, McLeod, and McCarty, through a series of chemical analyses, determined that the transforming material was most likely DNA (4). Since these early reports in *S. pneumoniae* a number of other gram-positive and gram-negative bacteria have been shown to transform (5), and although the ability to transform is not uniformly distributed among all the species of any bacterial genus, the breadth of genera with at least some transformable species is considerable (Table 12.1).

Once alternative mechanisms for genetic exchange were discovered (6), it became important to distinguish them. Natural transformation is distinguished from transduction, genetic transfer in bacteria mediated by bacteriophage infection, through the use of a DNase I sensitivity test. Since natural transformation involves the uptake of DNA from solution, that DNA is unprotected and can be degraded by addition of DNase I to the transformation mixture. The phage capsid protects transduced DNA from enzymatic attack, making transduction resistant to DNase I (60). Also, transduction is traditionally defined as being transfer mediated by phage, and consequently this property has been associated with a plaque-forming agent, whereas transformation is not associated with an infectious agent.

Conjugation is distinguished from transduction and transformation in that it involves transfer associated with an extrachromosomal element (6). The process involves a single-strand transfer from a donor to a recipient cell, requiring direct contact between the mating types, another feature that distinguishes it from the other transfer mecha-

TABLE 12.1   Genera of Bacteria
Capable of Natural Transformation

Genus	Reference
Achromobacter	75
Acinetobacter	29
Anacystis	83
Azotobacter	76
Bacillus	17
Haemophilus	5
Halobacterium	77
Methylobacterium	78
Micrococcus	79
Mycobacterium	80
Moraxella	75
Neisseria	14
Pseudomonas	30
Streptococcus	1
Streptomyces	81
Synechococcus	82
Vibrio	84

nisms. Finally, the conjugational mechanism protects the DNA from enzymatic attack; thus conjugation is resistant to DNase I, distinguishing it from transformation (6).

The distinguishing traits clearly defined the three exchange processes until recently, but studies in various systems have invalidated many of these distinguishing features. For instance, Marrs and colleagues reported a genetic exchange mechanism in *Rhodobacter capsulatus* (formerly *Rhodopseudomonas capsulata*), called gene transfer agent (GTA), that is similar in many respects to transduction. The process mobilized genes randomly, but in units of a fixed length. If one incubated a culture, then separated the cells from the supernatant, the supernatant retained a gene transfer activity that was DNase I resistant. Electron microscopic analyses revealed viruslike structures in the supernatant, yet no plaque formation was ever associated with the donor stains or the supernatant GTA preparations (7). Thus, the hallmark of plaque-forming units defining transduction was brought into question.

Recent reports on gene transfer in the genus *Streptococcus* have required a modification of the definition of conjugation. Certain strains of this genus have demonstrated donor properties that are in all respects identical to conjugation: they are DNase I resistant; they demonstrate a unidirectional, linked transfer of genes that corresponds to the gene order on the chromosome; they allow the transfer of reasonably large pieces of single-strand DNA; and they require direct contact between donors and recipients. This system differs from traditional conjugation systems, however, in that it is not associated with extrachromosomal plasmids. These strains derive their ability to transfer DNA to recipients from chromosomally linked conjugative elements (8). Therefore, the definition of conjugation as gene transfer through plasmid-facilitated exchange is no longer valid.

Even the concept of natural transformation may require redefinition. Natural transformation has been traditionally viewed as the uptake of soluble, extracellular DNA by competent recipient cells. This process has been viewed as a donor-independent process because studies of natural transformation have been limited to the laboratory and have been performed in most cases with purified DNA provided to the recipient in a soluble form. However, reports in *Pseudomonas stutzeri* indicate that natural transformation can occur at a reasonably high frequency between intact donor and recipient cells (9). This process remains DNase I sensitive, is random with respect to marker transferred, and occurs in the absence of detectable plasmid or phage. Reports in *Haemophilus influenzae* (10), and in marine isolates of *P. stutzeri,* and *Vibrio* species (84, 85) suggest that donor-mediated natural transformation may be a reasonably common property. Review of the literature on gene transfer in *Bacillus subtilis* (11) suggests that this donor role in transformation may occur in gram-positive as well as gram-negative naturally transformable bacteria. The significance of these reports with respect to the environmental impact of natural transformation is that the instability of exogenous DNA as an environment provides no assurance that natural transformation will not occur.

It is clear in the light of these recent reports that our traditional definitions of gene transfer mechanisms in bacteria must be viewed with a certain degree of skepticism, especially as we consider departure from the controlled conditions of the laboratory to more variable environmental simulations and field studies. However, our perception of the potential for genetic exchange by natural transformation in the environment can be guided to some degree by the knowledge gained from the past 50 years of laboratory studies. By considering what is known about the stability of DNA in the environment, coupled with the knowledge of the important physical properties of DNA that are likely to influence its activity and survival in the environment, it may be possible to predict something of the potential significance of natural transformation in the field, guiding the design of experiments to address environmental transformation potential. To that end this chapter will review current knowledge of the definition, diversity, and mechanisms of natural transformation and will explore some significant properties of DNA likely to influence its survival and transfer.

The traditional definition of natural transformation is that it is a property of bacteria that allows cells to acquire and express genes from the environment. This is a normal, physiological function of these bacteria that is reasonably complex, involving as many as ten gene products (12). The process differs from artificial transformation, which requires some physical, chemical, or enzymatic treatment of the cells (or some combination of these). Natural transformation is physiologically controlled and requires no specialized treatment. Natural and artificial transformation differ as well in their specificity. Artificial transformation is most effective with covalently closed, circular DNA (13), whereas natural transformation is reasonably inefficient for plasmid DNA unless a degree of homology is retained internally on the plasmid or is shared by the plasmid and the endogenote (5). Natural transformation is also a regulated function, that is, it is only expressed under certain conditions; the exception to this rule being *Neisseria gonorrhoeae,* in which competence is constitutively expressed (14).

Natural transformation has only been studied extensively in a limited number of systems. Our earliest reports of transformation were made with a *S. pneumoniae* model. This particular system has continued to contribute to our understanding of the mechanisms involved in this process. Transformation has also been studied in other streptococcal species, but only studies in *S. sanguis* have contributed significantly to our knowledge of the process beyond what has been learned in the pneumococci. Among the other gram-positive bacteria, only species of *Bacillus* have received serious attention. The bulk of our information about transformation in this genus results from studies of *Bacillus subtilis.* The *Bacillus* and *Streptococcus* models appear to be quite similar in many respects, as will become clear in the discussion on the mechanisms of natural transformation.

Although a number of gram-negative bacteria have been shown to transform naturally, our understanding of transformation in the gram-negative bacteria is based almost entirely on studies of transformation in *Haemophilus* species. *H. influenzae* has been most extensively evaluated, with a limited number of studies performed in *H. parainfluenzae.* A reasonable effort has been made to study transformation in *Neisseria gonorrhoeae,* and considerably less is known about the transformation systems of *Azotobacter, Acinetobacter, Pseudomonas,* and *Vibrio.* As more studies are performed on the mechanisms of natural transformation, the various systems appear more similar than they are different. This observation will be more extensively explored later in this chapter.

In addition to developing models for the mechanism of natural transformation from a limited number of species, experimental methods and sources of transforming DNA are quite limited. Virtually everything known about natural transformation results from experiments performed under optimal conditions in the research laboratory. Until recently scientists have only speculated as to the significance of natural transformation as an environmentally important process. Studies of the mechanisms for uptake of linear chromosomal DNA by recipient strains are extensive, but only recently have studies on the utility of the mechanism for uptake of plasmid DNA been performed. Both of these questions are especially important in light of new efforts to release GEMs.

Finally, although the process of natural transformation has been studied for 50 years, the question of the potential source of transforming DNA has never been satisfactorily addressed. Until recently, natural transformation was used as a tool in the laboratory. It was considered a "laboratory phenomenon," and since DNA could be readily prepared and applied, the question of a source of DNA for transformation was a moot one. Recent evidence of the potential of an active donor function in natural transformation raises the question of DNA release as a part of the process. Thus, studies that broaden our available pool of transformation model systems and those that address the release of DNA are critical and require additional effort.

Obviously, a discussion of the potential of natural transformation as an environmentally significant process would be incomplete without a discussion of the stability and reactivity of DNA in the environment. The susceptibility of DNA to enzymatic degradation, the impact of binding of DNA to sands, soils, and sediments, and the availability of DNA to competent cells are all factors that will influence the potential for gene transfer by natural transformation in environmentally significant situations.

Mechanism of Natural Transformation

The classical view of natural transformation has focused on the recipient role, that is, the DNA uptake and expression process. This mechanism is generally viewed as consisting of four discrete phases: development of competence, DNA binding, DNA uptake, and recombination and expression. Each of these phases has traits or properties that are specific. Each has been defined by mutant and physiological analysis. This four-stage scheme will be used to discuss the recipient function. Rather than discussing each bacterial model, the various model systems will be compared at each phase of the recipient process to identify the similarities and differences that exist.

Development of competence

Competence is a physiological state of transformable bacteria that allows the organism to bind and take up DNA. As stated earlier, with the exception of *Neisseria gonorrhoeae,* all naturally transformable bacteria regulate competence development (9). The means by which competence is controlled in bacteria varies, but the outcome appears to be the same; competence is triggered under physiological conditions that mimic those of unbalanced growth. *Streptococcus* demonstrates something of an exception to this rule in that it also demonstrates competence for a finite period of time during exponential phase (15). The development of competence has been characterized for two genera of gram-positive bacteria. Competence is controlled in both *Streptococcus* and *Bacillus* by a low-molecular-weight protein, competence factor (16, 17). In each case competence factor is synthesized and excreted constitutively as the culture grows. As the culture density increases, so does the concentration of competence factor. When the exogenous concentration of the protein is sufficiently high, it triggers a series of physiological changes in the cells (18). Competence factor appears to reach this critical concentration when the density of the culture approaches 10^8 cells per milliliter (15). This state is maintained as long as the competence factor concentration remains high. Removal of the competent cells from their growth medium followed by resuspension in fresh medium lacking competence factor results in rapid loss of competence (5). Conversely, one can resuspend non-competent cells in spent medium from a competent culture and induce competence at a rate significantly greater than the growth rate of the culture, suggesting that competence involves conversion of existing cells rather than growth into competence (19). However, there are detectable differences in competence development when one compares the *Streptococcus* and *Bacillus* systems. The shift to competence in *Bacillus* is a gradual process. It involves an eventual conversion of cells to the competent state over the duration of time for conversion of exponential- to stationary-phase cells (20). The *Streptococcus* model displays a much more united shift within the population in that the population shifts to maximum competence over a very short period of time (15). Because of the rapid and complete shift to competence displayed by *Streptococcus,* extensive studies have been performed on the control of competence in this system. As with other competence systems, a series of physiological changes are detected either coincidentally with competence induction or as a result of it. As competence increases, the chain length of clusters of cells increases by as much as eightfold (21). The growth rate of the competent cells in the culture decreases rapidly, and in many cases cell death has been associated with the uptake of DNA by competent cells (22). The synthesis of proteins other than those specific for transformation appears to be temporarily inhibited (23). As many as sixteen proteins have been associated with competence development in *S. sanguis* (24), and a large mRNA transcript encoding as many as 10 polypeptides has been detected (12). The transient nature of the competent state also seems to be a regulated phenomenon. A specific inhibitor is released from the cells coincidentally with the termination of competence (25). Chen and Morrison (26) have reported a biphasic modulation of competence development and the number of competence shifts were affected by the pH of the medium, with more alkaline conditions resulting in biphasic competence patterns, and lower pH resulting in only a single shift to competence. In the case of the biphasic regime, the first peak consisted of about 80 percent conversion of the population to competence, whereas only about 12 percent of the cells were converted to competence in the latter induction. The practical implication of these studies is that slight variations in growth conditions can greatly influence the potential of a population to take up and express DNA, an important consideration if one looks to field trials where growth conditions may be difficult to standardize.

Competence development varies in other model systems. With the exception of the *Bacillus* and *Streptococcus* systems discussed above, competence development has probably been most extensively studied in *H. influenzae*. Competence development in this system is internally regulated. There are a number of ways to induce competence in the organism. Shift of an exponential culture to a medium that will not sustain growth will cause induction of competence in almost 100 percent of the population (5). Addition of cyclic AMP (27) and blocking of cell division while protein synthesis proceeds (5) are alternative methods for competence induction in this bacterium. The competent state in this model system appears to differ somewhat from the *Streptococcus* model. While competence in the latter is a transient phenomenon, in the case of *Haemophilus* the competent state is reasonably stable (28). Competent cells can be converted to noncompetent ones by shifting to rich medium that supports rapid, balanced growth.

Development of competence in other gram-negative bacteria is not as well characterized. *Acinetobacter calcoaceticus* (29) and *Pseudomonas stutzeri* (30) are induced to competence at the onset of stationary phase. Starvation for iron is related to induction of competence in the nitrogen-fixing bacterium *Azotobacter vinelandii* (W. J. Page, personal communication), and *P. stutzeri*, strain ZoBell, a marine isolate, appears to be competent during exponential phase as well as during stationary phase (85). To summarize, with the possible exceptions of *S. pneumoniae, N. gonorrhoea,* and perhaps marine strains of *P. stutzeri*, most transformable bacteria only achieve the competent state when shifted to stationary phase of growth, or to some other state of unbalanced growth. When one considers the potential environments where transformable bacteria may occur, it is likely that these cells will not be presented with conditions that allow exponential growth; it is more likely that some nutrient limitation in the environment will result in an unbalanced state for the cell. Therefore, it is reasonable to infer that transformable bacteria might exist in a competent state in their natural environments.

It is interesting to note that reports suggest a possible correlation between the competent state and the potential for cells to release DNA. *B. subtilis* releases DNA from cells in a form that is transformable (31). The release of this DNA correlates with competence development as a result of a tendency of part of the population to be prone to lysis. The released DNA tends to be most closely associated with intact cells in the population. This bacterium has also demonstrated an ability to release DNA from cells by extrusion. The DNA is released from cells in an ordered fashion starting at or near the origin of chromosomal replication (32). The exogenous DNA resulting from this process is selectively enriched for markers known to be associated with membrane/DNA and cell wall/DNA complexes from the bacterium. These data suggest at least one system where coordinated DNA release/DNA uptake mechanisms may exist. There have also been reports in *P. stutzeri* (9) and *H. influenzae* (10) that suggest a specific mechanism for release of DNA.

DNA binding

DNA binding by competent bacteria appears to involve at least two distinguishable stages. These phases are distinguished by the relative ease for removal of the bound DNA prior to uptake. In both *S. pneumoniae* and *B. subtilis* uptake of DNA occurs rapidly after binding of DNA to the outer surface of the cell (5). One can retard the uptake process substantially by addition of EDTA. This does not inhibit the binding process, but it prevents uptake (33). Competent cells first bind DNA in a "loose" form. In this form it is possible to remove the DNA by ionic washes or by addition of competing DNA. As the binding progresses to the "tight" binding stage, the DNA is no longer removed by these processes but it is still sensitive to shearing and to DNase I digestion (34). There is a very strong preference for binding of double-stranded DNA, whereas single-stranded DNA, RNA, and DNA:RNA hybrids bind very poorly (5). In both gram-positive model systems there appears to be no preference for specificity for homologous DNA; any double-stranded DNA will bind at high efficiency (35). However, reports of DNA binding in *Haemophilus* suggest the possibility of two binding specificities: one for homologous DNA and another, nonspecific binding system (36). Studies of DNA binding in *P. stutzeri* indicate that efficiency of transformation can be reduced if homologous DNA is mixed with heterologous DNA, suggesting a nonspecific binding process in this model (30).

In all cases, DNA binding is mediated by specific proteins associated with the cell membranes of the competent bacteria. Transformation mutants of *Streptococcus* (37) and *Haemophilus* strains have been isolated that lack specific membrane proteins, and in the latter case, some of these mutants have been shown to shed membrane vesicles that contain DNA-binding proteins (38). In strains of *Streptococcus* the DNA binding proteins exist in a complex with a nuclease involved in processing the bound DNA (5).

There are a number of features of DNA binding in naturally transformable bacteria that may be significant with regard to environmental transformation. First of all, loose binding suggests an ionic association between the DNA and the DNA binding site. The fact that high salt washes can dissociate the DNA-cell associations allows the ionic strength of the environment to influence the potential for natural transformation under field conditions. However, the real significance of ionic disruption of nucleic acid–cell associations may be less important than laboratory studies of DNA binding suggest, since these studies are often done in the presence of EDTA, creating an artificially slow DNA binding period. In fact, in the presence of divalent cations (as would exist in many environments) it is likely that the DNA binding step in natural transformation would be transient, with rapid transport of the DNA into the competent cell.

DNA uptake

Until recently it was thought that DNA uptake was mechanistically different in gram-positive and gram-negative bacteria (5). Transformation studies in the gram-positive bacteria established that DNA taken up by the cell passes through a single-stranded intermediate (39). Since there is a requirement for double-stranded DNA for binding, transforming intermediates during the uptake stage were incapable of transforming other competent cells if reisolated from the original transforming cells prior to recombination (40). This stage in the transformation process is referred to as the eclipse phase, since the DNA has no apparent transforming activity. The basis for the gram-reaction dichotomy of uptake processes was based on the absence of an eclipse phase of transforming intermediates in *H. influenzae* (41). Later, competent cells of *H. influenzae* were shown to possess specialized membrane structures, called transformasomes, reportedly involved in the DNA uptake process (42). Transforming DNA contains an eight-base-pair recognition sequence bound to proteins located in these transformasomes. The DNA was sequestered in the transformasome in a double-stranded form for periods of as much as an hour before the DNA was released into the cytoplasm for recombination. More recent reports indicate that transforming DNA may actually involve a single-strand intermediate as the sequestered DNA exits the transformasome into the cytoplasm, but the single-stranded DNA appears to enter a recombination complex almost immediately (47). This short lifetime for the single-stranded intermediate precludes its detection as an eclipse form. Thus, entry of transforming DNA into the cytoplasm appears similar for both gram-positive and gram-negative bacteria in that entry into the cytoplasm involves only single-stranded DNA. The specificity for an uptake sequence appears in both *Haemophilus* and *Neisseria* species.

The process of DNA uptake appears similar in most models. At the time of or shortly following binding of DNA to the membranes of competent bacteria, transformation-specific nucleases nick the bound DNA. One strand of the bound DNA is then translocated across the cell membrane while the complementary strand is degraded (5). In the case of *Streptococcus* and *Bacillus,* the single-strand intermediate is protected from degradation by protein binding (45, 46). In *Haemophilus* the single strand appears to enter a recombination complex immediately upon entry into the cytoplasm, preventing degradation of the transforming DNA by cytoplasmic nucleases (47).

Recombination

The first three phases of the transformation process, competence development, DNA binding, and DNA uptake, appear to involve gene products unique to the transformation process. The final phase, however, seems to utilize the general homologous recombination machinery of the recipient cell. The presence of single-stranded DNA in bacterial cells is known to trigger the strand replacement activity of the *recA* gene product (48). Thus, entry of single-stranded transforming DNA into recipients would activate the ability of the cells to mediate strand exchange. In each model where the recombination function of transformation has been studied, the gene product involved in the recombination phase is very similar in nature to the *E. coli recA* gene product. Transformation mutants blocked in the ability to establish transforming DNA once it has entered the cell display the classic pleiotropic phenotype of a recombination mutant (49, 50). This process is a rapid one, mediating the single-strand replacement almost immediately following the formation of the recombination complex.

The importance of this dependence on homologous recombination is clear as one considers issues of biotechnology risk assessment. The necessity for substantial homology to mediate strand exchange for the recombination gene products (35) significantly limits the potential for soluble exogenous DNA to alter transformable bacteria. Recombination frequencies are low, even between closely related species (51). Thus, it is not enough that DNA exist in the environment; that DNA must be substantially similar in sequence to the genomes of the endogenous organisms before gene replacement can occur. However, genetic engineering involves the use of specialized nucleic acid sequences, including transposons and plasmids. These elements frequently contain highly conserved sequences, some of which encode resistance to various antibiotics and to certain toxic compounds that could provide direct selection for transformants were these compounds to stress the environment containing the transformed species. We should also recognize that methods for biological containment of genetically engineered species often involve mutations in essential genes that also may be highly conserved among bacteria. As a result, while the homologous recombination step in natural transformation limits the potential gene pool, it may allow repair of GEMs altered in these highly conserved functions. Thus, release of GEMs presents a twofold risk by transformation: the dissemination of recombinant molecules to native flora in the environment, and the repair and subsequent loss of control of the genetically debilitated GEM.

The demand for homology also poses limitations to screening field isolates for transformability. Broad screens for transforming bacteria are difficult because of the sequence specificity of the field isolates. One almost requires the strains in pure culture before transformation probes can be developed. Potential methods for generic screening for transforming bacteria in the environment will be discussed later in this chapter.

Properties of DNA that might influence environmental transformation. The complexity of the DNA molecule presents its own set of special considerations as one discusses the potential for transformation in the environment. The hydrophilic interactions of deoxyribonucleic acid have been an area of research for a number of years. The role of these hydrophilic interactions provides the basis for base pairing of DNA strands (52), for transcription (53), and for overall migration of DNA in an electric field (54). It is reasonable to infer that hydrophilic interactions may influence the binding and stability of DNA in the environment, especially when interacting with positively charged molecules and surfaces. More recently attention has been focused on the hydrophobic interactions of DNA (55). This chemical property greatly increases the potential for binding of DNA to materials and surfaces in the field.

The potential for exogenous DNA to impact the genome of a population will be influenced by two important processes. The importance of competence and natural transformation in the expression of exogenous genes has been discussed. One must also consider that DNA, while an extremely stable molecule in isolation, is susceptible to enzymatic degradation. There have been a number of studies that document the concentration of DNA in the environment (56) and the rate of DNA turnover in the environment (57). For transformation to be environmentally significant, the rate of uptake of DNA by competent cells must be competitive with the rate of DNA turnover. The role of binding of DNA to surfaces could significantly alter the relative rates of transformation as compared with turnover. Studies have shown that frequencies of transformation are reduced when exogenous DNA binds to the surface of sediments (58), however, the rate of turnover of that DNA by DNases is also significantly reduced (59). Therefore, while binding of DNA to surfaces may reduce the immediate rate of transformation by the population, this same process may significantly increase the residence time for DNA in the environment by protecting it from enzymatic turnover. Once a foreign gene or sequence is introduced to the native population, it may be stabilized and amplified by reproduction of the transformed cell and its progeny. Thus, while binding of DNA to sediments and other surfaces may reduce the immediate rate of transformation, it may stabilize the transformation potential of the environment, ultimately resulting in broader distribution of the transforming DNA in the mixed population. It will be most important, therefore, to determine the potential for recombinant gene stabilization by binding of the DNA to materials in the environment.

Finally, in light of evidence for cell contact transformation, the rare transformation event could allow entry of a recombinant molecule into the gene pool of the native flora. Once stabilized in the pool, natural transformation between cells or release of DNA from cells as a donor function could facilitate the dissemination of sequences within the population. Thus, even rare transformation events could potentially have a major impact on the total genome of the population.

Current practices

Virtually everything known about natural transformation has resulted from laboratory studies where variables could be precisely controlled and where competence could be induced to maximal levels.
With the exception of Griffith's first report, we have had no evidence
that natural transformation will be a significant process in the field
until recently.

Predictions can be made, however, of how environments may be conducive to competence development. Most transformable bacteria are
induced to competence when they are shifted from balanced, exponential growth to unbalanced growth (35), as often occurs at the onset of
stationary phase (60). Additionally, shifts in biosynthetic capacity can
trigger competence (61). Many environments are limited in an essential nutrient for bacterial growth; consequently, it is possible that
many of these bacteria exist in the field in the competent state. If this
is the case, one might expect transformation to be environmentally
significant. One must also consider, however, that exogenous DNA
will always be susceptible to degradation by nucleases. As a result,
the frequencies of transformation detected in the environment will
probably be less, perhaps significantly less, than those detected for
laboratory transformations. Thus, any means to measure natural
transformation in the field must be sensitive, and means that are
amplifiable would have definite advantages. We must also remember
that differences in transcription machinery might allow a gene to be
harbored by a recipient cell without expressing in that cell. Thus, one
cannot rely completely on measurement methods that require or depend on transcription-translation steps. It is safe to say that there is
no single assay for environmental transformation, but in fact, that a
number of approaches will be necessary before meaningful results are
obtained. The following discussion will address some of the methods
currently used for transformation studies and will explore their utility
for studies of natural transformation in the field.

Use of auxotrophic markers.   One of the most powerful screens for natural transformation in the laboratory involves the use of auxotrophic
markers. Auxotrophs are mutants that are defective in a gene for
some metabolic pathway. The result of this mutation is the inability
to grow on chemically defined media without incorporation of the particular growth factor synthesized by that pathway (62). One can utilize auxotrophs as recipients in a transformational cross. If one utilizes DNA from a wild-type donor cell and uses this DNA to transform
the auxotroph, the resulting transformant will repair the auxotrophic

lesion, allowing the recipient to grow on media without the growth factor. Thus, one can directly select transformants. The value of auxotrophic markers in field studies of natural transformation is questionable, however, because many isolates (especially aquatic isolates) fail to grow on chemically defined media (63). Since most complex media contain a rich assortment of biological molecules, it is difficult to select auxotrophy because the medium will contain the required growth factor. There are a few auxotrophic mutations that could be of value, however. If one utilized sources of carbon such as vitamin-free casamino acids, one could probably utilize some vitamin mutants, and perhaps some pyrimidine or purine mutants; other mutants, i.e., amino acid auxotrophs, would be especially limited in value. One could destroy particular amino acids in the medium through heat or enzymatic inactivation, but the effort required to prepare this medium would probably exceed the value of the approach. One might be able to utilize chromogenic substrates of some biosynthetic pathways in complex media. There are a number of metabolic intermediates for which chromogenic analogs exist. One could screen for acquisition or loss of a particular genetic lesion by natural transformation utilizing these analogs.

The use of antibiotic resistance and auxotrophic mutants is complicated by two factors. First, transformation in the environment appears to occur at low frequency (86), and second, point mutations revert at reasonable frequencies, frequencies that may approach the spontaneous rate for reversion. Deletion mutants may offer the cleanest assay for transformation in that they will not revert at an appreciable frequency (88). Thus, detection of gene conversion of a deletion mutant would indicate a genetic exchange event.

Chromosomal antibiotic resistance markers.   An alternative approach to the use of auxotrophs is the use of mutants resistant to growth-inhibiting or toxic agents. Many commonly employed antibiotics exert their effects by inhibiting necessary cellular functions. One can obtain mutations in the genes that encode proteins involved in these functions such that the mutant continues to live, but the antibiotic no longer inhibits the process. For instance, streptomycin is an inhibitor of protein biosynthesis in bacteria. One can isolate point mutations in the gene for the ribosomal protein where streptomycin inhibits (64). The result is mutants that are capable of synthesizing protein nearly as efficiently as are wild-type, but which are no longer inhibited by streptomycin. Similarly, one can isolate mutants in RNA polymerase that synthesize RNA but which are no longer inhibited by rifampin, and nalidixic acid-resistant mutants that result from mutations in DNA gyrase genes. If one prepares DNA from these resistant mutants, one can use that DNA to transform recipients to resistance. These antibiotics are effective in complex media and offer a means of direct selection of mutants.

Certain environmental conditions will reduce the effectiveness of certain antibiotics. For instance, trimethoprim resistance is difficult to screen in complex media, and streptomycin may be of limited value for studies in marine bacteria due to masking of the aminoglycoside binding sites on the ribosome by cations (65). Another limitation to this approach is the variability in inherent resistance levels for various antibiotics in different field isolates. One may find that the naturally resistant bacteria in the environment will mask the frequencies of transformation that result from addition of resistant DNA to the field sample. Nonetheless, the chromosomally linked antibiotic resistance mutations probably offer the best first approach to detection and characterization of transformable bacteria from the field.

Plasmid marker transformation.   While the use of chromosomal antibiotic resistance offers a powerful tool for detecting transformation, it has one major limitation. The process requires substantial homology between the endogenote and the exogenous DNA before the transformed DNA can be incorporated into the cell by homologous recombination (66). As a result, one must have a strain in isolated form so that one can generate antibiotic-resistant mutants before one can screen environmental samples for natural transformation. One can circumvent this problem by utilizing exogenous DNA that contains its own origin of replication. The plasmids offer an excellent tool for such a screen. To be effective as a generic screen for natural transformation, the plasmid should have a reasonably broad host range. Especially among the gram-negative bacteria, there are classes of plasmids that replicate reasonably well in many genera. These plasmids owe their promiscuous nature to an origin of replication that is recognized by this variety of organisms (67). Since a number of these plasmids have been characterized, there are a collection of different resistance markers and plasmid sizes that could be employed in a transformation screen.

Uptake of plasmids by natural transformation differs in mechanism from artificial transformation of plasmids. Artificial transformation in the enterics results in the preferential uptake of circular DNA (13). Natural transformation of plasmids requires linearization of the DNA, followed by uptake of a single strand of the plasmid into the cell. Since plasmid DNA must be circular for replication, one must convert the linearized DNA to a circular form. If the plasmid shares homology with the endogenote, the ends of the plasmid can be positioned together by base-pairing with their homologous sequence in the endogenote, allowing the ends to be ligated. Unfortunately, this requires that the transforming DNA be identical, at least over a small span of bases, with a sequence in the cell. Again one is faced with the problem of needing the transforming strain in pure culture before one can screen for transformation in the environment. An alternative approach allows one to contain the requisite homology for recircularization on the transforming fragment itself. If utilizing a multimeric plasmid, one containing more than one copy of the genes in a single circular molecule, the plasmid will contain homology to itself and will recircularize by recombination between identical sequences. While this results in loss of one or more copies of the gene sequence, it allows at least one copy to be maintained on a circular, self-replicating element. Thus, one can use multimeric plasmids to detect transformation without the need for a strain of the recipient prior to screening.

There are limitations to the use of plasmids to detect transformation. If the presence of plasmid is detected by phenotypic expression, the plasmids must not only replicate in the host, but they must also be transcribed and translated. It is possible that a plasmid could be stably maintained by a bacterium in the environment without being expressed. This bacterium could then serve as a reservoir for the plasmid, which could be transferred to another recipient at a later time. Also, the inherent level of antibiotic resistance for different bacteria in the environment could mask acquired plasmid resistances, just as it could for the chromosomal markers. Plasmid resistance, however, offers one advantage as compared to chromosomal antibiotic resistance markers. While chromosomal resistance markers are scattered over the bacterial chromosome, many plasmids contain closely linked multiple resistances. Thus, one can select more than one resistance if an appropriate plasmid is used in the transformation assay, perhaps circumventing some of the restrictions of inherent resistance.

Confirmation of natural transformation.   When one studies gene transfer in new systems, it is important to determine the mechanism of exchange. As discussed earlier, gene transfer in bacteria is traditionally reviewed as one of three general mechanisms. We can distinguish natural transformation from transduction and conjugation by two criteria. First, and perhaps most importantly, natural transformation can be inhibited if DNase I is added to the recipient-DNA mixture (5). Sensitivity to deoxyribonuclease provides the best evidence of an exogenous source of DNA in the transfer process. Secondly, since natural transformation requires no extrachromosomal etiologic agent to drive the exchange process, genetic exchange in the absence of plasmid or bacteriophage is a good indication of natural transformation. Although gene transfer agent of *Rhodobacter capsulatus* and the conjugative elements of *Streptococcus* are clearly exceptions to this rule, important information can be gained from analysis of transforming supernatants for the presence of phagelike particles, and from analysis of transfer agents for an extrachromosomal replicon.

Isotopic measurements of natural transformation. The variability in transcription or translation of a gene in an expression-based assay and the limitation of required homology for establishment of transforming DNA have already been discussed. Clearly, as long as recombination and expression of genes are required for detection of transformation, we will probably underestimate the number of transformable bacteria in the environment. We can measure the binding and uptake of DNA without gene expression if we utilize radioactively labeled DNA. The use of an isotopic tag allows us to measure the amount of label associated with recipient cells without concern for expression. One can distinguish DNA that is bound to the outer surface of the cell from that internalized by DNase I digestion of the transformation complex. Since most transforming systems are nonspecific for the type of double-stranded DNA they bind and take up, one can measure the potential for DNA binding and uptake in these bacteria with virtually any source of labeled DNA. The limitation to the use of labeled DNA for field trials will probably be the stability of the DNA in the environmental sample. If the sample contains substantial concentrations of deoxyribonuclease, the DNA would be turned over as rapidly as it is introduced. This is not exclusively a limitation of the label-detection approach, however, since DNA would also be turned over as rapidly as it is introduced. This is not exclusively a limitation of the label-detection approach, however, since DNA would also be turned over in cases where gene expression was used for detection of transforming bacteria. The variety of methods for labeling DNA (in vivo labeling, nick-translation, end-labeling, etc.) offer a number of ways to produce transforming DNA of high specific activity.

Trends in natural transformation

Recipient function. Although competence development, DNA binding and uptake, and recombination, the defined recipient functions, have received a great deal of study over the past 50 years, there is still much to be learned about this aspect of natural transformation. Although a number of bacteria have been identified that transform naturally, only a very small number have been extensively studied with regard to the mechanisms of recipient function. Most of the information about recipient function results from studies in three genera: *Streptococcus, Bacillus,* and *Haemophilus* (35). These genera contain medically significant species, and studies on the basic mechanism of natural transformation have been supported primarily from agencies with interests in medically important organisms. There are many other bacteria known to transform naturally. The mechanisms involved in recipient function in these bacteria have received little attention. Some of these bacteria, for example, species of *Pseudomonas, Acinetobacter, Azotobacter,* and *Vibrio,* are very important environmental and industrial organisms. The mechanisms for transformation in these bacteria may be similar to those more extensively studied, or they may be significantly different. As the question of the impact of gene transfer in the environment is raised, it will become increasingly important for agencies with environmental and industrial interests to support analysis of these and other transforming systems.

DNA release and stability. While some understanding of the mechanisms for DNA binding, uptake, and processing exists in at least a few models, the source of transforming DNA has yet to be addressed. Since natural transformation has been traditionally considered a laboratory phenomenon, the importance of the mechanism of DNA release has been minimal. However, now that the environmental significance of natural transformation is being addressed, the mechanism(s) for release of DNA are of equal importance as the mechanisms of uptake. It is likely that many transforming species have developed mechanisms for release of DNA. In *P. stutzeri* the ability to release DNA is DNA-replication-dependent and enhances the frequency of transformation between intact donor and recipient cells (9). While donor function has not been proposed in other species of transforming bacteria, there is evidence that such a process may occur. *B. subtilis* has been shown to release DNA from synchronized cells in an ordered fashion starting at or near the origin of chromosomal replication (32). This organism has also been shown to be "prone to lysis" at the onset of competence (31). *H. Influenzae* has been shown to transfer DNA between intact cells within a colony (68), and reports of genetic exchange that are similar in nature to cell contact transformation in *P. stutzeri* have been made (10). As we attempt to determine the importance of natural transformation as an environmentally significant phenomenon, we must address the means by which DNA becomes available for transformation in the field; therefore, studies on the mechanisms of donor function in natural transformation are essential.

Natural transformation in the environment. Studies on the mechanisms of donor and recipient function in natural transformation will require study of the processes in pure culture. It is also important to study the potential for transformation in the field. Initial studies have been made for transformation in aquatic (85, 86), soil (69, 87), and wastewater environments (70). It is, perhaps, too early to draw definite conclusions from these studies as to the importance of transformation in the field, but the preliminary studies suggest that naturally occurring organisms at least have the potential to transform in the environment, and that in field simulations natural transformation occurs. Wackernagel, Lorenz, and colleagues have shown that transformation of *B. subtilis* occurs when cells and DNA are associated with sand grains. They have further demonstrated that attachment of transforming DNA to sands provides protection to DNA from turnover by nucleases (88), yet still allows the DNA to be taken up and expressed by competent cells. In aquatic sediments transformation occurs in sterile and nonsterile sediment columns when chromosomal DNA from antibiotic-resistant strains of *P. stutzeri* and recipient strains are added (86). Stewart and Sinigalliano (85) have shown that transformation can occur with soluble DNA and by cell contact within and between soil and marine isolates of this bacterium. Bale, Day, and Fry have demonstrated transformation of nonconjugal plasmid DNA in the epilithon layer on river stones (M. J. Day, personal communication). Finally, Lee and Stotzky have demonstrated transformation of *B. subtilis* in sterile and nonsterile soils (87).

Because of the complexity of the field as compared to the laboratory, a great deal of preliminary information is required before one can speculate as to the significance of transformation for gene transfer in the environment. Studies on temperature, salinity, DNA stability, and surface-associated versus solution transfer are only a few of the questions to be addressed.

Specific methods for measuring environmental transformation— strain-specific assays

If a strain has been identified as transformable, there are a number of assays that can determine the efficiency and regulation of transformation. These methods may be quite simple, requiring minimal purification of DNA, or they may involve elaborate methods for obtaining highly purified DNA. In any case, perhaps the most convenient method for analyzing transformation in these bacteria involves the use of chromosomal markers. One can utilize strains of recipients defective in a particular gene for cellular growth, in other words, an auxotrophic strain (62). The limitation of this approach is that most auxotrophic markers must be screened on chemically defined, synthetic media; however, many field isolates fail to grow of chemically defined media. An alternative chromosomal screen involves the use of point mutations that result in antibiotic resistance. These mutations can be screened on either synthetic or complex media and offer the advantage of direct selection of transformants. Streptomycin (64), rifampin (71), and nalidixic acid (72) are three antibiotics for which chromosomal resistance can be selected.

Selection of chromosomal antibiotic resistance mutations in bacteria (30)

1. Determine the natural resistance level of the isolate on the antibiotics of choice by plating on plates with increasing concentrations of the antibiotic. One can screen a number of isolates on a single set of plates by streaking each in a sector of the agar plate.

2. Once the inhibitory concentration of the antibiotic is determined, prepare plates containing antibiotic at twice the concentration of the minimum inhibitory concentration.

3. Prepare an overnight culture of the wild-type strain. Collect cells by centrifugation and resuspend in $0.1\times$ the volume of the centrifuged culture.

4. Plate 0.1 ml of the concentrated cells suspension on selective plates. Dilute the preparation 1:100 to 10^{-6} and 1:10 to 10^{-7}. Plate 0.1 ml of the two most dilute solutions on media without antibiotic to determine the viable count of the strain.

5. Determine the frequency of spontaneous resistance by determining the concentration of resistant mutants per volume of culture and the concentration of total cells per volume. Frequency of spontaneous resistance will be the concentration of resistant cells per wild-type cell.

6. Resistant strains should be selected, purified twice on nonselective agar, and screened again for resistance. Strains can be stored for later use by freezing and storing cells in 20% glycerol at $-70°C$.

Juni lysis procedure (29)

1. Streak a donor strain onto an agar plate of appropriate media. Incubate the plate until colonies are approximately 2 to 3 mm in diameter.

2. Add 0.5 ml of Juni lysis buffer (0.05% SDS, 47 mM sodium acetate, 150 mM sodium chloride) to a sterile, 13-mm screw-capped tube. With a sterile loop pick a single colony from your donor plate and transfer into the Juni buffer. Resuspend the cells by vortexing.

3. Sterilize the crude DNA preparation by heating at 70°C for 2 hours. Confirm purity by streaking the preparation on an appropriate plate and incubating.

4. DNA can be stored for several weeks at 4°C.

Preparation of purified DNA

1. Inoculate 100 ml of fresh liquid medium to a Klett value of 20 units (or A540 of about 0.1) with an overnight culture of your donor strain. Incubate the culture as appropriate until it has reached a mid exponential culture (approximately 100 KU or 0.8 absorbance).

2. Collect cells by centrifugation. Wash once with an appropriate buffer. Resuspend cells in 30 ml of sucrose buffer (25% sucrose, 50 mM tris-HCl [pH 8.0]).

3. Add 30 mg of lysozyme and incubate on ice for 30 minutes.

4. Add 3.0 ml of 0.25 M EDTA and incubate on ice for 15 minutes.

5. Gently lyse cells by the addition of 3.3 ml of 10% sarkosyl (sodium lauryl sarcosine) in TES buffer (50 mM tris-HCl [pH 8.0], 5 mM EDTA, 50 mM NaCl).

6. Add 10 μl of heat-treated RNase I (5 mg/ml, heated to 65°C for 10 minutes). Incubate at room temperature for 30 minutes.

7. Add 2 mg of proteinase K and incubate at 37°C for 30 minutes.

8. Extract the cell lysate twice with equal volumes of phenol-chloroform (1:1), followed by extractions with TES-saturated diethyl ether until lysate is clear.

9. Precipitate DNA by the addition of two volumes of 95% ethanol (-20°C).

Determining DNA concentration (73)

1. Prepare a stock solution of DNA (calf thymus or other commercially available preparation) standard at 25 μg/ml. Confirm concentration by measuring absorbance of the solution at 260 nm (solution should have an absorbance of 0.5).

2. Prepare fluorescence buffer by adding 25 µl of a 5 mg/ml stock of Hoescht 33528 to 50 ml of fluorescence buffer (2 M sodium chloride, 0.05 M sodium phosphate, pH 7.4).

3. Dispense 2.5 ml of buffer-dye solution into each of several 10 × 60 mm disposable tubes.

4. Prepare a standard curve for DNA by adding 0, 5, 10, 15, 20, and 25 µl of your DNA stock solution to separate buffer-dye tubes.

5. Adjust the upper range of your fluorometer with the 25 µl DNA standard as per manufacturer's instructions. Adjust the lower end of your scale against the 0 µl DNA blank.

6. Measure fluorescence of each DNA standard.

7. Add 2 µl of each DNA preparation to separate buffer-dye tubes. Measure fluorescence. Add additional solution, or dilute preparation to obtain readings within the range of your standard curve.

8. Use your DNA standards to prepare a standard curve of fluorescence versus nanograms of DNA. Determine the amount of DNA in each measured unit of DNA solution. Calculate total DNA in each preparation.

**Plate transformation method
(qualitative assay) (29)**

1. Using a permanent marker, divide an appropriate nonselective agar plate into three sectors. Label "DNA only," "recipient only," and "DNA + recipient."

2. With a sterile loop spot your DNA preparation (Juni or purified) onto the DNA-only and the DNA + recipient sectors. Carefully wipe loop to remove DNA.

3. With a sterile loop spot a loopful of an overnight recipient culture on the recipient-only and the DNA + recipient sectors. Use your sterile loop to mix the DNA and recipient spots on the combined sector.

4. Incubate plate at an appropriate temperature overnight or until a patch of growth is seen on the two recipient sectors (DNA-only sector should be free of growth).

5. With a sterile loop pick from the patch of the recipient-only sector and streak on one-half of a selective plate. Repeat for the DNA + recipient patch.

6. Incubate plate at an appropriate temperature. Monitor for growth. Growth of the culture from the DNA + recipient patch, but not from the recipient-only patch indicates transformation.

Filter transformation assay
(quantitative assay) (9)

1. Prepare an overnight culture of the recipient in an appropriate medium.

2. Filter 0.5 ml of the overnight culture dropwise onto a 0.22 μm, 47-mm membrane filter, forming a spot of approximately 1 cm in diameter.

3. With sterile forceps transfer the filter cell side up onto an appropriate, nonselective agar plate.

4. Overlay the cell spot with your DNA preparation (50 μl of a Juni preparation or 1 μl of purified DNA). Allow the plate to absorb solution. Incubate the plate at an appropriate temperature overnight. (Controls should include a recipient-only filter and a DNA-cell filter treated with 0.1 ml DNase I (10 mg/ml) in 4.2 mM MgCl$_2$ solution, sterile).

5. Remove filter with sterile forceps and place in a sterile 125-ml flask containing 10 ml of an appropriate nonselective liquid medium. Resuspend cells from the filter by shaking gently at room temperature for 30 minutes.

6. Prepare dilutions of the resuspended cells in nonselective media. Dilute cell suspension to 10^{-7}. Plate 0.1 ml of the undiluted, 10^{-1}, and 10^{-2} dilutions on transformant selective media to determine total transformants. Plate 0.1 ml of the 10^{-6} and 10^{-7} dilutions on nonselective media to determine numbers of recipients. Express frequency of transformation as transformants per recipient. (*Note:* It may be necessary to modify dilutions and incubation times for each species.)

Generic transformation screen using
multimeric plasmid DNA

Prepare your plasmid sample by isolating plasmid DNA from a *recA* mutant of *E. coli*. Since these strains are defective for homologous recombination, they tend to accumulate multimeric forms when grown under continual antibiotic stress. This procedure is designed for isolation of plasmids 20 Kb or smaller in size (74).

1. Inoculate an appropriate volume of LB medium containing an appropriate antibiotic to an optical density of 0.1 (540 nm) with an overnight culture of your plasmid bearing strain (the volume of culture required will depend on whether your plasmid is amplifiable or not). Incubate the culture at 37°C with shaking un-

til an optical density of 0.8 is obtained. If your plasmid is amplifiable, add chloramphenicol to a final concentration of 100 ug/ml (chloramphenicol should be first dissolved in methanol). Continue to incubate the culture overnight. If you are dealing with a nonamplifiable plasmid proceed immediately to step 2.

2. Collect cells by centrifugation (8000 rpm, 15 minutes, GSA rotor). Wash cells in 10 ml of glucose buffer (50 mM glucose, 10 mM EDTA, 25 mM tris-HCl, pH 8.0) (per 250 ml bottle). Recentrifuge.

3. Resuspend cells from each bottle in 2 ml of glucose buffer. Transfer to 30 ml Corex tubes.

4. Add 0.5 ml of lysozyme solution (20 mg/ml in glucose buffer). Incubate tube on ice for 10 minutes. (*Note:* Lysozyme should be freshly prepared and kept on ice.)

5. Add 5 ml of lysis buffer (0.2 M NaOH, 1% SDS freshly prepared). Mix gently and incubate on ice for 10 minutes.

6. Add 4 ml of acetate buffer [29.4% (w/v) potassium acetate, 11.5% (v/v) glacial acetic acid]. Mix very gently. Incubate on ice for 10 minutes.

7. Remove precipitate by centrifugation at 12,000 Xg for 10 minutes.

8. Transfer supernatant to a clean tube.

9. From a stock of 5 mg/ml of RNase A (heated to 95°C for 5 minutes, then cooled slowly to room temperature) add sufficient RNase to achieve a final concentration of 20 μg/ml. Incubate at room temperature for 20 minutes.

10. Add an equal volume of phenol-chloroform (1:1). Mix by gently rocking the tube several times. Separate layers by centrifuging the tube for 10 minutes at 12,000 Xg. Transfer aqueous (upper) layer to a clean tube. Repeat extraction.

11. Extract aqueous phase with an equal volume of chloroform. Centrifuge 12,000 Xg for 2 minutes. Repeat

12. Precipitate DNA by addition of 0.1× volume of 3 M sodium acetate (pH 7.0) and 2.5× volumes of 95% ethanol. Hold on ice for 10 minutes. Collect by centrifugation at 12,000 Xg, 4°C for 10 minutes. Discard supernatant. Wash pellet with 10 ml of 70% ethanol, or dry under vacuum for a few minutes. Resuspend pellet in 1 ml of chromatography buffer (0.5 M NaCl, 50 mM tris-HCl, pH 7.4, 10 mM EDTA).

13. Separate DNA from RNA by column chromatography over a sepharose 4B column bed (use 1 × 25 cm column, column bed 20 cm long). Prewash column with chromatography buffer. After

buffer level is at or slightly below that of the gel bed add your DNA solution. Allow the solution to flow into the gel bed, then wash with 1 ml of buffer. Overlay the column with about 2 cm of buffer, then attach a flow system to the head of the column and wash with buffer. Collect about twenty 1-ml fractions. Confirm content of fractions by electrophoresis of 5 μl of each fraction on an agarose gel. Pool fractions containing plasmid DNA but not RNA.

14. Extract fractions with phenol-chloroform and chloroform as above.

15. Precipitate DNA with 0.1× volume of 3 M sodium acetate and 2.5× volumes of 95% ethanol. Hold in a dry ice/ethanol bath for at least 10 minutes, or overnight at −70°C. Collect by centrifugation at 12,000 Xg for 20 minutes.

16. Resuspend pellet in a minimal volume of TES buffer. Store at 4°C.

17. Once DNA preparation is obtained transformations can be performed as described above.

References

1. Griffith, F. 1928. The significance of pneumococcal types. *J. Hygiene* 27:113–159.
2. Zinder, N. D., and J. Lederberg. 1952. Genetic exchange in *Salmonella*. *J. Bacteriol.* 64:679–699.
3. Hayes, W. 1952. Recombination in *Bact. coli* K-12: Unidirectional transfer of genetic material. *Nature* 169:118–119.
4. Avery, O. T., C. M. MacLeod, and M. McCarty. 1944. Studies on the chemical nature of the substance inducing transformation in pneumococcal types. *J. Exp. Med.* 79:137–159.
5. Smith, H. O., D. B. Danner, and R. A. Deich. 1981. Genetic transformation. *Ann. Rev. Biochem.* 50:41–68.
6. Low, K. B., and R. D. Porter. 1978. Modes of gene transfer and recombination in bacteria. *Ann. Rev. Genet.* 12:249–287.
7. Wall, J. D., R. F. Weaver, and H. Gest. 1975. Gene transfer agents, bacteriophages, and bacteriocins of *Rhodopseudomonas capsulata*. *Arch. Microbiol.* 105:217–224.
8. Clewell, D. B., and C. Gawron-Burke. 1986. Conjugative transposons and the dissemination of antibiotic resistance in streptococci. *Ann. Rev. Microbiol.* 40:635–639.
9. Stewart, G. J., C. A. Carlson, and J. L. Ingraham. 1983. Evidence for an active role of donor cells in natural transformation in *Pseudomonas stutzeri*. *J. Bacteriol.* 156:30–35.
10. Albritton, W. L., J. K. Setlow, and L. Slaney. 1982. Transfer of *Haemophilus influenzae* chromosomal genes by cell to cell contact. *J. Bacteriol.* 152:1066–1070.
11. Ephrati-Elizur, E. 1968. Spontaneous transformation in *Bacillus subtilis*. *Genet. Res.* 11:82–96.
12. Raina, J. L., and A. W. Ravin. 1980. Switches in macromolecular synthesis during induction of competence for transformation of *Streptococcus sanguis*. *Proc. Natl. Acad. Sci. USA* 77:6062–6066.
13. Cohen, S. N., A. C. Y. Chang, and L. Hsu. 1972. Nonchromosomal antibiotic resistance in bacteria: Genetic transformation of *Escherichia coli* by R-factor DNA. *Proc. Natl. Acad. Sci. USA* 69:2110–2114.

14. Sparling, P. E. 1966. Genetic transformation of *Neisseria gonorrhoeae* to streptomycin resistance. *J. Bacteriol.* 92:1364–1371.
15. Tomasz, A., and R. D. Hotchkiss. 1964. Regulation of the transformability of pneumococcal cultures by macromolecular cell products. *Proc. Natl. Acad. Sci. USA* 51:480–486.
16. Tomasz, A. 1966. Model for the mechanism controlling the expression of competent state in Pneumococcus cultures. *J. Bacteriol.* 91:1050–1061.
17. Bott, K. F., and G. A. Wilson. 1967. Development of competence in the *Bacillus subtilis* transformation system. *J. Bacteriol.* 94:562–570.
18. Pakula, R., and W. Walczak. 1963. On the nature of competence of transformable streptococci. *J. Gen. Microbiol.* 31:125–133.
19. Tomasz, A. 1970. Cellular metabolism in genetic transformation in *Pneumococci:* Requirement for protein synthesis during induction of competence. *J. Bacteriol.* 101:860–871.
20. Dooley, D. C., C. T. Hadden, and E. W. Nester. 1971. Macromolecular synthesis in *Bacillus subtilis* during development of the competent state. *J. Bacteriol.* 108:668–679.
21. Grist, R. W., and L. O. Butler. 1981. The use of M1 medium in transformation of *Streptococcus pneumoniae. J. Gen. Microbiol.* 127:147–154.
22. Grist, R. W., and L. O. Butler. 1983. Effect of transforming DNA on growth and frequency of mutation of *Streptococcus pneumoniae. J. Bacteriol.* 153:153–162.
23. Tomasz, A., and J. L. Mosser. 1966. *Proc. Natl. Acad. Sci. USA* 55:58–66.
24. Morrison, D. A., B. Mannarelli, M. N. Vijayakumar. 1982. Competence for transformation in *Streptococcus pneumoniae:* An inducible high-capacity system for genetic exchange. In D. Schlessinger (ed.), *Microbiology—1982.* American Society for Microbiology, Washington, D.C., pp. 136–138.
25. Tomasz, A. 1973. Cell surface structures and absorption of DNA molecules during genetic transformation in bacteria. In L. Leive (ed.), *Bacterial Membranes and Walls.* Marcel Dekker, New York, pp. 321–355.
26. Chen, J.-D., and D. A. Morrison. 1987. Modulation of competence for genetic transformation in *Streptococcus Pneumoniae. J. Gen. Microbiol.* 133:1959–1967.
27. Wise, E. M., S. P. Alexander, and M. Powers. 1973. Adenosine 3'5'-cyclic monophosphate as a regulator of bacterial transformation. *Proc. Natl. Acad. Sci. USA* 70:471–475.
28. Barnhart, B. J., and R. M. Herriott. 1963. Penetration of deoxyribonucleic acid into *Haemophilus influenzae. Biochim. Biophys. Acta* 76:25–39.
29. Juni, E. 1972. Interspecies transformation of *Acinetobacter:* Genetic evidence for a ubiquitous genus. *J. Bacteriol.* 112:917–931.
30. Carlson, C. A., L. S. Pierson, J. J. Rosen, and J. L. Ingraham. 1983. *Pseudomonas stutzeri* and related species undergo natural transformation. *J. Bacteriol.* 153:93–99.
31. Sinha, R. A., and V. N. Iyer. 1971. Competence for genetic transformation and the release of DNA from *Bacillus subtilis. Biochim. Biophys. Acta* 232:61–71.
32. Borenstein, S., and E. Ephrati-Elizur. 1969. Spontaneous release of DNA in sequential genetic order by *Bacillus subtilis. J. Mol. Biol.* 45:137–152.
33. Seto, H., and A. Tomasz. 1976. Calcium-requiring step in the uptake of deoxyribonucleic acid molecules through the surface of competent pneumococci. *J. Bacteriol.* 126:1113–1118.
34. Seto, H., and A. Tomasz. 1974. Early stages in DNA binding and uptake during genetic transformation of pneumococci. *Proc. Natl. Acad. Sci. USA* 71:1493–1498.
35. Stewart, G. J., and C. A. Carlson. 1986. The biology of natural transformation. *Ann. Rev. Microbiol.* 40:211–235.
36. Scocca, J. J., R. L. Poland, and K. C. Zoon. 1974. Specificity in deoxyribonucleic acid uptake by transformable *Haemophilus influenzae. J. Bacteriol.* 118:369–373.
37. Lacks, S., B. Greenberg, and M. Neuberger. 1974. Role of deoxyribonuclease in the genetic transformation of *Diplococcus pneumoniae. Proc. Natl. Acad. Sci. USA* 71:2305–2309.
38. Cocino, M. F., and S. H. Goodgal. 1982. DNA-binding vesicles released from the

surface of competence-deficient mutants of *Haemophilus influenzae*. *J. Bacteriol.* 152:441–450.

39. Venema, G. 1979. Bacterial transformation. *Adv. Microb. Physiol.* 19:245–331.

40. Morrison, D. A. 1977. Transformation in *Pneumococci:* Existence and properties of a complex involving donor deoxyribonucleate single strands in eclipse. *J. Bacteriol.* 132:576–583.

41. Stuy, J. H. 1965. Fate of transforming DNA in the *Haemophilus influenzae* transformation system. *J. Mol. Biol.* 13:554–570.

42. Kahn, M. E., G. Maul, and S. H. Goodgal. 1982. Possible mechanism for donor DNA binding and transport in *Haemophilus. Proc. Natl. Acad. Sci. USA* 79:6370–6374.

43. Goodgal, S. H. 1982. DNA uptake in *Haemophilus influenzae. Ann. Rev. Genet.* 16: 169–192.

44. Mathis, L. S., and J. J. Scocca. 1982. *Haemophilus influenzae* and *Neisseria gonorrhoeae* recognize different specificity determinants in the DNA uptake step of genetic transformation. *J. Gen. Microbiol.* 128:1159–1161.

45. Raina, J. L., E. Metzer, and A. W. Ravin. 1979. Translocation of the presynaptic complex formed upon DNA uptake by *Streptococcus sanguis* and its inhibition by ethidium bromide. *Mol. Gen. Genet.* 170:249–259.

46. Pieniazek, D., M. Piechowska, and G. Venema. 1977. Characteristics of a complex formed by a non-integrated fraction of transforming DNA and *Bacillus subtilis* constituents. *Mol. Gen. Genet.* 156:251–261.

47. Kahn, M. E., F. Barany, and H. O. Smith. 1983. Transformasomes: Specialized membrane structures that protect DNA during *Haemophilus* transformation. *Proc. Natl. Acad. Sci. USA* 80:6927–6931.

48. Radding, C. M. 1982. Strand transfer in homologous genetic recombination. *Ann. Rev. Genet.* 16:405–437.

49. Cassulo, E., and C. M. Radding. 1971. Mechanism for the action of Lambda exonuclease in genetic recombination. *Nature New Biol.* 229:13–16.

50. Howard-Flanders, P. 1981. Inducible repair of DNA. *Sci. Am.* 245:71–80.

51. Rudner, R., G. Tackney, and P. Gottlieb. 1982. Variations of nucleotide sequences among related *Bacillus* genomes. In U. N. Streips, S. H. Goodgal, W. R. Guild, and G. A. Wilson (eds.), *Genetic Exchange: A Celebration and a New Generation.* Marcel Dekker, New York, pp. 339–351.

52. Marsh, R. E. 1968. Some comments on hydrogen bonding in purine and pyrimidine bases. In A. Rich and N. Davidson (eds.), *Structural Chemistry and Molecular Biology.* Freeman, San Francisco.

53. Chamberlin, M. 1982. Bacterial DNA-dependent RNA polymerases. In P. Boyer (ed.), *The Enzymes,* vol. 15, part B. Academic Press, New York, p. 61.

54. Rickwood, D., and B. D. Hames. 1982. *Gel Electrophoresis of Nucleic Acids: A Practical Approach.* IRL Press, London.

55. Freifelder, D. 1987. *Molecular Biology,* 2d ed. Jones and Bartlett, Boston.

56. DeFlaun, M. E., J. H. Paul, and D. Davis. 1986. Simplified method for dissolved DNA determination in aquatic environments. *Appl. Environ. Microbiol.* 52:654–659.

57. Paul, J. H., W. H. Jeffrey, and M. E. DeFlaun. 1987. Dynamics of extracellular DNA in the marine environment. *Appl. Environ. Microbiol.* 53:170–179.

58. Lorenz, M. G., B. W. Aardema, and W. E. Krumbein. 1981. Interaction of marine sediment with DNA and DNA availability to nucleases. *Mar. Biol.* 64:225–230.

59. Aardema, B. W., M. G. Lorenz, and W. E. Krumbein. 1983. Protection of sediment-absorbed transforming DNA against enzymatic inactivation. *Appl. Environ. Microbiol.* 46:617–620.

60. Hartl, D. L., D. Freifelder, and L. A. Snyder. 1988. *Basic Genetics.* Jones and Bartlett, Boston, p. 161.

61. Finn, C. W., and O. E. Landman. 1985. Competence-related proteins in the supernatant of competent cells of *Bacillus subtilis. Mol. Gen. Genet.* 198:329–335.

62. Guirard, B. M., and E. E. Snell. 1981. Biochemical factors in growth. In P. Gerhardt

(ed.), *Manual of Methods for General Microbiology.* American Society of Microbiology, Washington, D.C.

63. ZoBell, C. E. 1941. Apparatus for collecting water samples from different depths for bacteriological analysis. *J. Mar. Res.* 4:173–188.

64. Stoffler, G., and G. W. Tischendorf. 1980. Antibiotic receptor sites in *Escherichia coli* ribosomes. In J. Drews and F. E. Hahn (eds.), *Drug Receptor Interaction in Antimicrobial Chemotherapy,* Vol. 1, *Topics in Infectious Disease.* Springer-Verlag, New York.

65. Washington, J. A., II, R. J. Snyder, P. C. Kohner, C. G. Wiltse, D. M. Ilstrup, and J. T. McCall. Effect of cation content of agar on the activity of gentamicin, tobramycin, and amikacin against *Pseudomonas aeruginosa. J. Infect. Dis.* 137: 103–110.

66. Carlson, C. A., S. M. Steenbergen, and J. L. Ingraham. 1985. Natural transformation of *Pseudomonas stutzeri* by plasmids that contain cloned fragments of chromosomal DNA. *Arch. Microbiol.* 140:134–138.

67. Curtiss, R. 1969. Bacterial conjugation. *Ann. Rev. Microbiol.* 23:70–135.

68. Stuy, J. H. 1985. Transfer of genetic information within a colony of *Haemophilus influenzae. J. Bacteriol.* 162:1–4.

69. Graham, B. J., and C. A. Istock. 1978. Genetic exchange in *Bacillus subtilis* in soil. *Mol. Gen. Genet.* 166:287–290.

70. Gealt, M. A., M. D. Chai, K. B. Alpert, and J. C. Boyers. 1985. Transfer of plasmids pBR322 and pBR325 in wastewater from laboratory strains of *Escherichia coli* to bacteria indigenous to waste disposal systems. *Appl. Environ. Microbiol.* 49:836–841.

71. Davis, B. J. 1982. Contributions to biology form studies on bacterial resistance. In S. Mitsuhashi (ed.), *Drug Resistance in Bacteria.* Japan Science Society Press, Tokyo, pp. 327–332.

72. Inoue, S., J. Yamagishe, S. Nakamura, Y. Furutani, and M. Shimizu. 1982. Novel nalixic acid-resistance mutations relating to DNA gyrase activity. In S. Mitsuhashi (ed.), *Drug Resistance in Bacteria.* Japan Science Society Press, Tokyo, pp. 411–414.

73. Paul, J. H., and B. Meyers. 1982. Fluorometric detection of DNA in quatic microorganisms by use of Hoechst 33258. *Appl. Environ. Microbiol.* 43:1393–1399.

74. Davis, L. G., M. D. Dibner, and J. F. Battey. 1986. Large-scale alkaline lysis method: Plasmid purification. In *Basic Methods in Molecular Biology.* Elsevier, New York, pp. 99–101.

75. Juni, E., and G. A. Heym. 1980. Transformation assay for identification of psychotrophic Achromobacters. *Appl. Environ. Microbiol.* 43:1393–1399.

76. Page, W. J., and H. L. Sadoff. 1976. Control of competence in *Azotobacter vinelandii* by nitrogen catabolite repression. *J. Bacteriol.* 125:1088–1095.

77. Mevarech, M., and R. Werczberger. 1985. Genetic transfer in *Halobacterium volcanii. J. Bacteriol.* 162:461–462.

78. O'Conner, M., A. Wopat, and R. S. Hanson. 1977. Genetic transformation in *Methylobacterium organophilum. J. Gen. Microbiol.* 98:265–272.

79. Tigari, S., and B. E. B. Moseley. 1980. Transformation in *Micrococcus radiodurans:* Measurement of various parameters and evidence for multiple independently segregating genomes per cell. *J. Gen. Microbiol.* 119:287–296.

80. Norgard, M. V., and T. Imeda. 1978. Physiological factors involved in the transformation of *Mycobacterium smegmatis. J. Bacteriol.* 133:1254–1262.

81. Roelants, P., V. Konvalinkova, M. Mergeay, P. F. Lurquin, and L. Ledoux. 1976. DNA uptake by Streptomyces species. *Biochim. Biophys. Acta* 442:117–122.

82. Chauvat, F., C. Astier, F. Vedel, and F. Joset-Espardellier. 1983. Transformation in the cyanobacterium *synechococcus* R2: Improvement of efficiency; the role of the pUH24 plasmid. *Mol. Gen. Genet.* 191:39–45.

83. Golden, S. S., and L. A. Sherman. 1984. Optimal conditions for genetic transformation of the cyanobacterium *Anacystis nidulans* R2. *J. Bacteriol.* 158:36–42.

84. Jeffrey, W. H., J. H. Paul, and G. J. Stewart. 1990. Natural transformation of a marine *Vibrio* species by plasmid DNA. *Microb. Ecol.* 19:259–268.

85. Stewart, G. J., and C. D. Sinigalliano. 1989. Detection and characterization of natural transformation in the marine bacterium *Pseudomonas stutzeri* strain ZoBell. *Arch. Microbiol.* 152:520–526.

86. Stewart, G. J., and C. D. Sinigalliano. 1990. Detection of horizontal gene transfer by natural transformation in native and introduced species of bacteria in marine and synthetic sediments. *Appl. Environ. Microbiol.* 56:1818–1824.

87. Lee, G. H., and G. Stotzky. 1989. Transformation is a mechanism of gene transfer in soil. Wind River Conference on Genetic Exchange Abstracts. Estes Park, Colo.

88. Lorenz, M. G., and W. Wackernagel. 1987. Adsorption of DNA to sand and variable degradation rates of adsorbed DNA. *Appl. Environ. Microbiol.* 53:2948–2952.

13

Evaluating the Potential for Genetic Exchange in Natural Freshwater Environments

Dennis J. Saye

Susan B. O'Morchoe

Introduction

The scarcity of information on gene transfer in natural habitats has become an issue only in recent years. The developing technology of recombinant DNA spawned a movement to provide biotechnological answers to environmental problems. Concern over introduction of genetically altered bacteria into the environment has created the need for basic information to intelligently assess the risk versus benefit of deliberate releases of microorganisms into the environment. Three mechanisms of genetic transfer are recognized: conjugation, transduction, and transformation. Conjugation requires direct cell-to-cell interaction. DNA transfer via transduction is mediated by bacteriophage. In transformation, naked DNA is taken up by naturally competent bacteria directly from the surrounding environment. Available studies indicate that DNA transfer in freshwater environments occurs via both conjugation and transduction (21, 22, 36, 41, 42, 50, 51, 55, 63).

Conjugation is a widespread phenomenon. The high incidence of naturally occurring plasmids and the presence of antibiotic resistance genes in a diverse range of taxonomic groups implicate conjugation as a likely means of gene transfer in freshwater environments. Plasmids from indigenous organisms, transferred into released organisms, could alter the ability of released organisms to survive and spread in the environment. It has been shown that plasmids isolated from natural waters are capable of transfer. Grabow et al. (22) reported the transfer of 3 out of 10 R factors from *Escherichia coli* which were isolated from river water. Transfer occurred at low frequency in dialysis bags suspended in a river. Cooke (13) also showed transfer of R factors in *E. coli* that were isolated from river water. A study by Lundquist and Levin (34) demonstrated that it is possible for naturally occurring plasmids to be established and maintained in a population by infectious transfer alone. However, this transfer is primarily dependent on the population density of potential recipients. Naturally occurring bacteria concentrations of 10^3 to 10^4 colony-forming units (CFU) per milliliter are common. The density of introduced organisms can be adjusted to any desired level in studies using containment chambers, but concentrations of introduced cells would be low in an actual release situation due to their dilution in a large body of water.

Another consideration for determining the impact of conjugation on the spread of genetic material is the transfer of plasmids from introduced organisms into the indigenous population. Genthner et al. (20) studied the ability of indigenous organisms to act as recipients of DNA from introduced strains. Of 68 freshwater isolates tested, 26 (38 percent) were found to serve as recipients for the broad-host-range plasmid R68. Ten of these (15 percent) were found to be recipients for R1162 by R68 mobilization. These experiments were conducted under laboratory conditions. Schilf and Klingmüeller (55) tested indigenous organisms for recipient ability and found that the occurrence of plasmid transfer was extremely low, with only 17.3 percent of their isolates able to act as a recipient for the broad-host-range plasmids RP4 and pRD1. Walter et al. (67) determined that laboratory strains were more effective recipients than natural isolates. Studies of conjugation between laboratory strains of bacteria performed in situ showed a low frequency of transfer (21, 42).

Transduction as a mechanism of gene transfer has been described in some detail in Chapter 11. The transduction of *Pseudomonas aeruginosa* chromosomal genes in a freshwater environment has been demonstrated (41, 50), as well as the transduction of plasmid DNA (51). The potential of transduction as a potent form of gene transfer in the environment has largely been dismissed, however, for a number of reasons.

In natural aquatic habitats, bacteriophage are present in such low numbers 1.6×10^2, plaque-forming units (PFU) per liter to 6.8×10^4 PFU/L, that they are usually not detected by direct assay (44, 45). Most of these bacteriophage have a relatively narrow host range.

DNA transfer by transduction is actually carried out by defective phage, called transducing particles. These transducing particles contain host DNA mistakenly packaged into phage capsid and make up only a small fraction of the total number of phage produced following induction of the prophage to lytic growth or during primary lytic infection. When considering the vast dilution potential of a freshwater lake, it seemed unlikely that transduction would be significant as a mode of gene transfer.

However, Bergh et al. (9) have recently reported that they have found up to 2.5×10^8 virus particles per milliliter using a new technique for quantitative enumeration. They concluded that the measured concentration of phage and host cells in natural waters is high enough to result in a high phage-host interaction rate.

There is now considerable evidence that both transduction and conjugation have significant potential as mechanisms of gene dispersal (5, 41, 42, 50, 51).

Transformation in a natural freshwater environment has not been described. Possible methods for studying this mechanism will be discussed briefly.

Principles

Establishing model systems

Species and strain considerations.   The specific species and strains used in conducting environmental testing will be dependent on the objectives of the specific study. However, there are a number of considerations that must be addressed in designing a test system. One must decide whether the chosen microorganisms are representative of the indigenous community at the test site. If the species used are naturally found at the test site, a major consideration is the ability to differentiate the test strains from the other members of the indigenous microbial population. One approach is to utilize genetic markers that impart specific phenotypic traits which are either selectable or nonselectable but provide a phenotype unique to that strain. Antibiotic resistance carried on the chromosome or extrachromosomally, auxotrophies for specific growth factors, chromosomally encoded resistance to specific bacteriophage, and genes which encode enzymes involved in catabolic processes have all been successfully used as markers (41, 42, 50, 51).

It may also be desirable to use markers which occur in close linkage along with a distant marker. Strains of well-characterized genotype that carry multiple genetic markers or blocks of linked genetic markers are particularly useful when evaluating the transfer of chromosomal DNA. There is evidence that transduction of chromosomal DNA occurs in freshwater habitats at low frequencies (50). It had previously been shown by Morrison et al. (41) that transduction of a chromosomally encoded resistance to streptomycin could be demonstrated in flow-through environmental chambers, with transductants being recovered at frequencies of 5×10^{-6}, 1.2×10^{-2}, and 9.5×10^{-1} transductants per recipient at 1 hour, 4 days, and 10 days, respectively. The demonstration of chromosomal gene transfer may be improved by the use of linked genetic markers. An altered phenotype due to change at a single locus may occur from a simple reversion rather than the result of a gene transfer event. However, the cotransduction of neighboring genes would strongly support evidence of DNA transfer since reversion of two neighboring mutations would have an extremely low probability relative to that of a single reversion. The use of linked markers reduces the background of spontaneous revertants and would improve the confidence that transfer of DNA has occurred rather than reversion and subsequent growth of the revertants (50).

Plasmids replicate independently of chromosomal DNA and generally carry genes which are nonessential for the growth and metabolism of the bacterial cell. They often encode a number of specialized functions which must be considered in strain choice. When choosing a plasmid-containing strain for a test system, a number of factors must be carefully addressed. Is the plasmid self-transmissible (Tra$^+$), mobilizable (Mob$^+$), or does it possess chromosome-mobilizing ability (Cma$^+$)? The host range of the plasmid and the size of the plasmid are important parameters.

The pattern of antibiotic resistance imparted by a particular plasmid is useful in identifying cells containing an exogenous plasmid, whether formed by conjugation, transduction, or transformation. The presence of multiple markers on a test plasmid is advantageous, since a greater number of easily identified markers simplifies identification of test strains in the presence of the indigenous community. Multiple markers also improve the ability to screen for members of the natural community that may have served as recipients of a test plasmid, since the likelihood of encountering an identical phenotype among the indigenous population will decrease with the complexity of the phenotype imparted by the plasmid.

The use of a plasmid that is well characterized both genetically and molecularly is helpful in the identification of recipients and also in identifying rearrangements, deletions, and insertions. During their studies of conjugation in a freshwater lake, O'Morchoe et al. (42) isolated transconjugants that had undergone a rearrangement of the test plasmid during the in situ incubation. Evidence that rearrangement had taken place included an altered antibiotic resistance pattern and changes in the restriction pattern of plasmid DNA isolated from transconjugants. Loss of antibiotic resistance was observed in laboratory simulations, but at a frequency two logs lower. Changes in the restriction pattern of the plasmid DNA were not detected in the laboratory. The differences between the laboratory and in situ frequencies underscores an important consideration in performing in situ studies; the ability to define all variables acting on the test system incubated in the environment.

Defining variables in a complex system.   Defining variables in the environment is a formidable, if not impossible, task. Many factors affect the survival of organisms and the transfer of genetic material in situ. Physical and chemical factors such as pH, salinity, dissolved oxygen, particulate matter, temperature, and nutrient availability all exert an influence, as do biological factors such as competition with other organisms, presence of extrachromosomal DNA, and specific nutrient requirements. Many of these factors may act synergistically to help or hinder the establishment of a novel biotype in an ecosystem. Some of these properties, such as temperature, pH, conductivity, dissolved oxygen, and salinity are measured at each sampling and provide a rough approximation of the environmental conditions. Other properties are not readily measured and their effect on gene transfer is unknown. Factors which have been studied by other investigators include seasonal variation in bacterial production and the number of bacteria which are free-living or particle-bound (25, 33). Nutrient availability varies seasonally and spatially, with the level of organic nutrients decreasing with the distance from point sources of pollution such as sewage effluent and groundwater runoff (27). In general, freshwater systems are nutrient-poor. Many naturally occurring bacteria are adapted to conditions of low nutrient concentration. The implications are that laboratory strains may not successfully compete with the indigenous microbiota and also that cultivation of indigenous species may fail because of their inability to grow on rich media under conditions normally employed in the laboratory.

The effect of suspended particulates on gene transfer has not been demonstrated in freshwater habitats, but bacteria and bacteriophage have been found associated with clay particles. In laboratory studies the attachment of clay particles has been shown to affect the interaction of *E. coli* and bacteriophage (46). Colloidal clay has also been shown to inhibit conjugal transfer of R plasmids in microcosms and the effect of colloidal clay on conjugation in aquatic systems is inferred (59). There is evidence that bacteria associated with particulate matter are metabolically more active than unattached bacteria (28). Attachment to such solid supports may increase metabolic activity by providing a source of nutrients (66) or may buffer changes in pH (64).

Whether this implies an increased ability for genetic transfer, recombination, and expression of exogenous genes is purely speculative.

Laboratory simulations.   The use of microcosms for simulating environmental situations has been routinely practiced and may be the best available tool for gaining insight into the potential outcome of an in situ study. Simulations permit the researcher to develop the methodology, define the parameters, and identify inherent pitfalls before incurring the expense in time and money of the in situ field trials.

Since the environment does not allow true controlled conditions, interpretation of in situ data is based on laboratory results and assumes that the observed effects of known variables in the lab exert similar influences in a natural setting. Conditions can be controlled in the laboratory, making it possible to test a model where the recovery of transconjugants and transductants can be optimized by the use of standard conditions and subsequently to study the effects of known variables on the recovery of transconjugants and transductants. The influence of many known variables can be evaluated in the lab and anomalies which are observed in subsequent in situ studies may sometimes be resolved in terms of the influence of one or more known variables. These anomalies may often be the result of an as yet undefined or untested variable.

Laboratory simulations should be set up to resemble, as closely as practical, the experimental design of the system to be used in situ. Simulations performed in our lab were set up using lake water obtained at the in situ test site and transported to the laboratory under refrigeration (42, 50, 51). Sterile lake water may be obtained by autoclaving or membrane filtering through a 0.2-μm pore-size membrane (Millipore, Bedford, Massachusetts). Autoclaved water retains all particulate matter whereas filter-sterilized lake water does not. In preliminary work from our laboratory we were unable to demonstrate any significant differences in the recovery of transductants or transconjugants when performing simulations using either autoclaved or filter-sterilized lake water. To our knowledge, the influence of suspended particles or the potential of fluxes in available nutrients (resulting from autoclaving) has not yet been rigorously tested.

The manipulation of individual variables, such as temperature, agitation, nutrient availability, or other testable factors is appropriate in laboratory simulations. However, any departures from the basic experimental design of the in situ system should be avoided to preclude the possibility of inadvertently introducing a variable that could lead to spurious conclusions.

Test systems in situ

Site selection.   The selection of a freshwater test site will be governed by a number of considerations. Practical considerations include the availability of a sufficient database on the ecology of the test site. Background data concerning the bacterial density of the indigenous microorganisms, plasmid occurrence, amount of organic pollution, and the antibiotic resistance among the natural microbial population are particularly useful in characterizing the test site. Factors which influence the ecology of the test location, such as the proximity to a sewage treatment facility or an outflow of industrial waste should be noted. A history of the physical and chemical attributes of the native water should be available and these attributes will be monitored throughout the incubation period.

Many of the procedures used during the course of these studies are not conducive to on-site testing and will require the transport and timely processing of samples in a suitable laboratory. This is especially true of the microbiological techniques used for culturing and enumerating the donor and recipient strains as well as identification of the isolates exhibiting a phenotype acquired as the result of DNA transfer. For this reason it is imperative that the laboratory facility be promptly accessible.

The investigators must be aware of and in compliance with any ordinances governing the test location. For example, if the proposed in situ test site is in a navigable waterway, the tethering of containment chambers may be restricted or prohibited.

Containment.   Recently, Bale et al. (4, 5) demonstrated conjugal transfer of plasmid DNA in *Pseudomonas aeruginosa* on the surface of stones in a laboratory microcosm and in situ in a Welsh river. Amin and Day (2) have reported the bacteriophage-mediated transfer of plasmid DNA between strains of *P. aeruginosa* on membrane filters suspended in the River Taff (Wales). These experiments were performed in situ without containment of the test strains. This lack of containment is noteworthy since most, if not all, other studies of gene transfer in situ in aquatic habitats were performed in containment chambers. The types of chambers which have been used include membrane diffusion chambers (41), dialysis sacs (21, 22), chambers constructed of impermeable Teflon film (42, 51) and gas-permeable, flexible, plastic flasks (Lifecell, Baxter Healthcare Corp., Fenwal Division) (50).

One advantage of an enclosed system is the ability to examine in situ gene transfer in the absence of indigenous microbiota. The influence of the microbial community on gene transfer can be observed by incubating parallel chambers containing the test strains in sterilized lake water and nonsterile fresh lake water, respectively. Examples of containment systems are shown in Fig. 13.1.

Measuring physical and chemical variables. Temperature, pH, conductivity, and dissolved oxygen are measured initially and at each sampling throughout incubation of the in situ systems. These measurements are best performed on site. Portable instruments are

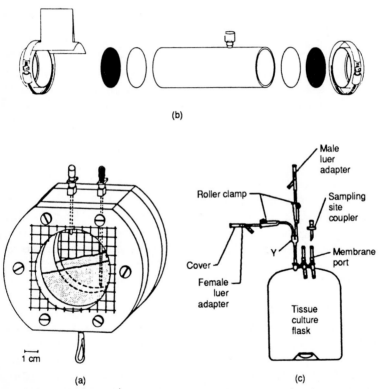

Figure 13.1 (*a*) Modified membrane diffusion chamber. Plexiglas construction has 0.4-nm-pore-size polycarbonate filters that are protected by stainless steel wire screens. Rubber septums cover the filling and sampling ports (18, 35). (*b*) Flow-through environmental chamber. A polycarbonate cylinder with a rubber septum–covered sampling port has 0.2-μm-pore-size polycarbonate membranes at each end. The membranes are protected by polyurethane foam covers and secured with a polyethylene ring and stainless steel band clamps (40). (*c*) Lifecell® (Baxter Healthcare Corp., Fenwal Division) tissue culture flasks are made of a flexible, gas-permeable plastic in 1-liter and 3-liter capacities (49).

commercially available that will provide accurate measurements of these parameters. Other factors which may be pertinent to these studies include salinity and turbidity. Physical and chemical measurements should be obtained for the samples as well as the surrounding waters at the time of each sampling. The weather conditions at the time of sampling should be noted as well as any observed changes in the characteristics of the surrounding waters.

Collection, transport, and storage of samples.   Collection of samples for microbiological assays is easily performed by transferring the sample to a sterile collection bottle and placing it on ice for transport to the laboratory. A second volume of sample is drawn for chemical and physical measurements (51). Both volumes must be drawn aseptically to prevent contamination of the test chambers.

Samples obtained for physical and chemical measurements must be carefully handled to prevent mixing. Agitation of the sample will cause the dissolved oxygen reading to be erroneously high. Measurements should be made immediately upon the collection of the sample.

The samples designated for microbiological assay are to be immediately placed on ice for transport to a suitable laboratory (51). Samples should be maintained on ice or under refrigeration until processed. Enumeration of the test microorganisms and subsequent workup must commence shortly after receipt by the laboratory. Although it is important to gather as much information as possible at each sampling, time and space restrictions must be kept in mind. Samples should not be stored for extended periods of time before analysis. Therefore, the sampling times should not be planned too frequently, without having adequate laboratory personnel available to promptly carry out the required analyses.

Enumeration of recombinant organisms

Culture and metabolic techniques.   The most widely used technique for enumeration of bacteria is plating the sample directly to specific media. Whether selective, differential, selective differential, or enrichment media are used, direct plating techniques are easy and inexpensive (38). However, enumeration of test strains is complicated by the use of selective media. Sensitivity to selective agents is enhanced when cells are compromised by environmental stresses. Injured cells are physiologically debilitated and are unable to grow and produce colonies under selection (11, 69). The use of indigenous organisms as potential donors and recipients presents special problems since many aquatic bacteria do not grow on media designed for laboratory organisms or have differing resistance levels to selective agents from that of the laboratory strains (27). In addition, bacteria isolated from natural aquatic habitats have been reported to have generation times as long as 20 to 200 h (26). Incubation times must be adjusted accordingly.

A primary criticism of the use of plate counts for the enumeration of aquatic bacteria is that they underestimate the number of viable cells in environmental samples. Large differences are consistently reported between the results of plate counts and direct microscopic counts (47, 48).

It must be kept in mind that all culturing techniques are somewhat selective (48). A bacterium's optimum growth rate is defined by its nutritional requirements, the pH, the temperature of incubation, and a host of other factors. Organisms which have requirements that fall outside the narrow range of conditions of the heterotrophic plate count either will not grow or will grow extremely slowly and be generally outcompeted by organisms that exhibit near-optimum growth under the conditions used for culturing. In addition, metabolic end products of one strain may accumulate and kill or adversely affect the growth of another organism. It is important to realize that in all culturing techniques some degree of selective pressure is inevitable. In spite of its limitations, however, a heterotrophic plate count performed on rich media without selective agents will provide data that are useful for comparison to results obtained on selective media.

The investigator may prefer to estimate total viable counts by direct counting techniques, but just as plate counts can underestimate the number of viable bacteria, direct counting can overestimate. Overestimation results from an inability to distinguish between living and nonliving cells. Direct counts are also subject to interference from particulate matter in the sample (29). An adaptation of the direct count technique combines the use of fluorescent antibody and the reduction of 2-(p-iodophenyl)-3-(p-nitrophenyl)-5-phenyltetrazolium chloride (INT reduction). This allows a count of viable organisms that is species-specific. This combination of fluorescent antibody and INT reduction (FAINT) was developed by Baker and Mills (3) to determine the number of respiring *Thiobacillus ferroxidans* cells in water samples. Using the FAINT technique, it is possible to enumerate a specific species that is actively respiring in a mixed aquatic population. This method is more sensitive than conventional cultural techniques. Two assumptions are made in the use of FAINT: (1) that the antibody preparation is specific and (2) that the INT reduction is an accurate reflection of viability. The main difficulty lies in the preparation of the specific antibody (12, 56). Enumeration of viable bacteria in ecosystems has been reviewed recently by Roszak and Colwell (47, 48).

Donor and recipient cells are enumerated on the basis of

chromosomally encoded phenotypes which can be directly selected (e.g., antibiotic resistance). The marker that is being selected should allow growth of test strains with minimal interference from the growth of indigenous microbes.

In many cases, the recombinant that results from gene transfer will also grow on the media used for enumerating the donor and the recipient strains. However, recombinant cells are usually present in numbers too low to significantly contribute to the total number of donors and recipients that are detected.

In general, recombinant strains are isolated by selecting for both the transferred markers and the markers unique to the recipient. Alternatively, selection for the transferred markers can be combined with counterselection of the donor strain. Morrison et al. (41) selected for transductants that had acquired streptomycin resistance by directly plating onto L-agar (Luria broth solidified with 1.5% Bacto-agar, Difco) containing 1000 μg/ml streptomycin and the Sm^r donor cells were eliminated by overlaying with a lysate of E79, a virulent *Pseudomonas* bacteriophage. The recipients carried a chromosomally encoded resistance to E79.

The prudent choice of markers to select and the stringency of the selection to be used for enumerating test strains and recombinants must be determined before the start of any experiment. O'Morchoe et al. (42) noted that the recovery of transconjugants was improved several hundredfold when Cb^r was used for the primary selection rather than G-418^r or Tc^r. All three antibiotic resistances are plasmid-encoded. The inability to detect the recombinant organisms by initial selection on G-418 or Tc-containing media does not necessarily reflect genetic instability, but may be a consequence of the specific mechanism of drug resistance. Environmental stresses to which the organisms are exposed may significantly alter their sensitivity to selective agents (11). Several lines of evidence indicate that short-term exposure in water causes cellular envelope damage in *E. coli*. This evidence includes a significant differential between growth on selective and nonselective media and also morphological changes observed in electron micrographs of injured cells (11, 69). Kropinski et al. (31) has described significant changes in the lipids, lipopolysaccharides, and outer membrane proteins of *P. aeruginosa* in response to growth at low temperature. Cells grown at 37°C were sensitive to bacteriophage E79 infection, whereas cells grown at 19°C were resistant, indicating an alteration in the cell-envelope-associated E79 receptor. The effects of temperature on the sensitivity of other phages and various antibiotics are largely unknown.

The sensitivity of detecting recombinants by direct plating to selective media is relatively low. The ability to detect small numbers of recombinants in a large sample volume is facilitated by the use of membrane filtration. The sample is filtered through a membrane of appropriate pore size (e.g., 0.45 μm pore size) and aseptically transferred to the surface of selective media. Bale et al. (5) enumerated transconjugants in this manner when demonstrating the in situ transfer of a naturally occurring plasmid to *Pseudomonas aeruginosa* PU21 in a Welsh river. The donor strain carried a chromosomally encoded resistance to streptomycin. PU21 is resistant to rifampicin. Putative transconjugants were recovered by membrane filtering 1-, 10-, or 100-ml volumes through a 0.45-μm membrane. The membranes were incubated on the surface of plate count agar (PCA) containing rifampicin and $HgCl_2$. Mercury resistance (Hg^r) is plasmid-encoded. Presumptive transconjugants were confirmed by verifying the PU21 phenotype using appropriately supplemented minimal glucose agar and also checked for sensitivity to streptomycin. Colonies showing the suitable phenotype were tested for the presence of the plasmid.

A similar procedure was used by Saye et al. (51) to recover nalidixic acid–resistant *P. aeruginosa* recipients which had acquired the plasmid Rms149 by transduction during incubation in situ. Rms149 encodes resistance to carbenicillin (Cb^r), streptomycin (Sm^r), gentamycin (Gm^r), and sulfonamides (Su^r). After filtration through a 0.45-μm membrane, the membrane was incubated on L-agar containing Nal (500 μg/ml) and Cb (500 μg/ml). Verification was accomplished by testing for the presence of an unselected chromosomal marker carried on the recipient chromosome, and verifying the plasmid identity by the pattern of antibiotic resistance and restriction analysis.

Membrane filtration and most probable number (MPN) techniques were compared for efficiency in recovering *P. aeruginosa* from water when present in low numbers. Membrane filtration was found to be superior to MPN techniques (15). The sensitivity of both methods decreases with increasing turbidity of the sample. One drawback to membrane filtration is that large amounts of particulate matter produce faulty filtrations with the suspended solids clogging the membrane filter. In addition, when cell densities are relatively high, the membranes become clogged with cells and it is difficult to complete filtration or wash the membrane. To circumvent this problem, a large sample volume can be divided into smaller volumes and each filtered and washed separately. The appropriate volumes are determined by the degree of turbidity present in each sample.

Molecular techniques. The identity of a transferred plasmid can be verified by isolating the plasmid DNA (37) and evaluating by restriction analysis. Plasmids from putative transductants or transconjugants are digested with restriction endonucleases and electrophoresed with a lane of authentic plasmid DNA. Comparison of banding patterns confirm the presence of the appropriate plasmid in the recipient (42, 51).

Hybridization techniques

Southern hybridization. Plasmid isolated from putative recombinants is hybridized to similarly restricted authentic plasmid DNA which has been labeled (60). If the plasmid has been transferred in total, all bands will "light up." This procedure could also be used to detect deletions and rearrangements of transferred DNA.

Colony hybridization. Methods for the direct detection of specific DNA sequences in bacteriophage plaques and in bacterial colonies have been described (23, 24). Colony hybridization using labeled DNA probes have been shown to be a useful tool for studying environmentally relevant problems (7, 54).

DNA:DNA colony hybridization techniques have been developed as an approach to determining the frequency, dispersal, and maintenance of specific sequences within a natural population. It is particularly useful in detecting poorly selectable phenotypes or genotypes that are not expressed well under routine culturing conditions. This protocol allows the detection of specific DNA sequences, without requiring gene expression or selection. This characteristic makes this method particularly attractive for following the transfer of exogenous DNA to indigenous organisms where a gene may persist without being expressed.

When detecting specific sequences among the indigenous population, probe specificity is of primary importance. Interference may be a problem when using a probe derived from a broad-host-range plasmid since there may be homology to other naturally occurring sequences. When using an appropriately specific probe, the sensitivity of detecting low-copy-number plasmid DNA by colony hybridization is high, i.e., 1 colony in 10^6 (54) to 10^8 colonies (35) of nonhomologous background. It may be possible to detect high-copy-number plasmids with even greater sensitivity after appropriate amplification (24, 54). A limiting parameter of colony hybridization is the specific activity (curies per mole) of the probe. When using a very small probe, detection requires the use of very high specific activity (52). DNA gene probes have been developed for detecting mercury resistance genes in gram-negative communities (7), for detecting catabolic genotypes in environmental samples (54), and for specifically detecting species of the *Pseudomonas fluorescens* group (18).

Samples are spread inoculated directly onto agar plate surfaces or onto hybridization supports overlaid on the respective medium. If inoculation is onto the agar surface, colonies which develop are transferred to hybridization supports (e.g., nitrocellulose or Gene Screen Plus) and subjected to hybridization without further incubation of the transferred cells. If inoculated onto an overlaid support, the hybridization support can be removed after colony development and immediately processed (54). Colony hybridizations will be discussed in greater detail in Chap. 19.

Polymerase chain reaction (PCR).   DNA:DNA colony hybridizations still require both the recovery and the growth of the organism prior to detection of specific sequences. The difficulties involved in recovering bacteria from the environment by culturing techniques have been discussed (48). The inability to recover the "viable but nonculturable" bacteria from environmental samples will also limit the sensitivity of the DNA:DNA colony hybridization technique (43, 48, 61). The PCR technique, originally developed for clinical diagnosis, has been adapted for detection of probe-specific sequences in environmental samples and does not require direct culturing of the organism (61).

DNA isolated from the environmental samples is melted and short oligomer primers are annealed to sequences which flank the target sequence. DNA polymerase and free deoxyribonucleotides are added to the mixture and DNA is extended from the primers across the target region. Repeating this process amplifies the target sequence exponentially with amplification being approximately 2^n, where n is the number of cycles of melting and primer extension. The use of a heat-stable DNA polymerase from *Thermus aquaticus* (Taq) permits many cycles to be performed without the need to add new polymerase after each melting (49, 61).

Steffan and Atlas (61) were able to enhance the detection of specific sequences by three orders of magnitude using PCR with dot-blot hybridization techniques. A target sequence in the pesticide degrading bacterium *Pseudomonas cepacia* AC110 was amplified to detectable levels after 25 cycles of PCR in a sample that contained as little as 0.3 pg of target sequence in a background of nonspecific DNA isolated from sediments. This detection level represents less than 10^4 cells of *P. cepacia* AC110 against a background of 10^{11} cells without the need to first grow the organism. Refinement of the PCR technique may significantly improve our ability to detect or track specific sequences in natural environments.

Effect of environmental parameters

Physical and chemical.   Stress is induced in a substantial proportion of bacteria introduced into aquatic environments such that they can no longer be detected on selective media due to injury or death. The characteristics of survival, injury, or recovery of *Escherichia coli* and *Streptococcus faecalis* in a freshwater river habitat have been described by Bissonette et al. (11). Using membrane diffusion chambers immersed at eight Montana stream sites, they observed considerable variation in survivability as a function of exposure time for both species. Four sets showed the highest fatality with less than 1 percent surviving after four days. The proportion of the survivors exhibiting nonlethal injury varied from 3 to 100 percent. The proportion of injured cells among the survivors increased with time of incubation. Enrichment using a rich nonselective media prior to selective plating improves recovery of stressed cells.

Natural microbial community.   Decline in the numbers of donors and recipients in situ has been observed by other investigators and found to be greatly accelerated by the inclusion of the natural microbial population (39, 42, 51). Mortality as a consequence of grazing by indigenous microorganisms has been studied (17, 25, 58). Servais et al. (58) removed grazers by filtration through a 2-μm Nucleopore membrane. The mortality rates were significantly lower in the filtered sample than in the unfiltered control. Grazing rates ranged from 0.005 to 0.021/h in water from the Meuse River, Belgium. Grazing pressure is greater for bacteria attached to particulate matter than for freeliving bacteria (17).

Methods

Conjugation in fresh water

Bacterial strains and plasmids.   The choice of bacterial strains and plasmids for conjugation studies is limited by only a few factors. The plasmid being tested must encode for a selectable marker, otherwise, finding the rare transconjugants in a huge background of donor, recipient, and indigenous organisms becomes time consuming and expensive. It is also desirable for the plasmid to have a secondary marker for confirmation of its presence in putative transconjugants by quick and inexpensive methods.

Promiscuous laboratory plasmids, such as those of the Inc P1 group, are more likely to find recipients among the resident population than are narrow-host-range plasmids. However, narrow-host-range plasmids and nontransmissible plasmids used as vectors for recombinant DNA will need to be tested because an undefined factor in the environment could change its transfer characteristics. These plasmids should be tested with and without the inserted DNA.

The chromosomal composition of donor and recipient strains must be such that they are distinguishable from each other as well as from the indigenous population. A selectable marker in the recipient organism aids in the primary enumeration of transconjugants by providing a means of eliminating the background of donor cells. Auxotrophic as well as prototrophic strains have been successfully used for in situ studies in fresh water (42). When using nalidixic acid to halt transfer in timed studies or to prevent transfer from occurring on a solid support during enumeration, the recipient strain must be resistant to nalidixic acid.

If indigenous organisms are to be used as recipients, a selectable marker may need to be introduced. By plating a suspension of the bacteria on media containing antibiotic, cells which have acquired resistance to the antibiotic by spontaneous mutation can be isolated. This method is the simplest way to obtain strains with selectable phenotypes and is commonly used. Alternatively, the insertion of a screenable gene for use in environmental studies, such as the gene encoding β-galactosidase, has been described (16). It is important to remember that the introduction of such a marker, whether for selection or for screening, could alter the characteristics of the strain in such a way that its ability to serve as a recipient may be either decreased or increased.

Factors to consider when examining conjugal transfer of a plasmid are: (1) the energy burden imposed on the host by production of nonessential gene products, (2) the potential benefit to the strain from plasmid-encoded traits such as resistance to antibiotics or heavy metals, (3) the rate of plasmid replication and transfer relative to the rate of vegetative segregation, and (4) the cell density of potential recipients in the population.

In situ conjugation.   Cells for conjugation are grown in the laboratory to midlog, harvested by centrifugation, and washed with sterile lake water. The bacterial density of these suspensions is estimated by reading the optical density and comparing it to a standard growth curve for that strain. The concentration of cells is adjusted to the desired levels. A high cell density of organisms makes recovery of transconjugants easier, since smaller sample volumes may be used. However, in natural habitats concentrations of bacteria will be low. Therefore, if high cell densities are used in contained experiments, it must be demonstrated that the results are also valid when cell densities are low. Appropriate volumes of diluted or undiluted washed cells are transported on ice in sterile syringes to the test site. Inoculation takes place at the test site into previously prepared chambers (42).

The initial and subsequent samples are aseptically drawn on site and the physical and chemical parameters are measured. Samples are then transported on ice to the laboratory facility for microbiological assay. Inoculation, incubation, and sampling procedures are similar to those described by Saye et al. (51).

Enumeration of transconjugants.   The primary method for enumeration of transconjugants from environmental samples is by plating to selective media. The technique used is identical to traditional laboratory methods for the isolation of transconjugants, although factors unique to the environmental situation must be considered. The first factor to consider is that the cell density of the introduced bacteria may decline rapidly, eventually falling below the limits of detection. The number of transconjugants is likely to be very small and large sample volumes are often needed to allow detection. Small numbers of recombinants can be found in large volumes by filtering the sample through a membrane and aseptically transferring the membrane to the surface of selective media. A problem often encountered with membrane filtration of environmental samples is clogging due to particulate matter or high cell density. Also, a buildup of cells on the surface of the membrane may limit diffusion of selective agents from the media, allowing overgrowth of the membrane surface during incubation. These problems are alleviated by dividing the sample into smaller volumes, which are filtered separately and incubated as previously described. Colonies arising on these membranes are counted and the numbers pooled.

A potential problem with the use of membrane filtration in conjugation studies is that conjugation can occur between donors and recipients after being brought together on the membrane rather than in the test chamber. Conjugation can be stopped by using nalidixic acid to inhibit transfer of the DNA (6, 42). The concentration of nalidixic acid necessary to halt DNA transfer without killing the recipient varies for different species and strains. O'Morchoe et al. (42) used a nalidixic acid–resistant strain of *P. aeruginosa* as a recipient to circumvent this problem.

Evaluation of conjugation potential.   At each sampling, the total viable count should be determined as well as enumeration of donor, recipient, and transconjugant cells. Most conjugal transfer will occur in the first 24 to 48 hours. During this time the numbers of introduced bacteria may decline significantly. The sampling schedule should be planned accordingly.

Plated samples should be incubated for at least one to two days. Longer incubation times may be necessary for strains with long generation times. Colonies which develop are counted as putative transconjugants and confirmed as true transconjugants by patching them onto media that will test each plasmid marker individually, including those markers which were not used for primary selection. Chromosomal markers, both selected and unselected, that are specific for the recipient are also scored. The presence of the plasmid and its identity can be verified by using small-scale plasmid preparation (37) with subsequent restriction analysis of the isolated DNA.

Conjugal transfer of DNA has been studied in lakes, ponds, and in rivers. O'Morchoe et al. (42) studied the transfer of two laboratory plasmids between *P. aeruginosa* donor and recipient strains in lake water both with and without the indigenous population. They showed that the potential for conjugal transfer in a freshwater environment exists, but that the ability to detect transconjugants is significantly reduced by inclusion of the indigenous microbial population. Using sterile pond water, Gowland and Slater (21) demonstrated the potential for conjugal transfer of laboratory plasmids between *E. coli* strains. The cells were placed in dialysis tubing which was suspended in the pond. Transfer occurred at very low frequencies and transconjugants isolated only after extended mating times of 92 and 306 hours.

Uncontained *P. aeruginosa* PU21 was the recipient of the naturally occurring plasmid QM1 in a study performed in a Welsh river by Bale et al. (5) and found to transfer at frequencies of 3.3×10^{-1} and 6.8×10^{-9} transconjugants per recipient. In a subsequent study, Bale et al. (4) were able to demonstrate the conjugal transfer of QM1 between *P. aeruginosa* that were first incorporated into the epilithon of river stones. Without containment and without increasing the natural densities of the bacteria, plasmid transfer was shown to occur at frequencies between 2.2×10^{-1} and 2.5×10^{-6} transconjugants per donor and appeared to be more dependent on donor to recipient ratio (489:1 to 0.0047:1) than on water temperature. These and other studies indicate that conjugal transfer of genetic information occurs in a freshwater environment (13, 20, 22, 55).

The potential for conjugal transfer of both introduced plasmids into indigenous bacteria and of naturally occurring plasmids into introduced bacteria has been demonstrated in natural freshwater habitats. However, the impact of conjugal transfer on the environment is yet to be determined.

Transduction in fresh water

Bacterial strains, plasmids, and bacteriophage.   The strains of bacteria used in the transduction studies will need to be readily distinguished from each other and from native microorganisms. Saye et al. (51) used donor and recipient strains which were prototrophic and utilized the plasmid-encoded antibiotic resistances to enumerate the donors. Recipients were enumerated by selecting for chromosomally encoded resistance to nalidixic acid and verified by the presence of an unlinked, unselected chromosomal marker (10, 14, 51). The use of prototrophs is not essential. Strains exhibiting a number of auxotrophies have been successfully used for in situ investigations (42, 50). An important consideration in strain choice is the convenience of using strains with directly selectable phenotypes.

In transduction studies it is imperative that the test plasmid is not self-transmissible (Tra$^+$), is not mobilizable (Mob$^+$), and does not possess chromosome-mobilizing ability (Cma$^+$). If plasmids are to be transduced, the size of the plasmid may be significant. Plasmids of molecular weights comparable to the phage genome were shown to transduce more frequently than smaller plasmids (51).

Some virulent as well as temperate bacteriophage have been shown to carry out generalized transduction (30, 40). A virulent phage must be introduced into the test system as a lysate. A temperate transducing phage, however, may be introduced either as a lysate or by introducing a lysogenic bacteria into the test system and is discussed in the following section. A transducing bacteriophage exhibiting a relatively wide host range would certainly be desirable for this type of study; however, bacteriophage generally exhibit a relatively narrow host range. For this reason, it is important that all test strains be appropriately sensitive. Prior to the field studies, the ability of the recipient to be transduced by a particular phage should be demonstrated under controlled conditions using laboratory transduction protocols.

Source of transducing bacteriophage. The transducing bacteriophage in the test system may be provided from a number of sources. Morrison et al. (41) added lysates of the *Pseudomonas* bacteriophage F116 to a flowthrough chamber containing the recipient strain. The lysate had been prepared on a donor strain which carried a chromosomally encoded resistance to streptomycin. Alternatively, transducing phage may be provided in the form of lysogens. The lysogens release free bacteriophage to the system upon induction of the prophage to lytic growth.

The introduced lysogen may be the donor (41) or the recipient (51). Three different paradigms for the source of transducing phage have been evaluated in situ by Saye et al. (51). The greatest number of transductants were recovered when a lysogenic strain served as both the source of the transducing phage and as the recipient of the plasmid DNA. The donor was nonlysogenic. Alternatively, the lysogen which serves as the source of the transducing phage could be neither the donor nor the recipient, being introduced solely as the phage source. To our knowledge, transduction in a natural habitat has not been demonstrated using this model.

Transfer of chromosomal markers has been demonstrated between two *P. aeruginosa* strains lysogenic for the same temperate transducing phage in microcosms and also in situ (50). The ability of lysogens to serve as recipients has been established (8, 51). It has been suggested that the immunity of the lysogen to superinfection following transduction enhances its ability to persist in the environment (51).

In situ transduction. Test strains are grown in the laboratory to midlog. Cultures are harvested by centrifugation and the supernatant is discarded. Cells are washed with sterile lake water (39). The washes prevent nutrient levels from being elevated as a result of culture media being transferred to the chambers along with the test bacteria. More extensive washing is necessary when using lysogenic strains to prevent the addition of significant levels of free bacteriophage (51). Free phage are present in midlog laboratory cultures at concentrations as high as 10^6 PFU/ml as the result of the spontaneous induction of prophage in a fraction of the growing bacteria. The cell density of the washed cultures should be estimated by measuring the optical density. A standard plate count of each strain is routinely performed to verify cell numbers. Appropriate volumes of the washed cultures are drawn up into labeled sterile syringes and transported on ice to the in situ field site (2, 42).

The containment chambers are presterilized and filled with a known volume of sterilized lake water or alternatively with fresh lake water if the presence of the indigenous microbiota is desired. The containment chambers are inoculated with the test strains, mixed, and immediately sampled to obtain the 0 time data. All physical and chemical parameters are measured. The 0 time sample is handled the same as all other samples to be taken. This establishes a reference point from which changes in microbial number and changes in physical and chemical attributes can be measured. At each sampling, physical and chemical measurements are made of the surrounding waters.

Saye et al. (51) found that transductants were recovered at the greatest frequency in the first 72 hours of incubation in situ. Morrison et al. (41) observed that the transduced cells increased in numbers over the first four days, then decreased at a rate comparable to that observed for the untransduced recipient. For this reason, sampling periods should be relatively frequent during the first four days of incubation. The intervals between sampling then increase as the time of incubation becomes lengthier.

The multiplicity of infection (MOI) is an important variable in laboratory transduction protocols. It seems reasonable that the concentration of phage in the test system, relative to cell density, may significantly affect the potential for transduction in situ. It is therefore necessary to obtain cell-free filtrates at the time of each sampling and determine the titer of the bacteriophage in each containment chamber.

When drawing a sample for microbiological assay it is convenient to use a large volume sterile syringe. The syringe can be fitted with a 0.45-μm membrane syringe filter and 1 to 5 ml of filtrate collected in a sterile tube for use in titrating the bacteriophage. The cell-free filtrate is obtained immediately at the time of sampling to preclude the possibility of decreasing the phage titers as the result of adsorption to cell debris or increasing the phage titer due to induction of the prophage by a variable inadvertently introduced during sample handling.

Determination of bacteriophage titers.   The sample which was filtered through the 0.45-μm filter is serially diluted. Each dilution is mixed with a midlog culture of a sensitive strain and tempered (45°C) top agar. An overlay is then made on a rich nonselective media and incubated for 18 to 24 hours. Phage are enumerated by the formation of plaques (51).

Enumeration of transductants.   Donors and recipients are enumerated on the appropriate selective media as described earlier. Total viable counts are obtained by serially diluting the sample and plating on a rich media such as YEPG agar (53). Bissonette et al. (11) found that use of phosphate buffer diluent injured cells and prevented their being detected on selective media. The use of a gelatin phosphate buffer improved the recovery of bacteria which had been incubated in situ. Transduction in situ is likely to occur at a relatively low frequency; therefore, sample size should be quite large and be expected to yield a small number of transductants. Detection of transductants is simplified by membrane-filtering the sample and incubating the membrane on the surface of media which is selective for the transduced phenotype (15, 32, 51).

The possibility that transduction can occur on the membrane must be considered. In order to preclude transduction from occurring on the membrane surface, Saye et al. (51) added nalidixic acid to the sample prior to filtration. At appropriate concentrations, the DNA metabolism of a sensitive donor will be halted, precluding infection and lysis of donor cells by the bacteriophage. After addition of the nalidixic acid, preformed transducing particles could still effect transduction of a recipient cell trapped on the membrane. Therefore, extensive washing of the membrane is performed to remove unadsorbed bacteriophage.

After filtering and washing, the membranes are placed directly on the surface of media containing agents selective for both the recipient markers and the transduced phenotype. Incubation should be allowed for at least two to three days. Isolates which appear on the membrane surface should be streaked for purity on selective media and subsequently patched onto media that selects each marker individually. The combination of markers expressed by an isolate is used to identify transductants.

Evaluation of transduction potential.   The transductants are verified by checking the individual phenotypes expected of a transduced cell. Plasmid DNA is isolated from the putative transductants and its identity confirmed as described.

In laboratory protocols for transduction the results are traditionally reported as the number of transductants per ingoing plaque-forming unit (i.e., transductants per PFU). Although this is a convenient way to express the transducing potential of a particular transducing bacteriophage, it may not be appropriate in environmental testing for the following reasons: (1) Complex interactions of phage, donor cells, and recipient cells take place in the test system. (2) New phage and transducing particles are introduced into the system by induction of donor lysogens (41) or by primary infection of nonlysogenic donors (51). (3) Changes in the number of donor and recipient cells and changes in their relative proportions are likely to occur over the course of the study.

Results need to be expressed in a manner which will allow meaningful comparisons between data obtained by different investigators. In the freshwater environmental studies cited here, results have been reported in terms of transductants per recipient or transductants per donor (41, 51).

Saye et al. (51) found that when incubating a nonlysogenic donor with a lysogenic recipient, the number of transductants per donor remained relatively constant regardless of the initial ratio of donors to recipients as long as donors were in excess. The numbers of transductants per recipient on the other hand increased with the increasing donor-to-recipient ratio and leveled off after 36 hours. This correlated well with the increase in bacteriophage titer observed in the chambers. Transductants were detected in chambers containing nonlysogenic plasmid donors and lysogenic recipients.

Morrison et al. (41) reported the F116-mediated transduction of *P. aeruginosa* increasing from 5×10^{-6} to 9.5×10^{-1} transductants per recipient in a chamber containing recipient bacteria and a cell-free lysate over 10 days of incubation in situ. They also observed transduction increasing from 1.4×10^{-5} to 8.3×10^{-2} in a chamber consisting of a lysogenic donor strain and a nonlysogenic recipient over the same period. Changes in bacteriophage titer were not measured by Morrison et al. in their demonstration of the F116-mediated transduction of *Pseudomonas aeruginosa*.

It is important to understand that these studies measured the frequency of recovering transductants in the test chambers and not the frequency of transduction. This distinction must be made. It cannot be determined in the above experiments whether the increasing number of transductants per recipient resulted from increasing numbers of transduction events or from the outgrowth of a small number of transduced cells.

The relationship of bacteriophage titer and cell density and its influence on the potential for gene transfer in situ is not known. In laboratory simulations in lake water and sewage, Wiggins and Alexander (68) concluded that bacteriophage do not exert an effect on populations of bacteria in natural ecosystems when the host species is present in numbers below a threshold density of 10^4 CFU/ml.

Effect of environmental factors.   The persistence of introduced species in environmental systems is a significant factor for in situ investigations. There is evidence that the recovery of the introduced species is significantly improved when incubating in the absence of the natural microbiota (51). Bacteriophage titers remain large in a sterile lake water environment, but decline rapidly in nonsterile lake water (51, 68).

Seeley and Primrose (57) recognized three physiological classes of aquatic bacteriophage based on differences in their efficiency of plating (e.o.p.) at high, medium, and low temperature. The maximum and minimum plating temperatures are stable phage properties and correlate well with the temperature of the environment from which they originate. High-temperature (HT) phages have a low efficiency of plating below 35°C and are presumed to have their origins from the gut of warm-blooded animals. The low-temperature (LT) phages, whose natural habitat is an aquatic environment, plate poorly above 30°C. Primrose et al. (45) found that only HT phages could be isolated when water temperature was low, but as the water temperature increased LT phages predominated. In aquatic habitats, where the average temperature is relatively low, transducing bacteriophage normally used in laboratory protocols may function poorly compared to temperature-adapted phage that are capable of mediating genetic exchange.

Transformation in a freshwater environment

Natural competency is a property of some environmentally important species of bacteria and is encoded in the bacterial chromosome. It is a normal physiological function in these species and allows uptake of DNA directly from the surrounding milieu (62). Many bacteria do not exhibit a natural competency for the uptake of naked DNA.

It is reasonable that in situ evaluation of transformation in a freshwater habitat could be accomplished by employing many of the features described for test systems of transduction and conjugation. However, since there are no reported examples of transformation being demonstrated in freshwater habitats, we will not discuss it further.

Conclusion

There are many limitations in evaluating genetic transfer among microorganisms in a natural freshwater environment. The use of containment chambers certainly imposes a barrier to the normal dynamics of a freshwater habitat and this limitation has been recognized. Only a few studies have demonstrated transfer of DNA between strains in an unenclosed system (4, 5). Demonstrations of DNA exchange in enclosed systems and in the presence of a natural community may be overestimates of the potential that exists in nature. The methodology presented here provides a starting point for investigating the potential for gene dispersal in freshwater environments by the three recognized mechanisms.

Acknowledgments

This work was supported by cooperative agreements No. CR812494 and CR815234 from the United States Environmental Protection Agency, Gulf Breeze Laboratory. The contents of this report do not necessarily reflect the views of the Environmental Protection Agency nor does mention of trade names or commercial products constitute endorsement or recommendation for use.

References

1. Altherr, M. R., and K. L. Kasweck. 1982. *In situ* studies with membrane diffusion chambers of antibiotic resistance transfer in *Escherichia coli. Appl. Environ. Microbiol.* 44:838–843.

2. Amin, M. K., and M. J. Day. 1988. Donor and recipient effects on transduction frequency *in situ.* Abst. No. 2, REGEM 1 Program. Cardiff, Wales, p. 11.
3. Baker, K. H., and A. L. Mills. 1982. Determination of the number of respiring *Thiobacillus ferroxidans* in water samples by using combined fluorescent antibody-2-(*p*-iodophenyl)-3-(*p*-nitrophenyl)-5-phenyltetrazolium chloride staining. *Appl. Environ. Microbiol.* 43:338–344.
4. Bale, M. J., M. J. Day, and J. C. Fry. 1988. Novel method for studying plasmid transfer in undisturbed river epilithon. *Appl. Environ. Microbiol.* 54:2756–2758.
5. Bale, M. J., J. C. Fry, and M. J. Day. 1987. Plasmid transfer between strains of *Pseudomonas aeruginosa* on membrane filters attached to river stones. *J. Gen. Microbiol.* 133:3099–3107.
6. Barbour, S. D. 1967. Effect of nalidixic acid on conjugational transfer and expression of episomal *Lac* genes in *Escherichia coli* K12. *J. Mol. Biol.* 28:373–376.
7. Barkay, T., D. L. Fouts, and B. H. Olson. 1985. Preparation of a gene probe for detection of mercury resistance genes in gram negative bacterial communities. *Appl. Environ. Microbiol.* 49:686–692.
8. Benedik, M., M. Fennewald, and J. Shapiro. 1977. Transposition of a beta-lactamase locus from RP1 into *Pseudomonas putida* degradative plasmids. *J. Bacteriol.* 129:809–814.
9. Bergh, Ø., Y. Børsheim, G. Bratbak, and M. Heldal. 1989. High abundance of viruses found in aquatic environments. *Nature* 340:467–468.
10. Betz, J. L., J. E. Brown, P. A. Clarke, and M. Day. 1974. Genetic analysis of amidase mutants of *Pseudomonas aeruginosa. Genet. Res. Camb.* 23:335–359.
11. Bissonette, G. K., J. J. Jezeski, G. A. McFeters, and D. G. Stuart. 1975. Influence of environmental stress on enumeration of indicator bacteria from natural waters. *Appl. Microbiol.* 29:186–194.
12. Bohlool, B. B., and E. L. Schmidt. 1968. Nonspecific staining: Its control in immunofluorescence examination of soil. *Science* 162:1012–1014.
13. Cooke, M. D. 1978. R-factor transfer in river and sea-water. Abstracts of New Zealand Microbiological Society Annual Conference. Nelson, New Zealand.
14. Cuskey, S. M., and P. V. Phibbs, Jr. 1985. Chromosomal mapping of mutations affecting glycerol and glucose catabolism in *Pseudomonas aeruginosa* PAO. *J. Bacteriol.* 162:872–880.
15. deVicente, A., J. J. Borrego, A. Francisco, and A. Romero. 1986. Comparative study of selective media for enumeration of *Pseudomonas aeruginosa* from water by membrane filtration. *Appl. Environ. Microbiol.* 51:832–840.
16. Drahos, D. J., B. C. Hemming, and S. McPherson. 1986. Tracking recombinant organisms in the environment: β-galactosidase as a selectable marker for fluorescent Pseudomonads. *Bio/Technology* 4:439–444.
17. Fenchel, T., and D. Jørgensen. 1977. Detritus food chains of aquatic environments: The role of bacteria. *Adv. Microb. Ecol.* 1:1–58.
18. Festl, H., W. Ludwig, and K. H. Schleifer. 1986. DNA hybridization probe for the *Pseudomonas fluorescens* group. *Appl. Environ. Microbiol.* 52:1190–1194.
19. Fliermans, C. B., and R. W. Gorden. 1977. Modification of membrane diffusion chambers for deepwater studies. *Appl. Environ. Microbiol.* 33:207–210.
20. Genthner, F. J., P. Chatterjee, T. Barkay, and A. W. Bourquin. 1988. Capacity of aquatic bacteria to act as recipients of plasmid DNA. *Appl. Environ. Microbiol.* 54: 115–117.
21. Gowland, P. C., and J. H. Slater. 1984. Transfer and stability of drug resistance plasmids in *Escherichia coli* K12. *Microbiol. Ecol.* 10:1–13.
22. Grabow, W. O. K., O. W. Prozesky, and J. S. Burger. 1975. Behavior in a river and dam of coliform bacteria with transferable or non-transferable drug resistance. *Water Res.* 9:777–782.
23. Grünstein, M., and D. S. Hogness. 1975. Colony hybridization: A method for the isolation of cloned DNAs that contain a specific gene. *Proc. Natl. Acad. Sci. USA* 72: 3961–3965.
24. Hanahan, D., and M. Meselson. 1980. Plasmid screening at high colony density. *Gene* 10:63–67.

25. Iriberri, J., M. Inanue, I. Barcina, and L. Egea. 1987. Seasonal variation in population density and heterotrophic activity of attached and unattached free-living bacteria in coastal waters. *Appl. Environ. Microbiol.* 53:2308–2314.
26. Jannasch, H. W. 1969. Estimations of bacterial growth rates in natural waters. *J. Bacteriol.* 99:156–160.
27. Jones, J. C., S. Gardener, B. M. Simon, and R. W. Pickup. 1986. Factors affecting the measurement of antibiotic resistance in bacteria isolated from lakewater. *J. Appl. Bacteriol.* 60:455–462.
28. Kirchman, D., and R. Mitchell. 1982. Contribution of particle-bound bacteria to total microheterotrophic activity in five ponds and two marshes. *Appl. Environ. Microbiol.* 43:200–209.
29. Kogure, K., S. Ushio, and N. Taga. 1979. A tentative direct microscopic method for counting living marine bacteria. *Can. J. Microbiol.* 25:415–420.
30. Kokjohn, T. A. 1989. Transduction: Mechanism and potential for gene transfer in the environment. In R. V. Miller and S. B. Levy (eds.), *Gene Transfer in the Environment.* McGraw-Hill, New York.
31. Kropinski, A. B., V. Lewis, and D. Berry. Effect of growth temperature on lipids, outer membrane proteins, and lipopolysaccharides of *Pseudomonas aeruginosa* PAO. *J. Bacteriol.* 169:1960–1966.
32. Krueger, C. L., and W. Sheikh. 1987. A new selective media for isolating *Pseudomonas* spp. from water. *Appl. Environ. Microbiol.* 53:895–897.
33. Lovell, C. R., and A. Konopka. 1985. Seasonal bacterial production in a dimictic lake as measured by increases in cell numbers and thymidine uptake. *Appl. Environ. Microbiol.* 49:492–500.
34. Lundquist, P. D., and B. R. Levin. 1986. Transitory derepression and the maintenance of conjugative plasmids. *Genetics* 113:483–497.
35. Maas, R. 1983. An improved colony hybridization method with significantly increased sensitivity for detection of single genes. *Plasmid* 10:296–298.
36. Mach, P. A., and D. J. Grimes. 1982. R-plasmid transfer in a wastewater treatment plant. *Appl. Environ. Microbiol.* 44:1395–1403.
37. Maniatis, T., E. F. Fritsch, and J. Sambrook. 1982. *Molecular Cloning: A Laboratory Manual.* Cold Spring Harbor Laboratory, Cold Spring Harbor, N.Y., pp. 364–373.
38. McCormick, D. 1986. Detection technology: The key to environmental biotechnology. *Bio/Technology* 4:419–422.
39. McFeters, G. A., and D. G. Stuart. 1972. Survival of coliform bacteria in natural waters. *Appl. Environ. Microbiol.* 24:805–811.
40. Morgan, A. F. 1979. Transduction of *Pseudomonas aeruginosa* with a mutant of bacteriophage E79. *J. Bacteriol.* 19:137–140.
41. Morrison, W. D., R. V. Miller, and G. S. Sayler. 1978. Frequency of F116 mediated transduction of *Pseudomonas aeruginosa* in a freshwater environment. *Appl. Environ. Microbiol.* 36:724–730.
42. O'Morchoe, S., O. Ogunseitan, G. S. Sayler, and R. V. Miller. 1988. Conjugal transfer of R68.45 and FP5 between *Pseudomonas aeruginosa* strains in a freshwater environment. *Appl. Environ. Microbiol.* 54:1923–1929.
43. Pettigrew, C., and G. S. Sayler. 1986. The use of DNA:DNA colony hybridizations in the rapid isolation of 4-chlorophenyl degradative bacterial phenotypes. *J. Microbiol. Methods* 5:205–213.
44. Primrose, S. B., and M. Day. 1977. Rapid concentration of bacteriophage from aquatic habitats. *J. Appl. Bacteriol.* 42:417–421.
45. Primrose, S. B., N. D. Seeley, K. B. Logan, and J. W. Nicholson. 1982. Methods for studying aquatic bacteriophage ecology. *Appl. Environ. Microbiol.* 43:694–701.
46. Roper, M. M., and K. C. Marshall. 1978. Effect of clay particle size on clay-*Escherichia coli*-bacteriophage interactions. *J. Gen. Microbiol.* 106:187–189.
47. Roszak, D. B., and R. R. Colwell. 1987. Metabolic activity of bacterial cells enumerated by direct viable count. *Appl. Environ. Microbiol.* 53:2889–2983.
48. Roszak, D. B., and R. R. Colwell. 1987. Survival strategies of bacteria in the natural environment. *Microbiol. Rev.* 51:365–379.

49. Saiki, R. K., S. Scharf, F. Faloona, K. B. Mullis, G. T. Horn, H. A. Ehrlich, and N. Arnheim. 1985. Enzymatic amplification of β-globin genomic sequences and restriction site analysis for the diagnosis of sickle cell anemia. *Science* 230:1350–1354.
50. Saye, D. J., O. Ogunseitan, G. S. Sayler, and R. V. Miller. 1990. Transduction of linked chromosomal genes between *Pseudomonas aeruginosa* strains during incubation in situ in a freshwater habitat. *Appl. Environ. Microbiol.* 56:140–145.
51. Saye, D. J., O. Ogunseitan, G. S. Sayler, and R. V. Miller. 1987. Potential for transduction of plasmids in a natural freshwater environment: Effect of donor concentration and a natural microbial community on transduction in *Pseudomonas aeruginosa*. *Appl. Environ. Microbiol.* 53:987–995.
52. Sayler, G. S., C. Harris, C. Pettigrew, D. Pacia, A. Breen, and K. M. Sirotkin. 1987. Evaluating the maintenance and effects of genetically engineered microorganisms. *Devel. Ind. Microbiol.* 27:135–149.
53. Sayler, G. S., L. C. Lund, M. P. Shiaris, T. W. Sherrill, and R. E. Perkins. 1979. Comparative effects of Arochlor 1254 (polychlorinated biphenyls) and phenanthrene on glucose uptake velocities by freshwater microbial populations. *Appl. Environ. Microbiol.* 37:878–885.
54. Sayler, G. S., M. S. Shields, E. T. Tedford, A. Breen, S. W. Hooper, and K. M. Sirotkin. 1985. Application of DNA-DNA hybridization to the detection of catabolic genotypes in environmental samples. *Appl. Environ. Microbiol.* 49:1295–1303.
55. Schilf, W., and W. Klingmüeller. 1983. Experiments with *Escherichia coli* on the dispersal of plasmids in the environment. *Recomb. DNA Tech. Bull.* 6:101–102.
56. Schmidt, E. L. 1973. Fluorescent antibody technique for the study of microbial ecology. *Bull. Ecol. Res. Commun.* 17:67–76.
57. Seeley, N. D., and S. B. Primrose. 1980. The effect of temperature on the ecology of aquatic bacteriophage. *J. Gen. Virol.* 46:87–95.
58. Servais, P., G. Billen, and J. V. Rego. 1985. Rate of bacterial mortality in aquatic environments. *Appl. Environ. Microbiol.* 49:1448–1454.
59. Singleton, P. 1983. Colloidal clay inhibits conjugal transfer of R-plasmid R1*drd19* in *Escherichia coli*. *Appl. Environ. Microbiol.* 46:756–757.
60. Southern, E. M. 1975. Detection of specific sequences among DNA fragments separated by gel electrophoresis. *J. Mol. Biol.* 98:503–517.
61. Steffan, R. J., and R. M. Atlas. 1988. DNA amplification to enhance detection of genetically engineered bacteria in environmental samples. *Appl. Environ. Microbiol.* 54:2185–2191.
62. Stewart, G. J., and C. A. Carlson. 1986. The biology of natural transformation. *Ann. Rev. Microbiol.* 40:211–235.
63. Stotzky, G., and H. Babich. 1986. Survival of and genetic transfer by, genetically engineered bacteria in natural environments. *Adv. Appl. Microbiol.* 31:93–138.
64. Stotzky, G., and L. T. Rem. 1966. Influence of clay minerals on microorganisms: I. Montmorillonite and kaolinite on bacteria. *Can. J. Microbiol.* 12:547–563.
65. Strauss, H. S., D. Hattis, G. Paage, K. Harrison, S. Vogel, and C. Caldart. 1986. Genetically engineered organisms: II. Survival, multiplication and genetic transfer. *Rec. DNA Tech. Bull.* 9:69–88.
66. VanEs, F. B., H. J. Laanbroek, and H. Veldkamp. 1984. Microbial ecology: An overview. In G. A. Codd (ed.), *Aspects of Microbial Metabolism and Ecology.* Academic Press, New York, pp. 1–33.
67. Walter, M. V., A. Porteous, and R. J. Seidler. 1987. Measuring genetic stability in bacteria of potential use in genetic engineering. *Appl. Environ. Microbiol.* 53:105–109.
68. Wiggins, B. A., and M. Alexander. 1985. Minimum bacterial density for bacteriophage replication: Implications for significance of bacteriophages in natural ecosystems. *Appl. Environ. Microbiol.* 49:19–23.
69. Zaske, S. K., W. S. Dockins, and G. A. McFeters. 1980. Cell envelope damage in *Escherichia coli* caused by short-term stress in water. *Appl. Environ. Microbiol.* 40:386–390.

Measurement of Conjugal Gene Transfer in Terrestrial Ecosystems

Michael V. Walter

Ramon J. Seidler

Introduction and Historical Perspective

Currently, over 10,000 public and private laboratories are involved in biotechnology on a worldwide basis (24). Much of this biotechnology deals with genetically engineered microorganisms (GEMs) intended for use in agriculture and pollution control and requires their release into the natural environment. Concern has been expressed regarding possible adverse effects their long-term survival may have on natural biological processes (9, 18, 20, 23, 28) and the undesirable spread of recombinant genes to members of the indigenous microflora that inhabit a particular environment (23, 25). Furthermore, identification and enumeration procedures developed for counting living recombinant microbes will not detect recombinant genes once they are transmitted to new species.

Much of the knowledge regarding genetic exchange has been obtained under controlled laboratory conditions using techniques and principles derived from the mobilization of R^+ plasmids. Although useful in explaining mechanisms of genetic transfer, it would be naive to extrapolate these data from the laboratory and apply them to the natural environment (22). Conjugal transfer of DNA in soil, water, and sewage has been demonstrated, but when observed, transfer frequencies are generally lower than those observed under laboratory conditions (8).

Environmental parameters and their mechanisms for affecting conjugation are not well understood. Observations have revealed that pH (27), temperature (26, 32), nutrient availability (31), and the physical environment (4) quantitatively affect DNA transfer by conjugation.

Plasmids have been shown to be present in numerous genera of soil microorganisms (2, 19, 21, 30). However, plasmid function and transfer in soil have been difficult to demonstrate (8). This has forced investigators to rely upon the addition of suitably tagged donor and recipient strains to evaluate in situ gene transfer in the environment. Often, autochthonous species have been used, e.g., *Escherichia coli* donors and recipients added to soil. These studies have relied heavily on the transmission of antibiotic-resistant markers on R^+ plasmids. This series of approaches has probably provided a narrow perspective on the significance of conjugal DNA transfer in natural ecosystems because such approaches rarely consider the natural evolution, maintenance, and transfer of indigenous plasmids. The fate (i.e., expression and replication) of plasmids newly introduced into a given ecosystem are also legitimate ecological issues which have been largely ignored.

Current Practices

Natural soil is an extremely complex environment. Since little is known about the microbial ecology of soil, and the environmental conditions that are most influential on conjugation and survival of transconjugants in this environment, initial investigations regarding gene exchange in soil have used sterilized soil, amended soil, or soil slurries. Such soils are valuable in obtaining preliminary information regarding conjugation by GEMs and identifying the factors that may affect conjugation in the environment. Results from such experiments are useful as estimates of what may be true in natural soils. However, these results cannot be extrapolated directly to the natural environment.

Using both prototrophic and auxotrophic strains of *E. coli*, Weinburg and Stotzky (38) demonstrated conjugal transfer in autoclaved soil collected from New York and Lima, Peru. Transconjugants were isolated from experiments when donors and recipients were inoculated in the same area, and when they were inoculated 10 to 15 mm apart in petri dishes. This indicates that both donors and recipients were capable of growth and migration since transconjugants were isolated in the area between the points of inoculation (38). In addition, Weinberg and Stotzky demonstrated that both the growth rate of donor and recipients and numbers of transconjugants were influenced by the physical components of the soil. Greater numbers of donors, recipients, and transconjugants were isolated from soil amended with the clay mineral montmorillonite. The pH of the soil was also shown to have an influence on conjugation.

Escherichia coli transconjugants were isolated only after the pH was adjusted from 5.5 to 6.9 (38).

The physicochemical and biological properties affecting conjugation in soil were further studied by Wamsley (36). Again, using strains of *E. coli,* he found that pair formation in soil was temperature-dependent. Pairs which did not form at 24°C were found to produce aggregates between 30 and 41°C. Trevors and Oddie (31) reported that nutrient availability could influence temperature dependence on conjugation. They found that, while conjugation occurred in nutrient-amended soil at 15°C, it failed to occur in non-nutrient-amended soil at the same temperature.

The effect of the indigenous microbial population on conjugation was studied by Stotzky and Krasovsky (28). The presence of *Pseudomonas fluorescens* at 5×10^9 cells per gram of soil and *Rhodotorula rubra* at 3×10^7 cells per gram of soil did not impair conjugation between strains of *E. coli* K-12. However, in the presence of the indigenous microbial population (total population of about 10^8 cells per gram of soil), the frequency of gene transfer was lower.

Conjugation in the rhizosphere has been investigated by Talbot et al. (29). Radish seeds, inoculated with antibiotic-sensitive *Klebsiella* recipients, followed by inoculation with 10^4 colony-forming units (CFU) per gram of antibiotic-resistant donor *Klebsiella,* demonstrated the presence of transconjugants during the first week. Conjugal frequencies ranged from 10^{-6} to 10^{-7} transconjugants per donor. Kooykaas et al. (12) demonstrated conjugation using bacteria of the genus *Rhizobium,* which fixes nitrogen in leguminous plants. The plasmid pRtr5a contained in *Rhizobium trifolii,* which encodes nitrogen fixation, was transferred in situ to *Rhizobium leguminosarum.* Transconjugant *R. leguminosarum* were able to fix nitrogen not only in pea and vetch, their normal symbionts, but in clover as well. *R. trifolii* was capable of undergoing conjugation with *Agrobacterium tumefaciens* that had been cured in its Ti-plasmid.

Conjugation on plant surfaces has also been investigated. Watson et al. (37) isolated transconjugant *Agrobacterium* strains five weeks after inoculating plants with antibiotic-sensitive, virulent donors containing the Ti-plasmid, and antibiotic-resistant avirulent recipients. Conjugal frequencies ranged from 10^{-4} to 10^{-2} transconjugants per donor over a three- to five-week period. The *in planta* transfer by *Pseudomonas* species of plasmid RP1 was reported by Lacy and Leary (15). Immature lima bean pods and leaflets of trifoliate leaves were inoculated with donor *E. coli* strains and recipient *Pseudomonas glycinea* or *Pseudomonas phaseolicola.* Transfer frequencies on leaf surfaces between *E. coli* and *P. glycinea* were lower by about a factor of 10 than in vitro transfer frequencies (4.9×10^{-3} and 5.8×10^{-2}, respectively). Lacy and Leary (15) reported transfer of RP1 between

Erwinia herbicola and *P. syringae,* pv. *syringae* to *Erwinia amylovera* on pear blossoms. They reported more efficient transfer of plasmid RP1 *in planta* than in vitro with frequencies as high as 10^{-1} transconjugants per donor.

Specific Methods to Evaluate Conjugal DNA Transfer

Laboratory techniques

The screening of GEMs for ability to undergo conjugation has been carried out under highly controlled laboratory conditions. The literature is filled with different techniques designed to measure conjugation. To determine whether one method is superior, and allow for standardization, Walter et al. (33) compared four common techniques for measuring conjugal DNA transfer: colony cross streak (CCS); broth mating (BM); combined spread plate (CSP); and membrane filtration (MF). Each technique was compared for sensitivity, reliability, technical ease, and cost. The transfer of five plasmids, two Inc W (pSa, R388::Tn1721), two Inc P (RK2, pRK2013), and one Inc C (R40a), were evaluated using the four techniques above; laboratory strains and isolates recently obtained from soil and plants were used as recipients.

Each of the four techniques were capable of demonstrating transconjugants. The MF technique produced transconjugants most consistently except when laboratory strains were used as recipients in matings transferring plasmid R40a. The effectiveness of the other three techniques varied with the plasmid and recipient used.

In general, laboratory strains were found to be better recipients than environmental isolates, except when plasmid R40a was transferred using BM or MF. The reliability of the four techniques varied with plasmid, technique, and recipient.

Conjugal frequencies were highly variable, even within the same plasmid group. Highest frequencies were observed in matings with laboratory strains as recipients using MF. Frequencies averaged 3.3×10^{-1} transconjugants per donor with pRK2013, but were only 1.4×10^{-9} using plasmid R40a. For the most part, the highest frequencies were observed with MF and the lowest with BM. Differences in conjugal frequencies between laboratory strains and environmental isolates ranged from 10-fold, as in plasmid RK2 using BM (8.2×10^{-8} transconjugants per donor and 3.2×10^{-9} transconjugants per donor for laboratory and environmental isolates, respectively), to more than 1000-fold as with R388::Tn1721 using CSP (3.5×10^{-4} laboratory strains versus 2.7×10^{-7} for environmental isolates).

No single mating technique proved superior with all plasmids when laboratory and environmental isolates were used as recipients. Therefore, an integrated technique which incorporates CCS, BM, CSP, and MF into a single process was designed. This procedure has been designated the "combined mating technique" (Fig. 14.1). It is designed to use a single source of washed cells for use in BM, CSP, and MF. The

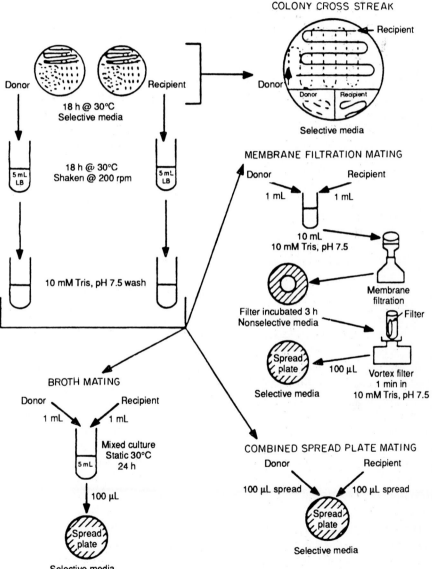

Figure 14.1 Combined mating technique. A standardized procedure for evaluating conjugal transfer of DNA using common laboratory procedures (from reference 33).

combined mating technique also provides a solid and liquid environment for matings. This is important, since some plasmids transfer
preferentially on solid medium and the chance of detecting transconjugants is increased (4).

Soil slurries

The combined mating technique gives no information to indicate
whether a GEM will be able to undergo conjugation in the natural environment or under specific environmental conditions. A soil slurry
assay was developed to give preliminary information about survival
and genetic stability in terrestrial environments (35). The soil slurry
assay consists of mixing 5 g of soil with 18 ml of a liquid (buffer, nutrients, etc.) and placing the mixture into a test tube. The addition of
the liquid phase allows alteration of chemical conditions within the
slurry; it greatly simplifies sampling. Parameters such as pH, nutrient content, and mineral content can easily be adjusted to meet
specific conditions and provide for an even dispersal of donors and recipients.

Tests of this procedure were evaluated using two strains of
Pseudomonas cepacia and the plasmid R388::1721. Experiments were
carried out in sterile and nonsterile commercial potting soil and a representative agricultural soil (Hazelaire complex) collected from a local
farm.

When incubated at 30°C for 24 h with LB broth as the liquid phase,
an average of 9×10^5 transconjugants per milliliter of soil slurry were
observed. This corresponded to a mating frequency of 1.7×10^{-2}
transconjugants per donor (initial donor populations averaged
8.5×10^7 CFU/ml soil slurry). Replacement of LB broth with sterile
buffer, either 10 mM Tris (pH 7.3) or 1.2 mM phosphate (pH 7.2), resulted in a 100-fold decrease in the number of transconjugants
observed, and a decreased mating frequency of 5.7×10^{-5} transconjugants per donor. No significant difference was observed in the
number of transconjugants when using either of the two buffers. No
significant difference ($p < .05$) was observed between the final donor
and recipient populations from LB broth-amended and buffer-
amended soil slurries. Therefore, the reduction in mating frequency
cannot be accounted for by decreases in population levels of donors
and recipients. It is probable that the decrease in transconjugants was
due to a physiological change in the microorganisms within the
buffer-amended microcosms, responding to decreased nutrients.

Temperature is one of the more variable parameters that GEMs entering the environment would encounter. Temperature has dramatic
effects on transconjugant formation. For example, increasing incuba-

tion temperatures from 25 to 35°C during mating experiments caused a 100-fold increase in both the number of transconjugants observed and in the conjugal frequency (2.8×10^{-3} to 4.8×10^{-1} transconjugants per donor, respectively) (35). Final donor and recipient populations were not affected by incubation temperature, thus eliminating any changes in population densities as a cause for increased transconjugant formation. These results are similar to those reported by others (3, 36) using sterile soil culture.

The effect of soil pH on conjugation was also investigated using the soil slurry technique. Increasing slurry pH with LB broth from 5.5 to 7.5 increased the numbers of transconjugants and conjugal frequency 1000-fold. Increasing the pH from 7.5 to 8.5 decreased the number of transconjugants. However, unlike nutrient and temperature experiments, donor populations at pH 5.5 and 8.5 were about 20 percent of those at 7.5. Compared with the baseline data obtained at neutral pH, a comparable decrease in donor population had less than a 10 percent effect on the number of observed transconjugants, that is pH per se had no effect on final donor populations and the 1000-fold decrease cannot be attributed to decreases in donor populations. These observations are also in agreement with results obtained in mating experiments conducted in sterile soil culture (28).

Mating experiments conducted in nonsterile soil slurries at 30°C with LB broth as the liquid phase did not demonstrate a significant effect on conjugal frequencies. The final population of the indigenous microflora averaged 5×10^8 CFU/ml slurry after incubation at 30°C for 20 h. Conjugal frequencies were 8.7×10^{-3} transconjugants per donor for nonsterile potting soil and 1.1×10^{-3} transconjugants per donor for nonsterile agricultural soil.

A test protocol designed to measure conjugal activity should be capable of accommodating diverse microorganisms and test conditions. The soil slurry assay is convenient, as it can be adjusted to simulate a wide range of environmental conditions. Conjugal experiments using this system gave results similar to experiments using sterile soil culture. This technique can be used with any type of soil. Therefore, soil from the areas intended for field studies could easily be included in a screening protocol. While not intended as a replacement for more realistic microcosm experiments, the soil slurry assay provides preliminary data on genetic stability and survival.

Terrestrial microcosms

A more complete understanding of conjugation in terrestrial environments will require the use of microcosms that bridge the gap between artificial laboratory experiments and the natural environment. A microcosm has been defined as controlled reproducible laboratory system that attempts to simulate a portion of the environment (10). Within the confines of the research laboratory, microcosms enable investigations of the impact of GEMs on the natural environment.

One microcosm has recently been described for use in investigating GEM fate or survival and genetic stability in the root rhizosphere ecosystem (1, 14). The design of the microcosm chambers was described in detail by Gillett and Witt (10). The chambers consisted of a box $1 \times 1.25 \times 0.75$ m, composed of glass, plexiglass, and polyethylene (Fig. 14.2). The microcosms were designed for control of temperature, light, airflow, and relative humidity (RH).

In the studies conducted at the Environmental Research Laboratory at Corvallis, airflow through the chambers was 400 L/min, tempera-

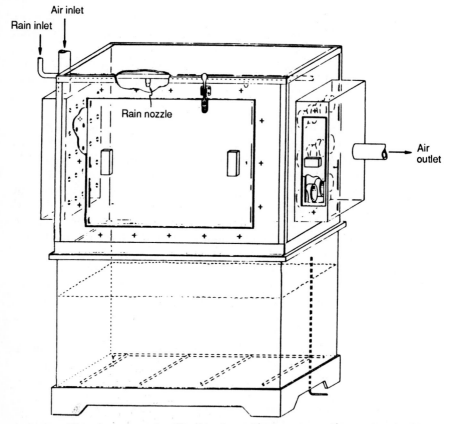

Figure 14.2   The microcosm unit utilized in the conjugal gene transfer experiments. Air flow, light duration, relative humidity, temperature, and soil moisture are controlled and electronically monitored. The unit would contain potted plants during a typical experiment.

ture was maintained at 28 to 32°C during the day and 22 to 24°C at night, RH was kept at 30 percent, and light intensity was maintained at 3500 foot-candles. Plants were kept on an 18-h photoperiod. Plants were propagated in double plastic pots 15 × 10 cm. The upper pot contained soil and plants were kept moist by plant nutrient solution absorbed by polyester "wicks" threaded into the lower pot.

Mating experiments conducted in this microcosm have used either radish (*Rhaphanus sativus,* cultivar Cherry Belle) or barley (*Hordeum vulgar,* cultivar Advance) to establish a root rhizosphere. Transconjugants were detected in the rhizosphere of both radish and barley seedlings, 2 to 3 days after planting (DAP) or at about the time of seed germination. The number of transconjugants detected peaked on freshly germinated plants at 9×10^2 transconjugants per gram of soil on barley and 8×10^2 transconjugants per gram on radish roots. Conjugal frequencies were 10^{-5} to 10^{-6} transconjugants per initial donor. However, after 5 to 7 days, the number of transconjugants decreased steadily to 1.1×10^2 on barley, and to below the level of detection (about 1×10^2) on radish.

The microcosm described above provided insights into the kinetics of transconjugant formation, as well as information on fate and survival of donors, recipients, and transconjugant populations. The microcosm approach is flexible and more "realistic" relative to the natural environment than mating experiments conducted on laboratory media. Experiments in excess of 100 days have evaluated the fate and survival of recombinant bacteria and transconjugants.

It is anticipated that many of the products of the biotechnology industry will be used in the phyllosphere. Therefore, a variation of the microcosm procedures described above was developed and used to evaluate conjugal gene transfer on the leaf surface (1). Plastic trays (25 × 40 cm) containing 5 cm of soil were kept in the microcosm chambers described above. Relative humidity (RH) was elevated by enclosing the trays in plastic bags. RH, temperature, and soil moisture were constantly monitored using sensors and a data logger. The RH inside the enclosed plastic bags was about 95 percent during the 8-h dark period; it decreased slightly to 85 percent during the 16-h light period. When the bags were opened, RH averaged between 25 to 30 percent. Conjugation experiments were carried out using either 3-week-old radish (*Raphanus sativus,* cultivar Cherry Belle) or bean plant (*Phaseolus vulgaris humilis,* cultivar Bus Blue Lake).

In situ conjugation experiments were initiated by spraying either radish or bean leaves with buffer washed donors and recipients until runoff. Samples collected consisted of 2 to 6 g of radish or 5 to 12 g of bean leaves. Leaves were blended in phosphate buffer and dilutions were inoculated into selective media contained in 96-well microtiter trays. Populations were estimated using an eight-tube most probable number (MPN) by counting the number of turbid wells and comparing counts with an MPN table (14). Counts were monitored over a 14-day period.

Transconjugants were detectable within 24 h after inoculation of donors and recipients on both bean and radish leaves. Transconjugant populations increased rapidly, peaking at about 1×10^3 transconjugants per gram within 2 to 3 days after inoculation. Transconjugant populations on the bean declined only slightly from a high of about 1×10^3 per gram to 2.5×10^2 per gram, while on the radish, transconjugants declined from 1×10^3 per gram on day 2 to about 2×10^1 per gram by day 14. Donor and recipient populations also declined more rapidly on radish leaves, implying some apparent preference of the *P. cepacia* for bean leaf surfaces.

Experiments where recipients were inoculated onto beans 24 h prior to donors also demonstrated a rapid rise in transconjugants. However, the transconjugant population remained steady at about 1×10^3 per gram for the duration of the experiment, while donor and recipient populations remained at or about 5×10^6 per gram.

The microcosms described above have allowed insights into conjugation in the natural environment, and have provided data about the genetic stability of GEMs under consideration for field releases. However, regardless of the experimental design, isolation and enumeration of transconjugants requires a selective medium. Since donors and recipients as well as transconjugants are present in samples, it is probable that conjugation would occur on the surface of selective agar plants, or in broth, especially when low dilutions containing high cell densities of donors and recipients are plated. Any "plate matings" would lead to artifacts by exaggerating the number of transconjugants detected. Since differentiation of transconjugants formed in situ from those formed on selected media would be impossible, these plate matings could make results from conjugation experiments of questionable value.

Nalidixic acid (Nal) has been shown to inhibit conjugation (6, 7, 11), possibly by acting on DNA gyrase, and inhibiting DNA replication (5). Therefore, including Nal in a selective medium should eliminate artifacts resulting from plate matings.

To experimentally determine the extent of plate matings, conjugation experiments were conducted using *P. cepacia* pR388::Tn1721, and a recipient *P. cepacia* resistant to 500 µg/ml of Nal. Conjugation experiments were conducted in broth, in the soil slurry, and the rhizosphere microcosms as described above. The number of transconjugants obtained on agar selective for transconjugants was compared with the numbers observed on the same medium supplemented with 500 µg/ml of Nal. In addition, in each of the mating techniques used, donor and recipient *P. cepacia* populations were inoculated and incubated separately. Prior to plating for transconjugants, donor and recipient were kept separate during all stages of sample preparation. After dilutions were prepared, equal volumes of donor and recipient were mixed and applied immediately to selective medium with and without Nal (Fig. 14.3). Any transconjugants observed from these experiments would only be produced by matings that occurred on plates (plate matings) and not those that occurred in broth or microcosms.

Following overnight incubation in broth, the numbers of transconjugants growing on selective media without Nal (1.6×10^5) were not significantly different from those that were detected after the plate mating procedure (1.4×10^5). Therefore, distinguishing between transconjugants formed in the broth and those that formed on the plate after mixing donor and recipient was not possible. However, when the plate-mating procedure was conducted in the presence of Nal, the number of transconjugants decreased to 4×10^2 per milliliter. Transconjugants were not entirely eliminated, as might have been expected. This low level of transconjugant formation may reflect the rare mating event that occurs on the agar before Nal exerts its effects. Transconjugants detected from both matings plated in the presence of Nal decreased from 1.6×10^5 to 2×10^4 CFU/ml. The number of transconjugants produced by the plate-mating procedure decreased by three orders of magnitude when Nal was present and decreased only one order of magnitude from broth matings plated with Nal present, indicating that the 2×10^4 transconjugants per milliliter were actually produced in broth.

The number of transconjugants detected from soil slurry matings plated without Nal averaged about 7.7×10^5 transconjugants per milliliter. Plate matings in the absence of Nal were significantly lower ($p < .05$), averaging 4×10^4 transconjugants per milliliter. The number of transconjugants from soil slurry matings plated in the presence of Nal averaged 4×10^5 transconjugants per milliliter while no transconjugants resulted from the plate mating procedure in the presence of Nal. These results indicate that when matings occur in a soil slurry, subsequent plate-mating artifacts are not a significant problem. It is possible that the soil particles present in serial dilutions interfered with conjugation, by either physically damaging pili or coating donors and recipients.

In rhizosphere microcosms, the number of transconjugants detected when selective agar lacked Nal were two orders of magnitude higher than transconjugants detected from plates containing Nal (8×10^4 and 8×10^2 transconjugants per gram, respectively). In addition, transconjugants were detected on plates lacking Nal from samples plated immediately after placing of donors and recipients in mi-

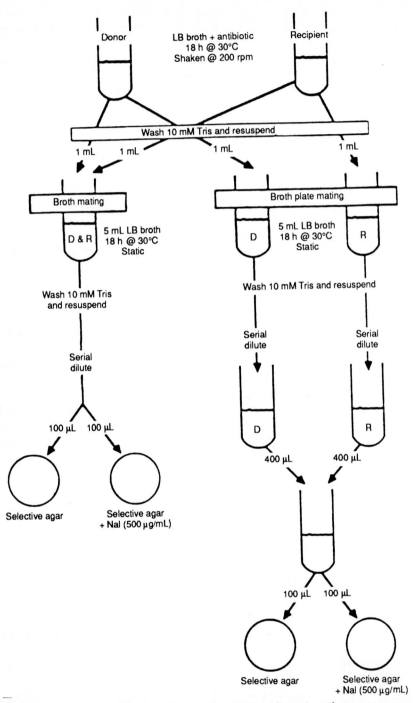

Figure 14.3 Schematic of the procedure used to distinguish matings that occur on agar plates (by plate mating procedure) and those that occur during overnight co-incubation of donors and recipients in broth.

crocosms. However, no transconjugants were detected on plates containing NaI, suggesting that the transconjugants resulted from plate matings only.

Results from these experiments indicate that conjugation can take place on media lethal to donors and recipients, and selective for transconjugants. Such continued conjugation takes place after incubation of donors and recipients in broth, soil slurries, and in the rhizosphere. These plate matings can be inhibited by including NaI in the medium used to select transconjugants. The failure to take such plate matings into account will lead to questions of whether transconjugants formed in situ or on the plate. We recognize that the inclusion of a NaI-resistant marker imposes important limitations on gene transfer experiments involving transfer into the indigenous microflora.

Target microbial populations, donors, recipients, and transconjugants which are used in the microcosms described above are detected by their resistance to certain antibiotics. High concentrations of nalidixic acid (500 μg/ml) or rifampicin (100 μg/ml) were required to achieve a level of detection of 1×10^1 CFU/g. While effective in eliminating the indigenous population (which ranges from 10^6 to 10^8 CFU/g), populations of transconjugants detected in soil peaked at only 9×10^2 transconjugants per gram in the rhizosphere microcosm soil (to be published) and about 1×10^3 transconjugants per gram in soil (14).

Conjugal frequencies ranged from 4.5×10^{-6} to 5×10^{-7} transconjugants per initial donor in the rhizosphere of barley and from 1.1×10^{-5} to nondetectable in radish. Conjugal frequencies in soil ranged from 1×10^{-5} to nondetectable (14). Comparable frequencies were reported by Talbot (29) using *Klebsiella,* who obtained conjugal frequencies of about 10^{-6} transconjugants per donor in the rhizosphere of the radish.

Conjugal frequencies in the phyllosphere were not much greater. Knudsen et al. (14) reported conjugal frequencies of about 1×10^{-4} transconjugants per donor. Lacey et al. (16) reported conjugal frequencies that ranged from 10^{-1} to 10^{-8} transconjugants per donor.

The low conjugal frequencies, therefore, require large initial populations of donors and recipients, usually 10^7 to 10^8 per gram of each, in order to observe detectable numbers of transconjugants. The use of a large initial population of a donor or recipient would be realistic, as the population of a GEM intended for release could be expected to be high. However, inoculation of high numbers of both donors and recipients is artificial.

Summary and Conclusion

Regardless of the sophistication of microcosms, computer models, and experimental controls, results obtained within the laboratory need to be validated by experiments conducted in the field. Small-scale releases of wild-type donors must be conducted and data from such experiments carefully analyzed before adequate judgments about the risk of genetic exchange between released GEMs can be made.

Adequate assessments of environmental impacts of released GEMs require development of methods that are widely applicable and capable of providing reproducible and accurate data (17). The techniques developed to measure genetic stability in terrestrial environments must not only be capable of accommodating the wide variety of different organisms intended for release but also must be standardized to allow for comparisons of results. The three experimental approaches described in this chapter permit standardization and flexibility. Future efforts must now be directed toward developing microcosms that include a greater variety of species such as insects, invertebrates, and additional plant species. In addition, testing of microcosms must be undertaken with a wider variety of microorganisms and preferably under carefully controllable environmental conditions.

The ability to predict conjugation in the natural environment using computer simulation models is currently under investigation. Knudsen et al. (14) used a mass-action model to predict conjugation in both potting soil and the phyllosphere. More sophisticated models that can account for fluctuations of environmental parameters (e.g., temperature, soil moisture, and RH), need to be developed and evaluated.

Techniques must be developed that allow for detection of conjugation between an introduced microorganism and members of the indigenous microflora. Such techniques must account for the plate-mating phenomenon described above. However, most members of an indigenous population are not resistant to Nal. An obvious way to circumvent these experimental difficulties is to conduct laboratory experiments to evaluate quantitatively gene transfer to pure cultures isolated from the environment where the GEM will be released. Such pure cultures can be made Nal resistant and can be added back to a microcosm environment. Studies from our laboratory demonstrated that approximately 50 percent of the gram-negative indigenous microbes isolated from soil and the rhizosphere served as recipients of plasmid DNA of Inc groups C, P, and W (33). Another way of measuring DNA transfer into indigenous populations is through the use of marker genes. The *xylE* gene from the Tol pathway has been cloned into pBR322 and used to track target microorganisms in both soil and the phyllosphere (34). The yellow colony which forms following catechol application could be used to signal the presence of a transconjugant on media which counterselects the donor strain.

Acknowledgments

The authors wish to thank their colleagues who contributed technical material to this manuscript. Appreciation is extended to Dr. Frank Lyon for the review of this manuscript. Mention of trade names or commercial products does not constitute endorsement or recommendation for use.

References

1. Armstrong, J. L., G. R. Knudsen, and R. J. Seidler. 1987. Microcosm method to assess survival of recombinant bacteria associated with plants and herbivorous insects. *Curr. Microbiol.* 15:229–232.
2. Beringer, J. E., and P. R. Hirsch. 1984. The role of plasmids in microbial ecology. In J. H. Slater, R. Wimpenny, and J. W. T. Wimpenny (eds.), *Microbes in Their Environment.* 34th Symposium of Society for General Microbiology. Cambridge University Press, New York, pp. 63–70.
3. Bradley, D. E. 1980. Conjugative pili of plasmids in *Escherichia coli* K-12 and *Pseudomonas* species. In S. B. Levy, R. C. Clowes, and E. I. Koenig (eds.), *Molecular Biology, Pathogenicity, and Ecology of Bacterial Plasmids.* Plenum Press, New York.
4. Bradley, D. E., D. Taylor, and D. Cohen. 1980. Specification of surface mating among conjugative drug resistance plasmids in *Escherichia coli* K-12. *J. Bacteriol.* 143:1466–1470.
5. Bouk, N., and E. A. Adelberg. 1970. Mechanism of action of nalidixic acid on conjugating bacteria. *J. Bacteriol.* 102:688–701.
6. Burman, L. 1977. R-plasmid transfer and its response to nalidixic acid. *J. Bacteriol.* 131:76–81.
7. Fenwick, R. G., and R. Curtis III. 1973. Conjugal deoxyribonucleic acid replication by *Escherichia coli* K-12: Effects of nalidixic acid. *J. Bacteriol.* 116:1236–1246.
8. Freter, R. 1984. Factors affecting conjugal transfer in natural bacterial communities. In M. J. Klug and C. A. Reddy (eds.), *Current Perspectives in Microbial Ecology.* American Society for Microbiology. Washington, D.C., pp. 105–114.
9. Gealt, M. A., M. D. Chai, K. B. Alpert, and J. C. Boyers. 1985. Transfer of plasmids pBR322 and pBR324 in wastewater from laboratory strains of *Escherichia coli* to bacteria indigenous to the waste disposal system. *Appl. Environ. Microbiol.* 49:836–841.
10. Gillett, J. W., and J. M. Witt. 1978. Terrestrial microcosms. In J. W. Gillett and J. M. Witt (eds.), *Proceedings of the Symposium on Terrestrial Microcosms and Environmental Chemistry.* National Science Foundation. Washington, D.C., pp. 1–7.
11. Hane, M. 1971. Some effects of nalidixic acid on conjugation in *Escherichia coli* K-12. *J. Bacteriol.* 105:46–56.
12. Hooykaas, P. J., A. A. VanBrussel, H. den Dulk-Ras, G. M. vanSlogteren, and R. A. Schilperoort. 1981. Symplasmid of *Rhizobium trifoli* expressed in different rhizobiol species and *Agrobacterium tumefaciens. Nature* 291:351–353.
13. Inouye, S., A. Nakazawa, and T. Nakazawa. 1981. Molecular cloning of gene *xyl* S of the Tol plasmid: Evidence for positive regulation of the *xyl* DEFG operon by *xyl* S. *J. Bacteriol.* 148:413–418.
14. Knudsen, G. R., M. V. Walter, L. A. Porteous, V. J. Prince, J. L. Armstrong, and J. Seidler. 1988. A predictive model of conjugative plasmid transfer in the rhizosphere and phyllosphere. *Appl. Environ. Microbiol.* 54:343–347.
15. Lacey, G. H., and J. V. Leary. 1975. Transfer of antibiotic resistance plasmid RP1

into *Pseudomonas glycinea* and *Pseudomonas phaseolicola in vitro* and *in planta*. *J. Gen. Microbiol.* 88:49–57.

16. Lacey, G. H., V. K. Stromberg, and N. P. Cannon. 1984. *Erwinia amylovera* mutants and *in planta*-derived transconjugants resistant to oxytetracycline. *Can. J. Plant Pathol.* 6:33–39.

17. Levin, M., R. Seidler, R. Borquin, J. Fowle III, and T. Barkay. 1987. EPA developing methods to assess environmental release. *Bio/Technology* 5:38–45.

18. Mancini, P., S. Fertels, D. Nave, and M. Gealt. 1987. Mobilization of plasmid pHSV 106 from *Escherichia coli* HB101 in a laboratory scale waste treatment facility. *Appl. Environ. Microbiol.* 53:665–671.

19. Maniatis, T. E., F. Fritsch, and J. Sambrook. 1982. *Molecular Cloning: A Laboratory Manual.* Cold Spring Harbor Laboratory, Cold Spring Harbor, New York.

20. Milewski, E. 1985. Field testing of microorganisms modified by recombinant DNA techniques: Applications, issues, and development of "points to consider" document. *Recomb. DNA Tech. Bull.* 8:102–108.

21. Radford, A. J., J. Oliver, W. J. Kelley, and D. C. Reanney. 1981. Translocatable resistance to mercuric ions in soil bacteria. *J. Bacteriol.* 147:1110–1112.

22. Reanney, D. C., P. C. Gowland, and J. H. Slater. 1984. Genetic interactions among microbial communities. In J. H. Slater, R. Whittenbury, and J. W. T. Wimpenny (eds.), *Microbes in Their Environment.* 34th Symposium of the Society for General Microbiology. Cambridge University Press, New York, pp. 380–419.

23. Rissler, J. 1984. Research needs for biotic environmental effects of genetically engineered microorganisms. *Recomb. DNA Tech. Bull.* 7:20–30.

24. Saftlas, H. B. 1984. Biotechnology, a sleeping giant stirs. Health care, hospitals, drugs, and cosmetics: Current analysis. *Std. Poors Ind. Survey* July 19, pp. H1–5.

25. Sinclair, J. C., and M. Alexander. 1984. Role of resistance to starvation in bacterial survival in sewage and lake water. *Appl. Environ. Microbiol.* 48:410–415.

26. Singleton, P., and A. E. Anson. 1981. Conjugal transfer of R-plasmid R1drd-19 in *Escherichia coli* below 22°C. *Appl. Environ. Microbiol.* 42:789–791.

27. Singleton, P., and A. E. Anson. 1983. Effect of pH on conjugal transfer at low temperature. *Appl. Environ. Microbiol.* 24:291–293.

28. Stotzky, G., and H. Babich. 1986. Survival of, and genetic transfer by, genetically engineered bacteria in natural environments. *Adv. Appl. Microbiol.* 31:93–108.

29. Talbot, H. W., D. Y. Yamamoto, M. W. Smith, and R. J. Seidler. 1980. Antibiotic resistance and its transfer among clinical and non-clinical *Klebsiella* strains in botanical environments. *Appl. Environ. Microbiol.* 3:97–104.

30. Trevors, J. T. 1985. Bacterial plasmid isolation and purification. *J. Microbiol. Methods* 3:259–271.

31. Trevors, J. T., and J. M. Oddie. 1986. R-plasmid transfer in soil and water. *Can. J. Microbiol.* 32:610–613.

32. Walter, M. V., and J. W. Vennes. 1985. Occurrence of multiple antibiotic-resistant enteric bacteria in domestic sewage and oxidation lagoons. *Appl. Environ. Microbiol.* 50:930–933.

33. Walter, M. V., A. L. Porteous, and R. J. Seidler. 1987. Measuring genetic stability in bacteria of potential use in genetic engineering. *Appl. Environ. Microbiol.* 53: 105–109.

34. Walter, M. V., R. H. Olson, V. Prince, F. L. Lyon, and R. J. Seidler. 1988. Use of catechol dioxygenase for the direct and rapid identification of recombinant microbes taken from environmental samples. In A. Balows, R. C. Tilton, and A. Turano (eds.), *Rapid Methods and Automation in Microbiology and Immunology.* Brixia Academic Press, Brescia, Italy.

35. Walter, M. V., L. A. Porteous, and R. J. Seidler. 1989. Evaluation of a method to measure conjugal transfer of recombinant DNA in soil slurries. *Curr. Microbiol.* 19: 365–370.

36. Wamsley, R. H. 1976. Temperature dependence of mating pair formation in *Escherichia coli. J. Bacteriol.* 126:222–224.

37. Watson, B., T. C. Curier, M. P. Gordon, M. D. Chilton, and E. W. Nester. 1975. Plasmid required for virulence of *Agrobacterium tumefaciens. J. Bacteriol.* 123:255–264.

38. Weinberg, S. R., and G. Stotzky. 1972. Conjugation and genetic recombination of *Escherichia coli* in soil. *Soil Biol. Biochem.* 4:171–180.

15

Gene Transfer
in Wastewater

Michael A. Gealt

Introduction

Microorganisms have been used in industry (including agriculture) in two basic modes: contained in vessels (for the production of beer, antibiotics, yogurt, etc.) or disseminated into the environment (as in the addition of *Rhizobium* onto seeds for improved nodule formation). Although they have been used successfully without health risk, the artificial alteration of their characteristics and capabilities through genetic engineering has prompted new doubts over efficacy and safety.

Several possible ways have been suggested in which the release of genetically engineered microorganisms (GEMs) might lead to environmental alterations, despite their low survival potential outside the laboratory or industrial setting. The most obvious is a potential change in the dynamics of ecosystem populations. Although the organisms initially engineered were deliberately debilitated so that there was virtually no chance for the successful establishment of accidentally released organisms (6), newer constructs have been developed specifically for release and should survive outside the laboratory for several weeks, if not months. If these organisms grow, their concentration will increase, thus limiting the amount of nutrients available for native species. The prolonged presence of GEMs also magnifies the chance for genetic interactions with indigenous strains. This genetic interaction may be either the transfer of genetic information from the modified strain into the indigenous, or the transfer of information from the indigenous strains into the GEM, so that it may acquire characteristics needed for long-term survival. Other alterations are also possible, such as acquisition of the potential for invasiveness (13, 14), which might lead to an increased level of pathogenicity. Alternatively, acquiring short DNA sequences could have the effect of disrupting a gene following a recombination event.

While the environmental use of GEMs would require a deliberate release of organisms, large-scale production facilities in pharmaceutical and other industries requires the use of GEMs contained in fermenters. While it is doubtful that sufficient numbers of bacteria might be released from a research laboratory to cause major problems, this is not necessarily the case with a production facility. Accidental release poses the problem of a large, rapid influx of organisms into a waste-treatment facility. These organisms are not likely to survive well outside the fermenter, but their immense number would suggest that sufficient numbers would remain viable long enough for possible genetic interaction with indigenous species.

At present we do not have the methods to adequately predict the result of interactions between microbes in the soil, water, etc. The survival period of plasmid-containing bacteria released into soil or water and the amount of genetic interaction is still under investigation (5, 8, 11, 33, 43, 44, 48, 49, 51, 53). The high cell concentrations and diversity of ecological niches make a treatment facility a reasonable model for analysis of genetic interactions which will occur, although probably at lower rates, in environments, both aquatic and terrestrial.

Gene Dissemination in Wastewater

The concept of gene dissemination is not a new one. Within recent history, antibiotic-resistance plasmids (R-plasmids) have spread from rare, early independent isolates to a general presence as a current medical hazard (24). The high incidence of use of antibiotics has increased the selective pressure for the maintenance of resistance to multiple antibiotics. This is despite the additional energy necessary for the replication of the plasmid, which is frequently, though not always, large enough to decrease the growth rates (5, 45). The occurrence of resistance to ampicillin in approximately 25 percent of all enterics found in wastewater from the Philadelphia Southwest Treatment Facility (31) suggests that there is a great number of resistant organisms in the natural environment of wastewater. In several cases these organisms were resistant to more than one antibiotic.

The transfer of resistance plasmids between wastewater bacteria has been demonstrated (2, 10, 25, 31). The transfer of R-plasmids may occur between genera. Thus, a gene from *Escherichia* may end up in *Salmonella*, *Proteus*, or *Klebsiella*. Very rapid spread of genes throughout a population can occur. Corliss (4) reported that the R1 resistance plasmid (chloramphenicol, ampicillin, kanamycin, streptomycin, and sulfonamide) can transfer from laboratory strains of *E. coli* to both *E. coli* and *Citrobacter freudii* isolated from human fecal samples. Mach and Grimes (25) have reported several interspecies transfers of conjugative plasmid material in a wastewater treatment plant. Alcaide and Garay (2) have reported R-plasmid transfer between *Salmonella* spp. isolated from wastewater and contaminated surface waters. We have observed the transfer of R100-1 from a laboratory strain of *E. coli* to raw wastewater isolates of both *E. coli* and *Enterobacter cloacae* (10).

Waste Treatment

Waste treatment attempts to remove particulate materials, including organisms, and degrade chemical contaminants, including hazardous materials, before the release of effluent back into the environment (3, 7, 22, 55). While the major microbiological objective is to decrease pathogen concentration, in the process there is generally an overall decrease in the total number of bacteria. Once in the treatment plant the GEM encounters an environment that may either allow survival or, at the least, provides for the genetic interaction with indigenous organisms. These conditions include a high titer of organisms (at least 10^6 per milliliter in the wastewater and several orders of magnitude higher in the sludge), adequate nutrition for viability, numerous bacteria containing conjugative plasmids, and physical surroundings sufficiently quiescent (at least in the sludge) to allow the cell-to-cell contact necessary for conjugation to occur.

Waste-treatment methods fall into two major categories: physical and biological. The physical approaches to microbial removal include settling of particulates and chemical precipitation, which aids the settling process. Chemical hazards are reduced by air stripping, ozonization, absorption on carbon, removal by ion-exchange chromatography, and membrane separation. Ozonization can also be antimicrobial. Biological approaches are primarily for decontamination of wastes and include degradation by aerobic, facultative anaerobes, or anaerobic organisms, especially bacteria and fungi. Organisms removed from the wastewater wind up in the sludge, from which they also reenter the environment. After drying, dewatering, heating, or chemical treatment, the sludge may be used for soil amendment. Placement in landfills or incineration are alternative fates for the sludge.

In municipal wastes most particulate materials are removed by settling (7, 32, 34, 55), frequently in tanks found as the initial site of treatment. Some small towns and cities use settling in lagoons as the major (and sometimes sole) method to treat all wastewater. Lagoons allow for a series of complex reactions to occur, including aerobic and anaerobic biodegradation in addition to settling.

A mixture of organisms resides in the settling tank and ponds. These organisms are capable of forming mats or filamentous nets which will also bind up additional organisms and provide domains of high bacterial titer with relatively low flow rates. Binding to these mats provides potential sites for genetic interaction between indigenous species and any incoming GEMs. Additional compounds may be needed to aid precipitation of solids by increasing the particle size (3). In some cases ionic compounds are added, e.g., calcium carbonate, ferric compounds, or alumina, while in others polymers are utilized. Some of the ionic compounds will decrease rates of bacterial conjugation (17).

In addition to precipitation, other methods have been used to remove particulates.

In ultrafiltration, wastewater is passed through a filter with pores small enough to retard most particulates. Reverse osmosis membranes are more selective in that they discriminate between compounds based primarily on chemical composition rather than size. The major problem with these membranes and columns is that they foul, requiring cleaning or regeneration (3). Carbon in activated charcoal may actually serve as nutritive material for bacterial growth. Fouling decreases flow rates and, therefore, increases the probability of bacterial genetic interactions.

Biological waste treatment allows organisms (either indigenous or those added during treatment, e.g., in activated sludge) in the wastewater to utilize wastes as nutritive sources or cometabolites (7, 28). Much of this biodegradation can be enhanced by selective modification of the return sludge from other portions of the treatment process. This natural selection of activated sludge capable of degradation is one method of enhancing the degradation processes. This kind of selection process has been suggested as a means of achieving a more natural type of genetic engineering (16, 18, 21, 23), utilizing processes dependent on indigenous conjugative plasmid containing cells and transducing phage. Genetic engineering has been suggested as a means of improving biodegradation during waste and wastewater treatment (19, 35, 38).

The organisms which are involved in the degradation processes are frequently found in biofilms, with both aerobic and anaerobic organisms in different layers, thus enabling them to carry out both type of conversions. The amount of gene transfer which may occur in the biofilm is currently unknown.

Plasmid Transmission in Wastewater

A high titer of bacteria in wastewater provides a major opportunity for different types of interactions. Besides the production of nutritive materials for use by other organisms, i.e., cross-feeding, and the cooperative cometabolism of toxic materials, indigenous wastewater organisms have an excellent opportunity for genetic interactions. In addition to interactions in wastewater, sludge removed from the treatment facility is also a potential site for genetic interaction.

Although it has been demonstrated that host *E. coli* cells containing plasmid vectors do not multiply in or colonize model (nonsterile) treatment facilities (15, 39, 40; Gealt and Selvaratnam, unpublished results), they are present, albeit in decreasing concentrations, for several days after entering the plant. Analysis of samples from wastewater has demonstrated the presence of several bacterial species which can serve as potential secondary hosts for plasmid vehicles (31).

Analysis of gene transmission to indigenous organisms requires extremely sensitive detection methods because of the complex nature of wastewater which can mask bacterial interactions. Transmission of genetically engineered DNA sequences (GEDS) during waste treatment has been demonstrated in a series of simple, interpretable model systems including a laboratory-scale (20-L) treatment plant seeded with wastewater organisms. Initial analysis was restricted to the transmission of plasmids pBR322 and pBR325, two commonly used nonconjugative vectors in a test-tube microcosm (10). While neither of these plasmids could transfer their phenotypic characteristics from an *E. coli* host into another *E. coli* (laboratory strain χ1997) devoid of plasmids, nor into either of two naturally occurring wastewater bacterial isolates (an *E. coli* and an *Enterobacter cloacae*), transfer did occur when the incubation mixture was triparental, i.e., included the conjugative strain *E. coli* χ1784. This conjugative strain contained plasmid R100-1, which acted as a mobilizer for the recombinant plasmids. When coincubations were performed in test tubes with L-broth, a rich bacterial medium, between 10^1 and 10^4 transconjugants per milliliter were recovered after a 25-h incubation. Coincubations in sterilized wastewater exhibited transfer at approximately the same frequency. Plasmid mobilization also occurred if the coincubation took place on a Millipore filter incubated on a wastewater-saturated filter pad.

This system can be used to examine the effects of exogenous compounds on conjugation (17). Experiments with several salts and detergents (under conditions which did not result in a significant decrease in viability of either the plasmid donor or recipient) indicated that compounds can be divided into at least three major groups: (1) those with no effect on conjugation, e.g., Triton X-100, (2) those which markedly decreased the number of transconjugants even at very low concentrations, e.g., sodium dodecyl sulfate (SDS), zinc, and iron, and (3) those which decreased the number of transconjugants only at moderately high concentrations, e.g., magnesium, calcium, potassium, and sodium. For example, the effect of zinc was seen at a concentration of approximately 2 mM while sodium required almost a 1 M concentration for the same degree of decrease in conjugation. Precise reasons for the effects are unknown in most cases, although the SDS presumably affects the proteins of the pilus (1, 50) and zinc is known to bind to and block specific sites in F-plasmid conjugation (36, 37, 52). Ferric ions appear to affect promoter activity for the *traJ* gene of plasmid R100-1 (Khalil and Gealt, unpublished results).

While mobilization can be monitored by the recipient cell's acquisition of phenotypic characteristics, i.e., antibiotic resistance, this is only circumstantial evidence for plasmid transfer. Confirmation of plasmid acquisition requires analysis of isolated DNA from donor, mobilizer, and recipient cells (10, 31). Several different plasmid patterns are observed in the transconjugants. Some recipients contain material comigrating with the recombinant plasmid, some have DNA comigrating with the conjugative plasmid, and some show complex rearrangements of the plasmid material. This common observation of rearranged plasmid profiles in recipient cells indicates that the achievement of a stable plasmid complement sometimes requires much intracellular recombination between the plasmids and, perhaps, the chromosome. Similar rearrangements have been noted in potable water (42). The frequency of plasmid recombination following conjugation means that merely noting the absence of plasmids comigrating with the original recombinant plasmid is not sufficient to eliminate the possibility that GEDS acquisition has occurred.

Although GEDS could be transferred in wastewater using an exogenous mobilizer, the probability of such transfer in actual treatment facilities depends on the presence of indigenous mobilizer strains. Several mobilizer strains were isolated from wastewater by setting criteria of resistance to at least one antibiotic, having at least one high-molecular-weight plasmid, and being able to transfer these characteristics to a plasmidless recipient (26, 31). Several indigenous strains were isolated which met these criteria and were able to mobilize recombinant plasmids.

Mobilization in a Laboratory Waste-Treatment Facility (LWTF)

Because indigenous bacteria may have antibiotic resistance profiles similar to those commonly found in GEMs, it is difficult to enumerate GEMs based solely upon these characteristics. It is necessary to follow specific gene sequences with nucleic acid hybridization procedures. A technical problem encountered in these initial analyses was the occurrence of cross-reactive sequences in the plasmids of indigenous recipients. Therefore, a test strain of *E. coli* was constructed which contained a plasmid with a gene which was not of bacterial origin. We utilized pHSV106 [Bethesda Research Laboratories,

Inc. (BRL), Gaithersburg, Maryland] which contains the thymidine kinase (*tk*) gene of herpes simplex virus (HSV) (29, 30, 54) for our construction (26). Because there is no bacterial gene with high sequence similarity to *tk*, a molecular probe utilizing this gene sequence would not have a high probability of cross-reacting with indigenous bacterial DNA sequences. This plasmid also has a *amp*^{r} gene derived from the pBR322 into which the HSV sequences had been cloned, making enrichment possible when necessary by selection for this antibiotic resistance. One of the problems encountered with this plasmid was a spontaneous loss of part of the complete plasmid, resulting in a smaller plasmid which was still *amp*^{r}, but lacked the *tk* sequences. The need for constant verification of the full-size plasmid was obvious.

Mobilization of pHSV106 from *E. coli* HB101 (*recA*) was demonstrated not only *in vitro*, i.e., in a test tube, but also in a laboratory-scale waste-treatment facility [LWTF (see Fig. 15.1); approximately

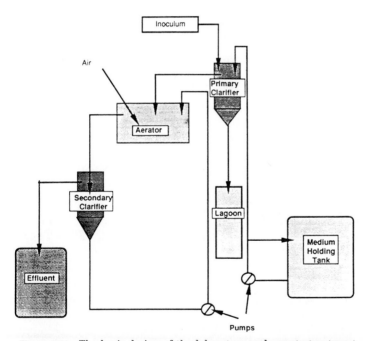

Figure 15.1 The basic design of the laboratory-scale waste treatment facility (LWTF). The medium is pumped from a large holding container to the primary settling tank. Flow from the primary to the aerator and on to the secondary settling tank is driven by gravity. The return activated sludge is pumped from the bottom of the secondary settling tank back to the aerator. The flow to lagoons is by gravity feed from the bottom of the primary settling tank. All effluent from the system passes through a hypochloride solution bath for disinfection. The LWTF in our laboratory consists of four identical systems using the same medium feed so that controls can be run simultaneously with experimental systems.

20 L total volume]. The LWTF models the actual occurrences in a waste facility without requiring the release of GEMs into the larger environment. This pilot plant can be run with complex media into which wastewater organisms are introduced or with authentic wastewater. Media was always prepared as needed in tap water (nonsterile). The LWTF consists of four major areas: the primary settling tank, the aerator, the secondary settling tank and lagoons. It is constructed of acrylic and PVC plastics with all joints sealed with epoxy cement. The settling tanks are acrylic cylinders, 16 cm (diameter) × 25 cm (high), each with a funnel at the bottom. The medium is pumped from a holding tank into the system at the primary settling tank at rates that are controllable and are usually operated at a flow of between 10 and 40 ml/min. The flow enters the primary settling tank via a standpipe. Beyond the primary, the wastewater is gravity fed into the aerator, a 15 cm by 16.5 cm by 30 cm acrylic container (total volume of 7.4 L). Gravity flow allows transfer of materials into the secondary settling tank. Besides the medium flow to the primary settling tank, the only movement not powered by gravity flow is that of the return activated sludge (RAS), which is pumped from the bottom of the secondary settling tank to the aerator. The aerator is fed by "house air," which is delivered through aquarium air stones at a rate rapid enough to provide constant motion in the tank. The dissolved oxygen in the primary settling tank is approximately 6.5 mg/L at the beginning of the experiments and drops rapidly to about 1.8 mg/L at 24 or 48 h. Similar decreases are observed in the secondary settling tank. The lagoons consist of stainless steel vessels (27.5 cm wide × 47.5 cm long × 15.0 cm deep) with lids. The lids have entry holes through which either wastewater or settled primary sludge can be delivered.

Experiments in the LWTF demonstrated that gene sequences of the *tk* gene could be detected in the DNA of the transconjugants produced by coincubation of GEM, mobilizer, and recipient (26). These transconjugants were selected initially because they grew on agar containing ampicillin, tetracycline, and nalidixic acid, which are selective agents for the GEDS of the pHSV106 plasmid, the R100-1 plasmid (of the mobilizer), and the recipient chromosome, respectively. Transconjugants were also observed when isolates of indigenous wastewater bacteria were used for the mobilizer and recipient inocula. Indeed, transfer into indigenous recipients by native mobilizers can be demonstrated even without the use of previously characterized strains (9; Gealt, Vettese, Cleaveland, and Tuckey, unpublished results). The levels of transconjugant production were the same regardless of nutrient composition, nutrient concentration, influent flow rate, or the origin of the parental cells. Most of the transconjugants were in the medium near the bottom of the primary settling tank and in the sludge lining the bottom of the settling tank. The localization of transconjugants at these sites may be a result of the cell distribution, which is dependent on flow rate characteristics. The presence of transconjugants was noted only when the titer of parental donor strains in the bottom of the primary settling tank and the sludge were at least 10^7 cells per milliliter, suggesting a dependence on donor cell number for the efficiency of plasmid mobilization. These numbers of cells are frequently exceeded in sludge from actual plants.

The general protocol of the LWTF experiments is shown in Fig. 15.2. Cells are removed from different sections at various time intervals after inoculation. In general we find that 24 to 48 h is more than enough time to obtain transconjugants. These cells are first screened by their antibiotic-resistant characteristics to increase the probability of finding recipients which have acquired gene sequences from pHSV106. The transconjugants are then transferred onto a membrane (Colony/Plaque Screen, NEN Research Products, Boston, Massachusetts) for hybridization (27).

Using the colony-blot methodology it is possible to detect the transfer and maintenance of the *tk* sequences in recipient cells (26). But, while the presence of *tk* in transconjugants did indicate that this nonselected viral gene was retained and did persist for at least several generations, the technique of colony hybridization is not optimal for large numbers of screenings because (1) it tests only a few cells and not an entire population and, therefore, may yield false negatives when the transfer occurs at low frequencies, (2) there is a possible difficulty in detecting the *tk* gene sequence in some transconjugants (because copy number may be low and, therefore, it may require a probe of higher specific activity for detection in such transconjugants), and (3) it requires colony growth on selective medium so that viable cells which are not culturable may be missed.

Solution of these problems required modification of the protocol to obtain DNA from a larger portion of the microbial population. Therefore, DNA from liquid cultures of transconjugants was used as the starting material for the hybridization (Fig. 15.3). Indigenous bacteria are generally used directly from wastewater primary effluent without any preselection or characterization, but may be grown overnight in rich medium so that a heavy inoculum can be used. After incubation, aliquots of approximately 50 ml were removed and grown in selective medium (to rule out growth of the *tk* donor), collected by centrifugation and lysed (26). There is no need to purify the plasmid away from the chromosome (and, indeed, this may not be desired as the GEDS may have recombined into the chromosome of the recipient cell). The DNA was applied to a hybridization membrane, e.g., Zeta-Probe (Bio-Rad, Rockville Centre, New York), for hybridization. Figure 15.4 is a

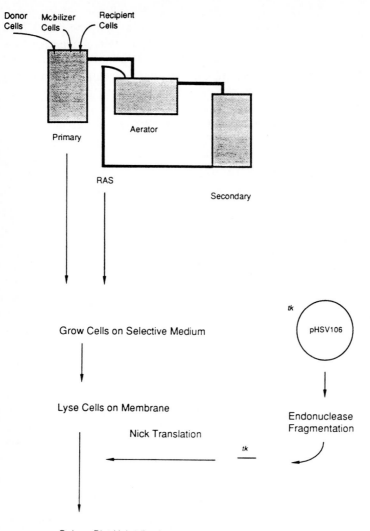

Figure 15.2 The experimental design used for the detection of transferred GEDS using colony-blot hybridization. The inoculum for growth on selective medium agar plates is obtained from the LWTF at any of several points, such as the primary sludge or return activated sludge. After growth on selective medium plates, the cells are transferred to a membrane, lysed, and hybridized with an appropriate nucleic acid probe. The probe is produced by restriction fragmentation of a recombinant plasmid followed by radiolabel addition by the nick translation method.

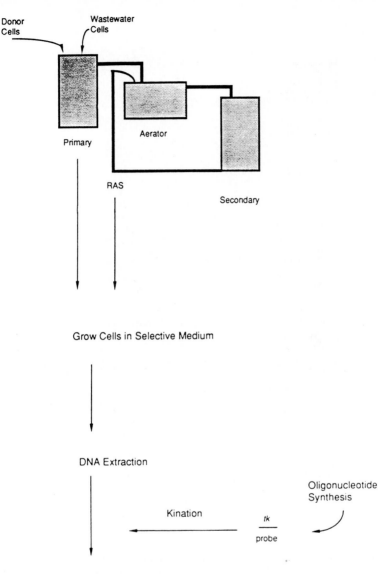

Figure 15.3 The experimental design used for detection of transferred GEDS using slot-blot hybridization. The inoculum for growth in liquid selective medium is obtained from the LWTF at any of several points in the system. After growth, the cells are collected by centrifugation, lysed, and the DNA applied to a membrane with a slot-blot apparatus for reaction with an appropriate nucleic acid probe. The radioactive probe was obtained by kination of ^{32}P to an in vitro synthesized oligonucleotide sequence.

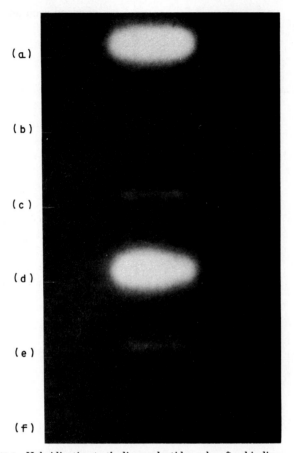

Figure 15.4 Hybridization to *tk* oligonucleotide probe after binding of DNA from LWTF samples to membrane in a slot-blot apparatus. (a) Positive control using DNA from LWTF system containing *E. coli* HB101(pHSV106), *E. coli* χ1784, and *E. coli* χ1997; (b) negative control from LWTF system containing *E. coli* χ1784 and *E. coli* χ1997; (c) experimental system containing *E. coli* HB101 (pHSV106), *E. coli* χ1784, and wastewater organisms; (d) purified plasmid pHSV106 DNA; (e) experimental system containing *E. coli* HB101(pHSV106) and wastewater organisms; (f) negative control containing salmon DNA.

DNA hybridization of the *tk* probe with DNA extracted from such cells. As Figs. 15.2 and 15.3 suggest, it is possible to use either a nick-translated restriction fragment or a kinated oligonucleotide sequence as a molecular probe. The probe used for hybridization was a 27-nucleotide sequence (located near the 5′ end of the *tk* gene) synthesized in vitro. DNA complementary for *tk* was found in transconjugants from the systems which contained only the GEM and uncharacterized wastewater organisms, as well as in the positive control LWTF systems (which included both a characterized recipient and a characterized mobilizer). The autoradiogram of the hybridization reaction was scanned with densitometer (e.g., LKB Ultrascan, LKB Biotechnology, Inc., Piscataway, New Jersey) and the area of each band compared as shown in Fig. 15.5. The cultures that contain transconjugant cells hybridized more of the probe than the wastewater cells alone (i.e., background), indicating that transfer had occurred. Although we grew our cells in liquid selective medium for signal enhancement (i.e., an increase in the number of transconjugant cells and elimination of donor parents), an alternative without any selective medium growth would be achieved by performing a DNA extraction of cells collected from large volumes of liquid, e.g., wastewater, primary effluent, etc., using methods described in the literature (20, 46).

While this experimental protocol does not distinguish between conjugation and other possible mobilization mechanisms (transduction and transformation), the utilization of containment chambers fitted with membranes for the donor bacteria which would allow viral, but not bacterial, passage would make it possible to determine the contribution of other mechanisms to the GEDS mobilization phenomenon.

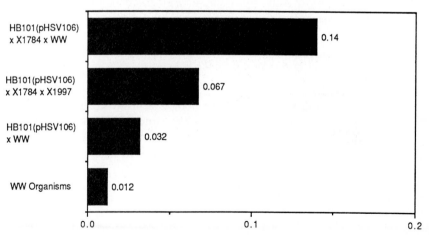

Figure 15.5 Quantification of slot-blot hybridization reaction autoradiogram results by scanning densitometry. The area units of density are plotted as a function of the amount of DNA applied to the slot-blot hybridization membrane. In an experiment using an oligonucleotide sequence probe for the *tk* gene, the wastewater (WW) organisms do contain some cross-reacting sequences, but considerably less than after incubation with the *tk* gene containing HB101(pHSV106). Addition of a mobilizer *E. coli* χ1784 (R100-1) and the recipient cells *E. coli* χ1997 or the wastewater organisms further enhances the mobilization.

Similar transfers of genetic material between introduced bacteria and indigenous strains have been noted in soil microcosms (12).

Conclusion

The release of genetically engineered organisms into the environment presents many opportunities for their interaction with indigenous species. The high titer of bacteria in wastewater, as well as the presence of conditions which allows mating to occur, suggests that a probability exists for transfer of GEDS within a waste-treatment facility. Genetic interactions due to conjugation between GEMs and indigenous wastewater organisms in a nonsterile laboratory-scale pilot plant resulted in the transfer of recombinant DNA sequences into the indigenous bacteria. Both selectable and nonselectable gene sequences were transferred. The presence of the nonselectable sequences in recipient organisms can be detected by nucleic acid hybridizations.

The ease with which GEDS can mobilize into indigenous species indicates that it is not sufficient, following release of a GEM, to monitor only the presence of the original host, but rather the modified DNA itself must be followed. The most sensitive method for doing this is nucleic acid hybridization. By extracting cell cultures directly, it has been possible to find and quantify the transfer of specific sequences into indigenous organisms. Possible future sensitivity enhancements include the use of sequence amplification by polymerase chain reaction (41, 47).

Acknowledgments

I would like to thank B. P. Sagik for his constant encouragement with this project and P. Cleaveland, S. Tuckey, and M. Vettese for their technical aid. This research was supported by grants R-810049 and R-812362 from the U.S. Environmental Protection Agency.

References

1. Achtman, M., G. Morelli, and S. Schwuchow. 1978. Cell-cell interactions in conjugating *Escherichia coli. J. Bacteriol.* 135:1053–1061.
2. Alcaide, E., and E. Garay. 1984. R-plasmid transfer in *Salmonella* spp. isolated from wastewater and sewage contaminated surface waters. *Appl. Environ. Microbiol.* 48:435–438.
3. Clifford, D., S. Subramonian, and T. J. Sorg. 1986. Removing dissolved inorganic contaminants from water. *Environ. Sci. Technol.* 20:1072–1080.
4. Corliss, T. L., P. S. Cohen, and V. J. Cabelli. 1981. R-plasmid transfer to and from *Escherichia coli* strains isolated from human fecal samples. *Appl. Environ. Microbiol.* 41:959–966.

5. Cruz-Cruz, N. E., G. A. Toranzos, D. G. Ahearn, and T. C. Hazen. 1988. In situ survival of plasmid-bearing and plasmidless *Pseudomonas aeruginosa* in pristine tropical waters. *Appl. Environ. Microbiol.* 54:2574–2577.
6. Curtiss, R., III, D. A. Pereira, J. C. Hsu, S. C. Hull, J. E. Clark, L. J. Marurin, R. Goldschmidt, R. Moody, M. Inoue, and L. Alexander. 1977. Biological containment: the subordination of *Escherichia coli* K-12. In R. F. Beers and E. G. Bassett (eds.), *Recombinant Molecules: Impact on Science and Society*. Raven, New York, pp. 45–56.
7. Forster, C. F. 1985. *Biotechnology and Wastewater Treatment*. Cambridge University Press, New York.
8. Fulthorpe, R. R., and R. C. Wyndham. 1989. Survival and activity of a 3-chlorobenzoate-catabolic genotype in a natural system. *Appl. Environ. Microbiol.* 55:1584–1590.
9. Gealt, M. A. 1988. Recombinant DNA plasmid transmission to indigenous organisms during waste treatment. *Water Sci. Technol.* 20:179–184.
10. Gealt, M. A., M. D. Chai, K. B. Alpert, and J. C. Boyer. 1985. Transfer of plasmids pBR322 and pBR325 in wastewater from laboratory strains of *Escherichia coli* to bacteria indigenous to waste disposal system. *Appl. Environ. Microbiol.* 49:836–841.
11. Genthner, F. J., P. Chatterjee, T. Barkay, and A. L. Bourquin. 1988. Capacity of aquatic bacteria to act as recipients of plasmid DNA. *Appl. Environ. Microbiol.* 54: 115–117.
12. Henschke, R. B., and F. R. J. Schmidt. 1990. Plasmid mobilization from genetically engineered bacteria to members of the indigenous soil microflora in situ. *Curr. Microbiol.* 20:105–110.
13. Isberg, R. R., and S. Falkow. 1985. A single genetic locus encoded by *Yersinia pseudotuberculosis* permits invasion of cultured animal cells by *Escherichia coli* K-12. *Nature* 317:262–264.
14. Isberg, R. R., D. L. Voorhis, and S. Falkow. 1987. Identification of invasin: A protein that allows enteric bacteria to penetrate cultured mammalian cells. *Cell* 50:769–778.
15. Kane, J. F., P. E. Jung, M. D. Hale, M. A. Heitkamp, and G. Bogosian. 1990. Fate of recombinant *Escherichia coli* in nonsterile water, soil, and sewage. *Abstracts of the Annual Meeting (Anaheim, Calif.)*. American Society for Microbiology, Abstract Q-157.
16. Kellogg, S. T., D. K. Chatterjee, and A. M. Chakrabarty. 1982. Plasmid-assisted molecular breeding: New technique for enhanced biodegradation of persistent toxic chemicals. *Science* 214:1133–1135.
17. Khalil, T. A., and M. A. Gealt. 1987. Temperature, pH, and cations affect the ability of *Escherichia coli* to mobilize plasmids in L broth and synthetic waste water. *Can. J. Microbiol.* 33:733–737.
18. Kilbane, J. J., D. K. Chatterjee, J. S. Karns, S. T. Kellogg, and A. M. Chakrabarty. 1983. Biodegradation of 2,4,5-trichlorophenoxyacetic acid by a pure culture of *Pseudomonas cepacia*. *Appl. Environ. Microbiol.* 44:72–78.
19. Kobayashi, H. A. 1984. Application of genetic engineering to industrial waste/wastewater treatment. In G. S. Omenn and A. Hollaender (eds.), *Genetic Control of Environmental Pollutants*. Plenum Press, New York, pp. 195–214.
20. Knight, I. T., S. Shults, C. W. Kaspar, and R. R. Colwell. 1990. Direct detection of *Salmonella* spp. in estuaries by using a DNA probe. *Appl. Environ. Microbiol.* 56: 1059–1066.
21. Kröckel, L., and D. D. Focht. 1987. Construction of chlorobenzene-utilizing recombinants by progenitive manifestation of a rare event. *Appl. Environ. Microbiol.* 53: 2470–2475.
22. Lawler, D. F. 1986. Removing particles in water and wastewater. *Environ. Sci. Technol.* 20:856–861.
23. Lehrbach, P. R., J. Zeyer, W. Reineke, H. J. Knackmuss, and K. N. Timmis. 1984. Enzyme recruitment *in vitro*: Use of cloned genes to extend the range of haloaromatics degraded by *Pseudomonas* sp. strain B13. *J. Bacteriol.* 158:1025.
24. Levy, S. B., R. W. Hedges, F. Sullivan, A. A. Medeiros, and H. Sosroseputro. 1985. Multiple antibiotic resistance plasmids in *Enterobacteriaceae* isolated from diarrhoeal specimens of hospitalized children in Indonesia. *J. Antimicrob. Chemother.* 16:7–16.

25. Mach, P. A., and D. J. Grimes. 1982. R-plasmid transfer in a wastewater treatment plant. *Appl. Environ. Microbiol.* 44:1395–1403.

26. Mancini, P., S. Fertels, D. R. Nave, and M. A. Gealt. 1987. Mobilization of plasmid pHSV106 from *Escherichia coli* in a laboratory waste treatment facility. *Appl. Environ. Microbiol.* 53:665–671.

27. Maniatis, T., E. Fritsch, and J. Sambrook. 1982. *Molecular Cloning.* Cold Spring Harbor Laboratories, Cold Spring Harbor, N.Y.

28. McCarty, P. L., and D. P. Smith. 1986. Anaerobic wastewater treatment. *Environ. Sci. Technol.* 20:1200–1206.

29. McKnight, S. L. 1980. The nucleotide sequence and transcript map of the herpes simplex virus thymidine kinase gene. *Nucleic Acids Res.* 8:5949–5964.

30. McKnight, S. L., and E. R. Gravis. 1980. Expression of the herpes thymidine kinase gene in *Xenopus laevis* ovocytes: An assay for study of deletion mutants constructed *in vitro. Nucleic Acids Res.* 8:5931–5948.

31. McPherson, P., and M. A. Gealt. 1986. Isolation of indigenous wastewater bacterial strains capable of mobilizing plasmid pBR325. *Appl. Environ. Microbiol.* 51:904–909.

32. Metcalf and Eddy, Inc. 1979. *Wastewater Engineering: Treatment, Disposal, Reuse.* McGraw-Hill, New York.

33. Morgan, J. A. W., C. Winstanley, R. W. Pickup, J. G. Jones, and J. R. Saunders. 1989. Direct phenotypic and genotypic detection of a recombinant pseudomonad population released in lake water. *Appl. Environ. Microbiol.* 55:2537–2544.

34. Nemerow, N. L. 1978. *Industrial Water Pollution.* Addison-Wesley, Reading, Mass.

35. Nicholas, R. B. 1987. Biotechnology in hazardous-waste disposal: An unfulfilled promise. *ASM News* 53:138–142.

36. Ou, J. T. 1972. Effect of Zn^{2+} on bacterial conjugation: Inhibition of mating pair formation. *J. Bacteriol.* 111:117–185.

37. Ou, J. T. 1973. Effect of Zn^{2+} on bacterial conjugation: Increase in ability of F^- cells to form mating pairs. *J. Bacteriol.* 115:648–654.

38. Rittmann, B. E. 1984. Needs and strategies for genetic control: Municipal wastes. In G. S. Omenn and A. Hollaender (eds.), *Genetic Control of Environmental Pollutants.* Plenum Press, New York, pp. 215–228.

39. Sagik, B. P., and C. A. Sorber. 1979. The survival of host-vector systems in domestic sewage treatment plants. *Recombinant DNA Tech. Bull.* 2:51–55.

40. Sagik, B. P., C. A. Sorber, and B. E. Moore. 1981. The survival of EK1 and EK2 systems in sewage treatment plant models. In S. B. Levy, R. C. Clowes, and E. L. Koenig (eds.), *Molecular Biology, Pathogenicity, and Ecology of Bacterial Plasmids.* Plenum Press, New York.

41. Sakai, R. K., D. H. Gelfand, S. Stoffel, S. J. Scharf, R. Higuchi, G. T. Horn, K. B. Mullis, and H. A. Erlich. 1988. Primer-directed enzymatic amplification of DNA with a thermostable DNA polymerase. *Science* 239:487–491.

42. Sandt, C., and D. Herson. 1990. Plasmid rearrangements associated with mobilization of the genetically engineered plasmid pHSV106 from *Escherichia coli* HB101 (pHSV106) to an environmental isolate of *Enterobacter cloacae. Abstracts of the Annual Meeting (Anaheim, Calif.).* American Society for Microbiology, Abstract Q-159.

43. Saye, D. J., and R. V. Miller. 1989. The aquatic environment: Consideration of horizontal gene transmission in a diversified habitat. In S. B. Levy and R. V. Miller (eds.), *Gene Transfer in the Environment.* McGraw-Hill, New York, pp. 223–259.

44. Scanferlato, V. S., D. R. Orvos, J. John Cairns, and G. H. Lacy. 1989. Genetically engineered *Erwinia carotovora* in aquatic microcosms: Survival and effects on function groups of indigenous bacteria. *Appl. Environ. Microbiol.* 55:1477–1482.

45. Seo, J.-H., and J. E. Bailey. 1985. Effects of recombinant plasmid content on growth properties and cloned gene product formation in *Escherichia coli. Biotechnol. Bioeng.* 27:1668.

46. Sommerville, C. C., I. T. Knight, W. L. Straube, and R. R. Colwell. 1989. Simple, rapid method for direct isolation of nucleic acids from aquatic environments. *Appl. Environ. Microbiol.* 55:548–554.

47. Steffan, R. J., and R. M. Atlas. 1988. DNA amplification to enhance detection of genetically engineered bacteria in environmental samples. *Appl. Environ. Microbiol.* 54:2185–2191.

48. Stewart, G. J. 1989. The mechanism of natural transformation. In S. B. Levy and R. V. Miller (eds.), *Gene Transfer in the Environment*. McGraw-Hill, New York, pp. 139–164.

49. Stotzky, G. 1989. Gene transfer among bacteria in soil. In S. B. Levy and R. V. Miller (eds.), *Gene Transfer in the Environment*. McGraw-Hill, New York, pp. 165–222.

50. Tomoeoda, M., M. Inuzuka, and T. Date. 1975. Bacterial sex pili. *Prog. Biophys. Mol. Biol.* 30:23–56.

51. Trevors, J. T., J. D. Van Elsas, M. E. Starodub, and L. S. Van Overbeek. 1989. Survival of and plasmid stability in *Pseudomonas* and *Klebsiella* strains introduced into agricultural drainage water. *Can. J. Microbiol.* 35:675–680. (Erratum: *Can. J. Microbiol.* 35:975.)

52. Tzagoloff H., and D. Pratt. 1964. The initial steps in infection with coliphages M13. *Virology* 24:372–380.

53. Van Elsas, J. D., J. T. Trevors, L. S. Van Overbeek, and M. E. Starodub. 1989. Survival of *Pseudomonas fluorescens* containing plasmids RP4 and pRK2501 and plasmid stability after introduction into two soils of different texture. *Can. J. Microbiol.* 35:951–959.

54. Wagner, M. J., J. A. Sharp, and W. C. Summers. 1981. Nucleotide sequence of the thymidine kinase gene of herpes simplex virus type 1. *Proc. Natl. Acad. Sci.* 78:1441–1445.

55. Weber, Jr., W. J., and E. H. Smith. Removing dissolved organic contaminants from water. *Environ. Sci. Technol.* 20:970–979.

Conjugal
Gene Transfer
on Plants

Stephen K. Farrand

Introduction and History

As we become more familiar with the bacteria that inhabit plants and their immediate environs, it becomes clear that these organisms have the capacity for genetic exchange. Many of these organisms are known to exchange experimentally introduced genetic elements (10, 14, 26, 30, 35, 37) and for some, experiments in culture have shown the presence of indigenous transferable factors (3, 7, 9, 16, 17, 25, 34, 38, 46). Given these findings, it is reasonable to assume that these organisms can and do exchange genetic information in their natural habitats, i.e., in, on, or around plants.

Surprisingly, there is very little direct evidence in the literature showing that such gene transfer takes place in situ. The evidence that does exist can be divided into two types. In the first, the presence of indistinguishable or highly conserved plasmids in isolates of the same or closely related bacterial species from diverse locations suggests that these extrachromosomal elements are transferred among these organisms. Perhaps the best example of this is the high conservation of the octopine-type Ti plasmids in various isolates of *Agrobacterium tumefaciens* (11, 39). In this system, although the plasmids are virtually indistinguishable, the characteristics of the bacteria themselves, presumably encoded by chromosomal determinants, can be very different. Similar observations have been made with *Xanthomonas* where plasmid conservation was found to be pathovar-specific (29), in *Pseudomonas syringae* pv. *glycinea* (8), and in *P. syringae* pv. *tomato*

(3). Finally, Schofield et al. (41) have presented a detailed molecular study suggesting that genetic exchange and recombination among *Rhizobium* Sym plasmids has taken place in natural environments. These are all retrospective studies and the proof of in situ transfer is made by inference only.

In the second class of experiments, plasmid transfer *in planta* has been demonstrated directly by controlled matings. These experiments fall into two groups; those using well characterized, highly conjugal model plasmids and those demonstrating transfer of plasmids indigenous to the donor strains. In the first group, broad-host-range R-plasmids of the IncP1 group usually have been employed. However, in both types of experiments, the mating conditions usually have not approximated conditions in the field. Most commonly, high titers of donors and recipients are inoculated onto the plant parts, or into the rhizosphere. For example, Lacy and Leary (27) demonstrated transfer of RP1 from *Pseudomonas glycinea* to *P. phaseolicola* in detached bean pods and on leaves. Similar results were reported for transfer of this plasmid between *Erwinia chrysanthemi* strains on maize (26) and from *E. herbicola* or *P. syringae* pv. *syringae* to *E. amylovora* on detached pear blossoms (28). In each case, the donors and recipients were inoculated as culture droplets or as spray mists with titers ranging from 10^6 to 10^9 cells per milliliter. Manceau, Gardan, and Devaux (33) showed that the closely related IncP1 R-plasmid, RP4, could be transferred between strains of *X. campestris* pv. *corylina* and from this organism to *E. herbicola* following inoculation into hazelnut trees. In experiments designed to test parameters of the system, these workers identified the one critical factor as the frequency in which the donors came into contact with recipient cells. Interestingly, they also detected transfer of the R-plasmid to two resident epiphytic isolates; a *P. fluorescens* and an *E. herbicola*.

Demonstrated *in planta* transfer of indigenous plasmids is almost exclusively confined to studies with the agrobacteria. In fact, the first demonstration of the conjugal nature of Ti plasmids came from *in planta* studies (24, 45). Panagopoulos, Psallidas, and Alivizatos (36) showed that a second *Agrobacterium* plasmid, pAgK84, can be transferred among strains in the field. However, as with the R-plasmid experiments, these analyses were all conducted by inoculating high titers of organisms onto experimentally wounded plants or plant parts.

Because of the artificial inoculation conditions, these experiments, while indicating that conjugal plasmids *can* transfer *in planta,* do not answer the question as to whether they *do* transfer on the plant under natural conditions. However, results from these experiments certainly are consistent with the hypothesis that such transfer does take place.

The question of conjugal plasmid transfer *in planta* under normal environmental conditions is important from a number of perspectives. First, as has been shown for *Agrobacterium* and certain *P. savastanoi* isolates, plasmids play a major and direct role in pathogenesis associated with these organisms. Even more intriguing is the recent observation by Stall, Loschke, and Jones (43) that an avirulence gene of *X. campestris* pv. *vesicatoria* is encoded on a conjugal 200-kb plasmid which also confers copper resistance. Thus plasmid transfer mechanisms are important in considerations of epidemiology, disease control, and evolution.

Second, in at least two cases, a plasmid-encoded determinant may be associated with resistance of the bacterial hosts to an economically important agricultural bactericide. Cooksey (6) has described a plasmid encoding copper resistance present in many isolates of *P. syringae* pv. *tomato*. Copper sprays are routinely used to control infections on tomato by this *Pseudomonas* pathovar. Significantly, this plasmid is conjugal and is retained in a highly conserved fashion by copper-resistant isolates of this organism. Similar observations have been made with copper-resistant isolates of *X. campestris* pv. *vesicatoria* (43). Thus, *in planta* transfer, under the selective pressure of the metal bactericide, may constitute a mechanism for genetic dissemination of the resistance plasmid.

Determining the conjugal nature of plasmids present in plant-associated bacteria may also provide information concerning the interactions between such bacteria and their hosts. For example, the conjugal transfer of Ti plasmids is normally repressed, but is induced by specific plant tumor-produced factors (17, 25). Thus, investigations into the conjugal nature of *Agrobacterium* Ti plasmids constituted one of the first examples of plant-microbe chemical signaling mechanisms.

Finally, assessment of *in planta* conjugal transfer is of major importance when considering biological containment of genetically engineered traits. It is crucial that transferability of traits engineered into organisms destined for environmental release be properly assessed. Such traits may be plasmid-borne or may be on replicons mobilizable by conjugal plasmids. This emphasizes the need for development of techniques for the critical evaluation of conjugal gene transfer in plant ecosystems.

General Considerations and Issues

Donors and recipients

As with any genetic cross, there are a number of parameters that must be considered in order to maximize gene transfer and the selective recovery of transconjugants. First, careful consideration should be given to the choice of donor and recipient organisms. It may well be that the donor is predetermined by the purposes of the experiment. For example, if a recombinant strain has been constructed and the purpose of the cross is to determine containment of the engineered trait, then the donor must be that strain itself. However, to maximize the level of detection for the cross, careful consideration should be given to choice of recipient. Ideally, the best recipient would be a strain isogenic to the donor except for the engineered trait and markers in the recipient required for counterselection. If this is not possible, or if one wishes to assess transfer to other, unrelated organisms, certain recipient characteristics must be considered. For example, the presence of restriction-modification systems within the recipient may result in undetectable or very low levels of transfer. Such low transfer frequencies may be interpreted as constituting a contained system. However, if the marker is transferred at a low frequency against the restriction barrier, once in the recipient, it will be modified, and then presumably will be capable of further transfer at higher frequencies.

Production by the donor or the recipient of antibiotics, bacteriocins, or bacteriophages active against the other member of the mating pair must also be considered. Such antagonistic characteristics should be assessed in culture by determining efficiencies of recovery for each member of the pair in mixed versus pure cultures. However, one must be cautious about extrapolating results from culture in artificial medium to events occurring *in planta*. It is conceivable that while such agents might not be detected under laboratory culture conditions, their production may be induced or substantially increased when the bacteria are inoculated onto plants. As a related concern, consideration should be given to production by the plant itself of compounds detrimental to the donor, the recipient, or both. It is well known that some bacteria act as elicitors, inducing plants to produce toxic defense compounds such as phytoalexins (1). Many such secreted factors have antibacterial activity. Finally, either the donor or the recipient may be unable to colonize and persist in the plant for a period sufficient to allow for mating pair formation and gene transfer. For all of these reasons it is prudent to determine differential titers for donors and recipients at the beginning and conclusion of all mating experiments be they *in planta* or on culture medium.

Consideration must also be given to parameters associated with the particular genetic elements. Host range limitation of the transferable element is an obvious factor. Again, care must be taken. Because a particular plasmid may be transmissible to one isolate of a bacterial species does not mean that it is transferable to all isolates of that species. Similarly, successful transfer of one plasmid to a particular recipient does not ensure that all derivatives of that plasmid will have the same host range. For example, RP4 is freely conjugal to *P. syringae* pv. *tabaci* strain BR2. However, the broad-host-range cloning vectors pRK290, pCP13 and pLAFR1, all derived from an indistinguishable plasmid, RK2, cannot be introduced into this phytopathogen (P. D. Shaw, personal communication).

One must also consider the plasmid complement of the recipient. The presence of a plasmid of the same incompatibility group in the recipient can reduce transfer frequencies below the level of detection by expression of entry exclusion and incompatibility functions. The presence of additional plasmids in the donor may also present a problem. Such plasmids, while being compatible with the element being tested for transmissibility, may exert fertility inhibition resulting in greatly depressed transfer frequencies. All of these factors should be monitored by crosses on laboratory media before *in planta* matings are attempted.

Markers for selection and counterselection

The success of any experiment in bacterial genetics depends in large part on the choice of markers for selections and counterselections. The element to be transferred must contain a marker which will allow for direct selection of a potentially small number of transconjugants in a background of a large excess of unconverted recipients. Transfer frequencies may range from 10^{-5} to 10^{-8}, illustrating the need for identifying a transconjugant among as many as 100 million recipients. Similarly, the recipients must contain markers allowing for counterselection of the donors. Such markers must allow for growth of a few transconjugants under conditions in which growth of a large excess of donors is suppressed.

A distinction must be made between selectable markers and screenable markers. For example, an antibiotic resistance trait present on the element being transferred but absent in the recipient is a selectable marker for transconjugants. One can select such progeny directly by plating the mating mix on medium containing the antibiotic at a concentration sufficient to completely inhibit the growth of unmated recipients. On the other hand, the *E. coli lacZ* gene coupled with X-gal indicator plates does not constitute a selective system. The marker does not confer a selectable growth advantage on transconjugants, and the rare blue colony would be lost in a background of confluent recipient growth. However, the *lacZ* gene can be used as a screening marker. Potential transconjugants, selected for resistance to an antibiotic, for example, can be verified on X-gal medium if the transferred element contains both the drug resistance determinant and the *lacZ* gene. Similarly, production of fluorescent pigment, while suitable as a screening marker, is not a selectable trait.

Markers for selection and counterselection must completely suppress the growth of the targeted member of the mating pair. In this light, a second factor, mutation to selection resistance, must also be considered. Thus, if antibiotics are used, they must not only effectively inhibit growth of the donor or recipient, but the sensitive organism should exhibit an acceptably low spontaneous mutation rate for resistance to the selective agent.

Susceptibility to an antibiotic can be determined by drug dilution experiments. Susceptibility levels are generally expressed as minimal growth inhibitory (MIC) or minimal bactericidal (MBC) concentrations. Techniques and protocols for determining these values are presented in Blair, Lannette, and Truant (5). The problem of spontaneous mutation can be assessed by spreading a high titer [10^8 to 10^9 colony-forming units (CFU)] of the organism to be tested onto medium containing the antibiotic at the proper inhibitory concentration. Colonies appearing following incubation at the appropriate temperature should be screened for stable resistance to the antibiotic. Generally, to be an effective selection agent, the spontaneous mutation rate for the antibiotic to be used should be less than one-tenth of the lowest transfer frequency one expects to detect.

Acceptable markers for selections and counterselections include not only antibiotic resistance traits, but also nutritional characteristics. For example, auxotrophies specific to the donor, coupled with the use of appropriate minimal media, can be used for counterselections as long as the mutations are stable and the nutritional deficiencies do not affect colonization or donor functions *in planta*. Similarly, differences in carbon utilization patterns can be used. For example, *E. coli* cannot utilize sucrose as a carbon and energy source. If the recipient organism can catabolize this sugar, a minimal medium containing sucrose as sole carbon source will effectively counterselect an *E. coli* donor. Similar schemes employing lactose can be used to counterselect *Pseudomonas* donors.

Nutritional markers encoded on the element to be transferred can also be used for selection of transconjugants as long as the trait is normally absent in the recipient. For example, relevant portions of the *E. coli* Lac operon contained on the element to be transferred, when combined with lactose-minimal medium could be used for transconjugant selection with Lac⁻ recipients such as soil pseudomonads (13). However, one must distinguish between organisms that are β-gal⁻ and those that are Lac⁻. For example, *Agrobacterium* biovar 1 strains lack β-galactosidase activity, and form white colonies on medium supplemented with X-gal (42). However, all can utilize lactose as sole carbon and energy source via a catabolic pathway not requiring β-galactosidase (2, 19). When the Lac genes of *E. coli* are introduced into such agrobacteria, they become β-gal$^+$ and form blue colonies on X-gal medium. Such Lac constructs are useful screening markers in *Agrobacterium,* but cannot be used for selection.

In addition to selectable markers, donors and recipients should contain other, differential markers. These are called unselected, or outside, markers, and play no role in the actual genetic selections imposed during the process of the cross. The purpose of these unselected markers is for screening and positive identification of putative transconjugants. Such markers are particularly important in cases where primary selections are not 100 percent effective in eliminating the donors or in selecting transconjugants from recipients. Screening potential transconjugants for the presence or absence of such markers can quickly verify their identity.

Suitable outside markers include all of those discussed above including antibiotic resistance traits, auxotrophies, and carbon utilization traits. In addition, as long as they are stable and characteristic, taxonomically associated traits can be used as unselected markers. For example, in crosses where one (but only one) member is a fluorescent pseudomonad, production of the characteristic fluorescent pigment can be used as an outsider marker. Similarly, in crosses involving a biovar 1-type *Agrobacterium,* production of 3-ketolactose (2) is a useful identificational marker. This emphasizes the importance of a thorough working knowledge concerning the characteristics of the organisms involved in the crosses.

It is also of value to have unselected markers associated with the element to be transferred. This is particularly important in cases where the primary selection for the transferred element is not completely effective. Such markers can include antibiotic resistance determinants, catabolism of carbon sources, or other, easily assayable properties. The latter might include susceptibility to plasmid-specific bacteriophages or sensitivity or resistance to ultraviolet light. A combination of screening for an unselected marker of the recipient (or absence of such a marker associated with the donor) along with an unselected marker on the transferred element can give almost complete assurance that colonies represent true genetic progeny.

Mating substrates

Although the primary aim is to assess gene transfer on plants, a number of parameters must still be considered. Chief among these is choice of the plant species to be used. Here two options exist. One can use plants that are convenient and suited to the laboratory, growth chamber, or greenhouse. This has the advantage of a ready supply of relatively homogeneous stock, allowing for control of substrate variability from one experiment to another. On the other hand, the experiments will address only the question of gene transfer on that plant species, and extrapolations to exchange events on other, perhaps more relevant, plant species may not be valid. This can present a problem if the organisms under study are adapted to or will be used with some other plant species. In such cases, transfer experiments may be influenced by the inability of either or both bacterial partners to adapt properly on the laboratory host plants.

The second option is the use of the relevant plant species or cultivars. While in many cases this might present no problem, in other instances, where the plants are not readily adaptable to controlled conditions, this option will set limitations on how the experiments can be performed. Examples of such problems might include use of plants not easily maintained in growth chambers or greenhouses in numbers suitable for experimental replication, or use of large plants such as shrubs or trees.

Under laboratory conditions, one can also choose between a number of variables for propagation of the plant host. Among these include plants grown in soil or in some other substrate including liquid (hydroponics), spray (aeroponics), or agar. If grown in soil, a choice can be made between sterilized medium or soil containing indigenous microflora. Although this might seem to be of concern only for crosses involving root-associated bacteria, such is not the case since aerial plant portions easily become contaminated by and colonized with soil organisms. The presence of indigenous microflora can present at least two problems. First, these organisms may be capable of growth on the selection media, thus presenting difficulties in isolating and enumerating transconjugants. Second, the indigenous microflora may outcompete or otherwise suppress the growth of the inoculated strains thus interfering with the formation of mating pairs.

Growth in liquid or on agar surfaces requires that the plant be rendered microorganism-free, otherwise fungi and bacteria normally colonizing the plant or seeds from which the plants germinate can contaminate the medium. This is not a critical problem with soil- or aeroponically-grown plants; however, if they are not decontaminated first, there exists a high probability that indigenous microflora will contaminate the root substrate.

If gene transfer is to be assessed in the absence of such competing microflora, the plants must be rendered axenic. This usually can be accomplished by raising plants from sterilized seeds and propagating the germinants in autoclaved soil or on sterile, artificial substrates as described above. Seeds can be decontaminated by treatment with mercuric salts, chlorine derivatives, or alcohols, or by regimes using two or more of these disinfectants. Procedures are described in detail in Evans et al. (15). Following such treatments, the seeds should be thoroughly rinsed with sterile water and germinated in sterilized substrate, either soil or sand, or on sterile supports such as moistened filter paper, water agar, or Hoagland's salts (21) solidified with agar. Care must be taken when autoclaving soil; the process has been known to generate by-products toxic to both plants and microorganisms.

As an alternative to whole plants, matings may be performed on excised plant parts. Virtually any plant part can be dissected and propagated for at least a short period of time on agar-based media. Especially amenable to this type of manipulation are leaves, cotyledons, and hypocotyls along with sections from fleshy roots such as carrots or sugar beets or from tubers such as potato. For example, disks cut from carrot roots, and maintained on moist filter paper or on water agar have been used to demonstrate *in planta* conjugal Ti-plasmid transfer between *Agrobacterium* strains (31). However, it must be remembered that such explants show altered physiology as compared to intact plants. This may include secretion or excretion of compounds having bacteriostatic properties. Extrapolating results from matings on such plant parts to events occurring on intact plants, especially those growing in the field, should be done with care.

Environmental conditions

Environmental conditions most likely play a critical role in the ability of bacteria to transfer genes among themselves. The effects of altered conditions are probably secondary; influencing the survival and propagation of one or both of the mating partners or the resulting transconjugants. Critical factors may include temperature, humidity, and in the case of root-associated organisms, soil matric potential. As an example, Tempé et al. (44) observed that conjugal Ti-plasmid transfer contains a temperature-sensitive component. Transfer frequencies drop below detectable levels when matings are conducted at temperatures of 33°C or higher.

For controlled experiments, matings should be performed on plants maintained in growth chambers or greenhouses. Temperature and humidity can be kept constant and soil matric potential monitored and maintained via soil probes (47). Field experiments, on the other hand, cannot be controlled with respect to these parameters. It therefore becomes important to monitor environmental conditions during the course of such experiments. At a minimum, continuous temperature and humidity readings should be obtained along with soil moisture values for crosses involving root-associated microorganisms. It is also important to monitor rainfall, as it has been shown for certain epiphytic bacteria that precipitation is the most valid prognosticator for microbial population increases on leaf surfaces (20). Rainfall may also be a particularly important parameter in gene transfer between root-associated bacteria; water percolation is known to be a major factor in the vertical movement of soil- and rhizosphere-associated microorganisms (18, 32).

Experiments to assess gene transfer under field conditions should also take into account the use of agricultural chemicals during preparation of the soil and subsequent planting and growth. Applications of bactericides, fungicides, herbicides and pesticides, along with fertilizers, may have direct or indirect effects on the capacity of the bacterial partners to establish themselves in the ecosystem or to transfer genetic elements. Although such treatments have not been tested for their effects on microbial gene transfer, many of these compounds are known to alter plant-associated microflora. For initial experiments, it would be wise to assess gene transfer frequencies in soils or on plants known not to have been treated with agricultural chemicals. Once such frequencies have been established, the influence of treatment regimes can be determined in subsequent experiments.

Questions of purpose

In the design of *in planta* gene transfer experiments two preliminary questions must be considered. For the first, one must decide whether the purpose is to determine if gene transfer *can* take place or if gene transfer *does* take place. The first possibility considers only the ability of organisms to transfer or accept genetic elements when incubated on plants. It sets no constraints on type of element, donors or recipients, or on the conditions of the matings. Mating parameters, especially with respect to input donor and recipient titers and inoculation methods, can be maximized to enhance the possibility of mating pair formation and recovery of transconjugants. The second option considers the actual phenomenon of conjugal gene transfer among bacteria under natural conditions of host plant, population dynamics, and enirvonment, i.e., under conditions of real-life ecosystems.

The second general problem concerns whether quantitative or qualitative results are required. If the latter, there is no need to determine population levels of donors, recipients, or transconjugants, and one must only consider the limits of detection for isolation of progeny. If quantitative results are desired, it becomes necessary to determine population titers, as well as detection sensitivity levels. This, in turn, requires an estimation of the fraction of organisms of each class that are recoverable from the plants.

Experimental Designs

The preceding discussion has covered some general considerations relevant to analysis of genetic transfer events *in planta*. Equal thought must be given to the design and execution of the mating experiments. What follows is a discussion of various steps in the performance of such experiments.

Preparation of inocula

It goes without saying that donor and recipient cultures should be clonally purified prior to their use in any mating experiment. Cultures for preparation of inocula should be propagated under conditions optimal for growth and selective for the markers involved in the matings. This is especially true of donors where the trait of interest is contained on an independently replicating plasmid. This will ensure that all members of the donor population contain the element being tested for transfer. However, if any of these traits necessitate growth in the presence of antibiotics, the bacteria should be harvested from the cultures, washed thoroughly, and resuspended in an appropriate medium lacking the antibiotic before proceeding with the mating. This will prevent the antibiotic from inhibiting the sensitive member of the mating pair. In general, cultures should be harvested for use during mid- to late-exponential growth as stationary phase bacteria are often poor donors or recipients.

Inoculations

Methods for inoculation of host plants will vary depending upon the purpose of the experiment, on the segment of the plant to be assessed as a substrate, and on the organisms being tested. Concerning the latter two points, choice of inoculation site should be dependent upon what part of the plant is normally colonized or invaded by the bacterial partners involved in the cross.

Roots may be inoculated by dipping into suspensions of the bacteria prior to planting in soil. Alternatively, the bacteria can be percolated into soil containing the plant, or can be mixed into the soil at the desired cell densities before planting. It may be a good idea to first resuspend the bacteria in some neutral medium such as isotonic NaCl to prevent any effects on the plants or the bacteria caused by growth medium carryover.

Aerial portions of the plant can also be inoculated by dipping into suspensions of the microorganisms. Under such conditions, it may be helpful to add a drop or two of Tween 80 to the bacterial suspensions to aid in complete wetting of the plant surfaces. Alternatively, various portions of the plant can be inoculated directly by pipetting or spraying suspensions of the cells. The latter may be especially effective in leaf or flower inoculations.

When microbial pathogens are involved, it may be important to use inoculating methods developed for enhancement of virulence. This may involve first wounding the plant and then placing the bacteria into the damaged sites. Stems can be wounded with sterile scalpels, toothpicks, or hypodermic needles. Leaves can be wounded by gentle rubbing with an abrasive such as carborundum. Methods have been developed for infiltrating microorganisms into stems or leaves using pressure or vacuum. A number of techniques and mechanical applicators for such purposes are described in Dhingra and Sinclair (12).

When only qualitative indications of gene transfer are required, relatively high-titer inocula should be used to ensure donor-recipient contacts and to maximize mating pair formation. However, when the purpose of the experiment is to determine whether gene transfer can take place under conditions similar to what exists in nature, more care must be taken in strategies for inoculation. Paramount here is to have some understanding of the normal sites of colonization and population levels characteristic of the organisms being tested. Once such information is available, inoculation procedures should be chosen which will lead to population distributions and levels mimicking those seen in the wild. If the question is to assess gene transfer between bacteria being tested for environmental release and indigenous microbial colonizers, the former should be introduced in a manner similar to that planned for full-scale field inoculations onto plants already colonized by their natural microflora.

Incubations

As discussed above, for controlled experiments, incubations should be conducted in greenhouses or preferably in growth chambers. This will ensure some control over environmental conditions from one repetition to another and for evaluation of such variables as donor and recipient titers and ratios, inoculation techniques, and incubation times. Furthermore, incubations in growth chambers allow for assessment of the impact on gene transfer by such environmental variables as light regime and intensity, humidity, and temperature. For experiments conducted in the field, there can be no control over these conditions.

However, as pointed out above, salient environmental parameters should be monitored and recorded.

Recovery of bacteria from inoculated plants

Success in detecting gene transfer, especially under conditions in which frequencies may be very low, is critically dependent upon the ability to recover microorganisms from plant parts. This is not a trivial problem, and recovery efficiencies will certainly contribute to setting the lower limits of detection for demonstration of conjugal gene transfer.

For crosses performed on leaf or flower surfaces, Bertoni and Mills (4) have published a simple procedure allowing for rapid and quantitative recovery of microorganisms. Although the technique was developed for monitoring growth, it should be equally effective in the recovery and titration of donors, recipients, and genetic progeny. The method is simple, inexpensive, and amenable to the processing of large numbers of samples.

Organisms inoculated onto surfaces of roots or stems can be recovered following vigorous vortexing of the excised plant part in buffers. Addition of salt or EDTA may help in releasing loosely adherent bacteria. In the case of roots, soil should first be removed by gentle shaking. Quantitative recovery may require repeated washes, especially for experiments involving organisms that attach tightly to plant surfaces. If this proves to be a problem, quantitative recovery may be better facilitated by grinding the tissue in a manner similar to that described in Ref. 4 or by the techniques described by James, Suslow, and Steinback (22).

Recovery following infiltrations into stems or roots requires maceration and extraction of the plant material. For fleshy roots and herbaceous stems grinding in buffer as described above (4, 22) should provide quantitative recovery. However, woody tissues present a more difficult problem. Such samples should be excised from the plant and ground in a blender or Polytron as described by Manceau, Gardan, and Devaux (33).

Selections

Following recovery, samples should be serially diluted in an appropriate buffer and plated onto selective media. For qualitative assessments, it is sufficient to plate samples onto a medium selective only for transconjugants. However, if quantitative results are required, it is necessary to titer donors, recipients, and transconjugants. Samples of dilutions should be plated onto sets of media selective for each of these populations.

Assuming that the selection markers are efficient, samples from axenic plants should present no major contamination problems. However, if inoculated plants have been grown in the greenhouse or in the field, or if soil is contaminated with indigenous microflora, selective media should also contain agents to prevent the growth of ancillary microorganisms. Fungicides such as benomyl can be included to suppress fungal growth. Additionally, medium selective for the organisms under study can be used. For example, if the cross involves pseudomonads, Kings B medium (40) can be used to suppress growth of other microorganisms. Such media can be supplemented with antibiotics or other selective agents depending upon the markers used in the cross. Selective or semiselective media are available for a range of plant-associated microorganisms including *Agrobacterium, Clavibacter* (previously *Corynebacterium*), *Erwinia, Pseudomonas,* and *Xanthomonas* (40).

Screens

To verify colonies as transconjugants, a representative number should be screened for unselected markers associated with the recipient and also with the genetic element being transferred. This is especially important if the selection schemes are not fully effective in excluding growth of the donor or unmated recipient, or if the plants are contaminated with indigenous microflora that are difficult to counterselect. The last point could result in unexpected but interesting results. Manceau, Gardan, and Devaux (33) were able to demonstrate *in planta* transfer not only to their intended recipients but also to indigenous epiphytic microorganisms colonizing the plant hosts.

When performing these screens it must be remembered that any colony appearing on primary selection medium may be mixed, containing a small but significant percentage of nongrowing donors and unconverted recipients. If directly transferred to screening medium not counterselective to these parents, they can easily regrow and be improperly scored. To correct this problem, colonies to be tested should first be purified by streaking for isolation on selective medium, or the screening medium can be supplemented with the agents used in the original selection regime.

Controls

Controls for any mating include platings of donor-alone and recipient-alone cultures. For *in planta* matings, these should be conducted from beginning to end in a fashion exactly the same as the mating mixes themselves. Colonies appearing on selective plates from such control inoculations should be identified as having derived from the donor or the recipient. If confirmed, their numbers will give some indication as to the effectiveness of the selection scheme. In addition, the donor-alone and recipient-alone titers should be determined on nonselective media and compared to those obtained from the *in planta* mating inoculations. Significant differences in titers of one or the other when compared to their levels in the mating mix could indicate antagonism between the members of the mating pair.

It is also important to consider the possibility of matings between donor and recipient taking place on the selective medium following recovery from the plant host. While this might seem to be unlikely, especially when antibiotics are used for the selections, a standard methodology for transferring plasmids among agrobacterial parents involves matings directly on antibiotic-containing selective media (14, 23). This problem can be assessed by sequentially plating aliquots from donor-alone and recipient-alone *in planta* extracts directly on the same plates of selective media. It is also conceivable that mating pairs could form following extraction but before plating on selective media. This can be minimized by working rapidly and not allowing extracts and dilutions to stand for long periods of time. The problem can be assessed and controlled for by mixing samples from the donor-alone and recipient-alone extracts before dilution and plating.

Enumeration and expression of mating frequencies

Donor, recipient, and transconjugant titers should be determined by standard statistical methods using data obtained from plate counts and dilution factors. Titers should be corrected for donor or recipient selection escapes and for any matings occurring in the extracts or on the selective media. Populations should then be expressed as colony-forming units (CFU) per volume or weight of tissue processed. Sufficient independent matings and plating replicates should be used to ensure statistical validity.

Mating frequencies are generally expressed as transconjugants per input donor or input recipient. However, this is meaningful only for short matings where bacterial growth and secondary matings do not occur. Such is certainly not the case for *in planta* crosses in which donors, recipients, and transconjugants may undergo complex interactions with each other and with the host plant. In addition, most *in planta* matings are conducted over periods of days to weeks, allowing for further, perhaps unequal growth of the various populations as well as additional, secondary matings. Assuming that the transconjugants have no selective advantage over their recipient counterparts, expressing mating frequencies as numbers of transconjugants per recovered recipient is probably more meaningful. However, even this does not take into account the likelihood of secondary mating events.

Summary and Conclusions

Assessment of gene transfer in the environment, be it in water, soils, or on plants, presents difficulties seldom encountered in the laboratory. The success of conjugal gene transfer experiments in general is dependent upon proper experimental design. Strategies for selection and counterselection are particularly important for the generation of unambiguous results. This is certainly the case for crosses performed on standard bacteriological media, but is even more critical when nontraditional conditions are imposed. Choice of selections and culture-handling practices will depend in large part on the organisms being studied. For this reason, a thorough understanding of the culture characteristics, physiology, and genetics of the bacterial species in question can be instrumental in the design and implementation of meaningful gene transfer experiments. There is no substitution for good bacteriological technique.

Acknowledgments

Work in the author's laboratory concerning conjugal transfer of *Agrobacterium* plasmids was supported by grant numbers R01-CA44051 from the National Cancer Institute and AG-87-CRCR-1-2353 from the United States Department of Agriculture.

References

1. Bailey, J. A., and J. W. Mansfield. 1982. *Phytoalexins*. John Wiley and Sons, New York.
2. Bennaerts, M. J., and J. DeLey. 1963. A biochemical test for crown gall bacteria. *Nature* 197:406–407.
3. Bender, C. L., and D. A. Cooksey. 1986. Indigenous plasmids in *Pseudomonas syringae* pv. *tomato*: Conjugative transfer and role in copper resistance. *J. Bacteriol.* 165:534–541.
4. Bertoni, G., and D. Mills. 1987. A simple method to monitor growth of bacterial populations in leaf tissue. *Phytopathology* 77:832–835.
5. Blair, J. E., E. H. Lennette, and J. P. Truant. 1970. *Manual of Clinical Microbiology*. ASM Press, Bethesda, Md.
6. Cooksey, D. A. 1987. Characterization of a copper resistance plasmid conserved in copper-resistant strains of *Pseudomonas syringae* pv. *tomato*. *Appl. Environ. Microbiol.* 53:454–456.
7. Coplin, D. L., R. G. Rowan, D. A. Chisholm, and R. E. Whitmayer. 1981. Characterization of plasmids in *Erwinia stewartii*. *Appl. Environ. Microbiol.* 42:599–604.
8. Curiale, M. S., and D. Mills. 1983. Molecular relatedness among cryptic plasmids in *Pseudomonas syringae* pv. *glycinea*. *Phytopathology* 73:1440–1444.

9. Currier, T. C., and M. K. Morgan. 1983. Plasmids of *Pseudomonas syringae*: No evidence of a role in toxin production or pathogenicity. *Can. J. Microbiol.* 29:84–89.

10. Datta, N., and R. W. Hedges. 1972. Host range of R factors. *J. Gen. Microbiol.* 70: 453–460.

11. De Vos, G., M. DeBeuckeleer, M. van Montagu, and J. Schell. 1981. Addendum. Restriction endonuclease mapping of octopine tumor-inducing plasmid pTiAch5 of *Agrobacterium tumefaciens. Plasmid* 6:249–253.

12. Dhingra, O. D., and J. B. Sinclair. 1985. *Basic Plant Pathology Methods*. CRC Press, Boca Raton, Fla.

13. Drahos, D. J., B. C. Hemming, and S. McPherson. 1986. Tracking recombinant organisms in the environment: β-galactosidase as a selectable non-antibiotic marker for fluorescent pseudomonads. *Bio/Technology* 4:439–444.

14. Ellis, J. G., A. Kerr, M. van Montagu, and J. Schell. 1979. *Agrobacterium:* Genetic studies on agrocin 84 production and the biological control of crown gall. *Physiol. Plant Pathol.* 15:311–319.

15. Evans, D. A., W. R. Sharp, P. V. Ammirato, and Y. Yamada. 1983. *Handbook of Plant Cell Culture, Volume I. Techniques for Propagation and Breeding*. Macmillan, New York.

16. Farrand, S. K., J. E. Slota, J.-S. Shim, and A. Kerr. 1985. Tn5 insertions in the agrocin 84 plasmid: The conjugal nature of pAgK84 and the locations of determinants for transfer and agrocin 84 production. *Plasmid* 13:106–117.

17. Genetello, C., N. van Larebeke, M. Holsters, A. De Picker, M. van Montagu, and J. Schell. 1977. Ti plasmids of *Agrobacterium* as conjugative plasmids. *Nature* 265: 561–563.

18. Hamdi, Y. A. 1974. Vertical movement of rhizobia in soil. *Zentralbl. Bakteriol. Parasitenkd. Infecktionskr. Hyg. Abt. 2* 129:373–377.

19. Hayano, K., and S. Fukui. 1970. α-3-Ketoglucosidase of *Agrobacterium tumefaciens. J. Bacteriol.* 101:692–697.

20. Hirano, S. S., and C. D. Upper. 1986. Temporal, spatial and genetic variability of leaf-associated bacterial populations. In N. J. Fokkema and J. Van den Heuvel (eds.), *Microbiology of the Phyllosphere*. Cambridge University Press, London, pp. 235–251.

21. Hoagland, D. R., and D. I. Arnon. 1950. *The Water-Culture Method for Growing Plants without Soil*. California Agricultural Experiment Station Circular No. 347.

22. James, Jr., D., T. V. Suslow, and K. E. Steinback. 1985. Relationship between rapid, firm adhesion and long-term colonization of roots by bacteria. *Appl. Environ. Microbiol.* 50:392–397.

23. Jones, D. A., M. H. Ryder, B. G. Clare, S. K. Farrand, and A. Kerr. 1988. Construction of a Tra⁻ deletion mutant of pAgK84 to safeguard the biological control of crown gall. *Mol. Gen. Genet.* 212:207–214.

24. Kerr, A. 1971. Acquisition of virulence by non-pathogenic isolates of *Agrobacterium radiobacter. Physiol. Plant Pathol.* 1:241–246.

25. Kerr, A., P. Manigault, and J. Tempé. 1977. Transfer of virulence *in vivo* and *in vitro* in *Agrobacterium. Nature* 265:560–561.

26. Lacy, G. H. 1978. Genetic studies with plasmid RP1 in *Erwinia chrysanthemi* strains pathogenic on maize. *Phytopathology* 68:1323–1330.

27. Lacy, G. H., and J. V. Leary. 1975. Transfer of antibiotic resistance plasmid RP1 into *Pseudomonas glycinea* and *Pseudomonas phaseolicola in vitro* and *in planta. J. Gen. Microbiol.* 88:49–57.

28. Lacy, G. H., V. K. Stromberg, and N. P. Cannon. 1984. *Erwinia amylovora* mutants and *in planta*–derived transconjugants resistant to oxytetracycline. *Can. J. Plant Pathol.* 6:33–39.

29. Lazo, G. R., and D. W. Gabriel. 1987. Conservation of plasmid DNA sequences and pathovar identification of strains of *Xanthomonas campestris. Phytopathology* 77: 448–453.

30. Levin, R. A., S. K. Farrand, M. P. Gordon, and E. W. Nester. 1976. Conjugation in *Agrobacterium tumefaciens* in the absence of plant tissue. *J. Bacteriol.* 127:1331–1336.

31. Liao, C. H., and G. T. Heberlein. 1978. A method for the transfer of tumorigenicity between strains of *Agrobacterium tumefaciens* in carrot root disks. *Phytopathology* 68:135–137.

32. Madsen, E. L., and M. Alexander. 1982. Transport of *Rhizobium* and *Pseudomonas* through soil. *Soil Sci. Soc. Am. J.* 46:557–560.

33. Manceau, C., L. Gardan, and M. Devaux. 1986. Dynamics of RP4 plasmid transfer between *Xanthomonas campestris* pv. *corylina* and *Erwinia herbicola* in hazelnut tissues, *in planta. Can. J. Microbiol.* 32:835–841.

34. Obukowicz, M., and P. D. Shaw. 1985. Construction of Tn3-containing plasmids from plant-pathogenic pseudomonads and an examination of their biological properties. *Appl. Environ. Microbiol.* 49:468–473.

35. O'Gara, F., and L. K. Dunican. 1973. Transformation and physical properties of R factor RP4 transferred from *Escherichia coli* to *Rhizobium trifolii. J. Bacteriol.* 116: 1177–1180.

36. Panagopoulos, C. G., P. G. Psallidas, and A. S. Alivizatos. 1979. Evidence of a breakdown in the effectiveness of biological control of crown gall. In B. Schippers, and W. Gams (eds.), *Soil-Borne Plant Pathogens.* Academic Press, London, pp. 569–578.

37. Panopoulos, N. J. 1981. Emerging tools for *in vitro* and *in vivo* manipulation of phytopathogenic *Pseudomonas* and other nonenteric gram negative bacteria. In N. J. Panopoulos (ed.), *Genetic Engineering in Plant Sciences.* Praeger, New York, pp. 163–186.

38. Poplawsky, A. R., and D. Mills. 1983. Conjugal transfer and incompatibility properties of a plasmid from a plant pathogenic pseudomonad. *Phytopathology* 73:826.

39. Sciaky, D., A. L. Montoya, and M.-D. Chilton. 1978. Fingerprints of *Agrobacterium* Ti plasmids. *Plasmid* 1:238–253.

40. Schaad, N. W. 1988. *Laboratory Guide for Identification of Plant Pathogenic Bacteria.* APS Press, St. Paul, Minn.

41. Schofield, P. R., A. H. Gibson, W. F. Dudman, and J. M. Watson. 1987. Evidence for genetic exchange and recombination of *Rhizobium* symbiotic plasmids in a soil population. *Appl. Environ. Microbiol.* 53:2942–2947.

42. Stachel, S. E., G. An, C. Flores, and E. W. Nester. 1985. A Tn3 *lacZ* transposon for the random generation of β-galactosidase gene fusions: Application to the analysis of gene expression in *Agrobacterium. EMBO J.* 4:891–898.

43. Stall, R. E., D. C. Loschke, and J. B. Jones. 1986. Linkage of copper resistance and avirulence loci on a self transmissible plasmid in *Xanthomonas campestris* pv. *vesicatoria. Phytopathology* 76:240–243.

44. Tempé, J., A. Petit, M. Holster, M. van Montagu, and J. Schell. 1977. Thermosensitive step associated with transfer of the Ti plasmid during conjugation. Possible relation to transformation in crown gall. *Proc. Natl. Acad. Sci. USA* 74: 2848–2849.

45. Watson, B., T. C. Currier, M. P. Gordon, M.-D. Chilton, and E. W. Nester. Plasmid required for virulence of *Agrobacterium tumefaciens. J. Bacteriol.* 123:255–264.

46. White, F. F., and E. W. Nester. 1980. Hairy root: Plasmid encodes virulence traits in *Agrobacterium rhizogenes. J. Bacteriol.* 141:1134–1141.

47. Wilkinson, H. T. 1986. An environmental cell to control simultaneously the matric potential and gas quality in soil. *Phytopathology* 76:1018–1020.

Chapter 17

Construction of Plasmids for Use in Survival and Gene Transfer Research

Gerben J. Zylstra

Stephen M. Cuskey

Ronald H. Olsen

Introduction and General Considerations

Experiments to assess the risk posed by releases of genetically engineered microorganisms to the environment are concerned, in part, with the possible dissemination of genes to the indigenous microbial community. Consideration has been given to both the self-transfer of recombinant plasmids or the mobilization of transfer-deficient plasmids by helper plasmids or bacteriophages (17, 22). Projects designed to evaluate gene transfer in the environment involve a wide variety of unrelated conjugal plasmids, cloning vectors, transducing bacteriophages, and transposons. As a result, experimental data derived from such diverse systems may be difficult to generalize. The data may reflect unique characteristics of the microorganisms indigenous to a microcosm (or mesocosm), the properties of the specific genetic element being utilized, or a combination of variables, some of which are unrecognized. Confusion may arise when data are compared from different laboratories, each using their own genetic system, for the purpose of establishing "typical" results. Confusion may also arise from the use of nonsterile environmental samples. Existing organisms may give false-positive readings in procedures employed for estimation of the survival of introduced microorganisms or the transfer of engineered DNA. Some of this confusion can be ameliorated if a series of benchmark plasmids are employed in studies conducted in different laboratories and in various ecosystems. The benchmark plasmids would share a common origin, but each would be designed for the study of different variables. These variables would include conjugation of the benchmark plasmid, mobilization of other replicons by the benchmark plasmid, mobilization of the benchmark plasmid by other conjugative replicons, survival and/or expression of cloned genes, and identification of organisms containing the genes of interest. The data between different experiments and different laboratories become more comparable if one uses the same basic set of plasmids to test all the variables. A series of both self-transmissible and non-self-transmissible (cloning vector) plasmids have been constructed for this purpose.

Benchmark Plasmids

The plasmid R388 has several attractive features useful for the construction of benchmark plasmids for environmental studies. R388 is a relatively small (33-kilobase pairs), broad-host-range, self-transmissible plasmid derepressed for transfer (7). R388 carries genes for resistance to the antibiotics trimethoprim and sulfonamide. There are relatively few cutting sites for common restriction enzymes on the plasmid (26). Many of these restriction sites are in or near the antibiotic resistance genes. The IncW group of plasmids has been reviewed recently (24).

In order to extend the utility of R388 several derivatives were constructed (Table 17.1). The first set of derivatives involved the transposition of a known genetic element to regions of R388 not involved in replication or self-transmissibility. R388 has been labeled with Tn*1* (carbenicillin resistance), Tn*904* (streptomycin resistance), Tn*1721* (tetracycline resistance), and Tn*501* (mercury resistance). These transposons have been well characterized in the literature (4, 15, 19, 23). The addition of the new antibiotic resistance genes allows the selection of the basic replicon in various microorganisms that may be intrinsically resistant to trimethoprim and/or sulfonamide. This provides a convenient selection for the acquisition of R388 by soil microorganisms in microcosm analyses. It may also be important to include in microcosm studies self-transmissible plasmids capable of mobilizing vector plasmids to mimic situations which can occur in nature. The transposed derivatives also function to mobilize other plasmids through transposition, cointegration, and conjugation to a recipient microorganism (5, 20). Therefore, the ability of particular transposons to mobilize other replicons can be assessed using the same self-transmissible benchmark plasmid (R388) as the mobilizing replicon.

TABLE 17.1   Relevant Properties of Benchmark Plasmids

Plasmid	Antibiotic resistance(s)	Self-transmission	Cloning vector
R388	Trimethoprim, sulfonamide	+	−
R388::Tn1	Trimethoprim, sulfonamide, carbenicillin	+	−
R388::Tn904	Trimethoprim, sulfonamide, streptomycin	+	−
R388::Tn1721	Trimethoprim, sulfonamide, tetracycline	+	−
R388::Tn501	Trimethoprim, sulfonamide, mercuric ions	+	−
pRO236	Trimethoprim, sulfonamide	−	−
pRO2313	Carbenicillin	−	+
pRO2316	Carbenicillin, tetracycline	−	+
pRO2317	Carbenicillin, tetracycline	−	+
pRO2318	Carbenicillin, trimethoprim	−	+
pRO2320	Carbenicillin, trimethoprim, tetracycline	−	+
pRO2321	Trimethoprim, tetracycline	−	+
pEPA74	Kanamycin, streptomycin	−	−
pEPA90	Ampicillin	−	−

A series of plasmids was constructed to be used as non-self-transmissible cloning vectors, all based on the R388 replication origin. The regions of R388 involved in conjugation, replication, and antibiotic resistance have previously been located (25). The IncW origin of replication of R388 was subcloned from pRO236, a spontaneous, nonconjugative, deletion derivative of R388. Tn1 was transposed randomly into pRO236, and the resulting derivatives pooled. The transposed derivatives were deleted with *Bam*HI and the smallest resulting plasmid was labeled pRO2313. Plasmid pRO2313 (Fig. 17.1) contains the R388 origin of replication and the Tn1 carbenicillin resistance gene. The Tn1 genes for transposition and the R388 genes for antibiotic resistance and conjugation have been deleted. Additional plasmid derivatives were constructed starting with pRO2313. Restriction maps of these vectors are shown in Fig. 17.1. Plasmid pRO2316 was constructed through the addition of pBR322 (3) to pRO2313. Plasmid pRO2317 is a deleted version of pRO2316 and is essentially pBR322 with the R388 IncW origin of replication inserted in place of the ColE1 origin of replication. Plasmid pRO2318 was constructed from pRO2313 and a *Bam*HI fragment of R388 containing the trimethoprim resistance gene. Plasmid pRO2320 was constructed from pRO2317 and pRO2318 and encodes resistances to tetracycline, carbenicillin, and trimethoprim. Plasmid pRO2321 is a Bal-31 deletion derivative of pRO2320 and encodes resistance only to tetracycline and trimethoprim.

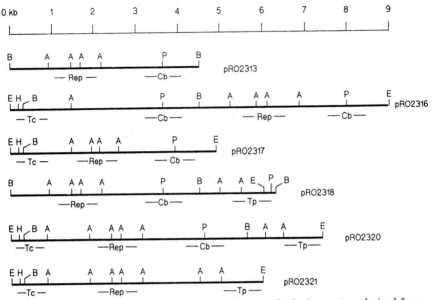

Figure 17.1 Functional and restriction maps of benchmark cloning vectors derived from R388. Abbreviations: A, *Ava*I; B, *Bam*HI; E, *Eco*RI; H, *Hin*dIII; P, *Pst*I; Rep, R388 origin of replication; Cb, Tc, and Tp, resistances to carbenicillin, tetracycline, and trimethoprim, respectively.

A more detailed description of the cloning vector construction has been described elsewhere (28). These cloning vectors allow a selection for particular antibiotic resistances, alone and in combination, and have unique restriction sites that can be used for cloning. The pRO2300 series constructs are small, non-self-transmissible, low-copy-number, broad-host-range, IncW cloning vectors analogous to many of the cloning vectors used in genetic constructions designed for potential environmental release.

Plasmids with a Unique DNA Sequence for Identification of Released Organisms and Indigenous DNA Recipients

A different concern has been addressed in the construction of plasmids containing noncoding regions of eukaryotic DNA. The use of gene probes and antibiotic resistance markers are the two most common means used to detect both introduced bacteria and recipient organisms in gene transfer experiments. Both methods are theoretically imprecise because of the possibility of false-positive readings from in-digenous bacteria. Intrinsic resistance to antibiotics runs over a wide range of concentrations and the DNA used in construction of probes, if prokaryotic, may be present in native bacteria. Inclusion of a frag-ment of eukaryotic DNA for use in DNA hybridization detection meth-ods may overcome these potential problems. Plasmid pRW022 contains both gram-positive and gram-negative origins of replication and noncoding regions of plant and animal DNA (R. Walter, unpub-lished results). Plasmid pEPA74 (Fig. 17.2) contains an approximately 350-base-pair *Eco*RI to *Sst*I fragment of plant DNA [part of a plant napin storage protein cloned in plasmid pN2, (6)] subcloned from pRW022 into the broad-host-range, IncQ plasmid pKT230. Plasmid pKT230 (2) also contains resistances to kanamycin and streptomycin. Plasmid pEPA90 contains the same 350-base-pair *Eco*RI to *Sst*I frag-ment of plant DNA cloned into the multiple cloning site of pUC18 (27). These plasmids can be utilized directly in environmental studies, but they are also useful for providing numerous restriction sites for subcloning the plant DNA to other genetic loci. Utilizing DNA hybrid-ization techniques organisms containing the unique fragment of plant DNA can be located without the background problems associated with screening for other cloned bacterial genes.

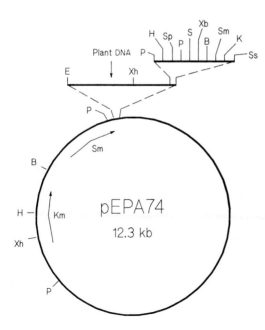

Figure 17.2 Restriction map of plasmid pEPA74. Abbreviations: B, *Bam*HI; E, *Eco*RI; H, *Hin*dIII; P, *Pst*I; Sp, *Sph*I; S, *Sal*I; Xb, *Xba*I; Sm, *Sma*I; K, *Kpn*I; Ss, *Sst*I; Xh, *Xho*I; Sm and Km, resis-tances to streptomycin and kana-mycin, respectively. The plant DNA insert with the pUC19 mul-tiple cloning site is not drawn to scale.

Other Vector Systems Useful for Environmental Testing

Several other plasmids have been designed as broad-host-range vectors. These systems have been the subject of many recent reviews (10, 11, 16, 18). Several cloning vectors have been constructed utilizing the pRO1600 replicon (21), the IncQ replicon (1, 2, 8), the IncP1 replicon (9, 13), the pVS1 replicon (12), and the IncW Sa replicon (14). However, the collection of plasmids derived from R388 described above allow the study of several parameters in environmental gene transfer. All of the plasmids share a common origin and thus will eliminate many of the variables associated with utilizing replicons from various sources.

Summary and Conclusions

The above are descriptions of plasmids designed to overcome the problems of reproducibility, confusion due to the use of different genetic structures with a wide range of capabilities, and background contamination when enumerating released organisms or recipients in gene transfer determinations. The inclusion of one or more of these plasmids in risk assessment experiments may, therefore, alleviate some of the ambiguity in analysis of data from different environmental milieus.

Acknowledgments

Portions of this work were supported by Environmental Protection Agency Cooperative Agreement no. CR812679 with the University of Michigan.

References

1. Bagdasarian, M., R. Lurz, B. Ruckert, F. C. H. Franklin, M. M. Bagdasarian, J. Frey, and K. N. Timmis. 1981. Specific-purpose plasmid cloning vectors II. Broad host range, high copy number, RSF1010-derived vectors, and a host-vector system for gene cloning in *Pseudomonas*. *Gene* 16:237–247.
2. Bagdasarian, M., and K. N. Timmis. 1982. Host:vector systems for gene cloning in *Pseudomonas*. *Curr. Top. Microbiol. Immunol.* 96:47–67.
3. Balbas, P., X. Soberon, E. Merino, M. Zurita, H. Lomeli, F. Valle, N. Flores, and F. Bolivar. 1986. Plasmid vector pBR322 and its special purpose derivatives—a review. *Gene* 50:3–40.

4. Bennett, P. M., J. Grinstead, C. L. Cjoi, and M. H. Richmond. 1978. Characterization of Tn501, a transposon determining resistance to mercuric ions. Mol. Gen. Genet. 159:101–106.
5. Crisona, N. J., J. A. Nowak, H. Nagaishi, and A. J. Clark. 1980. Transposon-mediated conjugational transmission of nonconjugative plasmids. J. Bacteriol. 142: 701–713.
6. Crouch, M. L., K. M. Tenbarge, A. E. Simon, and R. Ferl. 1983. cDNA clones for Brassica napus seed storage proteins: Evidence from nucleotide sequence analysis that both subunits of napin are cleaved from a precursor polypeptide. J. Mol. Appl. Gen. 2:273–283.
7. Datta, N., and R. W. Hedges. 1972. Trimethoprim resistance conferred by W plasmids in Enterobacteriacea. J. Gen. Microbiol. 72:349–355.
8. Davison, J., M. Heusterspreute, N. Chevalier, V. Ha-Thi, and F. Brunel. 1987. Vectors with restriction site banks V. pJRD215, a wide-host-range cosmid vector with multiple cloning sites. Gene 51:275–280.
9. Ditta, G., S. Stanfield, D. Corbin, and D. R. Helinski. 1980. Broad host range DNA cloning system for Gram-negative bacteria: Construction of a gene bank of Rhizobium meliloti. Gene 77:7347–7351.
10. Franklin, F. C. H., and R. Spooner. 1989. Broad-host-range cloning vectors. In C. M. Thomas (ed.), Promiscuous Plasmids of Gram-Negative Bacteria. Academic Press, New York, pp. 247–267.
11. Haas, D. 1983. Genetic aspects of biodegradation by pseudomonads. Experientia 39: 1199–1213.
12. Itoh, Y., and D. Haas. 1985. Cloning vectors derived from the Pseudomonas plasmid pVS1. Gene 36:27–36.
13. Keen, N. T., S. Tamaki, D. Kobayashi, and D. Trollinger. 1988. Improved broad-host-range plasmids for DNA cloning in Gram-negative bacteria. Gene 70:191–197.
14. Leemans, L., J. Langenakens, H. DeGreve, R. Deblaere, M. Van Montagu, and J. Schell. 1982. Broad-host-range cloning vectors derived from the W-plasmid Sa. Gene 19:361–364.
15. McCombie, W. R., J. B. Hansen, G. J. Zylstra, B. Maurer, and R. H. Olsen. 1983. Pseudomonas streptomycin resistance transposon associated with R-plasmid mobilization. J. Bacteriol. 155:40–48.
16. Mermod, N., P. R. Lehrbach, R. H. Don, and K. N. Timmis. 1986. Gene cloning and manipulation in Pseudomonas. In I. C. Gunsalus, J. R. Sokatch, and L. N. Ornston (eds.), The Bacteria, vol. 10. Academic Press, New York, pp. 325–355.
17. Miller, R. V. 1988. Potential for transfer and establishment of engineered genetic sequences. In J. Hodgson and A. M. Sugden (eds.), Planned Release of Genetically Engineered Organisms, Trends in Biotechnology/Trends in Ecology and Evolution special publication. Elsevier, Cambridge, pp. 523–527.
18. Morales, V., M. M. Bagdasarian, and M. Bagdasarian. 1990. Promiscuous plasmids of the IncQ group: Mode of replication and use for gene cloning in Gram-negative bacteria. In S. Silver, A. M. Chakrabarty, B. Iglewski, and S. Kaplan (eds.), Pseudomonas: Biotransformations, Pathogenesis, and Evolving Biotechnology. American Society for Microbiology, Washington, D.C., pp. 229–241.
19. Olsen, R. H., and J. Hansen. 1976. Evolution and utility of a Pseudomonas aeruginosa drug resistance factor. J. Bacteriol. 125:837–844.
20. Olsen, R. H. 1978. Evolution of Pseudomonas R-plasmids: Consequences of Tn1 insertion and resultant partial diploidy to chromosome and Tra⁻ R-plasmid mobilization. J. Bacteriol. 133:210–216.
21. Olsen, R. H., G. DeBusscher, and W. R. McCombie. 1982. Development of broad-host-range vectors and gene banks: Self-cloning of the Pseudomonas aeruginosa PAO chromosome. J. Bacteriol. 150:60–69.
22. Saye, D. J., O. A. Ogunseitan, G. S. Sayler, and R. V. Miller. 1990. Transduction of linked chromosomal genes between Pseudomonas aeruginosa strains during incubation in situ in a freshwater habitat. Appl. Environ. Microbiol. 56:140–145.

23. Schmitt, R., E. Bernhard, and R. Mattes. 1979. Characterization of Tn*1721*, a new transposon containing tetracycline genes capable of amplification. *Mol. Gen. Genet.* 172:53–65.
24. Valentine, C. R. I., and C. I. Kado. 1989. Molecular genetics of IncW plasmids. In C. M. Thomas (ed.), *Promiscuous Plasmids of Gram-negative Bacteria.* Academic Press, New York.
25. Ward, J. M., and J. Grinsted. 1978. Mapping of functions in the R-plasmid R388 by examination of deletion mutants generated in vitro. *Gene* 3:87–95.
26. Ward, J. M., and J. Grinsted. 1982. Physical and genetic analysis of the IncW plasmids R388, Sa, and R7K. *Plasmid* 7:239–250.
27. Yanisch-Perron, C., J. Vieira, and J. Messing. 1985. Improved M13 phage cloning vectors and host strains: Nucleotide sequences of the M13mp18 and pUC19 vectors. *Gene* 33:103–119.
28. Zylstra, G. J. 1987. Ph.D. thesis. University of Michigan.

Chapter 18

Lux and Other Reporter Genes

Clarence I. Kado

Introduction

As dictated by natural selection, microorganisms survive best in an environment in which they can effectively compete against others and intrinsically have protection against environmental stress factors. Thus, bacteria that reside on the foliage of plants are there because they are best and formidably equipped for survival in that particular niche. Likewise, bacteria that reside in soil are likely endowed with protective methods for their survival in that particular environment. Many epiphytic bacteria contain pigments that shield their genome from harmful ultraviolet irradiation of the sun. These bacteria along with other edaphophilic microorganisms may also produce defenses such as bacteriocins, capsular materials, and nucleases against competing bacteria and bacteriophages. They are also well equipped to utilize whatever nutrients that are released by the foliage or in the soil. Thus, for a given environment, whether plant, soil, or animal, bacteria equipped with protective cellular and physiological features that enable them to effectively occupy a given ecological niche should provide the most suitable material for the construction of genetically engineered microorganisms (GEMs) to be reintroduced into that niche. Once introduced, information on the fate of the GEM in the environment is generally needed.

Current techniques in following the fate of GEMs released in the environment rely heavily on antibiotic resistance to serve as a selectable marker. The antibiotic-resistant GEM is usually monitored by analyzing samples taken from the environment and neighboring environments by plating them on media containing the antibiotic. Problems can arise when other microorganisms that are naturally resistant to the antibiotic occupy the same niche, since they can either mask or complicate the accuracy of determining the fate of the GEM. Even when multiple antibiotic resistance is employed, particularly for GEMs released in soil, bacteria naturally resistant to many different antibiotics tend to obscure the analyses. Last, of some concern is the release of a GEM harboring an antibiotic resistance marker into the open environment. Nonetheless, antibiotic resistance markers remain useful in experimental environments because of their ease in use and simplicity in constructing antibiotic-resistant mutants.

To circumvent the problem of the lack of specificity and sensitivity, markers other than antibiotic resistance have been sought. This chapter provides some of the recent developments in molecular genetics that can be applied toward the monitoring GEMs in the environment. One of the most promising markers is the use of bioluminescence in tagging GEMs (Shaw and Kado, 1986, 1987), which enables the continued monitoring of the GEM without disruption of the niche of interest and may be adaptable to aerial surveillance (Shaw et al., 1987). These and other useful reporter genes are presented in this chapter.

Vector Construction

Because of the enormous diversity of GEMs that are likely to be developed in the near future, broad-host-range vectors are desirable for introducing genes that confer marker phenotypes. Such vectors are usually constructed from plasmids that have known broad-host-range properties. These plasmids belong to certain established groups based on plasmid incompatibility (Datta, 1975). The incompatibility groups P (Datta and Hedges, 1972), Q (Barth and Grinter, 1974), and W (Ward and Grinsted, 1982; Tait et al., 1983; Valentine and Kado, 1989), contain plasmids with broad-host-range properties. The following descriptions provide guidelines in the construction of vectors for the delivery of reporter genes to GEMs.

Basic requirements

Vectors can be constructed with relative ease from existing vector systems. However, many of the vectors described in the literature have not proved to be widely useful because of the diverse nature of bacteria in the environment. For successful vector construction, care should be taken in obtaining basic information on the origin of DNA replication (*ori*). This will include gathering information on the number, minimum size, copy control, partitioning, and stability functions often associated with the *ori* region. A particular *ori* needs to be selected when a plasmid with multiple *ori*'s is considered for vector construction. The broad-host-range characteristic of plasmids with more than one *ori* may be merely due to the presence of multiple *ori*'s and not due to a single broadly specific *ori*.

Besides *ori,* considerations should be given to the final size of the vector as well as the mobilization functions and genes involved in maintaining stability of the vector. Most useful vectors are nonconjugative by design and are generally equipped with mobility functions (*mob*) so that they can be efficiently transferred to the GEM recipient by the use of transfer functions encoded by a second plasmid often referred to as a "helper plasmid." Helper plasmids thus confer the transfer function *in trans,* and plasmid transfer operates well in this genetic arrangement in the donor strain.

The broad-host-range vector may be designed to undergo abortive replication once a desired gene set is integrated into the genome of a GEM recipient by generalized recombination. In this case, vectors lacking one or more partitioning genes (*par*) required for equal partitioning of daughter plasmids are generally used. The presence in *cis* of a *par* gene from either homologous or heterologous sources will stabilize the plasmid vector (Meacock and Cohen, 1980; Nordstrom et al., 1980; Gallie et al., 1985). Thus, vectors carrying a reporter gene may be fully stabilized when they carry a *par* gene. Vectors lacking *par* will undergo abortive replication and will be quickly lost by segregation. Such vectors bearing a desired gene flanked by DNA homologous to the recipient chromosome will be maintained long enough to permit insertion of the desired gene into the genome of the GEM by homologous recombination. Although viewed as a means to stabilize a reporter gene, integrated foreign genes are frequently not as stable as those carried extrachromosomally in a stabilized vector.

Broad-host-range vectors

Because GEMs will likely represent diverse types of bacteria, the construction of GEMs containing a reporter gene will depend on vectors that can efficiently deliver the gene. Besides the introduction of reporter genes, the construction of the GEM itself will also be dependent on a cloning vector that can shuttle useful genes from *Escherichia coli* to the GEM. Also, such a vector will be needed to remove existing undesirable genes from the GEM. Thus, vectors' broad-host-range characteristics have proved useful in transferring genetic traits that confer phenotypes for tracking the movement of the GEM.

The desirable properties of broad-host-range gene shuttle vectors have been reviewed (Kado and Tait, 1983; Gallie et al., 1987; Schmidhauser et al., 1987). They include the following:

1. The vector must replicate in diverse organisms to facilitate the isolation and characterization of genes.
2. The vector must be easily recognized by selectable markers.
3. The vector should be sufficiently small in size to accommodate DNA inserts.
4. Cloned genes should be easily detected.
5. Useful quantities of the vector must be easily obtained.
6. The vector must be stable, nonpathogenic, and non-stress-inducing.
7. Vectors must effectively deliver genetic information for stable maintenance in alternate desired recipients.
8. The introduced genetic information should be stably maintained as a new heritable determinant.

Vector mobilization

By design, vectors are preferably mobilized by donation rather than by conduction (Clark and Warren, 1979). Donation involves no physical association between a helper plasmid and the vector, but requires certain genes on the nonconjugative vector, usually a *trans*-acting *mob* protein and a *cis*-acting site. Conduction involves cointegration of the vector and helper plasmid at nonspecific sites through either a *rec*A-dependent homologous recombination or transposition. Transfer by conduction is not favored due to nonspecific rearrangements of the vector and cloned genes, and the frequency of transfer by conduction is less than conjugal transfer by donation.

Helper plasmids can be native conjugative plasmids or their derivatives. Examples of helper plasmids are listed in Table 18.1, together with their frequencies of mobilization. Vector plasmids usually contain *mob* genes that initiate transfer from their *ori*T site and are complemented for ancillary but necessary functions encoded by the conjugative helper plasmid that is specific for its own *ori*T.

If the use of a helper plasmid system is undesirable, there is the alternative of using *E. coli* donor strains that contain the transfer function integrated in their chromosome (Simon et al., 1983). The vector containing the reporter gene can be inserted into the transfer-proficient *E. coli* strain, which can then serve as the effective donor. The frequency of donation appears about the same as that for donors

TABLE 18.1   Mobilization of Nonconjugative Vectors by Helper Plasmids

Helper	Incompatibility group	Frequency of transfer		
		pBR325	pLAFR3	pUCD4
pRK2013	P	0.5	10^{-2}	5×10^{-4}
R46	N	2×10^{-6}	—	10^{-7}
pSa325	F1	10^{-3}	10^{-4}	10^{-4}
pSa	W	5×10^{-2}	—	—
R6K	X	5×10^{-4}	—	10^{-6}

NOTE: Cited plasmids are described elsewhere: pBR325 (Bolivar, 1978); pLAFR3 (Staskawicz et al., 1987); pRK2013 (Ditta et al., 1980); R46 (Hedges, 1972); pSa325 (Zaitlin, 1984; Shaw and Kado, 1987); pSa (Datta and Hedges, 1972); for review, see Valentine and Kado (1989); R6K (Kontomichalou et al., 1970).

carrying a helper plasmid and in some cases is higher when copy numbers of the helper plasmid are high.

Assessing Microbial Expression by Use of Reporter Genes

Although reporter genes are extremely useful in the study of gene regulation, many of them can also be used to track the fate of GEMs released into the environment. Obviously, reporter genes that disclose the unambiguous existence of the GEM are the most desirable. Because of the convenience of the assay system, GEMs equipped with antibiotic resistance markers have been popular. The merits and disadvantages on the use of antibiotic resistance as the reporter phenotype are discussed below. Reporter genes other than antibiotic resistance genes are becoming increasingly desirable.

Antibiotic resistance genes

Most antibiotic resistance genes are derived from R plasmids. They are particularly common in members of the Enterobacteriaceae, such as species of *Escherichia, Salmonella,* and *Shigella,* and in other bacteria such as staphylococci. Nearly all antibacterial antibiotics used in medicine have a corresponding resistance gene harbored on an R plasmid (Table 18.2). As shown, resistances to antibiotics conferred by R plasmids are generally due to antibiotic inactivation by enzymatic modifications. Resistance to these antibiotics can also be conferred by chromosomal genes, many of which confer resistance by preventing the uptake of the antibiotic or by modification of the antibiotic target site.

Resistance genes to certain antibiotics, such as rifampin, nalidixic acid, oxolinic acid, coumeromycin, novobiocin, and nitrofurans, which

TABLE 18.2   Antibiotic Resistance Genes on R Plasmids and Mechanism of
Antibiotic Resistance

Antibacterial antibiotic	Antibiotic mechanism	Resistance mechanism
Aminoglycosides: amikacin, gentamicin, kanamycin, neomycin, streptomycin	Mistranslation	Phosphotransferase, antibiotic modification
Cephalosporins	Cell wall synthesis	β-Lactamase
Chloramphenicol	Mistranslation	Transacetylation
Erythromycin	Mistranslation	Methylation of 23S rRNA
Lincomycin	Mistranslation	Methylation of 23S? rRNA
Penicillins: benzyl penicillin, ampicillin, carbenicillin, methicillin	Cell wall synthesis	β-Lactamase hydrolysis of β-lactam ring
Sulfonamides	Dihydropteroate synthetase competitor	Resistant form of dihydropteroate synthetase
Tetracycline	Mistranslation	Uptake inhibition
Trimethoprim	Dihydrofolate reductase inhibitor	Resistant form of dihydrofolate reductase

all affect nucleic acid metabolism in one way or another, i.e., polymerases and topoisomerases, have thus far only been found to be conferred by chromosomal genes. If such resistance genes are desired as a marker, the chromosomal resistance gene would have to be isolated and incorporated into the vector (Balganesh and Setlow, 1985).

R plasmids often contain genes for antibiotic resistance to more than one antibacterial drug. Genetic mapping studies have shown that these antibiotic resistance genes are clustered, and restriction enzyme sites are also heavily clustered among these genes (Tait et al., 1982; Ward and Grinsted, 1982; Valentine and Kado, 1989). The reason for the high density of restriction sites among antibiotic resistance genes is not known but may indicate a recent evolutionary accumulation of these genes on promiscuous plasmids. Whatever the explanation may be, these genes and the high frequency of differing restriction sites serve as excellent tools for the analysis of gene expression. In the case of GEMs, the antibiotic resistance genes are useful as reporters, providing that the background of naturally resistant strains is sufficiently low to provide confidence in the accurate assessment of the GEM population in the environment. Because of the ease of constructing spontaneous antibiotic-resistant mutants, many investigators have resorted to using these mutants. However, since very few workers have investigated the nature of the resistance itself, a common assumption is made that such spontaneous mutants will remain unaffected after they are released into the environment. Hence, uncharacterized antibiotic resistance mutants should be employed judiciously. Antibiotic genes contained on a number of well-characterized R plasmids and derivatives are excellent sources of reporter genes. Their mechanism of antibiotic resistance is known.

Chloramphenicol acetyltransferase. There are three types of chloramphenicol acetyltransferases (CAT). R plasmids of the W incompatibility group contain *cat* genes that encode type II CAT, which are immunologically and electrophoretically distinct from types I and III CATs. The type I *cat* genes are harbored by transposable elements such as Tn*9* (Alton and Vapnek, 1979), and type III enzymes are contained in a number of R plasmids such as pACYC177 (Chang and Cohen, 1978) outside of the IncW group. The type I *cat* gene has been used in promoter-probe vectors and in gene cartridges or cassettes, i.e., DNA fragments containing the *cat* gene flanked by linker sequences for easy cloning (Table 18.3).

Aminoglycoside phosphotransferase. Members of the aminoglycoside family of antibiotics such as streptomycin, neomycin, gentamicin, and kanamycin are inactivated by *O*-phosphorylation catalyzed by a phosphotransferase. Antibiotics of this family can also be inactivated by *N*-adenylation (Kawabe et al., 1979) and *O*-nucleotidylation. The most commonly used aminoglycoside antibiotic resistance genes are those derived from R plasmids and certain transposable elements such as Tn*5*. This latter transposable element encodes aminoglycoside phosphotransferase II, inhibiting the action of neomycin and kanamycin. Such genes are usually dissected from the R plasmid or transposon and fitted with linkers to make an antibiotic resistance

TABLE 18.3   Use of Type I Chloramphenicol Acetyltransferase as a Reporter of Gene Activity

Plasmid	Purpose	Reference
pCM1,4,7	Cartridge	Close and Rodriguez, 1982
pUCD206	Promoter selection	Hagiya et al., 1985
pUCD206B	Promoter selection	Close et al., 1985
pKK232-8	Promoter selection	Brosius and Lupski, 1987
pPUC29	Cosmid	Tandeau de Marsac et al., 1982
pDS1	Expression	Stueber and Bujard, 1982
pHV33	Deletion	Primrose and Ehrlich, 1981
pBW18	Induction	Byeon and Weisblum, 1985
pJD2000	Cloning	Hardesty et al., 1987
pBB2,3,5 & 6	Cloning	Bron et al., 1987
pSa4	Broad-host cloning	Tait et al., 1983
CTN	Promoterless cassette	Hirooka and Kado, 1986
pPR328	Cloning	Quigley and Reeves, 1987

cartridge (or cassette) (Table 18.3) that can be implanted in a desirable location of the chromosome of a particular GEM.

β-Lactamase. The principal β-lactam antibiotic is penicillin and its derivatives such as ampicillin, carbenicillin, and relatives such as the cephalosporin group. These penicillins are inactivated by β-lactamases (penicillinases) produced by a number of gram-negative and gram-positive bacteria. These enzymes are encoded by either the chromosome or plasmid. TEM β-lactamase [originally analyzed from plasmid R-TEM (now called R6K) thus the name TEM for a class of this enzyme] are encoded by pBR322 and related plasmids as a precursor molecule of 30,000 daltons (Achtman et al., 1979), which is excreted across the inner membrane into the periplasmic space, where it is processed into the active protein of 28,000 daltons (Dougan and Sherratt, 1977). The enzyme catalyzes the hydrolysis of the lactam ring of the penicillin. The TEM species of β-lactamase are encoded by transposons such as Tn3 harbored in gram-negative bacteria. In contrast to gram-positive sources of β-lactamase, which need to be induced for full expression, TEM lactamases are constitutively expressed.

It should be emphasized that the production of β-lactamase is not the only means of resistance to penicillins, since alterations in uptake, target proteins, and other forms of modification may play a role in conferring resistance. Noteworthy also is the fact that many bacteria, particularly members of the Pseudomonadaceae are naturally highly resistant to penicillins. β-lactamase genes are widely distributed among pseudomonads (Jacoby and Matthew, 1979). Thus, GEMs equipped with β-lactamase genes as reporters may be easily masked by such groups of penicillin refractory bacteria.

Other reporter genes

β-Galactosidase. The most frequently used reporter gene is the *lacZ* gene of *E. coli*. The *lacZ* gene of the lactose operon encodes β-galactosidase, an enzyme that has already proved extremely useful in molecular biology because of the ease by which its production can be detected on agar plates containing the indicator 3-chloro-4-bromo-indolyl β-galactoside (X-gal), which is hydrolyzed to give a visible blue color. GEMs endowed with a constitutive β-galactosidase gene can thus be detected on such agar by the blue colony color. This reporter gene has been recently employed in monitoring *Pseudomonas fluorescens* (Barry et al., 1986; Drahos et al., 1986; Drahos and Barry, Chap. 8). In this case, the *lacZY* genes were inserted between the termini of a transposase defective transposon of Tn7 and introduced into the chromosome of *P. fluorescens* by means of a helper vector system (Grinter, 1983).

β-Galactosidase is not ideal for all systems. A large number of biological systems bear endogenous β-galactosidase levels that make it difficult or nearly impossible to detect by current enzymatic methods. Systematic examination for β-gal activity in bacteria of the environment of interest will need to be made. Since bacterial populations are qualitatively different at different locations, these studies are imperative. Owing to their nonfermentative physiological mode, fluorescent *Pseudomonas* species are usually free of β-gal activity. Thus, the large number of soil pseudomonads that were tested were gal-negative (Drahos et al., 1986).

β-Glucuronidase.   β-Glucuronidase (GUS) is encoded by the *uidA* locus of *E. coli*. This enzyme catalyzes the cleavage of a wide variety of β-glucuronides, many of which are water-soluble and commercially available. The *uidA* structural gene has been isolated, sequenced, and used as a reporter gene (Jefferson et al., 1986). *E. coli* GUS is a very stable enzyme, with a broad pH optimum (pH 5.0 to 7.5), and is resistant to thermal inactivation at 55°C (half-life of 2 h). Unlike β-galactosidase, GUS has a monomer molecular weight of 68,200 daltons and is tolerant to many detergents and widely varying ionic conditions. It has no cofactors but is inhibited by copper and zinc ions, and therefore EDTA is required in the assay. Since many organisms do not have GUS activity, the *uidA* gene can nicely be used as a reporter gene by its incorporation into the chromosome of the GEM of interest.

Levansucrase.   Levansucrase is encoded by the *sacB* gene of *Bacillus subtilis* Marburg and is secreted from the bacterial cell into the culture medium after induction with sucrose. The enzyme catalyzes the transfer of fructosyl residues from donors such as sucrose to different acceptors such as water, alcohols, and sugars. This occurs by the hydrolysis of the donor and the formation of branched polymers of fructose known as levan. We have discovered that the production of levansucrase in many different gram-negative bacteria such as species of *Agrobacterium, Erwinia, Salmonella, Escherichia, Serratia, Shigella, Rhizobium, Pseudomonas,* and *Xanthomonas* is lethal in the presence of sucrose in agar medium. The *sacB* gene therefore has proved useful in the positive selection of insertion (IS) and transposable (Tn) elements (Gay et al., 1985). Gram-negative GEMs harboring *sacB* will be killed whenever they are exposed to sucrose. Thus, *sacB* serves as a suicide gene and an easily screenable phenotype.

Bioluminescence (lux), a novel reporter system.   The visible detection of GEMs as they disseminate in the environment would be the ideal assay system. This concept has already been put to use in our laboratory by using a constitutively expressed luciferase operon originating from *Vibrio fischeri* (Shaw and Kado, 1986, 1987; Shaw et al., 1985, 1987).

Bioluminescence occurs in nematodes, mollusks, insects, fish, diatoms, jellyfish, fungi, and bacteria. Marine bacteria such as members of the genus *Vibrio* are often bioluminescent, and the luciferases of *V. fischeri* and *V. harveyi* have been studied extensively (DeLuca, 1978; Nealson and Hastings, 1979; Van Dyke, 1985; Meighen, 1991). The *Vibrio* luciferase is composed of an α and β subunit of 40,000 and 38,000 daltons, respectively, which are encoded by two genes of a five-gene operon. The remaining three genes encode a fatty acid transacetylase and reductases (Engebrecht et al., 1983, 1984; Wall et al., 1984). The native luciferase operon is under the control of a complex autoregulation system involving two regulatory genes R and I. Constitutive expression is effected by the replacement of this regulatory unit with a constitutive promoter (Fig. 18.1). Aside from the regulatory unit, the remaining five genes are required for light production, which comprise two essential reactions:

(1) $RCHO + FMNH_2 + O_2 \rightarrow RCOOH + FMN + AMP + light$

(2) $RCOOH + NADPH_2 + ATP \rightarrow RCHO + NADP + AMP + PPi$

Reaction 1 is needed for the production of light, while reaction 2 is required for the recycling of fatty acid, which is the substrate for reaction 1. *N*-tetradecyl aldehyde, which is found naturally, may serve as this substrate. The fatty acid reductase reduces the fatty acid product of reaction 1, regenerating aldehyde for reuse (Wall et al., 1984). Thus, actively metabolizing bacteria will continuously produce light, while dead bacteria will not. The amount of light produced can vary between

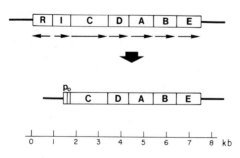

Figure 18.1   A derivative of the lux operon with the genes involved in autoregulation (R and I) removed and the sequences immediately upstream from gene C encoding fatty acid reductase, including the Shine-Dalgarno sequence, removed. This section was replaced with a constitutive promoter P and the Ω translational enhancer. Genes D and E encode a transferase (esterase) and acyl-protein synthetase, respectively. The α and β subunits of luciferase are encoded by genes A and B. The horizontal arrows indicate the direction of transcription.

strains of a given bacterium (Shaw and Kado, 1986), and therefore the brightest light-producing strains should be selected. The R and I regulatory genes of the luciferase gene set (*lux*) have been removed, and novel promoters were set in their place so that luciferase is expressed constitutively under the control of a constitutive promoter such as that derived from the tetracycline resistance gene. The constitutively expressed luciferase gene set has been placed in the broad-host-range vector pUCD5 (Close et al., 1984) to yield the luciferase plasmid pUCD607 (Shaw and Kado, 1986). pUCD607 has been introduced into a large number of different bacteria, and each has produced light at varying intensities (Table 18.4).

TABLE 18.4 Bacteria Made Bioluminescent by the Introduction of pUCD607 Containing the Constitutive Lux Gene Set

Agrobacterium radiobacter
Agrobacterium rhizogenes
Agrobacterium rubi
Agrobacterium tumefaciens
Azospirillum sp.*
Erwinia amylovora
Erwinia carotovora subsp. *carotovora*
Erwinia carotovora subsp. *atroseptica*
Erwinia chrysanthemi
Erwinia rubrifaciens
Erwinia herbicola
Erwinia nigrifluens
Erwinia salicis
Erwinia stewartii
Enterobacter agglomerans
Escherichia coli
Escherichia vulneris
Proteus vulgaris
Providentia stuartii
Pseudomonas fluorescens
Pseudomonas solanacearum
Pseudomonas syringae pv. *glumae*
Pseudomonas syringae pv. *glycinea*
Pseudomonas syringae pv. *phaseolicola*
Pseudomonas syringae pv. *savastanoi*
Pseudomonas syringae pv. *syringae*
Rhizobium meliloti
Rhizobium leguminosarum
Rhizobium fredii
Salmonella typhimurium
Serratia marcescens
Shigella flexneri
Xanthomonas campestris pv. *campestris*
Xanthomonas campestris pv. *oryzae*
Xanthomonas campestris pv. *phaseoli*
Xanthomonas campestris pv. *transluscens*
Zymomonas mobilis

*From R. K. Prakash, NPI, Inc., Salt Lake City, Utah, personal communication.

Variation in bioluminescence of bacteria harboring pUCD607 seems to be strain- and isolate-dependent. For instance, *E. coli* DH1 (pUCD607) is brighter than *E. coli* HB101 (pUCD607). Both strains contain pUCD607 and yet emit different levels of light. Such differences are even stronger at the species level. It appears that the physiology of the bacteria plays an important role in determining the amount of light produced. Cells in the stationary phase emit much less light per cell than those in the exponential stage of growth. The age, size, and density of cells on agar plates will affect luminescence levels. Plasmid copy number, promoter strength, and growth media will affect light production. Part of the reason for the physiological effects on bioluminescence may be due to energetic constraints where light production is dependent on a sufficient supply of reduction equivalents and ATP (DeLuca, 1978; Nealson and Hastings, 1979).

Optimization of Gene Expression

The efficient expression of a reporter gene is dependent on the transcription and translational start sequences. These sequences, which are normally upstream of the reporter gene, may function well in the organism from which the gene was derived; however, for a variety of reasons (DeBoer et al., 1982), these regulatory sequences may not operate well in a heterologous background. The available promoters on various expression vectors, which are now commercially available, have primarily been designed to operate in *E. coli*. Therefore, each reporter gene equipped with either its own promoter or a promoter from a heterologous source will need to be tested in the GEM of interest. Variations in reporter gene expression may be due to (1) weak promoter activity; (2) premature transcriptional termination; (3) inefficient translation; (4) reporter gene product instability or toxicity. One means of circumventing the problem of promoter efficiency is by the use of an endogenous promoter. Since many R-factor plasmids confer high levels of antibiotic resistance to GEMs, the promoter of an antibiotic resistance gene may suffice. Such an engineered reporter gene would still need testing since there are the other stated factors for inefficient expression.

Use of superpromoters

Superpromoters are promoters that enhance the expression of genes at maximal strength, where RNA chain initiation frequency is limited only by the rate of RNA chain elongation (DeBoer et al., 1982). These promoters are engineered either by using parts of existing promoters or by the construction of synthetic promoters. For example, the *lac*

Pribnow box and the *lac* operator joined to the −35 region of the *trp* promoter make up a superpromoter known as *tac* that is highly functional in *E. coli* at about the same level as a fully derepressed *trp* promoter (DeBoer et al., 1982). The *lac* promoter has a consensus Pribnow box sequence (TATAATG) (Pribnow, 1975) but no consensus −35 sequence, while the *trp* promoter has no consensus Pribnow box sequence, but it has a consensus −35 (TTGACA). Thus, the *tac* promoter contains both consensus sequences, the consequence of which elevates promoter efficiency.

Special care should be taken to assess hybrid promoters in the GEM relative to an endogenous promoter. Spacing between the Pribnow box and the −35 sequence is known to affect promoter strength (Jaurin et al., 1981). A given spacing that functions well in *E. coli* may not do so in the GEM since differences in RNA polymerase affinities relative to spacing constraints may differ.

Use of the Ω translation enhancer

The Ω enhancer sequence of tobacco mosaic virus has been shown to stimulate translational activity in eukaryotic and prokaryotic backgrounds (Gallie et al., 1987). Ω enhances translation efficiently even in the absence of a Shine-Dalgarno (S-D) sequence, and therefore promoter hybrid constructs can use Ω in lieu of S-D region (Gallie and Kado, 1989). However, Ω together with an S-D sequence provide optional translation efficiency. Translational enhancement with Ω is multifold (8×), a feature desirable for optimizing the detection of reporter gene activity. Care should be taken, however, to ensure that the protein product of the reporter gene does not poison the cell.

Methods and Applications

Antibiotic resistance

Construction of spontaneous antibiotic-resistant mutants. Spontaneously derived antibiotic-resistant mutants are the easiest to construct. A turbid culture of GEM cells is harvested by centrifugation, and the resulting pellet of cells is resuspended in <0.5 mL of buffer. Then 0.2 mL of the cell suspension is plated onto an appropriate agar culture medium supplemented with 5 μg of antibiotic per milliliter. Colonies that appear are suspended in buffer as above and plated onto medium containing 20 μg of the same antibiotic. This procedure is repeated several times with graded, increasing doses of antibiotic in the agar medium. This stepwise procedure will eventually produce GEM colonies resistant to high levels, e.g., 500 to 1000 μg of antibiotic per milliliter. Care should be taken to eliminate antibiotic-uptake-defective mutants since these will turn out to have high reversion frequencies.

Construction of antibiotic resistance gene cassettes.   The best sources of genes conferring antibiotic resistance, like those described in Antibiotic Resistance Genes in a previous section, are from well-characterized R plasmids and transposons. These elements have defined restriction maps and may contain unique restriction sites that flank a desirable antibiotic resistance gene. It is wise to test the activity of the gene first in the GEM before isolations are performed. The gene is isolated easily by electrophoretic fractionation of restricted fragments in agarose gels stained with 0.05 μg/mL ethidium bromide visualized by ultraviolet light. Microgel apparatus available commercially are not suitable for fragment isolation. The horizontal electrophoresis apparatus of McDonnell et al. (1977) is excellent for this purpose since gels that are 0.5 to 1.0 cm thick can be used. We have modified the original apparatus by adding a channeled Plexiglas section beneath the section where the gel is laid (Fig. 18.2). The desired DNA fragment is isolated by slicing out that portion of the gel containing it, which is then placed in a small dialysis bag prefilled with electrophoresis buffer (TE

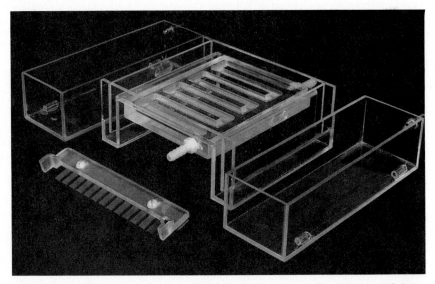

Figure 18.2 Horizontal water-cooled electrophoresis apparatus for isolation and detection of DNA fragments and cryptic plasmids. Melted buffered agarose is poured in the legs of the apparatus (center) and allowed to solidify. Additional agarose is then poured on the water-cooled plate to the desired thickness. The comb as shown is placed into the unsolidified agarose at one end of the apparatus (at the cathode side). The legs of the apparatus are placed in the rectangular reservoirs containing TE buffer (40 mM Tris-acetate, pH 7.9, 2 mM EDTA) and fitted with platinum electrodes. During electrophoresis at 120 to 140 V, the agarose gel is cooled by tap water that is circulated through the bottom of the plate. After electrophoresis, the gel is easily removed by cutting the two edges of the gel and inverting the apparatus over 300 mL of tap water containing 0.05 μg ethidium bromide per milliliter. The apparatus is a modification of one described by McDonnell et al. (1977).

buffer = 40 mM Tris-acetate [Trizma base (Sigma)] titrated with acetic acid to pH 7.9, 2 mM EDTA). Excess buffer is removed and the dialysis bag is tied carefully, avoiding air bubbles. The bag is immersed in TE buffer in a small plastic box containing platinum electrodes fixed at the bottom of opposite walls. The DNA is electroeluted by using 120 V for 1 to 2 h, after which time the polarity of the electric current is reversed for 2 to 3 min to release the DNA from the wall of the dialysis bag. The buffer surrounding the gel slice is recovered with a pipette, and the bag is rinsed with additional buffer. The pooled buffer is extracted twice with phenol (distilled)-chloroform (1:1, v/v) and twice with water-saturated diethyl ether. The DNA is precipitated by adding 20 μL of 3 M NH$_4$ acetate and two volumes of 95% ethanol. The precipitate is recovered by centrifugation in a microfuge and dried *in vacuo*.

The DNA fragment containing the antibiotic resistance gene must be flanked with chromosomal DNA of the GEM. This hybrid DNA is then equipped with linkers for a selected restriction site of a plasmid vehicle used to harbor the gene and to deliver it to the GEM, in which the gene is integrated into the chromosome by homologous recombination. The other means of incorporating the antibiotic resistance gene is by the use of a transposase defective transposon into which the gene has been inserted. This latter method requires a transposase gene on a helper plasmid as described earlier.

An alternative approach in the incorporation of antibiotic resistance genes into GEMs is simply to use an endogenous plasmid naturally present in the GEM. Such plasmids must be tested for self-transfer activity. Those that are nonconjugative are the most desirable. Conjugative plasmids would require modification by deletion of the transfer genes. Native plasmids are usually extremely stable and afford convenient locations for the insertion of a reporter gene.

The following plasmid miniscreening protocol (Kado and Liu, 1981) is useful in screening for plasmids in GEMs:

1. Cells are grown in 3 to 5 mL of L-broth in a screw cap culture tube (Pyrex 9826), overnight at optimum growth temperature.

2. The cells are collected by centrifugation of the same culture tube at 3600 × g, 10 min. The supernatant is discarded, and the tube is briefly drained on a paper towel.

3. The cell pellet is thoroughly resuspended in 1.0 mL of E buffer [40 mM Trizma base (Sigma) titrated with acetic acid to pH 7.9, 2 mM EDTA]. For gram-positive bacteria, lysozyme (50 to 100 μg/mL) is added and incubated for 20 min to facilitate lysis in the next step.

4. The cells are lysed by adding 2 mL of lysing solution (3% Na dodecylsulfate, 50 mM Trizma base titrated with NaOH to pH 12.6).

Immediately following adding the lysis solution, the mixture is shaken by hand for no more than 10 s.

5. The tube containing the mixture is placed in a water bath set at 65°C for 20 to 60 min, depending on the GEM. GEMs with higher GC contents should be incubated for the longer time.

6. Two volumes of phenol-chloroform mixture (1:1, v/v) are added to the mixture, and the solution is emulsified by shaking briefly by hand (no more than 10 s). The mixture is then centrifuged at 3600 rpm for 10 min in the same culture tube. Care should be taken to use phenol that has been distilled in an all-glass apparatus. Distilled phenol is stored in the freezer in 250-mL samples in polyethylene bottles (do not use nylon bottles since phenol will dissolve it).

7. The upper aqueous phase is carefully collected with a plastic pipette having a large orifice. All of the aqueous phase should not be collected since the lower one-quarter may contain debris. The aqueous phase may be stored at 4°C for long periods without appreciable degradation of the plasmid.

8. On a parafilm sheet, 50 μL of the aqueous phase is mixed with 10 μL of dye solution (0.25% bromocresol purple in 50 mM Tris-acetate, pH 8.0, 50% glycerol). This mixture is placed into a well of 0.7% agarose gel prepared for electrophoresis in E buffer (see above for description of apparatus). Electrophoresis is performed for 2 to 4 h at 12 V/cm. After electrophoresis the gel is stained for 20 min in 300 mL of tap water containing 0.05 μg/mL ethidium bromide, and viewed with ultraviolet light. The plasmid often can be visualized better in the photograph of the gel.

β-Galactosidase. In lieu of antibiotic resistance genes, the *lacZ* gene of the lactose operon of *E. coli* can be used. The transfer of *lacZ* is similar to that for antibiotic resistance genes. However, instead of positive selection of GEMs equipped with antibiotic resistance genes, *lacZ* requires screening for blue colonies on agar medium containing X-gal. The use of *lacZ* is described in detail by Drahos and Barry in Chap. 8.

β-Glucuronidase. GEM containing the GUS cassette, preferably under the control of a GEM promoter, may be plated onto medium containing 50 μg/mL 5-bromo-4-chloro-3-indolyl β-D-glucoronide (X-glu) (Research Organics, Cleveland, Ohio) (Jefferson et al., 1986). Blue colonies indicate GUS activity.

Luciferase (lux). Some vectors equipped with LUX are listed in Table 18.5. The lux cassette is comprised of five genes as described above and is free of sites for restriction enzyme *AccI*, *Bam*HI, *ClaI*, *Eco*RI, *KpnI*, *PstI*, *SacII*, *SalI*, and *SmaI*. Transfer of these vectors is mediated by a helper plasmid such as pRK2013 (see Table 18.1). Light can be detected in several ways—in order of increasing sensitivity—vision, 35-mm photographic film, x-ray film, and photoelectrically. Visual detection in a dark room is possible when the GEM containing pUCD607 produces relatively high light levels. Light may be produced either as colonies on agar medium or in culture flasks. Differences in the relative intensity of light produced by GEM colonies is expected, and those that produce the greatest amount of light can be selected. These "bright GEMs" can then be used in microcosm and field experiments. If visual detection is borderline, either 35-mm and/or x-ray film may be used. Photographs may be taken using either color or black-and-white film. Elapsed time exposures help, but reciprocity failure, i.e., variations in quality control for scientific photography, is common with conventional film. Instead, films hypersensitized with N_2 and H_2 (ASA 1000) used by astronomers will detect extremely faint light and withstand prolonged exposure. Such film can be purchased from Lumicon in Livermore, California. Photographs should be taken in a vibration-free area for obvious reasons. Cameras with mechanical shutters are recommended since batteries quickly deplete with prolonged exposures using cameras with electronic shutters. X-ray film is used to help in the selection of the brightest colonies on agar medium. The film is simply taped to the bottom of the petri plate, and the plate and film are wrapped in foil. Where many colonies appear on the medium, x-ray film wrapped in plastic (Saran Wrap) is laid directly on the colonies in order to improve resolution. Exposure times can be in minutes to several hours. Upon development of the film, exposed areas are compared to the distribution of colonies on the plate. Such methods are recommended for identifying GEMs taken from microcosm and field studies. This will enable one to quantify the number of GEMs recovered relative to the total number of bacteria that were screened.

TABLE 18.5   Lux Vectors

Vector	Properties	Selectable markers	Lux cassette	Reference
pUCD607	Broad-host-range	Ap, Km, Sp	Constitutive	Shaw and Kado, 1986
pUCD615	Broad-host-range	Ap, Km	Promoterless	Rogowsky et al., 1987
pUCD623	Transposon	Ap, Cm	Tc, Promoterless	Shaw et al., 1988

Ap = ampicillin, Cm = chloramphenicol, Km = kanamycin.

A more sophisticated approach entails the use of an X-Y scanner equipped with a photomultiplier-amplifier system that is connected to an electronic counter-translator. Such equipment (e.g., Hamamatsu Phototonics K. K.) is a two-dimensional photon counter that records images with very low light levels. Once an image is accumulated and stored in the random-access.frame memory, it can be displayed on a monitor and provide digital data translated into photons. Less expensive instruments can be developed using an end-on photocell attached to a counting device (e.g., luminometer, Beckman Instruments). Light readings can be taken directly from petri plates and flasks. In the latter culture vessel, nephelo flasks are desirable since the side arm of the flask is convenient for insertion in the photometer in a light-free box. Care should be taken to determine the linearity between light emitted by the GEM and light detected by the photomultiplier, especially at high light levels where the amount of light produced by the GEM is often underestimated. In such cases, neutral density filters (Kodak, wratten filters) are used to reduce incoming light. Such filters come in transmittance steps of 10 (0.01 percent, 0.10 percent, 1.00 percent, and 10 percent). The reduction of light by these filters is linear, and therefore the true value of light emitted is multiplied by the value of reduction by the filter. It is best to use filters rather than to dilute suspended bacterial cultures since dilution usually invigorates the bacteria to produce additional light and thus distort the true light value of the original culture.

Summary and Conclusions

Various reporter genes have been presented, and the advantages and disadvantages on the use of these reporters have been described. No reporter gene may be generally applicable since each has not been extensively tested in GEMs that are not related to bacteria used conventionally in research. Some common problems are reporter gene instability, expression, and background noise. Overall, the use of the LUX as the reporter is attractive since no special media, preparation of cells, or expensive chemicals (e.g., X-gal, X-glu) are required. In addition, bioluminescence, which can be quantified easily and reproducibly, indicates living cells (since dead cells do not produce light), and there is the potential for tracking them by photographic equipment (e.g., aerial surveillance).

Acknowledgments

The author is indebted to members of the Davis Crown Gall Group for their efforts on the construction of vectors and reporter genes. The studies summarized in this chapter were supported by NIH grants CA-11526 and GM-45550 from the National Cancer Institute and the Institutes of General Medicine, DHHS, and grants from the Competitive Research Grants Office, USDA.

References

Achtman, M., P. A. Manning, C. Edelbluth, and P. Herrlich. 1979. Export without proteolytic processing of inner and outer membrane proteins encoded by F sex factor *tra* cistrons in *Escherichia coli* minicells. *Proc. Natl. Acad. Sci. USA* 76:4837–4841.

Alton, N. K., and D. Vapnek. 1979. Nucleotide sequence analysis of the chloramphenicol resistance transposon Tn9. *Nature (London)* 282:864–869.

Balganesh, M., and J. K. Setlow. 1985. Effect of chromosome homology of plasmid transformation and plasmid conjugal transfer in *Haemophilus influenzae*. In D. R. Helinski, S. N. Cohen, D. B. Clewell, D. A. Jackson, and A. Hollaender (eds.), *Plasmids in Bacteria*. Plenum Press, New York/London, pp. 571–584.

Barry, G. F. 1986. Permanent insertion of foreign genes into the chromosomes of soil bacteria. *Bio/Technology* 4:446–449.

Barth, P. T., and N. J. Grinter. 1974. Comparison of the deoxyribonucleic acid molecular weights and homologies of plasmids conferring linked resistance to streptomycin and sulfanomides. *J. Bacteriol.* 120:618–630.

Bolivar, P. 1978. Molecular cloning vectors: Derivatives of plasmid pBR322. In H. W. Boyer and S. Nicosia (eds.), *Genetic Engineering*. Elsevier/North-Holland Biomedical Press, Amsterdam, pp. 59–63.

Bron, S., P. Bosma, M. van Belkum, and E. Luxen. 1987. Stability function in the *Bacillus subtilis* plasmid pTA1060. *Plasmid* 18:8–15.

Brosius, J., and J. R. Lupski. 1987. Plasmids for the selection and analysis of prokaryotic promoters. *Methods Enzymol.* 153D:54–68.

Byeon, W.-H., and B. Weisblum. 1985. Post-transcriptional regulation of chloramphenicol acety transferase. In D. R. Helinski, S. N. Cohen, D. B. Clewell, D. A. Jackson, and A. Hollaender (eds.), *Plasmids in Bacteria*. Plenum Press, New York/London, pp. 823–834.

Chang, A. C. Y., and S. N. Cohen. 1978. Construction and characterization of amplifiable multicopy DNA cloning vehicles derived from the p15A cryptic megaplasmid. *J. Bacteriol.* 134:1141–1156.

Clark, A. J., and G. J. Warren. 1979. Conjugal transmission of plasmids. *Annu. Rev. Genet.* 13:99–125.

Close, T. J., and R. L. Rodriguez. 1982. Construction and characterization of the chloramphenicol-resistance gene cartridge: A new approach to the transcriptional mapping of extrachromosomal elements. *Gene* 20:305–316.

Close, T. J., D. Zaitlin, and C. I. Kado. 1984. Design and development of amplifiable broad-host-range cloning vectors: Analysis of the *vir* region of *Agrobacterium tumefaciens* plasmid pTiC58. *Plasmid* 12:111–118.

Close, T. J., R. C. Tait, and C. I. Kado 1985. Regulation of Ti plasmid virulence genes by a chromosomal locus of *Agrobacterium tumefaciens*. *J. Bacteriol.* 164:774–781.

Datta, N. 1975. Epidemiology and classification of plasmids. In D. Schlessinger (ed.), *Microbiology—1974*. American Society for Microbiology, Washington, D.C., pp. 9–15.

Datta, N., and R. W. Hedges. 1972. Host ranges of R-factors. *J. Gen. Microbiol.* 70:453–460.

DeBoer, H. A., L. J. Comstock, D. G. Yansura, and H. L. Heyneker. 1982. Construction of a tandem *trp-lac* promoter and a hybrid *trp-lac* promoter for efficient and controlled expression of the human growth hormone gene in *Escherichia coli*. In R. L. Rodriguez and M. J. Chamberlain (eds.), *Promoters, Structure and Function*. Praeger Publishers, New York, pp. 462–481.

DeLuca, M. A. (ed.). 1978. Bioluminescence and chemiluminescence. *Methods Enzymol.* 57:125–226.

Ditta, G., S. Stanfield, D. Corbin, and D. R. Helinski. 1980. Broad host range DNA cloning system for gram-negative bacteria: Construction of a gene bank of *Rhizobium meliloti*. *Proc. Natl. Acad. Sci. USA* 77:7347–7351.

Dougan, G., and D. J. Sherratt. 1977. The transposon Tn1 as a probe for studying ColE1 structure and function. *Mol. Gen. Genet.* 151:151–160.

Drahos, D. J., B. C. Hemming, and S. McPherson. 1986. Tracking recombinant organisms in the environment: β-Galactosidase as a selective non-antibiotic marker for fluorescent pseudomonads. *Bio/Technology* 4:439–444.

Engebrecht, J., K. Nealson, and M. Silverman. 1983. Bacterial bioluminescence: Isolation and genetic analysis of functions from *Vibrio fischeri*. *Cell* 32:773–781.

Engebrecht, J., and M. Silverman. 1984. Identification of genes and gene products necessary for bacterial bioluminescence. *Proc. Natl. Acad. Sci. USA* 81:4154–4158.

Engebrecht, J., M. Simon, and M. Silverman. 1985. Measuring gene expression with light. *Science* 227:1345–1347.

Gallie, D. R., S. Novak, and C. I. Kado. 1985. Novel high- and low-copy stable cosmids for use in *Agrobacterium* and *Rhizobium*. *Plasmid* 14:171–175.

Gallie, D. R., P. Gay, and C. I Kado. 1987. Specialized vectors for members of the Rhizobiaceae and other gram-negative bacteria. In R. L. Rodriguez and D. T. Denhardt (eds.), *Vectors: A Survey of Molecular Cloning Vectors and Their Uses*. Butterworth Publishing, Stoneham, Mass., pp. 333–342.

Gallie, D. R., D. E. Sleat, J. W. Watts, P. C. Turner, and T. M. A. Wilson. 1987. A comparison of eukaryotic viral 5'-leader sequences as enhancers of mRNA expression in vivo. *Nucl. Acid. Res.* 15:8693–8711.

Gallie, D. R., and C. I. Kado. 1989. A translational enhancer derived from tobacco mosaic virus is functionally equivalent to a Shine-Dalgarno sequence. *Proc. Natl. Acad. Sci. USA* 86:129–132.

Gay, P., D. Le Coq, M. Steinmetz, T. Berkelman, and C. I. Kado. 1985. Positive selection procedure for entrapment of insertion sequence elements in gram-negative bacteria. *J. Bacteriol.* 164:918–921.

Grinter, N. J. 1983. A broad-host-range cloning vector transposable to various replicons. *Gene* 21:133–143.

Hagiya, M., T. J. Close, R. C. Tait, and C. I. Kado. 1985. Identification of pTiC58 plasmid-encoded proteins fro virulence in *Agrobacterium tumefaciens*. *Proc. Natl. Acad. Sci. USA* 82:2669—2673.

Hardesty, C., G. Colon, C. Ferran, and J. M. Dirienzo. 1987. Deletion analysis of sucrose metabolic genes from a *Salmonella* plasmid cloned in *Escherichia coli* K12. *Plasmid* 18:142–155.

Hedges, R. W. 1972. Resistance to spectinomycin determined by R factors of various compatibility groups. *J. Gen. Microbiol.* 72:407–409.

Hirooka, T., and C. I. Kado. 1986. Location of the right boundary of the virulence region on *Agrobacterium tumefaciens* plasmid pTiC58 and a host-specifying gene next to the boundary. *J. Bacteriol.* 168:237–243.

Jacoby, G. A., and M. Matthew. 1979. The distribution of β-lactamase genes on plasmids found in *Pseudomonas*. *Plasmid* 2:41–47.

Jaurin, B., T. Grundström, T-Edlund, and S. Normark. 1981. The *E. coli* β-lactamase attenuator mediates growth rate-dependent regulation. *Nature* 290:221–225.

Jefferson, R. A., S. M. Burgess, and D. Hirsh. 1986. β-glucuronidase from *Escherichia coli* as a gene-fusion marker. *Proc. Natl. Acad. Sci. USA* 83:8447–8451.

Kado, C. I., and S. T. Liu. 1981. Rapid procedure for detection and isolation of large and small plasmids. *J. Bacteriol.* 145:1365–1373.

Kado, C. I., and R. C. Tait. 1983. Bacterial-plant gene cloning shuttle vectors for genetic modification of plants. In P. F. Lurquin and A. Kleinhofs (eds.), *Genetic Engineering in Eukaryotes*. NATO Advanced Science Institute Series, vol. 61. Plenum Press, New York and London, pp. 103–110.

Kawabe, H., K. Fukasawa, S. Shimizu, T. Tanaka, K. Inoue, H. Umezawa, and S. Mitsuhashi. 1979. Biochemical mechanisms of R-mediated resistance to streptomycin and spectinomycin. In S. Mitsuhashi (ed.), *Microbial Drug Resistance*. University Park Press, Baltimore, pp. 257–261.

Kontomichalou, P., M. Mitani, and R. C. Clowes. 1970. Circular r-factor molecules controlling penicillinase synthesis, replicating in *Escherichia coli* under either relaxed or stringent control. *J. Bacteriol.* 104:34–44.

McDonnell, M. W., M. N. Simon, and F. W. Studier. 1977. Analysis of restriction fragments of T7 DNA and determination of molecular weights by electrophoresis in neutral and alkaline gels. *J. Mol. Biol.* 110:119–146.

Meacock, P. A., and S. N. Cohen. 1980. Partitioning of bacterial plasmids during cell division: A cis-acting locus that accomplishes stable plasmid inheritance. *Cell* 20: 529–542.

Meighen, E. A. 1991. Molecular biology of bacterial bioluminescence. *Microbiol. Rev.* 55:123–142.

Nealson, K., and J. Hastings. 1979. Bacterial bioluminescence: Its control and ecological significance. *Microbiol. Revs.* 43:496–518.

Nordstrom, K., S. Molin, and H. Aagaard-Hansen. 1980. Partitioning of plasmid R1 in *Escherichia coli*. II. Incompatibility properties of the partitioning system. *Plasmid* 4: 332–349.

Pribnow, D. 1975. Nucleotide sequence of an RNA polymerase binding site at an early T7 promoter. *Proc. Natl. Acad. Sci. USA* 72:784–788.

Primrose, S. B., and S. D. Ehrlich. 1981. Isolation of plasmid deletion mutants and study of their stability. *Plasmid* 6:193–201.

Quigley, N. B., and P. R. Reeves. 1987. Chloramphenicol resistance cloning vector based on pUC9. *Plasmid* 17:54–57.

Rogowsky, P. M., T. J. Close, J. A. Chimera, J. J. Shaw, and C. I. Kado. 1987. Regulation of the *vir* genes of *Agrobacterium tumefaciens* plasmid pTiC58. *J. Bacteriol.* 169: 5101–5112.

Schmidhauser, T. J., G. Ditta, and D. R. Helinski. 1987. Broad-host-range plasmid cloning vectors for gram-negative bacteria. In R. L. Rodriguez and D. T. Denhardt (eds.), *Vectors: A Survey of Molecular Cloning Vectors and Their Uses*. Butterworth Publishers, Stoneham, Mass., pp. 281–332.

Shaw, J. J., and C. I. Kado. 1986. Development of a *Vibrio* bioluminescence gene-set to monitor phytopathogenci bacteria during the ongoing disease process in a nondisruptive manner. *Bio/Technology* 4:560–564.

Shaw, J. J., and C. I. Kado. 1987. Direct analysis of the invasiveness of *Xanthomonas campestris* mutants generated by Tn4431, a transposon containing a promoterless luciferase cassette for monitoring gene expression. In D. P. S. Verma and N. Brisson (eds.), *Molecular Genetics of Plant-Microbe Interactions*. Martinus Nijhoff Publishers, Dordrecht, pp. 57–60.

Shaw, J. J., T. J. Close, J. Engebrecht, and C. L. Kado. 1985. Use of bioluminescence to monitor *Agrobacterium, Erwinia, Pseudomonas* and *Xanthomonas* in plants. *Phytopathology* 75:1288.

Shaw, J. J., P. M. Rogowsky, T. J. Close, and C. I. Kado. 1987. Working with bioluminescence. *Plant Mol. Biol. Reptr.* 5:225–236.

Shaw, J. J., L. G. Settles, and C. I. Kado. 1988. Transposon Tn4431 mutagenesis of *Xanthomonas campestris* pv. campestris: characterization of a nonpathogenic mutant and cloning of a locus for pathogenicity. *Mol. Plant-Microbe Interactions* 1:39–45.

Simon, R., U. Priefer, and A. Puhler. 1983. A broad host range mobilization system for in vivo genetic engineering: Transposon mutagenesis in gram negative bacteria. *Bio/Technology* 2:784–791.

Staskawicz, B., D. Dahlbeck, N. Keen, and C. Napoli. 1987. Characterization of cloned avirulence genes from race 0 and race 1 of *Pseudomonas syringae* pv. *glycinea*. *J. Bacteriol.* 169:5789–5794.

Stueber, D., and H. Bujard. 1982. Transcription from efficient promoters can interfere with plasmid replication and diminish expression of plasmid specified genes. *EMBO* 1:1399–1404.

Tait, R. C., R. C. Lundquist, and C. I. Kado. 1982. Genetic map of the crown gall suppressive IncW plasmid pSa. *Mol. Gen. Genet.* 186:10–15.

Tait, R. C., T. J. Close, R. C. Lundquist, M. Hagiya, R. L. Rodriguez, and C. I. Kado. 1983. Construction and characterization of a versatile broad host range DNA cloning system for gram-negative bacteria. *Bio/Technology* 1:269–275.

Tandeau de Marsac, N., W. E. Borrias, C. J. Kuhlemeier, A. M. Castets, G. A. van

Arkel, and C. A. M. J. J. van den Hondel. 1982. A new approach for molecular cloning in *Cyanobacteria:* Cloning of an *Anacystis nidulans met* gene using a Tn901-induced mutant. *Gene* 20:111–119.

Valentine, C. R. I., and C. I. Kado. 1989. Molecular genetics of IncW plasmids. In C. Thomas (ed.), *Promiscuous Plasmids of Gram-Negative Bacteria.* Academic Press, New York, pp. 125–163.

Van Dyke, K. (ed.). 1985. *Bioluminescence and Chemiluminescence: Instruments and Applications,* vol. 1. CRC Press, Boca Raton, Fla.

Ward, J. M., and J. Grinsted. 1982. Physical and genetic analysis of the Inc-W group plasmids R388, Sa, and R7K. *Plasmid* 7:239–250.

Wall, L., D. Byers, and E. Meighen. 1984. In vivo and in vitro acylation of polypeptides in *Vibrio harveyi:* Identification of proteins involved in aldehyde production for bioluminescence. *J. Bacteriol.* 159:720–724.

Zaitlin, D. 1984. Genetic characterization of *Agrobacterium tumefaciens* virulence suppression encoded on the IncW R-plasmid pSa. Ph.D. thesis, Univ. of Calif., Davis.

Practical Considerations of Nucleic Acid Hybridization and Reassociation Techniques in Environmental Analysis

Thomas C. Dockendorff

A. Breen

O. A. Ogunseitan

J. G. Packard

G. S. Sayler

Introduction

DNA and RNA hybridization and DNA reassociation techniques are being applied to ecological problems in microbiology, allowing researchers to quantify answers to pressing ecological questions and to ask new kinds of questions. Investigators now possess a better understanding of microbial population variability, genetic stability and transfer mechanisms, the occurrence and frequency of genetic exchange, the phenomenon of viable, but nonculturable environmental bacteria, and even the complexity of bacterial "community genomes." This better understanding may illuminate the habitat-dependent dynamics of specific populations.

Specific target bacteria have been monitored and recovered from natural habitats using selective enumeration media, antibiotic susceptibility and resistance scoring, and fluorescent antibody labeling

(17). The application of these techniques can be enhanced when complemented with nucleic acid hybridization technology (23). Probing unknown isolates with specific labeled DNA sequences to identify bacteria was initially developed by molecular biology and medical diagnostics researchers. Gene probe technology has now become widely accepted in the quality control, food, and biochemical industries. Nucleic acid hybridization techniques are currently used in microbiology for detection and systematics and appear to be more developed than techniques for characterizing in situ bacterial diversity and abundance in microbial communities (29).

Applying DNA hybridization techniques to environmental biotechnology practices yields powerful interpretations based on highly reproducible data. Conventional methods depend largely on the phenotypic expression of certain traits by bacteria; DNA probes allow direct observation of genotypes. The reliability of DNA hybridization data depends on the specificity of the nucleic acid sequence used as probe, which is in turn described by its uniqueness among a particular group of organisms. Procedural manipulations may be used to preset the level of specificity (stringency of hybridization) required for signal detection and quantification. The universal chemical structure of DNA makes DNA hybridization applicable across species, genus, and even kingdom boundaries.

In this chapter, we summarize current environmental practice and discuss practical considerations for using nucleic acid hybridization techniques in environmental biotechnology research. This chapter is a comparative analysis of available methodology; we discuss the molecular basis of each protocol, its possible limitations, and trends for future development. A summary of the parameters and practical considerations that will be discussed is given in both Fig. 19.1 and Table 19.1.

Molecular Basis of Hybridization Reactions

Nucleic acid hybridization takes advantage of the physical-chemical properties of double-stranded DNA. The two strands of DNA run antiparallel to each other and are held together through the hydrogen bonds between the bases of each strand. There are two hydrogen bonds for every A-T base pair and three hydrogen bonds for every G-C base pair. A double-stranded DNA molecule "melts" or dissociates into two single-stranded fragments when it is heated to a particular temperature range. Not all bonds between bases melt at the same temperature; there is a transition zone between fully double-stranded and completely dissociated DNA that can span 15°C. The point at which 50 percent of the DNA in solution is in single-stranded form is defined as

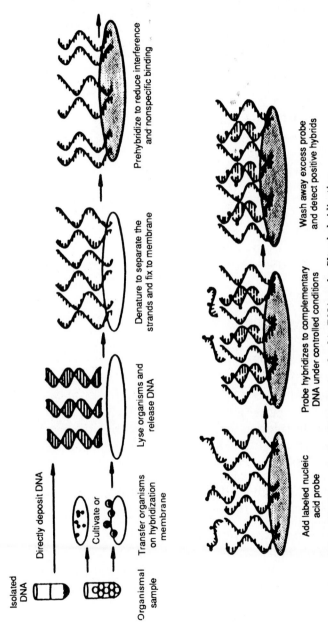

Figure 19.1 Schematic demonstrating the steps involved in DNA probe filter hybridization.

TABLE 19.1   Practical Considerations for Nucleic Acid Hybridization*

NA-probe sequence:
  Purity:
    Chromosomal contamination
    Cloning vector influence
    Restriction fragments
  Length:
    Greater length = greater sensitivity
    Shorter length = greater specificity
  Label
NA-target sequence:
  Adequate lysis
  Mass abundance
  Ancestry
Hybridization conditions:
  Saturation of nonspecific binding sites
  Cloning vector
  Hybridization temperature
  Single strandedness of probe
  Stringency of washing
  Appropriate negative-positive controls
Detection:
  Qualitative vs. quantitative
  Control
  Exposure
  Background
Postdetection:
  Need for confirmation

*Discussions of these considerations appear in the text.

the melting point, or "Tm." A strand's melting temperature depends on the base composition of the DNA molecule and on the ionic strength of the solution in which it is held. High temperatures and low salt concentrations favor dissociation of a double-stranded DNA molecule.

Dissociated DNA can re-form into double-stranded DNA. DNA reassociation can, under the proper conditions, be extremely precise, and it is this property that is taken advantage of during hybridization experiments.

Reassociation rates of single-stranded DNA to a double-stranded form depend on temperature and salt concentration. With a fixed salt concentration, reassociation depends on temperature, with the maximum rate of reassociation occurring at about 25°C below Tm. Varying salt (cation) concentrations at a fixed temperature affect the rate of reassociation. As the NaCl concentration changes from about 0.01 M to 0.1 M, reassociation between complementary single strands increases as negatively charged phosphate groups on the DNA backbone become shielded from one another. Once the NaCl concentration goes beyond 0.1 M, the rate of reassociation declines. Excessive shielding of phosphate groups allows imperfectly base-paired genes to form within a single strand of DNA. Such self-associated molecules are temporarily unavailable for specific reassociation with a complementary strand; therefore, hybridization rates decrease.

The length of the DNA molecules involved with the hybridization reaction affects the rate of reassociation. The reassociation process is thought to occur in two steps. The first step, the collision of two complementary strands, is the rate-limiting step. The second step is rapid; the two partially reassociated strands "zipper" to complete the hybridization. The rate at which reassociation occurs is approximately proportional to the square root of the length of the DNA molecule.

The rate of DNA reassociation also depends on the relative abundance of complementary strands of DNA. Sequences that are more abundant in a particular sample are more likely to collide with each other and hybridize. Thus, it is possible to use DNA reassociation kinetics as a measure of the number of copies of a specific sequence that are present in a mixed community.

DNA Hybridization Procedures

DNA hybridization is a technique that enables an investigator to determine whether an unknown sequence of DNA (target DNA) has any significant homology with a labeled probe DNA of known structure and/or function. Target DNA is bound to a support, usually a filter disk or membrane made of nylon or nitrocellulose. Probe DNA is added and allowed to hybridize. Unbound probe is washed off and hybrids are detected. Commonly used procedures for DNA hybridization include colony hybridization, dot-slot hybridization, and Southern hybridization.

Colony hybridization

Colony hybridization (Fig. 19.2) is a technique developed by Grunstein and Hogness (13) for detecting cloned sequences directly from bacterial colonies. The method involves lysis of bacteria that have been taken up by a nylon or nitrocellulose disk placed on the surface of a plate containing the bacterial colonies of interest. The released DNA is then fixed to the disk, and prehybridization, hybridization, and washing steps follow. The colony hybridization technique has been used successfully for the rapid detection of genes that code for the catabolism of environmental contaminants (24, 27, 28), for the

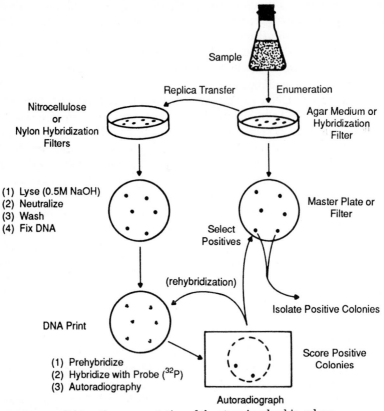

Figure 19.2   Schematic representation of the steps involved in colony hybridization.

identification of certain *Rhizobium* strains (14), and for the detection of *Salmonella* species (8). Sayler et al. (27) have used this technique and demonstrated that one organism of interest can be detected against a background of 10^6 organisms harboring nonhomologous DNA.

Organisms that have a positive signal can be recovered by matching an autoradiogram to colonies on a replica plate from which a positive signal has been detected. The organisms can then be isolated for further analysis. If a particular organism is selectable on a given type of medium, colony hybridization can rapidly determine the percentage of selected bacteria harboring the gene of interest. A major limitation to this technique is that only culturable bacteria harboring the sequence will be detected. Nonculturable organisms containing the sequence will not be seen. Therefore, this technique cannot be used to accurately calculate the total number of sequences present in a sample.

Also, environmental isolates can be difficult to lyse, so rigorous lysis protocols that include lysozyme and sodium dodecyl sulfate (SDS) may be necessary.

Dot-slot hybridization

Dot-slot hybridization is a semiquantitative method for determining the concentration of a particular DNA sequence from an environmental sample. The technique involves binding target DNA to a filter and conducting ordinary prehybridization, hybridization, and washing steps. Anderson and Young (2) give a mathematical treatment of the kinetics of such hybridizations, and their article is recommended reading for anyone wishing to use dot-slot methods.

This hybridization protocol can be used to accurately determine the level at which a particular DNA sequence is present in an environmental sample. Unlike colony hybridization, dot-slot hybridization does not require cultivation of the organisms; rather, total DNA must be extracted from the sample. (Methods for DNA extraction are discussed below.) The investigator spots known amounts of homologous DNA and target DNA onto a filter and estimates the level of homologous sequences in the target DNA sample by comparing the intensity of the signal on the filter to the signal given by the positive controls. Densitometry can be used to give a quantitative value. Using commercially made vacuum manifolds, dot-slot hybridization can make analysis of many samples a relatively quick job.

Southern hybridization

Developed by E. M. Southern in 1975 (30), Southern hybridization involves the transfer of DNA from an agarose gel to a nylon or nitrocellulose membrane. DNA transfer is carried out by capillary action, electroblotting or vacuum blotting. Hybridization is then carried out using standard procedures.

The gel can be run in a vertical or horizontal mode. However, vertical gels used for Southern hybridization are thinner than horizontal gels, and thus transfer of the DNA from the agarose to the membrane is faster and more efficient. After electrophoresis is completed, gels are stained with ethidium bromide and photographed.

There are published protocols for treatment of the gel prior to blotting (19). If large fragments of DNA [greater than 5 kilobase pairs (kb)] are to be transferred, it is recommended that the gel be soaked in dilute (0.25 N) HCl for 15 min to acid depurinate the DNA. Soaking facilitates transfer of large DNA fragments from gel to membrane. The transfer of DNA should be quantitative, and by restaining and

UV light visualization, the gel should be examined for the presence of any remaining DNA.

Southern hybridization is useful for determining whether a gene of interest is located on the plasmid or chromosome of the organism being screened. Gene location can be determined quickly by using one of the rapid plasmid extraction protocols that work well on environmental isolates (1, 18). Plasmid and chromosomal DNA can be separated on an agarose gel, transferred to the filter, and hybridized. If a genome has been physically mapped through restriction analysis, one can localize the gene of interest to a particular restriction fragment.

Common Hybridization Protocol Steps

Most hybridization procedures include steps for binding DNA to a filter, prehybridization, hybridization, and washing. A general outline for each step appears below. Commercial vendors of filter materials supply complete, detailed protocols, including reagent preparation, with their products. It is recommended that a vendor's protocol for filter treatment be followed; protocols vary, and what works for one brand of filter may not work for another.

Binding of target DNA

Binding of the target DNA simply involves introducing the DNA to the filter. A denaturation step, which allows probe DNA to bind, is followed by a neutralization step.

DNA binds to nylon or nitrocellulose filters in a noncovalent manner. Supercoiled plasmid DNA should be linearized by a restriction enzyme or by heating, because supercoiled DNA will quickly reassociate during the neutralization step.

Prehybridization

In prehybridization, the filter is incubated in a solution that precoats all areas of the filter where target DNA is not bound, so that probe DNA does not bind and cause high background levels on the autoradiogram. Denhardt's solution is commonly used for this purpose. It contains Ficoll, polyvinylpyrrolidone, and heterologous DNA, usually salmon sperm DNA. Denhardt's solution, other buffers, and the filters are placed in a resealable polyethylene pouch. Prehybridization fluid is added in a ratio of 0.08 mL of fluid to 1 cm^2 of filter material. The DNA in the prehybridization fluid must be denatured before it is added to the filter. The solution should be prewarmed to the temperature of the incubation bath before it is used.

Hybridization

Hybridization involves adding the probe DNA to the sample and incubating to allow establishment of DNA hybrids. Hybridization can be carried out in either an aqueous solution or a 30 to 50% solution of formamide. Anderson and Young (2) note that using a formamide solution allows the incubation temperature to be lowered. Lower incubation temperatures lead to greater probe stability, and bound DNA is less likely to be lost from the filter. Lower temperatures are also less harsh on the membrane, which is desirable if the membrane is to be used again for reprobing.

Hybridization can also be carried out in an aqueous solution at varied elevated temperatures. We are not aware of any advantages of hybridization in aqueous solutions over those using formamide and lower temperature.

Washing the filter

Washing is done to remove both unbound probe and probe DNA that has bound weakly to nonhomologous sequences. The washing step sets the stringency level of the hybridization procedure, as determined by the level of mismatching allowed between hybridizing strands. Stringency is a function of both the salt concentration in the wash solutions and the temperature of the wash. High incubation temperatures during the wash (65°C) favor closely matched hybrids. High salt concentrations tend to stabilize poorly matched hybrids.

The melting temperature (Tm) of two hybridized strands of DNA has been mathematically described:

$$Tm = 81.5 + 16.6(\log M) + 0.41(\% \text{ G} + \text{C}) - 0.72(\% \text{ formamide})$$

where M = molarity of the monovalent cation.

The equation holds for perfectly matched hybrids, but the melting temperature is lowered by mismatched and unevenly distributed base pairs. If all mismatches are concentrated in one area, and the rest of the hybrid is perfectly matched, the Tm will still be high. If the mismatches are evenly distributed, however, the hybrid will probably be unstable and Tm will be lowered. It has been shown during solution hybridization that a 1 percent mismatch reduces Tm by an average of 1°C, depending on the G + C content (2).

Investigators are free to vary stringency levels as needed. For instance, the degeneracy of the genetic code may not allow the probe and target to hybridize at stringencies requiring 90 percent homology, yet the probe and target could code for the identical amino acid sequence. Carrying out hybridization and washing steps at 65°C and low salt concentrations will allow about a 5 percent mismatch. There is no empirical formula to determine the level of mismatching allowed under a given set of hybridization and washing conditions. If a range of hybridization and washing conditions are to be used, the investigator should closely monitor hybridization levels against a negative control.

Amount of probe used

Probe concentrations of 50 to 100 ng/mL are used, unless dextran sulfate is added to the hybridization solution. Dextran sulfate is a large polymer that excludes DNA from the volume it occupies, in effect increasing the concentration of DNA in a hybridization mix. When dextran sulfate is added, probe concentrations of greater than 10 ng/mL are unnecessary and would lead to high backgrounds.

Incubation time

Existing protocols call for a wide range of incubation times. At first glance, it might appear that a long incubation time would ensure more complete hybridization, but this is not necessarily so. If hybridization is carried out for a long time in an aqueous solution (68°C), sequences can be lost from the filter, reducing sensitivity. Also, probe DNA will reassociate to double-stranded form over time. Flavell et al. (9) have noted that no more than 80 percent of probe DNA appears in hybrids. Anderson and Young (2) conclude that for double-stranded probes, incubation past $3 \times C_0t\frac{1}{2}$ is unnecessary. C_0 is defined as the concentration of single-stranded DNA in a solution; t is time. $(C_0t)\frac{1}{2}$ is the C_0t value at which 50 percent of the original solution has reassociated. Maniatis et al. (19) provide the following guideline for determining $C_0t\frac{1}{2}$:

$$N = \left(\frac{1}{x}\right)\left(\frac{y}{5}\right)\left(\frac{z}{10}\right)x2$$

where x = weight of the probe, µg
   y = complexity
   z = volume of the reaction, mL
   N = hours needed to achieve $C_0t\frac{1}{2}$

Single- vs. double-stranded probes

Probe DNA can be in either single- or double-stranded form. (Double-stranded DNA must, of course, be denatured before use.) Double-stranded probe may be partially lost when it reassociates in solution with a complementary self strand. Using single-stranded probe requires extra preparation, but the probe does not reassociate to itself in solution. Single-stranded probe can be produced using strand-separating gels (19), by having the probe sequence cloned into M13 phage vectors (16), or by producing synthetic oligonucleotides.

Length of probe

A probe can be a short oligonucleotide of 10 to 20 bases, or it can be a longer polynucleotide prepared by nick translation of a large DNA fragment. Longer probes (400 to 800 nucleotides) contain larger amounts of radioactivity and have been described as more sensitive than shorter probes; oligonucleotide probes have been described as more specific. Longer probes have the disadvantage of having lower levels of diffusion to the filter. Also, if the probe is a long, denatured double strand, reassociation rates of the probe to itself can be high.

Using and synthesizing an oligonucleotide probe requires that part of the DNA sequence be known. When using these probes, controls should ensure that the known sequence is not part of a conserved domain (found in some regulatory proteins) or that it is not a repetitive sequence (unless these are desired targets for hybridization).

Hybridization and washing procedures differ for oligonucleotide probes owing to their shorter length. Prehybridization and hybridization salt buffers differ; prehybridization salt is composed of 3 X SSC/ 0.1% SDS, and hybridization buffer is composed of 6 X SSC/0.05% pyrophosphate. Hybridization and washing temperatures are much lower for oligonucleotides than for longer probes, and vary for oligonucleotide probe length. Recommended temperatures are given in Table 19.2 (conditions may vary for filter brand). The length of time required for hybridization ranges from 14 to 48 h.

TABLE 19.2   Recommendations for Hybridization and Washing Temperatures for Oligonucleotide Probes

Length of oligonucleotide, in bases	Temperature for hybridization, °C
14	Room temp.
17	37
20	42
23	48

Length of oligonucleotide, in bases	Temperature for washing, °C
14	37
17	48
20	55
23	60

SOURCE: Adapted from Duby (7).

Detection and quantitation

Detection of radiolabeled hybrids can be done by autoradiography or by scintillation counting. Dot blots can be punched from the filter and placed into a scintillation cocktail, yielding a quantitative value of the amount of hybridization taking place. Alternatively, autoradiography coupled with densitometry can be used to quantitate hybrids. Development of the autoradiogram requires wrapping the filter in plastic wrap (Saran Wrap), loading the filter and x-ray film (Kodak X-OMAT AR is recommended) into a cassette, and placing the cassette in a freezer at -70 to $-90°C$. Using intensifying screens to enhance the film image can shorten exposure times. Approximately 1000 to 5000 counts per minute (cpm) of ^{32}P in a band 1 cm in width will give a visible image in 12 to 16 h. A review of autoradiography is given by Voytas (34).

Other Considerations

Filter choice

The two most commonly used filter materials for binding DNA are nylon and nitrocellulose. Nylon filters are sturdier, perform better in target-DNA binding, and can be reused for hybridization with another probe. DNA binding to these filter materials is noncovalent; thus DNA can be lost from the filter during prolonged or repeated hybridization or when the filter is exposed to high incubation temperatures.

DNA can be covalently bound to nylon filters through UV cross-linking (5). Cross-linking with nitrocellulose filters is not recommended because of the possibility of fire. Cross-linking DNA to a filter results in potentially greater probe sensitivity because smaller amounts of target DNA leach from the filter. Procedures have been developed for covalent bonding of DNA to filters through chemical means, but Anderson and Young (2) claim that the binding capacity of the filter is lowered considerably when these methods are employed.

There are a number of commercial suppliers of nylon and nitrocellulose membranes. Each supplier has its own recommended treatment procedures; the supplier's protocol for DNA binding, prehybridization, and hybridization should be followed.

Reprobing

In some cases it is desirable to reprobe a filter with different probes. Nylon filters are more durable than nitrocellulose and would be the filter material of choice if reprobing were to be done frequently. The loss of noncovalently bound target DNA may be a problem and must be considered when analyzing data.

One must ensure that all probe material is removed from the filter before reprobing. Anderson and Young (2) suggest the following procedure:

1. Wash filters twice for 10 min at room temperature in 50 mM NaOH.
2. Wash and neutralize the filter at room temperature by doing five 5-min washes in TE buffer, changing the buffer each time.
3. Perform autoradiography to ensure that all probe has been removed.

It is vital that the filters not be allowed to dry out, as this will cause irreversible binding of the probe.

MPN-DNA Hybridization

DNA hybridization has been used in conjunction with microtiter most-probable-number (MPN) for detecting and enumerating specific bacterial genotypes in environmental samples (12). An environmental sample is mixed with a medium selected to be optimal for incubation of the organism to be enumerated. A serial dilution series of the mixture is prepared in a sterile microtiter plate. For example, a five-replicate 10-fold dilution series may be used, and the microtiter plate is incubated for 3 to 4 days at the optimum incubation temperature for the organism.

To conduct a dot blot on the DNA of the organisms in the dilution series, the total solution volume from each well is transferred to a corresponding well in a filtration manifold. The sedimented particles in the bottom of the wells are not disturbed during the transfer. The solution from the wells is then filtered onto a hybridization membrane. The cells are lysed and the DNA is denatured and fixed on the membrane according to standard methods. Prehybridization and hybridization of the membrane are conducted using a labeled probe specific to the organism to be enumerated. After autoradiography of the membrane, positive signals that indicate the presence of target DNA are scored. The MPN value is calculated from a table appropriate for the number of replicates and the type of dilution series used. The number of organisms in the original sample can then be calculated. A positive control—a mixture containing the organism to be enumerated—and a negative control—a mixture of organisms other than the one to be enumerated—should be run alongside the environmental samples. The detection limit for the MPN-DNA hybridization procedure varies.

The interaction between the species to be enumerated and the soil or sediment particles is crucial; any target organisms adhering to the colloidal particles are not available for enumeration.

Probe and Target Isolation and Purification

DNA isolation protocols have been evolving since Meishner first isolated DNA in 1869. Avery and coworkers at Rockefeller University introduced many of the now routine steps in DNA isolation, such as chloroform extraction of proteins and alcohol precipitation of nucleic acids (21). In 1961, Marmur (20) refined the older techniques into a protocol that has been widely used, though often with modifications. Using detergent and lysozyme to lyse the culture, buffered phenol is employed to remove proteins. Then chloroform-isoamyl alcohol (24:1) is used to further extract proteins and phenol. This step is usually done at least twice because traces of phenol may damage DNA or inhibit enzymes that may later be used on the DNA. During these extractions, DNA always remains in the aqueous phase. The nucleic acid is precipitated from the aqueous phase with cold ethanol or isopropanol. Ethanol will precipitate DNA and RNA and evaporate relatively quickly. Isopropanol will precipitate DNA and minimize the occurrence of RNA and polysaccharides.

The inactivation of nucleases is critical to all DNA isolation protocols; for this reason EDTA is incorporated into most procedures. The methods of lysis vary from sample to sample. Often, lysozyme treatment followed by treatment with a detergent such as SDS at 65°C is sufficient. A salt wash prior to lysis will help remove exopolysaccharides that often copurify with nucleic acids. For some samples, such as for many gram-positive bacteria, a more rigorous procedure is needed. Further steps may include additional enzymatic treatments or physical disruption, such as French pressure cell lysis or ballistic disruption with glass beads. In general, bacteriolytic enzymes have a temperature optimum between 35 and 40°C. Many of these enzymes are very specific for their substrate, and some require that their target be pretreated with a detergent (this is particularly true of the gram-negative bacteria in which the peptidoglycan is protected by the outer cell membrane) (3). It is important to note that physical disruption of cells often results in DNA shearing; hence, these methods are not appropriate when high molecular weight DNA is desired.

Cesium chloride (CsCl) density centrifugation is another method of separating DNA from RNA and proteins. It is most commonly employed in the isolation of plasmid DNA during cesium chloride–ethidium bromide equilibrium centrifugation (19). The use of CsCl in the preparation of DNA was believed by some to subject the DNA to less shearing in preparation and thus yield DNA of a higher molecular weight. Under the proper density conditions, CsCl centrifugation will pellet RNA on the bottom of a tube while proteins float on top and DNA forms a band in the center of the tube. Marmur's experiments show DNA isolated by CsCl to be almost identical to DNA isolated by chloroform phenol extraction followed by alcohol precipitation (20).

An alternative to solvent extraction is to use hydroxyapatite (HA), a modified form of crystalline calcium phosphate, to isolate DNA. A crude lysate can be loaded onto an HA column equilibrated with 0.24 M sodium phosphate in 8 M urea. Elution with this buffer will allow most proteins and RNA to pass through the column. DNA will bind to the column and can then be eluted with 0.4 M sodium phosphate. The phosphate groups of nucleic acids are believed to adsorb to HA calcium ions. DNA can be desorbed by increasing the molarity of sodium phosphate buffer—the buffer outcompetes the DNA for HA binding sites. DNA in double-stranded form binds more strongly to HA because the phosphate groups of the double-stranded form molecules are more accessible to the HA binding sites (4). Hydroxyapatite varies greatly from batch to batch and must be checked to determine which molarity buffers should be used.

Direct Recovery of DNA from Environmental Samples

The tools of molecular biology have added a new level of sophistication to the field of environmental microbiology. Probing nucleic acids from environmental samples with a battery of function-specific gene probes can be done rapidly and reliably.

A number of techniques have emerged for isolating nucleic acids from environmental samples without cultivation of the microorganisms present (15, 22, 33; Fig. 19.3). These methods may eliminate the bias inherent in the cultivation of environmental isolates on artificial media. The extracted DNA, representative of the entire microbial population in situ, may be analyzed by a number of DNA-DNA hybridization techniques. As might be expected, the methods for isolating DNA from the environment vary from sample to sample. Nonetheless, all procedures have certain requirements: DNase must be eliminated or neutralized; and lysis must be universal to ensure that the DNA is representative of the environmental population; DNA must be sufficiently pure to allow accurate estimates of yields to be made.

An investigator must begin the extraction process with a sufficient amount of biomass. At the present time, extraction procedures should yield at least 10 µg of DNA. A small amount of DNA is all that is needed for a single hybridization, but in the environment, genes of in-

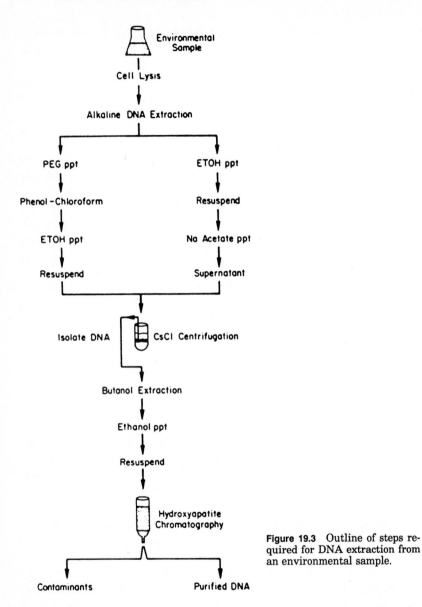

Figure 19.3 Outline of steps required for DNA extraction from an environmental sample.

terest may be present in low copy. Naturally, with more DNA more experiments can be conducted. Some samples, such as activated sludge or fecal material, are relatively easy to work with because they are naturally concentrated materials. Environments such as the open ocean or oligotrophic lakes may require a considerable sample concentration process to yield sufficient biomass for DNA extraction. DNA extracted from environmental samples can also be used for target gene amplification using polymerase chain reaction techniques (PCR) (see Part 1 for PCR).

Two very successful procedures for sample concentration are filtration concentration and continuous centrifugation. Filtration concentration can disrupt the microorganisms if pump pressures become too high; however, this is generally a very suitable means for concentrating biomass from a water column. Continuous centrifugation is somewhat more gentle (but is also more time consuming). Using a centrifuge on a boat may be impractical; a gimble is needed to keep the centrifuge balanced. Continuous centrifugation should not be used in rough waters.

Soil and sediments have high levels of biomass but are difficult matrices from which to extract DNA efficiently. Most procedures for DNA extraction from soil or sediments involve separation of the bacterial fraction from particulate matter with a series of high salt washes and centrifugation. In general, sodium pyrophosphate is the extractant of choice. The extracted biomass is lysed, usually by lysozyme treatment at 37°C for 15 to 30 min followed by SDS treatment at 65°C for 1 h. The sample is then checked microscopically to ensure that lysis is complete. In many cases, more vigorous disruption methods must be employed. French pressure cell treatment is effective, but care must be taken to ensure that gritty particulates in the sample do not damage the French press. Ballistic disruption using glass beads in a "bead beater" is another efficient way to ensure cell lysis. Both methods are used following an SDS pretreatment at 65°C for 1 h, and both result in DNA shearing down to the 10- to 0.5-kb range.

Following lysis, a clearing centrifugation step is used to remove cellular debris, particulates, and some humic materials. It is relatively easy to extract DNA from samples with little or no organic content; two or three phenol extractions followed by two chloroform extractions and an ethanol precipitation may yield high-quality DNA. In most cases, however, a more rigorous purification is needed. Yields can be increased by extracting the pellet after the clearing spin. Liberated DNA can bind to soil particulates as an anion over a broad pH range, and at low pH it can bind as the cation to negatively charged soil particles. Thus sodium pyrophosphate, which is an amphoteric molecule, is one of the most effective extractants for the removal of DNA from soil.

Torsvik (33) used ion exchange chromatography and hydroxyapatite on soils with high humic content. Her procedure is summarized below:

1. The bacterial fraction from 60 to 90 g of wet soil is washed with 100 mL 0.1 M sodium pyrophosphate (pH 7.0) and then with 0.15 M NaCl, 1 mM EDTA (pH 8.0), and is then resuspended in the same solution.

2. The cells are lysed by first adding lysozyme (1 mg/mL) to the solution and incubating at 37°C for ½ h, then by adding SDS to 1.0 percent and incubating the solution at 60°C for 1 h.

3. Urea and KCl are added to the solution to final concentrations of 8 M urea in 0.7 M KCl. The solution is kept at 0°C overnight and then centrifuged at 1000 g for 20 min.

4. The material is resuspended in 20 mL of cold 8 M urea in 0.7 M KCl, homogenized, and centrifuged as described in step 3.

5. The material is loaded on a DEAE-Sepharose CL-6B column equilibrated with 0.7 M KCl, 10 mM Tris HCl (pH 7.3). (This step may be omitted if the sample is light brown or yellow rather than dark brown.)

6. Hydroxyapatite chromatography is carried out in a three-step elution. The sample is loaded on the column and eluted first with 0.24 M sodium phosphate (pH 6.8) in 8 M urea, then with 14 mM sodium phosphate (pH 6.8) and then with 0.4 M sodium phosphate (pH 6.8).

The Torsvik protocol yields high-quality DNA in quantities as high as 1.5 mg from 90 g of wet soil. The low EDTA concentration (1 mM) of the lysing buffer increases the efficiency of humic removal during centrifugation. Column chromatography greatly expands the volume of the solution; hence, a concentration step such as alcohol precipitation is required to bring the DNA to a usable concentration.

A second protocol for extracting DNA from soils with high humic content homogenizes the sample material in a solution containing a dilute Winogradsky salt solution (0.2 M sodium ascorbate) (15). Homogenizing is done in a household blender in the presence of acid-washed polyvinylpolypyrrolidone. The bacterial fraction undergoes detergent, enzymatic, and heat treatments. The lysate is cleared by centrifugation, and the supernatant is centrifuged with cesium chloride–ethidium bromide. The protocol yields DNA in a concentrated form, but soils with high humic content will retain some level of humic contamination.

The protocols thus far described depend on efficient extraction of microorganisms from soil particulates. To eliminate the need to extract microorganisms, Ogram et al. (22) have developed a procedure, optimized for sediments with high humic and clay content, for in situ lysis of microorganisms and subsequent extraction of the liberated DNA. One percent SDS in 0.12 M sodium phosphate (pH 8) is added to the environmental sample, which is then incubated at 70°C and homogenized in a "bead beater." Repeated extractions with the alkaline phosphate buffer desorb DNA from the particulates and make a concentration step necessary. Both polyethylene glycol (PEG) and ethanol are effective for sample concentration. PEG has the advantage of not greatly increasing the sample volume during the precipitation process. Extracting the resuspended precipitates to remove the polyethylene glycol or ethanol also removes additional humic contaminants. Bringing the resuspended ethanol precipitate to 0.5 M potassium acetate and storing on ice for 2 h precipitates a large percentage of the humic materials.

The final step in DNA purification is cesium chloride–ethidium bromide centrifugation followed by hydroxyapatite chromatography, or, if the sample is relatively clean, hydroxyapatite chromatography without centrifugation. The standard buffers for hydroxyapatite chromatography are 0.24 M sodium phosphate (pH 6.8) in 8 M urea to remove humic contaminants; 0.014 M sodium phosphate (pH 6.8) to remove the urea; and 0.4 M sodium phosphate (pH 6.8) to elute the DNA.

A common problem in DNA isolation, especially when using environmental samples, is expansion of the sample extract volume. In an expanded sample the purified DNA is extremely dilute, ethanol precipitation becomes inefficient and cumbersome, and very large volumes of ethanol are required for precipitation. Centricon tubes (Amicon, Danvers, Massachusetts) are commonly used to concentrate DNA or change the buffer solution of a DNA preparation. The tubes feature an upper and lower chamber separated by a membrane (maximum 10,000 molecular weight porosity). The sample to be concentrated is put in the upper chamber and centrifuged; DNA concentrates in the upper chamber.

The procedure chosen for nucleic acid extraction depends on the type of sample to be processed. As a general rule, the higher the organic content of a sample, the more difficult the DNA extraction will be. An idea of the amount of biomass present is useful in determining sample size. The procedure used should provide a DNA preparation representative of the community in situ, not biased against microorganisms that are easily lysed or easily dislodged from particulate matter. Table 19.3 summarizes the characteristics of site environments that determine appropriate DNA extraction procedures.

TABLE 19.3 Characteristics of Sample Site in Regard to DNA Extraction

Sample	Characteristics
Water column	Usually low organic content; low biomass levels require a large volume of water processing
Sediment	Usually a large amount of biomass present; organics, particularly humic substances, are extremely difficult to separate from DNA
Soil	Same as sediment
Activated sludge	Very high biomass, does not require large sample processing; some organic contaminants may slow DNA isolation

Quantitation

Accurate quantitation of DNA requires a highly pure sample; however, hybridization can occur with minimal purification (e.g., in colony hybridization). The most widely used method to evaluate both the quantity and quality of DNA is an ultraviolet absorption scan of the DNA sample from 240 to 280 nm. An optical density (OD) value of 1 at 260 nm corresponds to a DNA concentration of 50 μg/μL. The OD 260- to 280-nm ratio provides a check on the quality of DNA (19). Pure DNA will have a ratio of 1.8:1 to 2.0:1. Contamination will reduce the ratio and make accurate quantitation impossible. When dealing with minute amounts of DNA (< 0.3 μg), an alternate method of quantitation is desirable. The agarose plate method works well in these instances (19). The sample is spotted onto a petri plate containing a 1.0% agarose gel with an ethidium bromide concentration of 0.5 μg/mL. Fluorescence of the sample is compared with a standard DNA dilution series. The sample should be rigorously purified because some contaminants may also fluoresce.

Isolation of Probe DNA

Probe DNA is generally in one of two forms—either the gene or genes of interest are cloned onto a plasmid or the probe is an oligonucleotide synthesized on an automated machine. The form of the probe varies with the experimental design.

Plasmid DNA can be isolated in a procedure similar to that used to extract sample DNA. Typically, cells in which the plasmid of interest are maintained are grown to late log phase and harvested. Many cloning vectors are amplifiable, and treatment with chloramphenicol greatly increases plasmid DNA yield (19). Plasmid isolation procedures feature a denaturation step (either alkali or heat treatment) that permanently dissociates the chromosomal DNA. It is then possible to use a clearing spin to remove much of the chromosomal DNA and the cellular debris. The supernatant may be precipitated with ethanol or isopropanol and checked for the presence of plasmid DNA with agarose gel electrophoresis. Although many protocols claim sufficiently "clean" (free of chromosome) yields at this stage, it is generally necessary and advisable to subject the preparation to cesium chloride–ethidium bromide ultracentrifugation. The intercalation of ethidium bromide into supercoiled, closed circular DNA gives the plasmid a different buoyant density than the linear chromosomal DNA. This results in separate banding of the chromosome and plasmid in the gradient. The plasmid DNA will make up the lower band and can be extracted from the gradient with the aid of a syringe and long-wave ultraviolet light. The ethidium bromide is extracted with *sec*-butanol, and the DNA is precipitated with ethanol. The DNA can then be resuspended in the appropriate buffer and quantitated.

As a general rule, small plasmids (< 20 kb) are more easily isolated than large plasmids. Many interesting plasmids, such as catabolic plasmids and Ti plasmids, cannot be isolated by typical procedures. A number of isolation protocols have been optimized for the more difficult plasmids. Among them are the Currier and Nester (6) procedure for Ti plasmid DNA, the Anderson and McKay (1) procedure for streptococcal plasmids, and the Franklin and Williams (11) procedure for the TOL plasmid. All of these protocols are effective for the isolation of large, low-copy-number plasmids.

When using genes that have been cloned into a vector plasmid, care must be taken to ensure that the vector itself does not add ambiguity to the experiment due to its hybridizing with homologous DNA. To avoid ambiguity, gene(s) of interest can be separated from the vector by digesting the plasmid with the appropriate restriction endonuclease and extracting the fragment from agarose using chemical means or electroelution (32). Correct salt concentrations are critical to the activity of restriction enzymes; hence, a dialysis cleanup of the plasmid after the cesium–ethidium bromide centrifugation may be necessary. Generally, large DNA fragments (> 10 kb) are difficult to recover from agarose; however, fragments of 1 kb or less are easily and efficiently purified.

An alternative to isolating a particular fragment from a plasmid is using an unlabeled cloning vehicle in a prehybridization step. Unlabeled DNA saturates homologous sites, thus eliminating the ambiguity resulting from using a labeled cloning vehicle. This procedure may be preferable if a large DNA probe is being used and extraction from agarose would be difficult.

Oligonucleotide probes are useful as very specific probes. They have been used to discern subsets of genes and to isolate clones from very low frequency mRNAs. In a novel approach to microbial ecology, oligonucleotide rRNA probes have been used to describe microbial community structure in environmental populations (31). Obviously, knowledge of the sequence is required for probe construction, and preparation is time-consuming. The cost is generally $100 per high-pressure liquid chromatography (HPLC)–purified probe, and probes usually must be further purified by polyacrylamide gel electrophoresis.

Probe Labeling

To detect a specific DNA sequence, one must have a method for detecting hybrids formed during the hybridization assay. This is most commonly done by labeling the probe DNA with some type of easily detectable tag.

A number of methods exist for the in vitro labeling of DNA. The choice of labeling procedure depends on the experimental design and form of the probe (gene, genes, or oligonucleotide). The "nick translation" procedure makes use of the DNA repair enzyme, DNA polymerase I, to incorporate radiolabeled nucleotides into double-stranded DNA. In the reaction, DNase I is used to introduce "nicks" (single-stranded breaks) into the DNA. DNA polymerase I recognizes the nicks and initiates the incorporation of nucleotides, one of which is radiolabeled, into the DNA molecule (19). The reaction is carried out at 15°C for 1 h and produces radiolabeled probe fragments of approximately 0.5 kb in length. Kits for running the nick translation reaction are available commercially.

A schematic of the nick translation process appears in Fig. 19.4. The probe is separated from unincorporated nucleotides by G-50 Sephadex chromatography. The radiolabeled probe will elute as a peak before the nucleotides. The radioactive isotope of choice is usually ^{32}P-dCTP. Specific activities of 1.0 to 5.0 × 10^7 cpm/μg DNA can routinely be obtained. Autoradiography can be carried out rapidly due to the high energy of the emitted beta particle. ^{32}P does not have a long shelf life, and probes are subject to radiochemical decay over time; hence, the probe is best used shortly after nick translation. An alternative to ^{32}P is ^{35}S, generally in the form deoxyadenosine 5'-[-thio-^{35}S]triphosphate (25). ^{35}S has a number of desirable characteristics. It has a much longer half-life (87 days) than does ^{32}P (14 days). ^{35}S also has a lower emission energy than ^{32}P and thus produces somewhat less diffuse bands during autoradiography. ^{35}S, however, cannot be monitored with a Geiger counter; hence, dextran blue, which comigrates with the probe, is added to the nick translation mixture before application to Sephadex. Also, ^{35}S-dATP is incorporated more slowly than ^{32}P-labeled nucleotides, and so nick translation reactions are run overnight.

Recently, nonradioactive DNA labels have become popular because of the instability, hazard, and cost of radioactive probes. The most commonly used system involves the use of biotin, which is incorporated into the probe in the form of biotinylated UTP (10). Biotin can be attached to the pyrimidine ring of the nucleotide, which is then used in place of dTTP. Biotin derivitized nucleotides can be incorporated by nick translation. Hybridization is detected with antibiotin

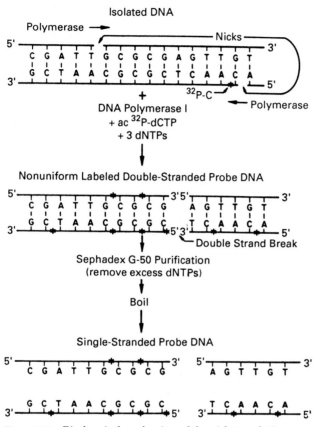

Figure 19.4   Biochemical mechanism of the nick translation reaction and probe preparation.

antibodies labeled fluorescently or enzymatically with streptavidin. Biotin-labeled probe is separated from unincorporated nucleotides by ethanol precipitation or by G-50 Sephadex, once again using dextran blue as a tracer. Unfortunately, nonspecific binding of these probes is a problem that often results in unacceptably high levels of background.

Oligonucleotide probes are labeled by the end-labeling process, rather than by nick translation, using bacteriophage T4 polynucleotide kinase (19). This enzyme incorporates the gamma phosphate of ATP to the 5'OH terminus of DNA (or RNA). Note that the gamma phosphate is labeled, whereas in the nick translation reaction the alpha phosphate of the NTP is labeled. The reaction is carried out at 37°C for 30 min, after which the sample is extracted with phenol and chloroform and, finally, ethanol-precipitated. Labeled

DNA is separated from unincorporated nucleotides by G-50 Sephadex column chromatography.

Applications for Solution Hybridization and DNA Reassociation Kinetic Analysis

Solution hybridization of a radioactive probe to DNA extracted from an environmental sample gives information similar to that obtained from dot blots, but the procedure used is quite different. An outline of the steps required is given in Fig. 19.5. In the solution hybridization

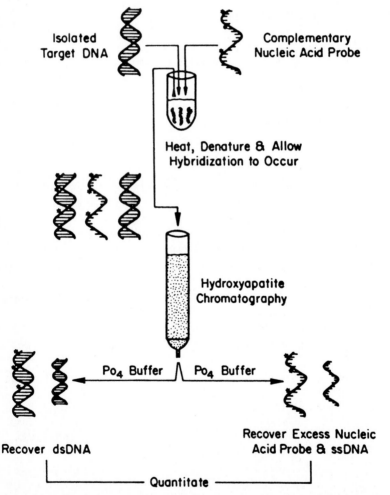

Figure 19.5 Protocol for solution hybridization and separation of single- and double-stranded DNA molecules.

procedure, the target DNA is not immobilized on a filter; it is in a solution to which the radioactive probe is added. Hybridization between the probe and the homologous sequences from the sample DNA occurs in the solution. Double-stranded DNA can be separated from single-stranded DNA with hydroxyapatite chromatography. Scintillation counting of both fractions determines the radioactivity in each fraction. The percentage of DNA in the sample that is homologous to the probe can then be calculated.

The reassociation kinetics of DNA are determined by a type of solution hybridization in which the total DNA is sheared, denatured, and allowed to reassociate. A small subsample of the total DNA is radioactively labeled in order to detect, over time, the percentage of DNA that is reassociated. The process produces a graph of the kinetics of the reassociation. The theory, interpretation, and technique of DNA reassociation kinetics are well known because of their use in many areas of genetics. An excellent review of DNA reassociation kinetics is provided by M. L. M. Anderson and B. D. Young in *Nucleic Acid Hybridization: A Practical Approach* (2).

The application of reassociation kinetics to DNA that has been extracted from environmental samples is currently being investigated (23, 29). In a process that is analogous to the use of reassociation kinetics to determine the complexity of genomes of single eukaryotic species, DNA reassociation kinetics could theoretically be used to determine the complexity of the bacterial community genome in an environmental example. Fractions of the reassociated DNA can be probed with a DNA sequence of interest to determine if the probe sequence is in a high, medium, or low copy number within the community genome. Reassociation kinetics techniques are more sensitive than dot-blot hybridization when they are used to locate a sequence that is in a very low copy number. Good results are possible if the DNA that is extracted is primarily bacterial. Feasibility also depends on relatively unique bacterial genomes, since genomes of individual species reassociate during the procedure.

Practical problems restrict the application of reassociation kinetics to the very complex DNA obtained from environmental samples. For instance, reassociation of the DNA takes a prohibitively long time using normal DNA concentrations. Higher DNA concentrations should speed up the reassociation, but extracting enough DNA to obtain higher DNA concentrations requires large environmental samples. Our experience has shown that for a complex community (i.e., greater than 100 genotypes), the mass of DNA required is in excess of 1 mg. The rate of reassociation of DNA is directly proportional to the square of the DNA concentration. Thus a twofold dilution of the DNA concentration would result in a fourfold reduction in the reassociation rate.

Therefore, a large amount of sample DNA (1 to 5 mg) must be isolated in order to conduct the reassociation reaction in a reasonable time (i.e., a 1-week incubation at 65°C in 0.12 M sodium phosphate).

For high C_0t values, ^{32}P is not an appropriate label for the probe because its high-energy beta emission is strong enough to break the phosphodiester bonds of the DNA backbone. If C_0t readings take place over more than 2 days, ^3H would be a more appropriate label for the probe. An alternative to using radioactivity is monitoring the hyperchromic shift of DNA. Single-stranded DNA will absorb more strongly at 260 nm than will double-stranded DNA. Thus the rate of reassociation can be measured spectrophotometrically.

Conclusion

The use of nucleic acid hybridization techniques in the study of environmental microbiology will facilitate the analysis of microbial community structures and the monitoring of their dynamics. When combined with techniques for extracting nucleic acids from environmental samples, hybridization makes it possible to gauge the "potential" a community has for performing some physiological function (i.e., catabolism of toxic chemicals, nitrogen fixation, etc.). The proliferation and spread of a gene within a community can be monitored by such techniques; this is important if considering the release of a recombinant organism. Knowledge of a community's potential, coupled with data on the level of expression, will enable investigators to deduce what types of environmental parameters influence the expression of a particular genotype.

References

1. Anderson, D. G., and L. L. McKay. 1983. Simple and rapid method for isolating large plasmid DNA from lactic streptococci. *Appl. Environ. Microbiol.* 46:549–552.
2. Anderson, M. L. M., and B. D. Young. 1985. Quantitative filter hybridization. In B. D. Hames and S. J. Higgins (eds.), *Nucleic Acid Hybridization: A Practical Approach.* IRL Press, Washington, D.C., pp. 73–111.
3. Andrews, B. A., and J. A. Asenjo. 1987. Enzymatic lysis and disruption of microbial cells. *Trends Biotech.* 5:273–277.
4. Brooks, T. L. 1981. Hydroxyapatite. *Calbiochem. Behring,* San Diego, Calif.
5. Church, G. M., and W. Gilbert. 1984. Genomic sequencing. *Proc. Natl. Acad. Sci. USA.* 81:1991–1995.
6. Currier, T. C., and E. W. Nestor. 1976. Isolation of covalently closed circular DNA of high molecular weight from bacteria. *Anal. Biochem.* 76:431–441.
7. Duby, A. 1987. Using synthetic oligonucleotides as probes. In F. M. Ausubel, R. Brent, R. E. Kingston, D. D. Moore, J. G. Seidman, J. A. Smith, and K. Struhl (eds.), *Current Protocols in Molecular Biology.* Greene Publishing Assoc., New York.
8. Fitts, R., M. Diamond, C. Hamilton, and M. Neri. 1983. DNA-DNA hybridization

assay for detection of *Salmonella* spp. in foods. *Appl. Environ. Microbiol.* 46:1146–1151.

9. Flavell, R. A., E. J. Birfelder, J. P. M. Sanders, and P. Borst. 1974. DNA-DNA hybridization on nitrocellulose filters 1. General considerations and non-ideal kinetics. *Eur. J. Biochem.* 47:534–543.

10. Forster, A. C., J. L. McInnes, D. C. Skingle, and R. H. Symons. 1985. Nonradioactive probes prepared by chemical labeling of DNA and RNA with a novel reagent, photobiotin. *Nucl. Acids Res.* 13:745–761.

11. Franklin, F. C. H., and P. A. Williams. 1980. Construction of a partial diploid for the degradative pathway encoded by the TOL plasmid (pWWO) from *Pseudomonas putida* mt-2: Evidence for the positive nature of the regulation by the *xylR* gene. *Mol. Gen. Genet.* 177:321–328.

12. Fredrickson, J. K., D. F. Bezdicek, F. J. Brockman, and S. W. Li. 1988. Enumeration of Tn5 mutant bacteria in soil by using a most probable number—DNA hybridization procedure and antibiotic resistance. *Appl. Environ. Microbiol.* 54:446–453.

13. Grunstein, M., and D. S. Hogness. 1975. Colony hybridization: A method for the isolation of cloned DNAs that contain a specific gene. *Proc. Natl. Acad. Sci. USA* 72: 3961–3965.

14. Hodgson, A. L. M., and W. P. Roberts. 1983. DNA colony hybridization to identify *Rhizobium* strains. *J. Gen. Microbiol.* 129:207–212.

15. Holben, W. E., J. K. Jansson, B. K. Chelm, and J. M. Tiedje. 1988. DNA probe method for the detection of specific microorganisms in the soil bacterial community. *Appl. Environ. Microbiol.* 54:703–711.

16. Hu, N., and J. Messing. 1982. The making of strand specific M13 probes. *Gene* 17: 271–277.

17. Jain, R. K., R. S. Burlage, and G. S. Sayler. 1988. Methods for detecting recombinant DNA in the environment. In *CRC Critical Reviews in Biotechnology,* vol. 8 (1). CRC Press, Boca Raton, Fla., pp. 33–84.

18. Kado, C. I., and S. T. Liu. 1981. Rapid procedure for detection and isolation of large and small plasmids. *J. Bacteriol.* 145:1365–1373.

19. Maniatis, T., E. F. Fritsch, and J. Sambrook. 1982. *Molecular Cloning: A Laboratory Manual.* Cold Spring Harbor Laboratory Press, Cold Spring Harbor, N.Y.

20. Marmur, J. 1961. A procedure for the isolation of deoxyribonucleic acid from microorganisms. *J. Mol. Biol.* 3:208–218.

21. McCarty, M. 1985. *The Transforming Principle.* W. W. Norton and Co., N.Y.

22. Ogram, A., G. S. Sayler, and T. Barkay. 1987. The extraction and purification of microbial DNA from sediments. *J. Microbiol. Meth.* 7:57–66.

23. Ogram, A., and G. S. Sayler. 1988. The use of gene probes in the rapid analysis of natural microbial communities. *J. Indus. Microbiol.* 3:281–292.

24. Pettigrew, C. A., and G. S. Sayler. 1986. The use of DNA: DNA colony hybridization in the rapid isolation of 4-chlorobiphenyl degradative bacterial phenotypes. *J. Microbiol. Meth.* 5:205–213.

25. Priefer, V. 1984. Characterization of plasmid DNA by agarose gel electrophoresis. In A. Puhler and K. N. Timmis (eds.), *Advanced Molecular Genetics.* Springer-Verlag, New York.

26. Radford, A. J. 1983. Comparisons using ^{35}S and ^{32}P labeled DNA for hybridization on nitrocellulose filters. *Anal. Biochem.* 134:269–271.

27. Sayler, G. S., M. S. Shields, E. T. Tedford, A. Breen, S. W. Hooper, K. M. Sirotkin, and J. W. Davis. 1985. Application of DNA-DNA colony hybridization to the detection of catabolic genotypes in environmental samples. *Appl. Environ. Microbiol.* 49: 1295–1303.

28. Sayler, G. S., R. K. Jain, A. Ogram, C. A. Pettigrew, L. Houston, J. Blackburn, and W. S. Riggsby. 1988. Applications for DNA probes with biodegradation research. In *Proceedings of the 4th International Symposium on Microbial Ecology.*

29. Sayler, G. S., and G. Stacey. 1985. Methods for evaluation of microorganism properties. In J. Fiksel and V. T. Covello (eds.), *Biotechnology Risk Assessment: Issues and Methods for Environmental Introductions.* Pergamon Press, New York.

30. Southern, E. M. 1975. Detection of specific sequences among DNA fragments separated by gel electrophoresis. *J. Mol. Biol.* 98:503.

31. Stahl, D. A. 1986. Unity in variety. *Bio/Technology.* 4:625–628.
32. Stroop, W. G. 1988. Simultaneous electroelution and concentration of DNA fragments from agarose gels. *Anal. Biochem.* 169:194–196.
33. Torsvik, V. L., and J. Goksoyr. 1978. Determination of bacterial DNA in soil. *Soil Biol. Biochem.* 10:7–12.
34. Voytas, D. 1987. Autoradiography. In F. M. Ausubel, R. Brent, R. E. Kingston, D. D. Moore, J. G. Seidman, J. A. Smith, and K. Struhl (eds.), *Current Protocols in Molecular Biology.* Greene Publishing Assoc., New York.

Part 3

Fate and Transport

20

Overview:
Fate and Transport
of Microbes

William E. Holben

Mary A. Hood

James M. Tiedje

Demonstrations that microorganisms can be transported in the environment in a viable form capable of reproduction are ample in the public health literature. For example, the transport of disease-causing bacteria in water supplies is well documented. That viable infectious agents can be transported by air is evidenced by the ready transmission of flu and common cold organisms via this vector. Further evidence for the transport of microorganisms by air is readily found in the plant pathology literature.

This part focuses on methods, experimental design, and facilities for monitoring the fate of microorganisms and their transport by biological and abiological means from their immediate environment. The methods described herein should be generally suitable for monitoring not only introduced genetically engineered microorganisms (GEMs) but also introduced wild-type organisms and indigenous populations. A main consideration in monitoring the fate and transport of microorganisms in the environment is the requirement for the detection and enumeration of a specific population of interest against a very complex background of similar organisms.

The early decades of environmental microbiology largely focused on describing microbial populations and communities by using plate and most-probable-number (MPN) counting procedures. In the late 1960s and early 1970s this approach gave way to methods focused more on measuring the processes microorganisms carried out in their natural communities with little regard to understanding the populations responsible. An exception was the fluorescent antibody procedure, pioneered for ecological work by Schmidt and coworkers (6). This technique allows the populations of important groups such as rhizobia and nitrifiers to be monitored without prior culturing as required by the classical methods. It has now become more important to determine the fate of individual populations of microorganisms, and even of specific DNA sequences in these populations, in the environment. The ability to monitor a population at this level is confounded by the complexity of the microbial community and the poorly understood interactions between microbial populations and other biological and abiological components of their environment.

This need to monitor specific types of bacteria within communities has triggered a new wave of methods development which has greatly expanded the capabilities of the microbial ecologist and soil microbiologist both in terms of basic research and for risk assessment analyses of microbial pest control agents (MPCAs) and GEMs. Many of these methods are refinements or extensions of previously used methods such as selective plating (5, 9, 19) and MPN analyses (17). Many others derive from techniques of molecular biology which were developed largely for laboratory studies of the genetics of pure cultures of organisms but have been shown to have applications to studies of populations of organisms in mixed communities in the environment (22, 23, 32, 36). Often, techniques for monitoring the fate and transport of microbes involve combining sampling instrumentation with traditional or new methods of microbial detection and/or identification and appropriate sampling schemes.

The fate of microbial populations added to an environment consists of two components: the physiological features that determine survival and growth of the population at the original site, and the transport and subsequent fate of the portion of the organisms that leave the original site. These two components of fate and the key features within each component are diagrammed in Fig. 20.1 for the agricultural field which is the only habitat in which requests for field testing of GEMs have so far been submitted. The chapters in Part 3 amplify on these components of fate.

It is helpful to visualize these fates because we can then evaluate how easily each can be controlled or measured. For most agricultural field tests, the horizontal transport vectors (Fig. 20.1) can be controlled by choice of site (slope, windbreaks, soil type) and management (dikes, fences, controlled access, surrounding vegetation). The vertical transport vector, leaching, is more difficult to control since one doesn't have control over heavy rainfall, but one may be able to choose a site

SURVIVAL and TRANSPORT

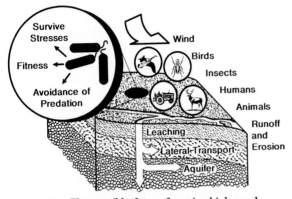

Figure 20.1 The possible fates of a microbial population introduced into an agricultural field. The fate of the portion of the population that remains at the introduction site is determined by physiological features of the genotype such as fitness and ability to withstand stresses and predation. Another portion of the population might also be transported from the site by various horizontal and vertical vectors that deposit organisms at new sites, some of which may support colonization.

where such factors as clay lenses, depth to water table and lateral flow phenomena would minimize this means of dispersal.

A number of possible fates exist for a population of microorganisms whether introduced or indigenous. *Persistence* of a population of microorganisms usually results from the steady-state rates of cell growth and death producing a constant number of cells per gram of soil. Conversely, persistence need not involve growth of cells and replication of DNA but may simply result from stress-survival mechanisms which allow the cells to withstand desiccation, starvation, and other stresses that might result in cell death (11, 25, 33, 35). Methods capable of detecting specific populations of microorganisms may not always distinguish between these mechanisms of persistence but can determine whether or not the organism is present in the sample above the level of detection of that method. Physical factors that affect the persistence and survival of microorganisms include relative humidity, ultraviolet radiation, temperature, pH, nutrient availability, and oxygen tension. Biological factors affecting persistence include competition from other organisms and predation by bacteriovorous organisms. Predictive laboratory experiments to assess the significance of these parameters are difficult since many of the factors (both biological and abiological) may interact to produce synergistic effects and different factors may predominate in the environment at different times of the day or year.

R. Y. Morita's contribution to this part (Chap. 21) considers the survival of microorganisms in the environment. Environmental stresses and competition in the environment and the response of added populations to these stresses will ultimately determine whether or not these populations will persist following introduction. It is noted that each population or strain of bacteria will likely respond differently to environmental factors, and so questions regarding fitness and survival may well require case-by-case studies at least until a database which reasonably represents the many different types of bacteria has evolved. According to Morita, nutrient availability is perhaps the most important among the stresses encountered by organisms in the environment. The available energy in soil and many aquatic systems is generally very limited and any population of organisms added to the environment must immediately adjust to these conditions. Measurements of available energy are confounded by the fact that total organic carbon measurements are not necessarily a true measure of the energy that is actually available to microorganisms (bioavailable carbon). The concept that only a small part of the microbial community is active at any time is explored, as are bacterial strategies for surviving environmental stresses. Of particular interest is the ability of non-spore-forming bacteria to form ultramicrocells (30, 33). These ultramicrocells are generally considered to represent a dormant stage analogous to the formation of endospores that is observed with some gram-positive bacteria. Presumably these ultramicrocells are better able to withstand the stresses of the environment such as nutrient deprivation and desiccation. It is likely that ultramicrocells are physiologically different than the "giant" cells observed in actively growing populations. These physiological differences might be expected to complicate methods of detection and enumeration that require metabolic activity, particularly since some bacteria are known to be able to exist in a viable but nonculturable stage where they cannot be cultured on normal laboratory media without first going through resuscitative steps (5, 11).

Other stresses encountered by bacteria in the environment are also discussed. Among these are water potential and temperature which in terrestrial ecosystems can vary drastically in a 24-h period as well as seasonally. The hydrogen ion concentration (pH), redox potential, and ionic environment are also considered. These parameters can vary far beyond the optima for growth in a number of environments and thus can play a large part in determining the persistence of a population. Morita also considers the role of ultraviolet light and visible light as environmental stresses. As mentioned elsewhere, these parameters have their greatest impact on the survival of bacteria in aquatic environments and during transport through the air as aerosols or particles. Also mentioned is the potential impact of antibiotics produced by other microorganisms in the ecosystem, but a clear demonstration of the competitive importance of these compounds in the environment is still wanting. It is suggested that the methods used to experimentally assess survival and to prepare bacterial inocula for introduction into the environment should be considered carefully (e.g., prestarvation for extended periods before addition to soil) so that the prevailing conditions in the open environment are approximated.

Another possible fate of a population of microorganisms is *die-off* resulting in its loss from the community. Die-off might result from a lack of competitiveness or fitness in the case of an introduced organism (8, 10) or from grazing by protozoa (26) or simply from abiotic stresses resulting from the prevailing environmental conditions. A potential point of confusion that may require clarification is that persistence of a population at a level below the limit of detection of the method being used may be construed as die-off of that population (Fig. 20.2, and also Ref. 37). Therefore statements regarding these phenom-

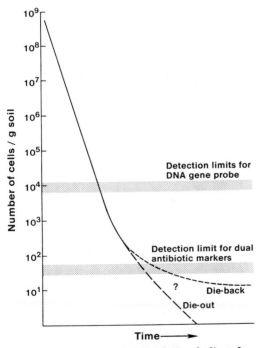

Figure 20.2   The pattern of population decline often observed for bacterial populations inoculated into nonsterile soil. The lower limits of the detection methods do not allow the distinction between complete die-out of the organism and die-back to undetectable levels.

ena should always be qualified as die-off to levels below the limit of detection and not simply as total loss of the population from the community. There are also cases where a population of organisms that was undetectable by certain methodologies was demonstrated to be present and viable when alternative methodologies or recuperative steps to recover stressed cells were employed (11; W. E. Holben, unpublished observations).

A population of microorganisms being monitored may experience a net *growth* in the size of the population if the environmental conditions favor the organism. For example, an indigenous organism or one engineered to mineralize a particular xenobiotic compound may exhibit net growth to densities far greater than the initial inoculation density by virtue of being the sole metabolizer of the novel growth substrate (the xenobiotic compound) in the bacterial community present at a spill site (21). Following depletion of the substrate, the population may demonstrate persistence or die-off to a lower or undetectable level (24).

In the above scenario, the introduced organism could be said to have produced an *effect* on the environment, namely the depletion of an undesirable compound. This would be considered a positive effect of the presence of the organism. It is also possible for populations of organisms to produce deleterious or negative effects on the environment such as pathogenicity, depletion of a nontarget growth substrate, or loss of other populations of microorganisms due to competition from an added or invading organism. The necessity of assessing the potential for negative effects resulting from planned or accidental introduction of microorganisms to a particular environment and the heightened awareness of the commercial and scientific importance of understanding the diversity of microbial communities are the driving forces behind the new wave of methods development in microbial ecology.

If the fate of genetic information is of concern, and not the organism, another fate for consideration is the *transfer of genetic information* between similar or dissimilar organisms. There have been reasonable demonstrations that gene exchange occurs between organisms in the environment (4, 18, 31). It is also possible that genetic information released by cell death or even from viable cells into the environment may be protected from degradation (1, 13) and be available for transfer into new populations via the process of transformation. These phenomena will be particularly important where genetically engineered organisms are involved. Methods which monitor the engineered organism by virtue of an unrelated phenotypic trait have proved useful for tracking an engineered organism (15) but would not be sufficient to monitor the engineered genetic information itself, if that is of ultimate concern. Toward this end, the techniques of molecular biology are being modified and moved from studies of pure cultures of laboratory strains to studies involving environmental samples. With these methods complex mixtures of nucleic acids are isolated from all of the component populations in microbial communities and, using appropriate probes, the engineered genetic information itself is detected (22, 32, 36). Molecular techniques can also be used to detect similar, but not identical, target DNA or RNA sequences which serve as signatures for organisms (e.g., among indigenous populations) that represent unique populations of interest, to identify other populations that perform a similar gene-encoded function, or even to differentiate between specific strains of a bacterial species.

Another potential fate of microorganisms is *transport* from one location to another (e.g., from the site of application of an engineered organism to another, unintended site). Although some bacteria are motile, transport to other geographic locations will generally involve some type of vector. Some sites of initial dispersal may provide adequate conditions for colonization and growth which can then result in secondary transport or dispersal from this site. Assuming the dispersal of some microorganisms to a new site, successful *colonization* at this new site generally would require significant outgrowth of the transported population from what would be an initial low-density inoculum to a population level high enough to be detected. Once a transported microorganism is deposited and becomes established in a new locale, all of the considerations of possible fates again come into play.

Transport may occur by either *biotic* or *abiotic* modes. Biotic vectors include insects, invertebrates such as earthworms, and vertebrates such as birds and humans. It also seems likely that some microorganisms can be transported through soil by other vectors. An example would be the transport of rhizobia through soil in association with the roots of living plants. J. L. Armstrong, who contributes a chapter to this part (Chap. 24), outlines the design of terrestrial microcosms and methods for on-line monitoring of environmental parameters. Among the controllable environmental parameters built into these sophisticated microcosms are light intensity and duration, temperature, relative humidity, rainfall, oxygen and water tension, and the ability to contain both micro- and macroorganisms. Precise directions for the construction of contained microcosms appropriate for monitoring the survival of inoculated bacteria on plants and in soil as well as the transport of these bacteria via insect vectors are given, as are methods for the containment of GEMs or MPCAs being tested in these microcosms. Armstrong also provides insight into appropriate sampling strategies to enumerate bacteria on plants, in soil and insect guts, as well as on insect surfaces using these microcosms. These microcosms have been used to monitor the growth and transport of introduced bacteria via insect vectors (2).

Abiotic transport media or vectors include air, surface water, and ground water. Transport of microorganisms through the air as aerosols and particles by air currents has been demonstrated (3, 7, 28, 29). The residence time of the organisms in the air appears to be inversely related to persistence such that the more time spent in transport, the lower the survival rate. Yet there are reports of the transport of microbial spores over hundreds of miles from the Black Sea to Sweden (7). The time required for *Bacillus* spores to travel this distance was estimated to be 36 h.

The cultivation of soils and harvesting of crops can facilitate transport of microorganisms as particles or when adhered to dust (27). Rainfall can also facilitate the movement of bacteria via aerosolization of the splashes. There are two chapters in this part that address aerial transport of microorganisms. The first, by L. D. Stetzenbach, B. Lighthart, R. J. Seidler, and S. C. Hern (Chap. 22), considers factors influencing the dispersal and survival of aerosolized microorganisms. The focus is on those factors that have implications when developing determinative particle-type models for the dissemination of bacteria via air. Both geometrical and gaussian plume models to describe the transport of particles are described. Among the environmental factors that affect particle distribution are wind, relative humidity, solar radiation, and temperature. The main mechanical consideration is droplet size, which is determined largely by the type of sprayer used; several are discussed. The composition of the droplets (other than aqueous suspension) is also important as this can affect viability. Although the viability of organisms in aerosols is affected by most of the other factors already discussed as well, most existing models for aerial transport fail to account for these parameters.

The second chapter on the topic of aerial dispersal is authored by L. D. Stetzenbach, S. C. Hern, and R. J. Seidler (Chap. 27). This chapter explores experimental methods and field trial strategies appropriate for the detection of airborne organisms. Air monitoring for microorganisms includes assessing both the extent and the distribution of inoculum drift. Four parameters for air monitoring are given, namely, (1) microbial source strength, (2) meteorological conditions, (3) characteristics particular to the target site that would affect dispersal and survival of microorganisms, and (4) the type and placement of sampling equipment. In particular, the various types of samplers are described and their relative advantages and disadvantages compared. Each of these four parameters is considered separately with the overall goal of developing a sensible, effective strategy to monitor the airborne dispersal of microorganisms from the site of application.

It is also known that water can play a large role in the dispersal of microorganisms in aquatic as well as in terrestrial systems. Widespread movement of bacteria in aquatic habitats via currents and aerosolization is well documented. M. A. Hood presents a chapter in this part (Chap. 25) that describes and compares experimental methods for studies of the fate and transport of microorganisms in aquatic systems. As each aquatic system has particular environmental factors which affect the biological components of the system, appropriate sampling strategies must be designed to accommodate these factors in the system being tested. Hood also points out that, even when considering only microbes, there are several parameters to consider in experimental design. For example, there are planktonic microorganisms as well as particle-associated organisms and among these there are the epilithic and epiphytic organisms. The relative importance of these modes of existence can vary from system to system, e.g., planktonic organisms predominate in open water while attached organisms would be of primary importance in aquatic systems characterized by shallow, moving water. Presented in this chapter is a table summarizing the most common field sampling procedures and equipment for aquatic field sampling. Another important consideration for field sampling is sample storage and transport. Storage effects are particularly important in aquatic samples (e.g., compared to soil samples) and must be considered in experimental design. Several examples of storage effects on aquatic samples are cited and recommendations for appropriate transportation and storage prior to detection or enumeration are given.

The use of aquatic microcosms as systems for assessing the fate and transport of microorganisms has also been expanding. Microcosms which better mimic the prevailing environmental conditions at the actual site of interest have been developed. The chapter by C. R. Cripe and P. H. Pritchard (Chap. 23) provides both an overview of important environmental parameters which must be accounted for, as well as providing design features for aquatic microcosms. Considerations of size, scaling, sampling strategies, enumeration methods and temperature and light control are included. Aquatic microcosms can be examined in the laboratory in the absence of environmental variables that can make the interpretation of field data difficult. Another important topic considered in this chapter is the extrapolation and validation of data derived from microcosm studies to the field. Totally contained microcosms which can be shown to provide data similar to what is obtained in the field can potentially be used in lieu of field studies—an important low-risk prelude to field release of GEMs and MPCAs. Data interpretation is key to the successful use of microcosms and the utility of modeling to compensate for factors unique to the field site or artifacts of the system. The authors of this chapter point out that each aquatic microcosm, by design, is site-specific such that baseline tests must be run and that these tests, when conducted in concert with field tests using the same parameters, constitute field calibration. Baseline calibration tests should also highlight the strengths and weaknesses of the system.

It is also known that water plays a large role in the transport of bacteria in terrestrial settings. Bacteria may be moved in the horizontal plane by surface runoff of water following rainfall or the irrigation of agricultural land. For example, the movement of fecal coliform bacteria by runoff from rainfall and snowmelts has been demonstrated (12, 14). Runoff that reaches streams and rivers might well result in transport of microorganisms to great distances from the source inoculum. Bacteria may also be transported by water in the vertical plane by percolation through the soil profile, but it appears that such movement is largely restricted, because soil serves as an effective biological filter (16). Significant vertical movement of particles such as bacteria through soil requires continuous water channels (i.e., saturated soil) which does not normally occur under typical agricultural conditions (20, 34). Bacterial movement through soil is best represented by the movement of particles rather than the movement of chemicals (Scott Smith, personal communication). Thus, much of what is known about chemical movement through soils does not extrapolate well to considerations of microbial transport through soils. Groundwater adds another component for the lateral movement of bacteria in subsurface soils.

The contribution to this part by C. Hagedorn (Chap. 26) outlines strategies for the appropriate design of field studies to monitor the fate and transport of microorganisms in terrestrial systems. Recommendations on choosing field sites to minimize the effects of soil heterogeneity and fertility gradients are given. Plot design is also considered and parameters involved in the proper choice of plot size and shape, block size and shape, and the degree of replication required are explored. Competition effects are also discussed. These are most easily visualized as differences in plant growth but are sure to play a role at the microbial level as well, especially in agroecosystems where much exploration on the use of plant-associated microbes to enhance plant growth and yield is taking place. Hagedorn also provides insight into sampling strategies that provide for ease of identification and measurement while providing high precision and low cost. Much of this discussion is geared toward the sampling of plants in field studies, but the same principles apply to sampling and monitoring microbial populations.

References

1. Aardema, B. W., M. G. Lorenz, and W. E. Krumbein. 1983. Protection of sediment-absorbed transforming DNA against enzymatic inactivation. *Appl. Environ. Microbiol.* 46:417–420.

2. Armstrong, J. L., L. A. Porteous, and N. D. Wood. 1989. The cutworm *Peridroma saucia* (Lepidoptera: Noctuidae) supports growth and transport of pBR322-bearing bacteria. *Appl. Environ. Microbiol.* 55:2200–2205.

3. Bausum, H. T., S. A. Schuab, K. F. Kenyon, and M. J. Small. 1982. Comparison of coliphage and bacterial aerosols at a wastewater spray irrigation site. *Appl. Environ. Microbiol.* 43:28–38.

4. Bender, C. L., and D. A. Cooksey. 1986. Indigenous plasmids in *Pseudomonas syringae* pv *tomato*: Conjugative transfer and role in copper resistance. 1986. *J. Bacteriol.* 165:534–541.

5. Bissonnette, G. K., J. J. Jegeski, G. A. McFeters, and D. G. Stuart. 1975. Influence of environmental stress on enumeration of indicator bacteria from natural waters. *Appl. Environ. Microbiol.* 29:186–194.

6. Bohlool, B. B., and E. L. Schmidt. 1968. Nonspecific staining: Its control in immunofluorescence examination of soil. *Science* 162:1012–1014.

7. Bovallius, A., B. Bucht, R. Roffey, and P. Anas. 1978. Long range air transmission of bacteria. *Appl. Environ. Microbiol.* 35:1231–1232.

8. Bromfield, E. S. P., D. M. Lewis, and L. R. Barran. 1985. Cryptic plasmid and rifampin resistance in *Rhizobium meliloti* influencing nodulation competitiveness. *J. Bacteriol.* 164:410–413.

9. Burr, T. J., and B. Katz. 1982. Evaluation of a selective medium for detecting *Pseudomonas syringae* pv *populans* and *P. syringae* pv *syringae* in apple orchards. *Phytopathology* 72:564–567.

10. Buttner, M. P., and P. S. Amy. 1989. Survival of ice nucleation-active and genetically engineered non-ice-nucleating *Pseudomonas syringae* strains after freezing. *Appl. Environ. Microbiol.* 55:1690–1694.

11. Colwell, R. R., P. R. Brayton, D. J. Grimes, D. B. Roszak, S. A. Huq, and L. M. Palmer. 1985. Viable but non-culturable *Vibrio-cholerae* and related pathogens in the environment: Implications for release of genetically engineered microorganisms. *Biotechnology* 3:817–820.

12. Culley, J. L. B., and P. A. Phillips. 1982. Bacteriological quality of surface and subsurface run-off from manured sandy clay loam soil. *J. Environ. Qual.* 11:155–158.

13. DeFlaun, M. F., J. H. Paul, and D. Davis. 1986. Simplified method for dissolved DNA determination in aquatic environments. *Appl. Environ. Microbiol.* 52:654–659.

14. Doran, J. W., and D. M. Linn. 1979. Bacteriological quality of run-off water from pastureland. *Appl. Environ. Microbiol.* 37:985–991.

15. Drahos, D. J., B. C. Hemming, and S. McPherson. 1986. Tracking recombinant organisms in the environment: b-galactosidase as a selectable non-antibiotic marker for fluorescent Pseudomonads. *Biotechnology* 4:439–444.

16. Edmonds, R. L. 1976. Survival of coliform bacteria in sewage sludge applied to a forest clearcut and potential movement in groundwater. *Appl. Environ. Microbiol.* 32:537–546.

17. Fredrickson, J. K., D. F. Bezdicek, F. E. Brockman, and S. W. Li. 1988. Enumeration of Tn5 mutant bacteria in soil by MPN-DNA hybridization and antibiotic resistance. *Appl. Environ. Microbiol.* 54:446–453.

18. Genthner, F. J., P. Chatterjee, T. Barkay, and A. W. Bourquin. 1988. Capacity of aquatic bacteria to act as recipients of plasmid DNA. *Appl. Environ. Microbiol.* 54:115–117.

19. Gould, W. D., C. Hagedorn, T. R. Bardinelli, R. M. Zablotowicz. 1985. New selective media for enumeration and recovery of fluorescent pseudomonads from various habitats. *Appl. Environ. Microbiol.* 49:28–32.

20. Hagedorn, C., D. T. Hansen, and G. H. Simonson. 1978. Survival and movement of fecal indicator bacteria in soil under conditions of saturated flow. *J. Environ. Qual.* 7:55–59.

21. Holben, W. E., B. M. Schroeter, V. G. Matheson, V. O. Biederbeck, A. E. Smith, and J. M. Tiedje. 1990. Gene probe hybridization and most probable number (MPN) analyses of soil microbial populations following treatment with 2,4-D: Correlation between hybridization and MPN data. (abstracts) 82nd Annual Meeting of the A.S.A., C.S.S.A. and S.S.S.A., San Antonio, Texas.

22. Holben, W. E., J. K. Jansson, B. K. Chelm, and J. M. Tiedje. 1988. DNA probe method for the detection of specific microorganisms in the soil bacterial community. *Appl. Environ. Microbiol.* 54:703–711.
23. Holben, W. E., and J. M. Tiedje. 1988. Applications of nucleic acid hybridization in microbial ecology. *Ecology* 69:561–568.
24. Kilbane, J. J., D. K. Chatterjee, and A. M. Chakrabarty. 1983. Detoxification of 2,4,5-trichlorophenoxyacetic acid from contaminated soil by *Pseudomonas cepacia*. *Appl. Environ. Microbiol.* 45:1697–1700.
25. Kurath, G., and R. Y. Morita. 1983. Starvation-survival and physiological studies of a marine *Pseudomonas* sp. *Appl. Environ. Microbiol.* 45:1206–1211.
26. Liang, L. N., J. L. Sinclair, L. M. Mallory, and M. Alexander. 1982. Fate in model ecosystems of microbial species of potential use in genetic engineering. *Appl. Environ. Microbiol.* 44:708–714.
27. Lighthart, B. 1984. Microbial aerosols: Estimated contribution of combine harvesting to an air shed. *Appl. Environ. Microbiol.* 47:430–432.
28. Lighthart, B., and J. Kim. 1989. Simulation of airborne microbial droplet transport. *Appl. Environ. Microbiol.* 55:2349–2355.
29. Lighthart, B., and A. J. Mohr. 1987. Estimating downwind concentrations of viable airborne microorganisms in dynamic atmospheric conditions. *Appl. Environ. Microbiol.* 53:1580–1583.
30. Morita, R. Y. 1985. Starvation and miniaturization of heterotrophs, with special emphasis on the maintenance of the starved viable state. In M. Fletcher and G. Floodgate (eds.), *Bacteria in the Natural Environment: The Effect of Nutrient Conditions*. Academic Press, New York, pp. 111–130.
31. Morrison, W. D., R. V. Miller, and G. S. Sayler. 1978. Frequency of F116-mediated transduction of *Pseudomonas aeruginosa* in a freshwater environment. *Appl. Environ. Microbiol.* 36:724–730.
32. Ogram, A., G. S. Sayler, and T. Barkay. 1987. The extraction and purification of microbial DNA from sediments. *J. Microbiol. Methods* 7:57–66.
33. Olsen, R. A., and L. R. Bakken. 1987. Viability of soil bacteria: Optimization of plate counting technique and comparison between total counts and plate counts within different size groups. *Microb. Ecol.* 13:59–74.
34. Rake, T. M., C. Hagedorn, E. L. McCoy, and G. F. Kling. 1978. Transport of antibiotic-resistant *Escherichia coli* through western Oregon hillslope soil under conditions of saturated flow. *J. Environ. Qual.* 7:487–494.
35. Roszak, D. B., and R. R. Colwell. 1987. Survival strategies of bacteria in the natural environment. *Microbial. Rev.* 51:365–379.
36. Steffan, R. J., J. Goksyr, A. K. Bej, and R. M. Atlas. 1988. Recovery of DNA from soils and sediments. *Appl. Environ. Microbiol.* 54:2908–2915.
37. Tiedje, J. M. 1987. Environmental monitoring of microorganisms. In J. W. Gillett (ed.), *Prospects for Physical and Biological Containment of Genetically Engineered Organisms*. Proceedings, Ohio Shackleton Point Workshop on Biotechnology Impact Assessment, Ecosystems Research Center Report no. 114, pp. 115–128.

Chapter 21

Survival and Recovery of Microorganisms from Environmental Samples

Richard Y. Morita

Introduction: Brief Review of Major Problems

The research on survival of microorganisms exposed simultaneously to the various environmental factors found in nature and their subsequent recovery has not been the subject of many investigations by microbial ecologists, mainly because it is impossible to duplicate all these environmental factors and fluxes in microcosms. Nevertheless, some headway has been made over the last decade. The complexity of an ecosystem is enormous and this chapter can only touch on this complex nature. This complexity in relation to the microbes is generally not well understood or appreciated. Even with the wide and extreme diversity of ecosystems, microorganisms are found in virtually all environments. The indigenous microflora of any specific environment have adapted to survive the changes that occur in that environment. In addition to foregoing difficulties, we still have to answer the question concerning the term "viable" as it applies to microbes (61), especially when we are now faced with the concept of "viable but not culturable" (67, 79). For instance, Roszak and Colwell (66), in a table, list 47 different methods for the differentiation of living from dead cells as well as a discussion on the viable state of microorganisms.

When laboratory-cultured microbes are placed in the various ecosystems by humans, their survivability and fate come into question. Their ability to cope with new stresses, including other microorganisms and predators, imposed by the new environment into which they are exposed must be investigated further. When there is a variety of environmental factors with their fluctuations, then the realization of the difficulties encountered in the study of survival, fate, and growth or nongrowth of microorganisms makes the problem a difficult task in the laboratory. To elucidate the introduced microbes into an environment, generally viable plate counts are employed. The use of viable plate counts brings into question as to whether the best medium is being employed or the correct medium concentration. If the introduced organism has somehow become "viable but nonculturable," the recovery of the organism becomes more complicated. Unless the organism introduced into the environment has specific antibiotic marker(s) and/ or has a specific unique substrate(s) for growth, it may be very difficult to distinguish between the introduced organisms from the indigenous microflora that have the ability to grow on the same medium. Even in a microcosm, it is difficult to duplicate the numerous environmental factors and their fluctuations as they affect the survival of microorganisms in environmental situations. Each species will respond differently to the various environmental factors and their fluctuations, hence no general rule can be laid down. Furthermore, the spores of various bacteria will survive differently than their vegetative counterpart. Any introduced microorganism, including genetically engineered microorganisms (GEMs), will cause a perturbation in the ecosystem. Is the perturbation transitory or permanent? If transitory, how long do the perturbations last? The stability of the engineered DNA of GEMs introduced into the various environments also becomes another question. In other words, when the GEM dies, does the recombinant DNA of the GEM survive and become involved in the genetics of other organisms in the ecosystem?

In a review, Reanney et al. (63) state the following:

> Studies in microbial genetics have uncovered a wealth of mechanisms that allow defined DNA sequences to migrate among replicons inside and between cells. An extensive understanding of the mechanisms by which DNA may be mobilized, transferred and relocated with the same species or across diverse genetic boundaries is now available. A naive extrapolation of these data from the precise, high manipulated test tube environments to natural habitats, could easily generate the impression the prokaryotic genes are in a perpetual state of flux and that the identity of a given bacterial species is often compromised by the uptake of nonhomologous DNA. Superficially, the scene may appear to be set for a jamboree of unfettered movement of genes, particularly as the microbial flora responds to changing or novel environmental conditions. On the other hand, the repeated isolation of virtually identical biotypes from specific ecological niches, such as the mammalian or the rhizosphere, suggests that bacteria genomes remain stable in spite of the potentially

high rates of gene transfer which can undoubtedly occur or be made to occur under appropriate conditions.

Environmental Factors that Determine Survivability

Availability of nutrients and energy

Soil and sediment. Introduced organisms into any ecosystem must compete for a share of the nutrients. Nearly all ecosystems are oligotrophic. Nutrient availability (including energy) is the most important environmental factor for the growth of any organism, and, in many cases, the energy is needed to offset the stresses of other factors. Competitive growth between various populations of organisms must also be taken into consideration, especially when the nutrient supply is limited. As early as 1923, Waksman and Starkey (74) concluded that nutrient availability was the chief factor affecting microbial numbers and survival in soil. There is little doubt that the primary limiting factor in the control and distribution of microorganisms is the availability of substrate (64). Because of nutrient limitations, Gray and Williams (26) concluded that the bulk of the soil population is inactive. To reaffirm this notion, McLaren (50) concluded that it has not been shown that there is any continually active biomass in soil. After surveying the literature, Williams (75) concluded that soil was grossly oligotrophic. According to Wagner (73), observations of microorganisms in soil only reveal existing structures, and it is difficult to determine whether or not active microbial growth is occurring. Cells seen by direct microscopy may be dormant or persistent dead structures that merely indicate past growth. Thus, in oligotrophic environment, there is a lack of nutrients, mainly energy, for cell growth as well as for the replication of DNA within the cell. However, in the rhizosphere, a variety of low-molecular-weight substrates are known to be released by the living root system. These include simple sugars, oligosaccharides, and amino acids.

Soil microbiologists have known for a long time that many of the soil bacteria isolated by the plating techniques were present in the dormant condition (27). This dormancy, in most instances, is brought about by the lack of available nutrients. The formation of the endospores by *Bacillus* and *Clostridium* and the formation of cysts of *Azotobacter* is also due to the lack of available nutrients. In most instances, the formation of spores in fungi and actinomycetes is induced by the lack of nutrients in the system. Many of the colonies on plates are the result of endospores (26) and this is also true for fungal spores (44). Evidence for the dormant nature of soil microflora being related to the available energy supply for growth is provided by Gray and

Williams (27) and Lockwood and Filonow (44). Microorganisms have developed a range of survival mechanisms that enable them to persist in soil (25). When the subject of microbiology was first initiated, the common practice was to preserve cultures in sterile soil. Preservation or storage of the bacteria, and not growth, was the reason for using soil. In other words, survival takes place well in soil. More recent data by Anderson and Domsch (5) indicate that maintenance energy requirements by microbes in soil, in most cases, appear to consume far more energy than is available in the soil substrate. This could result when microbes are utilizing endogenous substrate during the period when starvation starts. Nevertheless, the situation only illustrates that the available energy in soil is very limited, especially when compared to laboratory culture media, and any organism added to the soil must adjust immediately to this situation. Fungi use about four-fifths of the available energy in soil and bacteria use about one-fifth (4), probably due to the fact that the bacteria cannot withstand the lower water activity when soil dries periodically (see also "Water Activity").

However, when soil is perturbed, many changes take place, including an increase in the availability of nutrients. Through tillage, insolation, irrigation, and fertilization, the soil environment becomes more homogeneous, warmer, with improved texture, humidity, and gas exchange (23). Both Gochenaur (23) and Macfadyen (46) discuss the effects of perturbation of soils in relation to the microbial activity. Likewise, when marine sediments are perturbed, it brings about a short-term change, which, in turn, results in greater microbial growth rates (21).

When cells are deprived of nutrients, many of the species of bacteria will form ultramicrocells (51), which are much smaller than their counterpart growing on rich medium. Generally, these cells are less than 1.0 μm but have been defined as cells having a size of 0.3 μm or less (72). In soil, 60 to 80 percent of the bacteria are ultramicrocells (58), whereas in the marine environment (both nearshore and open ocean), most of the bacteria are between 0.8 and 0.4 μm (48), indicating that these environments are nutrient-poor.

Unfortunately, the amount of nutrients in natural environments is not taken into consideration when transduction, transformation, and transconjugation studies are undertaken in the laboratory. Not only must the physiological state of the GEMs be taken into consideration but also the physiological state of the recipient in the natural environment. For instance, it takes approximately 3 weeks for the metabolic activities of a *Pseudomonas* sp. to become adapted to the starved state (39). Thus, it appears that if one wishes to extrapolate laboratory experiments, one must use recipients that have been starved for at least a few weeks. In most environments, the recipient would be in some stage of starvation due to the lack of energy in the ecosystem. In the laboratory, as soon as the recipient and donor are mixed together they are plated out on rich media, whereas in the ecosystem, such a rich nutrient source does not exist. Genetic transfer is promoted by high population densities (reflection of nutrient availability) and surfaces (site of adsorption of nutrients) where bacteria tend to concentrate in nature. The high numbers of organisms (donor and recipient) used in many studies certainly do not duplicate nature. The extrapolation of these data to a natural condition warrants careful analysis. Furthermore, the GEMs would not become dominant in the ecosystem when there is a limited energy source, especially when one recognizes that most of the organic carbon is recalcitrant. The average generation times of bacteria in the habitat may be over a hundred times longer than those observed in the laboratory (29). The GEMs may survive, but still the question is whether they will become dominant if and when nutrients become available. This becomes an important question. Survival does not necessarily mean that the GEMs will be a hazard in any ecological situation, but the possibility of transferring the engineered DNA into indigenous microflora remains a possibility.

Growth conditions, including the composition of the medium, can affect the survival properties of bacteria. Death rate of exponential-phase cells is greater than cells in the resting phase of growth.

The measurement of organic carbon is not a true measure of the amount of energy that is available to microorganisms, and this will be discussed later. The usable and nonusable energy source in various ecosystems will vary quite remarkably. In addition, when one adds a readily utilizable energy source to soil experiments or aquatic samples, it does not reflect the in situ activity but the potential activity, mainly because the amount of energy added is, in most cases, greater than the indigenous level of the energy source. Furthermore, the more readily utilizable energy sources are released by the root structure of the plants and do not reflect the bulk phase of the soil.

The various photosynthetic bacteria, including the cyanobacteria, use light as the energy source. If any of these organisms are genetically altered, the survival and fate of the photosynthetic bacteria must also be investigated in terms of the influence of light energy.

The studies of Cochran et al. (14), employing ^{14}C-labeled straw and glucose-amended soils, indicate that if a readily utilizable energy source is added to soil, the microbial biomass derived from the readily available carbon rapidly consumes the available carbon and dies. Meanwhile, a different microbial biomass is growing on the less-available carbon components of straw. Its growth is slower but is sustained longer. Thus, microbes themselves deplete the environment of available energy.

Each type of soil is different. Most of the energy in terrestrial systems comes from the cellulose, hemicellulose, or lignin of plants. On a dry-weight basis of the plant, cellulose ranges from 10 to 30 percent, hemicellulose ranges from 10 to 30 percent, and lignin ranges from 5 to 30 percent (2). When plants die and their cellulose is decomposed by the cellulose digesters, the majority of the bacteria must rely on syntrophy in order to obtain their energy. The more readily decomposable material from plants is taken care of by the zymogenic bacteria, but they make their appearance (in terms of numbers) very rapidly and then die off. This leaves the humic substances, tannins and lignins, as well as complexed organic substances. These compounds are recalcitrant and form complexes with the more utilizable organic compounds rendering them unavailable for use by the microbes. In addition, phenolic substances are also formed and they may inhibit the growth of some indigenous bacteria (9). Clay minerals also adsorb some of the organic matter (71).

Generally, measurement of energy is conducted in terms of organic carbon analysis which, in a good loam, is about 25 to 30 mg per gram of soil. However, about 2 to 3 percent of this organic carbon is available for use by microorganisms, probably due to the fact that the more recalcitrant material is slowly being degraded. Readily utilizable carbon is severely limited in soil (13, 43, 44). To illustrate the fact that a chemical measurement is not what is available, Christensen and Blackburn (12) determined that in marine sediment 900 nM of the dissolved alanine was biologically unavailable, and a more realistic free dissolved pool would be 10 nM. Two pools of acetate in sediment were identified by Parkes et al. (60), one being unavailable to *Desulfobacter* sp. The concept that only a small part of the microbial community is active at any given time appears to be attracting more attention (55). Thus, the range of microbial biomass carbon turnover in different soils ranged from 0.4 (36) to 5.5 (49) times per year. Nedwell and Gray (55) compiled a table showing that in soils, generation times ranged from 30 to 15,168 h, and in marine sediments, from 3.8 to 3049 h.

Within the rhizosphere, various sugars, amino acids, organic acids, fatty acids and sterols, growth factors, nucleotides, flavonones, enzymes, and other compounds which form the exudates from the root system of plants have been identified. These exudates were identified from plants grown in microbe-free environments. Naturally, the qualitative and quantitative nature of the exudates will vary according to the type of plant. Further characterization of these compounds is summarized by Rovira (68). However, most organic molecules are not free but bound in polymers (2). Most of the microbes in the rhizosphere closely surround the roots, and these exudates are their main source of nutrients. GEMs added to the soil will have to be able to make their way to the area adjacent to the roots in order to benefit from these nutrients. In addition, they will have to compete with the indigenous microflora already present. Although there may be more organic matter in the rhizosphere than the surrounding soil, the number of bacteria is much larger and competition for the nutrients is still as great or greater than the surrounding soil.

Aquatic environments. The amount of nutrients will depend on the freshwater system in question since there are oligotrophic to eutrophic systems. In many of the eutrophic situations, industrial sewage and animal pollution add to the organic content of the water or cause the eutrophic condition.

Generally, the ocean is considered oligotrophic with nearshore waters having a higher organic matter content. In nonpolluted areas of the aquatic environment, there is approximately 10 mg of organic carbon per liter. Approximately 60 to 70 percent of the organic carbon in seawater is not available to organisms because of its recalcitrant nature. Nevertheless, pollutants can change the systems in localized areas. The availability of energy in ecosystems for bacteria is discussed by Morita (52).

Response to nutrient deprivation. In general there are four responses to the lack of nutrients (energy) by bacteria (Fig. 21.1) (1, 51). Microorganisms represented by line C of Fig. 21.1 have been studied the most. Organisms representing lines A and C have been studied by Amy and Morita (1) and line B by Jones and Morita (35). The increase in cell number in the early phase of starvation, as indicated by line C of Fig. 21.1, is due to fragmentation but depends on the concentrations of cells introduced into the starvation menstruum (56). This

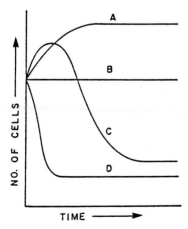

Figure 21.1 Starvation-survival patterns of microorganisms. [*After Morita (45).*]

increase in the number of cells also occurs in *Pseudomonas* as well as in vibrios. In all likelihood, GEMs grown in rich media and placed in certain environments will also increase in numbers due to the fragmentation of the large "giant" cells. Many of the organisms, when deprived of nutrients, form ultramicrocells (45, 56). [The terms minibacteria, nanobacteria, picobacteria, and dwarf cells are not employed for the reasons stated by Torrella and Morita (72).] As stated previously, ultramicrocells are found in soil as well as aquatic environments. However, cells do not necessarily form ultramicrocells when starved. One strain of nitrifying bacteria does not increase or decrease in size or numbers upon ammonia deprivation (35).

The large surface-volume ratio of ultramicrocells is an advantage for the cell to obtain nutrients as well as to avoid predation. Studies in this laboratory have shown that when cells are grown at a low dilution rate in a chemostat, the cells are ultramicrocells and that ultramicrocells reproduce ultramicrocells (53, 54). Thus in any oligotrophic environment, ultramicrocells reproduce ultramicrocells and do not go through the large cell stage seen when grown under ideal laboratory-rich media. The slow growth of organisms, including GEMs, may aid their ability to survive under the various ecosystems. However, GEMs may be at a disadvantage since they must undergo the starvation process due to the lack of energy in the ecosystem as well as compete with the native microflora and predation by protozoans. Cells resulting from slow growth rate are probably better suited for long-term survival than cells grown at a faster growth rate. The ability of Ant-300 (a vibrio) to survive hydrostatic pressure results when the cells are first starved (57).

There are other differences between the starved ultramicrocell and the unstarved parent cell. Drastic decreases in the total lipid and phospholipid content of *Vibrio cholerae* occur when the organism is starved (34) as well as the trans-to-cis fatty-acid ratio (30). When Ant-300 cells are starved, it is more difficult to break the cells to make membrane preparations (M. Slominski, personal communication). In cells deprived of nutrients, it is known that a high-affinity uptake system for arginine results (22). Thus, it appears that membrane changes occur with cells deprived of nutrients, and these changes in membrane of an indigenous organism in poor nutrient environments may affect its ability to interact in transduction and conjugation with any introduced GEM.

When energy is depleted or is insufficient to provide for growth and metabolism, the cell enters metabolic arrest (or suspended animation). This physiological state permits the cells to retain their ability to grow and reproduce when all growth conditions are met in its environment. Metabolic arrest could easily happen in the deep sea, aquifers, and other ecosystems where utilizable energy is depleted. The genetically engineered portion of the DNA of GEMs would still be present but nothing will happen unless nutrients become available.

Water activity

The concentration of water in soil (except in waterlogged soil) is constantly undergoing a change due to rain, irrigation, drainage by gravity, and dried by evaporation and extraction by plant roots; thereby, changes in the water potential will vary accordingly. Most microcosm experiments dealing with the survival of GEMs in soil are conducted at 100 percent of the soil's water-holding capacity or greater, hence the extrapolation of the data to natural ecosystems should be done with great caution.

Water potential does play a great role on microbial growth and activity (29, 31, 68, 76, 77) and acts as selective factor (29). The physical configuration of water in soil which related directly to water potential may influence the motility of certain microorganisms as well as diffusion of gases, nutrients, and exudates to and away from sites of biological activity (59). Soil water potential is the sum of matric, osmotic or solute, and pressure potentials, but the latter component under most conditions can be ignored. If the water potential is over -1 bar, the effects of the osmotic or solute potential are negligible because of the extreme dilution (29). Growth of plants and microorganisms often respond to changes in water potential of several bars or less (15, 16).

The permanent wilting point is around -15 bars. Under field conditions, plants can often dry a soil to potentials well below -15 bars (59). Below -1 bar, the water content–water potential relationship of soil is dominated by surface adsorption effects (59). When the thickness of adsorbed water films is reduced to six or eight molecular layers of water, plants are virtually unable to extract the water and growth ceases (65). Many bacterial activities in soil have been shown to decrease sharply as matric potential falls to between -0.5 and -3 bars. With decreasing water potentials, the activity of bacteria appear to be limited first, mainly because the decreasing portion of water-filled pores results in reduced mobility of bacterial cells and limited availability of nutrients (69). As the matric potential declines to between -3 to -6 bars, bacterial respiration falls rapidly, and at -20 bars was negligible (78). However, total microbial respiration showed little decline between -8 and -30 bars (matric potential) and most of this was probably attributable to fungi. Cook and Papendick (15) stressed that there is need to take into consideration the heterogeneous nature of water potential in field soil. In surface soil, water potential may be less than -100 bars, while that in the rhizosphere at greater depth may be only -30 bars.

A consistent feature in comparison of microbial activity under osmotic- and matric-induced water stress is that activity is reduced substantially more by lowering the matric potential than by lowering osmotic potential. Solute diffusion to and from microorganisms is more severely limited to systems with matric-induced stress than in osmotic systems with higher water content (31). Thus, reduced activity results more from an indirect effect on substrate availability and transport of metabolic products rather than a direct effect of water potential. Since the water content of the microorganisms stays almost constant, the major contribution of mass flow to solute uptake would come from solutes moving the water that flows past the microorganisms in the normal course of water infiltration and distribution in soil, especially when microorganisms are restricted to a microsite. Therefore, the mass flow is generally unimportant in supplying nutrients to bacteria. Mass flow is important, however, in replenishing nutrient in the bulk soil solution. Bacterial movement becomes negligible in soils drained to matric potentials between -0.2 and 1 bar (78). Under conditions where the rate of substrate supply cannot meet the substrate requirements for growth, the microbial population becomes energy-starved, and survival becomes a primary physiological concern (31). The ability to survive desiccation is also a function of temperature as well as the species in question (25).

The water potential of marine environment, with a salinity of 35 percent, is about -28 bars and halophiles normally grow at water potential between -274 to 422 bars. These latter organisms are found in salt lakes.

The method for measuring the water potential in soil is by use of a membrane pressure plate which is reasonably accurate, but this is time consuming when a large number of soil samples need to be measured. The simpler method to measure the effect of water potential is to use polyethylene glycol to create different water potentials. With polyethylene glycol, a vapor osmometer can be employed to measure the water potential.

If spray application of GEMs is used in agriculture, the spray is subject to desiccation. Once airborne, the microbes are rapidly desiccated to an extent depending on the temperature and humidity of the atmosphere. When it hits the soil, further desiccation may take place. Both these desiccation processes will have a deleterious effect on the GEM. At certain relative humidities, oxygen is toxic to airborne bacteria (19, 32). Rehydration conditions affect the viability of populations of *Escherichia coli* from aerosols at certain critical humidities (17, 18).

Desiccation will also take place if GEMs are applied to seeds, unless the seeds are kept and planted in the wet state.

Temperature

The temperature flux must also be taken into consideration in the survival of GEMs, especially when cold injury to the GEM can take place during periods of frost or freezing. Different cooling and warming rates will produce different kinds of damage to cells (37). Microcosm studies at room temperature or at a set temperature generally do not undergo a temperature drop to 0°C or below.

Hydrogen ion concentration

The pH of soil is generally between 4.0 and 8.5, whereas in the aquatic environments it lies between 6.5 and 8.5. In the marine environment, it is usually between 8.0 and 8.3. Low-pH environments do occur in nature, such as in peat bogs and acid mine drainage. GEMs, cultured in laboratory media with pH values around 7.0, when introduced into an ecosystem, will probably experience a pH value not optimum for its growth. The adjustment to the new pH environment must also be made. Naturally, if the pH values are below the minimum or above maximum pH value of the GEM, it will not grow. Thus, a change from optimal pH to a suboptimal pH will be a further stress on the GEM affecting its growth and survival in the ecosystem into which it is introduced.

Redox potential

Most of the biosphere is aerobic due to the pressure of oxygen. All soil organisms respire, and this respiration changes the gases within the soil creating carbon dioxide concentrations 10 to 100 times greater than that in the atmosphere. Thus, there are gradients of both these important gases. Soil microorganisms under restricted aeration might compete with the roots of the higher plants for oxygen (70). In ecosystems where the diffusion of oxygen from the atmosphere is exceeded by the utilization of oxygen, anoxic conditions will result which can then result in an anaerobic environment. Microhabitats may have a different Eh value than the main bulk portion of the ecosystem.

Salinity and ionic environment

GEMs may encounter saline environments, such as the ocean and saline lakes, due to freshwater runoff from rain or irrigation. In turn, this runoff may enter streams, rivers, and finally into a saline lake or the ocean. How terrestrial microorganisms survive the saline environment becomes an important question. However, if marine GEMs are introduced into the marine environment, we are also faced with their tolerance to various salinities that they must face in nearshore environments. Most marine bacteria are stenohaline and require Na^+. There are many reports that marine bacteria can be grown on freshwater media, but it should be recognized that media constituents contain varying amounts of Na^+. Very few organisms can tolerate high salinity found in salt lakes where the water evaporation exceeds the water input (40). Many microorganisms can withstand or survive in high salinity but will not grow or multiply. Whether terrestrial GEMs can tolerate, survive, or grow in various salinities should also be addressed.

Pressure

Osmotic pressure. Osmotic pressure can be created by salts but will not be discussed further since the salinity and ionic environment are briefly presented above.

Hydrostatic pressure. In all likelihood, terrestrial GEMs will not be exposed to the deep sea unless they are dumped into the deep sea. Likewise, any GEMs will probably not experience the hydrostatic pressure in the deep sea unless the introduction of the GEM is near an area where convergence (downwelling) takes place. Nevertheless, when sewage is placed in a simulated deep-sea environment, many of the sewage microorganisms do survive (8).

Ultraviolet radiation and visible light

Solar radiation plays a major role in the temperature of the environment. Due to solar radiation, soils can reach temperatures above the maximal temperature for growth of many organisms. It also plays a role in the desiccation of soil.

Although the earth receives various types of radiation, the most lethal to microorganisms is ultraviolet radiation, especially wavelengths of 260 nm. Ultraviolet light, particularly the wavelengths 250 to 260 nm, is absorbed most strongly by nucleic acids. The main DNA photoproducts formed during ultraviolet radiation are the dimers between adjacent pyrimidines. The ability of bacteria to survive irradiation varies some 1000-fold from one species to another and the amount of water in the cell has a profound effect on the radiation sensitivity. Other factors that enter the picture are the oxygen tension, temperature, and medium used for recovery (10).

The amount of ultraviolet radiation that hits the earth's surface depends on the time of day and weather conditions (seasonal as well as local conditions). Thus, this factor also fluctuates. If experiments on the effect of ultraviolet radiation effects on GEMs are carried out in the laboratory, it must be remembered that the penetration of ultraviolet light is weak. Hence, glassware must not obstruct the penetration of the ultraviolet light. In some cases, GEMs may be sprayed onto plants. During this spray period, the GEMs are exposed to the ultraviolet radiation. This radiation coupled with desiccation will definitely have a deleterious effect on the GEMs.

Visible light also has a deleterious effect on microorganisms and it appears unclear as to the mechanisms by which photodynamic action occurs. The failure of bacteria to form colonies when exposed to visible light has been studied by many investigators (38).

Antibiotics

Although it is known that many soil microbes can produce antibiotics, large-scale effects of antibiotics in nature have not been demonstrated (20). There may be two reasons for this in soil: (1) they are not produced in nature, and (2) microbes can develop resistance to antibiotics by acquisition of the R-factor (plasmids). On the other hand, antibiotics are known to be produced on a small scale (microenvironment) and can have an effect (2).

Many aquatic microorganisms also produce antibiotics and these antibiotics have been attributed to the bactericidal action of seawater (6, 28, 37). However, other investigators attribute the bactericidal action of seawater to heavy metals.

What effect antibiotics in ecosystems will have on introduced GEMs remains questionable.

Bacterivory and predation

Bacterivory and predation are common in soil as well as aquatic ecosystems. GEMs introduced into an ecosystem are also subjected to bacterivory and predation. Excellent documentation exists concerning *Bdellovibrio bacterivorus* predatory action on most gram-negative bacteria. The myxobacteria also have the ability to prey on bacteria. Protozoans also have the capacity to prey on bacteria (11). Even organisms higher on the evolutionary scale have the ability to utilize bacteria as food. Most bacteria in soil are ultramicrocells (7, 58) as well as in the marine environment (48). However, when first applied to any ecosystem, GEMs are probably "giant" cells compared to most of the indigenous bacteria, mainly because the GEMs have been grown under optimal laboratory conditions. Predators can distinguish between cell size, and naturally the larger cells are preferred by the predators (24).

There have been many experiments dealing with bacterial predators where the predator has been isolated and fed the prey. Trying to design experiments in the laboratory in relation to the ecosystem dealing with prey-predator interactions is difficult. The main questions that arise when employing sterilized soil or water samples are (1) are the right predators present, (2) is the ratio of the various predators equal to that in the environment, (3) is the right prey present? In addition one also has to take into consideration all the physical and chemical factors of the ecosystem in question.

Predation also occurs when bacteria attach to surfaces. Filter-feeding organisms will intake particles with thigmotactic bacteria. If the GEM is a thigmotactic microorganism, it could also be consumed by a filter-feeding organism. Even if intense predation occurs, it will not eliminate the prey (3) so that it is, more than likely, that introduced GEMs would not be completely eliminated.

Seasonal variabilities

The seasons of the year are also important variables since they are associated with temperature, water potential (in soil), growth of plants and other organisms (especially those involved in predation of microbes), etc. Survival and fate of organisms must be taken into consideration in relation to the seasons of the year. This may demand that studies be carried out over a year in order to include this environmental factor.

Seasonal variability also brings into consideration the factor of time. If GEMs are placed in any ecosystem, then the questions are (1) how long will they survive in the ecosystem, (2) what is the physiological state of the organism with time, (3) how genetically stable are they with time, and (4) will the genetic material remain stable without the cell? Very few studies take into consideration long periods of exposure to the environmental elements.

Environmental Testing

There is no substitute for the field-testing of GEMs. Unfortunately, regulations do not permit field-testing of GEMs without permission.

Nevertheless, permission can be obtained, as evidenced by the field study conducted by Lindow et al. (42). Field tests take into consideration all the various environmental fluxes that occur in the ecosystem into which the GEMs are to be introduced.

Experiments to determine the effect of one or two environmental factors on microorganisms in the laboratory are rather easy to set up. To duplicate the environment in the laboratory is an impossible task. In lieu of field-testing, various types of microcosms can be designed to duplicate the ecosystem as closely as possible. Numerous types of microcosms and their limitations are discussed by Pritchard and Borquin (62). These authors also discuss the precautions that must be taken into consideration in the use of microcosms.

Recovery or Detection of Viable Microorganisms from Test Material

The use of DNA probes for detecting GEMs in natural systems is not a sensitive method. Holben et al. (33) recently reported that the sensitivity of their method was the ability to detect *Bradyrhizobium japonicum* at densities as low as 4.3×10^4 cells per gram of soil (dry weight). However, one has to realize that the extraction procedure will also extract DNA from dead cells that remain intact. Amplification of DNA from ecological samples also runs into the difficulty of not knowing whether the DNA comes from dead, dormant, or living cells. Thus, the dead and dormant cells are not contributing to the physiological processes that occur in soil. Dormant cells could be a risk factor when dormancy is broken and when, if ever, there is enough energy in the ecosystem to permit the GEMs to become a risk factor.

The spread-plate method on various selective media is still the most sensitive method compared to detection by fluorescent antibody, DNA gene probes, rRNA sequencing, and fingerprinting (41). If GEMs undergo some kind of injury when placed in a test system, then resuscitation measures may be taken. If resuscitation is necessary, are these GEMs a problem?

The main questions concerning the recovery or detection of GEMs by the spread plate method from any environment are the following:

1. Is the right medium being employed? If so, is the strength of the medium employed the best to use? In this latter situation, starved cells have a better ability to form colonies on dilute agar medium than regular-strength medium. Hence, how long should the plates be incubated before they are counted?

2. Is the right type of diluent (e.g., buffer, dilute medium, saline, etc.) being employed for plate counts?

3. Are the physical and chemical conditions of incubation correct for the GEMs?

Conclusion

It is nearly impossible to duplicate the various ecosystems into which GEMs may be placed, mainly because of the various environmental factors, their fluxes, and changes. It is quite impossible, also, to specify any microbe in precise terms of its growth, structure, function, survivability, or fate without specifying the environmental conditions (and its history) prevailing at the time when the system was tested.

If field-testing cannot be performed, then systems that duplicate nature as much as possible should be employed. This also includes the physiological state of the donor and recipients of recombinant DNA. Whatever system is employed, stabilization of the system should occur before the system receives any test sample. The main logical and practical way to test introduced GEMs into any ecosystem is to field-test the parent strains of the GEMs that have been genetically altered with an innocuous gene(s) so that it provides a means by which it can be recovered from the ecosystem.

References

1. Amy, P. S., and R. Y. Morita. 1983. Starvation-survival patterns of sixteen freshly isolated open-ocean bacteria. *Appl. Environ. Microbiol.* 45:109–115.
2. Alexander, M. 1977. *Introduction to Soil Microbiology*, 2d ed. Wiley, New York.
3. Alexander, M. 1981. Why microbial predators and parasites will not eliminate their prey and hosts. *Ann. Rev. Microbiol.* 35:113–133.
4. Anderson, J. P., and K. H. Domsch. 1975. Measurement of bacterial fungal contributions to respiration of selected agricultural and forest soils. *Can. J. Microbiol.* 21:314–322.
5. Anderson, T.-H., and K. H. Domsch. 1985. Maintenance carbon requirements of actively-metabolizing microbial populations under *in situ* conditions. *Soil Biol. Biochem.* 17:197–203.
6. Aubert, M., J. Aubert, M. Gauthier, and S. Daniel. 1966. Origine et nature des substances antibiotique présentes dans le mileau marin. Deuxième Partie: Méthodes et techniques d'études. *Rev. Int. Oceanogr. Med.* 1:27–34.
7. Bakken, L. R., and R. A. Olsen, 1987. The relationship between cell size and viability of soil bacteria. *Microb. Ecol.* 13:103–114.
8. Baross, J. A., F. J. Hanus, and R. Y. Morita. 1975. The survival of human enteric and other sewage microorganisms under simulated deep sea conditions. *Appl. Microbiol.* 30:309–318.
9. Benoit, R. E., and R. L. Starkey. 1968. Enzyme inactivation as a factor in the inhibition of the decomposition of organic matter by tannins. *Soil Sci.* 105:203–208.
10. Bridges, R. A. 1976. Survival of bacteria following exposure to ultraviolet and ionizing radiation. *Symp. Soc. Gen. Microbiol.* 26:183–208.

11. Casida, L. E., Jr. 1988. Minireview: Nonobligate bacterial predation of bacteria in soil. *Microb. Ecol.* 15:1–8.
12. Christensen, D., and T. H. Blackburn. 1982. Turnover of ^{14}C-labelled acetate in marine sediments. *Mar. Biol.* 71:113–119.
13. Clark, F. E. 1965. The concept of competition in microbial ecology. In K. F. Baker and C. W. Snyder (eds.). *Ecology of Soil-Borne Plant Pathogens.* University of California Press, Berkeley, pp. 339–345.
14. Cochran, V. L., K. A. Horton, and V. C. Cole. 1988. An estimation of microbial death rate and limitations of N or C during wheat straw decomposition. *Soil Biol. Biochem.* 20:293–298.
15. Cook, R. J., and R. I. Papendick. 1972. Influence of water potential of soils and plants on root disease. *Annu. Rev. Phytopathol.* 10:349–374.
16. Cook, R. J., and R. I. Papendick. 1978. Role of water potential in microbial growth and development of plant disease, with special reference to postharvest pathology. *Hort. Sci.* 13:559–564.
17. Cox, C. S. 1966. The survival of *Escherichia coli* atomized into air and into nitrogen from distilled water and from solution of protecting agents as a function of relative humidity. *J. Gen. Microbiol.* 43:383–399.
18. Cox, C. S. 1966. The survival of *Escherichia coli* in nitrogen under changing conditions of relative humidity. *J. Gen. Microbiol.* 45:283–288.
19. Cox, C. S., S. J. Gagen, and J. Baxter. 1974. Aerosol survival of *Serratia marcescens* as a function of oxygen concentration, relative humidity and time. *Can. J. Microbiol.* 20:1529–1534.
20. Doetch, R. N., and T. M. Cook. 1973. *Introduction to Bacteria and Their Ecobiology.* University Park Press, Baltimore.
21. Findlay, R. H., M. B. Trexler, and D. C. White. 1990. Response of a benthic microbial community to biotic disturbance. *Mar. Ecol. Prog. Ser.* 62:135–148.
22. Geesey, G. G., and R. Y. Morita. 1979. Capture of arginine at low concentrations by a marine psychrophilic bacterium. *Appl. Environ. Microbiol.* 38:1092–1097.
23. Gochenaur, S. E. 1981. Response of soil fungal communities to disturbance. In D. T. Wicklow and G. C. Carroll (eds.). *The Fungal Community.* Marcel Dekker, pp. 459–479.
24. Gonzalez, J. M., E. B. Sherr, and B. F. Sherr. 1990. Size-selective grazing on bacteria by natural assemblages of estuarine flagellates and ciliates. *Appl. Environ. Microbiol.* 56:583–589.
25. Gray, T. R. G. 1976. Survival of vegetative microbes in soil. *Symp. Soc. Gen. Microbiol.* 26:327–364.
26. Gray, T. R. G., and S. T. Williams. 1971. Microbial productivity in soil. *Symp. Soc. Gen. Microbiol.* 21:255–286.
27. Gray, T. R. G., and S. T. Williams. 1971. *Soil Microorganisms.* Oliver and Boyd, Edinburgh.
28. Greenberg, A. E. 1956. Survival of enteric organisms. *Pub. Health Rep.* 71:77–86.
29. Griffin, D. M. 1981. Water potential as a selective factor in microbial ecology of soils. *Soil Sci. Amer. Spec. Publ. No.* 9:141–151.
30. Guckert, J. B., M. A. Hood, and D. C. White. 1986. Phospholipid ester-linked fatty acid profile changes during nutrient deprivation of *Vibrio cholerae*: Increases in the *trans/cis* ratio and proportions of cyclopropyl fatty acids. *Appl. Environ. Microbiol.* 52:794–801.
31. Harris, R. F. 1981. Effect of water potential on microbial growth and activity. *Soil Sci. Amer. Spec. Publ.* 9:23–96.
32. Hess, G. E. 1965. Effects of oxygen on aerosolized *Serratia marcescens. Appl. Microbiol.* 13:781–787.
33. Holben, W. E., J. K. Jansson, B. K. Chelm, and J. M. Tiedje. 1988. DNA probe method for the detecting of specific microorganisms in the soil bacterial community. *Appl. Environ. Microbiol.* 54:703–711.
34. Hood, M. A., J. B. Guckert, D. C. White, and F. Deck. 1986. The effect of nutrient (carbon) deprivation on the levels of lipids, carbohydrates, DNA, RNA, and protein in *Vibrio cholera. Appl. Environ. Microbiol.* 52:788–793.

35. Jones, R. D., and R. Y. Morita. 1985. Survival of a marine ammonium oxidizer under energy-source deprivation. *Mar. Ecol. Prog. Ser.* 26:175–179.
36. Jenkinson, D. S., and J. N. Ladd. Microbial biomass in soil-measurement and turnover. *Soil Biochem.* 5:415–471.
37. Ketchum, B. H., J. C. Ayres, and R. F. Vaccaro. 1952. Progress contributing to the decrease of coliform bacteria in a tidal estuary. *Ecology* 33:247–258.
38. Krinsky, N. I. 1976. Cellular damage initiated by visible light. *Symp. Soc. Gen. Microbiol.* 26:209–240.
39. Kurath, G., and R. Y. Morita. 1983. Starvation-survival physiological studies in a marine *Pseudomonas* sp. *Appl. Environ. Microbiol.* 45:1206–1211.
40. Kushner, D. J. 1968. Halophilic bacteria. *Adv. Appl. Microbiol.* 10:73–97.
41. Levin, M. A., R. Seidler, A. W. Borquin, J. R. Fowle, III, and T. Barkay. 1987. EPA developing methods to assess environmental release. *Biotechnology* 5:38–45.
42. Lindow, S. E., G. R. Knudsen, R. J. Seidler, M. V. Walter, V. W. Lambou, P. S. Amy, S. Schmedding, V. Price, and S. Hern. 1988. Aerial dispersal and epiphytic survival of *Pseudomonas syringae* during a pretest for the release of genetically engineered strains into the environment. *Appl. Environ. Microbiol.* 54:1557–1563.
43. Lockwood, J. L. 1977. Fungistasis in soils. *Biol. Rev.* 52:1–43.
44. Lockwood, J. L., and A. B. Filonow. 1981. Responses of fungi to nutrient-limiting conditions and to inhibitory substances in natural habitats. *Adv. Microb. Ecol.* 5:1–61.
45. MacDonell, M. T., and M. A. Hood. 1982. Isolation and characterization of ultramicrobacteria from a Gulf Coast estuary. *Appl. Environ. Microbiol.* 43:566–571.
46. Macfadyen, A. 1970. Soil metabolism in relation to ecosystem energy flow and to primary and secondary production. In J. Phillipson (ed.), *Methods for the Study of Production and Energy Flow in Soil Communities*. Paris: Unesco.
47. MacLoed, R. A., and P. H. Colcott. 1976. Cold shock and freezing damage to microbes. *Symp. Soc. Gen. Microbiol.* 26:81–110.
48. Maeda, M., and N. Taga. 1983. Comparison of cell size of bacteria from four marine localities. *La Mer* 21:207–210.
49. McGill, W. B., H. W. Hunt, R. G. Woodmansee, and J. O. Reuss. 1981. Phoenix—a model of the dynamics of carbon and nitrogen in grassland soils. *Ecol. Bull. Stockholm.* 33:49–115.
50. McLaren, A. D. 1973. A need for counting microorganisms in soil mineral cycles. *Environ. Lett.* 5:143–154.
51. Morita, R. Y. 1985. Starvation and miniaturisation of heterotrophs, with special emphasis on the maintenance of the starved viable state. In M. Fletcher and G. Floodgate (eds.), *Bacteria in the Natural Environment: The Effect of Nutrient Conditions*. Academic Press, New York, pp. 111–130.
52. Morita, R. Y. 1988. Bioavailability of energy and its relationship to growth and starvation-survival in nature. *Can. J. Microbiol.* 34:436–441.
53. Moyer, C. L., and R. Y. Morita. 1989. Effect of growth rate and starvation-survival on the viability and stability of a psychrophilic marine bacterium. *Appl. Environ. Microbiol.* 55:1122–1127.
54. Moyer, C. L., and R. Y. Morita. 1989. Effect of growth rate and starvation-survival on the cellular DNA, RNA, and protein of a psychrophilic marine bacterium. *Appl. Environ. Microbiol.* 55:2710–2716.
55. Nedwell, D. B., and T. R. G. Gray. 1987. Soils and sediments as matrices for microbial growth. *Symp. Soc. Gen. Microbiol.* 41:23–54.
56. Novitsky, J. A., and R. Y. Morita. 1977. Survival of a psychrophilic marine vibrio undergoing long-term nutrient deprivation. *Appl. Environ. Microbiol.* 33:635–641.
57. Novitsky, J. A., and R. Y. Morita. Starvation induced barotolerance as a survival mechanism of a psychrophilic marine vibrio in the waters of the Antarctic Convergence. *Mar. Biol.* 49:7–10.
58. Olsen, R. A., and L. R. Bakken. 1987. Viability of soil bacteria: Optimization of plate-counting technique and comparison between total counts and plate counts within different size groups. *Microb. Ecol.* 13:59–74.

59. Papendick, R. I., and G. S. Campbell. 1981. Theory and measurement of water potential. *Soil Sci. Amer. Spec. Publ. No.* 9:1–22.
60. Parkes, R. J., J. Taylor, and D. Jorck-Ramberg. 1984. Demonstration using *Desulfovibrio* sp., of two pools of acetate with different biological availabilities in marine pore water. *Mar. Biol.* 83:271–276.
61. Postgate, J. R. 1967. Viability measurements and the survival of microbes under minimum stress. *Adv. Microb. Physiol.* 1:1–23.
62. Pritchard, P. H., and A. W. Borquin. 1984. The use of microcosms for evaluation of interactions between pollutants and microorganisms. *Adv. Microb. Ecol.* 7:133–215.
63. Reanney, D. C., P. C. Gowland, and J. H. Slater. 1983. Genetic interactions among microbial communities. *Symp. Soc. Gen. Microbiol.* 34:379–421.
64. Richards, B. N. 1987. *The Microbiology of Terrestrial Ecosystems.* Longman Scientific & Technical, Essex, England.
65. Richards, L. A. 1960. *Advances in Soil Physics. 7th Int. Congr. Soil. Sci.* I:67–79.
66. Roszak, D. B., and R. R. Colwell, 1987. Survival strategies of bacteria in the natural environment. *Microbiol. Rev.* 51:365–379.
67. Roszak, D. B., D. J. Grimes, and R. R. Colwell. 1984. Viable but nonrecoverable stage of *Salmonella enteritidis* is aquatic systems. *Can. J. Microbiol.* 30:334–338.
68. Rovira, A. D. 1965. Plant root exudates and their influence on soil microorganisms. In K. F. Baker and W. C. Snyder (eds.), *Ecology of Soil-Borne Pathogens.* University of California Press, Berkeley, pp. 170–186.
69. Sommers, L. E., C. M. Gilmour, R. E. Wildung, and S. M. Beck. 1981. The effect of water potential on decomposition processes in soils. *Soil Sci. Amer. Spec. Publ. No.* 9:24–96.
70. Stotzky, G. 1965. Microbial respiration. In *Methods of Soil Analysis.* Monogr. No. 9. American Society of Agronomy, Madison, Wis., pp. 1550–1569.
71. Stotzky, G. 1986. Influence of soil mineral colloids on metabolic processes, adhesion, and ecology of microbes and viruses. SSS Spec. Publ. No. 17. pp. 305–428.
72. Torrella, F., and R. Y. Morita. Microcultural study of bacteria size changes and microcolony formation by heterotrophic bacteria in seawater. *Appl. Environ. Microbiol.* 41:518–527.
73. Wagner, G. H. 1975. Microbial growth and carbon turnover. In E. A. Paul and A. D. McLaren (eds.). *Soil Biochemistry,* Vol. 3. Marcel Dekker, New York, pp. 269–305.
74. Waksman, S. A., and R. L. Starkey. 1923. Partial sterilization of soil, microbiological activities and soil fertility: 1. *Soil Sci.* 16:137–156.
75. Williams, S. T. 1985. Oligotrophy in soil: Fact or fiction? In M. Fletcher and G. Floodgate (eds.). *Bacteria in the Natural Environment: The Effect of Nutrient Conditions.* Academic Press, New York, pp. 89–110.
76. Wilson, J. M., and D. M. Griffin. 1975. Water potential and respiration of microorganisms in the soil. *Soil Biol. Biochem.* 7:269–274.
77. Wilson, J. M., and D. M. Griffin. 1975. Respiration and radial growth of soil fungi at two osmotic potentials. *Soil Biol. Biochem.* 7:199–204.
78. Wong, R. T. W., and D. M. Griffin. 1974. Effect of osmotic potentials on streptomycete growth, antibiotic production, and antagonism to fungi. *Soil Biol. Biochem.* 6:319–325.
79. Xu, H.-S., N. Roberts, F. L. Singleton, R. W. Attwell, D. J. Grimes, and R. R. Colwell. 1982. Survival and viability of non-culturable *Escherichia coli* and *Vibrio cholerae* in estuarine and marine environments. *Microb. Ecol.* 8:313–323.

Factors Influencing the Dispersal and Survival of Aerosolized Microorganisms

Linda D. Stetzenbach

Bruce Lighthart

Ramon J. Seidler

Stephen C. Hern

Aerosolized particles provide an effective mode of passive transport for a wide range of minute particles of dust, minerals, trace elements, and microorganisms attached to the surface or incorporated into the droplet (18). The fate and transport of microorganisms in the atmosphere has recently received renewed attention owing to concern over the release of genetically engineered microorganisms (GEMs) and microbiological pest control agents (MPCAs) to the environment. The inoculation of plants in the field with GEMs and MPCAs to enhance agricultural productivity often involves an aerosol mist application to the leaf surface. Consequently, airborne organisms may be disseminated to nontargets. Additional factors that contribute to the transport of organisms via the air include resuspension from plant surfaces (38, 39), aerosolization resulting from raindrop impaction (24), agricultural processing (32), and wind-blown soil (23). The increased use of GEMs and MPCAs then will likely lead to increased requirements for environmental monitoring to assess their dispersal and fate.

Incidence of disease resulting from airborne pathogenic microorganisms has been discussed for over one hundred years since Tyndall presented a theory of miniature clouds of germs floating above open wounds. More recently, concern has focused attention on the aerosolization of pathogens such as Legionnaire's disease transmission via shower heads and faucets (8). While the aerosolization of pathogens is a serious problem in hospitals and nursing facilities, the spread is relatively easy to identify and confine. Outdoor aerosolization, however, may result in the dissemination of suspended material far from its source as illustrated by the transport of bacterial spores from the Black Sea to Sweden (9). Detection of airborne microorganisms 1.2 km from sewage treatment plants (1) has prompted interest in the release of enteric pathogens to the atmosphere from sewage treatment facilities and reuse water (3, 7a, 7b, 29, 47, 50, 53, 54). Infection of crops by airborne plant pathogens from neighboring fields has also been discussed (24, 55).

The influence that GEMs and MPCAs transported via air may have on the environment has prompted renewed efforts to understand the factors that affect the dispersal and survival of airborne organisms.

Dispersal

An aerosol cloud consists of discrete droplets suspended in air ranging in size from 0.001 to 100 μm in diameter (12, 18, 49) and is passively transported. Droplets larger than 100 μm may be generated by mechanical spray nozzles, but they generally settle quickly. Since 1 cm^3 of a liquid divided into 1-μm particles would change the surface area from 6 cm^2 to approximately 3×10^8 cm^2 (21), the dispersal of organisms via aerosols may have an impact over a large area.

Microbial dispersion models have been used either to predict dispersion of microorganisms prior to release or to determine the extent of dispersion following a release (27). Predictive models in the past have been gaussian-type models used for large-scale systems. However, the focus of more recent research efforts is with determinative particle-type models for use in both small- and large-scale systems.

Models can employ empirical or geometric approaches. Empirical curve-fitting uses predetermined settling rates of inert particles in Pasquill diffusion equations (26, 45). Unfortunately, variations in settling or deposition rate of the airborne particles can greatly affect the accuracy of the predicted travel distance. In the geometric approach, the volume of air in which a particle travels increases as the cube of the distance traveled, or the surface of the ground where the particle could fall increases in size with the square of the distance traveled. In a strict sense, the distance traveled is directly proportional to the velocity at which the particle is moving. A simple geometric model has been prepared to estimate the microbial load generated during agricultural combining of crops (32). Aerosolized microorganisms, however, do not necessarily travel in a straight line and the settling rate can be affected by environmental conditions.

Gaussian plume models, derived to predict the transport of particles downwind from a point source, describe the concentration of inert, nonsettling droplets that are normally distributed about multiple axes of a plume (34). At any downwind point there is a maximum particle concentration along the mean axis position and the concentration decreases symmetrically crosswind of this axis. Microorganism-containing droplet models are being developed for predictive and determinative uses that will address point as well as nonpoint sources (33) and models incorporating environmental parameters that affect dispersal are forthcoming.

Among the predominant environmental factors (Table 22.1) that affect an aerosol distribution are wind, temperature, humidity (18, 37) and characteristics of the aerosol (42).

Environmental considerations

Wind is the primary meteorological factor involved in the spread of particles in the atmosphere (Table 22.1; 26). Assumptions on dispersal, however, cannot be made on the velocity of the wind alone. Zones of turbulence and quiescence, which vary through eight orders of magnitude and may occur in wind eddies (26), have been called the most important characteristic in the atmosphere with regard to dispersion (13). Turbulence has three effects on dispersion based on the relative size of the particles and the eddy zone (13, 26). If the eddy is larger than the cloud of particles, the particles are carried as a unit on the local wind field. If the eddy is smaller than the particle cloud, the particles are carried forward laterally in a straight line. If, however, the cloud and the eddy are similar in size, turbulence may occur and the particles may not move in a straight line at all.

Size and density of the liquid droplets are significant factors in transport of the aerosol, and these characteristics may change after release (18). Temperature- and humidity-induced changes in particle

TABLE 22.1   Factors Affecting the Dispersal of Airborne Microorganisms

Parameter	Effect	References
	Environmental Factors	
Electrostatic forces	Release of cells from plant leaves	39
Humidity	Changes in droplet size washout with rain	12, 18, 39
Temperature	Changes in droplet size	12, 18, 37
Wind	Altered course, distance upward flux	13, 26, 27, 39
	Mechanical Factors	
Particle size-density	Distance traveled	18, 25, 30, 37
Sprayer characteristics	Droplet size selection	48, 54

size of sewage effluent have been reported (37). Effluent may evaporate almost instantly, leaving droplet nuclei containing a single bacterium (30). Conditions of low humidity and elevated temperature often cause such variations of droplet size (12).

The influence of changing temperature and humidity on dispersal (Table 22.1) was demonstrated by Lindemann and Upper (39) with the upward flux of bacteria from bean plants. They reported greater numbers of airborne bacteria during periods of elevated temperature and wind shearing stress. These conditions result in unsettled air, which aids in the release of bacteria from leaves and keeps them in suspension above the plant canopy. Electrostatic forces (Table 22.1) have been proposed as another factor in the retention or release of organisms from leaf surfaces as positive charges were present on leaves during sunny dry conditions when upward flux of organisms was observed (39). On the other hand thermal inversion or stagnant air was shown to reduce airborne bacterial numbers.

Mechanical considerations

Transport of organisms to the target site is more easily predicted with mechanically generated aerosols since droplets with a predetermined size range and density can be released (Table 22.1). Distribution of the organisms within the liquid suspension medium is assumed to be random following Poisson distribution from the spray nozzle to the surrounding environment.

The Minute Man Jet Fog Sprayer (International Industries, Chicago, Illinois) has been used for dispersal studies owing to its high aerosolization efficiency (54). This sprayer was reported to disperse droplets of 18 to 50 μm in diameter, resulting in a plume 1.5 m above the ground when held at a 1-m height. Field studies with GEMs in California used a low-pressure hand-held Hudson sprayer or a sprayer with a Tee-Jet nozzle adjusted to produce large droplets (48). The Tee-Jet nozzle sprayer was reported to produce 8×10^6 droplets per second with 40 percent of those measuring 100 to 250 μm in diameter.

Survival

Historically, models to predict the transport of airborne particles considered only the physical effects of environmental conditions on dispersal (27). These models failed to address the factors that affect survival of the organisms being transported. Viability parameters are essential elements in predicting the successful dissemination of GEMs (51) because released organisms must be able to survive and colonize once they settle onto a surface (52).

Many of these factors have been studied with effluent. Sewage effluent contains 10^3 to 10^5 viruses per liter (7b) and greater than 10^2 coliforms per milliliter (49, 50), and it is reported that 50 percent of the droplets dispensed from spray equipment at sewage plants contain viable bacteria in the respirable size range (1.0 to 5.0 μm in diameter). Attempts to correlate proximity to sewage treatment plants with increased enteric disease (11, 22), however, have not shown the aerosolized organisms to be the causative agent (29). Kenline and Scarpino (30) reason that this is due in part to the fact that the average half-life of such organisms is only 14 s, and therefore many of the aerosolized bacteria that arrive downwind are no longer viable. They report that a short half-life such as 14 s could result in inactivation of 90 percent of the initial concentration of microorganisms within the first 100 ft downwind. Both environmental and experimental conditions result in this loss of viability (Table 22.2).

TABLE 22.2   Environmental Factors Affecting the Survival of Airborne Microorganisms

Parameter	Effect	References
Carbon monoxide	Dependent on humidity	31
Humidity:		
Low	Flavin-linked damage by free radicals	6
Midrange	Decreased viability	6, 14, 17, 20, 26, 28
	Increased death with CO	31
	Increased death with SO_2	36
High	Stress-instability	14–16
	Protective against CO	31
	Resistance to SO_2	36
	Instability in phosphate buffer	14
	Disruption of sRNA synthesis	6
Ozone	Dependent on humidity	18, 44
Nitrogen oxides	N_2 fixation decreased with nitric oxide exposure	43
	Modifies sensitivity to radiation effects	43
Radiation:		
UV	Dimerization of DNA	5, 18
Visible	Aid in repair of DNA dimers	4, 5
	Cell damage by singlet oxygen	4
Ionizing	Random breaks in DNA	18
	Enzyme damage and altered membrane permeability by free radicals	5, 18
Gamma	Low, not a hazard	5
X-ray	Low, not a hazard	5
Sulfur dioxide	Decreased viability related to humidity	36
Temperature	Increased death rate with increasing temperature	5, 46

Environmental factors

Settling terms (45) and microbial death rate (34) were added to dispersal models in an effort to address die-off of airborne microbes. Camann (10) included a "microorganism impact factor," which was to aid in the estimate of organisms disseminated via spray irrigation. A microbial death rate constant was developed by Teltsch et al. (54) in an effort to improve transport estimates, but a model for dispersal that included several environmental parameters has only recently been developed (35). Relative humidity (RH), temperature, and solar radiation are among the environmental factors affecting the dispersal of viable bacteria (Table 22.2) that are being incorporated into that model.

Rapid cell death has been reported for aerosolized *Escherichia coli* with RH values in the 60 to 80 percent range (14) and midrange RH values have also been reported to adversely affect cell viability for several other microbial genera (15, 28). The effects of low RH have been linked to free radical damage of the flavin-linked enzymes in the bacteria, and those of high RH may be the result of effects on RNA synthesis (6). Low and high RH values are lethal for microorganisms depending on the wet or dry state of the aerosolization (15). Cox and Goldberg (20) concluded that water leaves bacteria aerosolized from a wet state in order to establish equilibrium with the air. This water loss results in the loss of viability of cells with minimum survival occurring at 50 to 55 percent RH for a *Pasteurella* sp. (17). Survival at minimal RH values was attributed to collection methods and the process of rehydration of aerosols generated in a dry state (15–17). The native habitat of the bacteria also seems to be an important consideration in their tolerance of changes in humidity. Terrestrial bacteria appear to be more tolerant of low RH values when compared to genera isolated from aquatic habitats (18). Relative humidity is also a factor in the lethal effects of atmospheric pollutants such as carbon monoxide (31) and sulfur dioxide (36) and is linked to effects due to temperature. Temperature tolerance ranges for microorganisms are related to water activity with increased desiccation generally occurring with increasing temperature (46). This relationship has been reported for airborne bacteria at RH of greater than 30 percent, although little effect on survival was noted below 20 percent or above 80 percent RH (26).

Solar radiation is also important to the survival of airborne microorganisms (Table 22.2). Ultraviolet radiation is strongly germicidal at 250 to 260 nm, leading to dimerization in the bacterial DNA (18). This dimerization can be repaired, however, if bacterial growth rates are slowed. Visible light may aid in the repair mechanism (5) but can also lead to the formation of chemically reactive singlet oxygen (4). Yellow, orange, and red carotenoid pigments help to prevent the formation of singlet oxygen, and organisms with these pigments are more resistant to exposure to sunlight (4). While a component of solar radiation, infrared radiation has low energy and penetrating power and, therefore, is discussed in the context of temperature effects on microorganisms. Ionizing radiation causes random breaks in bacterial DNA (18) and has also been associated with the production of free radicals that inactivate microbial enzymes and alter membrane permeability (5, 18). Levels of gamma and x-ray radiation in the environment, however, are low and not a hazard for airborne organisms (5).

The effects of radiation on bacteria have been modified by exposure to nitrogen oxides (43). Radiation sensitivity was reportedly increased following a 0.5% nitric oxide exposure, but the effects decreased at higher concentrations. The reduction of molecular nitrogen to ammonia by bacteria is affected by exposure to nitrogen oxides at concentrations of 0.039% and 0.0025% and nitrogen fixation by a *Clostridium* sp. ceased with a 0.01% nitric oxide concentration (43).

Ozone as an air pollutant has also shown antibacterial activity. Cox (18) cites several studies in which the toxic effects of ozone were demonstrated when ozone was present with other photochemical oxidants commonly associated with automobile exhaust, petroleum vapors, and olefins. The toxic effects of ozone are increased when the relative humidity is elevated. A 90 percent kill within 30 min of exposure to a 0.025-ppm ozone concentration was reported for an RH of 70 percent, but at a 45 percent RH and 1 percent ozone concentration, little bacterial death was reported (44).

Experimental factors

Characteristics of the aerosol itself, residence time in the aerosol, characteristics of the microorganisms, cultural techniques and storage of the organisms prior to aerosolization, and sampling, recovery, and enumeration methods are experimental considerations that can affect the survival of airborne organisms (Table 22.3; 28).

Characteristics of the aerosol include size and composition of the droplets. Droplet size not only affects dispersal settling terms but also is an important factor in viability in the airborne state (42). Larger droplet sizes have been reported to contain higher numbers of viable *E. coli* over time in an aerosol than smaller particle sizes (41). After 1 min, only 13 percent of the organisms were viable in smaller size ranges (less than 7 μm in diameter) compared to 30 percent for droplets greater than 7 μm. After 3 min approximately 3 percent of drop-

TABLE 22.3 Experimental Factors Affecting the Survival of Airborne Microorganisms

Parameter	Effect	References
Aerosol:		
Droplet size	Increased viability with larger droplet	25, 41
Suspending medium	Protective agents increase survival	4, 15, 18, 28
Residence time	Decreased recovery with time	19, 30
	Increased CO toxicity with time	31
Microorganism:		
Genera	Gram-positive genera more aerostable	25
	Differences in survival between strains	14, 15
Culture	Log phase cells less aerostable	18
Storage	Freeze with liquid N_2 and $-70°$ storage	18
Sampling design	Varying efficiencies of field samplers	2, 18, 40, 56

lets less than 4.7 μm in diameter were viable compared to greater than 20 percent for larger droplets. Green and Green (25) also noted decreasing viability with decreasing droplet sizes with an aerosolized *Proteus mirabilis*. The *P. mirabilis* recorded an approximately 22 percent loss in viability with droplets greater than 7 μm but suffered greater than 60 percent losses in droplets less than 1 μm in diameter.

Composition of the fluid suspension of organisms prior to aerosolization is also important (28). Suspensions of water produce a droplet of different properties than one formed with organic and inorganic ingredients such as sewage effluent or culture medium. After aerosolization, the droplet will lose water and shrink. Droplets formed from fluid consisting of washed cells and water will equilibrate with the relative humidity. If, however, the fluid contains nonvolatile components, these agents will increase in concentration as the droplet equilibrates. This increase in concentration may subject the cell to harmful levels (28) and set up a hypertonic state forcing more water out of the cell (4). The addition of protective agents to the suspending fluid prior to aerosolization or to the collection medium may improve the recovery of viable organisms (15, 28).

Residence time in the aerosol also appears to be important to survival with greater recoveries at 30 min than at 2 h for airborne *E. coli* (19). Residence time was also a factor in the lethal effects of carbon monoxide-exposed airborne bacteria (31).

Differences in survival with aerosolization among bacterial genera have also been shown. Green and Green (25) compared a *Staphylococcus aureus* and a *Proteus mirabilis* for survival and noted that the *S. aureus* suffered no loss in viability, while the *P. mirabilis* declined by 20 to 60 percent. Varying recoveries of the same genus using different sampling methods have also been shown (2, 18, 40, 56).

References

1. Adams, A. P., and J. C. Spendlove. 1970. Coliform aerosols emitted by sewage treatment plants. *Science* 169:1218–1220.
2. Andersen, A. A. 1958. New sampler for the collection, sizing, and enumeration of viable airborne particles. *J. Bacteriol.* 76:471–484.
3. Applebaum, Y., N. Guttman-Bass, M. Lugten, B. Teltsch, B. Fattal, and H. I. Shuval. 1984. Dispersion of aerosolized enteric viruses and bacteria by sprinkler irrigation with wastewater. *Monographs in Virology* 15:193–201.
4. Atlas, R. M. 1984. *Microbiology: Fundamentals and Applications.* Macmillan, New York.
5. Atlas, R. M., and R. Bartha. 1981. *Microbial Ecology: Fundamentals and Applications.* Addison-Wesley, Reading, Mass.
6. Benbough, J. E. 1967. Death mechanisms in airborne *Escherichia coli. J. Gen. Microbiol.* 46(3):325–333.
7a. Bitton, G. 1980. Fate of viruses in sewage treatment plants. *Introduction to Environmental Virology.* John Wiley and Sons, New York, pp. 121–152.
7b. Bitton, G. 1980. Fate of viruses in land disposal of wastewater effluents. *Introduction to Environmental Virology.* John Wiley and Sons, New York, pp. 200–240.
8. Bollin, G. E., J. F. Plouffe, M. F. Para, and B. Hackman. 1985. Aerosols containing *Legionella pneumophila* generated by shower heads and hot-water faucets. *Appl. Environ. Microbiol.* 50:1128–1131.
9. Bovallius, A., B. Bucht, R. Roffey, and P. Anas. 1978. Long range air transmission of bacteria. *Appl. Environ. Microbiol.* 35:1231–1232.
10. Camann, D. E. 1980. A model for predicting dispersion of microorganisms in wastewater aerosols. In *Wastewater Aerosols and Disease.* U.S. Environmental Protection Agency. PB81-169864. Health Effects Res. Lab., Cincinnati, Ohio.
11. Camann, D. E., and M. N. Guentzel. 1984. The distribution of bacterial infections in the Lubbock infection surveillance study of wastewater irrigation. In *Future of Waste Reuse.* Water Reuse Symposium III, San Diego, Calif.
12. Chatigny, M. A., R. L. Dimmick, and J. B. Harrington. 1979. Deposition. In R. L. Dimmick (ed.), *Aerobiology: The Ecological Systems Approach.* Dowden, Hutchinson, and Ross, Inc., Stroudsburg, Pa.
13. Chatigny, M. A., R. L. Dimmick, and C. J. Mason. 1979. Atmospheric transport. In R. L. Dimmick (ed.), *Aerobiology: The Ecological Systems Approach.* Dowden, Hutchinson, and Ross, Inc., Stroudsburg, Pa.
14. Cox, C. S. 1966. The survival of *Escherichia coli* sprayed into air and into nitrogen from distilled water and from solutions of protecting agents, as a function of relative humidity. *J. Gen. Microbiol.* 43:383–399.
15. Cox, C. S. 1968. The aerosol survival and cause of death of *Escherichia coli* K12. *J. Gen. Microbiol.* 54:169–175.
16. Cox, C. S. 1970. Aerosol survival of *Escherichia coli* B disseminated from the dry state. *J. Appl. Microbiol.* 19:604–607.
17. Cox, C. S. 1971. Aerosol survival of *Pasteurella tularensis* disseminated from the wet and dry states. *J. Appl. Microbiol.* 21:482–486.
18. Cox, C. S. 1987. *Aerobiological Pathways of Microorganisms.* John Wiley and Sons, New York.
19. Cox, C. S., and F. Baldwin. 1964. A method for investigating the cause of death of airborne bacteria. *Nature* 202(4937):1135.
20. Cox, C. S., and L. J. Goldberg. 1972. Aerosol survival of *Pasteurella tularensis* and the influence of relative humidity. *Appl. Microbiol.* 23:1–3.
21. Dimmick, R. L. 1969. Production of biological aerosols. In R. L. Dimmick, and A. B. Akers (ed.), *An Introduction to Experimental Aerobiology.* Wiley-Interscience, New York.
22. Fattal, B., M. Margalith, H. I. Shuval, and A. Morag. 1984. Community exposure to wastewater and antibody prevalence to several enteroviruses. In *Future of Water Reuse.* Water Reuse Symposium III, American Water Works Assoc., Denver, Colo.
23. Gillette, D. A., I. H. Blifford, Jr., and C. R. Fenster. 1972. Measurements of aerosol size distribution and vertical fluxes of aerosols on land subject to wind erosion. *J. Appl. Meteorol.* 11:977–987.

24. Graham, D. C., C. E. Quinn, and L. F. Bradley. 1977. Quantitative studies on the generation of aerosols of *Erwinia carotovora* var. *atroseptica* by raindrop impaction on blackleg infected potato stems. *J. Appl. Bacteriol.* 43:413–424.
25. Green, L., and G. Green. 1968. Direct method of determining viability of a freshly mixed bacterial aerosol. *Appl. Microbiol.* 16:78–81.
26. Gregory, P. H. 1973. In N. Polunin (ed.), *The Microbiology of the Atmosphere*, 2d ed. Leonard Hill, London.
27. Hanna, S. R., G. A. Briggs, and R. P. Hosker, Jr. 1982. *Handbook on Atmospheric Diffusion*. DOE/TIC-11223. Technical Information Center, U.S. Department of Energy, Washington, D.C.
28. Hatch, M. T., and H. Wolochow. 1969. Bacterial survival consequences of the airborne state. In R. L. Dimmick and A. B. Akers (eds.), *An Introduction to Experimental Aerobiology*. Wiley-Interscience, New York, pp. 267–295.
29. Jakubowski, W. 1983. Wastewater aerosol health effects studies and the need for disinfection. In *Proceedings of the Second National Symposium on Municipal Wastewater Disinfection*. EPA-600/9-83-009. U.S. Environmental Protection Agency, Cincinnati, Ohio.
30. Kenline, P. A., and P. V. Scarpino. 1972. Bacterial air pollution from sewage treatment plants. *Am. Ind. Hyg. Assoc. J.* 33:346–352.
31. Lighthart, B. 1973. Survival of airborne bacteria in a high urban concentration of carbon monoxide. *Appl. Microbiol.* 25:86–91.
32. Lighthart, B. 1984. Microbial aerosols: Estimated contribution of combine harvesting to an airshed. *Appl. Environ. Microbiol.* 47:430–432.
33. Lighthart, B. 1988. Transport modelling of airborne microorganisms—A short review and recent progress. *Proceedings of U.S. EPA Workshop "Integration of Research and Model Development in Biotechnology Risk Assessment."* Breckenridge, Colo., January 11–15.
34. Lighthart, B., and A. S. Frisch. 1976. Estimation of viable airborne microbes downwind from a point source. *Appl. Environ. Microbiol.* 31:700–704.
35. Lighthart, B., and A. J. Mohr. 1987. Estimating downwind concentrations of viable airborne microorganisms in dynamic atmospheric conditions. *Appl. Environ. Microbiol.* 53:1580–1583.
36. Lighthart, B., V. E. Hiatt, and A. T. Rossano, Jr. 1971. The survival of airborne *Serratia marcescens* in urban concentrations of sulfur dioxide. *J. Air Pollut. Control Assoc.* 21:639–642.
37. Lighthart, B., J. C. Spendlove, and T. G. Akers. 1979. Factors in the production, release, and viability of biological particles. In R. L. Edmonds (ed.), *Aerobiology: The Ecological Systems Approach*. US/IBP synthesis series 10. Dowden, Hutchinson, and Ross, Inc., Stroudsburg, Pa., pp. 11–22.
38. Lindemann, J., H. A. Constantinidou, W. R. Barchet, and C. D. Upper. 1982. Plants as sources of airborne bacteria, including ice nucleation-active bacteria. *Appl. Environ. Microbiol.* 44:1059–1063.
39. Lindemann, J., and C. D. Upper. 1985. Aerial dispersal of epiphytic bacteria over bean plants. *Appl. Environ. Microbiol.* 50:1229–1232.
40. Lundholm, I. M. 1982. Comparison of methods for quantitative determinations of airborne bacteria and evaluation of total viable counts. *Appl. Environ. Microbiol.* 44:179–183.
41. May, K. R. 1966. Multistage liquid impinger. *Bacteriol. Rev.* 30:559–570.
42. May, K. R. 1972. Assessment of viable airborne particles. In T. T. Mercer, P. E. Morrow, and W. Stoeber (eds.), *Assessment of Airborne Particles: Fundamentals, Applications, and Implications to Inhalation Toxicology*. Charles C. Thomas, Springfield, Ill.
43. National Research Council Criteria Document. 1977. *Medical and Biologic Effects of Environmental Pollutants: Nitrogen Oxides*. T. T. Crocker, Committee Chair. National Academy of Sciences, Washington, D.C.
44. National Research Council Criteria Document. 1977. *Medical and Biologic Effects of Environmental Pollutants: Ozone and Other Photochemical Oxidants*. S. K. Friedlander, Committee Chair. National Academy of Sciences, Washington, D.C.

45. Peterson, E. W., and B. Lighthart. 1977. Estimation of downwind viable airborne microbes from a wet cooling tower including settling. *Microb. Ecol.* 4:67–79.
46. Poon, C. P. C. 1968. Viability of long-storaged airborne bacterial aerosols. *J. Sanit. Eng. Div.* (Am. Soc. Civil Eng.) 94(6):1137–1146.
47. Randall, C. W., and J. O. Ledbetter. 1966. Bacterial air pollution from activated sludge units. *Am. Ind. Hyg. Assoc. J.* 27:506–519.
48. Seidler, R. J., and S. C. Hern. 1988. *EPA Special Report: The Release of Ice Minus Recombinant Bacteria at California Test Sites.* ERL-Corvallis, Oreg.
49. Sorber, C. A., and B. P. Sagik. 1980. Indicators and pathogens in wastewater aerosols and factors affecting survivability. In *Wastewater Aerosols and Disease*, PB81-169864. EPA Health Effects Res. Lab., Cincinnati, Ohio.
50. Sorber, C. A., H. T. Bausum, S. A. Schaub, and M. J. Small. 1976. A study of bacterial aerosols at a wastewater irrigation site. *J. Water Pollut. Control Fed.* 48: 2367–2379.
51. Stotzky, G., and H. Babich. 1986. Survival of, and genetic transfer by, genetically engineered bacteria in natural environments. *Adv. Appl. Microbiol.* 31:93–138.
52. Strauss, H. S., D. Hattis, G. Page, K. Harrison, S. Vogel, and C. Caldart. 1986. Genetically engineered microorganisms: II. Survival multiplication and genetic transfer. *Recomb. DNA Tech. Bull.* 9(2):69–88.
53. Teltsch, B., and E. Katzenelson. 1978. Airborne enteric bacteria and viruses from spray irrigation with wastewater. *Appl. Environ. Microbiol.* 35:290–296.
54. Teltsch, B., H. I. Shuval, and J. Tadmor. 1980. Die-away kinetics of aerosolized bacteria from sprinkler application of wastewater. *Appl. Environ. Microbiol.* 39:1191–1197.
55. Vinette, J. R., and B. W. Kennedy. 1975. Naturally produced aerosols of *Pseudomonas glycinea. Phytopathology* 65:737–738.
56. Zimmerman, N. J., P. C. Reist, and A. G. Turner. Comparison of two biological aerosol sampling methods. *Appl. Environ. Microbiol.* 53:99–104.

Chapter 23

Site-Specific Aquatic Microcosms as Test Systems for Fate and Effects of Microorganisms

C. R. Cripe

P. H. Pritchard

Introduction

Environmental studies designed to increase the understanding of ecological processes or to assess the risk associated with the introduction of toxic chemicals or novel genetic material are limited by the complex and dynamic nature of the environment. While field studies provide important information about a particular ecosystem, they often have many drawbacks. For example, separate variables (i.e., temperature, nutrients) cannot be controlled to evaluate their significance; meteorological events may disrupt the studies; and the introduction of toxic chemicals or genetically engineered organisms (GEMs) to the field for the purpose of research may be unacceptable until adequate data are available. For these reasons laboratory studies are frequently used to examine ecological processes. Laboratory systems offer a further advantage in that they can be duplicated and manipulated in the laboratory and yet may be complex enough to contain important ecosystem processes found in the field.

Microcosms are laboratory systems that have been used by researchers to obtain environmentally relevant ecological information

(64). The term "microcosm" has been applied to almost any laboratory test system, including a container of site water or media incubated with an inoculum (53). Our definition of a microcosm study is narrower and refers to an attempt to bring an intact, minimally disturbed piece of an ecosystem into the laboratory for study (64). An intact piece of an ecosystem, contained in the laboratory within certain physical and chemical constraints, offers the best way to study ecological processes under controlled experimental conditions (lighting, temperature, mixing, and so forth). This type of microcosm is site-specific, since it reflects ecological processes found in the particular aquatic environment from which it is taken. It is unclear at this time how similar aquatic environments must be to extend extrapolations to other sites.

A variety of generic test systems using structured or synthetic communities (also termed "model ecosystems") have been reported for studying the fate of a xenobiotic compound in an aquatic system (31, 43) as well as the effects on a community (38, 83). An artificial, defined community is created by placing water and microbiota in a container according to an established protocol. These multispecies tests were designed to provide reproducible environments to evaluate and rank toxic chemicals according to their fate or effects. Because they generally focus on the interaction of fewer species, generic systems may aid in the examination of an effect observed in a more complex site-specific system. However, generic microcosms differ significantly from site-specific microcosms in that they cannot routinely be calibrated with a field site, and thus it is difficult to relate effects observed in these systems to the environment. Since extrapolation to a field site is of primary concern in risk assessment, site-specific systems are generally more appropriate for this purpose. Generic microcosms are discussed in more detail in Chaps. 31 and 32.

Calibration

It is assumed that segments of an ecosystem sampled with minimal disturbance and maintained in the laboratory contain many of the complex ecological processes found in the ecosystem from which they were taken. This assumption requires that various structural and functional ecological parameters measured in the field and in microcosms be compared to determine how well and how long microcosm measurements simulate those from the field. We call this comparison "calibration." Since a major objective of microcosm research is the extrapolation of information obtained in the laboratory to the field, the concept of calibration is an important aspect of these test systems. Even if some parameters do not indicate close agreement between lab and field systems, the calibration process provides important information concerning how well a microcosm reflects ecological processes in the natural environment.

The diversity of parameters available for microcosm-field calibration is quite large. Perez et al. (54) used structural and functional characteristics such as phytoplankton and zooplankton enumeration, chlorophyll *a*, ammonia, oxygen uptake, and primary productivity in their simulation of Narragansett Bay. For comparison of seagrass microcosms with the field, Morton et al. (44) determined the density of dominant macroinvertebrates, concentrations of chlorophyll *a* in *Thalassia* leaves and in epiphytes, and epiphyte biomass. Portier (58) used microcosms from both a freshwater swamp and a salt marsh, and monitored both field and microcosm for density of bacteria, actinomycetes, fungi, and yeasts, as well as enzyme activity (phosphatase, dehydrogenase), and microbial ATP. Studies of Livingston et al. (41) and Federle et al. (18) relied on infaunal macroinvertebrates and microbial lipids for comparison of microcosm and field communities from the Apalachicola Bay system. Kroer and Coffin (37) compared the behavior of water-sediment core microcosms designed for testing GEMs with the field using the following parameters: phytoplankton biomass and productivity, microbial density and production, grazing on bacteria, pH, dissolved oxygen, and wall growth.

Parameters that are useful in the lab-field calibration process may also play an important role as indicators of microcosm perturbation (see Biomass and Functional and Structural Endpoints in this chapter). In addition, Pritchard (63) suggests that field calibration enhances the ability to extrapolate microcosm data to a field site. Further studies, however, involving field verification (i.e., establishing qualitatively and quantitatively that an observation observed in a microcosm also occurs in the field) are necessary to demonstrate the feasibility of extrapolating GEM persistence and effects information to the field (see Field Verification in this chapter).

Role of microcosms in risk assessment of GEMs

A primary reason for the use of microcosms in GEM risk assessment is the concern that standard single-species tests might fail to accurately predict the persistence or effects of a GEM. Single-species aquatic tests generally focus on algae, invertebrates, and fish. Although these organisms might be directly affected by a GEM, the tests provide no information about GEM persistence in aquatic habitats (other than those associated with the test organisms) or effects on important ecological processes controlled by natural microbial assemblages. Microcosms, on the other hand, can be used to examine the possible effects of introduced organisms on the structural and functional aspects of complex communities. Because of their complexity, microcosms may indicate indirect trophic level effects. Thus, a GEM that overproduces cellulase (compared to naturally occurring cellulose degraders) may not only increase leaf litter decomposition but, due to increased nutrient availability, may also decrease oxygen concentration and thereby indirectly affect other organisms.

Microcosms whose ecological processes have been calibrated with those of a field site could potentially be used as surrogates for field testing (see Field Verification in this chapter), particularly in cases where experimental field application would require great effort, cost, or risk. Containment is much easier with microcosms than with outdoor field studies, and the cost savings make microcosms attractive for research purposes.

Little information exists on the use of microcosms to monitor the fate of microorganisms. While chemostat-type systems may be applied to study the population dynamics of microorganisms suspended in the water column, they do not aid in understanding the growth of attached cells. The understanding of microbial populations on surfaces (i.e., sediment, plants, or walls of microcosms) is poor compared to the knowledge on pelagic microbial populations, yet the magnitude of sediment activity may, in shallow systems, greatly exceed that of the water column. Thus, site-specific aquatic microcosms may be used to provide information on three primary areas of concern for biotechnology risk assessment: (1) persistence (survival and colonization), (2) effects on ecological structure and function of an aquatic system, and (3) transfer of genetic material from a GEM to indigenous organisms in the aquatic environment. We will discuss each of these areas below.

GEM persistence. Persistence in a particular aquatic habitat is primarily a function of survival within the habitat and transport between habitats. A variety of physical, chemical, and biological factors may influence microorganism survival in an aquatic habitat (see Table 23.1; 78). Factors affecting survival are addressed in detail in Chap. 21, while transport is covered in Chap. 22. We will consider the role of microcosms as a tool for the further examination of microorganism persistence.

While the effects of individual parameters on survival could be quantified in a flask study, such simple test systems are not amenable for assessing the interactions of several parameters. For example, the effect of nutrient stimulation on an axenic GEM culture may be well-suited for a flask study, but a microcosm test would be required to de-

TABLE 23.1 Factors Affecting Survival and Transport

Factors affecting survival			Transport processes	
Chemical	Physical	Biological	Physical	Biological
pH	Light	Predation	Turbulence	Bioturbation
Eh	Surfaces	Competition	Water flow	Chemotaxis
Nutrients	Temperature		Turbidity	
			(sorbed)	
Toxicants				
Salinity				

termine GEM survival in the presence of nutrients while competing with indigenous organisms in a variety of aquatic habitats. Careful selection of microcosm systems and intact sampling of a particular field site produces realistic environments for observing these interactions. Habitats that may be included in microcosms for studying persistence include potential surfaces for attachment such as sediment particles (suspended and settled), rocks, and benthic macrophytes. The complex redox gradient established by biogeochemical processes in sediment provides a variety of habitats for GEM colonization; homogeneous test systems (i.e., shake flasks) cannot provide the redox gradient characteristic of site-specific, intact microcosm cores. Finally, microcosms may be used to verify information (persistence, effects, and genetic exchange) observed in simpler systems.

Another example of a study involving complex interactions that would require a microcosm includes transport processes (see Table 23.1) such as bioturbation. Reichardt (69) reported that the bioturbation activities of a marine worm impacted several aspects of the microbiology of intertidal sediments, including bacterial production, heterotrophic activity, and hydrolytic enzymes. Microbes are important sources of food to some benthic organisms (89), and water movement and sediment reworking by benthic organisms could distribute GEMs to habitats deep in the sediment. The distribution of xenobiotic compounds sorbed to sediment particles has been shown to be affected by such processes (47, 48).

Potential GEM effects. Many parameters used to monitor the responses of aquatic ecosystems to perturbations can be used for assessing the effects resulting from the introduction of GEMs. Odum (46) suggests a number of general changes (e.g., energetics, nutrient cycling, community structure) that might be expected in stressed ecosystems. Biomass and Functional and Structural Endpoints, in this chapter, lists a variety of structural and functional parameters that could be considered endpoints or effects parameters upon which a risk assessment decision might be based. Many of these represent ecological processes that require the complexity of a microcosm for proper evaluation.

The selection of appropriate endpoints to measure indirect effects may be aided by the use of a positive control designed to investigate the sensitivity of the parameter. Although selected microorganisms (with or without recombinant DNA) would represent the most appropriate positive controls for purposes of biotechnology risk assessment, the application of chemicals chosen to perturb a particular aspect (i.e., specific organisms, communities, or ecological process) may suffice. This approach may also aid our understanding of microcosm interactions. A hypothetical approach for a seagrass microcosm community is given in Table 23.2. For example, the addition of an antibiotic may directly affect bacteria involved in nutrient (i.e., nitrogen) cycling and detritus decomposition. However, other components of the system, such as the benthic fauna, may also be affected indirectly due to trophic level interactions. The result should be a better assessment of the potential ecological effects of the introduction of a GEM into an aquatic environment because of the ability to monitor unexpected secondary effects. Although laborious, this type of basic microcosm research is necessary to avoid errors in the interpretation of results and to use the microcosm to its fullest potential.

Gene transfer. Transfer of recombinant DNA to indigenous microbes presents several risk considerations. These include (1) detection methods must be developed to detect not only the presence of the GEM but also the novel DNA, (2) the indigenous recipient may be better suited to survive in the environment than in the GEM, and (3) the novel DNA may present a greater potential ecological risk when expressed by the indigenous recipient than by the GEM.

The concentration of active microbial communities on interfaces (i.e., air-water, water-sediment) appears to favor gene exchange, par-

TABLE 23.2 **Hypothetical Purturbants and Response Parameters Measurable in a Seagrass Microcosm Community**

Direct response parameters	Perturbator					
	Antibiotic	Predator	Grazer	GEM	Insecticide	Herbicide
Nutrient cycling	X		X	X		
Benthic fauna		X			X	
Detritus decomposition	X		X	X		
Epiphytes			X			X
Macrophyte growth						X

ticularly the presence of solid interfaces, which stimulate gene transfer through conjugation (84). The microcosm can be used to provide a variety of environmental surfaces (sediment, benthic macrophytes, surface and digestive tract of small fish and invertebrates) to assess the potential of transfer under relatively natural conditions. The topic of gene transfer is covered in more detail in Chap. 7.

Design Features for Sediment-Water Microcosms

Size

Because of the difficulty in sampling, containing, and transporting segments of the aquatic environment for laboratory testing, a microcosm will only characterize a relatively small part of a field site. For any risk assessment study, the selected size and design will depend primarily on the sampling volume and frequency, scaling (see Scaling in this section), and the trophic levels and habitats required to answer whatever questions are asked of the system. GEM risk assessment may place substantial restrictions on microcosm size since all components (water, sediment, etc.) may have to be sterilized by the end of the test. Thus the systems should be as small as possible and still allow for adequate sampling and trophic level representation.

With this size constraint in mind, we will briefly describe two systems as examples of relatively small microcosms. Each system was designed or modified to answer specific questions concerning the fate of xenobiotic compounds. Selection of any particular system design for GEM risk assessment would depend on the aspects of GEM persistence, ecological effects, or gene transfer to be addressed. Ecocore I (see Ecocore I in next section) has a volume of about 175 mL, and Ecocore II holds approximately 3 liters. Ecocore I functions as both corer and microcosm. Thus, many replicate cores can be taken in the field in a very short time. Ecocore II microcosms involve the use of a separate coring device and thus require more effort to core and extrude an intact plug. The limited volume capacity of Ecocore I, however, restricts sampling. The larger Ecocore II microcosms, on the other hand, will also accommodate macrophytes. The relatively small size of both microcosms maximizes their advantages (i.e., operation within a controlled laboratory environment, replicability, ease of establishment, and containment), and still accommodates many important ecological processes found in natural aquatic ecosystems. Although inclusion of large invertebrate or vertebrate predators may not be appropriate in microcosms of only a few liters, effects on microbiota and benthic infauna can be determined in relatively small systems (44, 71). In such microcosms, size is primarily limited by the amount of water required for samples, physical constraints (i.e., size of monitoring probes and mixing apparatus), and the sediment surface-to-volume ratio. Sampling may further affect this ratio.

Scaling

Because a microcosm consists of only a part of an aquatic environment, the magnitude of physical, chemical, and biological processes occurring in the microcosm may take place on a different scale than in the field. Perez et al. (54) examined a number of scaling parameters, including water turnover, turbulence, incident radiation, and ratio of water volume to sediment surface area, all of which significantly affected microcosm behavior. Each parameter is discussed below.

Sediment serves as source and sink for both microorganisms and chemicals in the overlying water (24) and consequently serves as an important consideration in scaling. Because of the importance of sediment as a potential habitat for GEMs, as well as the role of sediment-associated microorganisms in water quality, both microcosm designs discussed here include intact sediment cores. However, extrapolation of data related to processes controlled by sediment-associated microorganisms from microcosms to the field must take into account the water column volume to sediment surface area ratio. For example, sediment-associated processes will be accentuated in shallow bodies of water (i.e., tidal marshes and swamps) but may be insignificant compared to water column processes in deeper water. In general, the difference in sediment contribution between a microcosm and a field site may be accommodated by reducing the surface area of the intact core and, if necessary, increasing the water volume.

Microcosm construction materials

Careful consideration should be given to the composition of all materials in contact with microcosm water. To avoid potential problems from extraneous contamination (i.e., phthalate plasticizers from vinyl tubing), we suggest that only inert fluorocarbons, silicone rubber, or glass be used. Metals, various plastics, and rubber can be toxic to aquatic organisms. For example, Price et al. (61) found that latex tubing was quite toxic to phytoplankton, zooplankton, and bacteria, and that both latex and Tygon* tubing affected ^{3}H-thymidine incorporation by bacteria and ^{14}C incorporation into phytoplankton. If stoppers must be used, those made of silicone appear to be less toxic to some aquatic organisms than

---

*Mention of trade names or commercial products does not constitute endorsement by the U.S. Environmental Protection Agency.

other materials, such as neoprene (P. Borthwick, personal communication).

Light

Light quality and quantity are among the most important controllable parameters in microcosms. Cool white, fluorescent lights provide the least expensive choice, and their spectra is similar to sunlight. Their intensity is very low, however, which may reduce photosynthesis. We have used metal halide lamps, which provide both high intensity as well as wavelengths needed by benthic macrophytes. Multi-Vapor, 400-W lamps (General Electric, MVR400/U) were found to be sufficient for maintaining *Thalassia testudinum* plants in the laboratory for short periods (10).

Photoperiod is typically controlled by timers and may be fixed at a ratio such as 12:12 (light-dark, L-D), 14:10, or maintained at field conditions. This ratio directly affects the amount of primary production and also may be important for regulating seasonal physiological changes (i.e., reproductive or growth cycles) of benthic macrophytes and larger fauna.

Temperature control

Rates of most biological processes are dependent on temperature; therefore, regulation of temperature is important. Although water has a relatively high specific heat capacity, external factors, such as artificial lighting, may alter the temperature of aquatic microcosms. Temperature may be maintained by placing microcosms in a chamber with constant air temperature, but a circulating water bath or jacket (Fig. 23.1) is recommended.

Water replacement

Microcosms may be maintained in three modes: flow-through, static, or static renewal of a fixed volume (i.e., 10 percent of the microcosm water volume per day). Although flow-through and static renewal modes may be more difficult to operate and may require more time and effort to sterilize effluents, they may avoid nutrient deficiencies and accumulation of metabolic waste products typical of closed systems. Both static renewal and flow-through allow the addition of microorganisms, that is, introductions of new species as they become seasonally available at the field site and also recolonization of those that may have been eliminated from the microcosm at some point.

Water may be supplied to flowing-water microcosms by a pump or headbox and siphon arrangement. All parts of the pump and delivery tubing that come in contact with supply water should be inert (see Mi-

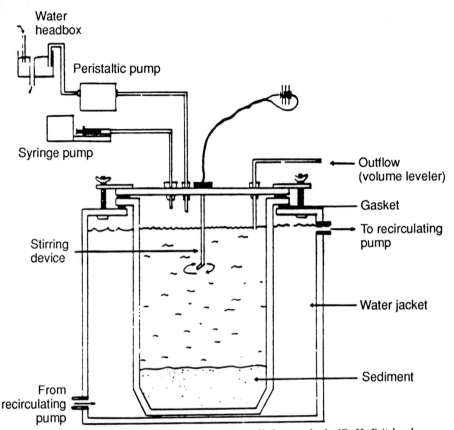

Figure 23.1 Ecocore II microcosm in temperature-controlled water bath. [*P. H. Pritchard, C. A. Monti, E. J. O'Neill, J. P. Connolly, and D. G. Ahearn, Movement of Kepone (chlordecone) across an undisturbed sediment-water interface in laboratory systems. Environ. Toxicol. Chem. 5:647–657. Reprinted with permission. Copyright 1988, Pergamon Press.*]

crocosm Construction Materials in this section). Peristaltic pumps and silicone tubing are acceptable. Care should be taken that the delivery tubing is as short as possible to reduce biofilm formation that might affect microbial distribution and remove nutrients.

Mixing

Mixing is required to prevent stratification of microorganisms, nutrients, and oxygen in the water column. Water column homogeneity is important for sampling and proper application of mathematical models. In addition, the amount of turbulence has been found to affect the composition of phytoplankton assemblages in enclosures (16). Mixing can be accomplished with pumps, aerators, or stirrers. Stirring must be gentle so that the sediment-water surface is not disturbed, since Wainwright (87) found that suspended marine sediments added to seawater stimulated suspended bacteria and protozoa. A disadvantage of pumps is their use of tubing with its associated surface area for microorganism attachment (biofilm formation). Mixing by aeration may produce aerosols that contain GEMs, particularly if they are present in large numbers at the air-water interface, creating another transport route. Aeration is generally not required for oxygenation, since photosynthesis by the phytoplankton community should produce sufficient oxygen. Glass or Teflon stirrers and small motors appear most desirable for mixing (see Ecocore II in the next section).

Coring

For microcosms consisting of sediment and a water column, it is important to obtain sediment cores that are intact, thus preserving the structural integrity of the sample, including redox gradient and the benthic community. For Ecocore I (see Ecocore I in the next section), the corer physically becomes the microcosm. In the case of Ecocore II microcosms (see Ecocore II in the next section), a sediment core is extruded into the microcosm vessel. A simple, effective coring device (Fig. 23.2) can be made of clear acrylic tubing (12.5 × 56 cm length) serrated along the bottom for cutting through plant roots. A PVC pipe fastened (silicon glue) through the top and perpendicular to the corer provides a handle. The top is sealed with an acrylic disk that contains a hole and stopper. The corer is inserted into the sediment to a depth (approximately 8 cm) marked on the corer with tape. The hole on top is plugged with a stopper, and the corer is raised and inserted in the microcosm vessel. A partial vacuum, produced as the corer is raised, helps maintain sediment integrity. The stopper is removed and the corer is lifted from the microcosm, leaving the sediment plug intact.

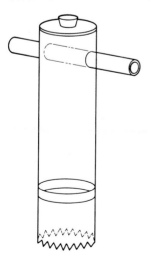

Figure 23.2 Clear acrylic corer (12.5 × 56 cm length) used to obtain intact sediment cores for Ecocore II microcosms.

Examples of Site-Specific Microcosm Designs

Ecocore I

Ecocore I microcosms (62, 75) consist of a glass tube (3.5 cm diameter × 40 cm) inserted vertically into sediment to a depth of about 8 cm. The top is sealed with a silicone stopper, the tube is removed from the sediment, and the bottom silicone stopper is inserted (Fig. 23.3). Coring may require Scuba divers for sampling sites in relatively deep water. After cores are carefully transported to the laboratory, water volume is adjusted to approximately 175 mL with site water. An acrylic stand with upper and lower holes drilled for Ecocores makes a convenient rack for transportation and use. A long, 18-gauge hypodermic needle (Owens and Minor Co., Harahan, Louisiana) is inserted through the upper silicone stopper to within 3 cm of the sediment, and a gentle stream of air is introduced for mixing (see Coring in preceding section for caution on use of aeration). A short length of glass tubing through the upper stopper provides an outlet for gases.

Ecocore II

The Ecocore II microcosm (48, 66) consists of a Pyrex reaction kettle bottom (Corning 6947; Fig. 23.1). A larger version, based on a 27-liter

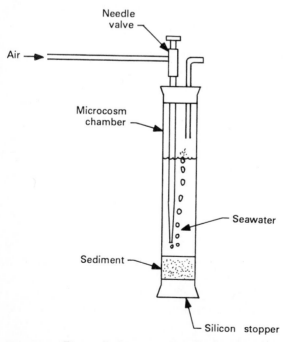

Figure 23.3   Ecocore I microcosm. (*Pritchard et al., 1979.*)

jar (Corning 6942-27L), can be utilized to obtain larger sediment cores and provides a greater capacity for water sampling. After a sediment core is added in the field, the microcosm is carefully transported to the laboratory. Since the coring process and transport from the field often cause some suspension of sediment into the overlying water column, a carboy of site water is also collected. At the laboratory, the water column is siphoned off without disturbing the water-sediment interface, and new water from the carboy is gently added with a volumetric pipette.

Test systems are covered with clear acrylic tops drilled and fitted with silicone stoppers to allow sampling and continuous monitoring by various probes. Reaction kettles are housed in a clear acrylic water bath and maintained at field (or another experimental) temperature. Acrylic or glass tops can be sealed with a gasket and stopcock grease and then bolted firmly in place to the top of the water bath if an airtight seal is necessary. In a sealed system, air is supplied above the surface of the water to maintain a constant volume of water, and an exit glass tube of small diameter is positioned at the air-water interface so that incoming air removes a volume of water equal to that of the incoming water. Microcosm effluent is collected in a vessel that can be sterilized, if necessary. Flow-through microcosms may be operated unsealed with a constant-level siphon used to maintain water at a preset level.

Mixing is accomplished with a small, 300-rpm constant revolution motor (Hurst, Princeton, Indiana) mounted on the microcosm cover. A glass rod, 0.5 × 8 cm long with a 120° angle in the lower 1 cm, provides effective mixing without disturbing the water-sediment interface.

This microcosm design was used with minor modifications to study the effects of bioturbation and seagrass on the distribution of Kepone (47), and, in conjunction with a mathematical model, to describe the movement of Kepone within the test system (65). Other studies included scaling and field verification tests on the fate of p-nitrophenol (76), comparison of the fate of p-chlorophenol in water and sediment from two salt-marsh sites (66), and biodegradation of fenthion (48).

Biomass and Functional and Structural Endpoints

Adequate assessment of ecological hazards of GEMs introduced into an aquatic environment requires sensitive techniques to determine fluctuations in GEM populations and also ecological changes in the community. Aquatic environments offer a variety of diverse habitats for GEM colonization. For example, GEMs may remain suspended in the water column either alone or attached to particulates or be transported to other environments by currents. Sedimentation of larger particles or attachment to interfaces (e.g., sediment-water, air-water, animal, or plant surfaces) may result in colonization of new habitats. Ingestion by aquatic invertebrates or vertebrates may further increase the number of available sites.

Ecological communities may be characterized by the biomass of their constituent populations, structural diversity (in terms of biomass of species or group of species, and numbers of species or groups comprising the community), and by the functions that these organisms perform, especially biogeochemical processes. Once introduced, a GEM may interact with and possibly alter populations of indigenous organisms or affect ecological processes of an aquatic ecosystem. Traditionally, risk assessment with larger organisms has primarily focused on biomass (i.e., acute and chronic toxicity, effects on growth or reproduction) or structure (i.e., diversity indices), while microbial measurements have often been related to function. Recent developments have resulted in other methods to measure the effects of stress, such as the application of guild approaches (14) to group invertebrates according to how they function within their communities, or the characterization of microbial communities by lipid analysis (18).

The structural and functional measures listed below may be useful in the microcosm calibration process and serve as endpoints to assess the effects of GEM introduction. Microbial ecological methods were reviewed by Karl (33). The selection of these parameters will depend on the aquatic site chosen for the microcosm study and characteristics of the GEM under study. For example, a sulfate-reducing community would be more appropriate for a salt-marsh habitat containing highly organic, anoxic sediment than for a habitat composed of well-oxygenated sand grains low in organic matter. Similarly, carbon cycling processes would be preferred to sulfate-reducing processes for examining the potential effects of a GEM with enhanced cellulase production capabilities.

Biomass enumeration

Evaluation of the fate and effects of a GEM in aquatic habitats depends on specific, sensitive enumeration techniques to establish trends in populations of GEMs and indigenous microorganisms. An introduced GEM may affect or be affected by an indigenous species or group of organisms either directly (i.e., predation, pathology, etc.) or indirectly (competition, alterations of chemical or physical environment, etc.). Enumeration of the most significant species and groups forms the basis for determining GEM effects on community structure (see Structural Parameters in this section). Functional endpoints (see Functional Parameters in this section) rely on enumeration to normalize the estimates of activity per organism to separate effects on absolute activity from those resulting from biomass changes.

Sampling and enumeration for GEM persistence. Most standard techniques for the enumeration of specific microbes, such as selective media, immunoassays (5), and gene probes (3), can be used with microcosms. Enumeration techniques for introduced GEMs are covered in detail in Chap. 2.

Water samples should be collected from all microcosms at dosing time (after an appropriate mixing period) and periodically thereafter, according to increases or decreases in GEM population.

The air-water interface surface film may be an important area for monitoring GEM concentrations. This hydrophobic microlayer (51, 86) may contain 1000-fold more bacteria on a volume basis than the water column (40). A number of techniques have been described and evaluated for sampling this zone, many of which use filters or metal or plastic mesh screens (40, 86).

The collection of sediment may significantly disturb the integrity of the remaining core in small microcosms. Destructive sampling at selected intervals avoids this problem if replicate systems are set up and tested serially. Prior to sediment sampling, the overlying water column should be carefully siphoned so that the water-sediment interface is not disturbed. The sediment then can be cored, extruded, and sliced into segments, and the center of each segment sampled by aseptic techniques.

The process of bioturbation may draw GEMs from the overlying water column or attached to sediment deep into the sediment bed where they may interact with benthic organisms. A variety of techniques can be used to measure these movements and compare the migration of water and particles into the sediment with the depths to which GEMs are distributed. Some approaches use tracers, such as tritiated water (19, 66) to examine diffusion. Although Harvey et al. (26) found that bromide and chloride ions or fluorescent microspheres may not provide ideal surrogates for modeling the movement of bacteria through an aquifer, these tracers may still be useful in modeling water movement and quantifying sediment reworking activity.

In addition to sediment, planktonic particles, and leaf and microcosm wall surfaces, benthic invertebrates should be examined for the presence of GEMs both on their surface and in their gut, since such habitats may be suitable for GEM colonization.

Enumeration of indigenous microorganisms. Since it is possible that an interaction of GEMs with indigenous microbes may result in the increase or decrease in the biomass of the indigenous organisms, periodic sampling for total numbers of microbes may provide one endpoint for assessing the effects on an aquatic community. Direct counting techniques using epifluorescence methods and stains, such as acridine orange (28) and DAPI (56), are commonly used for enumeration of bacteria. Microflagellates may be enumerated by epifluorescence microscopy with a primulin stain (8). The same stain can also be used for determining ciliate densities. See Chap. 5 for a discussion of methods of enumerating indigenous microbes in aquatic habitats and their sensitivity.

Direct count techniques have two limitations: they do not indicate the fraction of the microbial population that is metabolically active, and they represent only the standing biomass of a population that may be rapidly turning over. The latter problem is addressed in Bacterial, later in this section.

Several techniques address the problem of distinguishing metabolically active and inactive populations. The Kogure technique (36) utilizes nalidixic acid to prevent division of cells while incubating with yeast extract. The bacteria then are stained with acridine orange, and cell elongation is used to identify actively growing organisms. In microautoradiography, microbes are incubated with a radiolabeled (^{3}H-, ^{14}C-, etc.) substrate (i.e., acetic acid, thymidine, glucose, amino acid), filtered, transferred to a slide to which a photographic emulsion is applied, and allowed to develop. Active cells are identified by the presence of silver grains. Tabor and Neihof (82) combined epifluorescence microscopy with microautoradiography to enumerate both active and total microorganisms.

Microbial populations on surfaces are more difficult to estimate than those in the water column, although epifluorescence has proved useful to enumerate bacteria on particles (35). Peroni and Rossi (55) measured active microbial biomass in the sediment by monitoring reduction of resazurin by the electron transport system.

Phytoplankton biomass. The most useful chemical method for estimating phytoplankton biomass (at least in seawater) is through the determination of chlorophyll, either by spectrophotometry or fluorometry (52).

Structural parameters

Structural parameters are used to determine potential community composition changes as a result of GEM introduction. Assessment of structural changes requires enumeration of all or part of the members of a population to determine how they change with respect to other populations. Some structural endpoints identify the nature of the community composition change, while others (e.g., diversity indices) do not.

Diversity indices. Diversity is a measure of a community's composition in terms of the number of species and number of individuals within each species. High diversity measurements in microcosms approximating those of the field suggest that the laboratory system is reasonably healthy. Diversity changes may be observed in both water column (plankton, bacterial) and benthic communities (invertebrate, bacterial). Since they do not identify the specific organisms affected, diversity indices provide only a relative measure of perturbation.

Atlas (2) discussed the application of diversity indices to microbial communities and recommended the Shannon index or, if the community is small, Brillouin's index. Kaesler et al. (32) also examined the use of diversity indices for studies of stressed communities and suggested that the Brillouin index is superior to other diversity measures for applied aquatic ecology.

Diversity indices have been criticized (32, 50) for such inadequacies as insensitivity, too simplistic to reflect complex ecosystems, not comparable from one region to another, and information is lost when numbers of species and individuals are reduced to a single number. Thus, diversity indices should be used for risk assessment applications with some caution, but they are still considered useful tools to summarize long taxonomic lists describing a community if their limitations are recognized (32).

Molecular taxonomy. Application of molecular taxonomy allows additional characterization by DNA base composition, and DNA/DNA reassociation. Molecular genetics techniques appear promising for enumerating the large fraction of microbial communities that cannot readily be isolated in pure cultures. Species-specific and group-specific 16S rRNA hybridization probes have been used to characterize natural microbial ecosystems (77).

Bacterial lipid analysis. Analysis of lipids extracted from a microbial community has been used to characterize the community (18). The technique has been shown to be sensitive to stress; it was used to detect differences in sedimentary microbiota with and without epibenthic predators (17).

Guild approach. Diaz et al. (14) tried various grouping approaches to characterize benthic community changes observed in microcosms, given the numerous fluctuations in individual species populations that obscured details of community responses. Grouping by taxonomic level was abandoned because some closely related species sometimes performed different functional roles within a community. They settled on a guild or "ecological types" approach, based on how a guild used resources, lived and moved in sediments, and reproduced.

Benthic community colonization.   Changes in macrobenthic community composition have been used as an endpoint for toxicity tests. Silica sand is exposed to various concentrations of a toxicant while being colonized with larval invertebrates from unfiltered seawater. When the test systems are harvested, sediments are treated to relax the organisms, which are then preserved and stained prior to sorting (44). This test represents certain advantages over single-species toxicity tests (25). For example, it uses early developmental stages of organisms, which are often more sensitive than older stages and may indicate sensitive species that are not normally tested in single-species tests. In addition, effects on more complex interactions may be suggested, such as the unexpected increase in tunicates in communities exposed to Aroclor 1254, possibly due to decreases in predator or competitor species (25).

Meiofauna, generally removed from sediments by a density separation technique (73), has also been examined in benthic colonization toxicity tests (25). A review by Raffaelli (67) examined the usefulness of a nematode-to-copepod index for evaluating pollutant stress.

Another type of colonized community has been described and used for assessing the effects of toxic chemicals (59, 60). Microbial communities (bacteria, protozoa, algae, and microinvertebrates) that colonize polyurethane foam substrates can be evaluated for structural (protozoan species number, biomass) and functional (colonization rate, oxygen production) endpoints.

Functional parameters

Functional parameters are a result of the activity of microbial communities. Since each process may be the product of many individual chemical reactions, several methods can be used to measure different aspects of each process and thereby provide a means of detecting perturbation. Since all of these endpoints could be considered part of nutrient cycling, we have organized them into the following two categories:

Carbon cycling.   Carbon cycling activities amenable to microcosm studies consist primarily of those carried out by primary producers or associated with bacterial cycling of organic material. Productivity-respiration studies attempt to assess ecosystem or microcosm energetic processes in one assay; productivity (P) and respiration (R) are estimated from changes in measurements of dissolved oxygen concentrations under light and dark conditions (21, 42). A P/R ratio reflects the health of an ecosystem and its ability to sustain itself; a ratio of 1 indicates an energy balance characteristic of mature communities (45).

Other assays were designed to look at specific aspects of populations involved with carbon cycling:

Primary producers. Phytoplankton photosynthetic activity is commonly measured by the uptake of ^{14}C-labeled sodium carbonate or bicarbonate, a technique described by Parsons et al. (52). The productivity of benthic macrophytes may be determined by a similar ^{14}C uptake technique or leaf-biomass procedure whereby leaves are marked and growth over time is determined by actually weighing the new leaf material; both methods are compared by Bittaker and Iverson (4).

Bacterial. The measurement of turnover of organic carbon (heterotrophic activity) by bacteria generally requires the use of radiolabeled substrates such as carbohydrates or amino acids and is expressed as the amount of substrate assimilated or mineralized per unit time. These radiolabeled compounds are generally added at concentrations well below those of the naturally occurring substrate to avoid stimulating microbial activity. With certain assumptions, Michaelis-Menten kinetics may be used to examine the effect of substrate concentration on assimilation and mineralization. Measurement of bacterial heterotrophic activity through D-[U-^{14}C] glucose uptake has been described by Parsons et al. (52). Changes in activity may be a function of either biomass or physiological state and nutrient availability; therefore, it is important to determine biomass (see Diversity Indices in this section) each time a heterotrophic activity measurement is taken. This parameter has been used to monitor the effects of pollutants (1, 49, 70). Bacterial heterotrophic growth may be measured by the incorporation of thymidine[methyl-^{3}H] into DNA, a technique described by Parsons et al. (52).

Bacterial production and microflagellate grazing on bacteria can be measured simultaneously by inhibiting grazing on bacteria, either with selective filtration (88) or an antibiotic (11). Bacterial production is determined as the change in bacterial density in the treated samples over time, and grazing rate is the difference between this rate and the change in bacterial density in an untreated sample.

Carbon cycling may also be measured as a function of leaf litter decomposition (6, 44, 79), based on the significance of allochthonous leaf matter (for freshwater streams) or autochthonous benthic macrophyte matter (for seagrass beds) to the carbon pools of each ecosystem. Since this relatively large quantity of carbon is made available to other members of the ecosystem only through the actions of a relatively small group of organisms capable of mechanically and biochemically breaking down cellulose and lignin material, the process limits the rate of production of labile carbon material that might be utilized by other members of the ecosystem. Weighed leaf material is placed in mesh bags suspended in the water column, and periodic measurements of weight loss are made as the leaves are decomposed. An introduced GEM could theoretically affect this rate of decomposition by interacting with the indigenous degrading organisms, thereby slowing the rate, or by producing enzymes that accelerate this process. Either change in rate may deleteriously affect an aquatic ecosystem by causing detrimental changes in the quantity of available carbon at a given point in time.

Microbial carbon cycling potential may also be estimated by assessing enzymatic activity. Since each nutrient cycling process performed by microorganisms is generally the result of many metabolic reactions, the potential for carrying out a process can be monitored by assaying certain key enzymes, such as protease, chitinase, and cellulase. They may also be used to examine the mechanism by which a nutrient cycling activity has been affected. A common approach to enzyme assays uses fluorogenic dyes attached to substrates that are cleaved by specific enzymes. Examples include fluorescein diacetate for assaying esterase (81), Azocoll for protease (7), and methylumbelliferyl substrates for phosphatase, and α- and β-D-glucosidase (29). A number of studies have been reported that used microbial enzymes to monitor the effects of xenobiotic compounds; enzymes assayed included phosphatase and dehydrogenase (58), chitinase (57), and chitobiase, and polysaccharide hydrolase (23).

Nutrient cycling.   Microbes play important roles in the biogeochemical cycling of nutrients in ecosystems and thus provide some of the most obvious functional endpoints for biotechnology risk assessment. Even relatively small changes may be detrimental, resulting in eutrophication or a reduction of biomass.

Nitrogen cycling in sediment-water systems has been reviewed by Keeney (34). Nitrogen fixation is the conversion of inorganic nitrogen to a more oxidized state. It is typically measured by the acetylene reduction method (15) and has been used as an endpoint for pesticide effects studies (68). One aspect of nitrogen cycling that has been generally overlooked as an effects endpoint is denitrification. This process eliminates utilizable forms of nitrogen, such as nitrate, by a reduction to gaseous nitrogen and may spare some aquatic systems receiving relatively high nitrogen loads from eutrophication. Methods for assessing denitrification are described by Sorensen (74) and Chan and Knowles (9).

Phosphorous concentrations have been shown to be limiting to phytoplankton growth in many freshwater communities (27), and thus relatively small fluctuations of this nutrient might be expected to significantly affect primary productivity in some environments. Phosphate in the water column is typically measured spectrophotometrically after reacting it with molybdate (52).

Sulfur cycling, which plays an important role in protein metabolism, is most commonly measured in marine anaerobic sediments. The typical assay procedure (30) involves monitoring the reduction of $^{35}SO_4^{2-}$ to $^{35}S^{2-}$ (and sometimes pyrite and elemental sulfur).

Data Interpretation

A microcosm, like the community from which it is taken, is site-specific, and care should be taken when extrapolating data derived form microcosms beyond that community. Extrapolation to the field is greatly enhanced by mathematical models that can compensate for factors unique to a field site that cannot be incorporated into microcosm design (i.e., turbulence, water flow, and water volume to a sediment surface area). Most environmental model development effort has been associated with determining the fate of xenobiotic compounds in the aquatic environment. Models and concepts associated with the persistence and transport of wastewater microbes (90) appear relevant to the study of GEMs, but their application must be tested.

We urge caution in interpreting the various microcosm responses for risk assessment. Some parameters in a given microcosm may prove relatively insensitive to stress, while others may prove so sensitive that significant differences between control and experimental microcosms have no environmental relevance. We suggest that a variety of biomass, structural, and functional parameters be selected for baseline tests that use specific perturbators as "positive control" agents to determine the sensitivity and significance of each parameter. The most sensitive endpoints would be monitored for possible ecological effects due to the presence of GEMs.

Field Verification

Verification studies represent the last phase of microcosm testing. Ideally, both test site and microcosms are dosed simultaneously with the test organisms, and persistence, effects, and gene exchange are compared. For practical reasons (i.e., risk to the environment), initial field verification tests can be conducted with nongenetically engineered microorganisms selected as surrogates for GEMs. Several examples of field verification studies have been reported for fate and/or effects studies with xenobiotic compounds (12–14, 20, 22, 58, 76). After field verifications of microcosm-derived survival and effects are observed with a variety of microorganisms, it may be possible to exclude the costly field study.

Reported Microcosm Studies for Biotechnology Risk Assessment

Although aquatic microcosms are used frequently to address the fate and effects of xenobiotic compounds, they only recently were proposed for microbial persistence and ecological effects studies (80) and microbial genetic interaction studies (85). A few studies do exist, however, and more are expected. Liang et al. (39) observed the persistence of seven species of microbes (selected for their potential for use in genetic engineering) in shake flasks containing sewage and lake water. Fisher and Sheeran (Abstract, 8th Annual Meeting Soc. Environ. Toxicol. Chem. 1987) used aquatic microcosms to determine if persistence of *Bacillus thuringiensis* was affected by the presence of sediment and indigenous microbes, factors which also influence gene transfer. Shannon et al. (Abstract, 12th Symposium Aquat. Toxicol. Hazard Assess., Amer. Soc. Test. Mater., 1988) also reported on *B. thuringiensis* persistence, using mixed flask cultures. Finally, Scanferlato et al. (72) examined the persistence of genetically engineered *Erwinia carotovora* and its effects on indigenous bacteria performing various nutrient cycling (proteolytic, pectolytic, and amyollytic) functions.

Conclusions

Microcosms, especially those "calibrated" with the field site from which they are taken, offer several attractive features for GEM risk assessment over the use of actual field tests, specifically, control of certain environmental variables and elimination of potential risk to the environment. Actual research topics for which microcosms would be appropriate include GEM persistence, potential ecological effects, and gene transfer.

A number of important design features were presented, including size, scaling, construction materials, light, temperature, water replacement, mixing, and coring. Several examples of site-specific aquatic microcosms were described, each with intact sediment cores and ranging in volume from 175 mL to 27 liters.

Successful use of microcosms for studying potential ecological effects depends on the selection of appropriate endpoints. A variety of endpoints concerned with biomass enumeration, and numerous structural and functional parameters should be considered. Use of "positive control" agents to perturb a microcosm may aid in determining the sensitivity of the various effects parameters.

Mathematical models may aid in the extrapolation of data derived from microcosms to a field site, but it is too early to determine how far the extrapolation can be taken (i.e., to similar ecosystems or dissimilar ecosystems). Much additional basic microbial ecology research will be necessary to properly address this problem, although direct approaches involving field verification of microbial persistence, effects, and gene transfer will also be helpful.

The greatest contribution of microcosms to environmental risk assessment to date has been in the area of chemical fate testing. When coupled with an approach utilizing existing mathematical models, a relatively strong case can be made for the use of microcosms. Unfortunately, the number of microcosm studies with GEMs is still very small, and it is still difficult to determine how significant a role they will play in GEM risk assessment. The potential risks involved with taking GEMs directly to the field for testing instead of performing lower level tests in microcosms appears to be too great to ignore this test approach.

References

1. Albright, L. J., and E. M. Wilson. 1974. Sublethal effects of several metallic salts-organic compound combinations upon the heterotrophic microflora of a natural water. *Water Res.* 8:101–105.
2. Atlas, R. M. 1984. Diversity of microbial communities. In K. C. Marshall (ed.), *Advances in Microbial Ecology*, vol. 7. Plenum Press, New York, pp. 1–47.
3. Barkay, T., and G. Sayler. 1988. Gene probes as a tool for the detection of specific genomes in the environment. In W. J. Adams, G. A. Chapman, and W. G. Landis (eds.), *Aquatic Toxicology and Hazard Assessment*, 10th vol., ASTM STP 971. American Society for Testing and Materials, Philadelphia, Pa., pp. 29–36.
4. Bittaker, H. F., and R. L. Iverson. 1976. *Thalassia testudinum* productivity: A field comparison of measurement methods. *Mar. Biol.* 37:39–46.
5. Bohlool, B. B., and E. L. Schmidt. 1980. The immunofluorescence approach in microbial ecology. *Adv. Microb. Ecol.* 4:203–241.
6. Burton, T. M., R. M. Stanford, and J. W. Allan. 1985. Acidification effects on stream biota and organic matter processing. *Can. J. Fish. Aquat. Sci.* 42:669–675.
7. Caplan, J. A., and J. W. Fahey. 1980. A rapid assay for the examination of protein degradation in soils. *Bull. Environ. Contam. Toxicol.* 25:424–426.
8. Caron, D. A. 1983. Technique for enumeration of heterotrophic and phototrophic nanoplankton, using epifluorescence microscopy, and comparison with other procedures. *Appl. Environ. Microbiol.* 46:491–498.
9. Chan, Y. K., and Knowles, R. 1979. Measurement of denitrification in two freshwater sediments by an in situ acetylene inhibitor method. *Appl. Environ. Microbiol.* 37:1067–1072.
10. Clark, J. R., and J. M. Macauley. Comparison of the seagrass. *Thalassia testudinum* and its epiphytes in the field and in laboratory test systems. In W. Wang, J. W.

Gorsach, and W. R. Lower (eds.), *Plants for Toxicity Assessment,* ASTM STP 1091, American Society for Testing and Materials, pp. 59–68.

11. Coffin, R. B., and J. H. Sharp. 1987. Microbial trophodynamics in the Delaware Estuary. *Mar. Ecol. Prog. Ser.* 41:253–266.

12. Crossland, N. O., and D. Bennett. 1984. Fate and biological effects of methyl parathion in outdoor ponds and laboratory aquaria. I. Fate. *Ecotoxicol. Environ. Safety* 8:471–481.

13. Crossland, N. O. 1984. Fate and biological effects of methyl parathion in outdoor ponds and laboratory aquaria. II. Effects. *Ecotoxicol. Environ. Safety* 8:482–495.

14. Diaz, R. J., M. Luckenbach, S. Thornton, M. H. Roberts, Jr., R. J. Livingston, C. C. Koenig, G. L. Ray, and L. E. Wolfe. 1987. Project Report. Field validation of multi-species laboratory test systems for estuarine benthic communities. EPA/600/3-87/016. U.S. Environmental Protection Agency, Environmental Research Laboratory, Gulf Breeze, Fla.

15. Dicker, H. J., and D. W. Smith. 1980. Acetylene reduction (nitrogen fixation) in a Delaware, USA salt marsh. *Mar. Biol.* 57:241–250.

16. Estrada, M., M. Alcaraz, and C. Marrase. 1987. Effects of turbulence on the composition of phytoplankton assemblages in marine microcosms. *Mar. Ecol. Prog. Ser.* 38:267–281.

17. Federle, T. W., R. J. Livingston, D. A. Meeter, and D. C. White. 1983. Modifications of estuarine sedimentary microbiota by exclusion of epibenthic predators. *J. Exp. Mar. Biol. Ecol.* 73:81–94.

18. Federle, T. W., R. J. Livingston, L. W. Wolfe, and D. C. White. 1986. A quantitative comparison of microbial community structure of estuarine sediments from microcosms and the field. *Can. J. Microbiol.* 32:319–325.

19. Feijtel, T. C., Y. Salingar, C. A. Hordijk, J. P. R. A. Sweerts, N. van Breemen, and T. E. Cappenberg. 1989. Sulfur cycling in a Dutch moorland pool under elevated atmospheric S-deposition. *Water, Air, Soil Pollut.* 44:215–234.

20. Franco, P. J., J. M. Giddings, S. E. Herbes, L. A. Hook, J. D. Newbold, W. K. Roy, G. R. Southworth, and A. J. Stewart. 1984. Effects of chronic exposure to coal-derived oil on freshwater ecosystems: I. Microcosms. *Environ. Toxicol. Chem.* 3:447–463.

21. Giddings, J., and G. K. Eddleman. 1978. Photosynthesis/respiration ratios in aquatic microcosms under arsenic stress. *Water Air Soil Pollut.* 9:207–212.

22. Giddings, J. M., P. J. Franco, R. M. Cushman, L. A. Hook, G. R. Southworth, and A. J. Stewart. 1984. Effects of chronic exposure to coal-derived oil on freshwater ecosystems. II. Experimental ponds. *Environ. Toxicol. Chem.* 3:465–488.

23. Griffiths, R. P., B. A. Caldwell, W. A. Broich, and R. Y. Morita. 1982. Long-term effects of crude oil on microbial processes in subarctic marine sediments: Studies on sediments amended with organic nutrients. *Mar. Pollut. Bull.* 13:273–278.

24. Gunnison, D. 1986. New frontiers in applied sediment microbiology. *Adv. Appl. Microbiol.* 31:207–232.

25. Hansen, D. J., and M. E. Tagatz. 1980. A laboratory test for assessing impacts of substances on developing communities of benthic estuarine organisms. In J. G. Eaton, P. R. Parrish, and A. C. Hendricks (eds.), *Aquatic Toxicology,* ASTM STP 707. American Society for Testing and Materials, Philadelphia, Pa., pp. 40–57.

26. Harvey, R. W., L. H. George, R. L. Smith, and D. R. LeBlanc. 1989. Transport of microspheres and indigenous bacteria through a sandy aquifer: Results of natural- and forced-gradient tracer experiments. *Environ. Sci. Technol.* 23:51–56.

27. Hecky, R. E., and P. Kilham. 1988. Nutrient limitation of phytoplankton in freshwater and marine environments: A review of recent evidence on the effects of enrichment. *Limnol. Oceanogr.* 33:796–822.

28. Hobbie, J. E., R. J. Daley, and S. Jasper. 1977. Use of Nucleopore filters for counting bacteria by fluorescence microscopy. *Appl. Environ. Microbiol.* 33:1225–1228.

29. Hoppe, H. -G. 1983. Significance of exoenzymatic activities in the ecology of brackish water: Measurements by means of methylumbelliferyl-substrates. *Mar. Ecol.* 11:299–308.

30. Howarth, R. W., and B. B. Jorgensen. 1984. Formation of ^{35}S-labelled elemental

sulfur and pyrite in coastal marine sediments (Limfjorden and Kysing Ford, Denmark) during short-term $^{35}SO_4^{-2}$ reduction measurements. *Geochim. Cosmochim. Acta* 48:1807–1818.

31. Isensee, A. R., and N. Tayaputch. 1986. Distribution of carbofuran in a rice-paddy-fish microecosystem. *Bull. Environ. Contam. Toxicol.* 36:763–769.
32. Kaesler, R. L., E. E. Herricks, and J. S. Crossman. 1978. Use of indices of diversity and hierarchical diversity in stream surveys. In K. L. Dickson, J. Cairns, Jr., and R. J. Livingston (eds.), *Biological Data in Water Pollution Assessment; Quantitative and Statistical Analyses*, ASTM STP 652. American Society for Testing and Materials, Philadelphia, Pa., pp. 92–112.
33. Karl, D. M. 1986. Determination of *in situ* microbial biomass, viability, metabolism, and growth. In J. S. Poindexter and E. R. Leadbetter (eds.), *Bacteria in Nature*, Vol. 2, *Methods and Special Applications in Bacterial Ecology*. Plenum Press, New York, pp. 85–176.
34. Keeney, D. R. 1973. The nitrogen cycle in sediment-water systems. *J. Environ. Qual.* 2:15–29.
35. Kirchman, D., and R. Mitchell. 1982. Contribution of particle-bound bacteria to total microheterotrophic activity in five ponds and two marshes. *Appl. Environ. Microbiol.* 43:200–209.
36. Kogure, K., U. Simidu, and N. Taga. 1979. A tentative direct microscopic method for counting living marine bacteria. *Can. J. Microbiol.* 25:415–420.
37. Kroer, N., and R. B. Coffin. Microbial trophic interactions in aquatic microcosms developed for testing genetically engineered microorganisms: A field comparison. In review.
38. Leffler, J. W. 1984. The use of self-selected, generic aquatic microcosms for pollution effects assessment. In H. H. White (ed.), *Concepts in Marine Pollution Measurements*. Maryland Sea Grant College, University of Maryland, College Park, Md., pp. 139–157.
39. Liang, L. N., J. L. Sinclair, L. M. Mallory, and M. Alexander. 1982. Fate in model ecosystems of microbial species of potential use in genetic engineering. *Appl. Environ. Microbiol.* 44:708–714.
40. Liss, P. S. 1975. Chemistry of the sea surface microlayer. In J. P. Riley and G. Skirrow (eds.), *Chemical Oceanography*. Academic Press, London, pp. 193–244.
41. Livingston, R. J., R. J. Diaz, and D. C. White. 1985. Project Report. Field validation of laboratory-derived multispecies aquatic test systems. EPA/600/4-85/039. U.S. Environmental Protection Agency, Environmental Research Laboratory, Gulf Breeze, Fla.
42. McIntire, C. D., R. L. Garrison, H. K. Phinney, and C. E. Warren. 1964. Primary production in laboratory streams. *Limnol. Oceanogr.* 9:92–102.
43. Metcalf, R. L., G. K. Sangha, and I. P. Kapoor. 1971. Model ecosystem for the evaluation of pesticide biodegradability and ecological magnification. *Environ. Sci. Technol.* 5:709–713.
44. Morton, R. D., T. D. Duke, J. M. Macauley, J. R. Clark, W. A. Price, S. J. Hendricks, S. L. Owsley-Montgomery, and G. R. Plaia. 1986. Impact of drilling fluids on seagrasses: An experimental community approach. In J. Cairns, Jr. (ed.), *Community Toxicity Testing*, ASTM STP 920. American Society for Testing and Materials, Philadelphia, Pa., pp. 199–212.
45. Odum, E. P. 1969. The strategy of ecosystem development. *Science* 164:262–270.
46. Odum, E. P. 1985. Trends expected in stressed ecosystems. *Bioscience* 35:419–422.
47. O'Neill, E. J., C. A. Monti, P. H. Pritchard, A. W. Bourquin, and D. G. Ahearn. 1985. Effects of lugworms and seagrass on Kepone (chlordecone) distribution in sediment/water laboratory systems. *Environ. Toxicol. Chem.* 4:453–458.
48. O'Neill, E. J., C. R. Cripe, L. H. Mueller, J. P. Connolly, and P. H. Pritchard. 1989. Fate of fenthion in salt-marsh environments. II. Transport and biodegradation in microcosms. *Environ. Toxicol. Chem.* 8:759–768.
49. Orndorff, S. A., and R. R. Colwell. 1980. Effect of Kepone on estuarine microbial activity. *Microb. Ecol.* 6:357–368.
50. Oviatt, C. A., H. H. White, and A. Robertson. 1984. Summary and synthesis. In

H. H. White (ed.), *Concepts in Marine Pollution Measurements*. Maryland Sea Grant College, University of Maryland, College Park, Md., pp. 725–735.

51. Parker, B., and G. Barsom. 1970. Biological and chemical significance of surface microlayers in aquatic ecosystems. *Bioscience* 20:87–93.

52. Parsons, T. R., Y. Maita, and C. M. Lalli. 1984. *A Manual of Chemical and Biological Methods for Seawater Analysis*. Pergamon Press, New York.

53. Paul, J. P., and A. W. David. 1989. Production of extracellular nucleic acids by genetically altered bacteria in aquatic-environment microcosms. *Appl. Environ. Microbiol.* 55:1865–1869.

54. Perez, K. T., G. M. Morrison, N. F. Lackie, C. A. Oviatt, S. W. Nixon, B. A. Buckley, and J. F. Heltshe. 1977. The importance of physical and biotic scaling to the experimental simulation of a coastal marine ecosystem. *Helgolander will. Meeresunters.* 30:144–162.

55. Peroni, C., and G. Rossi. 1986. Determination of microbial activity in marine sediments by resazurin reduction. *Chem. Ecol.* 2:205–218.

56. Porter, K. G., and T. S. Feig. 1980. The use of DAPI for identifying and counting aquatic microflora. *Limnol. Oceanogr.* 25:943–948.

57. Portier, R. J., and S. P. Meyers. 1981. Chitin transformation and pesticide interactions in a simulated aquatic microenvironmental system. *Dev. Ind. Microbiol.* 22: 543–555.

58. Portier, R. J. 1985. Comparison of environmental effect and biotransformation of toxicants on laboratory microcosm and field microbial communities. In T. P. Boyle (ed.), *Validation and Predictability of Laboratory Methods for Assessing the Fate and Effects of Contaminants in Aquatic Ecosystems*, ASTM STP 865. American Society for Testing and Materials, Philadelphia, Pa., pp. 14–30.

59. Pratt, J. R., N. J. Bowers, B. R. Niederlehner, and J. Cairns, Jr. 1988. Effects of atrazine on freshwater microbial communities. *Arch. Environ. Toxicol.* 17:449–457.

60. Pratt, J. R., and J. Cairns, Jr. 1988. Use of microbial colonization parameters as a measure of functional response in aquatic ecosystems. In J. Cairns, Jr., and J. R. Pratt (eds.), *Functional Testing of Aquatic Biota for Estimating Hazards of Chemicals*, ASTM STP 988. American Society for Testing and Materials, Philadelphia, Pa., pp. 55–67.

61. Price, N. M., P. J. Harrison, M. R. Landry, F. Azam, and K. J. F. Hall. 1986. Toxic effects of latex and Tygon tubing on marine phytoplankton, zooplankton and bacteria. *Mar. Ecol. Prog. Ser.* 34:41–49.

62. Pritchard, P. H., A. W. Bourquin, H. L. Fredrickson, and T. Maziarz. 1979. System design factors affecting environmental fate studies in microcosms. In A. W. Bourquin, and P. H. Pritchard (eds.), *Microbial Degradation of Pollutants in Marine Environments: Proceedings of the Workshop*. EPA-600/9-79-012. U.S. Environmental Protection Agency, Environmental Research Laboratory, Gulf Breeze, Fla., pp. 251–272.

63. Pritchard, P. H. 1981. Model ecosystems. In R. A. Conway (ed.), *Environmental Risk Analysis for Chemicals*. Van Nostrand Reinhold, New York, pp. 257–353.

64. Pritchard, P. H., and A. W. Bourquin. 1984. The use of microcosms for evaluation of interactions between pollutants and microorganisms. *Adv. Microb. Ecol.* 7:133–215.

65. Pritchard, P. H., C. A. Monti, E. J. O'Neill, J. P. Connolly, and D. G. Ahearn. 1986. Movement of Kepone (chlordecone) across an undisturbed sediment-water interface in laboratory systems. *Environ. Toxicol. Chem.* 5:647–657.

66. Pritchard, P. H., E. J. O'Neill, C. M. Spain, and D. G. Ahearn. 1987. Physical and biological parameters that determine the fate of *p*-chlorophenol in laboratory test systems. *Appl. Environ. Microbiol.* 53:1833–1838.

67. Raffaelli, D. 1987. The behavior of the nematode/copepod ratio in organic pollution studies. *Mar. Environ. Res.* 23:135–152.

68. Ray, R. C. 1983. Toxicity of the pesticides hexachlorocyclohexane and benomyl to nitrifying bacteria in flooded autoclaved soil and in culture media. *Environ. Pollut.* (series A). 32:147–155.

69. Reichardt, W. 1988. Impact of bioturbation by *Arenicola marina* on microbiological parameters in intertidal sediments. *Mar. Ecol. Prog. Ser.* 44:149–158.

70. Sayler, G. S., J. B. Waide, M. C. Waldron, T. W. Sherrill, and R. E. Perkins. 1979. Comparative effects of Aroclor 1257 (polychlorinated biphenyls) and phenanthrene on glucose uptake by freshwater microbial populations. *Appl. Environ. Microbiol.* 37:878–885.

71. Sayler, G. S., R. E. Perkins, T. W. Sherrill, B. K. Perkins, M. C. Reid, M. S. Shields, H. L. Kong, and J. W. Davis. 1983. Microcosm and experimental pond evaluation of microbial community response to synthetic oil contamination in freshwater sediments. *Appl. Environ. Microbiol.* 46:211–219.

72. Scanferlato, V. S., D. R. Orvos, J. Cairns, Jr., and G. H. Lacy. 1989. Genetically engineered *Erwinia carotovora* in aquatic microcosms: Survival and effects on functional groups of indigenous bacteria. *Appl. Environ. Microbiol.* 55:1477–1482.

73. Schwinghamer, P. 1981. Extraction of living meiofauna from marine sediments by centrifugation in a silica sol—sorbitol mixture. *Can. J. Fish. Aquat. Sci.* 38:476–478.

74. Sorensen, J. 1978. Capacity for denitrification and reduction of nitrate to ammonia in coastal marine sediment. *Appl. Environ. Microbiol.* 35:301–305.

75. Spain, J. C., P. H. Pritchard, and A. W. Bourquin. 1980. Effects of adaptation on biodegradation rates in sediment/water cores from estuarine and freshwater environments. *Appl. Environ. Microbiol.* 40:726–734.

76. Spain, J. C., P. A. Van Veld, C. A. Monti, P. H. Pritchard, and C. R. Cripe. 1984. Comparison of *p*-nitrophenol biodegradation in field and laboratory test systems. *Appl. Environ. Microbiol.* 48:944–950.

77. Stahl, D. A., B. Flesher, H. R. Mansfield, and L. Montgomery. 1988. Use of phylogenetically based hybridization probes for studies of ruminal microbial ecology. *Appl. Environ. Microbiol.* 54:1079–1084.

78. Stotzky, G., and H. Babich. 1986. Survival of, and genetic transfer by, genetically engineered bacteria in natural environments. *Adv. Appl. Microbiol.* 31:93–138.

79. Stout, R. J., and W. E. Cooper. 1983. Effect of *p*-cresol on leaf decomposition and invertebrate colonization in experimental outdoor streams. *Can. J. Fish. Aquat. Sci.* 40:1647–1657.

80. Suter, G. W., II. 1985. Application of environmental risk analysis to engineered organisms. In H. O. Halvorson, D. Pramer, and M. Rogul (eds.), *Engineered Organisms in the Environment: Scientific Issues.* American Society for Microbiology, Washington, D.C., pp. 211–219.

81. Swisher, R., and G. C. Carroll. 1980. Fluorescein diacetate hydrolysis as an estimator of microbial biomass on coniferous needle surfaces. *Microb. Ecol.* 6:217–226.

82. Tabor, P. S., and R. A. Neihof. 1982. Improved microautoradiographic method to determine individual microorganisms active in substrate uptake in natural waters. *Appl. Environ. Microbiol.* 44:945–953.

83. Taub, F. B., A. C. Kendig, and L. L. Conquest. 1987. Interlaboratory testing of a standardized aquatic microcosm. In W. J. Adams, G. A. Chapman, and W. G. Landis (eds.), *Aquatic Toxicology and Hazard Assessment,* 10th vol., ASTM STP 971. American Society for Testing and Materials, Philadelphia, Pa., pp. 384–405.

84. Trevors, J. T., T. Barkay, and A. W. Bourquin. 1987. Gene transfer among bacteria in soil and aquatic environments: A review. *Can. J. Microbiol.* 33:191–198.

85. Trevors, J. T. 1988. Use of microcosms to study genetic interactions between microorganisms. *Microbiol. Sci.* 5:132–136.

86. Van Vleet, E. S., and P. M. Williams. 1980. Sampling sea surface films: A laboratory evaluation of techniques and collecting materials. *Limnol. Oceanogr.* 25:764–770.

87. Wainwright, S. C. 1987. Stimulation of heterotrophic microplankton production by resuspended marine sediments. *Science* 23:1710–1712.

88. Wright, R. T., and R. B. Coffin. 1984. Measuring microzooplankton grazing on planktonic marine bacteria by its impact on bacterial production. *Microb. Ecol.* 10: 137–149.

89. Wavre, M., and R. O. Brinkhurst. 1971. Interactions between some tubificid oligochaetes and bacteria found in the sediments of Toronto Harbour, Ontario. *J. Fish. Res. Bd. Can.* 28:335–341.

90. Yates, M. V., and S. R. Yates. 1988. Modeling microbial fate in the subsurface environment. *CRC Crit. Rev. Environ. Control* 17:307–344.

24

Persistence of Recombinant Bacteria in Microcosms

John L. Armstrong

Overview of Historical and Current Perspectives

Interest in microcosm research has been renewed recently (6, 8, 15) by all the attention given to the release of genetically engineered microorganisms (GEMs) into the environment (11, 13, 16). As with toxic-chemical testing, it is assumed that GEMs can be tested in microcosms before field release to clarify (1) survival in soil, on plants, and in association with animals; (2) transport into higher trophic levels; and (3) effects on the chemistry and biology of the targeted ecosystem.

The general objectives of microcosm-based research with toxic chemicals are analogous to those of GEMs. Therefore, methods developed for chemical studies are pertinent to the newer applications of microcosms with GEMs. Hammons's review of the diversity of methods developed for studies of toxic chemicals (9) is a thorough summary, including aquatic and terrestrial test systems, mathematical models, population interactions, and ecosystem effects.

Gillett, Witt, and Wyatt's review (8) includes worthwhile information on the chief characteristics of several terrestrial microcosm types that have been applied to effects research with hazardous chemicals. Topics that are central to all microcosm experiments are discussed. They recommend using (1) a mixed soil containing representative components from a field site, (2) invertebrate species from at least three phyla, and (3) at least one grassy and one broadleafed plant. They also consider the use of a graded series in microcosm complexity, and express concern over the general lack of validation (precision) and verification (accuracy) of microcosm systems.

Parkes (14) concentrates more on the applications of microcosms to microbiological problems. Although the bacteria used in the reviewed studies were nonrecombinant strains, the basic issues are relevant to research with GEMs. In addition to the size and composition of the microcosm, he discusses (1) the stabilization period that must precede collection of reliable data, (2) the source of the microbial inoculum and method of inoculation, and (3) the applicability of results to environmental questions.

Van Voris (19) presents a terrestrial soil-core microcosm design which encloses a "piece" of intact soil containing indigenous assemblages of biota in a large polyethylene tube fitted with a Buchner funnel to allow leachate collection. The study of relatively undisturbed soil has advantages for testing effects on natural ecological processes that may be disrupted if the soil is mixed.

Few publications, however, are available that specify microcosms as tools for the study of GEMs. The review by Fredrickson et al. (6) considers the benefits of microcosms for analysis of the fate and risks of disseminating chemicals and GEMs into the environment. Fredrickson et al. (5) added nonrecombinant *Azospirillum lipoferum,* a root-colonizing organism, to soil-core microcosms to determine if the methods would permit detection of changes in nutrient dynamics that might be caused by engineered bacteria.

To develop microcosm methods that can be applied to the study of the survival and environmental effect of GEMs, Armstrong et al. (1) used a simple microcosm consisting of plants and cutworms in nonsterile potting soil and assessed bacterial survival in cutworms. These investigators demonstrated the appearance of *Pseudomonas cepacia* (R388::Tn*1721*) in the foreguts of larvae feeding upon sprayed leaves, but not in their fecal pellets. The value of this microcosm design was also demonstrated by others who noted: (1) pBR322-carrying strains of *Enterobacter cloacae, Klebsiella planticola,* and *Erwinia herbicola* differed in their relative persistence on plants, in soil, and in the digestive tracts of cutworm larvae (2); (2) transconjugation occurred between bacteria in the digestive tracts of insects (3); and (3) a predictive model could be verified for the conjugative transfer of plasmids in the phyllosphere (10).

In line with the general methods theme of this book, the objective of this chapter is to contribute further to the effective use of microcosms as tools for GEM studies. Therefore, an emphasis is placed on materials and methods, supplemented with several examples of applications. Included are descriptions of two microcosms with different levels of methodological complexity and microbiological techniques for obtaining and processing microcosm samples.

Laboratory Methods

Design of microcosms

Trays containing plants and soil (described below) were housed in a chamber (Fig. 24.1; Ref. 4) consisting of a 1.0 × 0.75 × 0.75 m box with Pyrex* glass top and acrylic or glass walls, which rested on a polyethylene box measuring 1.0 × 0.75 × 0.55 m. Each chamber was irradiated with a 1000-W Sylvania metal halide lamp (positioned 55

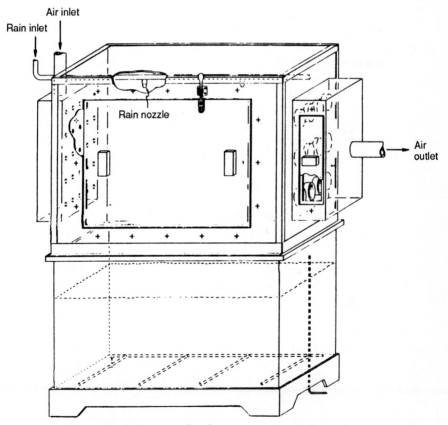

Figure 24.1 Schematic of microcosm chamber.

---

*Mention of trade names or commercial products does not constitute endorsement or recommendation for use.

cm above the box) on an 18/6-h light-dark cycle. Air was forced through the chamber and exhausted through a Hepa filter (Astrocel, American Air Filter, Louisville, Kentucky). Room temperature was maintained at 22°C (± 1°C) during the dark period and 23°C (±1°C) during the light period.

A simple microcosm (lacking environmental monitoring and control; Fig. 24.2) consisted of a plastic tray (27 × 55 × 5 cm) lined with a polyethylene bag containing potting mix (1 part Sunshine Mix to 1 part perlite) as described by Armstrong et al. (1). In experiments with cutworm larvae (see below), the bag was clamped near the top of a wire frame to stop their escape. During experiments, trays were placed on a metal rack in the chamber described above, supported 50 cm above the floor by bricks.

A second microcosm design was called "complex" since the environment was monitored and partially controlled and the soil was collected from a field site and not treated. This microcosm consisted of two wooden nursery trays (47 × 37 × 7 cm) lined with polyethylene bags, each holding about 6 to 7 cm soil (classified as Hazelair-Veneta series; collected near Corvallis, Oregon). These trays were placed on metal racks in the chamber as described above. Each chamber had a humidifier (#76850, Conviron Products of America, Pembina, North Dakota) below the rack. Water accumulating on the chamber floor during ex-

Figure 24.2 Schematic of simple microcosm.

tended periods of humidification was suctioned through a tube and collected in a 5-gal container attached to a vacuum line and was disinfected with bleach. At the end of each experiment, soil and plants were autoclaved for 1 h 15 min, and the inner surfaces of the chambers were disinfected with a dilute bleach solution.

Environmental monitoring

Environments of complex microcosms were monitored using a 21× micrologger (Campbell Scientific, Inc., Logan, Utah). Probes monitored air temperature and relative humidity (RH; probe #207), soil temperature (thermistor #107), soil moisture (gypsum block #227), and light (pyranometer LI200S).

Instrument probes were calibrated as follows. Each RH probe was wrapped in a moist paper towel and sealed in a plastic bag for 2 h to calibrate to 100 percent RH. Correction factors were determined so that all probes would read 95 (equivalent to 100 percent RH according to the manufacturer) in the bags. Temperature probes were calibrated by immersion in a water bath. Correction factors were determined to within 0.1°C of a National Bureau of Standards calibrated thermometer. Soil moisture probes were placed in saturated soil (100 percent water holding capacity, WHC) which was weighed as the soil dried (a moisture reading of −1 bar was equivalent to 62 percent WHC). Each chamber contained one temperature/RH probe (hanging 25 cm above the soil surface), two soil temperature probes (each buried 3 to 4 cm in the center of a tray), and two soil moisture probes (each near the soil surface in the center of a tray). Microloggers were programmed to collect environmental readings every minute and store hourly means.

Bacteria and culture conditions

A strain of *Enterobacter cloacae* was isolated from the feces of a variegated cutworm (*Peridroma saucia* Hübner) collected in a peppermint field near Albany, Oregon. The isolate was identified with the API 20E identification system (Analytab Products, Plainview, New York). A spontaneous rifampicin (Rif) resistant mutant isolate was transformed (12) with plasmid pBR322, which encodes oxytetracycline (Otc) resistance. Bacteria were cultured in LB broth or agar-based media (12) amended with three antibiotics (Rif, 100 μg/ml; Otc, 15 μg/ml; cycloheximide, 25 μg/ml; Sigma Chemical Co.), which effectively inhibited growth of all bacteria and fungi indigenous to the leaves and in soil. To enumerate colonies, samples were serially diluted in phosphate buffer (1.2 mM KH_2PO_4, pH 7.2), spread in duplicate on LB agar plus antibiotics, and incubated 20 h at 30°C. Colony counts from a pair of plates were averaged for calculations of colony-forming units (CFU).

Plant growth conditions

In the simple microcosm, 50 bean seeds (cultivar Blue Lake Bush) were planted in five rows per tray. For studies with complex microcosms, two seeds were planted in each of twelve holes per tray and one plant was thinned from each pair after one week. Plants were raised under 40-W fluorescent lamps (Sylvania Cool White, F40/CW) on a 18/6-h light-dark cycle. They were watered and fertilized with 2-L nutrient solution (North Carolina State University phytotron nutrient solution lacking hampol copper and sequestrene cobalt; see Table VI of Ref. 3) per tray every 48 to 72 h. After the first appearance of flowers (about 4 weeks), plants were used in experiments. Three days before an experiment, trays were placed in a microcosm chamber to expose the plants to the experimental environment.

Insect growth conditions

Variegated cutworm larvae were collected from fields near Corvallis, Oregon, and raised from eggs in petri dishes containing diet 9000 (Bioserve Inc., New Jersey, Ref. 17) which had been sterilized by autoclaving for 15 min. To coordinate maturation of larvae and plants for experimental use, larval growth was slowed or quickened by incubation at temperatures between 12 to 30°C. Fifth- and sixth-instar larvae were transferred to fresh medium every 48 h. Late sixth-instar larvae were then transferred to medium 9000 in 1-oz plastic containers, where they pupated. Pupae were put in petri dishes inside a wire mesh cage (enclosed in a plastic bag with a damp sponge to maintain high humidity) and incubated at 25°C with an 18/6-h light-dark cycle. Emergent moths deposited eggs on paper towels in the cage. Egg masses were transferred to petri dishes and incubated at 25 to 30°C for 4 to 6 days. Within 18 h of emergence, first-instar larvae were placed on diet 9000. Bactericidal chemicals were removed from the insects by feeding them fresh plants for 3 days prior to microcosm studies.

Inoculation of microcosm contents with bacteria

A spray inoculum was prepared from bacteria cultured for 15 to 18 h at 30°C with shaking in LB broth containing Otc and Rif. Cells were pelleted by centrifugation (10 min at $3500g$ at 4°C), suspended in 1.2 mM phosphate buffer, centrifuged again, and diluted 100-fold in phosphate buffer to about 2 to 5×10^7 CFU/ml. Microcosm contents were sprayed within the first 4 h of a light period. To contain escape of aerosolized bacteria into the room, sprays were done inside a negative-airflow hood with a Hepa filter on the exhaust port. Plants were sprayed to runoff (about 100 ml of suspension per tray) using a plastic misting bottle (rinsed with 95% ethanol followed by sterile phosphate buffer). In studies with insects in simple microcosms, about 40 larvae were added to the microcosm contents 2 h after spraying.

Sample collection and processing

From simple microcosms, three samples each of plant, soil, foregut, and frass were collected on each sampling day from the simple micro-cosm. From complex microcosms, six plant samples (three per tray) and six soil samples (three per tray) were taken on a sampling day from each complex microcosm. On day 0, plant samples were collected 4 to 6 h after the spray, when the leaf surfaces were dry. Plant samples (weighing about 1.5 to 2.5 g) consisted of one cotyledon, one first trifoliate, and one second trifoliate. These were aseptically collected at random using scissors and forceps, put in sterile plastic bags holding 20 ml phosphate buffer, and blended for 1 min in a Stomacher blender (Model 80, Tekmar Co., Cincinnati, Ohio). Using a spatula, soil samples of about 0.7 to 1.5 g each were aseptically collected at random from the top centimeter of soil. These were vortex-mixed for 1 min in 18 × 150 mm screwcapped tubes containing 5 ml phosphate buffer and seven glass beads (5 mm diameter). To sample a larval foregut, the head was removed with scissors and the foregut (weighing about 0.1 to 0.2 g) pulled out with forceps (1). The sample was vortex-mixed for 1 min in an 18 × 150 mm screwcapped tube containing glass beads and 5 ml of phosphate buffer. A frass sample was aseptically collected with forceps as a pellet was extruded by an insect (1). Each pellet (about 5 to 15 mg) was placed on parafilm, weighed, and dispersed for 30 sec with a Pellet Pestle (Kontes, Vineland, New Jersey) in a 1.5-ml Eppendorf tube containing 0.05 ml phosphate buffer, which was fol-lowed by addition of 0.95 ml of phosphate buffer and vortex-mixing for another 30 sec. After an experiment, soil and plants were autoclaved and the inner surfaces of the microcosm boxes disinfected with dilute bleach.

Data analysis

Bacterial survival was plotted as the mean of the logarithm (log) CFU per gram fresh weight for all samples obtained on each sampling day (bars about each mean represent the standard error of the mean, S.E.). Statistical analyses of pairs of experiments (each consisting of samplings from two complex microcosms) were performed using a two-way factorial, completely randomized block design with microcosm and day as the two factors (18). Slopes (units = log CFU/g/day) for data from days 5 to 21 were computed by regressing a least squares analysis of log CFU/g of samples versus day. Calculations were done with the General Linear Methods procedure with SAS software.

Methods Applications and Discussion

Use of the simple microcosm to study ingestion of bacteria by insects

The simple microcosm was useful in exploratory development of methods for inoculation of plants and soil with bacteria and sample collection, despite the possible disadvantages of a shallow soil bed, the type of soil used, and the absence of sensors to record environmental parameters. Figures 24.3 and 24.4 illustrate examples of the utility of simple microcosms for the study of ingestion of recombinant bacteria by insects. *E. cloacae* (pBR322) was sprayed on bean plants and soil and then tracked in the digestive systems of cutworms feeding on these plants. Figure 24.3 shows how the populations associated with soil and leaves persisted at 10^5 to 10^6 per gram of sample for 14 days. During this time, these bacteria were also detected in the digestive tract and frass of the cutworms (Fig. 24.4). These large populations of *E. cloacae* (pBR322) in foregut and frass samples indicated that this species survived passage through the gut. This contrasts with the *P. cepacia* (R388::Tn*1721*) strain studied earlier (1), which was detected in foreguts, but not in feces. Other possible applications of these mi-

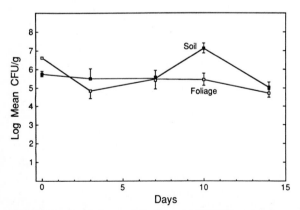

Figure 24.3 Persistence of *E. cloacae* (pBR322) sprayed on bean plants and soil in simple microcosm containing cutworms.

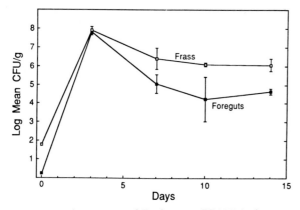

Figure 24.4   Appearance of *E. cloacae* (pBR322) in foreguts and frass of cutworms transferred to simple microcosm containing bean plants and soil sprayed with *E. cloacae* (pBR322).

crocosm methods include (1) studies of other herbivorous insects, (2) site-to-site transport of bacteria associated with insects when infected insects are transferred to uninoculated plants, (3) survival of recombinant bacteria through larval pupation and adult emergence, and (4) regrowth of bacteria in frass deposited on leaves and in soil.

Use of the complex microcosm to study bacterial survival.   The second microcosm was termed complex since it (1) had probes for environmental monitoring, (2) was set up for partial control of environmental conditions, and (3) contained natural soil from the field. This microcosm design has been conveniently used in experiments with chambers in pairs for comparative studies. Its utility was illustrated by an experiment in which survival data were obtained from two chambers with closely controlled environments. Such experiments were considered necessary to establish information on the reproducibility of survival curves between chambers. Figure 24.5 depicts the environmental data collected from the two microcosms and shows the similarity between the environments. Figure 24.5a displays the changes in RH, which were achieved by programming each micrologger to turn on a humidifier as needed to maintain RH near 95 percent during the 8-h dark period. Since humidifiers were off during the light period, RH dropped to 45 to 55 percent. In microcosm 1 and microcosm 2, respectively, the mean hourly RH values were as follows: dark period, 90 percent (standard deviation, SD = 6 percent) and 88 percent (SD = 8 percent); light period, 57 percent (SD = 7 percent) and 56 percent (SD = 7 percent). The soil moisture (data not shown) was between 62 to 100%

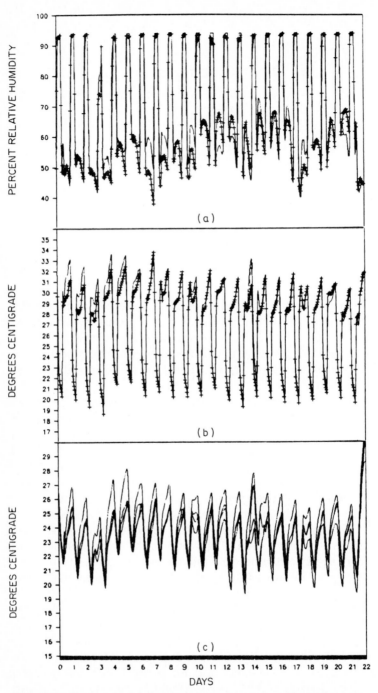

Figure 24.5 Environmental data from two microcosms with similar environments. (a) Percent relative humidity, (b) air temperature, (c) soil temperature. _____ , microcosm 1; + + + + + + + , microcosm 2.

WHC for 73 percent of the time in microcosm 1 and 60 percent of the time in microcosm 2. Figure 24.5b illustrates the typical cyclic pattern of chamber air temperature, fluctuating between 19 to 34°C during the daily cycle with respective day and night mean hourly values of 30°C (SD = 1°C) and 22°C (SD = 1°C) in both microcosms. These changes were due to heat from the lamps. Figure 24.5c shows the soil temperatures recorded by thermistors in the four trays. Since differences between the readings of soil temperature probes depended on thermistor depth in the soil and the overlying foliar layer, the tracings for the probes are not identified in the graph. Rather, they are presented here only to illustrate the quality and extent of daily variations. During the dark period, soil temperature typically approached the air temperature, but during the light period, the soil was about 5°C cooler than the air. The qualitative similarity in all the environmental data from the paired microcosms is indicative of their usefulness for comparative tests. These results demonstrate the similarity of environments in a pair of chambers, and the reproducibility of environmental monitoring between chambers.

E. cloacae (pBR322) was sprayed on plants and soil in two complex microcosm chambers and the populations were monitored. Figure 24.6 presents the number of *E. cloacae* (pBR322) recovered from plants in both microcosms. The curves depict the biphasic die-off that was typical for epiphytic populations of *E. cloacae* (pBR322). During the first 5 days after the spray, the detectable numbers decreased more rapidly than during days 5 to 21. Only 1 percent of the initial population was recovered 5 days after a spray (slopes for days 0 to 5: microcosm 1,

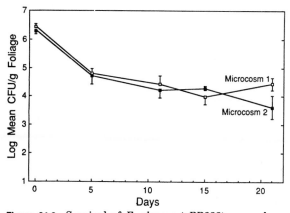

Figure 24.6 Survival of *E. cloacae* (pBR322) sprayed on bean plants in two microcosms with similar environments.

−0.33; microcosm 2, −0.32), contrasting with the slower die-off between days 5 to 21 (slopes: microcosm 1, −0.03; microcosm 2, −0.06).

Figure 24.7 shows the changes in *E. cloacae* (pBR322) populations in soil samples. About 10 percent of the initial soil population could be detected on day 5 (slopes for days 0 to 5: microcosm 1, −0.16; microcosm 2, −0.21). By day 21, only 10 percent of the day 5 population was detected (slopes for days 5 to 21: microcosm 1, −0.06; microcosm 2, −0.05).

The qualitative similarities between environmental data and the slopes of survival curves point to the utility of the microcosms for comparative studies of bacterial persistence. To substantiate these similarities with statistical analyses, the above experiment with two complex microcosms with similar environments was repeated two more times (data not shown). When the analysis of variance was performed on all plant and soil data from days 0, 5, 11, 15, and 21, in all cases, the "day effect" (i.e., mean of plant or soil sample data from all days combined) was highly significant ($p \leq .005$), due to the marked difference between the day 0 values versus those for days 5 to 21. This result was confirmed by an analysis of the data with day 0 omitted, which showed the day effect to be insignificant ($p > .30$ for plant sample data; $>.10$ for soil sample data). Table 24.1 summarizes the probabilities from analyses of variance that were done for the three pairwise combinations of the three experiments for data from days 5 to 21. No differences between microcosms were observed at the $p = .05$ level. This pairwise analysis of the three experiments (with each experiment consisting of two microcosms with similar environments) was developed for use in studies that contrasted bacterial survival in two chambers that differed in treatment. To eliminate possible effects

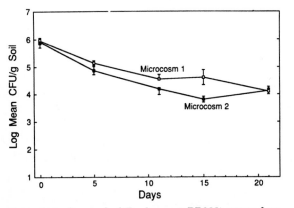

Figure 24.7 Survival of *E. cloacae* (pBR322) sprayed on soil in two microcosms with similar environments.

TABLE 24.1   Probability Values from Analysis of Variance of Bacterial Survival.*

Experiments compared†	Sample type	Probability
1 and 2	Foliage	0.53
1 and 2	Soil	0.85
1 and 3	Foliage	0.23
1 and 3	Soil	0.94
2 and 3	Foliage	0.50
2 and 3	Soil	0.90

*Bacterial populations on leaves and in soil were enumerated on days 5, 11, 15, and 21 after spraying microcosm contents with *Enterobacter cloacae* (pBR322). Each experiment consisted of a pair of microcosm chambers inoculated with similar quantities of these bacteria. The environments of both chambers were then maintained as closely as possible. The experiment was performed three different times. Survival curves were then statistically compared to determine the similarity of results from each pairwise combination of experiments.

†Graphic form of results from experiment 1 are presented in Figs. 24.6 and 24.7. Data for experiments 2 and 3 are not presented.

caused by the chambers, a crossover experimental design should be used. The treatment used in one chamber during the first experiment would be applied to the other chamber during the repeat of the experiment. This reversal ensures that each treatment will be studied in each microcosm box. Based on this experimental approach, the probability values in Table 24.1 are indicators of the limits of resolution that must be exceeded to distinguish differences between survival curves obtained from chambers with different treatments.

The crossover experimental design is clarified by considering its applications. For example, it might be of interest to study the effect of elevated RH on survival of bacteria on leaves. Two experiments would be performed. In the first, the humidifier in chamber 1 would be programmed to produce a high RH, while the RH in chamber 2 is kept low. Bacterial survival would be assessed. Then, during the second experiment, the control of RH conditions would be reversed between chambers. If bacterial survival on the foliage is independent of RH, there would be no significant difference (depending on the significance level chosen) between results for high versus low RH treatments during the two experiments. When survival is dependent on RH, there would be a significant difference (depending on the significance level chosen) between results for high versus low RH treatments during the two experiments. When applied to this example, Table 24.1 indicates that the effect of RH on bacterial survival would be significant only when $p < .23$ for survival on leaves and $p < .85$ for survival in soil.

The biphasic characteristic of the survival curves, especially for *E. cloacae* (pBR322) on foliage, that distinguishes days 0 to 5 from days 5 to 21, suggests that the suspension of bacteria at the time of spraying may have contained a small proportion (about 1 percent of the total) of biochemically and/or physiologically different cells better able to colonize the leaf surfaces and soil. It may also reflect a limited availability of leaf surface sites for bacterial colonization.

As shown here, microcosms can serve as useful preliminary indicators of viability and die-off of recombinant bacteria in lieu of field-based studies. Further useful applications might include: (1) long-term (months or longer) persistence of recombinant bacteria, (2) effects of low and high relative humidity and soil moisture on survival, (3) uptake of bacteria into the digestive tracts of cutworms and transfer into yet higher trophic levels, (4) site-to-site transfer of recombinant bacteria in cutworms as carriers, (5) comparisons of recombinant bacteria populations on plants alone versus populations on plants fed upon by cutworms, and (6) factors that determine the effect on the colonization of bacteria on the leaf surface after the spray.

Acknowledgments

For their assistance in various aspects of this research, I thank Dr. Ramon Seidler, L. Arlene Porteous, Dr. Guy Knudsen, and Robert DeLongchamp.

References

1. Armstrong, J. L., G. R. Knudsen, and R. J. Seidler. 1987. Microcosm method to assess survival of recombinant bacteria associated with plants and herbivorous insects. *Curr. Microbiol.* 15:229–232.
2. Armstrong, J. L., L. A. Porteous, and N. D. Wood. 1989. The cutworm *Peridroma saucia* (Lepidoptera: Noctuidae) supports growth and transport of pBR322-bearing bacteria. *Appl. Environ. Microbiol.* 55:2200–2205.
3. Armstrong, J. L., N. D. Wood, and L. A. Porteous. 1990. Transconjugation between bacteria in the digestive tract of the cutworm *Peridroma saucia. Appl. Environ. Microbiol.* 56:1492–1493.
4. Downs, R. H., and H. Hellmers. 1975. *Environment and the Experimental Control of Plant Growth.* Academic Press, New York.
5. Fredrickson, J. K., H. Bolton, Jr., S. A. Bentjen, K. M. McFadden, S. W. Li, and P. Van Voris. 1990. Evaluation of intact soil-core microcosms for determining potential impacts on nutrient dynamics by genetically engineered microorganisms. *Environ. Toxic. Chem.* 9:551–558.
6. Fredrickson, J. K., P. Van Voris, S. A. Bentjen, H. Bolton, Jr. 1990. Terrestrial microcosms for evaluating the environmental fate and risks associated with the release of chemicals or genetically engineered microorganisms to the environment. In J. Saxena (ed.), *Hazard Assessment of Chemicals.* Hemisphere Publishing, New York, pp. 157–202.
7. Gile, J. D. 1983. 2,4-D—Its distribution and effects in a ryegrass ecosystem. *J. Environ. Quality* 12:406–412.
8. Gillett, J. W., J. M. Witt, and C. J. Wyatt (eds.). 1979. *Symposium on Terrestrial Microcosms and Environmental Chemistry.* NSF/RA 79-0026, June 1977, Corvallis, Oreg.
9. Hammons, A. S. (ed.). 1981. *Methods for Ecological Toxicology, A Critical Review of Laboratory Multispecies Tests.* Ann Arbor Science Publishers, Ann Arbor, Mich.
10. Knudsen, G. R., M. V. Walter, L. A. Porteous, V. J. Prince, J. L. Armstrong, and

R.J. Seidler. 1988. Predictive model of conjugative plasmid transfer in the rhizosphere and phyllosphere. *Appl. Environ. Microbiol.* 54:343–347.

11. Liang, L. N., J. Sinclair, L. Mallory, M. Alexander. 1982. Fate in model ecosystems of microbial species of potential use in genetic engineering. *Appl. Environ. Microbiol.* 44:708–714.

12. Maniatis, T., E. F. Fritsch, and J. Sambrook. 1982. *Molecular Cloning, A Laboratory Manual.* Cold Spring Harbor Laboratory, Cold Spring Harbor, N.Y.

13. Milewski, E. A. 1985. Field testing microorganisms modified by recombinant DNA techniques: Applications, issues, and development of "points to consider document." *Recomb. DNA Tech. Bull.* 8:102–108.

14. Parkes, R. J. 1982. Methods for enriching, isolating, and analyzing microbial communities in laboratory systems. In A. T. Bull and J. H. Slater (eds.), *Microbial Interactions and Communities,* Vol. 1. Academic Press, New York.

15. Pritchard, P. H., and A. W. Bourquin. 1984. The use of microcosms for evaluation of interactions between pollutants and microorganisms. In K. C. Marshall (ed.), *Advances in Microbial Ecology,* Vol. 7. Plenum Press, New York, pp. 133–215.

16. Rissler, J. F. 1984. Research needs for biotic environmental effects of genetically engineered microorganism. *Recomb. DNA Tech. Bull.* 7:20–30.

17. Shorey, H. H., and R. L. Hale. 1965. Mass rearing of 9 noctuid species on a simple artificial medium. *J. Econ. Entomol.* 58:522.

18. Steel, R. G. D., and J. H. Torrie. 1980. *Principles and Procedures of Statistics, A Biometrical Approach,* 2d ed., McGraw-Hill, New York.

19. Van Voris, P., D. A. Tolle, and M. F. Arthur. 1985. Experimental terrestrial soil-core microcosm test protocol. PB85-213338, EPA/600/3-85/047. U.S. Environmental Protection Agency, Corvallis, Oreg.

Chapter 25

Experimental Methods for the Study of Fate and Transport of Microorganisms in Aquatic Systems

Mary A. Hood

Introduction

Until recently, genetically engineered microorganisms (GEMs) have not been intentionally released into the environment. Thus, little direct data are available concerning their fate and transport. Beringer and Bale (8) in a review of the survival and persistence of GEMs state "released GEMs are highly likely to enter freshwater or marine environments via agricultural run off,...drainage of industrial effluent, or fecal contamination by birds, animals or man...." They make this statement on the basis of studies of the survival of enteric bacteria such as the various indicator organisms.

Since few examples of released GEMs are available, the methods and procedures used to determine the fate and transport of microorganisms in general must be applied.

This chapter discusses the sampling methods used for studying the fate and transport of microorganisms in aquatic systems. The basic approaches used today by microbial ecologists to study fate and transport employ field, microcosm, and laboratory studies. The advantages and limits of each are briefly presented, and the procedures used in sampling are discussed. These include methods of sampling, handling of samples, and statistical applications.

Selection of Study System

Field study

Field studies are probably the best approach to answering practical fate, survival, and transport questions for the obvious reasons. Field studies are conducted in the "real" environment. However, the field study presents certain limitations, the most obvious being the risk factor. The release of any organism to the natural environment may present a potential risk. For example, in 1950, as part of a germ-warfare study, the U.S. Army released aerosols of the gram-negative bacterium *Serratia marcescens* around the San Francisco Bay area (10, 11). *S. marcescens* is not as harmless as originally believed, for it is now recognized that the organism is an important nosocomial agent in the United States causing urinary tract, pulmonary, and other infections and diarrhea. Five months after the release, a number of patients in local hospitals in the Bay area developed infections from *S. marcescens*. Although it could not be positively confirmed that this was the same strain the Army had originally released, the occurrence of these infections (never before reported) at the time of release strongly suggested that there was some correlation. This incident illustrates that if inappropriate organisms are released to the environment in inappropriate ways, they may have negative effects.

In contrast to this example, *Bacillus thuringiensis* (and other related microorganisms) have been used as microbial pesticides for more than 20 years (2) and have been released onto a variety of plants (and in numerous environments) with no apparent negative impact to humans, crops, soil, or the environment in general. Furthermore, the nitrogen-fixing bacteria *Rhizobium* and *Frankia* have been used for nearly a hundred years and have not adversely affected the soils into which they have been introduced (42). These few examples of large concentrations of wild-type (or genetically unaltered) microorganisms released to the environment suggest that microorganisms may have a wide and variable range of risk potentials, ranging from zero risk to possible human health risks.

Field studies *cannot* be conducted using microorganisms (either GEMs or wild types) that present any risk to human health or to the environment. However, determining whether a particular microorganism presents a risk to the health of the environment is a difficult problem to assess and much has been written concerning this critical issue. Beringer and Bale (8) categorize the questions of risk into four groups: (1) will the microorganism survive in the environment; (2) will the microorganism grow in the environment; (3) will the microorganism displace or compete with the natural microorganisms; and (4) will the microorganism directly harm humans, animals, or plants? If the answers to questions 3 and 4 can be determined by examining the historical data about a particular microorganism and it can be determined that that microorganism presents little risk of harm to either humans or the environment, a field study or a contained field study may be considered. Certainly the field study is the most valuable approach to understanding the fate and transport of GEMs.

Another limitation of the field study is its complexity, which will inevitably be expressed in terms of cost. Every water body has its own special characteristics (49). Freshwater lakes, ponds, and reservoirs (lentic or standing waters) are usually described microbiologically by limnological profiles, and the samples depend on the size and depth pattern. However, lotic (or running) waters, such as streams, creeks, and rivers where the current depends upon the slope and bottom of the riverbed are described microbiologically in horizontal terms. Head waters may be very different from waters that empty into a bay, gulf, or ocean. Marine waters such as estuaries and near-shore waters fluctuate widely in many physical and chemical parameters because of salinity and nutrient changes, and these waters probably reflect the greatest diversity of environmental conditions. In contrast, ocean waters are more constant and are influenced primarily by light, temperature, and pressure. Since each water body has its own set of environmental factors which influence the biological factors, samples must be taken to reflect the variations in these influences.

In any field study, several distinct microbial populations must be examined (15). The planktonic microorganisms are those which are free-living in the water column. Particle-associated microorganisms are obviously those associated with some type of particle suspended in the water column, while epilithic and epiphytic are microorganisms attached to nonliving substrates (such as rocks) and living organisms (such as aquatic plants). There are also microorganisms associated with aquatic sediments that are much like the epiphytic organisms. Furthermore, each environment and its microbial population has its own special relevance within the aquatic system. For example, epilithic and epiphytic microorganisms play a more significant role in flowing streams, while in large deep-water bodies, planktonic microorganisms may be more important.

Sampling represents the critical step in a field study and must include all the different habitats. Table 25.1 presents a summary of the most common field sampling procedures and gear. Included is the reference that describes the method and a brief description or character-

TABLE 25.1   Types of Samplers and Methods Used for Aquatic Field Samples

Environment	Sampler or method	Description
Surface films	Collapsed bottle (26) Volumetric pipette (57, 58)	Collects too much subsurface water.
	Continuous sampler (28, 51)	Not practical in rough waters.
	Screen sampler (20, 25, 35, 46)	Probably most practical and efficient of all surface samplers.
	Glass plate sampler (19, 29, 30)	Fails to release some of the surface film.
	Teflon plate sampler (37, 39)	No studies done to determine efficiency.
	Polycarbonate filter (sterile) (18)	Only method that collects sterile sample. No studies done to determine efficiency.
Water column	J-Z sterile sampler (and various modifications) (63)	Cannot be used at great depths, only samples of 500 ml can be taken.
	Niskin sterile water sampler (43)	Difficult to operate and can develop leaks but collects *sterile* sample; Is most common of all marine water samplers.
	Friedinger, Rutterner, Niskin, Nassen, Van Dorn bottles (49, 52)	Samplers that do not collect sterile samples.
Aquatic plants, Animals and other substrates	Collecting/scrapping/washing procedures (24)	Bacteria on macrophytes.
	(60)	Microorganisms on leaves and wood.
	(36)	Review of methods for sampling microorganisms on a variety of substrates.
	(3)	Review of methods for sampling microorganisms on epiphytes.
	(59)	Review of methods for sampling microorganisms on aquatic invertebrates.
	Direct observations (16)	Review.
	(14)	Review.
Aquatic sediment	Grab samplers (1, 49)	
	Peterson grab (1, 49)	Good for shallow and deep waters with hard bottoms and swift currents.
	Ponar grab (49)	Medium to deep rivers, lakes of sand, gravel, small rocks with mud.
	Van Veen grab (1, 49)	Better leverage to operate, good in firm sandy soil, sometimes releases prematurely in deep water.

TABLE 25.1   Types of Samplers and Methods Used for Aquatic Field Samples (*Continued*)

Environment	Sampler or method	Description
Aquatic sediment	Smith-McIntyre grab (53)	Very heavy, good for continental shelf sediments, triggers only after grab is on floor.
	Ekman grab (49)	Good for sampling silt, mulch, sludge in water with little current.
	Okean (49)	Designed for deep-sea samples.
	Shipek grab (48)	For marine water and in land water bodies, easy to operate, gives a scoop sample.
	Reinek box corer (49)	Very heavy and cumbersome (750 kg).
	Orange peel grab (49)	Good for marine waters and deep lakes with sandy substrates; prevents washout of sample.
	Gravity corers (The disadvantage of all gravity corers is possible washout of sample on ascent and compression of sample.)	
	Emery and Dietz corer (21) and other modified simple gravity corers (34)	Designed for short cores of marine sediment.
	Gravity corer with external retaining devices (22, 40)	Designed to avoid disturbance of sediment.
	Sholkovitz gravity corer (47)	Designed for deep cores (3 m).
	Knutson suction corer (33)	Designed for larger diameter core and sucks up sample.
	Bouma box corer (9)	Designed for larger cores in sandy sediments.
	Frame-mounted corers Jenkins corer (44)	Designed for soft silts.
	Craib corer (17)	Marine version of Jenkins corer.
	Multiple corer (6)	Up to 12 cores.

istics of the method. There are also a number of reviews and summaries which provide excellent details and diagrams (1, 3, 12, 31, 38, 49, 52). Obviously, not all sampling gear or methods can be presented, so only the most common ones are listed.

Field studies have also included diffusion chambers, membrane-filter chambers, and dialysis bags (8). Briefly, these studies involve suspending microorganisms in a chamber or bag, and placing the chamber in situ so that the organism is exposed to the water column.

The microorganism's survival is then measured by determining its presence over time. Studies using these methods are limited to predicting the survival rates of microorganisms in the planktonic stage. However, they do not speak to the issue of transport.

Microcosms

The use of microcosms to simulate the natural aquatic environment is another approach to answering questions concerning the fate, survival, and transport of microorganisms. The advantages and disadvantages of microcosms are reviewed in Chaps. 23 and 24. With respect to sampling, most of the same principles used in field studies can be applied to microcosms, but since the microcosm is a smaller, generally simpler system, the sampling regime is far simpler. As an example, in a number of microcosm studies as reviewed by Prichard and Bourquin (48), the parameter of temperature is kept constant. Thus, seasonality is eliminated and sampling does not have to account for this variable.

The sampling regime in a microcosm generally depends on the complexity of the system. If the microcosm is composed of only one component such as water, then only the water column would be sampled. However, if the microcosm consists of water, sediment, aquatic plants, and animals all these components must be sampled.

Specific sampling methods in microcosms include relatively simple procedures because the microcosm is housed in the laboratory. For example, sampling the water column of a microcosm can often be accomplished using a sterile pipette. Since the size of a microcosm is by definition small, smaller samples must be taken. For example, in one recent study of freshwater and estuarine microcosms, three replica samples (of 1- and 10-ml volumes) were taken at several-day intervals in the water column to enumerate bacteria (viable counts, AODC, and chitinoclastic bacteria) (O'Neill et al., 1988, ASM abstracts). Sediment sampling in microcosms can often be accomplished with a sterile blunt-ended pipette as a core, although some microcosm designs are modified to accommodate different types of sediment collection (see Chap. 23).

Laboratory cultures

Obviously, the simplest (but most limited) means of answering questions concerning the survival, fate, and transport of microorganisms in the aquatic environment is the use of laboratory studies. It is a relatively simple procedure to expose an organism in the laboratory to conditions expected in the field. Beringer and Bale (8) using *Rhizobium* as a model, list the following factors as affecting survival and persistence: (1) antagonism and stimulation by other organisms, (2) competition with other strains, (3) moisture, (4) organic matter, (5) pH, (6) presence of host and other plants, (7) predation, (8) soil properties, (9) temperature, and (10) others. For aquatic environments, this list could be expanded to include salinity, pressure, and sunlight.

Many studies have used the pure culture or flask culture approach to determine an organism's response to a particular aspect of the environment (for review see 4, 54, 55). The critical step is to extrapolate the results obtained in the laboratory to what is likely to occur in the environment. For instance, if the pH of the water of a particular lake ranges from 5.8 to 6.2 and a particular microorganism does not grow or survive well at pH's below 6.2 in culture (or in flasks containing the lake water), the conclusion might be that the organism would not survive well in the aquatic environment of the lake. However, what actually occurs in the environment may not be so simple. The organism may survive well and even grow in this aquatic system when it is in the gut of certain fish; or it may adhere to certain aquatic plant surfaces and survive quite well; or it may go into a viable but nonculturable state (13) and be cultured only when the pH is altered; or it may go into a resting or dormant state (41). Studies which use laboratory responses to predict the survival of a microorganism are especially common in starvation-survival experiments (see Chap. 21).

Exposing microorganisms to environmental waters in the laboratory may give a crude indication of that organism's survival capabilities, but such studies alone can never completely answer the questions. However, combinations of several approaches such as laboratory studies and microcosm studies may give more accurate or appropriate information.

Sample Procedures and Statistical Needs

Approaches and constraints

The approaches and constraints that affect sampling procedures are a function of the type of study undertaken. If the fate and transport of microorganisms are examined using field studies, the experimental design must include choice of sampling sites, the type of sampling methodology, and the logistics of returning the samples to the laboratory.

Care must be taken in choosing appropriate sampling sites because local variations may result in data that do not represent the overall environment. The choice of sample sites should represent the entire range of variability in the environment. This requires background data or a baseline survey if such background data are unavailable. After the survey, the appropriate number and positions of sample sites can be determined which represent the variabilities within that environment.

Sampling gear should fit the sample. For example, it is unnecessary to use a (rather expensive) Niskin sterile water sampler for collecting water from a shallow lake when a "lab-made modified" J-Z sterile sampler would do. Likewise, a J-Z sterile sampler would not work as well as the Niskin sterile water sampler for deeper ocean-water samples. Thus, decisions must include the proper sampling instruments.

Finally, the logistics of returning samples must be considered. If it takes more than a few hours to get samples back to the laboratory, either in situ measurements must be considered or other sample sites chosen. Furthermore, if there are too many samples to analyze immediately and samples remain stored too long, the delays will result in questionable data.

If microcosms or laboratory studies are used, the constraints are far less difficult. Microcosms and laboratory studies can generally be sampled with simple procedures and the analyses can be conducted immediately since these systems are housed in the laboratory.

Experimental design, sampling periodicity, sample size, and time constraints

Most microbiological field studies include at least a one-year seasonal cycle, although data are often collected for two or more years. The reason for at least a one-year cycle is to ensure that temperature-related or seasonal factors are included. Most commonly, microbiological samples are taken every month or twice a month. Routine procedure generally requires that samples be taken at the same time of day, usually in the morning (49). Diurnal fluctuations can be determined to establish the endpoints or range of variations expected within the day. The number of sample sites are designed to represent the variability in the environment and must be kept within the limitation of how many samples can be processed within the appropriate time. Likewise, the number of samples taken from each site is also a function of handling and processing logistics, but it is generally recognized that three replica samples represent the minimum. This, however, depends on the type of sample. For instance, in the water column, microorganisms are probably fairly evenly distributed but in sediments or on aquatic plants or animals, they are not uniformly distributed and far more replica samples must be taken. While small samples are required for actual bacterial analysis, it is generally recognized that the larger the container in which the sample is collected, the less the impact of "containerization" or the "bottle effect." The general rule of thumb in considering sample size is to remember "The objective of sampling is to collect a portion of material small enough in volume to be transported conveniently and handled in the laboratory while still accurately representing the material being sampled" (1).

Statistical procedures

An important component of a field, microcosm, or laboratory study is the appropriate application and use of statistics. There are many statistical packages available for use. Probably the most commonly used by microbial ecologists is SAS (Statistical Analysis System; 7). SAS includes most of the statistical manipulations a microbiologist would be expected to perform or calculate including mean, standard deviation, range, confidence interval, analysis of variance, correlation analysis, and clustering. The best review of basic statistical tools and experimental design for the microbial ecologist has been done by Atlas and Bartha (3). Several general texts for statistics and experimental design for the biologist include those by Green (27), Winer (61), Ott (45), and Schefler (50). For a review of precision and accuracy of methods see (1).

Sample storage

Time and temperature considerations.   It is generally recognized that samples should be analyzed as soon as possible after collection, because changes take place very quickly in both total number and the proportion of groups and species of microorganisms (1, 12, 49). Zobell (63) first reported that in marine water samples, there were initial decreases in total numbers followed by rapid increases as well as changes in the relative proportions of different microorganisms when the marine samples were stored. In a recent study of the bottle effect on the bacterial communities in seawater, Ferguson et al. (23) demonstrated that within 16 to 32 h at 25°C, significant changes in seawater occurred in bacterial total numbers, culturable numbers, cell volume, turnover rates of amino acids, and predominant species. Factors such as time, temperature, and light seem to especially affect the bacterial community. In a study by Taylor and Collins (56), it was demonstrated that increases in bacterial numbers occurred much more rapidly when seawater samples were stored in the dark at higher temperatures.

Early studies on the effect of storage of freshwater samples also showed changes in the microbiological community. Whipple (62) demonstrated that the number of culturable bacteria increased dramatically with storage. Recent microbial studies of low-organic ground water revealed that the storage of groundwater samples at 4 and 9°C in the dark resulted in up to a 17-fold increase in cell counts and loss of diversity within 24 h. *Vibrio* and *Spirilla* blooms were observed during this period of time. With additional storage, the cell numbers decreased slowly (32). In a study of the bacterial community structure of a slightly acidic organic freshwater lake in Florida, Hood et al. (manuscript in preparation) found that certain bacterial species "disappeared" after as short a time as 1 h of storage. However, diversity indices did not reveal significant changes until after 12 h of storage. There were also differences in the effect of storage temperature on the bacterial community. For example, samples stored on ice for 1 to 6 h contained fewer types (species) than samples kept at the original in situ temperatures (25°C). However, after 12 h, the number of species in the cold stored sample leveled off and remained relatively constant for 72 h.

The studies cited here illustrate the importance of analyzing samples as soon as they are collected. Sometimes, this is not possible, but the time between collection and analysis should be kept at a minimum. If there are no means of preventing long time periods of storage, studies should be conducted to determine the effects of these storage conditions.

Most standard methods (1, 49) recommend that samples arrive at the laboratory no later than 1 to 3 h after collection. If the outside temperature is hot, samples should be transported in double-walled containers packed with ice or cooling mixtures such that the temperature of the sample does not exceed 4°C. However, if the air temperature is freezing, sacs of warm water should be placed in the transport box to maintain as close to in situ temperature as possible. Finally, samples should be transported in containers that are sealed or will not leak.

Clearly, the aim of sample storage is to ensure that samples change as little as possible (or if change occurs, to be able to predict it) between the time of collection and analysis. In handling microbial samples, this goal should always be kept in mind.

In summary, it can be concluded that the purpose of appropriate sampling methodology for the study of fate and transport of GEMs, like all microbial sampling methodology, is to provide an adequate representation of what is there. For the microbial ecologist, this is especially relevant. Whether microcosm, laboratory, or field studies are employed in fate and transport studies, the collection of samples, the storage and handling of these samples and the application of appropriate statistical tools must follow protocols designed to minimize change in the samples and to reflect the "natural" environment.

Sometimes proper sampling procedures are minimized or not stressed. Yet if sampling is done improperly, it may produce misleading results.

References

1. American Public Health Association. 1985. *Standard Methods for the Examination of Water and Wastewater,* 16th ed. American Public Health Association, Washington, D.C.
2. Aronson, A. I., W. Beckman, and P. Dunn. 1986. *Bacillus thuringiensis* and related insect pathogens. *Microbial Rev.* 50:1–24.
3. Atlas, R. M., and R. Bartha. 1987. *Microbial Ecology: Fundamentals and Applications,* 2d ed. Benjamin/Cummings Publishing, Menlo Park, Calif., pp. 487–513.
4. Babich, H., and G. Stotzky. 1980. Environmental factors that influence the toxicity of heavy metals and gaseous pollutants to microcosms. *CRC Crit. Rev. Microbiol.* 8:99–145.
5. Baker, J. H. 1988. Epiphytic bacteria. In B. Austin (ed.), *Methods in Aquatic Bacteriology.* Wiley and Sons, New York, pp. 171–191.
6. Barnett, P., J. Watson, and D. Connelly. 1984. A multiple corer for taking virtually undisturbed samples from shelf, bathyal and abyssal sediments. *Ocean. Acta,* 7: 399–406.
7. Barr, A. J., J. H. Goodnight, J. P. Sall, and J. A. Helwig. 1976. *Statistical Analysis Systems.* SAS Institute Inc., Raleigh, N.C.
8. Beringer, J. E., and M. J. Bale. 1988. The survival and persistence of genetically-engineered microorganisms. In M. Sussman, C. H. Collins, F. A. Skinner, and D. E. Stewart-Tull (eds.), *The Release of Genetically-Engineered Microorganisms.* Academic Press, London, pp. 29–46.
9. Bouma, A. H. 1969. Large box sampler. In *Methods for the Study of Sedimentary Structures.* Wiley, New York, pp. 332–345.
10. Cano, R., and J. S. Colone. 1986. *Microbiology.* West Publishing, St. Paul, Minn. p. 882.
11. Cole, L. A. 1988. *Clouds of Secrecy: The Army's Germ Warfare Tests Over Populated Areas.* Rowman and Littlefield, N.J.
12. Collins, V. G., J. G. Jones, M. S. Hendrie, J. M. Shewan, D. D. Wynn-Williams, and M. E. Rhodes. 1973. Sampling and estimation of bacterial populations in the aquatic environment. In R. G. Board and D. W. Lovelock (eds.), *Sampling—Microbiological Monitoring of Environments.* Society for Applied Bacteriology Technical Series No. 7, Academic Press, London, pp. 1–35.
13. Colwell, R. R., P. R. Brayton, D. J. Grimes, D. B. Roszak, S. A. Hug, and L. M. Palmer. 1985. Viable but nonculturable *Vibrio cholerae* and related pathogens in the environment: Implications for the release of genetically engineered microorganisms. *Bio. Technology* 3:817–820.
14. Costerton, J. W., J. C. Nickel, and T. I. Ladd. 1986. Suitable methods for the comparative study of free-living and surface-associated bacterial populations. In J. S. Poindexter and E. R. Leadbetter (eds.), *Bacteria in Nature, Vol. 2, Methods and Special Applications in Bacterial Ecology,* Plenum Press, New York, pp. 49–85.
15. Costerton, J. W., and G. G. Geesey. 1979. Which population of aquatic bacteria should we enumerate. In J. W. Costerton and R. R. Colwell (eds.), *Native Aquatic Bacteria: Enumeration, Activity and Ecology,* American Society for Testing and Materials, STP, 695, pp. 7–18.
16. Costerton, J. W. 1984. Direct ultra-structural examination of adherent bacterial populations in natural and pathogenic ecosystems. In M. J. Klug and C. A. Reddy (eds.), *Current Perspectives in Microbial Ecology,* American Society for Microbiology, Washington, D.C., pp. 115–123.
17. Craib, J. S. 1965. A sampler for taking short undisturbed marine cores. *J. Cons. Perm. Int. Explor. Mer.* 30:34–39.

18. Crow, S. A., D. G. Ahearn, and W. L. Cook. 1975. Densities of bacteria and fungi in coastal surface films as determined by a membrane-adsorption procedure. *Limnol. Oceanogr.* 20:644–646.
19. Dietz, A. S., L. J. Albright, and T. Tuominen. 1976. Heterotrophic activities of bacterioneuston and bacterioplankton. *Can. J. Microbiol.* 22:1699–1709.
20. Duce, R. A., J. G. Quinn, C. E. Olney, S. R. Piotrowicz, B. J. Ray, and T. L. Wade. 1972. Enrichment of heavy metals and organic compounds in the surface microlayer of Narragansett Bay, Rhode Island. *Science* 176:161–163.
21. Emery, K. O., and R. S. Dietz. 1941. Gravity coring instrument and mechanics of sediment coring. *Bull. Geol. Soc. Am.* 52:1685–1714.
22. Fenche, T. 1967. The ecology of microbenthos. I. Quantitative importance of ciliates as compared with metazoans in various types of sediments. *Ophelia* 4:121–137.
23. Ferguson, R. L., E. N. Buckley, and A. V. Palumbo. 1984. Response of marine bacterioplankton to differential filtration and confinement. *Appl. Environ. Microbiol.* 47:49–55.
24. Fry, J. C., and N. C. B. Humphrey. 1978. Techniques for the study of bacterial epiphytes on aquatic macrophytes. In D. W. Lovelock and R. Davies (eds.), *Techniques for the Study of Mixed Populations*, Academic Press, New York, pp. 1–29.
25. Garrett, W. D. 1965. Collection of slick-forming materials from the sea surface. *Limnol. Oceanogr.* 10:602–605.
26. Goering, J. J., and D. W. Menzel. 1965. The nutrient chemistry of the sea surface. *Deep-Sea Res.* 12:839–843.
27. Green, R. H. 1979. *Sampling Design and Statistical Methods for Environmental Biologists.* Wiley, New York.
28. Harvey, G. W. 1966. Microlayer collection from the sea surface: A new method and initial results. *Limnol. Oceanogr.* 11:603–618.
29. Harvey, G. W., and L. A. Burzell. 1972. A simple microlayer method for small samples. *Limnol. Oceanogr.* 17:156–157.
30. Hatcher, R. F., and B. C. Parker. 1974. Laboratory comparisons of four surface microlayer samplers. *Limnol. Oceanogr.* 19:162–165.
31. Herbert, R. A. 1988. Sampling methods. In B. Austin (ed.), *Methods in Aquatic Bacteriology*, Wiley, New York, pp. 3–25.
32. Hirsch, P., and E. Rades-Rohkohl. 1988. Some special problems in the determination of viable counts of groundwater microorganisms. *Microb. Ecol.* 16:99–113.
33. Holme, N. A. 1964. Methods of sampling the benthos. *Adv. Mar. Biol.* 2:171–260.
34. Holme, N. A., and A. D. McIntyre. 1971. Methods for the study of the marine benthos. In *IBP Handbook*, vol. 16, Blackwells, Oxford.
35. Jarvis, N. L., W. D. Garrett, M. A. Scheiman, and C. O. Timmons. 1967. Surface characterization of surface-active material in seawater. *Limnol. Oceanogr.* 12:88–96.
36. Jones, J. G. 1977. The study of aquatic microbial communities. In F. A. Skinner and J. M. Shewan (eds.), *Aquatic Microbiology*, Academic Press, New York, pp. 1–30.
37. Larsson, K., G. Odham, and A. Sodergren. 1974. On lipid surface films on the sea. I. A simple method for sampling and studies of composition. *Mar. Chem.* 2:49–57.
38. MacIntyre, F. 1974. The top millimeter of the ocean. *Sci. Am.* 230:62–77.
39. Miget, R., H. Kator, C. Oppenheimer, J. L. Laseter, and E. J. Ledet. 1974. New sampling device for the recovery of petroleum hydrocarbons and fatty acids from aqueous surface films. *Analyt. Chem.* 46:1154–1157.
40. Mills, A. A. 1961. An external core-retainer. *Deep-Sea Res.* 7:294–295.
41. Morita, R. Y. 1985. Starvation and miniaturization of heterotrophs, with special emphasis on maintenance of the starved viable state. In M. Fletcher and G. Floodgate (eds.), *Bacteria in the Natural Environments: The Effect of Nutrient Conditions*, Academic Press, New York, pp. 111–130
42. National Academy of Sciences. 1987. *Introduction of Recombinant DNA-Engineered Organisms into the Environment: Key Issues.* National Academic Press, Washington, D.C.

43. Niskin, S. J. 1962. A water sampler for microbiological studies. *Deep-Sea Res.* 9: 501–503.
44. Ohnstad, F. R., and Jones J. G. 1982. The Jenkin surface-mud sampler user manual, Occasional Publication No. 15. Freshwater Biological Association. The Ferry House, Ambleside, Cumbria, LA 22 OLP.
45. Ott, L. 1984. *An Introduction to Statistical Methods and Data Analysis.* Duxbury Press, Boston.
46. Parker, B. 1978. Neuston sampling. In A. Sournia (ed.), *A Phytoplankton Manual.* UNESCO, Paris.
47. Pederson, T. F., S. J. Malcolm, and E. R. Sholkovitz. 1985. A lightweight gravity corer for undisturbed sampling of soft sediments. *Can. J. Earth Sci.* 22:133–135.
48. Prichard, P. H., and A. W. Bourquin. 1984. The use of microcosms for evaluation of interactions between pollutants and microorganisms. In K. Marshall (ed.), *Advances in Microbial Ecology,* Vol. 7, Plenum, New York, pp. 133–215.
49. Rodina, A. G. 1972. Marine ecology. In R. R. Colwell and M. S. Zambruski (eds.), *Methods in Aquatic Microbiology.* University Park Press, Baltimore.
50. Schefler, W. C. 1979. *Statistics for the Biological Sciences.* Addison-Wesley, Reading, Mass.
51. Schlieper, C. (ed.). 1972. *Research Methods in Marine Biology* (English translation). University of Washington Press, Seattle.
52. Sieburth, J. M. 1979. *Sea Microbes.* Oxford, New York.
53. Smith, W., and A. D. McIntyre. 1954. A spring-loaded bottom-sampler. *J. Mar. Biol. Ass. UK.* 33:257–264.
54. Stotzky, G. 1972. Techniques to study interactions between microorganisms and clay minerals in vivo and in vitro. In T. Rosswall (ed.), *Modern Methods in the Study of Microbial Ecology,* Swedish Natural Science Research Council, Stockholm, pp. 17–28.
55. Stotzky, G. 1972. Activity, ecology and population dynamics of microorganisms in soil. *CRC Crit. Rev. Microbiol.* 2:59–137.
56. Taylor, C. B., and U. G. Collins. 1949. Development of bacteria in waters in glass containers. *J. Gen. Microbiol.* 3:32–42.
57. Tsyban, A. V. 1967. On an apparatus for the collection of microbiological samples in the near-surface micro-horizon of the sea. *Hydrobiology* 3:84–86.
58. Tsyban, A. V. 1971. Marine bacterioneuston. *J. Oceanogr. Soc. Japan* 27:56–66.
59. West, P. A. 1988. Bacteria of aquatic invertebrates. In B. Austin (ed.), *Methods in Aquatic Bacteriology.* Wiley and Sons, New York, pp. 143–170.
60. Willoughby, L. G. 1978. Methods for studying microorganisms in decaying leaves and wood in freshwater. In D. W. Lovelock and R. Davies (eds.), *Techniques for the Study of Mixed Populations,* Academic Press, New York, pp. 31–50.
61. Winer, B. J. 1971. *Statistical Principles in Experimental Design.* McGraw-Hill, New York.
62. Whipple, C. E. 1901. Changes that take place in the bacterial content of waters during transportation. *Technol. Q. Proc. Soc. Arts.* 14:21–24.
63. Zobell, C. E. 1946. *Marine Microbiology.* Chronica Botanica, Waltham, Mass.

Chapter 26

Experimental Methods for Terrestrial Ecosystems

Charles Hagedorn

The literature on agricultural field research contains many journal articles, handbooks, and statistical textbooks that encompass the procedures involved in conducting a field experiment. It is possible to quite readily obtain information, once an experimental objective has been established, that will guide the researcher through the process of selecting an appropriate experimental design, placing the actual test in the field, collecting data and observations from the test, and analyzing and interpreting the data. This chapter does not provide a stepwise progression through any of these processes because such information is already available. However, a sampling of appropriate references is included that can provide details of the techniques and procedures employed in field plot research.

This chapter does provide a description of those aspects of field research that are of particular importance to microbiological studies. In analyzing a field test where some of the treatments are an expression of a microbiological component (e.g., population counts) it has been historically difficult to obtain significance as a result of any one treatment, or to identify correlations between treatments, and error factors are usually high. Large sources of variation can also be anticipated to occur frequently in tests with genetically engineered organisms as well. This chapter details those sources of variation such as soil heterogeneity and competition effects that, while accountable in tests where crops are the treatments, often render microbial treatments to insignificance. The consideration of plot design, the duration of field tests, site selection, and data collection are also reviewed in the context of conducting field plot research with microorganisms.

Site Selection

Heterogeneity in soil

Adjacent research plots, developed simultaneously to the same treatment and maintained as similarly as possible, will differ in as many characters as one would care to measure. The causes for these differences are numerous, but the most obvious, and probably the most important, is soil heterogeneity. Experience has shown that it is almost impossible to obtain an experimental site that is totally homogeneous (5, 15).

To choose an experimental site that has minimal soil heterogeneity, a researcher must be able to identify the features that magnify soil differences. Fertility gradients are generally most pronounced in sloping areas, with lower portions more fertile than high areas. This is because soil nutrients are soluble in water and tend to settle in lower areas. An ideal experimental site, therefore, is one that has no slope. If a level area is not available, an area with a uniform and gentle slope is preferred because such areas generally have predictable fertility gradients, which can be managed through the use of proper experimental design.

Different treatments used in experimental planting usually increase soil heterogeneity. Thus, areas previously planted to different crops, fertilized at different levels, or subjected to varying cultural managements should be avoided, if possible. Otherwise, such areas should be planted to a uniform variety and fertilized heavily and uniformly for at least one season before conducting an experiment. Another source of soil heterogeneity is the presence of nonplanted alleys that are common in field experiments. Plants grown in previously nonplanted areas tend to perform better. Nonplanted areas should be marked so that the same areas are left as alleys in succeeding plantings. Grading usually removes topsoil from elevated areas and dumps it in the lower areas of a site. This operation, while reducing the slope, results in an uneven depth of surface soil and at times exposes infertile subsoils. These differences persist for a long time. Thus, an area that has had any kind of soil movement should be avoided. If this is not possible, it is advisable to conduct a uniformity trial to assess the pattern of soil heterogeneity so that a suitable remedy can be achieved.

Avoid areas surrounding permanent structures. Such areas are usually undependable because of the shade they create, and the probability of some soil movement during their construction. These factors could contribute to unexplained variation in a test. A productive crop is an important prerequisite to a successful experiment. Thus, an area with poor soil should not be used, unless the experiment is set up specifically to evaluate such conditions (12).

An adequate characterization of soil heterogeneity in an experimental site is a good guide, and at times even a prerequisite, to choosing an appropriate experimentation technique. Based on the premise that uniform soil when cropped uniformly will produce a uniform crop, soil heterogeneity can be measured as the difference in performance of plants grown in a uniformly treated area. Uniformity trials involve planting an experimental site with a single crop variety and applying all cultural and management practices as uniformly as possible. All sources of variability, except that due to native soil differences, are kept constant. The planted area is subdivided into small units of the same size from which separate measurements of productivity, such as yield, are made. Yield differences between these units are taken as a measure of the area's soil heterogeneity. The size of the unit is governed mostly by available resources. The smaller the unit, the more detailed is the measurement of soil heterogeneity. Reducing heterogeneity effects is especially important in microbial field tests where plot size is usually small and large numbers of treatments are frequently employed.

Plot design

Once the fertility pattern of an experimental area is described, several options are available for reducing the effect of soil heterogeneity. Three options that are commonly used involved the proper choice of plot size and shape, block size and shape, and number of replications. These options can be inexpensive, involving only a change of plot or block orientation, but at times, the option may involve enlarging the experimental area or increasing the total number of plots (7).

The contribution of soil heterogeneity to experimental error stems from differences in soil fertility between plots within a block. The smaller this difference is, the smaller is the experimental error. The choice of suitable plot size and shape, therefore, should reduce the differences in soil productivity from plot to plot within a block and consequently reduce experimental error. Two major considerations are involved in choosing plot size, namely, practical considerations and the nature and size of variability. Practical considerations generally include ease of management in the field. The nature and size of variability is generally related to soil heterogeneity. From the empirical relationship between plot size and between-plot variance, it can be seen that while variability becomes smaller as plot size becomes larger, the gain in precision decreases as plot size becomes increasingly large. Furthermore, higher costs are involved when large plots are used. Hence, the plot size that a researcher should aim for is one that balances precision and cost.

Once the optimum plot size is determined (15), the choice of plot shape is governed by several considerations. Long and narrow plots should be used for areas with distinct fertility gradient, with the length of the plot parallel to the fertility gradient of the field. Plots should be as square as possible whenever the fertility pattern of the area is spotty or not known, or when border effects are large.

Block size is governed by the plot size chosen, the number of treatments tested, and the experimental design used. Once these factors are fixed, only the choice of block shape is left to the researcher. The primary objective in choosing the shape of blocks is to reduce the differences in productivity levels among plots within a block so that most of the soil variability in the area is accounted for by variability between blocks. Information on the pattern of soil heterogeneity in the area is helpful in making this choice. When the fertility pattern of the area is known, orient the blocks so that soil differences between blocks are maximized and those within the same block are minimized. For example, in an area with a unidirectional fertility gradient, the length of the block should be oriented perpendicular to the direction of the fertility gradient. On the other hand, when the fertility pattern of the area is spotty, or is not known to the researcher, blocks should be kept as compact, or as nearly square, as possible. Because block size, for most experimental designs, increases proportionately with the number of treatments and because it is difficult to maintain homogeneity in large blocks, a researcher must also be concerned with the number of treatments. If the number of treatments is so large that uniform area within a block cannot be attained, incomplete block designs may be used (5).

The number of replications that is appropriate for any field experiment is affected by

- The inherent variability of the experimental material
- The experimental design used
- The number of treatments to be tested
- The degree of precision desired

Because experimental variability is a major factor affecting the number of replications, soil heterogeneity clearly plays a major role in determining the number of replications in field experiments. In general, fewer replications are required with uniform soil. However, larger numbers of replications are often necessary where microbial tests are concerned because of the potential for cross-contamination of plots by treatments. Unlike plants, microbes can move between treatments (e.g., by wind, water, insects, roots) and add to the heterogeneity of the test (10).

Competition Effects

Types of competition effects

For plants, growth is affected greatly by the size and proximity of adjacent plants. Those surrounded by large and vigorous plants can be expected to produce less than those surrounded by less vigorous ones. Plants a greater distance from one another generally produce more than those nearer to each other. Interdependence of adjacent plants because of their common need for limited sunshine, soil nutrients, moisture, carbon dioxide, oxygen, and so on, is commonly referred to as competition effects. Because field experiments are usually set up to assess the effects on crop performance of several management factors, or genetic factors, or both, experimental plots planted to different varieties and subjected to different production techniques are commonly placed side by side in a field. As a consequence, border plants have an environment different from those in the plot's center: plants in the same plot are exposed to differing competitive environments. Competition effects between plants within a plot should be kept at the same level to ensure that the measurement of plant response really represents the condition being tested and to reduce experimental error and sampling error (11, 15).

For a given experiment, the significance of any competition effect depends primarily on the treatments being tested and the experimental layout. Nonplanted borders are areas between plots or around the experimental areas that are left without plants and serve as markers or walkways. These areas are generally wider than the area between rows or between plants in a row, and plants adjacent to these nonplanted borders have relatively more space. They are, therefore, exposed to less competition than plants in the plot's center. In trials involving different varieties of a given crop, adjacent plots are necessarily planted to different varieties. Because varieties generally differ in their ability to compete, plants in a plot will be subjected to different environments depending upon location relative to adjacent plots. Plants generally affected by varietal competition effect are the ones near the plot's perimeter. The size of the difference in plant characters between the varieties included in the trial plays an important role in determining the extent of varietal competition effects. The larger the varietal difference, the greater is the expected disadvantage. The disadvantage of one plot is usually accompanied by a corresponding advantage to the adjacent plot (3).

Fertilizer competition effect is similar to the varietal competition effect except that adjacent plots receive different levels of fertilizer instead of being planted to different varieties. Here the competition effect has two sources. First, plots with higher fertilizer application will be more vigorous and can probably compete better for sunshine and carbon dioxide. Second, the fertilizer could spread to the root zone of an adjacent plot, putting the plot with higher fertilizer at a disadvantage. Because these two effects are of different direction, their difference constitutes the net competition effect. In most instances, the effect of fertilizer dispersion is larger than that due to the difference in plant vigor, and the net advantage is usually with the plot receiving lower fertilizer.

Experimental plots frequently contain spots where a living plant is supposed to be but is absent because of poor germination, insect or disease damage, or physical mutilation. Because of the numerous factors that can kill a plant, even the most careful researcher cannot be assured of a complete stand for all plots in an experiment. A missing plant causes the plants surrounding its position to be exposed to less competition than the other plants. These plants, therefore, usually perform better than those surrounding a living plant. In microbial field tests, competition effects need to be considered when selecting those plants for sampling that will be used to determine microbial survival or colonization. Another type of competition (between the introduced organism and the indigenous population) is very difficult to quantify because "baseline" data on soil microbial populations is simply not reliable, even for rhizobial studies (6, 10).

Evaluating competition effects

Because the type and size of the competition effects can be expected to vary considerably from crop to crop and from one type of experiment to another, competition effects should be evaluated separately for different crops grown in different environments. Measurements of competition effects can be obtained from experiments planned specifically for that purpose or from those set up for other objectives.

Experiments can be set up specifically to measure competition effects by using treatments that simulate different types of competition. At least two treatments, representing the extreme types of competitor, should be used. For example, to evaluate fertilizer competition effects, use a no-fertilizer application and high fertilizer rate to represent the two extremes. For varietal competition, use a short and a tall variety, a high-tillering and a low-tillering variety, or an early-maturing and a late-maturing variety. When resources are not limited, intermediate treatments can be included to assess the trend in the effects under investigation. For example, to find out whether border effects are affected by the width of an unplanted alley, test several sizes of unplanted alley. Too many treatments should be avoided, however, to cut costs and to simplify the design and the interpretation of results. Experiments specifically set up to measure competition effects have two distinctive features:

1. Because the competition effects are usually small relative to the treatment effects in the regular experiments, the number of replications is usually large.

2. The plot size that is optimum for the regular experiments may not necessarily be optimum for experiments to measure competition effects. First, the unit of measurement is usually the individual rows, or individual plants, rather than the whole plot. Second, the plot must be large enough to ensure sufficient number of rows (mostly in the center of the plot) that are as free as possible of competition effects.

Long-term experiments

The use of crop production practices such as fertilization, insect and disease control, and cropping pattern, will generally change important physical and biological factors in the environment. This process of change can take many years. Consequently, the productivity of some types of technology must be evaluated by a series of trials repeated over time. Such trials are generally referred to as long-term experiments. Their distinguishing features are

1. Change over time is the primary performance index. Even though the average performance over time remains as an important measure of productivity, the change over time, in either crop or environmental traits, or both, is the more critical parameter. Obviously, an increasing rather than a decreasing productivity trend is an important feature of a desirable technology. In addition, although the initial productivity may be high, the buildup of pests or the depletion of soil nutrients resulting from continuous use of the technology can be serious enough to cause the technology to be abandoned.

2. The experimental field, the treatments, and the plot layout remain constant over time. Randomization of treatments to plots is done

only once during the first crop season and the layout in all subsequent cropping seasons exactly follows the initial layout.

Some examples of long-term experiments are

1. Long-term fertility trials that are designed to evaluate changes in soil properties and nutrients as a consequence of the application of some soil amendments over time.
2. Maximum-yield trials, which are designed to measure crop yields and change over time, in both physical and biological environments under intensive cropping and best management.
3. Pest-control trials, which are designed to measure the change in pest or disease pressure over time following different types of control measures. Most microbial field tests fall into this category, including evaluations of genetically engineered organisms (4).

The first step in the analysis of a long-term experiment is to identify one or more characters to be used as an index of crop productivity. Because random fluctuation can mask changes over long time periods, a good index is one that is least sensitive to, or is least affected by, the random changes over time with respect to environmental factors such as climate and pest incidences. Crop yield and soil properties are commonly used indexes. Although crop yield is an excellent index of productivity, it is greatly influenced by the changes in both the climate and biological environments over time. Soil properties, on the other hand, are less affected by environments but may not be as closely related to productivity as yield. Thus, more than one index is generally analyzed in a long-term experiment. With microbial field tests, long-term experiments usually result in more difficult interpretation because the microbes may be transported over other blocks and, as populations tend to decline over time, lower numbers make detection of amended strains difficult (4, 6).

Sampling Experimental Plots

Sampling concerns

Plot size for field experiments is usually selected to achieve a prescribed degree of precision for measurement of the character of primary interest. Because the character of primary interest (usually economic yield such as grain yield for grain crops or forage yield for forage crops) is usually the most difficult to measure, the plot size required is often larger than that needed to measure other characters.

Thus, expense and time can be saved if the measurements of additional characters of interest are made by sampling a fraction of the whole plot. For example, make measurements for plant height from only 10 of the 200 plants in the plot; for tiller number, count only 1 m^2 of the 15 m^2 plot; and for leaf area, measure from only 20 of the approximately 2000 leaves in the plot (11).

There are times, however, when the choice of plot size may be greatly influenced by the management practices used or the treatments tested. In an insecticide trial, for example, relatively large plots may be required to minimize the effect of spray drift or to reduce the insect movement caused by insecticide treatments in adjacent plots. This is also the situation encountered with microbial tests, especially if the treatments are applied as a foliar spray (1, 9). In such cases, plot size would be larger than that otherwise required by the character of primary interest. Consequently, even for the primary character such as grain yield, it may still be desirable to sample from a fraction of the whole plot.

An appropriate sample is one that provides an estimate, or a sample value, that is as close as possible to the value that would have been obtained had all plants in the plot been measured—the plot value. The difference between the sample value and the plot value constitutes the sampling error. Thus, a good sampling technique is one that gives a small sampling error (5, 11, 15).

Components of sampling

For plot sampling, each experimental plot is a population. Population value, which is the same as the plot value, is estimated from a few plants selected from each plot. The procedure for selecting the plants to be measured and used for estimating the plot value is called the plot sampling technique. To develop a plot sampling technique for the measurement of a character in a given trial, the researcher must clearly specify the sampling unit, the sample size, and the sampling design (2, 5).

The sampling unit is the unit on which actual measurement is made. Where each plot is a population, the sampling unit must necessarily be smaller than a plot. Some commonly used sampling units in replicated field trials are a leaf, a plant, a group of plants, or a unit area. The appropriate sampling unit will differ among crops, among characters to be measured, and among cultural practices. Thus, in the development of a sampling technique, the choice of an appropriate sampling unit should be made to fit the requirements and specific conditions of the individual experiments (11).

The important features of an appropriate sampling unit are

1. *Ease of identification:*   A sampling unit is easy to identify if its boundary with the surrounding units can be easily recognized. For example, a single hill is easy to identify in cotton because each hill is equally spaced and is clearly separated from any of the surrounding hills. In contrast, plant spacing in row cotton is not uniform and a single hill is, therefore, not always easy to identify. Consequently, a single hill may be suitable as a sampling unit for one planting technique but not another.

2. *Ease of measurement:*   The measurement of the character of interest should be made easy by the choice of sampling unit. For example, in drill-planted wheat, counting plants from a 2 × 2 sampling unit can be done quite easily and can be recorded by a single number. However, the measurement of plant height for the same sampling unit requires independent measurements of selected plants, the recording of many numbers and, finally, the computation of an average of those numbers.

3. *High precision and low cost:*   Precision is usually measured by the reciprocal of the variance of the sample estimate; while cost is primarily based on the time spent in making measurements of the sample. The smaller the variance, the more precise the estimate is; the faster the measurement process, the lower the cost is. To maintain a high degree of precision at a reasonable cost, the variability among sampling units within a plot should be kept small. For example, in transplanted rice, variation between single-hill sampling units for tiller count is much larger than that for plant height. Hence, although a single-hill sampling unit may be appropriate for plant height, it may not be so for tiller count.

The number of sampling units taken from the population is sample size. In a replicated field trial where each plot is a population, sample size could be the number of plants per plot used for measuring plant height, the number of leaves per plot used for measuring leaf area, or the number of hills per plot. The required sample size for a particular experiment is governed by the size of the variability among sampling units within the same plot (sampling variance), and the degree of precision desired for the character of interest.

In practice, the size of the sampling variance for most plant characters is generally not known to the researcher. The desired level of precision can, however, be prescribed by the researcher based on experimental objective and previous experience. The usual practice is for the researcher to prescribe the desired level of precision in terms of the margin of error, either of the plot mean or of the treatment mean. For example, the researcher may prescribe that the sample estimate should not deviate from the true value by more than 5 or 10 percent. With an estimate of the sampling variance, the required sample size can be determined based on the prescribed margin of error, of the plot mean, or of the treatment mean (2, 5, 8, 11).

The sample size for a simple random sampling design that can satisfy a prescribed margin of error of the plot mean is computed as

$$n = \frac{(Z_a^2)\,(V_s)}{(d^2)\,(\overline{X^2})}$$

where n is the required sample size, Z_a is the value of the standardized normal variate corresponding to the level of significance a, V_s is the sampling variance, $\overline{X}$ is the mean value, and d is the margin of error expressed as a fraction of the plot mean.

The information of primary interest to the researcher is usually the treatment mean (the average over all plots receiving the same treatment) rather than the plot mean (the value from a single plot). Thus, the desired degree of precision is usually specified in terms of the margin of error of the treatment mean rather than of the plot mean. In such a case, sample size is computed as

$$n = \frac{(Z_a^2)\,(V_s)}{r\,(D^2)\,(X^2) - (Z_a^2)\,(V_p)}$$

where n is the required size, Z_a and V_s are as defined in the previous equation. V_p is the variance between plots of the same treatment (i.e., experimental error), and D is the prescribed margin of error expressed as a fraction of the treatment mean. Take note that, in this case, additional information on the size of the experimental error (V_p) is needed to compute sample size (5, 15).

A sampling design specifies the manner in which the n sampling units are to be selected from the whole plot. There are five commonly used sampling designs in replicated field trials: simple random sampling, multistage random sampling, stratified random sampling, stratified multistage random sampling, and subsampling with an auxiliary variable. In a simple random sampling design, there is only one type of sampling unit and, hence, the sample size n refers to the total number of sampling units to be selected from each plot consisting of N units. The selection of the n sampling units is done in such a way that each of the N units in the plot is given the same chance of being selected. In plot sampling, two of the most commonly used random procedures for selecting n sampling units per plot are the random-number technique and the random-pair technique (5). This detailed discussion of sampling is included because the approach to sampling microbial field tests is often poorly conceived and large error factors often result (10). The detail provided is meant to demonstrate that there are procedures used in soil sampling (2) that are applicable to microbial tests and attention to sampling design can reduce experimental variation and contribute directly to interpreting the results.

Technology Development Experiments

Test site selection

The farmer's field provides a convenient and economical way to sample a wide array of physical and biological conditions in the generation of technology. Procedures for on-farm technology generation trials are similar to those at research stations, and most of the experimental procedures discussed in the previous sections can be used. Consequently, it is necessary to concentrate on modification of existing procedures to allow researchers to cope with features of the farm that are distinctly different from that of the research station. The test site for a technology generation trial is selected to provide a set of physical and biological conditions under which the trial is to be conducted or for which the technology is to be developed. Thus, the method of site selection is deliberate, rather than at random (12, 14). The steps in the selection procedure are

1. Clearly specify the desired test environment in terms of the specific physical and biological characteristics such as soil, climate, topography, landscape, water regime, etc.

2. Classify each of the specified environmental characteristics according to
   a. The relative size of contiguous area in which homogeneity of a given characteristic is expected. For example, areas with the same climate will be larger than those having the same landscape or water regime.
   b. The availability of existing information, or the relative ease in obtaining the information, on the desired characteristics. For example, climatic data are usually more readily available than information on landscape, water regime, and cropping pattern. The latter is usually obtained through farm visits by the researchers.

3. Select a large contiguous area that satisfied those environmental features that are usually homogeneous over a wide area; and, within that area, identify subareas (or farms) that satisfy those environmental conditions that are more variable. For example, a large contiguous

area can be first selected to satisfy the required climate and soil. These can be based on weather-station records and a soil map. Within the selected area, farm visits and interviews of selected farmers will help identify farms that have the topography, landscape, water regime, and cropping pattern that most closely approximate the required test environment.

If more than one farm is found to satisfy the specified test environment, select those most accessible, have more available resources, and are managed by cooperative farmers.

4. For each selected farm, choose an area or a field that is large enough to accommodate the experiment and has the least soil heterogeneity (through visual judgment, based on information of past crops, etc.). If no single field is large enough to accommodate the whole experiment, select the smallest number of fields that can accommodate it.

Experimental design

The design of experiments in farmers' fields must aim at keeping the size of experiment small. This is done by keeping the number of treatments and number of replications at the minimum. To reduce the number of treatments, a fractional factorial design may be used. To determine the number of replications, two contrasting considerations should be examined:

1. Errors can be expected to increase in on-farm trials because the fields are less accessible and more difficult for researchers to supervise. In addition, damage by vehicles, stray animals, vandalism, and theft are more apt to occur in on-farm trials and will increase experimental error.
2. The generally low insect and disease pressure and the more uniformly managed farmers' fields that are free of residual effects from previous treatments may, on the other hand, result in less experimental error.

Thus, the choice of the number of replications to be used depends on the relative importance of these two conflicting features. For example, if the chance for increase in experimental error overshadows the chance for less error, the number of replications should be greater than that used in research-station trials. Experience in such research has indicated that, with proper management, experimental error for on-farm technology generation experiments can be smaller than that in a research station; and, subsequently, the number of replications need not be larger than that used at research stations (5, 14).

For plot layout in a farmer's field, the techniques used in research stations generally apply. However, the following considerations should be used in laying out plots:

1. Plot shape may have to be adjusted to suit the irregular shape of the field and the manner in which the leveling of land has been done.
2. If a distinct parcel of land is not enough to accommodate the whole experiment, each parcel must accommodate at least one whole replication.

Data collection and analysis

All data normally collected in research-stations trials should be collected in an on-farm technology generation trial. In addition, data such as those on weather, soil, and history of plot management that are usually available in research stations but not for farmers' fields must be collected. Data collection must be flexible enough to handle unexpected incidents such as vehicle damage, stray cattle, or theft (3, 5).

Because a primary objective of technology generation trials in farmers' fields is to sample a wide array of environments, it is common for a technology generation trial to be on more than one farm. To effectively assess the interaction effect between treatment and farm environment, a uniform set of data should be collected on all farms (14).

The objective of technology generation trials remains the same whether on research stations or in farmers' fields. Consequently, data analysis on a per-trial basis is the same. When the trial is on more than one farm, procedures for combining data over locations should be employed. Because test farms represent different environments, emphasis should be placed on explaining the nature of the interaction effect between treatment and farm in terms of the relevant environmental factors—to gain a better understanding of the influence of those factors on the effectiveness of the treatments tested.

Analysis over sites

Technology adaptation experiments are designed to estimate the range of adaptability of new production technologies, where adaptability of a technology at a given site is defined in terms of its superiority over other technologies tested simultaneously at that site. The primary objective of such a trial is to recommend one or more new practices that are an improvement upon, or can be substituted for, the currently used farmers' practices. With the development of genetically engineered microbes, there is now much emphasis on using biological agents in agricultural practices and this will require adaptation of field techniques to adequately evaluate novel organisms (16). Thus, a technology adaptation experiment has three primary features:

1. The primary objective is the identification of the range of adaptability of a technology. A particular technology is said to be adapted to a particular site if it is among the top performers at that site. Furthermore, its range of adaptability includes areas represented by the test sites in which the technology has shown superior performance.

2. The primary basis for selecting test sites is representation of a geographical area. The specific sites for the technology adaptation experiments are purposely selected to represent the geographical area, or a range of environments, in which the range of adaptability of technology is to be identified. Such areas are not selected at random. In most cases, these test sites are research stations in different geographical areas. However, when such research stations are not available, farmers' fields are sometimes used as test sites.

3. The treatments consist mainly of promising technologies. Only those technologies that have shown excellent promise in at least one environment (e.g., selection from a preliminary evaluation experiment) are tested. In addition, at least one of the treatments tested is usually a control, which represents either a no-technology treatment (such as no fertilizer application or no insect control) or a currently used technology (such as local variety).

Two common examples of technology adaptation experiments are crop variety trials or a series of fertilizer trials at different research stations in a region or country. For the variety trials, a few promising newly developed varieties of a given crop are tested, at several test sites and for several crop seasons, together with the most widely grown variety in a particular area. The results of such trials are used as the primary basis for identifying the best varieties as well as the range of adaptability of each of these varieties. For fertilizer trials, on the other hand, several fertilizer rates may be tested at different test sites and for several crop seasons—in order to identify groups of sites having similar fertilizer responses (14).

Because technology adaptation experiments are generally at a large number of sites, the size of each trial is usually small and its design simple. If factorial experiments are used, the number of factors generally does not exceed 2. Thus, the two most commonly used designs are a randomized complete block and a split-plot design (12).

Technology adaptation experiments at a series of sites generally have the same set of treatments and use the same experimental design, a situation that greatly simplifies the required analysis. Data from a series of experiments at several sites are generally analyzed together at the end of each crop season to examine the treatment X site interaction effect and the average effects of the treatments over homogeneous sites. These effects are the primary basis for identifying the best performers, and their range of adaptability, among the different technologies tested (5, 7).

The applications of new microbial technologies will require evaluations that are quite different from the traditional field research approach, especially in attempts to characterize resident microbial populations. It may be that population dynamics used in ecology studies will be needed to adequately assess many microbial parameters (13). However, there are many field procedures that are perfectly adaptable to microbial tests, and have only to be applied to improve the quality of the evaluation (4).

References

1. Bishop, D. H. L., P. F. Entwistle, I. R. Cameron, C. J. Allen, and R. D. Possee. 1988. Field trials of genetically-engineered baculovirus insecticides. In M. Sussman, C. H. Collins, F. A. Skinner, and D. E. Stewart-Tull (eds.), *The Release of Genetically-Engineered Micro-organisms*. Academic Press, London, pp. 143–180.
2. Brown, J. R. (ed.), 1987. *Soil Testing: Sampling, Correlation, Calibration, and Interpretation*. Soil Science Society of America, Inc., Madison, Wis.
3. Dixon, W. J. 1986. Extraneous values. In A. L. Page (ed.), *Methods of Soil Analysis*, 2d ed., American Society of Agronomy, Madison, Wis., pp. 83–90.
4. Drahos, D. J., G. F. Barry, B. C. Hemming, E. J. Brandt, H. D. Skipper, E. L. Kline, D. A. Kluepfel, T. A. Hughes, and D. T. Gooden. 1988. Pre-release testing procedures: US field test of a lacZY-engineered soil bacterium. In M. Sussman, C. H. Collins, F. A. Skinner, and D. E. Stewart-Tull (eds.), *The Release of Genetically-Engineered Micro-organisms*. American Press, London, pp. 181–192.
5. Gomez, K. A., and A. A. Gomez. 1984. *Statistical Procedures for Agricultural Research*. 2d ed., Wiley, New York.
6. Hagedorn, C. 1986. Role of genetic variants in autecological research. In R. L. Tate III (ed.), *Microbial Autecology: A Method for Environmental Studies*. Wiley, New York, pp. 61–74.
7. John, P. W. M. 1971. *Statistical Design and Analysis of Experiments*. Macmillan, New York.
8. Kempthorne, O., and R. R. Allmaras. 1986. Errors and variability of observations. In A. L. Page (ed.), *Methods of Soil Analysis*, 2d ed. American Society of Agronomy, Madison, Wis., pp. 1–30.
9. Lindow, S. E. 1990. Use of genetically altered bacteria to achieve plant frost control. In J. P. Nakas and C. Hagedorn (eds.), *Biotechnology of Plant-Microbe Interactions*. McGraw-Hill, New York, pp. 85–111.
10. Nutman, P. S. 1976. IBP field experiments on nitrogen fixation by nodulated legumes. In P. S. Nutman (ed.), *Symbiotic Nitrogen Fixation in Plants*. Cambridge University Press, London, pp. 211–238.
11. Petersen, R. G., and L. D. Calvin. 1986. Sampling. In A. L. Page (ed.), *Methods of Soil Analysis*, 2d ed. American Society of Agronomy, Madison, Wis., pp. 33–50.
12. Petersen, R. G. 1976. Experimental designs for agricultural research in developing areas. Oregon State University Bookstores, Corvallis, Oreg.
13. Pielou, E. C. 1977. *Mathematical Ecology*. Wiley, New York.

14. Silva, J. A. 1981. Experimental designs for predicting crop productivity with environmental and economic inputs for agrotechnology transfer. Dept. Pap. 49. Hawaii Institute of Tropical Agriculture and Human Resources, University of Hawaii, Honolulu.
15. Steel, R. G. D., and J. H. Torrie. 1980. *Principals and Procedures of Statistics: A Biometrical Approach*, 2d ed. McGraw-Hill, New York.
16. Sumner, M. E. 1987. Field experimentation: Changing to meet current and future needs. In J. R. Brown (ed.), *Soil testing: Sampling, Correlation, Calibration, and Interpretation*. Soil Science Society of America, Madison, Wis.

Chapter 27

Field Sampling Design and Experimental Methods for the Detection of Airborne Microorganisms

Linda D. Stetzenbach

Stephen C. Hern

Ramon J. Seidler

The release of genetically engineered microorganisms (GEMs) and microbiological pest control agents (MPCAs) by spray application has renewed interest in field sampling strategies to detect airborne organisms. The development of appropriate experimental design to track the transport and survival in the environment of these agents is becoming increasingly important. The following chapter will present important points to consider when designing a field monitoring plan for aerosolized organisms and will discuss many of the currently available field sampling devices for airborne bacteria.

Study System

The spray application of GEMs or MPCAs in the field not only results in the inoculation of organisms to a specific target crop but also produces airborne cells that may be transported to surrounding areas. Air monitoring for the released organisms, therefore, includes assessment of the distribution of the released cells within the target site and the extent of inoculum drift. Analysis of organisms resuspended from the soil or plant material into the atmosphere and the redistribution of those cells may also be necessary to achieve monitoring goals.

During the field release of GEMs onto strawberry plants in California in the spring of 1987, 99.99 percent of the organisms sprayed out were deposited on the target plants and soil within the test plot (40). Approximately 0.0001 percent of the spray inoculum (1.2×10^6 cells) settled onto a 15-m uncultivated zone surrounding the target site established to reduce passive inoculation. Viable bacterial cells were also recovered at the outside edges of the buffer zone indicating drift of the GEMs beyond that designated area. Recombinant cells were similarly detected beyond a 30-m wide buffer zone in the release studies onto potato plants at Tulelake, California (40). The drift of cells off these two sites was later estimated at 19.2 m and 35.2 m, respectively. Increased plot dimensions, the concentration of bacterial cells released, and the size of the droplet(s) sprayed from the applicator were cited as possible reasons for the increased drift at Tulelake. These factors may have provided a longer period of viability in the aerosol for the organisms at the potato release plot and increased the likelihood for colonization beyond the buffer zone (40).

In addition to parameters inherent to all field releases (i.e., characteristics of the organism), the field study system for air monitoring should consider four parameters that may significantly impact the aerial field release. These parameters are (1) the microbial source strength, (2) the on-site weather conditions, (3) the characteristics of the target site that would affect dispersal and survival of the organism, and (4) the types and location of the air samplers to be used.

Source strength

The inoculum source strength required for successful colonization of the target crop should be established during prerelease greenhouse studies. These prerelease studies must be conducted under the same conditions as the actual aerial release. Since droplet size is important in estimations of aerosol drift (10, 22, 27) and survival of airborne organisms (17, 32), the sprayer to be used in the field release should be used during the prerelease studies. A CO_2-pressurized Tee-Jet 8004 hand-held sprayer operated at 2800 gs/cm^2 pressure was used in the application of organisms to oat plants (28) while a Hudson hand-held pump-feed sprayer was used for field application of GEMs to strawberries (40). These sprayers were selected because they produced large droplets that would settle out quickly and minimize drift.

Weather conditions

The weather conditions on the release site at the time of the spray event and in the days to follow are a major concern in the design of air sampling strategies. A wind speed of 0 to 5 mph (0-26.7 m/s) has been suggested as acceptable conditions for spray release of GEMs to maximize the inoculation of the target crop while minimizing drift off site (40). Wind direction is also important in the dispersal of airborne bacteria as demonstrated in the release of organisms on oat plants where winds were light but variable (0.73 to 1.5 m/s) resulting in an asymmetrical dispersal pattern (28). If the wind speed is greater than 5 mph or if the direction of the wind is widely variable over the spray time period, attempts to spray the GEMs may result in extensive drift of the organism beyond the buffer zone to nontarget plants. Sampling schedules and the placement of sampling platforms are also affected by on-site weather patterns. It may be more difficult to select representative sites for weather stations and biological samplers when unpredictable weather patterns exist and require several platforms throughout the plot to account for these often subtle changes.

The time of the day that the organisms are sprayed may also influence the drift of the bacterial aerosol. Slower air movement, common earlier in the day, was cited as a possible reason for more concentrated counts within the target site boundaries during a morning release on oat plants (28). Similarly, increased airborne bacterial numbers have been attributed to upward flux of air from plant leaves during hotter daytime hours (29).

Site characteristics

Features such as the geographical profile of the area including elevation changes and the placement of permanent buildings, shrubs or trees can subtly alter wind currents or air temperature within the test plot creating significant changes in transport and survival. Placement of temporary research stations and laboratories may also affect the normal site conditions and should be carefully considered. Previous releases in open fields did not encompass major buildings or vegetation within the target sites (28, 40), but a greater than 10°C temperature drop was detected between the meteorological sampling station on a bluff and the canopy height within the target plot (40).

Sampling stations and times

Sampling stations can be located using dispersal models based on expected wind speed and direction and on data obtained from tracking nonrecombinant organisms during release trials at the site. Stations can also be placed by simply arranging a grid pattern throughout the site.

The vertical location of the sampling stations will be determined based on the goals of the study. Several heights may be required for tracking the dispersal of a released organism while canopy height may be all that is necessary for confirmation of application to the target plant.

Sampling prior to release of the GEM or MPCA should be used to detect an indigenous background level of the released organism at the site. A general time frame for monitoring during and after the spray event can be established based on greenhouse studies prior to the release. The greenhouse studies should detail the expected colonization of the test crop and the survival of the released organism under conditions projected during the release and provide estimates of survival potential. The results of initial monitoring after the spray event can be used in conjunction with the greenhouse data to estimate the extent of long-term monitoring.

Liberation of organisms from plant leaf surfaces was cited as the source for a midday diurnal peak resulting from wind-generated upward flux (29), while dual peaks were noted in an urban setting with a peak in the morning and another 12 h later (34). These studies indicate that optimal sampling times may depend on conditions at the site and sampling events should be scheduled accordingly.

The goals of the project will dictate the selection of air sampler types, location, and sampling time, while the selection of the number of samplers and number of replicate samples should be based on the predicted variability of the data. Variability can be estimated using prerelease studies with nonrecombinant parental organisms under meteorological conditions similar to expected test conditions.

In the 1987 California trials, bacteria were released within the test sites in a systematic fashion consisting of a Latin square or a randomized block design (40). Since the monitoring goals included tracking the dispersal of the organisms, a combination of air samplers was placed at each site at various heights above the plant canopy surrounding the estimated limits of the aerosol plume both horizontally and vertically on all four sides. Discussion of several field samplers follows in the next section.

Field Sampling Methods

Sample collection of viable aerosolized bacteria requires apparatus that will entrain the desired organism while minimizing sampling stress on vegetative cells. The air sampler(s) must be valid for the detection of the released GEM or MPCA in the aerosolized state and be appropriate for the data quality objectives of the project. Several parameters, therefore, should be considered in the selection of a sampling method. These include the ability of the sampler to collect the organism of interest, the sensitivity and detection limits of the system, and the operational constraints and physical requirements for use of the equipment in the field. In addition to providing qualitative data, the method may need to allow for the enumeration per unit volume of air or detect the size of the aerosolized particles.

Techniques have been developed to detect airborne bacteria using radioactive tracers and enzymes (4) but culture methods are more commonly used. Culture methods rely on growth of the organisms and subsequent enumeration of typical colonies (2). These methods assume that the organism grow on artificial medium and produce a recognizable colonial morphology within a specified time period. Airborne microorganisms, however, are stressed during transport and may not respond to classic culture techniques (23). In addition, genetically altered bacteria have been shown to be sensitive to conventional selective media (25) and the use of selective media for the recovery of airborne bacteria has been criticized (11, 23).

Recent reports on the presence of viable but nonculturable organisms resulting from environmental stresses have prompted recommendations for the use of total (direct) count enumeration procedures for GEMs (9). Early direct count methods relied on the uptake of acridine orange by cells to enhance enumeration (19) and viability determinations using direct count techniques have recently been developed using amendments of 0.002% nalidixic acid and 0.025% yeast extract (24). The addition of fluorescently labeled antibodies for increased selectivity can also provide a direct viable count method useful for studies of a specific organism (39), and the application of the most-probable-number technique to direct counts has added statistical considerations (38). While these methods were developed for use in aquatic and terrestrial microbiology, applications to aerobiology and the enumeration of airborne GEMs are evident.

Numerous air samplers have been designed for use in aerobiology. These air samplers can be categorized as either passive or active depending on their use of forced air flow. Passive techniques rely on the settling out of particles from the air onto a sampling platform such as open sampling dishes with culture medium or glass slides covered with a sticky film (Table 27.1). This method assumes that the organisms of interest will settle onto the surfaces under gravity and efficiency is dependent on particle size, wind velocity, and turbulence (7). The sampling surfaces are exposed for a preset period of time to allow for inoculation and then returned to the laboratory for processing. Passive settling has been widely used for pollen and large particulates (14, 16, 18) but minute particles (i.e., bacterial cells) may remain suspended for long periods of time resulting in false negatives (13, 14). In addition, passive sampling methods do not permit quantitation of particles per unit volume of air (7).

TABLE 27.1   Field Air Samplers for the Detection of Microorganisms

Sampler type	Collection surface	Air, m^3/min	Use*	Special concerns	References
Dish or slide	Agar or sticky tape	NQ	S	Simple, settling, inexpensive	10, 14, 16, 18, 40
Filter	Filters	0.001	A	Viability less	8, 14, 35
Electrostatic	Fluid	0.4–15	B, V	Complex, difficult to field-sterilize, O_3 generated	8, 14, 33, 41, 42
Centrifugal					
Cyclone	Fluid	0.001–0.4	V	Subject to humidity, autoclavable	14, 43
XM2	Fluid	1.05	B, V	Prototype, difficult to sterilize	6
Reuter	Agar	0.04	B	Aggregates >3 μm, easy to field disinfect	14, 31, 36
Impactor					
Andersen	Agar	0.028	B	Particle sizing, "positive-hole," plates overloaded	3–5, 10, 12, 15, 20, 21, 30, 40
Casella	Agar	0.03	B	Time data	1, 14, 30
Burkard	Agar	0.01	S	24-h/7-day, indoor use	7, 37
Impinger					
AGI-30	Fluid	0.0125	B	Increased recovery	10, 14, 40, 44
May	Fluid	0.0125	B	Gentler air flow than AGI-30, particle sizing	10, 14, 32, 44

*A = antigen, B = bacteria, V = virus, S = spores.
†NQ = not quantitative.

To enhance the detection of microorganisms in air, active samplers with a mechanically induced air flow were designed to force air through or over sampling surfaces. Active aerosol samplers suitable for field research with microorganisms include filters, electrostatic high-volume samplers, centrifugal samplers, impactors, and impingers (Table 27.1; 10).

Filter samplers

Collection of microorganisms by filtration, common in water micro-biology (2), is not widely used for airborne cells due to the loss of via-bility through desiccation (Table 27.1; 8, 14). Direct counts and size distribution of bacteria and fungi, however, has been reported for in-door monitoring of microbial antigens where viability of the cells is not a concern (35).

Electrostatic precipitation samplers

Electrostatic precipitation samplers (Table 27.1) are primarily used in collecting particles under 1 μm in diameter. These systems operate by creating an electrical charge on particles passing through a collection tube (33) and are not affected by mass loading (42). The Litton Model M large-volume sampler (LVAS) and the liquid electrostatic aerosol precipitator (LEAP) use the principle of electrostatic precipitation. These samplers are very efficient in sampling situations where low concentrations of bacteria or virus are expected, in part due to their high sampling rate (8). Unfortunately, ozone is generated as a result of electrical arcing which reduces the viability of many microorgan-isms (33, 42). Ultraviolet light and nitrogen oxides may also be present and similarly reduce microbial viability (42). Electrostatic samplers also require more extensive training for the operators and rigorous monitoring efforts than other field samplers (14).

Centrifugal samplers

Cyclone centrifugal samplers (Table 27.1) are useful in monitoring airborne virus but recovery of culturable bacteria using this method has been quite variable (14). These samplers collect organisms as a re-sult of tangential impingement and collection efficiency is dependent on particle size. The particles are impinged onto a fine fluid but fluid loss due to evaporation and foaming of the collection medium is a problem (14). The addition of antifoaming agents can reduce foaming but these substances may be bactericidal (43) and must be tested prior to use. The cyclone scrubbing samplers are easier to operate than the electrostatic high volume samplers but they require monitoring and adjustments of the fluid level constantly during operation (14).

The XM2 prototype sampler, a newly developed biological sampler, combines scrubbing and impingement of deposit organisms into a col-lection liquid (6). This sampler processes large volumes of air without ozone generation but is difficult to sterilize and requires a trained op-erator (Table 27.1).

The Reuter centrifugal sampler (RCS) recovers organisms by impaction onto an agar strip rather than into a fluid (Table 27.1). While other centrifugal samplers are not recommended for bacterial sampling, the RCS has shown good recovery of organisms with particle sizes, under 3 μm in diameter (8, 36) and is portable and easily sterilized.

Impactor samplers

Impactor samplers are among the earliest methods of aerosol sampling (37). Particles are collected onto an agar surface as air is drawn through an orifice at high velocity. Since these samplers are dependent on the growth of microbial colonies for enumeration, the selection of the agar medium to support the growth of the organisms of interest is critical.

The Andersen multistage sampler is an impactor field sampler design recommended for assay of viable airborne microorganisms (5) and is portable, easy to operate and sterilize, and relatively efficient (Table 27.1). These samplers are cascade samplers designed to direct air flow through a series of stacked stages depositing organisms onto an agar plate at the stage corresponding to their particle size (Fig. 27.1a; 3). Single-stage (21), two-stage (12, 20), six-stage (15, 21, 30), and eight-stage (12) Andersen samplers have been used for assay of airborne bacteria. The number of culturable organisms is determined af-

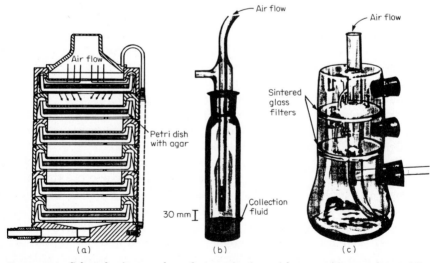

Figure 27.1 Selected air samplers for monitoring airborne microorganisms. (a) Andersen six-stage impactor air sampler. (b) All glass impinger AGI-30. (c) May three-stage impinger sampler.

ter the plates are removed from the sampler and incubated. Unfortunately, aggregates of organisms may be deposited as a unit onto the agar and multiple deposits through the same hole are counted as a single unit. This may result in lower counts and requires adjustment of the data for this "positive-hole" effect (3). In addition, the plates may become overloaded in situations of high bacterial counts and an accurate enumeration would not be possible (10). The use of the Andersen, therefore, should be limited to those situations when the expected counts are below 350 colonies per plate (8). Conversely, Andersen samplers are not recommended in low-density sampling situations owing to the low sensitivity of this device (20). Care should also be taken in the operation of Andersen samplers to ensure the proper particle size distribution estimates. Glass Petri dishes were recommended to eliminate electrostatic interactions of the particles and the dish edges for older Andersen sampler issues (3) but newer models can accommodate disposable plates. The amount of agar in each dish, however, must be controlled to provide proper flow between the Petri dish and the impactor surface (10).

The slit sampler, unlike the Andersen, is an impactor sampler primarily used to record the presence of microorganisms at a specific time and does not provide size discrimination (Table 27.1; 14). Air is directed through a slot in the slit sampler and particles are deposited onto a rotating agar plate (1). The air flow is comparable with the Andersen sampler and the agar plates also must be cultured prior to evaluation. The plate from the slit sampler, however, must be periodically removed and replaced with a noninoculated plate during the sampling operation in the field. The Reynier (40) and Casella (30) are slit samplers that have been used for previous field studies.

The Burkard spore trap is a slit sampler design that uses a rotating drum to enable continuous 24-h or 7-day studies (Table 27.1; 7, 37). While this device is not recommended for vegetative bacterial cells, it is often used for indoor sampling of fungal spores and pollen (7).

Impinger samplers

Impingers were originally designed to simulate air flow through nasal passages and operate by drawing air through a curved tube and into a liquid medium (10). The liquid collection fluid then can be used as an inoculum for culture media and/or examined by total count techniques. Impinger samplers will break apart aggregates to give a more accurate enumeration of individual cells but they may subject the organisms to damaging velocities, thereby reducing viability.

The AGI-30 is an all-glass impinger designed with a jet held 30 mm above the collection medium to reduce the effects of sonication (Fig.

27.1*b*). This sampler has been used as a reference for respirable sized particles (0.8 to 15 μm) since the First International Symposium on Aerobiology in 1963 (10). The May three-stage impinger sampler (Fig. 27.1*c*) uses sintered glass filters to separate particles for size determination (Table 27.1; 32). In addition to size characterization, the air flow through this sampler is gentler than the AGI-30 and it has been suggested as the best air sampler for fragile bacteria (10).

Both the AGI-30 and the May impinger samplers provide a liquid collection medium that can be sampled for microscopic counts and diluted prior to culturing. In this way the problem of overinoculation in high-density situations can be avoided. Subsamples can also be plated onto several media to enhance the recovery of injured or fastidious organisms. The impinger samplers do, however, require strict adherence to aseptic techniques to avoid contamination during the preparation, sampling, and subculturing steps (44). In addition, the sampling fluid can evaporate quickly and freeze in cool dry air. Restrictions on the sampling times are necessary, therefore, to reduce the chance of altering the pH, volume, or osmotic pressure of the collecting fluid (14).

Sampler comparisons

A linear relationship over a concentration range of 10^3 to 10^5 colony-forming units (CFU) per cubic meter of air has been demonstrated for the Andersen sampler (26). A similar relationship was shown in the same study for the all-glass impingers over a range of 10^3 to 10^7 CFU/m^3. A high degree of variability, however, was observed in the data and an increase in the number of samplings may be necessary to improve precision.

In previous releases to strawberry and potato plants a combination of sampler types including gravity plates, AGI-30s and Andersen impactor samplers, Reynier samplers, and oat or bush bean plants were used (40). While all the sampler types showed useful features, the gravity plates and the AGI-30s were the samplers of choice owing to satisfactory efficiencies for the data quality objectives, ease of use, and low cost. In addition, gravity plates recorded the presence of organisms resulting from resuspension when the mechanical methods did not.

Several researchers have compared the efficiencies of air samplers in field situations with a variety of organisms (Table 27.2). These data are not directly comparable but demonstrate the variability of results experienced in air monitoring studies.

TABLE 27.2   Relative Efficiencies of Selected Microbial Air Samplers

Sampler*	Efficiency	Compared to	Organisms	Reference
May	82%	Andersen	*E. coli*	44
LEAP	100%	Andersen	Coliforms	41
LEAP	40–70%	AGI-30	Not specified	8
Cyclone	63%	AGI-30	*B. subtilus* spores	14
S-T-A	< recovery	Reuter	Bacterial counts	36
Andersen	> recovery	Casella	*S. epidermidis* *S. marcescens*	30
Andersen	100%	Casella	Sewage bacteria	30
Andersen	> recovery	All-glass impinger	*S. epidermidis* *S. marcescens*	30
Andersen	> recovery	Filtration	*S. epidermidis* *S. marcescens*	30
Casella	98%	Andersen	Total bacteria	37
Casella	> recovery	All-glass impinger	*S. epidermidis*	30
Casella	> recovery	Filtration	*S. epidermidis*	30
Filtration	1–2%	Not specified	*S. marcescens*	14

*LEAP = liquid electrostatic aerosol precipitator; S-T-A = slit-to-agar.

Sentinel plants

Sentinel plants can be an additional means to evaluate dispersal and colonization by placing target host plants or nontarget plants susceptible to colonization in areas beyond the test plot. The sentinel plants would be colonized by viable organisms that have drifted off the target site and could signal the follow-up testing of other vegetation in the area.

Oat plants were positioned up to 33 m from target site in the Tulelake release study and recombinant cells were detected on their leaves up to 27 m away immediately after the release (28). Cayuse oat plants and bush beans were grown in trays prior to placement on the target site in the California studies and similarly detected the presence of released organisms (40). These plants were particularly useful in the detection of resuspended cells that were not detected by the mechanical samplers (40).

While the above-described field monitoring design components have provided useful information on airborne microorganisms, further research on the survival and transport of airborne microorganisms and retrieval techniques is needed to develop standardized protocols for the monitoring of GEMs in the environment. Recent developments in biotechnology (i.e., gene probes) will be providing new detection methodologies applicable to field monitoring and aerobiology. These methods will need to be compared in the field with the traditional sampling and analysis and may result in enhanced field monitoring designs.

References

1. Akers, A. B., and W. D. Won. 1969. Assay of living, airborne microorganisms. In R. L. Dimmick and A. B. Akers (eds.), *An Introduction to Experimental Aerobiology.* Wiley-Interscience, New York.
2. American Public Health Assoc., American Water Works Assoc., Water Pollution Control Federation. 1985. *Standard Methods for the Examination of Water and Wastewater.* American Public Health Assoc., Washington, D.C.
3. Andersen, A. A. 1958. New sampler for the collection, sizing, and enumeration of viable airborne particles. *J. Bacteriol.* 76:471–484.
4. Anderson, J. D., G. T. Crouch. 1967. A new principle for the determination of total bacterial numbers in populations recovered from aerosols. *J. Gen. Microbiol.* 47:49–52.
5. Brachman, P. S., R. Erlich, H. F. Eichenwald, V. J. Gabelli, T. W. Kethley, S. H. Madin, J. R. Maltman, G. Middlebrook, J. D. Morton, I. H. Silver, and E. K. Wolfe. 1964. Standard sampler for assay of airborne microorganisms. *Science* 144:1295.
6. Brenner, K. P., P. V. Scarpino, and C. S. Clark. 1988. Animal viruses, coliphage, and bacteria in aerosols and wastewater at a spray irrigation site. *Appl. Environ. Microbiol.* 54:409–415.
7. Burge, H. A., and W. R. Solomon. 1987. Sampling and analysis of biological aerosols. *Atm. Environ.* 21:451–456.
8. Chatigny, M. A. 1983. Sampling airborne microorganisms. In P. J. Lioy (ed.), *Air Sampling Instruments for Evaluation of Atmospheric Contaminants.* American Conference of Governmental Industrial Hygienists, Cincinnati, pp. E1–E9.
9. Colwell, R. R., P. R. Brayton, D. J. Grimes, D. B. Roszak, S. A. Huq, and L. M. Palmer. 1985. Viable but nonculturable *Vibrio cholerae* and related pathogens in the environment: Implications for the release of genetically engineered microorganisms. *Biotechnology* 3:817–820.
10. Cox, C. S. 1987. *Aerobiological Pathways of Microorganisms.* Wiley, New York.
11. Crawford, G. V., and P. H. Jones. 1979. Sampling and differentiation techniques for airborne organisms emitted from wastewater. *Water Res.* 13:393–399.
12. Curtis, S. E., R. K. Balsbaugh, and J. G. Drummond. 1978. Comparison of Andersen 8 stage and 2 stage viable air samplers. *Appl. Environ. Microbiol.* 35:208–220.
13. Dimmick, R. L. 1969. Production of Biological Aerosols. In R. L. Dimmick and A. B. Akers (eds.), *An Introduction to Experimental Aerobiology.* Wiley-Interscience, New York.
14. Fannin, K. F. 1980. Wastewater aerosols and disease. U.S. Environmental Protection Agency. PB81-169864. Health Effects Research Laboratory, Cincinnati.
15. Gillespie, V. L., C. S. Clark, H. S. Bjornson, S. J. Samuels, and J. W. Holland. 1981. A comparison of two-stage and six-stage Andersen impactors for viable aerosols. *Am. Ind. Hyg. Assoc.* 42:858–864.
16. Gillette, D. A., I. H. Blifford, Jr., and C. R. Fenster. 1972. Measurements of aerosol size distribution and vertical fluxes of aerosols on land subject to wind erosion. *J. Appl. Meteorol.* 11:977–987.
17. Green, L., and G. Green. 1968. Direct method of determining viability of a freshly mixed bacterial aerosol. *Appl. Microbiol.* 16:78–81.
18. Gregory, P. H. 1973. In N. Polunin (ed.), *The Microbiology of the Atmosphere,* 2d ed. Leonard Hill, London.
19. Hobbie, J. E., R. J. Daley, and S. Jasper. 1977. Use of Nuclepore filters for counting bacteria by fluorescence microscopy. *Appl. Environ. Microbiol.* 33:1225–1228.
20. Jones, B. L., and J. T. Cookson. 1983. Natural atmospheric microbial conditions in a typical suburban area. *Appl. Environ. Microbiol.* 45:919–934.
21. Jones, W., K. Morring, P. Morey, and W. Sorenson. 1985. Evaluation of the Andersen viable impactor for single stage sampling. *J. Am. Ind. Hyg. Assoc.* 46: 294–298.
22. Kenline, P. A., and P. V. Scarpino. 1972. Bacterial air pollution from sewage treatment plants. *J. Am. Ind. Hyg. Assoc. J.* 33:346–352.
23. Kingston, D. 1971. Selective media in air sampling: A review. *J. Appl. Bacteriol.* 34:221–232.

24. Kogure, K., U. Simidu, and N. Taga. 1984. An improved direct viable count method for aquatic bacteria. *Arch. Hydrobiol.* 102:117–122.
25. LeChevallier, M. W., A. K. Camper, S. C. Broadaway, J. M. Henson, and G. A. McFeters. 1987. Sensitivity of genetically engineered organisms to selective media. *Appl. Environ. Microbiol.* 53:606–609.
26. Lembke, L., R. N. Kniseley, R. C. VanNostrand, and M. D. Hale. 1981. Precision of the all-glass impinger and the Andersen microbial impactor for air sampling in solid waste handling facilities. *Appl. Environ. Microbiol.* 42:222–225.
27. Lighthart, B., J. C. Spendlove, and T. G. Akers. 1979. Factors in the production, release, and viability of biological particles. In R. L. Edmonds (ed.), *Aerobiology: The Ecological Systems Approach.* US/IBP Synthesis Series 10. Dowden, Hutchinson, and Ross, Inc., Stroudsburg, Pa., pp. 11–22.
28. Lindow, S. E., G. R. Knudsen, R. J. Seidler, M. V. Walter, V. W. Lambou, P. S. Amy, D. Schmedding, V. Prince, and S. E. Hern. 1988. Aerial dispersal and epiphytic survival of *Pseudomonas syringae* during a pretest for the release of genetically engineered strains into the environment. *Appl. Environ. Microbiol.* 54:1557–1563.
29. Lindemann, J., and C. D. Upper. 1985. Aerial dispersal of epiphytic bacteria over bean plants. *Appl. Environ. Microbiol.* 50:1229–1232.
30. Lundholm, I. M. 1982. Comparison of methods for quantitative determinations of airborne bacteria and evaluation of total viable counts. *Appl. Environ. Microbiol.* 44:179–183.
31. Macher, J. M., and M. W. First. 1983. Reuter centrifugal air sampler: Measurement of effective airflow rate and collection efficiency. *Appl. Environ. Microbiol.* 45:1960–1962.
32. May, K. R. 1966. Multistage liquid impinger. *Bacteriol. Rev.* 30:559–570.
33. Morris, E. J., H. M. Darlow, J. F. H. Peel, and W. C. Wright. 1961. The quantitative assay of mono-dispersed aerosols of bacteria and bacteriophage by electrostatic precipitation. *J. Hyg., Camb.* 59:487–496.
34. Pady, S. M., and C. L. Kramer. 1967. Diural periodicity in airborne bacteria. *Mycologia.* 59:714–716.
35. Palmgren, U., G. Strom, G. Blomquist, and P. Malmberg. 1986. Collection of airborne microorganisms on Nuclepore filters, estimation and analysis-CAMNEA method. *J. Appl. Bacteriol.* 61:401–406.
36. Placencia, A. M., J. T. Peeler, G. S. Oxborrow, and J. W. Danielson. 1982. Comparison of bacterial recovery by Reuter centrifugal air sampler and slit-to-agar sampler. *Appl. Environ. Microbiol.* 44:512–513.
37. Rajhans, G. S. 1983. Inertial and gravitational collectors. In P. J. Lioy (ed.), *Air Sampling Instruments for Evaluation of Atmospheric Contaminants.* American Conference of Governmental Industrial Hygienists, Cincinnati, pp. Q1–Q40.
38. Roser, D. J., H. J. Bavor, and S. A. McKersie. 1987. Application of most-probable-number statistics to direct enumeration of microorganisms. *Appl. Environ. Microbiol.* 53:1327–1332.
39. Roszak, D. B., and R. R. Colwell. 1987. Metabolic activity of bacterial cells enumerated by direct viable count. *Appl. Environ. Microbiol.* 53:2889–2983.
40. Seidler, R. J., and S. C. Hern. 1988. EPA Special Report: The release of ice minus recombinant bacteria at California test sites. ERL-Corvallis, Oreg.
41. Sorber, C. A., H. T. Bausum, S. A. Schaub, and M. J. Small. 1976. A study of bacterial aerosols at a wastewater irrigation site. *J. Water Pollut. Control Fed.* 48:2367–2379.
42. Swift, D. L., and M. Lippmann. 1983. Electrostatic and thermal precipitators. In P. J. Lioy (ed.), *Air Sampling Instruments for Evaluation of Atmospheric Contaminants.* American Conference of Governmental Industrial Hygienists, Cincinnati, pp. R1–R15.
43. White, L. A., D. J. Hadley, D. E. Davids, and R. Naylor. 1975. Improved large-volume sampler for the collection of bacterial cells from aerosol. *Appl. Microbiol.* 29:335–339.
44. Zimmerman, N. J., P. C. Reist, and A. G. Turner. Comparison of two biological aerosol sampling methods. *Appl. Environ. Microbiol.* 53:99–104.

Ecosystems Effects

28

Overview: Identifying Ecological Effects from the Release of Genetically Engineered Microorganisms and Microbial Pest Control Agents

James K. Fredrickson

Charles Hagedorn

Introduction

Scientists, public policymakers, and others have expressed concerns about the potential for ecological impacts from the intentional release of recombinant microorganisms to the environment. Numerous scenarios have been presented that describe how the release of a recombinant microbe could result in negative ecological effects; however, there is essentially no practical experience in studying such an event. Johnston and Robinson (36) describe such a scenario, wherein nitrification genes are transferred to and expressed in heterotrophic bacteria to improve wastewater treatment systems. Autotrophic nitrifying bacteria are known to produce small quantities of N_2O during ammonia oxidation, and under certain conditions, the proportion of evolved N_2O can increase. The promiscuous transfer of these genes to indigenous heterotrophic bacteria in the environment, in conjunction with cellular environmental conditions that favor N_2O production during nitrification, could result in the production of N_2O in quantities that could influence global ozone levels.

A more subtle example of how a recombinant bacterium, when introduced into the environment, could result in a detrimental ecological effect was provided by Odum (46) using ice-minus *Pseudomonas syringae* as an example. This genetically engineered microorganism (GEM) is generally believed to be innocuous because the genetic manipulation involved deletion of genomic DNA rather than the insertion of rDNA. In this example, the ecological significance of ice nucleation in bacteria was postulated as being important in the formation of raindrops in clouds. A decrease in microbial ice nucleation could result in changing precipitation patterns, in contrast to the negative impact that ice nucleation can have on horticultural crops by promoting frost formation.

These scenarios, which involve considerable speculation despite considerable scientific information on these processes from laboratory studies, demonstrate how little is known about many of the microbially mediated processes that occur in the environment. Nevertheless, considerable experience exists regarding the actual introduction of organisms into the environment. Familiar examples include vaccines, agricultural crop varieties, and biological control agents (42, 54). The positive and negative lessons learned from these experiences provide a basis for the rational development of safeguards to prevent undesirable ecological effects. Nevertheless, many unknowns exist concerning such factors as systematics; community structure; factors governing the survival, growth, and spread of populations of introduced organisms; and techniques for monitoring the organisms once they are introduced into the environment. The development of novel techniques for manipulating the genome of microorganisms opens new possibilities for basic research involving the close collaboration among molecular biologists, ecologists, and evolutionists (16). The results of such studies will help scientists understand the genetic structure of populations and develop improved methods for studying natural microbial communities. Ecological and epidemiological perspectives may help to elucidate fundamental aspects of molecular biology, such as the population biology of plasmids and transposable elements.

Scientists generally agree that the risk from the release of GEMs into the environment is slight, citing numerous examples of the introduction of laboratory-cultured or uncharacterized microorganisms into the environment with no apparent detrimental effects on human health or the environment (13). Stringent environmental constraints (i.e., homeostasis) typically prohibit the establishment of new species in an ecosystem from which they were previously absent. For example, when strains of *Rhizobium* sp. that are improved in their N_2-

fixing abilities are introduced into soils that contain strains that nodulate the target host plant, the introduction is usually unsuccessful because the improved strains fail to compete with native populations. Alternatively, when strains are introduced into soils in which native rhizobia that nodulate a particular leguminous crop are absent, the new strains can establish and make introduction of more efficient N_2-fixing strains difficult (8). This microbial "competitive exclusion" has also been demonstrated for *P. syringae* establishment on the phylloplane (39, 40).

The introduction of exotic species to the environment has been used as an analogy to the introduction of GEMs. Although the overwhelming majority of known introductions of exotic species have been unsuccessful, well-documented cases exist in which such introductions have resulted in ecological change. Levin and Harwell (38) grouped GEMs into four classes based on their relationships to the target environment: (1) slightly modified forms of resident organisms, (2) organisms that exist in the target environment, but that require continual supplemental support, (3) organisms that exist naturally elsewhere in the environment, but that previously had not reached the target environment, and (4) genuine novel organisms. Arguments against using the exotic species (class 3) analogy for GEMs are based on the fact that the majority of GEMs intended for release to the environment will be identical, or at least similar to, species that already exist within that ecosystem (classes 1 and 2). However, because microorganisms are readily disseminated once released and can be transported to nontarget environments, potential ecological effects from microorganisms in classes 1 and 2 cannot be ruled out.

In determining whether a particular GEM introduction will become established or will spread, several important factors must be considered, including size, geographic scale, and frequency of introduction. Therefore, small-scale field testing involves different considerations than does large-scale (e.g., commercial) application. This does not mean that small-scale field testing should be exempt from examination for regulation, but simply that associated risks are likely to be reduced and more easily managed than those for larger applications (43). Likewise, not all large-scale applications will be problematic, as demonstrated by many examples of successful biological introductions in the field of agriculture.

Examples of GEMs intended for release to the environment include *P. syringae* ice-minus deletion mutants (39) and root-colonizing *Pseudomonas fluorescens* containing the *B. thuringiensis* subsp. *kurstaki* delta-endotoxin gene (67). Although the introduction of genetically engineered species into ecosystems in which similar or identical species already exist is unlikely to have an ecological effect, zero risk cannot be guaranteed. Alternatively, the risk to the environment from the release of the majority of carefully designed GEMs or microbial pest control agents (MPCAs) is likely to be low (63) and, in many cases, such releases may replace or reduce the use of more toxic chemicals.

Predictions of environmental impacts from the release of novel or recombinant bacteria are difficult to make with our current level of knowledge. In addition, the expanding use of recombinant DNA techniques in numerous disciplines within the agricultural and environmental sciences opens the door to increasing environmental applications of recombinant microorganisms. As the applications of this technology expand, the probability of associated environmental problems increases (2). Unfortunately, except for a few microorganisms (e.g., *Rhizobium*), little relevant scientific data exist for scientists and policymakers to use in making rational decisions regarding release. Although the Environmental Protection Agency and U.S. Department of Agriculture are considering exempting certain categories of organisms from regulatory scrutiny, current practice is to assess the risks of releasing novel or recombinant bacteria on a case-by-case basis (44, 64). A case-by-case assessment requires relevant tests to characterize and predict the microorganisms' survival, transport, and ecological effects. The scientific community is responsible for identifying the ecological parameters that require evaluation and the appropriate methods for making applicable measurements. Though some approaches currently used for assessing environmental impact of chemicals can be adopted for GEMs and MPCAs, new methods will also be required.

Ecological Effects and Risk Assessment

Ecology has been defined as the study of ecosystem structures and functions. Ecosystem structure involves three factors: (1) the composition of the biological community, including species, numbers, biomass, life history and spatial distribution of populations; (2) the quantity and distribution of the abiotic materials such as nutrients, water, etc.; and (3) the range, or gradient, of conditions of existence such as temperature, light, etc. Ecosystem function involves two factors: (1) the rate of biological energy flow through the ecosystem, that is, the biogeochemical cycles; and (2) biological or ecological regulation, including both regulation of organisms by environment and regulation of environment by organisms (45).

The traditional separation of ecosystems into structures and functions holds certain advantages for risk assessment. Some ecosystem structures tend to be more readily measured than functions and are more susceptible to change. Typically, ecosystem function will not be upset dramatically by the displacement or loss of an individual species because of the functional redundancy in most ecosystems. An exception may be ecosystems intensively managed by humans, such as agricultural ecosystems, which tend to be less diverse and more susceptible to functional impact from individual species changes. The structure-function approach to ecosystem analysis has the additional advantage of readily lending itself to models, which are necessary to integrate ecosystem functions and structures to determine system-level processes and changes (68).

Many ecosystem functions used for analysis of ecological impacts (i.e., those functions requiring protection) from the release of GEMs are similar to those used for evaluating the environmental effects of chemicals. Such ecological impacts include effects on microbial species and communities as well as processes related to biogeochemical cycling. An approach for evaluating the risks associated with the intended release of GEMs and MPCAs to the environment can therefore incorporate some of the same components used in chemical ecotoxicology testing (1).

The risks of introducing a specific GEM or MPCA must be weighed against the perceived benefits and risks of not making the introduction. The properties of the introduced organism and its target environment are the key features in risk assessment. Such factors as the demographic characterization of the introduced organism, its genetic stability, and the fitness of the organism relative to the physiological and biological environment are important in evaluating the potential for adverse or favorable ecological effects. The scale and frequency of the introductions are also important for both modified and unmodified organisms. In fact, for modified organisms, scale and frequency of the introduction apply independently of the techniques used to achieve modification. That is, the organism itself is the critical factor, rather than the method used for its construction.

Each proposed introduction must be evaluated on its own merits in the risk assessment process, but each introduction does not need to be considered *de novo*. As experience accumulates with particular kinds of introductions in specific environments, more genetic approaches to the evaluation of these introductions can be developed. The basis for evaluating the environmental release of GEMs or MPCAs should be refined continually, providing a set of criteria that will allow any proposed introductions judged to be innocuous to be carried out speedily, and those judged to be potentially problematic to be given appropriate attention. Note that generalizations developed for particular groups of organisms (e.g., microorganisms) cannot be extended automatically to other groups (e.g., plants), which may have very different genetic and demographic characteristics, dispersal and reproductive mechanisms, and trophic positions.

Microbially mediated processes

Analysis of the effects of GEMs on nutrient cycling processes must be a key component of biotechnology risk assessment. Ecological processes such as the cycling of carbon and other nutrients are mediated principally by microorganisms and are critical to the functioning of ecosystems. Many of the existing and proposed biotechnology applications involve the release of microorganisms that are altered in the way or rate at which they process organic and inorganic compounds. Examples include bacteria genetically altered to extend their substrate range (53) or enhance the rate at which they degrade environmental contaminants, *Rhizobium* enhanced in their ability to fix nitrogen in symbiosis with leguminous plants (64), and metal-leaching bacteria that are better able to survive and function under the extreme environmental conditions that can occur in metal-leaching systems (12).

The specificity of certain physiological groups of microorganisms (usually chemoautotrophs) for mediating specific nutrient processes such as nitrification and sulfur reduction is well documented (58, 61). By comparison, it has been demonstrated that microorganisms are involved in the oxidation-reduction and methylation reactions of numerous elements (e.g., iron, manganese, mercury, selenium), although scientists do not clearly understand to what extent these reactions occur in ecosystems or how the reactions are mediated by environmental conditions such as temperature and moisture. Despite numerous examples of nutrient cycling and mineral transformations where knowledge is incomplete, it is still necessary to determine the potential for released GEMs to cause perturbations in biogeochemical cycles or in the population dynamics of microorganisms that catalyze these processes.

It has been hypothesized that a change in the rate of degradation of organic matter in soil could result from the establishment in soil of an organism genetically engineered to extend its substrate range or to alter the rate at which it degrades organic compounds (52). In the only known study to demonstrate the potential for such an effect to occur, recombinant strains of *Streptomyces lividans* genetically engineered to enhance their lignin-degrading capabilities significantly increased soil carbon turnover rates in comparison to the wild type (52). This study emphasized the comparison of the wild type with the recombinant strain. The study also demonstrated the necessity of including the parent strains in assessment of the impacts of GEMs on ecological processes in order to discern those effects associated with the manipulated genome from those due to the parent strain.

Some of the microbially mediated nutrient cycling processes for which methods are available for their measurement include

- Primary productivity
- Respiration
- Mineralization of carbon, nitrogen, sulfur, and phosphorus
- Nitrification
- Denitrification
- Nitrogen fixation
- Soil ethylene production
- Litter decomposition
- Decomposition of specific carbon substrates (both natural and contaminant)
- Enzyme synthesis and activity

Considerable information on the effects of xenobiotics and available methods for measuring microbially mediated nutrient cycling processes is available from chemical toxicology research. The effects of pesticides and other toxic chemicals on soil microorganisms and microbial processes has been the subject of numerous reviews (6, 21, 28, 30, 66) and will not be discussed extensively here. These reviews offer valuable insight into potential processes and organisms in soils that can serve as indicators of ecological effects from the introduction of GEMs and MPCAs. The reviews also provide numerous references for methods.

No single parameter will likely be capable of providing an overall assessment of the impacts from the release or GEMs or MPCAs on ecological processes; a group of tests will be needed. For example, a battery of microbial functional and structural parameters was used in conjunction with multivariate statistical analysis to determine the impact of coal-coking effluent on sediment microbial communities (55). An important finding of this study was that multiple estimates of microbial function were necessary to effectively evaluate the effects of coal-coking wastewater on sediment microbial communities.

The multivariable approach has also been evaluated for the ecological impacts associated with the release of GEMs and MPCAs. A soil multitest system that monitors various soil microbiological functions including respiration, mineralization of various carbon sources, nitrogen transformations, species diversity, and soil enzyme activity has been developed (22) to assess the ecological effects from the release of GEMs. The methods in such test systems are well-established, reliable, and do not require expensive instruments. Although GEM ecological effects testing will need to be tailored for specific GEMs and their intended applications, the use of a standard array of tests would be complementary to this approach.

Microbial species and communities

Many ecological processes such as the metabolism of organic matter and the mineralization of nitrogen are functionally redundant in that they are catalyzed not by a single species of microorganism but rather by a large, diverse group. Because of this functional redundancy, many ecological processes are relatively resistant to upset simply because if one or several organisms are affected by the introduction of a chemical or GEM, the loss of their functional contributions does not affect the overall rate or extent of the process. In contrast, the dynamics of specific populations of microorganisms are usually more sensitive to environmental insult. In addition, certain important ecological functions are catalyzed by a relatively narrow range of microorganisms, such as nitrification. Therefore, evaluation of specific microbial populations, including species, functional groups, and communities, can play an important role in assessing the risks from the release of GEMs and MPCAs.

Scientific understanding and appropriate methodologies are lacking for examining the effects of GEMs on ecological structure and function. Studies dealing with the effects of adding a GEM to a natural ecosystem will usually require intensified microcosm simulations (4). Three research approaches have proved to be useful in this respect

1. Expanded use of microcosms as representations of various ecosystems that include appropriate design features, characterization of the most critical parameters, calibration with field studies (usually non-GEM), and standardization of protocols

2. Establishment of ecosystem baseline data for any particular habitat

3. Generation of positive controls to ensure appropriate test systems for evaluating potential ecological effects.

The size of specific microbial populations in environmental samples can be measured using a variety of techniques. Most techniques that use standard methods are dependent upon culturing on media that is either selective or can differentiate the population of interest. To continue the example used above, the process of nitrification in soil is catalyzed principally by a relatively narrow range of autotrophic bacteria that obtain their energy from the oxidation of NH_4^+ or NO_2^-. This property is exploited by the use of an inorganic medium to determine the most probable number of this functional group of bacteria in environmental samples (57). Many cultural methods have been devised for enumerating a variety of microbial functional groups. These groups include organisms with the ability to use specific electron donors (organic and inorganic) or electron acceptors (i.e., denitrifiers, sulfate-reducing bacteria, methanogens), and the ability to carry out specific metabolic functions (i.e., fix nitrogen).

Microfauna, an often-neglected component of terrestrial communities, have an important role in decomposition processes (51) and have been the subject of ecotoxicological studies with xenobiotic chemicals. The assessment of the impacts of GEMs and MPCAs on important taxa or protozoa, nematodes, oligochaetes, and microarthropods may be desirable because of their importance in decomposition processes. Standard methods are available for measuring populations of nematodes (72), protozoa (62), earthworms (37), and microarthropods (70).

Another approach for measuring the structure of microbial communities involves determining the species composition in terms of the abundance of species and the relative numbers of each type. A wide variety of microbial characteristics—including morphological, physiological, genetic, and associative relationship properties—can be used for differentiating individual species (5). Species diversity and richness have been used for calibrating microcosms with the field based on colony diversity of heterotrophic bacteria on a nonselective medium (48) and could also be used as a measurement to determine the effects of GEMs on heterotroph diversity.

The analysis of microbial lipids, directly extracted from sediments, has also been used as a community structure measurement and has the advantage of being independent of culturing. Lipids that are restricted to a specific group of microorganisms, termed signature or fingerprint lipids, have provided information on the composition of microbial communities in surface and subsurface sediments (69). Lipid analysis can also provide an estimation of the size of the microbial biomass.

The size of the soil microbial biomass is a fundamental property of terrestrial ecosystems and has been shown to be sensitive to a variety of disturbances (14). The soil biomass can be determined by a variety of well-established methods (50) that include determining the CO_2 released following soil fumigation or after the addition of a readily metabolizable carbon substrate, such as glucose and the analysis of specific biomass constituents such as ATP, muramic acid, hexosamines, or nucleic acids. Direct microscopic observation has also been used to estimate the soil microbial biomass. However, this technique requires assumptions regarding the density, water content, and carbon content of microbial tissues (35) and is relatively labor-intensive.

In addition to the size of the microbial biomass and the relative abundance of the individual species, the activity or productivity levels may be potential indicators of environmental stress. The synthesis of DNA has been used as one estimate of productivity of natural microbial communities by measuring the rate of incorporation to ^{3}H-thymidine into microbial cells (24). While this technique has been routinely used and its methods are reasonably well-established, certain precautions must be taken (34).

Microcosms

Single-species toxicity assays and other simple laboratory tests are necessary to determine the mechanisms by which microorganisms may exert a negative impact on the resident biota. However, because of the complex interactions between the biotic and abiotic components, the response of an ecosystem to introduction of GEMs and MPCAs will not likely be predicted from simple laboratory tests. Ecotoxicology, the science concerned with the ecosystem level effects from the release of toxic substances, has used microcosms extensively for evaluating the fate and effects of toxic chemicals. More recently, microcosms have been suggested as tools to evaluate certain aspects of the fate, survival, transport, and ecological effects from the release of GEMs and MPCAs (15, 20, 25).

Microcosms, or model ecosystems, are laboratory systems used to simulate some portion of the real environment (26). Although microcosms obviously cannot represent complete ecosystems, they can represent portions of the environment and, therefore, may be used to determine effects on specific ecosystem structures and functions. This information can then be integrated into system-level models to help identify effects on system-level processes. Although the effects of chemicals on ecosystem properties and functions have been tested in microcosms, little work has been conducted to evaluate terrestrial microcosms for determining the effects of releasing GEMs to the environment.

The terrestrial soil-core microcosm protocol developed by Van Voris (65) for evaluating the fate and effects from the release of chemicals to the environment was adapted to evaluate the fate and effects from the release of GEMs (7). These studies used intact soil cores to monitor the fate and effects from the introduction of transposon Tn5 mutants of *Azospirillum lipoferum* as a model GEM. *Azospirillum* sp. live in association with plant roots and are capable of fixing N_2 and stimulating plant growth, although the latter two factors are not necessarily directly related. The studies evaluated effects on populations of total heterotrophic bacteria, free-living N_2-fixing bacteria, *Rhizobium* sp., and autotrophic nitrifying bacteria in the rhizosphere and rhizoplane in inoculated and control microcosms. Though no deleterious effects were observed, it was concluded that the occurrence of measurable natural populations of these organisms in the intact soil cores, in conjunction with their key roles in the processing of carbon and nitrogen, make them candidates for indicators of ecological impact.

The rate at which nutrients are leached from soil has been used as a method for environmental impact assessment because it is a holistic measurement, representing the functions of numerous microbial groups as well as abiotic processes. In a study to evaluate nutrient leaching from microcosms as means for monitoring environmental impact, O'Neill et al. (49) found that effects from the addition of toxic chemicals to microcosms could be detected in nutrient export in leachate from the microcosms but not by changes in population or community parameters. A similar conclusion was drawn by Jackson et al. (33) in a study that measured the efflux of calcium and nitrate from soil-core microcosms amended with arsenic. In addition, Jackson and Levin (32) found that intact terrestrial grassland microcosms were more sensitive to arsenic impact than the grassland itself, likely due to greater dispersal in the field. In this study, the microcosms proved to be conservative screening tools for evaluating ecological effects from the dispersal of arsenic. Microcosms have also been used to evaluate the ecological effects from a genetically engineered kanamycin-resistant strain of *Erwinia carotovora* and the wild-type strain introduced into water and sediment microcosms (56). An increase in total heterotrophic and proteolytic bacteria was observed in the inoculated microcosms early in the study but by 32 days after inoculation, the introduced bacteria had no apparent effect on native bacterial populations. Also, in this study, the genetically engineered and wild-type strains declined at approximately the same rate. Microcosms may also be valuable in testing new techniques to monitor changes in microbial community composition or function (10) or to study the interactions among natural microbial populations. The latter is necessary because an assessment of effects on an ecosystem following release of a chemical or a GEM requires prior knowledge of the interactions of organisms and how those interactions affect the functioning of that ecosystem.

Malanchuk and Joyce (41) measured N_2-fixation using the acetylene reduction assay and CO_2 evolution in simple soil microcosms to determine the ecological effects of the pesticide 2,4-D. Though no statistically significant effects were observed, the investigators concluded that their microcosm studies required validation with real environmental data. One microcosm measured gaseous effluxes as well as nutrient leaching from soils (3). Although relatively simple in design, this microcosm could lend itself to monitoring process changes associated with a specific GEM. While it is desirable to monitor for effects from the introduction of GEMs that are rationalized on the basis of the known functions of an individual GEM and the environment into which they will be introduced, it is also necessary to design standard microcosm test systems for evaluating effects. Because of the variety of microorganisms and processes (i.e., survival, transport, gene transfer, ecological effects) likely to require evaluation, no single microcosm design can be used universally for biotechnology risk assessment. Rather, a variety of microcosms, including both natural and synthetic designs, will be required to address the issues.

Microcosms have some limitations for evaluating the release of GEMs because they cannot represent all the complexities of the actual environment or all the physical processes that act on the natural ecosystem. Recognition of these limitations will enable the identification of ecosystem structures and functions that can be represented in microcosms and can therefore be used to evaluate potential effects induced by the release of GEMs. A valuable component to any microcosm or field evaluation of the ecological effects from the release of GEMs is the concurrent evaluation of effects and population densities so that dose response curves can be constructed.

Though studies on the calibration or comparison of terrestrial microcosm behavior with field studies are lacking, calibration is critical in using microcosms for biotechnology risk assessment. Lack of field calibration can lead to erroneous conclusions and conceptions. It is not necessary that microcosms behave identically in the field to be useful for risk assessment; however, processes and functions must show similar trends to those in the field even if the microcosms differ significantly from the field in magnitude.

Innovative Approaches

A limitation to evaluating the effects of novel or recombinant bacteria on indigenous microbial populations and communities is the current lack of knowledge regarding the interactions between microorganisms and microbial communities, and methods for measuring these communities and their functions. However, a new discipline within microbial ecology—"molecular" microbial ecology—has led to new methods that may be used for analyzing impacts of GEMs and MPCAs on natural microbial communities and their associated functions. Sensitive methods involving molecular biology techniques can be used to monitor not only the introduced microorganisms, as described in earlier chapters, but indigenous microbial populations as well. Many techniques have been developed for the direct extraction of nucleic acids from soils, sediments, and water and their purification to the point where they are suitable for molecular manipulations such as hybridization analysis, polymerase chain reaction amplification, and digestion with restriction enzymes (31, 47, 59, 60).

Nucleic acids that have been directly extracted from environmental samples can be used in conjunction with DNA probes to identify various components of the indigenous microbial community. For example, DNA probes developed for specific groups of environmental microorganisms could, with the appropriate controls, be used to monitor changes in the microorganism's population in the microbial community. Specific DNA probes have already been developed for such diverse microorganisms as *Pseudomonas fluorescens* (23), *Thiobacillus ferrooxidans* (71), *Rhizobium meliloti* (9), and the fungal plant pathogen *Gaeumannomyces graminis* (29). The number of groups for which probes will be available will certainly increase. Probes have also been developed that can distinguish DNA from broad groups of microorganisms such as archaebacteria, eubacteria, and eukaryotes (17, 27). More recently, oligodeoxynucleotide probes have been conjugated with various fluorochromes and used as probes for direct microscopic analysis of microbial populations (19). In addition to the detection of specific and general groups of microorganisms, DNA that codes for specific proteins involved in nitrogen fixation and degradation of natural and xenobiotic organic compounds can be used as function-specific probes.

Although the extraction and purification of DNA from soils is typically labor-intensive, portions of the purified DNA can be analyzed using a variety of DNA probes. A major advantage to using DNA hybridization analysis of directly extracted DNA is that the method is independent of culturing. Changes in the composition of a microbial community, resulting from the introduction of a GEM or MPCA, may be detected that would otherwise be unnoticed because of the inability or difficulty in culturing the target population.

Overview of Part 4

Part 4 describes methods for monitoring impacts on biogeochemical and energy cycling and native microbial populations and describes various microcosm systems and their potential uses for monitoring microbial survival and ecological effects. Chapter 29 emphasizes methods for evaluating carbon and nitrogen cycling because of their universal importance in ecosystems and because they are principally biologically catalyzed cycles. Also described are procedures for evaluating nutrient-cycling processes in soil-core microcosms and the potential for use of DNA probes to determine the genetic potential for specific nutrient cycling functions in ecosystems. Chapter 30 describes methods for determining the presence, population size, and activity of indigenous microbial populations. Approaches for the use of molecular probes to identify specific indigenous microbial populations are discussed as are means for assessing in situ microbial activity. Methods developed for evaluating the effects of metals on natural microbial communities are addressed in Chap. 33. Direct and indirect approaches, both classical and novel, for detecting and enumerating natural microbial communities and for studying their associated activities, are discussed in this chapter. The methods discussed are appropriate for investigating the response of microbial communities to metal stress and are also applicable to studying potential effects from the introduction of microorganisms.

Few microcosm systems have been used to evaluate the fate and effects of microorganisms; however, in Chap. 31 an aquatic microcosm design is described which has been used to study the MPCA *B. thuringiensis*. The mixed-flask-culture microcosm design is described in detail, as are the ecosystem structural and functional parameters that are operational in this microcosm and can be used to monitor effects from the introduction of GEMs or MPCAs. A variety of synthetic aquatic microcosms varying in complexity are described in Chap. 32. The advantages of the discussed microcosm designs, ranging from single to multiple species, are discussed as are their potential applications in biotechnology risk assessment.

Conclusions and Research Needs

Perhaps the most critical issue of biotechnology risk assessment is that of ecological effects. Answers to key questions such as "what constitutes an ecological effect" and "which ecosystem properties need protection" will, in part, dictate the ecological parameters to be evaluated prior to and after the release of GEMs to the environment. Specific concerns, though discussed elsewhere (2, 13, 64), include toxicity to nontarget species such as microbiota and higher animals, displacement of indigenous microbial species that leads to changes in community structure, and deleterious impacts on the cycling of nutrients and energy. Although several methods are currently available for evaluating the effects of GEMs on ecological processes, scientists must develop additional methods and information for addressing the unique aspects of biotechnology risk assessment.

The planned release of GEMs into terrestrial and aquatic environments raises concerns about potential short- and long-term effects that GEMs could have on indigenous processes and populations. Important research issues include:

- Identification of effects of GEMs on various components of ecosystem structure and function

- Delineation of biogeochemical processes that could be most useful for describing ecological structure and function

- Performance of microcosm and/or field studies (e.g., verification of laboratory-derived methodologies and results) that estimate ecological responses under GEM-use conditions

- Verification of biological indicators developed in the laboratory on population, community, and ecological system response to GEM exposure

- Development of model systems to predict resiliency (impact and recovery) of populations, communities, and ecological systems exposed to GEMs

- Assessment of risks associated with release of GEMs by integration and interpretation of biological, chemical, and physical data on impacted ecological systems

Any research on the potential environmental impact of GEMs will need to consider multiple-year scenarios because effects may become evident only after an accumulation of GEMs resulting from multiple releases over time. In addition, released GEMs may only demonstrate specific effects if environmental conditions support a "bloom" or period of rapid growth of the GEM. It is also possible that a GEM may have no impact on the environment into which it is released but could be transported into a different environment where the potential for perturbations is greater (18).

Topics that warrant further investigation or development of methods for their study include:

- Microbial interactions in natural communities such as the rhizosphere and phylloplane, including antagonism of pathogens as well as beneficial microorganisms (i.e., mycorrhizae)

- Methods for assessing the structure of natural microbial communities using cellular components that can be directly extracted from environmental samples (i.e., DNA, phospholipids, proteins)

- Interactions between microorganisms and microfauna

- Ecosystem functional and structural interrelationships between laboratory, microcosm, and field systems
- Studies with varied inoculum levels to determine dose-response relationships
- Long-term studies to determine whether observed ecological effects are transient or irreversible
- Direct comparison of GEMs with the parent strain to assess the effects associated only with the genome

With the tremendous diversity in ecosystem populations and processes, it will be a major research undertaking to attempt to attribute changes in any ecological parameters to the presence or activities of a GEM. By applying a suitable modeling program, conducting experiments in a representative microcosm, and focusing on the most important processes, it may be possible to determine the potential of a GEM for causing ecological changes. Microcosms, because of ease of replication, ability to achieve a high degree of containment of the test microorganism, and representation of a portion of the natural environment, will likely play an important role in biotechnology risk assessment. Unfortunately, few studies with naturally derived microcosms have attempted to describe the structures and functions within the model system, much less recognize a distinction between them. Future developments in the use of microcosms for evaluating the ecological effects from the release of GEMs and MPCAs must include the measurement of ecosystem structural and functional parameters in the microcosm concurrent with the field to calibrate the model system. Functional and structural parameters used for calibration can also be used to assess ecological effects, many of which are described in this part.

Finally, numerical analysis of data assumes a uniquely important aspect in ecological studies because of the large variation inherent in the components of natural systems and the difficulty in establishing cause-and-effect relationships (11). It will be necessary to attempt to define normal ranges of variability and function for specific ecological populations and/or processes and naturally occurring levels of change. Only then can ecological perturbations be detected and linked to the presence and/or activities of a GEM.

A variety of methods exist for evaluating the structural and functional properties of ecosystems. Many of these methods have been shown to be sensitive indicators of environmental insult from chemicals and therefore can be appropriate tools for evaluating the ecological effects from the release of GEMs and MPCAs. In addition, specific, sensitive techniques being developed in molecular microbial ecology will enable the study and description of natural microbial communities at a previously unachievable level of detail. These methods will be useful in biotechnology risk assessment. The challenge will be to identify and/or develop methods to effectively evaluate the potential ecological effects from the release of microorganisms to minimize the environmental risks and yet enable the continuing development of this promising technology.

References

1. Alexander, M. 1984. Ecological constraints on genetic engineering. In G. S. Omenn and A. Hollaender (eds.), *Genetic Control of Environmental Pollutants*. Plenum Press, New York, pp. 151–168.
2. Alexander, M. 1985. Ecological consequences: Reducing the uncertainties. *Issues Sci. Technol.* 1:57–68.
3. Anderson, J. M., and P. Ineson. 1982. A soil microcosm system and its application to measurement of respiration and nutrient leaching. *Soil Biol. Biochem.* 14:415–416.
4. Armstrong, J. L., G. R. Knudsen, and R. J. Seidler. 1987. Microcosm method to assess survival of recombinant bacteria associated with plants and herbivorous insects. *Curr. Microbiol.* 15:229–232.
5. Atlas, R. M. 1984. Diversity of microbial communities. *Adv. Microb. Ecol.* 7:1–47.
6. Babich, H., and G. Stotzky. 1985. Heavy metal toxicity to microbe-mediated ecologic processes: A review and potential application to regulatory policies. *Environ. Res.* 36:111–137.
7. Bentjen, S. A., J. K. Fredrickson, S. W. Li, and P. Van Voris. 1989. Intact soil-core microcosms for evaluating the fate and ecological impacts from the release of genetically engineered microorganisms. *Appl. Environ. Microbiol.* 55:198–202.
8. Beringer, J. E., and M. J. Bale. 1988. The survival and persistence of genetically-engineered micro-organisms. In M. Sussman, C. H. Collins, F. A. Skinner, and D. E. Stewart-Tull (eds.), *The Release of Genetically-Engineered Micro-organisms*. Academic Press, San Diego, pp. 29–46.
9. Bjourson, A. J., and J. E. Cooper. 1988. Isolation of *Rhizobium meliloti* strain-specific DNA sequences by subtraction hybridization. *Appl. Environ. Microbiol.* 54: 2852–2855.
10. Breen, A., B. Reynolds, L. Burtis, B. Bellew, and G. Sayler. 1988. Introduction and effects of the pSS50 degradative genotype in 4-chlorobiphenyl lake water microcosms, *Abstracts of the Annual Meeting of the American Society of Microbiology*, Q-159, p. 309.
11. Brezhnev, A. I., L. R. Ginzburg, and R. A. Poluektov. 1974. The structural optimization of exploited populations. *Econ. Math. Methods* 4:808–811.
12. Brierley, J. A. 1985. Use of microorganisms for mining metals. In H. O. Halvorson, D. Pramer, and M. Rogul (eds.), *Engineered Organisms in the Environment: Scientific Issues*. American Society for Microbiology, Washington, D.C., pp. 141–146.
13. Brill, W. J. 1985. Safety concerns and genetic engineering in agriculture. *Science* 227:381–384.
14. Brookes, P. C., and S. P. McGrath. 1986. Effects of heavy metal accumulation in field soils treated with sewage sludge on soil microbial processes and soil fertility. In V. Jensen, A. Kjoller, and L. H. Sorenson (eds.), *Microbial Communities in Soil*. Elsevier, Amsterdam.
15. Cairns, J., Jr., and J. R. Pratt. 1986. Ecological consequence assessment: Effects of bioengineered organisms. In J. Fiksel and V. T. Covello (eds.), *Biotechnology Risk Assessment*. Pergamon Press, New York, pp. 88–108.
16. Cairns, J., Jr., and J. R. Pratt. 1987. Ecotoxical effect indices: A rapidly evolving system. *Water Sci. Technol.* 11:1–12.
17. Chen, K., H. Neimark, P. Rumore, and C. R. Stewinbman. 1989. Broad range DNA

probes for detecting and amplifying eubacterial nucleic acids. *FEMS Microbiol. Lett.* 57:19–24.

18. Corapcioglu, M. Y., and A. Haridas. 1985. Microbial transport in soils and groundwater: A numerical model. *Adv. Water Resources* 8:188–200.

19. DeLong, E. F., G. S. Wickham, and N. R. Pace. 1989. Phylogenetic stains: Ribosomal RNA-based probes for the identification of single cells. *Science* 243:1360–1363.

20. Domsch, K. H., A. J. Driesel, W. Goebel, W. Andersch, W. Lindenmaier, W. Lotz, H. Reber, and F. Schmidt. 1988. Considerations on release of gene-technologically engineered microorganisms into the environment. *FEMS Microbiol. Ecol.* 53:261–272.

21. Domsch, K. H., G. Jagnow, and T.-H. Anderson. 1983. An ecologic concept for the assessment of side-effects of agrochemicals on soil microorganisms. *Residue Rev.* 86: 65–105.

22. Doyle, J., R. Jones, M. Broder, and G. Stotzky. 1988. Effects of genetically engineered microbes on microbe-mediated ecological processes in soil. *Abstracts of the Annual Meeting of the American Society for Microbiology*, Q5:285.

23. Festl, H., W. Ludwig, and K. H. Schleifer. 1986. DNA hybridization probe for the *Pseudomonas fluorescens* group. *Appl. Environ. Microbiol.* 52:1190–1194.

24. Findlay, S. E. G., J. L. Meyer, and R. T. Edwards. 1984. Measuring bacterial production via rate of incorporation of ^{3}H-thymidine into DNA. *J. Microbiol. Methods* 2:57–72.

25. Fisher, S. W., and J. D. Briggs. 1988. Environmental and ecological problems in the introduction of alien microorganisms in soil. *Agric. Ecosyst. Environ.* 24:325–335.

26. Gillett, J. W., and J. M. Witt. 1979. Terrestrial microcosms. NSF-79-0024, National Science Foundation, Washington, D.C.

27. Giovannoni, S. J., E. F. DeLong, G. J. Olsen, and N. R. Pace. 1988. Phylogenetic group-specific oligodeoxynucleotide probes for identification of single microbial cells. *J. Bacteriol.* 170:720–726.

28. Greaves, M. P. 1982. Effect of pesticides on soil microorganisms. In R. G. Burns and J. H. Slater (eds.), *Experimental Microbial Ecology*, Blackwell, Oxford, pp. 613–630.

29. Henson, J. M. 1989. DNA probe for identification of the take-all fungus, *Gaeumannomyces graminis. Appl. Environ. Microbiol.* 55:284–288.

30. Hicks, R. J., G. Stotzky, and P. Van Voris. 1989. Review and evaluation of the effects of xenobiotic chemicals on microorganisms in soil. *Adv. Appl. Microbiol.* 35. In press.

31. Holben, W. E., J. K. Jansson, B. K. Chelm, and J. M. Tiedje. 1988. DNA probe method for the detection of specific microorganisms in the soil bacterial community. *Appl. Environ. Microbiol.* 54:703–711.

32. Jackson, D. R., and M. Levin. 1979. Transport of arsenic in grassland microcosms and field plots. *Water Air Soil Pollut.* 11:3–12.

33. Jackson, D. R., C. D. Washburne, and B. S. Ausmus. 1977. Loss of Ca and NO_3-N from terrestrial microcosms as an indicator of soil pollution. *Water Air Soil Pollut.* 8:279–284.

34. Jeffrey, W. H., and J. H. Paul. 1988. Underestimation of DNA synthesis by [3H]thymidine incorporation in marine bacteria. *Appl. Environ. Microbiol.* 54: 3165–3168.

35. Jenkinson, D. S., and J. N. Ladd. 1981. Microbial biomass in soil: Measurement and turnover. In E. A. Paul and J. N. Ladd (eds.), *Soil Biochemistry*, Vol. II, Marcel Dekker, New York, pp. 415–471.

36. Johnston, J. B., and S. G. Robinson. 1984. *Genetic Engineering and Pollution Control.* Noyes Publications, Park Ridge, N.J.

37. Karnak, R. E., and J. L. Hamelink. 1982. A standardized method for determining the acute toxicity of chemicals to earthworms. *Ecotoxicol. Environ. Safety* 6:216–222.

38. Levin, S. A., and M. A. Harwell. 1986. Potential ecological consequences of genetically engineered organisms. *Environ. Manag.* 10:495–513.

39. Lindow, S. E. 1985. Ecology of *Pseudomonas syringae* relevant to the field use of ice-deletion mutants constructed in vitro for plant frost control. In H. O. Halvorson, D. Pramer, and M. Rogul (eds.), *Engineered Organisms in the Environment: Scientific Issues.* American Society for Microbiology, Washington, D.C., pp. 23–35.

40. Lindow, S. E. 1987. Competitive exclusion of epiphytic bacteria by ice-*Pseudomonas syringae* mutants. *Appl. Environ. Microbiol.* 53:2520–2527.
41. Malanchuk, E. A., and K. Joyce. 1983. Effects of 2,4-D on nitrogen fixation and carbon dioxide evolution in a soil microcosm. *Water Air Soil Pollut.* 20:181–189.
42. Milewski, E. A. 1985. Field testing of microorganisms modified by recombinant techniques: Applications, issues, and development of "points to consider." *Recomb. DNA Tech. Bull.* 8:102–108.
43. Minogue, K. P., and W. E. Fry. 1983. Models for the spread of disease: Model description. *Phytopathology* 73:1168–1173.
44. National Academy of Sciences. 1987. Introduction of recombinant DNA-engineered organisms into the environment: Key issues. National Academy Press, Washington, D.C.
45. Odum, E. P. 1962. Relationships between structure and function in ecosystems. *Jpn. J. Ecol.* 12:108–118.
46. Odum, E. P. 1985. Biotechnology and the biosphere. *Science* 229:1338.
47. Ogram, A., G. S. Sayler, and T. Barkay. 1987. The extraction and purification of microbial DNA from sediments. *J. Microbiol. Methods* 7:57–66.
48. O'Neill, E. O., M. Hood, C. Cripe, and P. Pritchard. 1988. Field calibration of aquatic microcosms. *Abstracts of the Annual Meeting of the American Society for Microbiology,* Q-89, p. 297.
49. O'Neill, R. V., B. S. Ausmus, D. R. Jackson, R. I. Van Hook, P. Van Voris, C. Washburne, and A. P. Watson. 1977. Monitoring terrestrial ecosystems by analysis of nutrient export. *Water Air Soil Pollut.* 8:271–277.
50. Parkinson, D., and E. A. Paul. 1982. Microbial biomass. In A. L. Page, R. H. Miller, and D. R. Keeney (eds.), *Methods of Soil Analysis, Part 2, Chemical and Microbiological Properties,* 2d ed. American Society of Agronomy, Madison, Wis., pp. 821–830.
51. Petersen, H., and M. Luxton. 1982. A comparative study of soil fauna populations and their role in decomposition processes. *Oikos* 39:287–422.
52. Rafii, F., D. H. Crawford, B. H. Bleakley, and Z. Wang. 1988. Assessing the risks of releasing recombinant *Streptomyces* in soil. *Microbiol. Sci.* 5:358–361.
53. Ramos, J. L., A. Wasserfallen, K. Rose, and K. N. Timmis. 1987. Redesigning metabolic routes: Manipulation of TOL plasmid pathway for catabolism of alkylbenzoates. *Science* 235:593–596.
54. Rissler, J. 1984. Research needs for biotic environmental effects of genetically engineered microorganisms. *Recomb. DNA Tech. Bull.* 7:20–30.
55. Sayler, G. S., T. W. Sherrill, R. E. Perkins, L. M. Mallory, M. P. Shiaris, and D. Pedersen. 1982. Impact of coal-coking effluent on sediment microbial communities: A multivariate approach. *Appl. Environ. Microbiol.* 44:1118–1129.
56. Scanferlato, V. S., D. R. Orvos, J. Cairns, Jr., and G. H. Lacy. 1989. Genetically engineered *Erwinia caratovora* in aquatic microcosms: Survival and effects on functional groups of indigenous bacteria. *Appl. Environ. Microbiol.* 55:1477–1482.
57. Schmidt, E. L., and L. W. Belser. 1982. Nitrifying bacteria. In A. L. Page, R. H. Miller, and D. R. Keeney (eds.), *Methods of Soil Analysis, Part 2, Chemical and Microbiological Properties,* 2d ed. American Society of Agronomy, Madison, Wis., pp. 1027–1042.
58. Silver, M., H. L. Ehrlich, and K. C. Ivarson. 1986. Soil mineral transformations mediated by soil microbes. In P. M. Huang and M. Schnitzer (eds.), *Interactions of Soil Minerals with Natural Organics and Microbes.* Soil Science Society of America Special Pub. No. 17. Madison, Wis., pp. 497–514.
59. Sommerville, C. C., I. T. Knight, W. L. Straube, and R. R. Colwell. 1989. Simple, rapid method for direct isolation of nucleic acids from aquatic environments. *Appl. Environ. Microbiol.* 55:548–554.
60. Steffan, R. J., J. Goksoyr, A. K. Bej, and R. M. Atlas. 1988. Recovery of DNA from soils and sediments. *Appl. Environ. Microbiol.* 54:2908–2915.
61. Stevenson, F. J. 1986. *Cycles of Soil: Carbon, Nitrogen, Phosphorus, Sulfur, and Micronutrients,* Wiley, New York.

62. Stoudt, J. D., S. S. Bamforth, and J. D. Lousier. 1982. Protozoa. In A. L. Page, R. H. Miller, and D. R. Keeney (eds.), *Methods of Soil Analysis, Part 2, Chemical and Microbiological Properties,* 2d ed. American Society of Agronomy, Madison, Wis., pp. 1103–1120.
63. Tiedje, J. M., R. K. Colwell, Y. L. Grossman, R. E. Hodson, R. E. Lenski, R. N. Mack, and P. J. Regal. 1989. The release of genetically engineered organisms: A perspective from the Ecological Society of America. *Ecology* 70:298–315.
64. United States General Accounting Office. 1988. Managing the risks of field testing genetically engineered organisms. Report to the Chairman, Subcommittee on Oversight and Investigations, Committee on Energy and Commerce, House of Representatives. GAO/RCED-88-27. U.S. General Accounting Office, Gaithersburg, Md.
65. Van Voris, P. 1988. Standard guide for conducting a terrestrial soil-core microcosm test. Standard No. E-1197-87. Annual book of ASTM Standards 1104:743–755.
66. Wainwright, M. 1978. A review of the effects of pesticides on microbial activity in soils. *J. Soil Sci.* 29:287–298.
67. Watrud, L. S., F. J. Perlak, M. T. Tran, K. Kusabno, E. J. Mayer, M. A. Miller-Wideman, M. G. Obukowicz, D. R. Nelson, J. P. Kreitinger, R. J. Kaufman. 1985. Cloning of the *Bacillus thuringiensis* subsp. *kurstaki* delta-endotoxin gene into *Pseudomonas fluorescens:* Molecular biology and ecology of an engineered microbial pesticide. In H. O. Halvorson, D. Pramer, and M. Rogul (eds.), Engineered Organisms in the Environment: Scientific Issues, American Society for Microbiology, Washington, D.C., pp. 40–46.
68. Weigert, R. G. 1986. Ecosystem structural and functional analysis. In J. Fiksel and V. T. Covello (eds.), *Biotechnology Risk Assessment.* Pergamon Press, New York, pp. 129–143.
69. White, D. C., J. S. Nickels, J. H. Parker, R. H. Findlay, M. J. Gehron, G. A. Smith, and R. F. Martz. 1984. Biochemical measures of the biomass, community structure, and metabolic activity of the ground water microflora. In C. H. Ward, W. Geiger, and P. L. McCarty (eds.), *Ground Water Quality.* Wiley, New York, pp. 307–329.
70. Wooley, T. A. 1982. Mites and other soil microarthropods. In A. L. Page, R. H. Miller, and D. R. Keeney (eds.), *Methods of Soil Analysis, Part 2, Chemical and Microbiological Properties.* 2d ed. American Society of Agronomy, Madison, Wis., pp. 1131–1142.
71. Yates, J. R., J. H. Lobos, and D. S. Holmes. 1986. The use of genetic probes to detect microorganisms in biomining operations. *J. Indust. Microbiol.* 1:129–135.
72. Yeates, G. W. 1981. Nematode populations in relation to soil environmental factors: A review. *Pedobiologia* 22:312–338.

Chapter 29

Methods for Evaluating the Effects of Microorganisms on Biogeochemical Cycling

James K. Fredrickson

Harvey Bolton, Jr.

Guenther Stotzky

Introduction

The potential impacts of genetically engineered microbes (GEMs) on the structure and function of environments into which the GEMs are introduced are major aspects of the concern about the survival of, and genetic transfer by, GEMs in these environments. If a GEM survives in the habitat into which it is introduced, carries out the function for which it was intended, and if the novel gene(s) is transferred to indigenous microbes, there should be little cause for concern unless the novel gene(s), either in the introduced GEM or in an indigenous recipient(s), results in some unexpected impacts on the habitat. Although this concept is easy to state, it is difficult to translate into an effective experimental design. What effects (i.e., environmental perturbations) should be sought, especially if the novel gene(s) codes for a limited function(s) and the GEM has been selected or altered for poor survival in that habitat? Considering the current state of the art and the paucity of data on detection, enumeration, survival, growth, and transfer of genetic information (both intraspecifically and interspecifically) by GEMs in natural habitats, the detection, measurement, and evaluation of potential effects of an introduced GEM on ecologic processes is analogous to finding "a needle in a haystack." However, as the effects of GEMs on ecologic processes is a major concern, more studies on this aspect must be considered. It is important to emphasize that inasmuch as insufficient basic data are available about the fate of GEMs in natural habitats, data from studies on the ecologic effects of GEMs must be interpreted and applied cautiously to avoid far-reaching and long-lasting policies, criteria, and regulations that may be based on incomplete and erroneous data.

Processes that involve the cycling of nutrients and energy are fundamental properties of ecosystems and are critical for their continued function. Dysfunction in the rate of processing nutrients and energy is known to have adverse impacts on ecosystems (65). Because of the functional redundancy of many microbially mediated ecologic processes, a detrimental impact on a single species or taxa may not necessarily result in a direct impact on an ecologic process. However, the opposite is not true. Alterations in the processing of nutrients and energy in an ecosystem are likely to affect the structure of the biological community (56). Therefore, in addition to effects on individual species and communities, the impact of recombinant or novel microorganisms on nutrient and energy cycling processes can complement the assessment of the environmental risks resulting from their release.

The majority of nutrient cycling processes are mediated by microorganisms and may be susceptible to change from the introduction of microorganisms, particularly those microorganisms that are altered in their ability to process nutrients. Numerous recombinant microorganisms have already been developed that are altered in the way they process nutrients. These microorganisms include those with enhanced N_2-fixation capabilities (108) and improved ability to degrade organics, either natural compounds (73) or contaminants (74). Technology is also available to amplify cloned genes to give stable high-copy numbers in microorganisms such as *Streptomyces* (1). This provides the GEMs with the capacity to produce high levels of specific enzymes and, hence, the potential for increased rates of nutrient processing in some GEMs. It is likely that the deliberate release of microorganisms with specialized or enhanced abilities to process nutrients will increase in the future.

In addition to evaluating the potential effects of GEMs on ecologic processes related to their intended function, scientists should be alert to the possible occurrence of unanticipated effects that cannot be predicted from the information encoded on the novel DNA (i.e., pleiotropic effects). For example, the acquisition of a plasmid harboring the genes for N_2-fixation and antibiotic resistance by various species of phytopathogenic bacteria apparently resulted in a spectrum of unrelated, unanticipated, and unpredicted biochemical and physiological alterations (54, 99). Other pleiotropic effects have also been observed and reported (98). If pleiotropic effects are indicated, the battery of tests performed on GEMs to determine ecologic effects should be extended before the microorganisms are released to the environment because unanticipated alterations could affect ecologic processes in soil and other environments. Furthermore, the growth rates of various GEMs, as well as of the homologous host bacteria, and their ability to compete with indigenous microbes in soil should be determined.

Nutrient and energy cycles are intimately linked, and an effect on one process often results in changes in a number of other processes. For example, an increase in the rate of mineralization of organic C in soil could also result in an enhanced release of N (i.e., mineralization) or a decrease in inorganic N (i.e., immobilization). Therefore, methods need to be developed that can evaluate the effects of microorganisms on a variety of nutrient cycling processes concurrently. These studies should be conducted initially in the laboratory, because of the potential risks associated with the release of an untested GEM to the environment. A variety of terrestrial microcosms that purportedly simulate field conditions have been developed. These microcosms range from extremely simple systems that inoculate a microorganism into sterile soil in test tubes, for example (112); to nonsterile soil in a test tube, flask, jar, or other container (102); to multiple containers of nonsterile soil enclosed within a larger container (96); to more complex systems. These complex systems may consist of undisturbed soil cores of varying size that are brought into the laboratory with minimum disturbance of the structure and biotic composition of the soil (9, 29, 40, 109), or may contain undisturbed or mixed soils that are cropped and maintained within chambers that enable the control of temperature, relative humidity, light and dark cycles, and other environmental variables (4, 34, 52). Examples of microcosms with varying degrees of complexity and the rationales for their use have been discussed in previous reports (5, 31, 35, 38, 72, 101). Many techniques that are available for measuring ecologic processes related to nutrient and energy cycling have been developed and/or refined as a result of their use to evaluate the effects of xenobiotics on these same processes. A considerable amount of information is available on the toxicity of heavy metals toward microbe-mediated ecologic processes such as primary productivity, C mineralization, various N transformations, mineralization of organic P and S, and litter decomposition (8). These same processes are also subject to impact from toxic organic chemicals, such as pesticides (23). Procedures that may be used for evaluating the effects of pesticides on microbial processes in soil have been outlined by Greaves (37). These processes vary in their sensitivity to impacts from xenobiotics and in their reproducibility (40), but well-developed methods are available for their measurement. Unfortunately, relatively few studies have assessed the effects of introduced GEMs on such processes. However, as these processes are fundamental to the functioning of ecosystems, they provide a relevant basis for evaluating the influence of the products of microbial biotechnology on ecologic processes.

Although microorganisms mediate the transformation of a variety of nutrients in biogeochemical cycles (e.g., C, N, P, S, trace elements), the focus of this chapter will be on specific methods for measuring C and N cycling processes, as these processes are most often cited as the major factors that control ecosystem productivity. Also discussed will be nutrient export from microcosms as indicators of ecologic effects of introduced microorganisms, and the potential application of the techniques of molecular microbial ecology in conjunction with function-specific gene probes.

General Considerations

Primary productivity is the process by which some members of the biota use light or chemical energy to fix CO_2 during photosynthesis or chemoautotrophy and incorporate nutrients and water into biomass. In the terrestrial environment, higher plants are mainly responsible for this function, whereas in aquatic systems, planktonic species are the major photosynthesizers. These biomass sources are the major energy source for heterotrophic organisms, where the higher trophic levels include various animals. Both plants and animals are eventually decomposed by aerobic and anaerobic microorganisms, and their C is converted principally to microbial biomass, recalcitrant humus, and CO_2. Soil organic matter and marine humus account for approximately 2.5×10^{15} kg and 3×10^{15} kg, respectively, of the global C while atmospheric reservoirs of CO_2 account for 7×10^{14} kg (93). A simplified version of the C cycle is shown in Fig. 29.1. The rate of microbial decomposition of organic matter is an important ecosystem characteristic and any change in the rate of this process would certainly be regarded as a perturbation.

Other than C, no single nutrient element is as intimately associated with the biota than N. Although a number of transformations in the overall N cycle are abiotic, the vast majority of the transformations are mediated by microorganisms. Because of the importance of these processes, that is, the potential benefit that can be derived from enhancing or altering of these properties through genetic manipulation, there is a need to consider the effects that introduction of such microorganisms may have on N transformations. Fortunately, because of the importance of N in plant growth and because it is often the primary nutrient limiting plant growth in many ecosystems, N transformations have been well studied, especially in agricultural systems. The major processes and pools in the N cycle are shown in Fig. 29.2. A

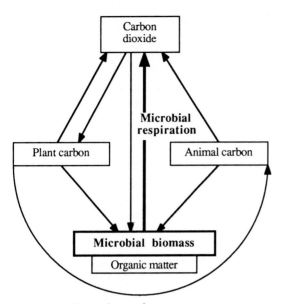

Figure 29.1   The carbon cycle.

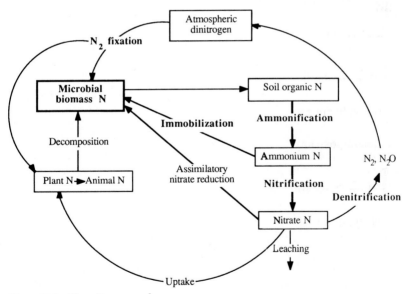

Figure 29.2   The nitrogen cycle.

number of methods have been developed to evaluate both the rate of biological processing of N as well as the direct evaluation of some of the important functional groups of microorganisms that catalyze these transformations. Numerous other nutrients including S, P, and some trace elements also undergo microbial transformations as part of their biogeochemical cycling. Methods are also available to evaluate the microbial transformation of these nutrients (69) but these will not be addressed in this chapter.

It is important to emphasize that the cycling of nutrients is a dynamic process that is influenced by a variety of abiotic factors, such as moisture, temperature, pH, Eh, and surfaces. Thus, for many nutrient cycling processes, the rate of processing, rather than single endpoints, should be measured. In many cases this requires repeatedly measuring the size of the various pools. In some instances the pool size may not change, but rates of transformation into and out of the pool are quite rapid. In these cases rate measurements can also be made by using isotopes of C and N. It is also important to include appropriate controls in the experimental design, as normal fluctuations in the rates of nutrient cycling as the result of environmental variation can be large and may dampen any effects caused by the addition of microorganisms. In addition, effects may be transient, therefore, long-term experiments should include multiple samplings to determine the length of time for which a specific process may be impacted. These variables need to be considered when choosing a method of analysis and the length of the experiment.

It is also important to keep in mind that each particular method has limitations with respect to specific applications and locations. For example, a number of methods designed for laboratory studies can be readily adapted to microcosms, but do not lend themselves well to the field. This is an important point, as certainly the need for evaluating the effects of GEMs on nutrient cycling processes in the field is likely to increase. Alternatively, a number of methods have been developed specifically for field applications, and others can be adapted with little difficulty to the field.

Although a particular process may be affected by the introduction of microorganisms into a particular ecosystem, the information needed to properly assess the risks is only partially complete. In conjunction with measuring a particular function or process (e.g., CO_2 respired, N mineralized), the population size and/or activity of the introduced microorganisms must be determined to establish relationships between a particular level of response and the concentration of the introduced organisms. The concept of ecologic dose (EcD), developed by Babich et al. (7) could be applied to analyzing the effects of GEMs on ecologic processes. (The ECD is the concentration of a toxicant that inhibits a process by some percentage.) If the introduction of a GEM results in an effect on a particular process, then dose-response curves can be constructed to identify the particular concentration of microbes or activity levels that are necessary to exert a particular effect (i.e., reduction or enhancement). Of course the EcD will apply only to the particular environment being studied because physicochemical environmental factors have a profound impact on the extent to which a particular process is affected. An example of the influence a physicochemical environment can exert on microbe-mediated effects is the variation in inhibition of microbial respiration in soil by Pb (22). The differences in inhibition of respiration between various Pb-containing soils were attributed to variations in the cation exchange capacities of the soils.

In contrast to measuring the actual flux of nutrients through various pools, techniques now exist that could be used to measure the genetic potential for the processing of nutrients. This may be achieved by hybridization analysis of microbial DNA using function-specific DNA probes for gene sequences coding for enzymes or pathways that are involved in the processing of nutrients. This type of analysis can provide complementary information to the actual measurement of nutrient cycling processes. If an effect is observed in a cycling process, hybridization analysis using function-specific probes allows for determining whether the genes involved in the function are absent, or present and declining or increasing in copy number. This also provides information on the potential for the ecosystem to recover the function.

Available Methods

Carbon cycling

In soil, the decomposition of organic matter is often equated with respiration, a process for which a number of methods of measurement are available. Some of these methods are automated and, therefore, ideal for obtaining rate information. Respiration from soil and sediments can be readily measured using a variety of techniques that monitor either CO_2 evolution or O_2 consumption. These methods probably provide the best and most easily measured index of the cumulative activity of mixed microbial populations in soil (2, 94, 95, 97). Respiration can be measured by a variety of means, including an apparatus described by Nordgren (66) that can continuously monitor the respiration of a large number of soil samples for long periods of time and is based on the measurement of changes in the conductivity of a hydroxide solution as it accumulates carbonate ions from respiring soil. This system is controlled by a microcomputer that also accumulates the data and calculates the respiration rate. Brookes and Paul (16) also developed a computer-controlled automated system to measure soil respiration. Their design allows for 36 samples to be monitored for CO_2 evolution by flushing the head space gas into a gas chromatograph for CO_2 analysis. Soil CO_2 evolution can be monitored every 75 minutes, or at even longer time intervals. Electrolytic respirometry (61) is also a convenient way to measure respiration in soil and the units are easily constructed and operated. The "master jar" system (95) enables the removal of soil subsamples during extended incubations for various analyses (e.g., transformation of substrates, species diversity, enzyme activities, and gene transfer) without disturbing the remainder of the soil and, thereby, eliminating artifactual peaks in CO_2 evolution that can result from the physical disturbance caused by sampling (100, 101). The soils are incubated under the desired conditions, and the amount of CO_2, trapped in NaOH is determined, after precipitation of the CO_2 with $BaCl_2$, by automatic potentiometric titration with HCl. Additional methods that involve the measurement of O_2 and CO_2, some of which can be used in the field, are described in detail elsewhere (2, 96).

Dehydrogenase activity has also been used as a general indicator of microbial activity in soil (64), as it is an enzyme involved in microbial respiratory processes necessary for the metabolism of C compounds. Dehydrogenase activity in soil has been shown to be a sensitive indicator of the ecological impacts of smoke (57), heavy metals (70), and other ions (76). Several procedures for measuring the level of dehydrogenase activity in soil are available (50, 103), that are based on the reduction of 2,3,5-triphenyltetrazolium chloride (TTC) by dehydrogenases under anoxic conditions to the colored 2,3,5-triphenylformazan (TPF). The TPF is extracted with methanol and its concentration is determined spectrophotometrically by comparison with a standard curve. Other enzyme activities that can be evaluated in soil include acid and alkaline phosphatases, arylsulfatases, and urease. The activities of phosphatases can be used to estimate the mineralization of organic phosphates by measuring colorimetrically the release of ρ-nitrophenol from ρ-nitrophenylphosphate: a buffer at pH 6.5 is used to assay the acid phosphatase activity (104), and a buffer at pH 11 is used to assay the alkaline phosphatase activity (26). The activity of arylsulfatases, which catalyze the hydrolysis of arylsulfate anions, can be used to estimate the mineralization of organic S by measuring colorimetrically the release of ρ-nitrophenol from potassium ρ-nitrophenol sulfate (105). The measurement of urease activity in soil, which catalyzes the hydrolysis of urea to CO_2 and NH_3, is based on the determination of NH_4^+ in soil incubated with tris (hydroxylmethyl) aminomethane (THAM) buffer, urea, and toluene (103). Evaluation of activities of additional enzymes, e.g., amylases (21), cellulases (85), proteases (18), nucleases [to provide information on the possibility of transforming DNA persisting in soil, see (103)], should also be considered. The activities of these and other hydrolytic enzymes should be evaluated when appropriate substrates (e.g., starches, celluloses) are added to soil. Specific enzyme activities in different soils can be compared by using the formula (1-B/A) 100, where A is the activity in the control soils and B is the activity in soils inoculated with a GEM or homologous host.

The metabolic activity of specific components of the soil microbiota can be evaluated by the addition of specific substrates (e.g., cellulose, starches, lipids, proteins, aldehydes) whose mineralization is dependent on the ability of these microbial components to synthesize the appropriate enzymes. The measurement of the decomposition of these materials can be greatly enhanced by the use of ^{14}C-labeled material. In laboratory or microcosm incubations, the mineralization of the amended material is readily determined by trapping respired CO_2 in alkali and then measuring the amount of ^{14}C by liquid scintillation counting. Methods and vessels for conducting such incubations have been described by Anderson (2). A convenient flask system for evaluating the degradation of ^{14}C-labeled compounds in soil is described by Loos et al. (58). Labeled materials may also be used in field experiments. If the CO_2 is not trapped in alkali the soil must be subsampled and the total C oxidized to CO_2, and ^{14}C measured to determine the amount of parent material remaining. Because all or portions of the ^{14}C remaining in the soil may be metabolites or incorporated into microbial biomass (depending on the material amended), methods for extraction and analysis of the parent material by direct means should also be used.

Methods are also available for determining the anaerobic degradation of labeled compounds. The simplest and most widely used methods for the cultivation of anaerobic bacteria employ sealed serum bottles (42). Shelton and Tiedje (82) developed a method that utilized serum bottles to test the susceptibility of organic chemicals to anaerobic degradation in sewage sludge by measuring gas production. This method could be adapted to evaluate the effects of GEMs on anaerobic biodegradative processes.

The microbial decomposition of natural C compounds, such as cellulose, have been used in past ecotoxicologic evaluations of the effects of chemicals on microbially mediated processes (28). Measurements of microbial decomposition could be readily adapted to evaluate the effects that may result from the introduction of microorganisms. In addition to the metabolism of naturally occurring C compounds, the ability of a microbial community to process organic contaminants or pesticide residues is also an important ecologic function that can be evaluated. In particular, soils could be amended with the specific substrate (e.g., toluene, xylenes, 2,4-dichlorophenoxyacetic acid) of the gene product of an introduced GEM transforms. This approach would determine whether the novel gene provides a specific advantage to the GEM and the effects on both nonspecific and specific metabolic activities and other microbe-mediated ecologic processes (25). When the GEM contains genes that confer resistance to heavy metals, the soils can be amended with the appropriate metal to determine whether such a stress (simulated "worst-case scenario") confers an ecologic advantage on the GEM and whether this advantage, in turn, affects that activity and population dynamics of the indigenous microbiota. The presence or survival of the introduced GEMs and of the novel genes (i.e., in the event of transfer to indigenous soil microorganisms) should be evaluated in the subsamples of soil from the incubation vessels or "master jars" (e.g., by phenotypic characterization on selective media, DNA fingerprinting, DNA probes) to establish whether changes in metabolic activity and other processes are, in fact, related to the presence of the GEMs or the novel genes.

The soil microbial biomass is the living fraction of the soil organic matter, excluding roots and soil animals larger than approximately 1 mm^3 (46). Because the microbial biomass has a critical role in the cycling of energy, carbon, and other nutrients, different methods for measuring biomass have been developed. Briefly, methods to measure microbial biomass include direct microscopic observation (48, 81), soil ATP content (71, 110), respiratory response to amendment with a readily metabolizable C source (3), and chloroform fumigation (48, 71). Detailed discussion of these techniques and their advantages and disadvantages are presented in several review articles (46, 86).

The chloroform fumigation technique (47) to measure microbial biomass is especially amenable to biogeochemical cycling studies and has been used to quantitate both the C and N contents of the microbial biomass. This technique involves fumigating a soil with chloroform to kill the majority of the microbial biomass, removing the chloroform, incubating the soil for 10 days, and measuring the CO_2 evolved (or N mineralized) from the killed biomass by cells that survive the fumigation. The initial amount of biomass C or N is calculated by subtracting a control that consists of the CO_2 evolved (or N mineralized) from nonfumigated soil incubated for 0 to 10 days (71), nonfumigated soil incubated for 10 to 20 days (47), or no control (111) from the amount of C or N mineralized following fumigation. Microbial biomass, as measured by the chloroform fumigation technique, has been shown to be sensitive to the addition of some chemicals and to fluctuations in environmental conditions; for example, metals (14), tillage (24), low-input agricultural practices (11), and seasonal variations (11) can significantly affect the size of the microbial biomass in soil. The effect of additions of microorganisms on the soil microbial biomass could be used as an indicator of ecosystem effects. An introduced microorganism that significantly impacts the microbial biomass could also impair the cycling of C and other nutrient processes in soil, as the result of the integral role the microbial biomass plays in these cycles.

Nitrogen cycling

Many of the key processes of the N cycle are catalyzed by microorganisms. These processes include the fixation of atmospheric N_2 by bacteria, some of which form symbiosis with plant roots; the mineralization of organic forms of N to NH_4^+ (ammonification); the oxidation of NH_4^+ to NO_2 and NO_3^- (nitrification); the reduction of NO_3 to gaseous products including N_2, NO, and N_2O (denitrification); and the incorporation of N into microbial biomass (immobilization, or if NO_3^-, assimilatory NO_3 reduction) (Fig. 29.2). Specific methods to evaluate the effects of GEMs are available for most of these processes. For a more in-depth discussion of the N cycle, the reader is referred to Stevenson (92).

Many bacteria, e.g., cyanobacteria (blue-green algae), actinomycetes, and symbiotic and free-living bacteria, which occupy a broad spectrum of niches ranging from the rhizosphere to water and sediments, have the capacity to fix N_2 to a form that is biologically available. These bacteria are physiologically diverse and range from aerobes to microaerophiles to anaerobes. It is estimated that each year the total amount of N_2 converted to NH_3 through biological fixation is 1.39×10^{11} kg (93). Approximately 65 percent of this amount is from the fixation by bacteria (*Rhizobium* sp.) in symbiotic association with leguminous plants. Rhizobia, and their genes coding for nitrogenase, the enzyme system responsible for N_2 fixation, are obvious candidates for genetic manipulation to improve their capacity for N_2 fixation or infectivity and nodulation of plants, and there is considerable scientific information regarding the genetics, physiology, and ecology of these bacteria.

A sensitive assay has been developed for evaluating N_2 fixation, based on the reduction of acetylene (C_2H_2) to ethylene (C_2H_4) by the nitrogenase enzyme system. The K_m for acetylene is considerably less than for N_2, simplifying this assay considerably as there is no need to conduct the assay in the absence of N_2. The theoretical conversion factor for moles of C_2H_2 reduced to moles of N_2 fixed is 3.0; however, a number of factors can influence this factor making it necessary to use the fixation of $^{15}N_2$ for calibration. The reduction of $^{15}N_2$ to $^{15}NH_4^+$ has also been used to assay for N_2 fixation (17). However, this method is considerably less sensitive than the C_2H_2 reduction technique and requires mass or optical emission spectroscopic analysis rather than the more easily used gas chromatography analysis required for ethylene. Because of the relatively high sensitivity of this method, it can be used for pure cultures and environmental samples; however, the reliability of this technique for field studies has been questioned. The C_2H_2-reduction technique can be used in conjunction with the most-probable-number procedures for specific functional groups of N_2-fixing bacteria (51), and intact soil cores or plant roots. A continuous-flow method has also been devised that enables the measurement of N_2-fixation directly in the field based on C_2H_2 reduction (39). Soil can either produce or utilize C_2H_4 and appropriate controls must be included in experiments utilizing the C_2H_2-reduction technique (13).

The process whereby organic N is converted to NH_4^+ is termed ammonification and is mediated by heterotrophic microorganisms, both aerobic and anaerobic. Ammonification combined with nitrification (the oxidation of NH_4^+ to NO_2^- and NO_3^-), which is mediated primarily by autotrophic bacteria, is termed mineralization (93). The incorporation of inorganic N into microbial biomass is termed immobilization. Simple determinations of changes in concentrations of NH_4^+ and NO_3^- in soil will not allow for calculations of mineralization or immobilization rates but only provides a "net" mineralization or immobilization value depending on whether the concentrations of NH_4^+ or NO_3^- are increasing or decreasing. The analyses of mineralization-immobilization reactions in soils and sediments have been aided by the use of ^{15}N as a tracer. If the soil inorganic pool (NH_4^+ or NO_3^-) is labeled with ^{15}N, immobilization can be demonstrated by the appearance of ^{15}N in the organic pools and can aid in the determination of N transformation rates. Conversely, mineralization can be demonstrated if the organic pool is labeled with ^{15}N and ^{15}N appears in the inorganic pools (45). Without the use of ^{15}N as a tracer it is possible only to distinguish net mineralization from net immobilization.

One or more pools in the N cycle (Fig. 29.2) can be labeled with ^{15}N, and N transformations can be followed over time to derive rates of transformations between pools, as pool sizes are measured and their ^{15}N content calculated. Thus, ^{15}N-labeled organic compounds; ^{15}N-labeled microbial biomass, $^{15}NO_3^-$, or $^{15}NH_4^+$ can be added to the soil and the various N pools assayed over time for total N and ^{15}N content. This allows for the sensitive measurement of the rate of N transformations and changes in sizes of the various N pools.

The microbial biomass not only catalyzes many N transformations in soil, but it is also a major pool of soil N that can exhibit a rapid turnover. The chloroform fumigation technique has gained wide acceptance as a method for measuring soil microbial biomass C, as the N and ^{15}N contents of the soil microbial biomass can be determined at the same time (47), as described earlier. Several direct extraction techniques for determining biomass N are available (15, 87), but these procedures require that the fumigation-incubation procedure be used for calibration.

Methods to determine the mineralization of soil organic N, or the ability of a soil to supply N for plant growth, have been developed by agricultural scientists to estimate the amount of fertilizer N required for a particular crop. Several of these methods may have utility for determining the effects that the addition of GEMs may have on the ability of a soil or sediment to transform organic N to inorganic forms. A number of biological methods based on short-term incubations (7 to 25 days) under aerobic or anaerobic conditions have been developed that are suitable for assessing the ability of a soil to provide N for plant growth (40). It should be stressed that only "net" mineralization will be measured by these methods and that all forms of inorganic N should be measured (NH_4^+, NO_2^-, and NO_3^-) (88). Stanford et al. (89) developed a procedure to determine the labile soil N pool or the N mineralization potential (N_o) of a soil which represents N that will become available during the growing season and a rate constant (k) for the process. In this procedure, inorganic N is removed from the soil by leaching with 0.01 M $CaCl_2$ at selected time intervals for up to 30 weeks, although shorter time periods of 11 to 12 weeks have also been used (27, 83). The N_o and k are soil characteristics that theoretically represent the amount and rate of N that will be supplied over the long term and, therefore, are important ecosystem functional parameters that may be useful for evaluating the effects of GEMs on nutrient cycling processes.

The amount of N mineralized in soil may also be estimated by constructing a nutrient balance or budget for the system under study (32). The amount of mineral N initially present in the soil, the amount removed by plants (assimilation) during the experimental period, and the residual amount at the end of the plant growth period or experiment must be determined. The difference between the amount of N initially present and that present in the soil at the end of the experiment plus that removed by plants is the amount mineralized. A complete nutrient budget would also include inputs from N_2 fixation, atmospheric deposition, and precipitation and removals by leaching and denitrification, although experimental conditions can be manipulated to decrease these factors if desired. This nutrient budget approach and the measurement of N mineralization could be readily adapted to evaluate the fate and effects of microorganisms added.

The process of nitrification in soil is catalyzed principally by a few genera of autotrophic bacteria, although some heterotrophic bacteria and fungi have also been implicated in nitrification in soil, mainly under environmental conditions where autotrophic nitrifiers could not grow or function. The process of nitrification is important because it converts a cationic form of N (NH_4^+), which is adsorbed to predominantly negatively charged particulates in soils, to an anionic form (NO_3^-), which is more mobile in soil. In some soils this can increase the supply of N to plant roots; however, when the supply exceeds the ability of the plant community to remove it from soil, the excess is susceptible to leaching to ground water, undergo denitrification, or be immobilized into the soil microbial biomass.

Methods are available for both the enumeration of autotrophic nitrifying bacteria and for directly assaying the rate of nitrification in soils and sediments. The enumeration of autotrophic nitrifying bacteria is based on a most-probable-number (MPN) procedure that takes advantage of the chemoautotrophic nature of these bacteria (80). Defined mineral media with NH_4^+ for NH_4^+-oxidizing bacteria and NO_2^- for NO_2^--oxidizing bacteria are inoculated with a dilution series prepared from soil or other environmental samples. After incubation, the individual MPN tubes are tested using a colorimetric spot test for the presence of NO_2^- and/or NO_3^- as indicators of the presence of NH_4^+-oxidizing bacteria or for loss of NO_2^- for NO_2^--oxidizing bacteria, respectively. In contrast to measuring specific microbial populations that catalyze the oxidation of NH_4^+ to NO_3^-, the nitrifying potential of a soil can also be determined. In most soils, the concentration of NH_4^+ is relatively low, making it difficult to directly measure the rate of nitrification in isolation from other processes that influence the NH_4^+ and NO_3^- pools. Therefore, a nitrification potential is often determined to obtain information on the ability of a soil to convert NH_4^+ to NO_3^- over longer periods such as days or weeks. This procedure is relatively straightforward and involves the amendment of soil with a source of NH_4^+ such as ammonium sulfate, and is followed by analysis of the rate of NO_3^- formation. Methods for measuring both autotrophic nitrifying populations and nitrification potential in soil have been described by Schmidt and Belser (80). The ^{15}N pool dilution technique (53) can also be used to measure nitrification, but over short periods (e.g., 24 hours). This technique has the advantage that it can be used in situ if a low amount of total N, as NO_3^-, with a high ^{15}N enrichment is added so the soil NO_3^- pool is not substantially increased. After an incubation period (usually 24 hours), the dilution of ^{15}N in the NO_3^- pool is determined as well as the ^{15}N content of the other N pools. By knowing the initial content of the N pools before ^{15}N addition and the final N and ^{15}N contents of the N pools, nitrification can be quantified in situ. The data obtained can also be used to distinguish between immobilization of $^{15}NO_3^-$ into microbial biomass and assimilation by plant roots.

Denitrification is the process whereby facultative bacteria utilize NO_3^- as a terminal electron acceptor in the absence of O_2 during the metabolism of organic matter. True dissimilatory denitrification has N_2O and N_2 as the principal end products. A variety of genera of bacteria are capable of dissimilatory nitrate reduction (107). Advances in the methodologies for evaluating denitrification in soils and sediments have been motivated by the impacts associated with the loss from soil of N for plant growth and the potential effects of N_2O on the earth's ozone layer (60).

The most widely used procedure to measure denitrification in soil and sediments is the acetylene inhibition method, although methods that employ isotopes of N as tracers (e.g., $^{15}N_2$) have also been used in laboratory (62) and field (63) studies. Acetylene blocks the enzymatic reduction of N_2O to N_2 and, therefore, the rate of denitrification can be quantified using gas chromatography to measure the quantities of N_2O produced. Methods have been developed that employ this approach in conjunction with MPN enrichment cultures to enumerate denitrifying bacteria in laboratory incubations of soils or intact cores to directly evaluate the losses of N via denitrification and to determine the concentrations of enzymes involved in denitrification (106). The measurement of denitrification in soil cores in the laboratory using the acetylene inhibition technique requires incubation in a gastight chamber into which acetylene can be injected at a concentration of approximately 10 percent of the gas volume. The chamber headspace is then sampled periodically for N_2O. The acetylene reduction method has also been used in the field to measure denitrification (77), but the associated difficulties have limited its use.

The soil perfusion technique is an alternative technique for measuring nitrogen transformations in soil. It has been used successfully to evaluate the effects of SO_2, acid precipitation, heavy metals, clay minerals, and GEMs on N transformations in soil (10, 25, 55, 59). Soils, uninoculated or inoculated with GEMs or the homologous hosts, are amended with ammonium sulfate, simple N-containing organics (e.g., amino acids), or complex nitrogen-containing organics (e.g., protein or plant tissues), and net nitrification and nitrogen mineralization rates are determined. The perfusates are analyzed, after appropriate times, for different nitrogen fractions (i.e., NH_2, NH_4^+, NO_3^- and NO_2^-) and pH, and the soil can be analyzed for the presence of the GEMs or novel genes. Advantages of the soil perfusion technique include that it is highly sensitive, easy to sample, and, because it is a continuous system it can be used to obtain kinetic data.

Nutrient export from microcosms

The rate at which nutrients are leached from soil has been suggested as a method for assessment of environmental impacts from GEMs, as it is a holistic measurement that represents the functions of numerous microbial groups as well as abiotic processes (68). O'Neill et al. (68) found that changes in nutrient export in leachates from microcosms resulting from the addition of toxic chemicals could be detected, but that changes in population and community parameters were not evident. Similar results were reported by Jackson et al. (44) in a study that measured the efflux of Ca and NO_3^- from soil core microcosms amended with As. In addition, Jackson and Levin (43) found that intact terrestrial grassland microcosms were more sensitive to the effects of As than the field, probably the result of a greater dispersal of this element in the field. In this study, the microcosms proved to be conservative screening tools for evaluating potential ecologic effects from the dispersal of As. The export of organic carbon and inorganic N species leached from soil-core microcosms could also be used for microcosm analysis of ecosystem impacts because of the importance of C and N to ecosystem function and the extensive information available concerning the processing of these nutrients in ecosystems. The measurement of DOC concentrations has also been suggested as a means of calibrating microcosm function with the field (72).

A detailed protocol for evaluation of the environmental fate and ecologic effects of chemicals in a terrestrial soil-core microcosm has been developed by Van Voris (109) and has been adapted to evaluate the fate and effects of GEMs (9). The terrestrial soil-core microcosm was originally designed to yield information about a specific chemical in either a natural grassland ecosystem or an agricultural ecosystem planted with multiple-species crops. The microcosms consist of a 17-cm-diameter (or 25.5-cm-diameter) tube of Driscopipe containing an intact soil core (60 cm depth). The polyethylene pipe is an ultra-high-molecular-weight, high-density, nonplasticized material. It is impermeable to water, lightweight, tough, rigid, and highly resistant to acids, bases, and biological degradation. The intact soil core rests on a Buchner funnel that is covered by a thin layer of glass wool. As an alternative to glass wool, a degradation-resistant fabric such as polyester can be used to cover the bottom of the soil core. The Buchner funnels are washed with 0.1 M HCl before use and are reusable. Six to eight microcosms are typically contained in a movable cart, the dimensions of which are determined by the environmental chamber size or the maneuverability required in the greenhouse.

Soil cores can be extracted from soil in the region of interest or from a specific site where a field evaluation is planned. The intact core is extracted with a specially designed steel extraction tube (Fig. 29.3) and a backhoe. The steel extraction tube encases the polyethylene pipe to prevent the tube from warping and/or splitting under pressures created during extraction. Once the core is cut by the leading edge of the driving tube, it is forced up into the microcosm tube where it expands slightly and creates a relatively tight fit with the inside of the polyethylene pipe. A tight fit is necessary to limit movement of water along the interface between the soil core and the pipe during leaching. The soil core microcosm is removed as a single unit (soil and plastic pipe) from the extraction tube and taken to the laboratory.

Microcosms are watered as dictated by a predetermined water regime, usually established on the basis of site history, with either purified laboratory water (e.g., distilled, reverse osmosis) or with rainwater that has been collected, filtered, and stored at 4°C. Soil water potential, a key factor influencing microbial fate and survival in soil, should be carefully monitored so that the desired moisture regime can be maintained. Probes such as gypsum blocks can be used to monitor soil water content (33), and tensiometers (19) or thermocouple psychrometers (75) can be used to monitor soil water potential.

Microcosms should be leached at least once before adding microorganisms and at predetermined intervals thereafter. Natural rainfall data, however, should be used to guide selection of a leaching regime.

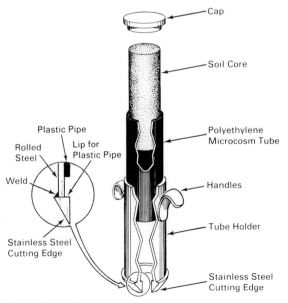

Figure 29.3   Diagram of intact soil-core microcosm coring device.

Leachate is collected in flasks washed with 0.1 M HCl after excess laboratory water or rainwater has been added to each microcosm. The 500-ml flask is attached to the Buchner funnel with tubing to facilitate leaching within a two-day period. When the microcosms are leached before planting, those that do not leach, leach too quickly, or require longer than two days to produce 100 ml of leachate after the soil has been brought to field capacity should be discarded.

Temperatures in environmental chambers and greenhouses are designed to approximate outdoor temperatures that occur during a typical growing season in the region of interest. If the experiment is not conducted in the greenhouse during the normal agricultural growing season, then lights that are suitable for plant growth and are controlled by timing devices should be used to simulate the photoperiod and intensity typical of the growing season in the area of interest. If the experiment is conducted in the greenhouse during periods when the photoperiod of the natural light is not long enough to induce flowering and seed set, then supplemental lighting should be supplied.

In a study by Fredrickson et al. (30), the utility of soil-core microcosms for evaluating the effects of microorganisms on nutrient cycling processes was determined by measuring nutrient export in the leachate. Genetically altered *Azospirillum lipoferum* was inoculated into microcosms consisting of intact soil cores (25.5-cm diameter by 60-cm deep and 17.8-cm diameter by 60-cm deep), from two different soil types, Burbank sandy loam (sandy, skeletal, mixed xeric Torriorthent) and Palouse silt loam (fine-silty, mixed, mesic Pachic Ultic Haploxerolls) and were then seeded with corn or wheat, respectively. After different times, the soil cores were collected and leachate was filtered to remove particulates and analyzed for PO_4^{3-}, SO_4^{2-}, NO_3^-, and NO_2^- by ion chromatography. Ammonium was determined with an ion-specific electrode, DOC by acidifying filtered leachate to pH 2 with phosphoric acid, sparging to remove inorganic C as CO_2, and injecting the sample into a carbon analyzer or gas chromatograph. Total carbon (TC) by direct injection into the carbon analyzer, and inorganic C was calculated as the difference between TC and DOC. The loss of purgable organics during the sparging of acidified aliquots was assumed to be negligible.

Inoculation with the genetically altered *A. lipoferum* had no significant effect on the levels of soluble nutrients in the microcosm leachates at days 33 or 67, except on day 33 the concentrations of TC and DOC were significantly higher in the controls than in the inoculated microcosms. This difference was probably the result of the addition of dead microbial biomass (i.e., heat-killed *Azospirillum* cultures) to the control microcosms. (See Tables 29.1 and 29.2.)

TABLE 29.1   Leachate Analyses (μg/ml) from Intact Soil-Core Microcosms 33 Days After Planting

Nu-trient	Maize, Burbank soil, A. lipoferum		Spring wheat, Palouse soil, A. lipoferum		Prob > F	
	Br17 Tn5[a]	Control[b]	RG20a Tn5[a]	Control	Soil	Inoculum
SO$_4$	41.02 (58.6)[c]	52.20 (62.2)	18.02 (37.9)	18.89 (34.9)	0.00	0.45
PO$_4$	1.90 (25.3)	2.34 (47.4)	0.83 (60.2)	0.96 (67.7)	0.00	0.12
NH$_4$	0.04 (50.0)	0.02 (50.0)	0.09 (200)	0.03 (33.3)	0.49	0.14
NO$_2$	0.04 (150)	0.02 (50)	0.04 (75.0)	0.05 (60.0)	0.43	0.14
NO$_3$	89.94 (28.2)	130.03 (47.6)	279.73 (44.2)	287.31 (50.0)	0.00	0.61
TC[d]	59.31 (25.3)	68.70 (12.9)	12.51 (39.6)	17.60 (42.4)	0.00	0.04
DOC[e]	16.36 (25.9)	21.42 (32.6)	5.37 (24.0)	6.91 (36.2)	0.00	0.05
DIC[f]	42.95 (29.2)	47.28 (74.3)	7.14 (60.5)	10.69 (50.7)	0.00	0.11

[a]Tn5 transposon mutants of *Azospirillum lipoferum*.
[b]Heat-killed *A. lipoferum* Tn5 mutants cells added to microcosms.
[c]() Coefficient of variation, $n = 6$.
[d]TC = total carbon.
[e]DOC = dissolved organic carbon.
[f]DIC = dissolved inorganic carbon.

TABLE 29.2   Leachate Analyses (μg/ml) from Intact Soil-Core Microcosms 67 Days After Planting

Nu-trient	Maize, Burbank soil, A. lipoferum		Spring wheat, Palouse soil, A. lipoferum		Prob > F	
	Br17 Tn5[a]	Control[b]	RG20a Tn5[a]	Control	Soil	Inoculum
SO$_4$	70.46 (77.6)[c]	23.50 (153)	22.04 (150)	3.57 (106.7)	0.07	0.09
PO$_4$	1.04 (40.4)	2.94 (40.1)	1.51 (65.6)	0.78 (43.6)	0.11	0.68
NH$_4$	0.15 (133)	0.04 (25.0)	0.04 (50.0)	0.11 (109)	0.68	0.93
NO$_2$	0.01 (100)	0.02 (50)	0.14 (136)	0.03 (66.7)	0.17	0.17
NO$_3$	0.77 (129)	0.53 (81.1)	25.98 (113)	2.24 (114)	0.12	0.10
TC[d]	86.05 (41.6)	93.25 (30.9)	28.10 (65.2)	17.00 (127)	0.00	0.70
DOC[e]	53.30 (55.6)	61.02 (41.3)	17.00 (74.4)	9.10 (129)	0.00	0.80
DIC[f]	32.75 (26.2)	32.23 (18.2)	11.11 (52.7)	7.89 (126.4)	0.00	0.56

[a]Tn5 transposon mutants of *Azospirillum lipoferum*.
[b]Heat-killed *A. lipoferum* Tn5 mutants cells added to microcosms.
[c]() Coefficient of variation, $n = 6$.
[d]TC = total carbon.
[e]DOC = dissolved organic carbon.
[f]DIC = dissolved inorganic carbon.

DNA probes for nutrient cycling functions

In contrast to measuring directly processes involved in nutrient cycling or the microbial populations that catalyze the transformation of nutrients, techniques are now available for measuring the genetic potential for such functions. If a GEM were to have a deleterious effect on nutrient cycling as determined by direct assay, it may be useful to determine whether there has been a loss of genetic potential for the particular functional attribute or whether it has simply been inhibited. The latter case would indicate that the potential for recovery of that particular function would be greater than in ecosystems where the genetic potential for it had been lost or diminished.

A number of genes that are associated with the cycling of C and N have been cloned and, therefore, can be used as DNA probes in the analysis of DNA extracted from environmental samples. DNA probes can be constructed from specific genes that code for individual proteins or catabolic operons involved in the processing of nutrients. One example are the *nif* genes that code for the structural components of the nitrogenase system. A *nif* DNA probe was used to determine if the introduction of a GEM into an aquatic microcosm had a specific effect on the presence of the DNA coding for nitrogenase (12). The *nif* genes are readily available in a number of different vectors from a number of different sources. For example, a *nifHD* probe from *Rhizobium meliloti* was used to determine the genetic diversity in *Bradyrhizobium japonicum* by determining the physical location of this gene in the genome of a number of field isolates (78).

A number of cloned genes are also available that have been cloned for enzymes involved in the catabolism of specific C compounds for use as DNA probes for evaluating the presence of these genes in environmental samples. Among these enzymes are endoglucanases (36) and xylanases (113) that catalyze the degradation of the natural C compounds, cellulose and xylan, and toluene monooxygenase (114) and toluene dioxygenase (115) that are involved in the catabolism of environmental contaminants such as toluene and some related aromatics. Due to the existence of functionally similar enzymes that differ in amino acid sequence, it may be necessary to identify highly conserved regions of these proteins in order to construct function-specific DNA probes.

Procedures for hybridization analysis of DNA by dot and Southern blots are now standard in molecular biology and details on their use can be found in a variety of sources (6). However, specific problems are associated with the quantitative extraction and purification of DNA from environmental samples. A number of methods are now available for extracting DNA molecules from soils, waters, and sediments in sufficient quantity and purity to allow for molecular analysis (41, 67, 84, 91). One of the disadvantages in using DNA probe methods is that they have limited sensitivity. The use of the *taq* polymerase chain reaction (PCR) amplification (79, 90) can increase the sensitivity of these methods by as much as a factor of 10^3 to 10^6, but the PCR method requires that at least part of the gene sequence be known so that appropriate primers can be constructed. Also, the available procedures for detecting specific DNA sequences in environmental samples are not quantitative and, therefore, do not provide information on the relative number of gene copies in a DNA sample.

This approach to the analysis of nutrient cycling processes assumes that the DNA of the particular gene being used as a function probe shares significant homology with DNA that codes for the same function in a particular environmental sample. For highly specific functions, such as N_2 fixation (nitrogenase), this is a relatively safe assumption. For example, nitrogenases from a variety of different sources have been shown to be structurally similar (20). However, for processes that may be catalyzed by a variety of enzymes such as lignin degradation, the assumption may not be valid.

The use of DNA probes for specific genes involved in the cycling of C and N is limited only by the availability of DNA sequences that contain the genes of interest. The relative ease and simplicity of most methods for the direct measurement of the cycling of N or C will probably make these the methods of choice for evaluating the effects of GEMs on nutrient cycling processes but the use of DNA probes is likely to increase in the future.

Conclusions and Recommendations

A variety of methods are currently available for evaluating the effects of GEMs on the cycling of nutrients and energy in ecosystems. The focus of this chapter was on the cycling of C and N because of the fundamental role of these elements in ecosystems and because most of their transformations are mediated by microorganisms. A majority of the methods described in this chapter were derived from the agricultural and ecotoxicologic sciences, as these methods have a history of successful use in studies designed to understand the microbial transformation of important plant nutrients and to determine the effects of chemicals on important ecologic processes. A detailed description of specific methods available for evaluating nutrient cycling processes was not the intent of this chapter; rather it was to provide an overview of the available methods, with a brief description of methods and key references that provide the detail necessary to conduct the measurements.

Redundancy, or the use of multiple analyses of ecologic function or structure, should be designed into evaluations of potential ecologic effects from GEMs. Redundancy not only provides internal controls for the validity and sensitivity of the different assays, but it will indicate those assays that best reflect the impact of an introduced GEM on microbe-mediated ecologic processes in soil. Such data should aid in the development of a battery of assays with which to evaluate the potential impacts of GEMs, both specifically and generally, on ecologic processes in soil and, perhaps, on other recipient natural habitats. The data obtained in these types of evaluations, whether positive or negative (i.e., whether or not an introduced GEM has an effect on the microbe-mediated ecologic process), should provide guidelines for the design of further investigations of introduced GEMs on the structure and function of natural or managed environments.

Microcosms, because they can simulate natural ecosystems and the nutrient cycling processes that occur in ecosystems, can be used to predict potential impacts of GEMs on nutrient cycling processes and perhaps provide insights on the underlying mechanism of the impacts before the release of GEMs to the environment. Microcosms also have the advantage that a number of techniques for measuring nutrient cycling processes in the laboratory, which may be difficult or cumbersome to adapt to the field, are readily applied in microcosms.

Although relatively new, the techniques of molecular biology, and more specifically DNA probes, offer means for determining the genetic potential for a specific process in an ecosystem, in contrast to its expression. This approach appears to be limited only by the availability of DNA sequences coding for important ecologic functions and will probably be used in conjunction with or following the application of more classical methods.

Acknowledgments

The preparation of this chapter and some of the studies discussed were supported in part by U.S. Environmental Protection Agency–Corvallis Environmental Research Laboratory under a Related Services Agreement with the U.S. Department of Energy under Contract DE-AC06-76RLO 1830 and Cooperative Agreements CR812484, CR813431, and CR813650 between the U.S. Environmental Protection Agency–Corvallis Environmental Research Laboratory to New York University. As this article has not yet been subjected to agency review and therefore does not necessarily reflect the views of the agency, no official endorsement should be inferred.

References

1. Altenburger, J., and J. Cullum. 1987. Amplification of cloned genes in *Streptomyces. Biotechnology* 5:1328–1329.
2. Anderson, J. P. E. 1982. Soil respiration. In A. L. Page, R. H. Miller, and D. R. Keeney (eds.), *Methods of Soil Analysis, Part 2, Chemical and Microbiological Properties,* 2d ed. American Society of Agronomy, Madison, Wis., pp. 831–871.
3. Anderson, J. P. E., and K. H. Domsch. 1978. A physiological method for the quantitative measurement of microbial biomass in soils. *Soil Biol. Biochem.* 10:215–223.

4. Armstrong, J. L., G. R. Knudsen, and R. J. Seidler. 1987. Microcosm method to assess survival of recombinant bacteria associated with plants and herbivorous insects. *Curr. Microbiol.* 15:229–232.

5. Atlas, R. M., and R. Bartha. 1981. *Microbial Ecology.* Addison-Wesley, Reading, Mass.

6. Ausubel, F. M., R. Brent, R. E. Kingston, D. D. Moore, J. A. Smith, J. G. Seidman, and K. Struhl. 1987. *Current Protocols in Molecular Biology.* Wiley, New York.

7. Babich, H., R. J. F. Bewley, and G. Stotzky. 1983. Application of the "ecologic dose" concept to the impact of heavy metals on some microbe-mediated ecologic processes in soil. *Arch. Environ. Contam. Toxicol.* 12:421–426.

8. Babich, H., and G. Stotzky. 1986. Heavy metal toxicity to microbe-mediated ecologic processes: A review and potential application to regulatory policies. *Environ. Res.* 36:111–137.

9. Bentjen, S. A., J. K. Fredrickson, S. W. Li, and P. Van Voris. 1989. Intact soil-core microcosms for evaluating the fate and ecological impacts from the release of genetically engineered microorganisms. *Appl. Environ. Microbiol.* 55:198–202.

10. Bewley, R. J. F., and G. Stozky. 1983. Effects of cadmium and simulated acid rain on ammonification and nitrification in soil. *Arch. Environ. Contam. Toxicol.* 12:285–291.

11. Bolton, Jr., H., L. F. Elliott, R. I. Papendick, and D. F. Bezdicek. 1985. Soil microbial biomass and selected soil enzyme activities: Effect of fertilization and cropping practices. *Soil Biol. Biochem.* 17:297–302.

12. Breen, A., B. Reynolds, L. Burtis, B. Bellew, and G. Sayler. 1988. Introduction and effects of the pSS50 degradative genotype in 4-chlorobiphenyl lake water microcosms. *Abstracts of the Annual Meeting of the American Society for Microbiology,* Q-159, p. 309.

13. Bremner, J. H. 1977. Use of nitrogen-tracer techniques for research on nitrogen fixation. In A. Ayannaba and P. J. Dart (eds.), *Biological Nitrogen Fixation in Farming Systems of the Tropics.* Wiley, New York, pp. 335–352.

14. Brookes, P. C., C. E. Heijnen, S. P. McGrath, and E. D. Vance. 1986. Soil microbial biomass estimates in soils contaminated with metals. *Soil Biol. Biochem.* 18:383–388.

15. Brookes, P. C., A. Landman, G. Pruden, and D. S. Jenkinson. 1985. Chloroform fumigation and the release of soil nitrogen: A rapid extraction method to measure microbial biomass nitrogen in soil. *Soil Biol. Biochem.* 17:837–842.

16. Brookes, P. C., and E. A. Paul. 1987. A new automated technique for measuring respiration in soil samples. *Plant Soil* 101:183–187.

17. Burris, R. H. 1972. Methodology. In A. Quispel (ed.), *The Biology of Nitrogen Fixation.* Elsevier North-Holland, New York, pp. 9–33.

18. Caplan, J. A., and J. W. Fahey. 1980. A rapid assay for the enumeration of protein degradation in soil. *Bull. Environ. Contam. Toxicol.* 25:424–426.

19. Cassell, D. K., and A. Klute. 1986. Water potential: Tensiometry. In A. Klute (ed.), *Methods of Soil Analysis. Part 1: Physical and Mineralogical Methods,* 2d ed. American Society of Agronomy, Madison, Wis., pp. 563–596.

20. Chen, J. S., J. S. Multani, and L. E. Mortenson. 1973. Structural investigation of nitrogenase components from *Clostridium pasteurianum* and comparison with similar components of other organisms. *Biochim. Biophys. Acta* 310:54–59.

21. Cole, M. A. 1977. Lead inhibition of enzyme synthesis in soil. *Appl. Environ. Microbiol.* 33:262–268.

22. Doelman, P., and L. Haanstra. 1979. Effect of lead on soil respiration and dehydrogenase activity. *Soil Biol. Biochem.* 11:475–479.

23. Domsch, K. H. 1984. Effects of pesticides and heavy metals on biological processes in soil. *Plant Soil* 76:367–378.

24. Doran, J. W. 1987. Microbial biomass and mineralizable nitrogen distributions in no-tillage and plowed soils. *Biol. Fertil. Soils* 5:68–75.

25. Doyle, J., K. Short, and G. Stotzky. 1989. Effects of *Pseudomonas putida* pPO301 (pRO103), genetically engineered to degrade 2,4-dichlorophenoxyacetic acid, on

microbe-mediated ecological processes in soil. *Abstracts of the Annual Meeting of the American Society for Microbiology*, Q-140, p. 283.

26. Eivazi, F., and M. A. Tabatabai. 1977. Effects of trace elements on urease activity in soils. *Soil Biol. Biochem.* 9:9–13.

27. El-Harris, M. K., V. L. Cochran, L. F. Elliott, and D. F. Bezdick. 1983. Effect of tillage, cropping, and fertilizer management on soil nitrogen mineralization potential. *Soil Sci. Soc. Am. J.* 47:1157–1161.

28. Federal Register. 1979. Toxic substances control act: Premanufacture testing of new chemical substances. *Fed. Reg.* 44:16240.

29. Fredrickson, J. K., S. A. Bentjen, H. Bolton, Jr., S. W. Li, and P. Van Voris. 1989. Fate of Tn5 mutants of root growth-inhibiting *Pseudomonas* sp. in intact soil-core microcosms. *Can. J. Microbiol.* 35:867–873.

30. Fredrickson, J. K., H. Bolton, Jr., S. A. Bentjen, K. M. McFadden, S. W. Li, and P. Van Voris. 1990. Evaluation of intact soil-core microcosms for determining potential impacts on nutrient dynamics by genetically engineered microorganisms. *Environ. Tox. Chem.* 9:551–558.

31. Fredrickson, J. K., P. Van Voris, S. A. Bentjen, and H. Bolton, Jr. 1989. Terrestrial microcosms for evaluating the environmental fate and risks associated with the release of chemicals or genetically engineered microorganisms to the environment. In J. Saxena (ed.), *Hazard Assessment of Chemicals-Current Development*, Vol. 7. Hemisphere Publishing, Washington, D.C., pp. 157–202.

32. Frissel, M. J. 1978. *Cycling of Mineral Nutrients in Agricultural Ecosystems*. Vol. 3 in *Developments in Agricultural Management of Forest Ecology*. Elsevier, Amsterdam.

33. Gardner, W. H. 1986. Water content. In A. Klute (ed.), *Methods of Soil Analysis, Part 1, Physical and Mineralogical Methods*, 2d ed. American Society of Agronomy, Madison, Wis., pp. 493–544.

34. Gile, J. D., J. C. Collins, and J. W. Gillett. 1982. Fate and impact of selected wood preservatives in a terrestrial model ecosystem. *J. Agric. Food Chem.* 30: 295–301.

35. Gillett, J. W. 1988. The role of terrestrial microcosms and mesocosms in ecotoxicologic research. In S. A. Levin, M. Harwell, J. Kelly, and K. D. Kimball (eds.), *Ecotoxicology—Problems and Approaches*. Springer-Verlag, New York, pp. 367–410.

36. Gilkes, N. R., D. G. Kilburn, M. L. Langsford, R. C. Miller, W. W. Wakarchuk, R. A. J. Warren, D. J. Whittle, and W. K. R. Wong. 1984. Isolation and characterization of *Escherichia coli* clones expressing cellulase genes from *Cellulomonas fimi. J. Gen. Microbiol.* 130:1377–1384.

37. Greaves, M. P. 1982. Effect of pesticides on soil microorganisms. In R. G. Burns and J. H. Slater (eds.), *Experimental Microbial Ecology*. Blackwell, Oxford, pp. 613–630.

38. Greenburg, E. P., N. J. Poole, P. H. Pritchard, J. Tiedje, and D. E. Corpet. 1988. Use of microcosms. In M. Sussman, C. H. Collins, F. A. Skinner, and D. E. Stewart-Tull (eds.), *Release of Genetically-Engineered Micro-organisms*. Academic Press, London, pp. 265–274.

39. Hardy, R. W. F., J. G. Criswell, and U. D. Havelka. 1977. Investigations of possible limitations of nitrogen fixation by legumes: (1) methodology, (2) identification, and (3) assessment of significance. In W. E. Newton et al. (eds.), *Nitrogen Fixation*. Academic Press, New York, pp. 451–467.

40. Hicks, R. J., G. Stotzky, and P. Van Voris. 1990. Review and evaluation of the effects of xenobiotic chemicals on microorganisms in soil. *Adv. Appl. Microbiol.* 35: In press.

41. Holben, W. E., J. K. Jansson, B. K. Chelm, and J. M. Tiedje. 1988. DNA probe method for the detection of specific microorganisms in the soil bacterial community. *Appl. Environ. Microbiol.* 54:703–711.

42. Hungate, R. E. 1968. A roll tube method for cultivation of strict anaerobes. In J. R.

Norris and D. W. Ribbons (eds.), *Advances in Microbiology,* vol. 3B. Academic Press, New York, pp. 117–132.

43. Jackson, D. R., and M. Levin. 1979. Transport of arsenic in grassland microcosms and field plots. *Water Air Soil Pollut.* 11:3–12.

44. Jackson, D. R., C. D. Washburne, and B. S. Ausmus. 1977. Loss of Ca and NO_3-N from terrestrial microcosms as an indicator of soil pollution. *Water Air Soil Pollut.* 8:279–284.

45. Jansson, S. L., and J. Persson. 1982. Mineralization and immobilization of soil nitrogen. In A. L. Page, R. H. Miller, and D. R. Keeney (eds.), *Methods of Soil Analysis. Part 2: Chemical and Microbiological Properties,* 2d ed. American Society of Agronomy, Madison, Wis., pp. 229–252.

46. Jenkinson, D. S., and J. N. Ladd. 1981. Microbial biomass in soil: Measurement and turnover. In E. A. Paul and J. N. Ladd (eds.), *Soil Biochemistry,* vol. V. Marcel Dekker, New York, pp. 415–471.

47. Jenkinson, D. S., and D. S. Powlson. 1976. The effects of biocidal treatments on metabolism in soil. V. A method for measuring soil biomass. *Soil Biol. Biochem.* 8:209–213.

48. Jenkinson, D. S., D. S. Powlson, and R. W. M. Wedderburn. 1976. The effect of biocidal treatment on metabolism in soil. III. The relationship between soil biovolume, measured by optical microscopy, and the flush of decomposition caused by fumigation. *Soil Biol. Biochem.* 8:189–202.

49. Keeney, D. R. 1982. Nitrogen-availability indices. In A. L. Page, R. H. Miller, and D. R. Keeney (eds.), *Methods of Soil Analysis. Part 2: Chemical and Microbiological Properties,* 2d ed. American Society of Agronomy, Madison, Wis., pp. 711–733.

50. Klein, D. A., T. C. Loh, and R. L. Goulding. 1971. A rapid procedure to evaluate the dehydrogenase activity of soils low in organic matter. *Soil Biol. Biochem.* 3: 385–387.

51. Knowles, R. 1982. Free-living dinitrogen-fixing bacteria. In A. L. Page, R. H. Miller, and D. R. Keeney (eds.)., *Methods of Soil Analysis. Part 2: Chemical and Microbiological Properties,* 2d ed. American Society of Agronomy, Madison, Wis., pp. 1071–1092.

52. Knudsen, G. R., M. V. Walter, L. A. Porteous, V. J. Prince, J. L. Armstrong, and R. J. Seidler. 1988. Predictive model of conjugative plasmid transfer in the rhizosphere and phyllosphere. *Appl. Environ. Microbiol.* 54:343–347.

53. Koike, I., and A. Hattori. 1978. Simultaneous determinations of nitrification and nitrate reduction in coastal sediments by a ^{15}N dilution technique. *Appl. Environ. Microbiol.* 35:853–857.

54. Kozyrovskaya, N. A., R. J. Gvozdyak, V. A. Muras, and V. A. Kordyum. 1984. Changes in properties of phytopathogenic bacteria affected by plasmid pRD1. *Arch. Microbiol.* 137:338–343.

55. Kunc, F., and G. Stotzky. 1980. Acceleration by montmorillonite of nitrification in soil. *Folia Microbiol.* 25:106–125.

56. Levin, S. A., and M. A. Harwell. 1986. Potential ecological consequences of genetically engineered organisms. *Environ. Manag.* 10:495–513.

57. Li, S. W., J. K. Fredrickson, M. W. Ligotke, P. Van Voris, and J. E. Rogers. 1988. Influence of smoke exposure on soil enzyme activities and nitrification. *Biol. Fert. Soils* 6:341–346.

58. Loos, M. A., A. Kontson, and P. C. Kearney. 1980. Inexpensive soil flask for ^{14}C-pesticide degradation studies. *Soil Biol. Biochem.* 12:583–585.

59. Macura, J., and G. Stotzky. 1980. Effects of montmorillonite and kaolinite on nitrification in soil. *Folia Microbiol.* 25:90–105.

60. McElroy, M. B., S. C. Wofsy, and Y. L. Yung. 1977. The nitrogen cycle: Perturbations due to man and their impact on atmospheric N_2O and O_3. *Philos. Trans. R. Soc. London, Ser. B* 277:159–181.

61. McGarity, J. W., C. M. Gilmour, and W. B. Bollen. 1958. Use of an electrolytic respirometer to study denitrification in soil. *Can. J. Microbiol.* 4:303–316.

62. Mulvaney, R. L. 1988. Evaluation of nitrogen-15 tracer techniques for direct measurement of denitrification in soil. III. Laboratory studies. *Soil Sci. Soc. Am. J.* 52:1327–1332.

63. Mulvaney, R. L., and R. M. Vanden Heuvel. 1988. Evaluation of nitrogen-15 tracer techniques for direct measurement of denitrification in soil. IV. Field studies. *Soil Sci. Soc. Am. J.* 52:1332–1337.

64. Nannipieri, P. 1984. Microbial biomass and activity measurements in soil: Ecological significance. In M. J. Keng and C. A. Reddy (eds.), *Current Perspectives in Microbial Ecology.* American Society of Microbiology, Washington, D.C., pp. 515–521.

65. Neuhold, J., and L. Ruggerio. 1975. Ecosystem processes and organic contaminants (NSF-RA-76008). National Science Foundation, Washington, D.C.

66. Nordgren, A. 1988. Apparatus for the continuous, long-term monitoring of soil respiration rate in large numbers of samples. *Soil Biol. Biochem.* 20:955–957.

67. Ogram, A., G. S. Sayler, and T. Barkay. 1987. The extraction and purification of microbial DNA from sediments. *J. Microbiol. Meth.* 7:57–66.

68. O'Neill, R. V., B. S. Ausmus, D. R. Jackson, R. I. Van Hook, P. Van Voris, C. Washburne, and A. P. Watson. 1977. Monitoring terrestrial ecosystems by analysis of nutrient export. *Water Air Soil Pollut.* 8:271–277.

69. Page, A. L., R. H. Miller, and D. R. Keeney (eds.). 1982. *Methods of Soil Analysis. Part 2: Chemical and Microbiological Properties,* 2d ed. American Society of Agronomy, Madison, Wis.

70. Pancholy, S. K., E. L. Rice, and J. A. Turner. 1975. Soil factors preventing revegetation of a denuded area near an abandoned smelter in Oklahoma. *J. Appl. Ecol.* 12:337–342.

71. Parkinson, D., and E. A. Paul. 1982. Microbial biomass. In A. L. Page, R. H. Miller, and D. R. Keeney (eds.), *Methods of Soil Analysis. Part 2: Chemical and Microbiological Properties,* 2d ed. American Society of Agronomy, Madison, Wis., pp. 821–830.

72. Pritchard, P. H. 1982. Model ecosystems. In R. A. Conway (ed.), *Environmental Risk Analysis for Chemicals.* Van Nostrand Reinhold, New York, pp. 257–353.

73. Ramachandra, M., D. L. Crawford, and A. L. Pometto. 1987. Extracellular enzyme activities during lignocellulose degradation by *Streptomyces* spp.: A comparative study of wild-type and genetically manipulated strains. *Appl. Environ. Microbiol.* 5:2754–2760.

74. Ramos, J. L., A. Wasserfallen, K. Rose, and K. N. Timmis. 1987. Redesigning metabolic routes: Manipulation of TOL plasmid pathway for catabolism of alkylbenzoates. *Science* 235:593–596.

75. Rawlins, S. L., and G. S. Campbell. 1986. Water potential: Thermocouple psychrometry. In A. Klute (ed.), *Methods of Soil Analysis. Part 1: Physical and Mineralogical Methods,* 2d ed. American Society of Agronomy, Madison, Wis., pp. 597–618.

76. Rogers, J. E., and S. W. Li. 1985. Effect of metals and other inorganic ions on soil microbial activity: Soil dehydrogenase activity as a simple test. *Bull. Environ. Contam. Toxicol.* 34:858–865.

77. Ryden, J. C., L. J. Lund, J. Letey, and D. D. Focht. 1979. Direct measurement of denitrification loss from soils. II. Development and application of field methods. *Soil Sci. Soc. Am. J.* 43:110–118.

78. Sadowsky, M. J., R. E. Tully, P. B. Cregan, and H. H. Keyser. 1987. Genetic diversity in *Bradyrhizobium japonicum* serogroup 123 and its relation to genotype-specific nodulation of soybean. *Appl. Environ. Microbiol.* 53:2624–2630.

79. Saiki, R. K., D. H. Gelfand, S. Stoffel, S. J. Scharf, R. Higuchi, G. T. Horn, K. B. Mullis, and H. A. Erlich. 1988. Primer-directed enzymatic amplification of DNA with a thermostable DNA polymerase. *Science* 239:487–491.

80. Schmidt, E. L., and L. W. Belser. 1982. Nitrifying bacteria. In A. L. Page, R. H. Miller, and D. R. Keeney (eds.), *Methods of Soil Analysis. Part 2: Chemical and Microbiological Properties,* 2d ed. American Society of Agronomy, Madison, Wis., pp. 1027–1042.

81. Schmidt, E. L., and E. A. Paul. 1982. Microscopic methods for soil microorganisms. In A. L. Page, R. H. Miller, and D. R. Keeney (eds.), *Methods of Soil Analysis. Part 2: Chemical and Microbiological Properties*, 2d ed. American Society of Agronomy, Madison, Wis., pp. 803–814.

82. Shelton, D. R., and J. M. Tiedje. 1984. General method for determining anaerobic biodegradation potential. *Appl. Environ. Microbiol.* 47:850–857.

83. Smith, J. L., R. R. Schnabel, B. L. McNeal, and G. S. Campbell. 1980. Potential errors in the first-order model for estimating soil nitrogen mineralization potentials. *Soil Sci. Soc. Am. J.* 44:996–1000.

84. Somerville, C. C., I. T. Knight, W. L. Straube, and R. R. Colwell. 1989. Simple, rapid method for direct isolation of nucleic acids from aquatic environments. *Appl. Environ. Microbiol.* 55:548–554.

85. Spalding, B. P. 1979. Effects of divalent metal chlorides on respiration and extractable enzymatic activities of Douglas-fir needle litter. *J. Environ. Qual.* 8:105–109.

86. Sparling, G. P. 1985. The soil biomass. In D. Vaughan and R. E. Malcolm (eds.), *Soil Organic Matter and Biological Activity. Developments in Plant and Soil Science*, vol. 16. Martinus Nijhoff/Dr. W. Junk, Dordrecht, pp. 223–262.

87. Sparling, G. P., and A. W. West. 1988. Modifications to the fumigation-extraction technique to permit simultaneous extraction and estimation of soil microbial C and N. *Commun. Soil Sci. Plant Anal.* 19:327–344.

88. Stanford, G. 1982. Assessment of soil nitrogen availability. In F. J. Stevenson (ed.), *Nitrogen in Agricultural Soils*. American Society of Agronomy, Madison, Wis., pp. 651–688.

89. Stanford, G., J. N. Carter, and S. J. Smith. 1974. Estimates of potentially mineralizable nitrogen based on short-term incubations. *Soil Sci. Soc. Am. Proc.* 38:99–102.

90. Steffan, R. J., and R. M. Atlas. 1988. DNA amplification to enhance detection of genetically engineered bacteria in environmental samples. *Appl. Environ. Microbiol.* 54:2185–2191.

91. Steffan, R. J., J. Goksoyr, A. K. Bej, and R. M. Atlas. 1988. Recovery of DNA from soils and sediments. *Appl. Environ. Microbiol.* 54:2908–2915.

92. Stevenson, F. J. (ed.). 1982. *Nitrogen in agricultural soils,* Agronomy Monograph No. 22. American Society of Agronomy, Madison, Wis.

93. Stevenson, F. J. 1986. *Cycles of Soil.* Wiley, New York.

94. Stotzky, G. 1960. A simple method for the determination of the respiratory quotient of soil. *Can. J. Microbiol.* 6:439–452.

95. Stotzky, G. 1965. Replica plating technique for studying microbial interactions in soil. *Can. J. Microbiol.* 11:629–636.

96. Stotzky, G. 1965. Microbial respiration. In C. A. Black et al. (eds.), *Methods of Soil Analysis. Part 2: Chemical and Microbiological Properties.* American Society of Agronomy, Madison, Wis., pp. 1550–1570.

97. Stotzky, G. 1972. Activity, ecology, and population dynamics of microorganisms in soil. *CRC Crit. Rev. Microbiol.* 2:59–137.

98. Stotzky, G. 1989. Gene transfer among bacteria in soil. In S. B. Levy and R. V. Miller (eds.), *Gene Transfer in the Environment.* McGraw-Hill, New York, pp. 165–222.

99. Stotzky, G., and H. Babich. 1986. Survival of, and genetic transfer by, genetically engineered bacteria in natural environments. *Adv. Appl. Microbiol.* 31:163–188.

100. Stotzky, G., and A. G. Norman. 1961. Factors limiting microbial activities in soil. I. The level of substrate, nitrogen, and phosphorous. *Arch. Mikrobiol.* 40:341–369.

101. Stotzky, G., and A. G. Norman. 1964. Factors limiting microbial activities in soil. III. Supplementary substrate additions. *Can. J. Microbiol.* 10:143–149.

102. Stotzky, G., M. A. Devanas, and L. R. Zeph. 1989. Methods of studying bacterial gene transfer in soil by conjugation and transduction. U.S. Environmental Protection Agency Protocol Document, EPA-600/3-89/042.

103. Tabatabai, M. A. 1982. Soil enzymes. In A. L. Page, R. H. Miller, and D. R. Keeney (eds.), *Methods of Soil Analysis. Part 2: Chemical and Microbiological Properties,* 2d ed. American Society of Agronomy, Madison, Wis., pp. 903–947.

104. Tabatabai, M. A., and J. M. Bremner. 1969. Use of ρ-nitrophenyl phosphate for assay of soil phosphatase activity. *Soil Biol. Biochem.* 1:302–307.

105. Tabatabai, M. A., and J. M. Bremner. 1970. Arylsulfatase activity of soils. *Soil Sci. Soc. Am. Proc.* 34:225–229.

106. Tiedje, J. M. 1982. Denitrification. In A. L. Page, R. H. Miller, and D. R. Keeney (eds.), *Methods of Soil Analysis. Part 2: Chemical and Microbiological Properties,* 2d ed. American Society of Agronomy, Madison, Wis., pp. 1011–1026.

107. Tiedje, J. M. 1988. Ecology of denitrification and dissimilatory nitrate reduction to ammonium. In A. J. B. Zehnder (ed.), *Biology of Anaerobic Microorganisms.* Wiley, New York, pp. 179–244.

108. U.S. Congress, Office of Technology Assessment. 1988. *New Developments on Biotechnology-Field–Testing Engineered Organisms: Genetic and Ecological Issues,* OTA-BA-350. U.S. Government Printing Office, Washington, D.C.

109. Van Voris, P. 1988. Standard guide for conducting a terrestrial soil-core microcosm test (Standard E-1197-87). In *1988 Annual Book of ASTM Standards,* vol. 11.04. American Society for Testing and Materials, Philadelphia, pp. 743–755.

110. Verstraete, W., H. Van de Werf, F. Kucnerowicz, M. Ilaiwi, L. M. J. Verstraeten, and K. Vlassak. 1983. Specific measurement of soil microbial ATP. *Soil Biol. Biochem.* 15:391–396.

111. Voroney, R. P., and E. A. Paul. 1984. Determination of k_c and k_n *in situ* for calibration of the chloroform fumigation-incubation method. *Soil Biol. Biochem.* 16:9–14.

112. Walter, M. V., K. Barbour, M. McDowell, and R. J. Seidler. 1987. A method to evaluate survival of genetically engineered bacteria in soil extracts. *Curr. Microbiol.* 15:193–197.

113. Whitehead, T. R., and R. B. Hespell. 1989. Cloning and expression in *Escherichia coli* of a xylanase gene from *Bacteroids rumincola* 23. *Appl. Environ. Microbiol.* 55:893–896.

114. Winter, R. B., K.-M. Yen, and B. D. Ensley. 1989. Efficient degradation of trichloroethylene by a recombinant *Escherichia coli. Biotechnology* 7:282–285.

115. Zylstra, G. J., R. McCombie, D. T. Gibson, and B. A. Finette. 1988. Toluene degradation by *Pseudomonas putida* F1: Genetic organization of the *tod* operon. *Appl. Environ. Microbiol.* 54:1498–1503.

30

Measurement of Microbial Population Dynamics: Significance and Methodology

Donald A. Klein

Introduction

Microbial growth and population dynamics in natural environments can be influenced by a variety of physical, physiological, and methodological factors. An understanding of these factors and how they influence microbial communities will be central to assessing the effects of added microorganisms.

Physical changes can affect microbial population dynamics independent of the organisms contained in a particular inoculum. The addition of water to a soil as the inoculum carrier may lead to the increased growth of indigenous microbes and carbon dioxide release (5, 6, 47). Physical disturbance (changes in aeration, mixing of stratified waters and heterogeneous soils) can also stimulate microbial population responses. The added organisms themselves also can influence the indigenous microbial communities without growth occurring. These added organisms can provide carbon, nitrogen, and other nutrients that can allow the indigenous microbial community to initiate or accelerate growth. Although inoculation levels might only be in the range of 10^6 to 10^8 per gram or milliliter of the environment, the bioavailability of nutrients from the inoculated microbes, especially in nutrient-limited oligotrophic aquatic or subsurface environments, may be able to stimulate microbial growth. In addition, these added organisms can serve as a source of genetic information, which can contribute to recombinant organism formation. Also, without growth occurring, the added microbes can serve as a food source for predatory microbes, insects, and animals, which can result in changes in the populations of these important groups of microbivores.

As a last level of interaction, the added microbes themselves can grow, leading to changes in their own populations and utilization of available resources in the environment. These changes can result in additional secondary physical effects and changes in indigenous microbial populations through competitive and commensalistic interactions.

The types of microbes used and their method of preparation also can affect the indigenous organisms and the recipient environment. For example, indigenous organisms can be added to an environment without laboratory culturing. This minimizes changes in the structure and physiology of the microbial community. The use of laboratory-grown organisms can produce markedly different results. The exposure of microbes from natural environments to higher nutrient levels in laboratory culture can lead to major changes in physiological characteristics. This can include differences in growth rates and substrate saturation characteristics, which has been observed with microbes from marine environments (79), as an example. These microbes will be more active physiologically and usually less able to compete with indigenous microbes (60).

The indigenous microbes, which will interact with these more physiologically active added organisms, are present in a spectrum of physiological states, ranging from substrate-responsive cells to ultramicro cells in a stasis condition (52, 53). For bacteria, this has been documented in terms of growth responses of these different bacterial groups (2). For fungi, variations in carbon and nitrogen allocation to cell walls and cytoplasm under different nutrient limitations (57) can play a critical role in the responses of this important part of the microbial community.

The addition of bacteria to an environment can result in a different physical distribution of these organisms in comparison with the indigenous bacteria. The presence of a complex physical matrix, particularly in soils and sediments, results in specific niches and micropores where the indigenous bacteria (especially the oligotrophs and ultramicrobacteria) are present at greater relative frequencies (30, 81) and where resources for growth must be in the immediate vicinity of the cells. The added bacteria will become associated with surfaces predominantly outside of these pores and microsites, where a different range of resources may be available. This can possibly lead to greater physiological stress on these introduced microbes, and a greater susceptibility to predation, especially when laboratory-grown cultures are used (24).

In contrast, fungi can have a different physical relationship to particulate materials found in their environment. The filamentous fungi develop to a greater extent on the surfaces of heterogeneous particles, such as soil aggregates, forming bridges between these structures, and the fungi can translocate nutrients across greater distances than can bacteria (54), making it possible to exploit more heterogeneously distributed nutrient resources (35). Fungi added to such environments can be expected to compete more directly with the indigenous fungi for available resources.

The levels of added microorganisms will also be important. With bacterial additions of less than 10^5 to 10^6 per milliliter or per gram of an environmental material, interactions with predators will be minimal (24), as well as opportunities for genetic exchange (72), especially in heterogeneous solid materials, such as soil or sediments. Even in aquatic environments, the presence of microbial flocs and organic growth on surfaces will not assure that the introduced microbes will be involved in predatory and genetic interactions to a greater degree, as the larger part of organisms in these flocs and surface growths are nonviable, especially in the interior of these structures (82).

The indigenous microbial community structure also can affect the ability to assess changes that occur with microorganism additions. Environments dominated by single species of microorganisms, such as extreme (pH, temperature, salinity), or energetically restricted environments (light characteristics, substrates), should be able to be studied with a greater degree of confidence, in comparison with more complex microbial communities.

Nutrient limitations are also critical, and added organisms can influence the ability of the indigenous microbes to deal with these limitations. As discussed by Tempest and Neijssel (74), organisms can use a variety of strategies to maintain physiological readiness, including overflow metabolism, ATP hydrolysis, and storage polymer accumulation. The addition of nutritionally rich microbial cells has the potential of triggering growth responses of these physiologically stressed indigenous populations by providing critical limiting nutrients, especially when higher populations might be added.

A large part of the recent work on microbial growth, turnover, and population dynamics of indigenous and added microbes has been carried out with aquatic systems; however, there is increasing interest in the analysis of soil and sediments, which, in general, are physically and chemically more complex.

Characterization of Added Microbial Populations

To assist in evaluating the growth responses of the indigenous and added microbes, and to allow interactions and secondary effects to be better assessed, information on organism characteristics, especially at the molecular level, will be critical. If the microbes to be added to the environment are cultured in the laboratory, it will be possible to characterize the organisms, to develop a range of detection and monitoring alternatives, and to identify the responses of these added microbes. The following capabilities and procedures can assist in separating the responses of the added versus the indigenous microbes:

1. Culture media, including selective media, to allow recovery of the added organism or genetic constructs.
2. Markers (Lux, antibiotic resistance, isotopic labeling, LacZY, etc.) to allow more efficient, sensitive, and rapid detection.
3. The availability of resuscitation media (49) or microviability assays to assess the occurrence and possible effects of stress on microbe growth potential (63). This phenomenon is important when organisms are added to the environment, in terms of assessing subsequent population dynamics and interactions with indigenous populations.
4. Specific genetic sequences contained in the organism can be characterized, making it possible to monitor the presence of the organism in terms of DNA-RNA recovery and detection by use of specific probes.
5. Appropriate controls will be able to be used to determine the actual cause of population changes that might occur. These can be due to physical, nutritional, genetic, or microbial growth-related processes or to a combination of these factors. The nonmodified parent organism can also be used on a comparative basis to more fully understand the possible effect of a specific genetic sequence on the indigenous microbial community.
6. Knowledge of microorganism distribution within the environment, if available, will assist in establishing experimental boundaries in terms of sample size and the possible effects of repeated sampling. This latter concern is particularly critical if sufficient replicates will not be available to allow destructive sampling for each desired assay time and experimental variable.
7. An ability to assess immigration and emigration, especially in aquatic environments, where physical mixing and microbial mobility may be major factors influencing microbial populations found at a particular site or region. As discussed by Brock (10, 12), immigration, emigration, and division in situ are critical considerations related to this concern.

Application of Techniques to Population Studies

Macroviability involving colony formation on solid media

One of the most common methods of assessing growth in natural environments is to measure colony-forming units at desired time intervals, and to then assume that increases observed between time a and b represent the growth rate. This approach has been used in studies of the rhizosphere (77, 78) and of microbial populations in soils and soil solutions (26, 34). In the latter study (34), periodic population increases were summed to give an all-over estimate of bacterial production over a summer season, an approach that could be integrated with the evaluation of effects of additions or particular microbes.

Depending on the time intervals used, population fluctuations may occur that will not be detected by such a general procedure:

- Population growth can be counteracted, to some extent, by turnover and loss of viability, and the actual growth rate can be higher.

- If the interval between the measurements is sufficiently long, the population may have exhibited additional changes including lags, or a higher population followed by decreases.

In these situations, the actual growth rate, growth potential, and population turnover may be markedly different than that shown by measurements at two particular time intervals.

An approach that is finding increasing use for evaluating the growth potential of a particular microbial community involves measuring the rate of colony appearance on a solid medium, or *lambda* (λ), the colony-forming potential (27–29), as noted in Fig. 30.1. The rate of visible colony formation on a solid medium can be considered to result from the development of two populations: (1) microcolonies originally present from the environment (curve I), which can continue growth that was in progress in the natural environment after transfer to a laboratory medium, and (2) individual cells that have not begun the process of active cell division (curve II).

By the use of such a differential graphing procedure, the relative occurrence of these two populations in different samples can be assessed. Colony appearance by these two populations which may be present in a sample can be treated mathematically in the following way:

$$\ln (N_o - N) = \ln N_o - \lambda T$$

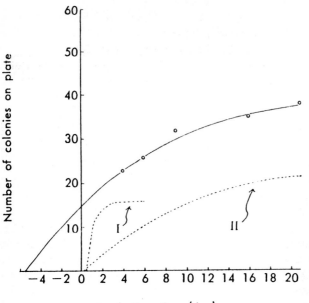

Figure 30.1   Measurement of subpopulations present in an environmental sample by analysis of colony formation curves. The subpopulations are noted as curves I and II. [*Hattori* (27).]

where N_o is the number of cells capable of forming a colony, N represents the number of colonies observed at a specific time, and λ is a probabilistic parameter.

By the use of this approach, the relative presence of dividing cells in different environments can be estimated. This procedure has been used to assess bacterial growth processes in the rhizoplane and rhizosphere of rape seedlings (1), as an example.

Jannasch (37) estimated microbial growth in natural waters by diluting samples at a series of different rates in a chemostat. In this approach, the natural water is diluted at a given rate, and the rate of culture loss was monitored by macroviability assays. Appropriate calculations were used to estimate the difference between the loss rate and the dilution rate to give the growth rate of the natural population. This approach could be used to detect interactions between indigenous microbes and marked added populations.

As a last, more general, concern with such macroviability approaches, population changes may simply reflect an increased ability to form such colonies on a given medium and not an actual increase in the viable microbial population itself. This is especially important in natural environments, where such macroviability-based population estimates are usually markedly lower than the populations derived from microscopic or microviability approaches. Additions of low levels of supplemental nutrients can increase the colony-forming potential of such nutritionally and/or environmentally stressed microbial populations (49, 58).

Microscopic and microviability approaches

A variety of microscopic and microviability-based approaches to the estimation of growth responses and the growth potential of added and indigenous microbes are available, as summarized in Table 30.1. These include measurements of staining and respiration, cell size and division potential in conjunction with staining, and the use of fluorescent antibody techniques. With nalidixic acid, which inhibits cell division, resulting in elongated viable cells (41–43), a variety of substrates and inhibitors (89, 90) can be used to give the assay functional and physiological specificity. These approaches have also been considered in a more general context by Van Es and Meyer-Reil (80).

Depending on the types of samples being examined, microscopic techniques may be subject to limitations that will influence the utility

TABLE 30.1 Basic Approaches Used to Evaluate Growth Responses and Growth Potentials Based on Microscopic and Microviability-Based Procedures

Method	Description	References
Microscopic estimates of bacterial numbers with varied times and treatments	Use of total and viable estimates	3, 33, 39
Microscopic measurement of colony growth or cell numbers on slides	Direct microscopic measurements Use of computer-enhanced images	8, 11 13, 14
Activity and respiration	AODC/Vmax ratio Staining with INT Staining with ethidium bromide FDA staining—fungi	58 89, 91 48, 73 68
Dividing cell frequency	Use in aquatic environment for growth rates and productivity estimation	55
Cell size measurements	Direct measurement of cell size ranges	2
Nalidixic acid responsive cells	Use with varied substrates and inhibitors	42, 89, 90
Fluorescent antibody technique	Requires growth of organisms for antibody production	7, 22, 62

of this approach, independent of the final microviability procedure that will be used. With aquatic samples, it is possible to work with a wide range of sample background population levels, and filtration is often used to concentrate organisms to minimize limitations when measuring dilute populations. For water samples, the main limitation is the ratio of the target populations to the background organisms in terms of staining and counting interferences, and appropriate dilution-concentration steps can be used to assure that countable fields are available.

With soil or sediment systems, background organism levels can be in the range of 10^9 to 10^{10} per gram of material. With the solid matrix in these samples, this can result in the need to dilute these samples to the point where a countable field can be obtained (65). If the added population levels are too low, the high background populations of indigenous organisms again will limit sensitivity, and it is necessary to achieve a balance between this concern and the need to minimize interference by particulate materials.

In contrast, in samples of subsurface materials the levels of indigenous bacteria can approach or be lower than the microscopic dilution factor, limiting the usefulness of microscopic techniques. When counting dilute populations of microorganisms in these samples, strict control of extraneous microbes on glass slides and filtration of all solutions is required to minimize background interference (W. C. Ghiorse, personal communication).

Use of molecular markers

A wide range of techniques are now available to detect indigenous and added microbial populations and to assess their growth potential, based on the ability to identify specific nucleic acid sequences. These approaches involve the use of molecular probes, which can be used to test colonies from plated bacteria grown on desired culture media for homology with known DNA sequences (22).

This colony blot approach has been complemented by procedures where nucleic acids are directly extracted from environmental samples. Techniques are available to recover nucleic acids from aquatic (56, 69) and soil environments (71, 75). Basic considerations with this approach involve whether direct DNA recovery will be used or whether microbes will first be recovered from the particular environment before extraction and DNA recovery. Steffan et al. (71) have compared these techniques and found that each approach provides

DNA with different characteristics, with direct recovery including eukaryotic DNA. These differences should be taken into account when developing a research design.

The availability of polymerase chain reaction (PCR) techniques (70) improves the potential to determine the presence and levels of specific genetic constructs even where present in microbial populations that cannot be recovered by plating, due to background organism interference. As few as 1 to 10 cells per gram of material containing a specific genetic sequence can be identified using this technique. The use of these molecular approaches will have major effects on the ability to assess microbial population dynamics, especially considering indigenous microbial populations for which specific nucleic acid sequences can be characterized, even if the organisms have not been grown.

A variety of techniques can be used to assess growth by measuring rates of nucleic acid synthesis (9, 38). These approaches do not measure growth or cell division directly, but they do measure the rate of radioactive precursor incorporation into macromolecules (51). Factors that can influence this process include the rates of precursor transport and incorporation by different organisms. Degradation and isotope dilution are also concerns. Although most work has been conducted with water and sediments, this approach has also been applied to sterilized soil systems (15).

A recent application of nucleic acid probes which shows a marked potential for monitoring dynamics of specific microbial populations is the use of ribosomal RNA-based probes for the identification of single cells (20). With the ability to identify nucleic acid sequences that are specific for different organisms, the potential to follow the dynamics of indigenous and added populations, based on phylogenetic and taxonomic factors involving ribosomal RNA, will be of increasing interest.

Analysis of cellular constituents and metabolism

Specific biochemical constituents that are only found in living cells, and that are assumed to degrade rapidly when the cell becomes nonviable or lyses, have potential for monitoring microbial population changes.

Among the cellular constituents that have been used for the analysis of microbial populations are muramic acid (50), glucoseamine in chitin (61), and phospholipid-linked ester fatty acids (25, 85). These and other biochemical signatures have been discussed in recent reviews (71, 83).

To be of maximum utility in assessing microbial population dynamics, the particular marker compound must degrade rapidly once the cell has lost viability, as has been suggested to occur for phosphorus release from diacyl phospholipids (85). If relatively short-term population dynamics are being assessed, degradation rates for the marker compound can be critical in assessing population dynamics. It is suggested that rigorous controls on survival, extractability, binding, and microbial interactions with the chemical signature compound be completed before applying these methods to any but the most general and longer-term population response studies.

Another method for following the effects of added organisms, in terms of responses of the indigenous microbial populations and the added organisms, is multilocus enzyme electrophoresis (66, 86). This technique, which has been used extensively in eukaryotic population genetics, is now also being applied to prokaryotic organisms.

Multilocus enzyme electrophoresis provides information on clonal population structure and clonal turnover in an environment based on the mobility variation (electromorphs) of given enzymes that correspond to alleles at specific structural gene loci. Although this process underestimates total genetic variation because of silent nucleotide substitutions and amino acid replacements that do not alter enzyme mobility (19, 44), this approach appears useful for estimating strain relationships and divergence among bacterial chromosomal genomes. Although this approach has been used primarily with microbes that can be cultured from the environment such as *Rhizobium* (88), the potential exists for carrying out direct extractions and analyses after physical recovery of microbial communities from a particular sample, as has been carried out with DNA (71).

Estimates of population changes and turnover based on nutrient and energy flows through particular habitats also have been considered. These include estimates of carbon added to particular environments in the process of primary production and its subsequent ability to allow growth and turnover of microbial populations (16, 67). Originally developed with aquatic systems (46), this approach has been applied to soils (31) to estimate the processing of organic substrates by organisms of different trophic levels, including the soil microfauna. Nutrient recycling within the bacterial cells and fungal hyphae is also considered to occur at specific rates (32), which can lead to difficulties in calculating responses at different trophic levels (18). In aquatic environments, whole-lake productivity can be assessed by measuring ^{14}C uptake rates from labeled carbon dioxide in a lake at specific depths (21). In this approach, specific growth rates (μ) have been calculated by using the following equation:

$$\mu = \ln\left(1 - \frac{P}{B}\right)$$

where P is photosynthetic production and B is biomass.

Calorimetry has also been used for measuring growth and population responses. Heat evolution from the degradation of several different carbon sources in soil and the effects of temperature on this process have been evaluated (87). This approach can be integrated with other experimental variables, including the addition of microbes, to determine if effects on all-over metabolic processes might have occurred.

Growth responses also can be estimated by stable isotope fractionation measurements, especially with simpler substrates (64). Comparative studies also can be completed, since inorganic carbon reduction by chemoautotrophs results in greater ^{13}C depletion than photosynthetic reduction (23). This approach can also be used with ^{13}C-labeled glucose and acetate to follow the synthesis or turnover of important nonnitrogen biochemical components such as phospholipid fatty acids and poly-β-hydroxybutyric acid (PHB) (84). These components also can indicate the physiological status of the microbial community and can thus be even more useful for evaluating the possible effects of added organisms on the structure and functional characteristics of the microbial community.

Mathematical approaches

A variety of mathematical and modeling approaches can be used to estimate the size, activity, and turnover of microbial populations (4). General models of microbial growth and decomposition have been developed for aquatic (17) and terrestrial (36) systems. These models are usually based on Monod-type treatments of substrate responses. Relatively slow microbial growth which occurs in natural environments can involve a greater allocation of available substrates for the maintenance of energy (D. E. Caldwell, personal communication; 59), which can have major effects on assumptions of substrate processing to microbial biomass versus metabolic products, and subsequent growth responses that might occur.

The resulting information can be used in models to assess the potential survival and growth of added organisms, including possible genetic exchange (40). Although in the early developmental stages (36, 45), modeling indigenous and introduced microorganism population dynamics and genetic interactions will provide increasingly valuable insights into the dynamics of indigenous and added populations in the future.

Summary and Future Challenges

A variety of techniques and approaches can be used to evaluate microbial populations and their dynamics, and to assess effects of added microorganisms. To use these approaches most effectively, a series of methodological and conceptual limitations must be considered. For macroviability approaches, medium characteristics, medium selectivity, and stress effects must be considered. Microviability approaches, in contrast, although allowing increased sensitivity and an ability to identify responses of individual organisms, are limited by the microscopic factors, background interferences, and the need to possibly concentrate cells from particular environments such as the subsurface, where microbial populations are much lower. High background populations found in surface soils and eutrophic aquatic environments can present additional concerns with regard to statistical analysis and sensitivity (69).

If used with direct DNA extraction, molecular probe approaches can be limited by the ability to quantitatively extract nucleic acids. This is especially critical when using a limited number of samples from a complex ecosystem, such as a lake or soil, which may have marked spatial and temporal differences in microbial (and DNA) distribution and extractability.

The use of biochemical signatures to estimate total and viable microbial populations is being carried out with increasing sensitivity. As the sensitivity of these methods is improved through analytical advances, the need to rigorously document the maintenance of a given chemical signature only in association with living cells will be of concern. This may limit the utility of this approach, especially in the analysis of short-term growth responses of specific microbial groups. Energy flow and modeling approaches continue to show increasing potential, although their use is only beginning to be addressed in depth.

Population response measurements are among the most difficult to make, especially when microbes will be added to complex natural ecosystems. The consideration of time as an experimental variable has the potential of providing much useful information and many experimental and conceptual challenges.

Acknowledgments

Prepared with partial support from the Department of Energy, Subsurface Science Program. The assistance of W. C. Ghiorse, Rhea

Garen, Nancy Tonso, Mark Radosevich, and Barbara Frederick with discussions during preparation of the manuscript is gratefully appreciated.

References

1. Baath, E., S. Olsson, and A. Tunlid. 1988. Growth of bacteria in the rhizoplane and the rhizosphere of rape seedlings. *FEMS Microb. Ecol.* 53:355–360.
2. Bakken, L. R., and R. A. Olsen. 1987. The relationship between cell size and viability of soil bacteria. *Microb. Ecol.* 13:103–114.
3. Barber, D. A., and J. M. Lynch. 1976. Microbial growth in the rhizosphere. *Soil Biol. Biochem.* 9:305–308.
4. Bazin, M. J. (ed.). 1982. *Microbial Population Dynamics.* CRC Press, Boca Raton, Fla.
5. Birch, H. F. 1958. The effect of soil drying on humus decomposition and nitrogen availability. *Plant Soil* 10:9–31.
6. Birch, H. F. 1959. Further observations on humus decomposition and nitrification. *Plant Soil* 11:262–286.
7. Bolool, B. B., and E. L. Schmidt. 1980. The immunofluorescence approach in microbial ecology. *Adv. Microb. Ecol.* 4:203–235.
8. Bott, T. L., and T. D. Brock. 1969. Bacterial growth rates above 90°C in Yellowstone hot springs. *Science* 164:1411–1412.
9. Brock, T. D. 1967. Growth rates in the sea by thymidine autoradiography. *Science* 155:81–83.
10. Brock, T. D. 1971. Microbial growth rates in nature. *Bacteriol. Rev.* 35:39–58.
11. Brock, T. D. 1978. *Thermophilic Microorganisms and Life at High Temperatures.* Springer Verlag, New York.
12. Brock, T. D. 1985. Procaryotic population ecology. In H. O. Halvorson, D. Pramer, and M. Rogul (eds.), *Engineered Organisms in the Environment.* Am. Soc. Microbiol. Washington, D.C., pp. 176–179.
13. Caldwell, D. E., and J. Germida. 1984. Evaluation of difference imagery for visualizing and quantitating microbial growth. *Can. J. Microbiol.* 31:35–44.
14. Caldwell, D. E., and J. Lawrence. 1986. Growth kinetics of *Pseudomonas fluorescens* microcolonies within the hydrodynamic boundary layers of surface microenvironments. *Microb. Ecol.* 12:299–312.
15. Christensen, H., D. Funck-Jensen, and A. Kjoller. 1989. Growth rate of rhizosphere bacteria measured directly by the titrated thymidine incorporation technique. *Soil Biol. Biochem.* 21:113–117.
16. Clark, F. E., and E. A. Paul. 1970. The microflora of grassland. *Advan. Agron.* 22:375–435.
17. Clesceri, L. S., P. Park, and J. Bloomfield. 1977. General model of microbial growth and decomposition in aquatic ecosystem. *Appl. Environ. Microbiol.* 33:1047–1048.
18. Coleman, D. C., C. P. P. Reid, and C. V. Cole. 1983. Biological strategies of nutrient cycling in soil systems. *Adv. Ecol. Res.* 13:1–55.
19. Coyne, J. A. 1982. Gel electrophoresis and cryptic protein variation. In M. C. Rattazzi, J. G. Scandalios, and G. J. Whitt (eds.), *Isozymes: Current Topics in Biological and Medical Research,* vol. 5. Alan R. Liss, New York, pp. 1–32.
20. DeLong, E. F., G. S. Wickham, and N. R. Pace. 1989. Phylogenetic stains: Ribosomal RNA-based probes for the identification of single cells. *Science* 243:1360–1363.
21. Folt, C. L., M. J. Weavers, M. P. Yoder-Williams, and R. P. Howmiller. 1989. Field study comparing growth and viability of a population of phototrophic bacteria. *Appl. Environ. Microbiol.* 55:78–85.
22. Ford, S. F., and B. H. Olson. 1988. Methods for detecting genetically engineered microorganisms in the environment. *Adv. Microb. Ecol.* 10:45–79.

23. Fuchs, G., R. Thauer, H. Ziegler, and W. Stichler. 1979. Carbon isotope fractionation by *Methanobacterium thermoautotrophicum*. *Arch. Microbiol.* 120:135–139.
24. Goldstein, R. M., L. M. Mallory, and M. Alexander. 1985. Reasons for possible failure of inoculation to enhance biodegradation. *Appl. Environ. Microbiol.* 50:977–983.
25. Guckert, J. B., and D. C. White. 1986. Phospholipid, ester-linked fatty acid analysis in microbial ecology. In F. Megusar and M. Gantar (eds.), *Perspectives in Microbial Ecology*, Proceedings, Fourth Int. Symp. Microb. Ecol., Slovene Soc. Microbiol., Ljublana, pp. 455–459.
26. Hartel, P., and M. Alexander. 1987. Effect of growth rate on the growth of bacteria in freshly moistened soil. *Soil Sci. Soc. Am. J.* 51:93–96.
27. Hattori, T. 1982. Analysis of plate count data of bacteria in natural environments. *J. Gen. Appl. Microbiol.* 28:13–22.
28. Hattori, T. 1983. Further analysis of plate count data of bacteria. *J. Gen. Appl. Microbiol.* 29:9–16.
29. Hattori, T. 1985. Kinetics of colony formation of bacteria: An approach to the basis of the plate count method. *The Reports of the Institute for Agricultural Research* 34: 1–36.
30. Hattori, T., and R. Hattori. 1976. The physical environment in soil microbiology: An attempt to extend principles of microbiology to soil microorganisms. *CRC Crit. Rev. Microbiol.* 4:423–461.
31. Heal, O. W., and J. Dighton. 1985. Resource quality and trophic structure in the soil system. In A. H. Fitter, D. Atkinson, D. J. Read, and M. B. Usher (eds.), *Ecological Interactions in Soil. Plants, Microbes and Animals*. British Ecological Society, Blackwell Scientific, Oxford, pp. 339–353.
32. Heal, O. W., and S. F. MacLean. 1975. Comparative productivity in ecosystems: Secondary productivity. In W. H. van Dobben and R. H. Lowe-McConnell (eds.), *Unifying Concepts in Ecology* W. Junk, The Hague, pp. 89–108.
33. Hendricks, C. W., E. A. Paul, and P. D. Brooks. 1987. Growth measurements of terrestrial microbial species by a continuous-flow technique. *Plant Soil* 101:189–195.
34. Hisset, R., and T. R. G. Gray. 1976. Microsites and time changes in soil microbe ecology. In J. M. Anderson and A. Macfadyan (eds.), *The Role of Terrestrial and Aquatic Organisms in Decomposition Processes*. Blackwell Scientific, Oxford, pp. 23–39.
35. Holland, E. A., and D. C. Coleman. 1987. Litter placement effects on microbial and organic matter dynamics in an agroecosystem. *Ecology* 68:425–433.
36. Hunt, H. W., D. C. Coleman, E. R. Ingham, E. T. Elliott, J. C. Moore, S. L. Rose, C. C. P. Reid, and C. R. Morley. 1987. The detrital food web in a shortgrass prairie. *Biol. Fertil. Soils* 3:57–68.
37. Jannasch, H. W. 1969. Estimation of bacterial growth in natural waters. *J. Bacteriol.* 99:156–160.
38. Karl, D. M. 1981. Simultaneous rates of ribonucleic acid and deoxyribonucleic acid syntheses for estimating growth and cell division of aquatic microbial communities. *Appl. Environ. Microbiol.* 42:802–810.
39. Klein, D. A., B. A. Frederick, and E. F. Redente. 1989. Fertilizer effects on soil microbial communities and organic matter in the rhizosphere of *Sitanion hystrix* and *Agropyron smithii. Arid Soil Res. Rehab.* 3:397–404.
40. Knudsen, G. R., M. V. Walter, L. A. Porteous, V. J. Prince, J. L. Armstrong, and R. J. Seidler 1988. Predictive model of conjugative plasmid transfer in the rhizosphere and phyllosphere. *Appl. Environ. Microbiol.* 54:343–347.
41. Kogure, K., U. Simidu, and N. Taga. 1979. A tentative direct microscopic method for counting living marine bacteria. *Can. J. Microbiol.* 25:415–420.
42. Kogure, K., U. Simidu, and N. Taga. 1984. An improved direct viable count method for aquatic bacteria. *Arch. Hydrobiol.* 102:117–122.
43. Kogure, K., U. Simidu, N. Taga, and R. R. Colwell. 1987. Correlation of direct viable counts with heterotrophic activity for marine bacteria. *Appl. Environ. Microbiol.* 53:2332–2337.
44. Kreitman, M. 1983. Nucleotide polymorphism at the alcohol dehydrogenase locus of *Drosophila melanogaster. Nature* 304:412–417.

45. Levin, B. R. 1986. The maintenance of plasmids and transposons in natural populations of bacteria. Banbury Report 24, *Antibiotic Resistance Genes: Ecology, Transfer and Expression.* Cold Springs Harbor Laboratory, New York, pp. 57–70.
46. Lindeman, R. L. 1942. The trophic-dynamic aspect of ecology. *Ecology* 23:399.
47. Lund, V. J., and J. Goksoyr. 1980. Effects of water fluctuations on microbial mass and activity in soil. *Microb. Ecol.* 6:115–123.
48. Marxsen, J. 1985. Investigations into the number of respiring bacteria in ground water of sandy and gravelly deposits. Preliminary results. *Verh. Internat. Verein. Limnol.* 22:2721.
49. McFeters, G. A., J. S. Kippen, and M. W. LeChevallier. 1986. Injured coliforms in drinking water. *Appl. Environ. Microbiol.* 51:1–5.
50. Millar, W. N., and L. E. Casida, Jr. 1970. Evidence for muramic acid in soil. *Can. J. Microbiol.* 16:299–304.
51. Moriarty, D. J. W. 1986. Measurements of bacterial growth rates in aquatic systems from rates of nucleic acid synthesis. *Adv. Microbiol. Ecol.* 9:245–292.
52. Morita, R. Y. 1982. Starvation-survival of heterotrophs in the marine environment. *Adv. Microb. Ecol.* 6:272–298.
53. Morita, R. Y. 1985. Starvation and miniaturisation of heterotrophs, with special emphasis on maintenance of the starved viable state. In M. Fletcher and G. Floodgate (eds.), *Bacteria in Their Natural Environments: The Effect of Nutrient Conditions.* Academic Press, New York, pp. 111–130.
54. Newman, E. I. 1985. The rhizosphere: Carbon sources and microbial populations. In A. H. Fitter, D. Atkinson, D. J. Read, and M. B. Usher (eds.), *Ecological Interactions in Soil, Plants, Microbes, and Animals.* British Ecological Society, Blackwell Scientific, Oxford, pp. 107–121.
55. Newell, S. Y., and R. R. Christian. 1981. Frequency of dividing cells as an estimator of bacterial productivity. *Appl. Environ. Microbiol.* 42:23–31.
56. Ogram, A., G. S. Sayler, and T. Barkay. 1988. DNA extraction and purification from sediments. *J. Microbiol. Meth.* 7:57–66.
57. Paustian, K., and J. Schnurer. 1987. Fungal growth response to carbon and nitrogen limitation: A theoretical model. *Soil Biol. Biochem.* 19:613–620.
58. Peele, E. R., F. L. Singleton, J. W. Deming, B. Cavari, and R. R. Colwell. Effects of pharmaceutical wastes on microbial populations in surface waters at the Puerto Rico dump site in the Atlantic Ocean. *Appl. Environ. Microbiol.* 41:873–879.
59. Powell, E. O. 1989. The growth rate of microorganisms as a function of substrate concentration. In E. O. Powell, C. G. T. Evans, R. T. Strange, and D. W. Tempest (eds.), *Microbial Physiology and Continuous Cell Culture.* Her Majesty's Stationary Office, London, pp. 34–55.
60. Reeve, C. A., P. Amy, and A. Matin. 1984. Role of protein synthesis in the survival of carbon-starved *Escherichia coli* and *Salmonella typhimurium. J. Bacteriol.* 160: 1041–1046.
61. Ride, J. P., and R. B. Drysdale. 1972. A rapid method for the chemical estimation of filamentous fungi in plant tissue. *Phys. Plant Pathol.* 2:7–15.
62. Robert, F. M., and E. L. Schmidt. 1982. Population changes and persistence of *Rhizobium phaseoli* in soil and rhizospheres. *Appl. Environ. Microbiol.* 45:550–556.
63. Roszak, D. B., and R. R. Colwell. 1987. Survival strategies of bacteria in the natural environment. *Microbiol. Rev.* 51:365–379.
64. Ruby, E. G., H. W. Jannasch, and W. G. Deuser. 1987. Fractionation of stable carbon isotopes during chemoautotrophic growth of sulfur-oxidizing bacteria. *Appl. Environ. Microbiol.* 53:1940–1943.
65. Schallenberg, M., J. Kalff, and J. B. Rasmussen. 1989. Solutions to problems in enumerating sediment bacteria by direct counts. *Appl. Environ. Microbiol.* 55:1214–1219.
66. Selander, R. K., D. A. Caugant, H. Ochman, J. M. Musser, M. N. Gilmour, and T. S. Whittam. 1986. Methods of multilocus enzyme electrophoresis for bacterial population genetics and systematics. *Appl. Environ. Microbiol.* 51:873–884.
67. Shields, J. A., E. A. Paul, and W. E. Lowe. 1973. Turnover of microbial tissue in soil under field conditions. *Soil Biol. Biochem.* 5:753–764.

68. Soderstrom, B., and S. Erland. 1986. Isolation of fluorescein diacetate stained hyphae from soil by micromanipulation. *Br. Mycol. Soc.* 86:465–468.
69. Somerville, C. C., I. T. Knight, W. L. Straube, and R. R. Colwell. 1989. Simple, rapid method for direct isolation of nucleic acids from aquatic environments. *Appl. Environ. Microbiol.* 55:548–554.
70. Steffan, R. J., and R. M. Atlas. 1988. DNA amplification to enhance detection of genetically engineered bacteria in environmental samples. *Appl. Environ. Microbiol.* 54:2185–2191.
71. Steffan, R. J., J. Goksoyr, A. K. Bej, and R. M. Atlas. 1988. Recovery of DNA from soils and sediments. *Appl. Environ. Microbiol.* 54:2908–2915.
72. Stotzky, G., and H. Babich. 1986. Survival of and genetic transfer by genetically engineered bacteria in natural environments. *Adv. Appl. Microbiol.* 31:93–138.
73. Swannell, R. P. J., and F. A. Williamson. 1988. An investigation of staining methods to determine total cell numbers and the number of respiring micro-organisms in samples obtained from the field and the laboratory. *FEMS Microbiol. Ecol.* 53:315–324.
74. Tempest, D. W., and O. M. Neijssel. 1978. Ecophysiological aspects of microbial growth in aerobic nutrient-limited environments. *Adv. Microb. Ecol.* 2:105–153.
75. Torsvik, V. L., and J. Goksoyr. 1978. Determination of bacterial DNA in soil. *Soil Biol. Biochem.* 10:7–12.
76. Tunlid, A., and G. Godham. 1986. Ultrasensitive analysis of bacterial signatures by gas chromatography/mass spectrometry. In F. Megusar and M. Gantar (eds.), *Perspectives in Microbial Ecology.* Proc., Fourth Int. Symp. Microb. Ecol., Slovene Soc. Microbiol., Ljublana, pp. 447–454.
77. Turner, S. M., and E. Newman. 1984. Fungal abundance on *Lolium perenne* roots: Influence of nitrogen and phosphorus. *Trans. Br. Mycol.* 82:315–322.
78. Turner, S. M., and E. Newman. 1984. Growth of bacteria on roots of grasses: Influence of mineral nutrient supply and interactions between species. *J. Gen. Microbiol.* 130:505–512.
79. Vaccaro, R. F. 1969. The response of natural microbial populations in seawater to organic enrichment. *Limnol. Oceanog.* 14:726–735.
80. Van Es, F. B., and L.-A. Meyer-Reil. 1982. Biomass and metabolic activity of heterotrophic marine bacteria. *Adv. Microb. Ecol.* 6:111–170.
81. Van Veen, J. A., and J. D. Van Elsas. 1986. Impact of soil structure and texture on the activity and dynamics of the soil microbial population. In F. Megusar and M. Gantar (eds.), *Perspectives in Microbial Ecology.* Proc. Fourth Int. Symp. Microb. Ecol., Slovene Soc. Microbiol., Ljublana, pp. 481–488.
82. Weddle, C. L., and D. Jenkins. 1971. The viability and activity of activated sludge. *Water Res.* 5:621.
83. White, D. C. 1983. Analysis of microorganisms in terms of quantity and activity in natural environments. In J. H. Slater, R. Whittenbury, and J. W. T. Wimpenny (eds.), *Microbes in Their Natural Environment.* Symp. 34, Soc. Gen. Microbiol. Cambridge University Press, Cambridge, pp. 37–66.
84. White, D. C., R. J. Bobbie, S. J. Morrison, D. Oosterhof, C. W. Taylor, and D. A. Meeter. 1977. Determination of microbial activity of estuarine detritus by relative rates of lipid biosynthesis. *Limnol. Oceanogr.* 22:1089–1099.
85. White, D. C., W. M. Davis, J. S. Nickels, J. D. King, and R. J. Bobbie. 1979. Determination of the sedimentary microbial biomass by extractible lipid phosphate. *Oecologia (Berl)* 40:51–62.
86. Whittam, T. S. 1989. Clonal dynamics of *Escherichia coli* in its natural habitat. *Ant. van Leeuwenhoek* 55:23–32.
87. Yamano, H., and K. Takahashi. 1983. Temperature effect on the activity of soil microbes measured from heat evolution during the degradation of several carbon sources. *Agric. Biol. Chem.* 47:1493–1499.
88. Young, J. P. W. 1985. *Rhizobium* population genetics: Enzyme polymorphism in isolates from peas, beans, and lucerne grown at the same site. *J. Gen. Microbiol.* 131:2399–2408.

89. Zelibor, J. L., Jr., M. W. Doughten, D. J. Grimes, and R. R. Colwell. 1987. Testing for bacterial resistance to arsenic in monitoring well water by the direct viable counting method. *Appl. Environ. Microbiol.* 53:2929–2934.
90. Zelibor, J. L., Jr., M. Tamplin, and R. R. Colwell. 1987. A method for measuring bacterial resistance to metals employing epifluorescence microscopy. *J. Microb. Meth.* 7:143–155.
91. Zimmerman, R., R. Iturriaga, and J. Becker-Birck. 1978. Simultaneous determination of the total number of aquatic bacteria and the number thereof involved in respiration. *Appl. Environ. Microbiol.* 36:926–935.

Chapter 31

Use of the Mixed Flask Culture Microcosm Protocol to Estimate the Survival and Effects of Microorganisms Added to Freshwater Ecosystems

Lyle J. Shannon

Richard L. Anderson

Introduction

The ability to manipulate an organism's genetic substance offers opportunities to benefit many aspects of human health and well-being. Coupled with this positive aspect of genetic engineering, however, is a concern about the potential adverse effects on human welfare and environmental quality. Some opponents of genetic engineering have been quick to paint images of new organisms running amok and taking over the planet. While this is unlikely, there is legitimate cause for concern over the effects of releasing novel organisms into natural systems. Experience with exotic species that have successfully exploited new niches underscores a need for caution.

New microorganisms are being developed for many purposes. The industrial applications for novel organisms appear boundless, ranging from manufacturing to pharmaceutical, to agricultural, to mining, to bioremediation. While many of these organisms are designed for use in processes where they can be controlled, others are targeted specifically for release into the environment. One such group, microbial pest control agents (MPCAs), have become the subject of intense development and serve as an example of the test requirements needed for the regulatory process.

Insect pest control has moved from the exclusive use of synthetic organic chemicals to an integrated program of chemicals and pathogenic microorganisms. Originally, naturally occurring pathogens such as *Bacillus thuringiensis, Bacillus popilliae,* and certain baculoviruses (5) were used. Low survival and virulence often offset the advantage of host specificity for natural MPCAs. Engineering techniques are being applied to a variety of entomopathogenic bacteria, fungi, and viruses in an effort to produce "new organisms" with enhanced pest control attributes missing from the natural organism.

The Environmental Protection Agency (EPA) has the responsibility for regulating distinct aspects of the production and use of these genetically engineered microorganisms (GEMs) under two legislative acts. The Federal Insecticide, Fungicide, and Rodenticide Act (FIFRA) directs the EPA to register MPCAs before they can be used in this country. The Toxic Substances Control Act (TSCA) provides EPA with jurisdictions over all chemicals already in commerce or new chemicals intended for commercial use that are not specifically covered by other regulatory authorities (22). Both acts give EPA the power to prescribe tests and data-reporting requirements for GEMs or MPCAs. Most tests currently available were developed to provide information for chemicals, however, and may not be directly applicable to predicting impacts of GEMs.

Both FIFRA and TSCA regulate through several procedures including an ecological assessment of the risk associated with using a material. For example, risk assessment for a new chemical pesticide would include some measure of its *toxicological hazard* determined through a tier testing process and an estimation of the potential *environmental exposure* of nontarget populations (24). Toxicological hazard data, for chemicals in fresh water, would include a suite of tests that produces acute 2- or 4-day exposure laboratory data, usually in the form of LC50 values, that is, a concentration affecting 50 percent of the test population. These tests might be followed with more complex life-cycle invertebrate and fish exposures to measure survival and reproduction. Such traditional LC50 values and reproductive studies may have less value in GEM or MPCA risk assessments because direct lethality and reproduction effects may not be the major environmental impacts. Environmental exposure is determined using a combination of the estimated amount of chemical in and available to organisms and the numbers, types, distribution, abundance, dynamics, and natural history of organisms that will be exposed to chemicals. Estimating environmental concentrations and predicting the effects of microorganisms on the structure and function of an ecosystem may be more difficult than for chemicals due to our limited knowledge of the fate and effects of microorganisms introduced into freshwater ecosystems.

Live microorganisms present a special problem when assessing their effects in natural systems. In any controlled testing paradigm the question of containment is critical. Since field tests release the test organism into the "wild," they should only be used as the final stage in hazard assessment. Some determination of the probability of the test organisms' survival in the wild must be made before field studies can be used. These considerations point to the need for an intermediate step, a test protocol that is more ecologically realistic than an acute or chronic single-species test but more contained than an outdoor experiment. Laboratory microcosms offer an appropriate solution.

Although many types of laboratory aquatic microcosms have been used for chemical testing (4, 6—8, 11, 23), their suitability for evaluating microbial effects has yet to be determined. For routine testing, a microcosm should contain a community capable of performing the important trophic interactions seen in the field. The system should be complex enough to allow for decomposition, nutrient cycling, and primary production, and contain at least one level of consumers. Yet the system must not be so complex as to be impossible to reproduce. While it is important to measure effects on both ecosystem function and structure, a balance must be struck between the information provided and the cost. Additional considerations include the suitability of the system for containment and decontamination. In general, small, static systems appear to be more easily contained and generate less waste than large or flow-through systems. Small systems also have the advantage of permitting a larger number of replicates, thereby increasing statistical power.

The size of the test system determines, in part, which species can be included. Most microcosms, even large ones, do not support a top predator (e.g., a fish). As the size of the system decreases, the size and numbers of organisms it can support also decreases. Systems containing only a few liters may be unable to support populations of large insect predators. Aquatic systems of less than a liter may be unable to sustain zooplankton populations. Test-tube-based systems might only support a microbial community. Each microcosm has advantages and disadvantages. This chapter describes a freshwater microcosm that we believe offers a good compromise on size and ecological complexity.

The Test System

The microcosm test procedures we have developed (15) are based on a "naturally derived" microcosm protocol originally proposed by Leffler (8). Although we have extensively modified a number of procedures and added many new ones, the philosophy behind this protocol and the basic approach are still those of Leffler. The test system is a generic one, predicated on the idea that many ecosystem processes (e.g., decomposition, nutrient cycling, primary production, and respiration) are independent of species composition. As long as suitable groups are represented in the microcosms, these processes will occur and a viable, self-sustaining community will develop.

Development of this protocol began in 1984 with a series of tests designed to compare the hazard rank of chemicals generated by community and ecosystem level tests with those generated by single-species tests (4, 26, 27). This work included 16 experiments with the Leffler protocol. During our work, other investigators completed eight experiments with this method and evaluated the developmental assumptions of the system (19, 20) (Table 31.1).

These experiments showed that Leffler's protocol produced viable, functioning aquatic communities. Modifications were necessary, however, to improve the replicability and repeatability of the test and enhance the procedures used for monitoring, data handling, and data analysis. In developing appropriate modifications, we completed four additional tests, which resulted in a new draft protocol for the mixed flask culture (MFC) microcosm test (15). Using the procedures outlined in this new protocol, we have completed three chemical tests and four tests with microorganisms (Table 31.1). The following sections briefly describe the procedures and provide examples of applications where the method has been used. The complete protocol is available from the senior author.

Design features

This protocol can provide data on the fate and effects of chemicals or survival and effects of microorganisms introduced into a freshwater environment. The system is designed as a generic model of small eutrophic ponds. It contains stable communities of primary producers, consumers, and decomposers, and changes are monitored at both the

TABLE 31.1   History of Tests Conducted in Mixed Flask Microcosms

Compound	No. of tests
Original Leffler protocol:	
Copper sulfate, $CuSO_4$	2
1-Octanol	2
1-Decanol	2
1-Hexanol	2
Aniline	1
2,6-Diisopropylaniline	1
4-Hexyloxyaniline	1
2,3,4,6-Tetrachloroaniline	2
2-(Octyloxy)acetanilide	1
Diuron	2
Salicyclanilide	2
Hercofloc 863	1
Hercofloc 872	1
Atrazine	4
Fluorene	4
Shannon (MFC) protocol:	
Dursban	2
Fenvalerate	1
Bacillus thuringiensis subsp. *israelensis*	3
Serratia marcescens	1

population and ecosystem levels. Measurements of pH and dissolved oxygen are used to estimate primary production and respiration. These functions are influenced by factors such as microbial decomposition activities, nutrient cycling, and grazing of protozoans, zooplankton, and other crustaceans and insects. As such they provide an integrated measure of effects. Although these functional measurements are sensitive indicators of perturbation, structural measurements (i.e., population densities) are also necessary to explain the cause of changes observed at the ecosystem level (Stay et al., 1989b).

Microcosms are constructed in 1-liter Pyrex beakers and are covered with a 150 × 15 mm plastic or glass petri dish cover. Because of the beaker lip, the system is not completely sealed and gas exchange with the atmosphere can occur. The cover reduces the probability of outside contamination and loss of the introduced agent to the outside. These microcosms contain a defined nutrient medium, a sand sediment, and a diverse community of bacteria, fungi, protozoa, algae, rotifers, planktonic, and benthic grazers. The starting community "seed" is derived from a variety of natural sources that are allowed to "coadapt" in the laboratory. The microcosms' small size permits a large number of them to be held in a small space. As many as five replicates in each of six different treatments (30 microcosms) can be managed by two people.

Test protocol

Development of a stock community. The stock culture is developed in a 40-liter aquarium held in an incubator or constant-temperature room at 20°C and a 12-h light-dark cycle with enough light to support algal growth. The aquarium is first acid-washed and carefully rinsed. Acid-washed white quartz sand is added to a depth of 2 to 3 cm, and the aquarium is filled with approximately 36 liters of T82 nutrient medium (Taub and Read, 1982). Samples taken from natural ecosystems are collected and added to the aquarium to start a stock culture. A variety of sites should be sampled in order to ensure a diversity of species in the stock cultures. Suitable sources include small ponds, lakes, marshes, vernal pools, or other standing water sources that have no known exposure to toxicants. Approximately 2 liters of material collected from natural ecosystems is used to inoculate the stock aquarium.

The stock community is allowed to mature for at least 3 months prior to use as an inoculum. This maturation period provides a relatively stable species assemblage since those species that are unable to coexist with other members of the community are eliminated, leaving a "coadapted" group of species. At least one new stock culture should be prepared every 6 months to ensure a healthy, growing inoculum. Inoculum from older stock cultures in addition to new "wild" material should be used to start a new culture. Stock cultures may be used for several years unless obvious changes suddenly occur in the system.

Before use, the stock culture should be microscopically examined to determine its species composition. The following minimum criteria must be met:

1. Two species of single-celled green algae or diatoms
2. One species of filamentous green alga
3. One species of nitrogen-fixing blue-green alga
4. One grazing macroinvertebrate
5. One benthic, detritus-feeding macroinvertebrate
6. Bacteria and protozoa species

In our experience, species diversity is typically far greater than this minimum.

Microcosm construction. Microcosm vessels must be scrupulously clean. We recommend that they be washed five times with tap water, five times with a solution containing 200 mg/liter hypochlorite, five times with 10% HCl, and five times with distilled water. They are then autoclaved for 15 min at 121°C. Microcosms are constructed by adding 50 mL of acid-washed quartz sand sediment and 900 mL of Taub 82 medium (supplemented with 15 μg $NaHCO_3$ per liter as an additional carbon source) to the microcosm vessel. Each microcosm is then inoculated with 50 mL of the stock community.

Test exposures usually consist of 4 treatment groups (a control and 3 test groups), each containing 5 replicate microcosms for a total of 20 beakers. Microcosm vessels are incubated at 20°C ± 1°C in a suitably sized growth chamber on a 12-h light-dark cycle. Each microcosm is randomly assigned to a treatment group, given a replicate letter, and positioned in the incubator according to a randomized block design. Each block is moved to a new shelf position twice a week to distribute any effects of shelf position. To simulate naturally occurring immigration and permit the reintroduction of extirpated species, all microcosms are reinoculated each week with 10 mL of the stock community.

Microcosms are allowed to mature for 6 weeks prior to treatment and are monitored through a 6-week exposure period. During the first 6 weeks, consistency among replicates is improved with a weekly cross-inoculation. This is accomplished by thoroughly mixing all microcosms and removing 100 mL from each. These 100-mL aliquots are combined in a 4-liter beaker, thoroughly mixed and redistributed among the vessels.

It is prudent to start more systems than needed for the experiment. The extra systems provide replacements for vessels that are accidentally broken or simply fail to develop. Just prior to treatment, beginning measurements are taken and each microcosm evaluated. To be acceptable for use in a test, a microcosm must (1) contain all the required functional groups, and (2) meet the minimum criteria of 4 mg/liter predawn dissolved oxygen and 11 mg/liter late-afternoon dissolved oxygen. Any not meeting these criteria are removed. If the remaining group of microcosms is larger than the number needed for a test, those microcosms showing the greatest deviation from the group means for oxygen and pH measurements are successively removed until an appropriate number remain. This process serves to reduce variability among the group of test microcosms.

Variables measured. During the exposure, monitoring is done on a schedule of decreasing frequency. Measurements are taken twice weekly for the first 2 weeks, and then weekly. Normally the exposure period lasts for 6 weeks, although longer exposure times can be used.

Functional variables. Dissolved oxygen (DO) and pH are used as indicators of effect. They are both easily measured and they reflect trophic interactions within the biogeochemical matrix of the ecosystem (14). By measuring oxygen content three times, at 7 A.M. and 4 P.M. on day 1 and at 7 A.M. on day 2, diel oxygen gain and loss can be calculated (10). Oxygen gain is a measure of net primary production, while oxygen loss is a measure of community respiration. From estimates of production (P) and respiration (R), a P/R ratio can be computed as a further indicator of the status of the system.

Dissolved oxygen concentrations are measured with a meter and polarographic oxygen electrodes. Separate electrodes are used for control and treated groups. Measurement of pH is done with a meter equipped with a junction box and four glass combination electrodes. Separate electrodes are used for each treatment group.

Structural variables. Populations of major taxa are determined according to the procedures outlined below. The level of taxonomy depends on the objectives of the study. In many cases, identification to order or major functional group is sufficient. In some tests, identification to species may be warranted. All sampling devices should be autoclaved before and after use.

Algae. Algal species are counted by transferring a small sample from a stirred microcosm to a Palmer-Maloney counting slide. A Pasteur pipette works well unless there are high densities of filamentous species present. These require a larger bore pipette. Counts are made at 40× magnification.

Zooplankton and macroinvertebrates. Macroinvertebrates (amphipods, insects, molluscs) are usually large enough to be counted with the naked eye. Normally they are present in relatively low numbers so the total number per microcosm can be counted. Zooplankton are normally too abundant for total counts and must be subsampled by removing 50-mL aliquots. These aliquots are counted under a dissecting microscope and then returned to the system.

Protozoa. Protozoa typically show an extremely clumped distribution that makes it difficult to obtain accurate counts. Best results are obtained by removing several (usually 10) 0.2-mL aliquots to form a composite sample. The composite is mixed on a Vortex mixer. A 0.02-

mL subsample is transferred to a clean glass slide and covered with a 10-cm^2 piece of plastic wrap (17). Counts are made under a phase contrast microscope at $40\times$ magnification.

Bacteria. Samples for determination of bacterial density are collected with an inverted 10-mL pipette fitted with a suction bulb on the tip. A combination sediment and water column sample is withdrawn from each microcosm by first collecting about 0.7 mL of sand-sediment from the bottom, then lifting the pipette while continuing to draw from the water column until the 10-mL mark on the pipette is reached. This results in a sample of about 13.2 mL of which 0.7 mL is sediment and 12.5 mL is liquid. The sample is transferred and temporarily stored in a sterile test tube. The choice of enumeration method will vary depending on the organism and the objectives of the test.

Data collection and analysis. Data collection and analysis have been automated through a series of programs developed for use on an IBM PC or compatible computer. Functional measurements are entered directly into the computer and echoed to a printer as they are collected. Structural measurements are first recorded on data sheets and later entered into the computer. At the conclusion of an experiment, all data are combined into a single ASCII file that can be readily imported into a variety of commercial programs. For routine analyses we have developed a program to calculate means, standard deviations, and Dunnett's significant difference (18). The Dunnett's value is used to determine which of the treatments are different from the control (at $p \leq 0.05$). This test is performed on each variable measured for each sampling day. It provides a determination of which variables are significantly affected and the duration of the effect.

From these calculations, the minimum observable and no-effect concentrations can be determined. From the periodic counts of the density of the microbial agent, determinations of the survival rate of the organism can also be reported. The calculations are visually summarized by plotting the means of each treatment group over time. An area corresponding to the upper and lower bounds of Dunnett's least significant difference ($p < 0.05$) is plotted around the line for the control mean (Fig. 31.1). This envelope defines a region where the treatments are not significantly different from the control. Points lying outside this shaded area are significantly different ($p < 0.05$) from the control. We developed a program that automatically creates these plots on a Postscript-equipped laser printer.

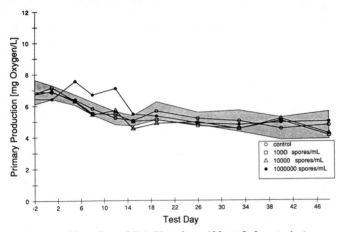

Figure 31.1 The effect of Bti (Vectobac, Abbott Laboratories) application on primary production (as mg/liter of O_2 per 12 h) in microcosm communities. Each point represents the mean of five replicate microcosms. The shaded envelope represents Dunnett's significant difference from the control ($p \leq 0.05$). Bti added on day 0.

Quality Assurance

The sensitivity of the variables measured in these microcosms was determined by calculating a minimum detectable difference (16). Structural and functional measures were found to differ widely in their sensitivity (15, 21). Functional variables were by far the most sensitive with average minimum detectable differences ranging from 3 to 19 percent. Structural variables were considerably less sensitive. For the broadest taxonomic groupings (where the number of individuals was large and the sampling variability was low) detectable differences were approximately 50 percent. Detectable differences for rare taxa were sometimes much higher, ranging well over 100 percent.

In tests with both chemicals and microorganisms, we have found good agreement among experiments (15, 27). Despite differences in community structure, the total densities of algae and zooplankton and the rates of primary production and community respiration attain similar levels in repeated experiments. Despite differences in species composition among repeated experiments, we have found that these microcosm systems develop in a predictable manner. The appropriate status of a microcosm ecosystem at the beginning of a test (after the 6-week development period) is summarized in Table 31.2. These values were derived from 150 microcosms used in our six most recent tests and should serve as a quality assurance guideline for determin-

TABLE 31.2   Mean Values for Major Microcosm Variables at the Sixth Week of the Development Phase

	AM pH	Oxygen gain	Oxygen loss	P/R	Total zooplankton per liter	Total algae per mL	Total protozoa per mL
Mean	8.49	4.40	4.61	0.97	255	2.1×10^8	35
SD	0.59	1.01	1.05	0.13	138	1.5×10^8	25

ing whether microcosms have developed properly. Microcosms falling below these guidelines should not be used in tests.

Applications

Although these microcosms have been used extensively for chemical testing, it is only recently that they have been adapted for evaluating survival and effects of introduced microorganisms. A series of tests with *Bacillus thuringiensis v. israelensis* (Bti) and *Serratia marcescens* indicated are suitable for monitoring both the survival and ecological effects of introduced microorganisms (15).

Survival of introduced microorganisms

Bti is a registered pest control agent used in mosquito (Culicidae) and blackfly (Simuliidae) control programs. When a commercial formulation (Vectobac, Abbott Laboratories) was added to microcosms, Bti spores persisted at their inoculation densities for the duration of a 6-week monitoring period (Fig. 31.2) (15). There were relatively few spore-formers among the "indigenous" microcosm microbial community, so Bti spores could be monitored by plating pasteurized samples of water and sediments on tryptic soy agar (TSA). Additional tests with another commercial formulation (Mosquito Attack, Reuters Laboratories) and preparations of washed Bti spores confirmed this pattern of persistence.

If a test MPCA survives in a microcosm community, it is important to determine how long it maintains its toxic or pathogenic properties. This can be accomplished either by periodically adding new target organisms to the microcosms and determining their survival, or by removing water and sediment samples from the microcosms and using them in bioassays against target organisms. Although Bti persisted in microcosms, its ability to kill target organisms quickly disappeared (15). Bioassays conducted using water and sediment samples with first instar mosquito larvae, *Aedes atropalpus,* consis-

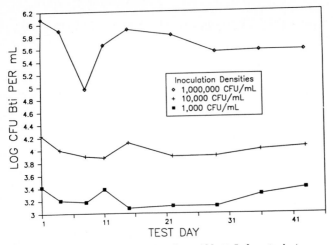

Figure 31.2 Survival of Bti (Vectobac, Abbott Laboratories) in microcosms. Counts are based on pasteurized samples plated on TSA agar.

tently showed a complete loss of toxicity between 7 and 14 days post-application. This was not an unexpected result since several investigators have noted a similar loss of Bti toxicity in natural systems (9, 12, 13, 25).

Serratia marcescens was used as a non-spore-forming organism to provide a contrast to the Bti study. *Serratia* was readily monitored in microcosms, but in contrast to Bti it did not persist (Fig. 31.3). This pattern of rapid decline is one of three patterns of bacterial survival noted in a recent review (Alexander, 1986) and is probably more typ-

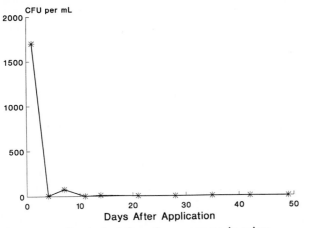

Figure 31.3 Survival of *Serratia marcescens* in microcosms.

ical of bacterial introductions than the constant spore densities maintained in the Bti tests.

Movement of the test organism into the food chain. In monitoring the distribution of an organism, it is necessary to examine its partitioning in water, sediments, and the biota. Our work in this area is continuing, but we have found such measurements are possible for most species. In Bti tests, for example, we removed amphipods, cladocerans, and copepods from treated microcosms on a weekly basis. These animals were sonified in 5 mL of sterile water, which was then pasteurized and plated on TSA agar. Bti accumulated rapidly in the animals and was gradually lost over the 4-week test period (Fig. 31.4; 15).

Predicting the distribution of an introduced microorganism requires detailed studies of the distribution of the organism among the abiotic and biotic components of an ecosystem. Because of their small size, these microcosm systems contain somewhat limited zooplankton populations (typically 200 to 600 individuals). Repeated sampling of animals for microbial content can severely deplete these populations and result in changes in population densities that affect rates of primary production, respiration, nutrient cycling, and other ecosystem processes. Distribution studies are probably best accomplished by using a large number of microcosms, and sacrificing complete systems at each sampling date. This will provide replication so that variability can be

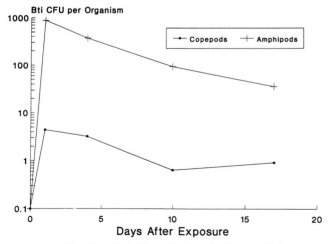

Figure 31.4 Uptake of Bti (Mosquito Attack, Reuters Laboratories) by crustacean zooplankton in microcosm communities. Each point represents the mean of 10 animals.

quantified and avoid the possibility of confusing sampling-derived effects with true effects.

Ecological effects

Bti does affect microcosm communities (15). At exposure levels of 10^3 spores per milliliter, all mosquito larvae and a few chironomids were killed within 24 h. At the high exposure (10^6 spores per milliliter) chironomids were totally eliminated, but zooplankton and amphipods remained unaffected. Some functional measures also showed significant effects at the high exposure rate. The most consistent of these were brief increases in primary production (see Fig. 31.1) and community respiration. These changes apparently were a response to the loss of grazing species and the addition of nutrients associated with the Vectobac formulation. Higher rates of community respiration probably reflected increased microbial activity in decomposing the carcasses of mosquitoes and chironomids and the organic materials used in the commercial formulation. Primary production increased in response to both a loss of grazer species and an increased nutrient pool. The relatively short duration of these responses (approximately 2 weeks) indicated that long-term ecological consequences of Bti application would likely be minimal.

In testing an MPCA it is important to demonstrate that the organism can function as a pest control agent under conditions found in the test system. If the test organism is not capable of operating under microcosm conditions, test results may have little relevance to natural systems where the agent is used. Our approach for MPCA testing is to incorporate a target animal into the test scheme. Aquatic target animals (e.g., mosquito larvae) can generally be added to the microcosm community. If these organisms are then killed by the test MPCA, it demonstrates that the test organism is capable of functioning in the system and that appropriate concentrations have been tested. The inclusion of target animals also allows for the detection of secondary effects arising from the loss of that species from the test community.

Interpretation

Statistical and ecological significance are not necessarily equivalent. As noted earlier, it is important to know the statistical properties of the variables being monitored. In these microcosms, functional measurements are quite sensitive and may show statistically significant differences for deviations from the control of as little as 5 to 10 percent. Such changes may be statistically significant, but their ecological significance is probably small. With less sensitive variables (e.g., measures of population density) where deviations of 50 percent or greater are necessary to show statistical significance, significant statistical effects are likely to be ecologically important. At the extreme end of the scale, some measurements have such a high degree of inherent variability (e.g., protozoan densities) that it is difficult to demonstrate statistical significance, even though there may have been large changes in populations and ecologically significant differences could have occurred. It is evident that both an understanding of the statistical properties of a system and a knowledge of ecological relationships will be necessary to predict whether an introduced microorganism will cause a significant ecological effect.

There is more to consider in making a judgment of whether an observed effect is ecologically significant. Beyond a determination of which variables were affected and the statistical properties of those variables, the ability and time required for a microcosm community to recover from perturbation should also be considered. A great advantage of microcosm tests is their ability to simulate immigration and recolonization, thereby affording the opportunity to demonstrate recovery. Disturbances, even major ones, from which microcosms can recover (i.e., return to control levels) may have less ecological significance than those from which they fail to recover. It is important that the ability of the microcosm community to recover (its resilience) be used in evaluating the ecological significance of observed effects. In the Bti tests, most affected variables recovered to control levels by the end of the test, providing evidence for a prediction that the ecological consequences of Bti application are relatively minor.

While these microcosms appear to work well, we need to establish their predictive capabilities. Lab-to-field comparisons with chemical tests have generally shown these microcosms to be good models of the small pond environment and good predictors of fate and effects (19, 21). Results of our own tests with the insecticide chlorpyrifos (Dursban) compared favorably with those observed in concurrent outdoor tests conducted in the littoral zone of a pond (3). The results showed that untreated (control) microcosms had many functional and structural similarities to the field sites, indicating that the microcosms were accurate models of shallow ponds. Both in the laboratory and in the field, amphipods and cladocerans were lost at 0.5 μg/liter of chlorpyrifos, while copepods and ostracods suffered less severe reductions. Algae were not directly affected in either system. Concurrent lab and field tests with Bti are currently under way.

Summary

Given the uncertainties concerning the fate and survival of microorganisms released into the environment, there is a clear need for a contained test system. While further lab-to-field calibration and validation are necessary, on the basis of current information, these microcosms can play a significant role in that regard. These microcosms have many structural and functional similarities to small ponds. Because of their simplicity, however, they are easy to construct and maintain. The behavior of these systems has been well studied and the predictive capabilities of the variables described. Their small size makes it possible to maintain large numbers of microcosms within a relatively small space. This not only facilitates containment but also enhances the statistical power of an experiment by increasing the numbers of replicates in a treatment group.

Acknowledgments

This work was largely sponsored by the U.S. EPA through cooperative agreements CR-812799 and CR-810741. Many people have been involved in the development and testing of this microcosm protocol. Dr. David Yount was instrumental in initiating this work and has contributed much to its success. Dr. Michael Harrass also provided major input into the early stages of this work. Much of the laboratory work was done by Carl Mach and Terry Flum with additional assistance provided by Charles Walbridge, Daniel Fitzsimmons, and Eric Mead.

References

1. Alexander, M. 1986. Potential impacts of environmental release of biotechnology products: Assessment, regulation, and research needs. 4. Fate and movement of microorganisms in the environment. Part 1: Survival and growth. *Environ. Manage.* 10:464–469.
2. Bourquin, A. W., R. L. Garnas, P. H. Pritchard, F. G. Wilkes, C. R. Cripe, and N. I. Rubenstein. 1979. Interdependent microcosms for the assessment of pollutants in the marine environment. *Int. J. Environ. Studies* 13:131–140.
3. Brazner, J. C., L. J. Heinis, and D. A. Jensen. 1989. A littoral enclosure for replicated field experiments. *Environ. Tox. Chem.* 8:1209–1216.
4. Flum, T. F., and L. J. Shannon. 1987. The effects of three related amides on microecosystem stability. *Ecotox. Environ. Safety* 13:239–252.
5. Fuxa, J. R. 1987. Ecological considerations for the use of entomopathogens in IPM. *Ann. Rev. Entomol.* 32:225–251.
6. Giddings, J. M. 1983. Microcosms for the assessment of chemical effects on the properties of aquatic ecosystems. *Hazard Assessment of Chemicals: Current Developments* 2:45–94.
7. Giesy, Jr., John P. 1980. *Microcosms in Ecological Research.* Coordinating ed., DOE Symposium Series; 52, CONF-781101, NTIS, U.S. Department of Commerce, Springfield, Va.
8. Leffler, J. W. 1984. The use of self-selected, generic aquatic microcosms for pollution effects assessment. In H. H. White (ed.), *Concepts in Marine Pollution Measurements.* Maryland Sea Grant College, University of Maryland, College Park, Md.

9. Margalit, J., and H. Bobroglo. 1984. The effect of organic materials and solids in water on the persistence of *Bacillus thuringiensis* var. *israelensis* serotype H-14. *Z. Angew. Entomol.* 97:516–520.

10. McConnell, W. J. 1962. Productivity relations in carboy microcosms. *Limnol. Oceano.* 7:335–343.

11. Metcalf, R. L. 1977. Model ecosystem approach to insecticide degradation: A critique. *Ann. Rev. Entomol.* 22:241–261.

12. Mulligan, F. S., C. H. Schaefer, and W. H. Wilder. 1980. Efficacy and persistence of *Bacillus sphaericus* and *B. thuringiensis* H. 14 against mosquitoes under laboratory and field conditions. *J. Econ. Entomol.* 73:684–688.

13. Ramoska, W. S., S. Watts, and R. E. Rodriguez. 1982. Influence of suspended particulates on the activity of *Bacillus thuringiensis* serotype H-14 against mosquito larvae. *J. Econ. Entomol.* 75:1–4.

14. Schindler, J. E., and J. B. Waide. 1980. 1. Theoretical Rationale. In J. P. Giesy (ed.), *Microcosms in Ecological Research*. DOE Symposium Series; 52, CONF-781101.

15. Shannon, L. J., T. E. Flum, R. L. Anderson, and J. D. Yount. 1989. Adaptation of mixed flask culture microcosms for testing the survival and effects of introduced microorganisms. In U. M. Cowgill and L. R. Williams (eds.), *Aquatic Toxicology and Hazard Assessment,* vol. 12. ASTM STP 1027. American Society for Testing and Materials, Philadelphia, pp. 224–239.

16. Snedecor, G. W., and W. G. Cochran. 1976. *Statistical Methods.* Iowa State University Press, Ames, Iowa.

17. Spoon, D. M. 1976. Use of thin, flexible plastic coverslips for microscopy, microcompression and counting of aerobic microorganisms. *Trans. Am. Micros. Soc.* 95(3):520–523.

18. Steel, R. G. D., and J. H. Torrie. 1960. *Principles and Procedures of Statistics.* McGraw-Hill, New York.

19. Stay, F. S., A. Katko, C. M. Rohm, M. A. Fix, and D. P. Larsen. 1988. Effects of fluorene on microcosms developed from four natural communities. *Environ. Toxicol. Chem.* 7:635–644.

20. Stay, F. S., A. Katko, C. M. Rohm, M. A. Fix, and D. P. Larsen. 1989a. The effects of Atrazine on microcosms developed from four natural plankton communities. *Arch. Environ. Contam. Toxicol.* 18:866–875.

21. Stay, F. S., T. E. Flum, L. J. Shannon, and J. D. Yount. 1989b. An assessment of the precision and accuracy of SAM and MFC microcosms exposed to toxicants. In U. M. Cowgill and L. R. Williams (eds.), *Aquatic Toxicology and Hazard Assessment,* vol. 12. ASTM STP 1027. American Society for Testing and Materials, Philadelphia, Pa., 189–203.

22. Stern, A. M. 1986. Potential impacts of environmental release of biotechnology products: Assessment, regulation, and research needs. 3. *Regulatory Aspects Environ. Manage.* 10:453–462.

23. Taub, F. B., and P. Read. 1982. *Standardized Aquatic Microcosm Protocol.* Final report, vol. 2. FDA Contract 223-80-2352.

24. Urban, D. J., and N. J. Cook. 1986. *Hazard Evaluation Division Standard Evaluation Procedure Ecological Risk Assessment.* U.S. Environmental Protection Agency, EPA-540/9-85-001.

25. Van Essen, F. W., and S. C. Hembree. 1982. Simulated field studies with four formulations of *Bacillus thuringiensis* var. *israelensis* against mosquitoes: Residual activity and effect of soil constituents. *Mosq. News.* 42:66–72.

26. Yount, J. D., and L. J. Shannon. 1987. Effects of aniline and three derivatives on laboratory microecosystems. *Environ. Toxicol. Chemistry* 6:463–468.

27. Yount, J. D., and L. J. Shannon. 1988. State changes in laboratory microecosystems in response to chemicals from three structural groups. In J. Cairns, Jr., and J. R. Pratt (eds.), *Functional Testing of Aquatic Biota for Estimating Hazards of Chemicals.* ASTM STP 988. American Society for Testing and Materials, Philadelphia, Pa., pp. 86–96.

Chapter 32

Synthetic Microcosms As Test Systems for Survival and Effects of Genetically Engineered Microorganisms

Frieda B. Taub

Introduction

The survival and functioning of species introduced into natural environments depend on the competitive balance between the introduced and indigenous species for limiting resources, and the balance between reproduction and mortality including predation. Interactions such as symbiosis or antibiotics also may affect growth and physiology. Therefore, the success of introduced species ultimately must be determined in mixtures of organisms. These statements are equally valid for genetically engineered microbes (GEMs) as for other exotic organisms.

Ecological studies begin and end with studies of natural communities, but there are times when a researcher needs intermediate, simpler test systems to reduce complexity. In these intermediate systems, the researcher sacrifices realism to gain analytical capability. Synthetic microcosms are generally used to minimize unknown or uncontrolled variables. They commonly are composed of chemically defined media and most, if not all, of the organisms are known. Synthetic microcosms are often used to avoid the seasonal and spatial variability of natural substrates and organism assemblages which, if undefined, cannot be repeated in subsequent experiments.

While some experimentation is served best by designing simple experiments with explicit limitations, and subsequently increasing the complexity from single species, through simplistic models toward natural assemblages, other research is better served by site-specific microcosms (12). The value of any microcosm study will be judged by its ability to help us understand and predict how the ecological system under consideration will respond to new organisms or chemicals. In some cases, predictions need to be made for a specific habitat, e.g., a specific freshwater pond, in other cases for generalized habitats (e.g., any freshwater pond) or, more broadly, for any aquatic community. The appropriateness of the test system depends on the type of prediction that needs to be addressed. The continuum from synthetic to naturally derived microcosms will be described among the examples and their relative use contrasted in the Discussion section of this chapter.

Examples of Synthetic Microcosms

The ecological study of GEMs is new, but methods can be adapted from earlier experiments that were useful in evaluating interactions among organisms under control and chemically stressed conditions. This paper offers a review of synthetic microcosm studies, from simple through complex. In some cases, applicable methods have yet to be applied to GEM research, and some potential uses are suggested.

Single species

The uses of axenic (pure) cultures and of synthetic media are widespread in microbial studies. Single-species studies are performed to ensure that the metabolic properties ascribed to a particular species are caused by that species and not another associated with it. Indeed, microbiology has suffered from the almost exclusive use of single-species studies with the result that species interactions have been ignored too often. Synthetic media are routinely used to test single substrates and to provide repeatable conditions for subsequent experiments.

The rates of segregation and selection of an *E. coli* B, transformed with DNA of the nonconjugative plasmid pACYC184, were studied in a simple glucose-limited continuous culture (37). By using a population dynamics model, Lenski and Bouma showed that both plasmid loss and negative selection were important processes in predicting the population dynamics of plasmid-free segregants. This is one of many studies in which the use of a single species can yield information on the stability of a GEM, isolated from competition or predation from other species, and in a medium with a known limiting factor. The simplicity of the system makes the data easier to interpret and makes the experiment capable of being repeated in other laboratories.

The study of microbial inclusions and their maintenance is important in GEMs, and the example above can be used for evaluating many kinds of organisms. The use of mathematical models to study phenomena such as genetic stability increases the amount of understanding that can be gained from the data and suggests ways to extrapolate from the test case to other cases.

Two or three species—studies of a process such as genetic exchange, mutualism, competition, grazing, or predation

It is easier to study species interactions in the absence of nonessential species, at least initially. For this reason, defined media, sterile soil or water are often inoculated with the species under investigation. The elimination of extraneous species is most helpful in genetic exchange experiments where many types of recombinants are possible, and the investigator wishes to know that all of the organisms are either the parental strains or new recombinants, without having to distinguish these from other organisms in a complex, natural ecosystem. Eventually, the information gained in two- or three-species studies needs to be integrated into our understanding of more complex species assemblages found in natural communities.

Numerous examples of genetic exchanges among two or three organisms in defined media or sterile soil have been summarized (55), including studies by Stotzky and his colleagues who showed that the clay montmorillonite increased the conjugation of prototrophic and auxotrophic stains of *E. coli.* 'Three organisms were used to demonstrate the effectiveness of the conjugative R100-1 plasmid from *E. coli* to transfer the plasmids pBR322 and pBR325 to indigenous wastewater organisms (20, 39, 42).

An example of mutualism was demonstrated in the adequacy of a consortium of three anaerobic microbes to degrade 3-chlorobenzoate, although the natural community from which they were derived contained at least nine organisms (75). Of the three, DCB-1 accomplishes reductive chlorination, using H_2 and releasing benzoate; strain BZ-2 grows only in the presence of a H_2-consuming bacteria, and uses benzoate; PM-1, a *Methanospirillum,* uses H_2 and produces methane. The reconstructed consortium consisting of only these three organisms grows as rapidly on 3-chlorobenzoate as did the original mixture of populations (14, 15, 75).

Many two- or three-species interaction studies are expanded to study survival and effects among more complex and less well defined natural biota; often the test organisms are tested in both sterile medium (either synthetic or naturally derived) and nonsterile medium containing the indigenous biota to test for survival under conditions that include competition and predation. For example, a *Pseudomonas* designed to degrade substituted benzoates, FR1 and FR2, and the parental strain B13 were tested in sterile synthetic sewage to which activated sludge organisms were inoculated. The parental and derived species were able to survive in similar numbers, and the derived strains were able to degrade the target compound (18). Wellington et al. demonstrated the transmission of a high-copy-number, self-transmissible plasmid, pIJ673, between *Streptomyces* species in sterile and nonsterile soil (77); Bej et al. tested a model suicide vector for GEMs in sterile and nonsterile soil (4); Ogunseitan et al. studied the interactions between *Pseudomonas* and bacteriophages in sterile and nonsterile lake water (44); and Saye et al. evaluated the effects of plasmid donor concentration on transduction of plasmids in sterile and nonsterile lake water (48). The survival of a GEM was tested in survival chambers that separated the test organism from the indigenous biota by 0.2-μm-pore filters; survival was also tested in direct contact with indigenous biota (1). These experiments are closely related to those in which GEMs have been tested in microcosms that contain indigenous microbial communities; e.g., Jain et al. tested the survival and stability of GEMs in groundwater aquifer microcosms (7) and Trevors et al. measured the vertical transport of GEMs through soil microcosms (76). In this series, there is no sharp demarcation between "synthetic" and naturally derived microcosm techniques.

Continuous cultures of two to three species have been used to study competition and predator (grazer)-prey interactions. Competition studies are often done to identify and select the more rapidly growing strain in a specific medium with a growth factor or toxicant that is being tested. Other studies involved simple competition to determine which species demonstrates competitive dominance in the presence and absence of a predator; in the absence of a predator, *E. coli* always eliminated *Azotobacter* in continuous culture, whereas in the presence of the predator *Tetrahymena*, the two bacterial species coexisted to the end of the 8-day experiment (32). Predation on bacteria or the equivalent process of grazing on algae has been studied, e.g., predation on *E. coli* by the protozoan *Colpoda* (17); persistence of *E. coli* in the presence of bacterivorous ciliates (47); predation on *Serratia* by the protozoan *Uronema* (25) or grazing on the alga *Chlamydomonas* by the protozoan *Tetrahymena* (5). These same algal-grazer (*Chlamydomonas-Tetrahymena*) data were analyzed by the application of Michaelis-Menten and continuous culture equations to describe the growth of the alga as a function of light and external nutrient (nitrate), and the growth of the protozoan as a function of the algal cell number, nitrogen content, and light intensity (high light intensity is detrimental to the protozoan) (D. H. McKenzie, Ph.D. thesis, University of Washington, Seattle, 1975). The model demonstrated that the protozoa became extinct at high dilution rates because the reproductive rate could not match the washout rate, thus explaining the apparent paradox of extinction in the presence of a nutritionally adequate food. It was possible to model the nutrient-algal-protozoan relationships over a wide range of nutrient, light, and dilution rates, and many of the results of the complete model were not immediately obvious if limiting factors were considered singly, or over narrower ranges (5, 66; F. B. Taub, Biological models of freshwater communities, EPA 600/3-73-008, 1973).

The analyses of these multispecies, continuous-culture studies benefited from mathematical models based on the well-described properties of continuous cultures (29, 33). Sambanis and Fredrickson provide an excellent bibliography of batch cultures, single-stage chemostats, and two-stage chemostat cascades (47).

An obvious use of a continuous culture would be to study competitive interactions between a parental strain and a GEM. Since most GEMs will have a common genome except for a few modified genes, one can hypothesize that these strains will be fiercely competitive. If the modified genetic material engenders a metabolic cost, it has often been assumed that the GEM will have reduced survival capability, except under conditions that provide a specific survival advantage for the GEM, e.g., having antibiotic resistance in the presence of the antibiotic or the ability to metabolize a substrate available to it, but not to other bacteria present. These hypotheses could be tested in continuous culture, in the presence and absence of the specific chemicals. In the case of a new metabolic capability, the concentration of the target chemical and competition with alternative substrates could be tested. Such studies could identify models of the conditions under which the parental or wild type might win, or conditions under which they might coexist. An analogous example has been developed by Landis (34) to predict competitive exclusion or coexistence in the presence of a toxic substance.

Gnotobiotic ecosystems—studies involving an interactive community

Slightly more complex species mixtures have been used to study the minimal assemblage of organisms that provide the three processes presumed necessary for self-sustaining ecosystems—(1) primary production, the conversion of inorganic chemicals, via light or chemical energy, into organic chemicals that can sustain heterotrophs; (2) grazing as a means of converting primary producers to heterotrophs; and (3) nutrient recycling for the reuse of primary producers. These communities have been termed gnotobiotic (16, 38) because all of the species were known.

A functional community was established with the single alga, *Chlamydomonas reinhardtii,* which was the source of fixed carbon; the protozoan *Tetrahymena vorax* and the rotifer *Philodina* sp., both of which grazed on the alga; and the bacteria *Aerobacter aerogenes* and *E. coli* (three additional species of bacteria were inoculated, but did not establish populations—*Cytophaga hutchinsonii, Cellulomonas gelida,* and *Cellvibrio fulvus*) (60). For comparison, less complex communities were established without the rotifer, without *Aerobacter,* and with only the alga. In all of the communities, the alga eventually established dense cultures; the grazers were able to delay, but not to prevent dense algal blooms, although the rotifers and protozoa were abundant. In the absence of *Aerobacter, E. coli* did not maintain measurable populations ($<10^2$ CFU/mL), whereas in the presence of *Aerobacter, E. coli* maintained populations above 10^4 CFU/mL. Contrary to the notion that simple communities would be unstable or unpredictable, these gnotobiotic communities were stable to the point of boredom.

Microcosms of the bacteria *Pseudomonas, Cytophaga,* and *Aerobacter,* the alga *Chlamydomonas,* and the protozoan *Tetrahymena* were studied in batch mode with multiple nutrient additions, and in sealed mode. The populations were stable, although there were physiological shifts observed by the yellowing of the algae and changes in the size and feeding vacuoles of the protozoa. Upon dilution by the addition of new inorganic medium, the organism densities declined, and after a slight lag regrew to a density higher than predilution, and then gradually declined back to the previous density (61). The protozoa persisted in the unsealed mode, but not after sealing from the atmosphere.

The properties of a continuous, gnotobiotic community with these same organisms, plus *E. coli* (and minus the *Cytophaga,* which never became established), were studied for properties of stability and yield at different light and dilution rates (59, 62). As in the simpler alga-protozoan studies described above, the grazer population was driven to extinction in the presence of a suitable food if the food density was reduced enough (by reducing the light intensity) to cause the grazer's reproductive rate to be less than the wash-out rate of the culture (determined by the dilution rate).

Gnotobiotic microcosms of a brine community were established using six bacteria, one alga (*Dunaliella viridis*), and the brine shrimp

Artemia salina (43). The alga-bacteria populations were stable, and oscillations did not occur in these simple systems, kept in a stable environment. The brine shrimp did not persist indefinitely, although populations persisted over 150 days in two of the experiments.

Synthetic communities of this level of complexity could be used to test a GEM within the context of ecosystem functions. Because all of the organisms are known (by definition in a "gnotobiotic ecosystem"), it is possible to follow all of the populations of microbes. Since the researcher determines the biota, organisms that cannot easily be distinguished from the test organism can be eliminated. Aside from contamination by extraneous organisms, the GEM and any new recombinants should be easy to detect. This makes such synthetic microcosms suitable for testing the effects of GEMs on other microbes, e.g., does a GEM replace an indigenous one, or alter dominance of the microbial community? All of the kinds of studies in paired systems could be studied here in more complex systems with greater potential for species competition, displacement, shifts in metabolic function, and so forth. A target chemical could be added as the initial substrate, but over time, other substrates would become available from the excretion of organisms. The continued use of the target substrate could then be measured in the presence of potentially competing, naturally produced, substrates.

A major limitation of gnotobiotic ecosystems is the relatively few grazers that can be easily grown axenically, or monoxenically (in the presence of one neighbor) (16, 38). *Tetrahymena vorax* is one of the few protozoa that can be grown axenically in organic media and can graze bacteria and algae as described in the studies above (5, 66), although it was not able to compete in more complex *standardized aquatic microcosms* (SAM) discussed below. The rotifer *Philodina* sp. can be grown monoxenically on a single bacterium, *Enterobacter aerogenes* (formerly called *Aerobacter aerogenes*). Even fewer predatory organisms were found to have been cultured gnotobiotically, and the few that we tested in more complex systems (SAM), e.g., *Didinium*, did not survive.

Synthetic ecosystems of greater complexity—multiple species on each trophic level

Microcosms of greater complexity have been developed as multispecies, multitrophic level communities by sacrificing the knowledge of all of the microbes present. General reviews of microcosms can be found in Cooke (11) and Giesy (23). Four related types of microcosms are discussed below as examples of synthetic systems that could be, or have been, adapted for GEM studies; each is described in Cairns (7) and has been compared by Taub (65). These communities contain numerous species of primary producers, grazers, and decomposers, but tend to lack large predators. Because these microcosms were developed to study the effects of chemicals on community function (estimated by oxygen and pH dynamics) and structure (based on algal and zooplankton populations or enzymes), the microbial populations have generally not been characterized, but methods could be adapted for this purpose.

The mixed flask culture (MFC) microcosms are based on the assumption that all ecosystems have common processes and that the particular species carrying out those processes are not important. The inoculum of the 1-liter microcosms is a mixed assemblage, from numerous sources of algae, zooplankton, and other microorganisms that have been allowed to develop for at least 3 months in a laboratory aquarium of synthetic medium (49, 50). The endpoint for this test system is the amount of deviation from controls established with the same organisms at the same initial conditions. A 6-week successional period with cross-inoculation helps ensure similar conditions in the replicates prior to treatment. Because the focus is on system-level properties, the precise species are not defined, but a minimum number of functional types are added if not present in the community: two species of unicellular green algae, one species of nitrogen-fixing blue-green algae, one species of filamentous green algae, one species of herbivorous grazer, one species of benthic detritivore, protozoa, and associated bacteria. Re-inoculation is done each week to simulate natural recolonization and allow recovery from temporarily toxic conditions. The ecosystem-level variables include oxygen change (net daytime production and night respiration), eH, and pH. The populations of some biota are also estimated. This methodology has been adapted to test the ecological effects of *Bacillus thuringiensis* (2). Additional details of the methodology can be found in J. Leffler, *Tentative Protocol of an Aquatic Microcosm Screening Test for Evaluating Ecosystem-Level Effects of Chemicals,* Battelle Columbus Subcontract T6411 (7197), Ferrum College, Ferrum, Virginia, 1981; and L. J. Shannon, T. E. Flum, and J. D. Yount, *Draft Protocol for a Mixed Flask Culture Aquatic Microcosm Toxicity Test and Support Document,* No. 7667A, University of Minnesota to U.S. EPA, Duluth, 1987.

A pond microcosm has been developed that simulates the shallow, rooted-vegetation-dominated communities common in much of the southern United States (19, 21). These communities were 80-liter aquaria filled with sediment and water from natural ponds. Each replicate received 100 g of rooted plants and associated biota; cross-seeding or transfer of larger organisms was undertaken prior to treatment to assure that replicates were similar before treatment with oil products. Crayfish had to be eliminated because they were capable of eating the rooted vegetation. The rooted plants filled most of the volume of liquid, and snails, oligochaetes, dipteran larvae, hydra, planeria, damselfly larvae, water mites, protozoa, nematodes, rotifers, gastrotrichs, copepods, and cladocerans were part of the community. The measurements included temperature, pH, specific conductivity, dissolved oxygen dynamics, alkalinity, hardness, and algal nutrients. Populations of small planktonic organisms were measured along with nontaxonomic aggregates (e.g., total phytoplankton chlorophyll, total bacterial density, and total ecosystem metabolism). The macrophyte biomass was determined at the end of the experiment. These aquaria microcosms responded similarly to oil as did outdoor ponds (22, 30).

A "generic aquatic ecosystem" test system was constructed of natural assemblages of planktonic organisms grown in synthetic medium (51, 52). Parent cultures from different lakes were tested to determine if they yielded similar rankings of toxic chemicals. Of the metabolic parameters, rankings based on electron transport system (ETS) activity were not consistent with the rankings based on other measurements (dissolved oxygen, ^{14}C-primary production, and ^{14}C-acetate uptake). Many of these methods could be adapted to the study of GEMs.

A "standardized aquatic microcosm" (SAM) was developed and tested to provide, at any season or location, microcosms of the same species composition (algae and animals) in similar numbers and culture state (9, 10, 63–65, 67–73). The microcosm medium and biota were selected to demonstrate ecologically important processes, much as the gnotobiotic ecosystems (primary production, secondary production, recycling), but with multiple species per trophic level. The procedures were designed to be less labor-intensive than the gnotobiotic microcosms because aseptic techniques were not required during sampling. The chemical medium and sediment are chemically defined; the limiting algal nutrients are nitrate (the N source) and phosphate. The biota, 10 species of algae, and 5 species of animals were selected for taxonomic diversity and different feeding habits. Weekly re-inoculation simulates immigration and allows recovery of populations if toxicity is eliminated by the transformation of a toxic chemical.

The sensitivity of the SAM to a variety of chemicals has been tested (24, 35, 54, 63–65, 70–73); the within-laboratory repeatability and between-laboratory reproducibility have been evaluated (10, 65, 71–73); results have been compared with naturally derived microcosms (F. J. Hardy, Ph.D. dissertation, University of Washington, Seattle, 1984); with the mixed flask cultures (53); and with ponds and lakes (27, 36). The results demonstrate that the standardized aquatic microcosms could be adapted to study the invasiveness of GEMs and the degradation of toxic chemicals. The ecological effects that could be detected include survival of the GEM, shifts in species abundance of the "indigenous" organisms and shifts in nutrient cycling, as the results of competition from the GEM, toxicity of GEM products, or detoxification of toxic chemicals.

A microcosm simulation model, MICMOD, was developed to explore biological responses to chemicals and to changes in initial conditions (46, 56–58; K. A. Rose, Ph.D. thesis, University of Washington, Seattle, 1985). The multispecies phytoplankton-zooplankton model includes eight groups of algae (some similar species were lumped), *Daphnia* (eggs and five size classes), amphipods, ostracods, protozoa, and rotifers. The algal submodel includes external and internal nutrients and calculated changes in photosynthesis by a variety of mechanisms; the animal submodel includes respiration, feeding, growth, and mortality (46). Using data derived from single and paired species, the model has been able to simulate the effects of streptomycin and copper in microcosm experiments (26, 46, 58; K. A. Rose, Ph.D. thesis, University of Washington, Seattle, 1985). The use of such ecological simulations, with examples from the microcosm experiments, have been published in a text (56).

A potential use of such simulation models for GEM research is to evaluate the growth requirements and conditions under which a GEM might be expected to successfully compete. For example, nitrate and phosphate are limiting nutrients in the SAM, and an organism with the ability to take up nutrients faster, or at lower concentration, than the indigenous species would have a competitive advantage. Simulations of conditions in which dissolved algal nutrients are available, such as early spring or where algae have been eliminated by grazing or toxic chemicals, might be more vulnerable to invasion by exotic algae than communities in which algal nutrients are scarce. The simulation model could be used to predict the community and organism conditions and growth parameters (nutrient uptake, temperature, light conditions) that might allow or prevent establishment of a novel organism. If chemicals are involved in the study, the distribution of the parent and degradation compounds could be modeled (6).

All of these "xenic" (that is, microbial community not known) microcosms could be used to test GEMs. The identification of bacteria from these microcosms will be as difficult as those of natural communities, because the bacteria are of multiple, unknown species. In addition to testing for the survival and potential competition of indigenous organisms with GEMs, these microcosms could be used to test for alterations in nutrient cycles and transformation of toxic chemicals. For example, the addition of a nitrogen fixer to a nitrogen-limited community might result in major shifts in total community metabolism, species dominance, and total abundances. In using these microcosms to test for the transformation of persistent toxic chemicals, a potential experimental design could be (1) control, (2) test chemical with control biota, (3) test GEM—to test for direct effects of the organism, and (4) test GEM and test chemical—to test for elimination of the toxicity associated with the test chemical. If the GEM is more effective than the control biota in degrading the test chemical to less harmful products, recovery should occur earlier in treatment (4) than in (2). Should there be any direct effects of the GEM (or its products), these could be detected by comparing treatments (1) with (3). Survival and population dynamics of the GEM in the presence and absence of the test chemical could be determined in treatments (3) and (4). Degradation or transformation of the test chemical should be confirmed by chemical analysis. Should the test chemical be transformed by the GEM to a biologically active product, its presence might be indicated by responses in treatment (4) as compared to (2) and (1). The SAM microcosm is being used to test the ability of a bacterium to degrade a recalcitrant chemical, dibenz-1,4-oxazepine (CR) (M. V. Haley, E. Vickers, T. C. Cheng, and W. G. Landis, Haley, M. V., E. Vickers, T. C. Cheng, and W. G. Landis. 1989. Naturally occurring organisms resistant to dibenz-1,4-oxazepine. In J. D. Williams, Jr. and M. D. Rausa (eds.), *Proceedings of the 1988 CRDEC Conference on Chemical Defense Research*, CRDEC-SP-013, pp. 1085–1090.

Discussion

Extrapolation

The value of a model system is vastly greater if it allows extrapolation to a broad range of natural ecosystems. The extent to which this can be done is still an open question. The general question of extrapolation of results needs intensive study. This problem arises every time the test community is different from the community to be protected. This question is equally valid when natural ponds are used to estimate the safety of organisms that may enter rivers and large lakes, and for the questions of extrapolation from field studies in remote sites to estimates of safety at other sites. Even at the same site, events at one season may be different than at another (74). Dean-Ross (13) expressed the concern that model ecosystems as test systems could not possibly measure all possible natural communities, but that the physical and chemical requirements of the test organisms could suggest appropriate test systems. Using the GEMs requirements may assist in some respects, such as selecting a freshwater vs. marine test system, but it will be far more difficult to know if the test system has appropriate competitive and predation pressures to those the GEM is likely to experience in the environment.

The source, potential use, and potential ecological effects of a GEM may aid in determining the appropriateness of a test system. If the GEM eliminates some organisms (e.g., the target of a microbial pesticide), we can expect an increase in the abundance of the target species' former prey or of its competitors, and potentially a decrease in the abundance of its predators. New pest control agents, such as those developed by transferring B. *thuringiensis* insecticidal activity to other bacteria (3, 78), may eliminate target and possibly some nontarget organisms. To fully test the ecosystem effects of these GEMs, the researcher may need to include (at least in conceptual models) effects on the food species, competitive species, and predators such as fish or birds. Some of these interactions are being studied using naturally occurring *Bacillus* (2). Nontoxic and nonpathogenic GEMs can compete with and replace indigenous organisms, and may produce structural and functional changes (13). Even without eliminating any indigenous biota, the successful colonization by an organism that altered the supply of a limiting nutrient, such as a nitrogen producer or a denitrifier in a nitrogen-limited community, or that could use an abundant resource that was unavailable to the indigenous species, such as lignin or cellulose, could have a marked ecological effect by modifying the limiting factors. The survival and ecological effects of such organisms may not be obvious from test systems with other limiting factors.

The problems of extrapolation from synthetic systems to natural ecosystems may be of no greater difficulty than extrapolation from one natural ecosystem to another. In synthetic microcosms, the species assemblages, nutrients, and substrates are better defined than in most natural systems. This information may provide some suggestions as to the types of natural communities that have some similar properties.

Complementary aspects of site-specific and synthetic microcosms

Some researchers perceive natural ecosystems as complex beyond understanding. Communities can differ from each other in a multitude of physical, chemical, and biological ways. In addition, most natural environments are heterogeneous in space and variable in time. The very complexity of chemical substrates and biological diversity of natural ecosystems is sometimes seen as so daunting that laboratory scientists, accustomed to experiments with a few variables, feel that natural communities are beyond rational study. Often it is difficult to define and repeat the initial conditions of a field experiment. If the results of successive experiments are similar, researchers are reassured, but if the results differ, under conditions where there may be an infinite number of potential differences between initial experimental conditions—the differences in results may result in frustration, rather than understanding.

Other researchers, in the search for "emergent properties" of ecosystems (those which could not be determined by studying individual components), state that only the study of natural communities contributes to new ecological understanding. The "real ecosystems = complexity = stability" paradigm has been responsible for the rejection of simple synthetic microcosms as ecological studies by some researchers. It had been hypothesized that complexity was a requirement of stability in ecosystems, but May (40, 41) has shown that complexity is not a mathematical requirement of stability, and Connell's field studies (8) have shown that the most complex ecological systems were not stable, but were in the process of changing to another state.

As a compromise between complexity and realism, some researchers have reduced the complexity of their study systems by working with relatively simple microbial communities that are adapted to temperature or chemical extremes, such as geothermal hot springs (45). Another useful natural assemblage of anaerobic organisms has been the rumen. Since these are specialized communities, we do not know the degree to which we can extrapolate their behavior to other systems.

The study of ecosystems is not qualitatively different from the study of other complex systems. Learning the interactions among subsystems is always useful in understanding the whole but may not be adequate for predicting responses to new organisms because some processes may cancel out others. In contrast to the belief that ecosystems are too complex to study, many properties become relatively easy to understand once the controlling factors are identified, for example, why a nitrogen-limited ecosystem is very responsive to changes in nitrogen availability, whereas a natural ecosystem with a different controlling factor would be less sensitive. This chapter strives to present experimental approaches that help bridge the gap between component and ecosystem understanding.

Ecology will be best served by complementary studies of general synthetic and site-specific microcosm studies. Each has unique strengths, and both are needed to provide convincing evidence that interactions among organisms are important in predicting the effects of releasing organisms or chemicals. Within each test system, the limitations and assumptions must be explicitly recognized. The use of microcosm data for regulatory decisions may encourage the use of ecological information beyond the currently used single-species toxicity and bioaccumulation. The use of microcosm data for regulatory decisions has been discussed by Harrass and Sayre (28).

Site-specific microcosms are intended to be limited to the specific community from which they were excised. Cripe and Pritchard (12) have stated, "A microcosm, like the community from which it is taken, is site specific, and care should be taken to avoid extrapolation beyond the community." Some scaling to larger systems may be possible by mathematical models. Naturally derived microcosms, like field studies, have the disadvantages of potential nonreproducibility for unknown reasons.

Synthetic microcosms are useful when reproducibility and analytical capability are important. Synthetic microcosms have virtue when the researcher wants to systematically vary conditions, e.g., with and without grazers, or with and without competing nutrient substrates, or in the presence or absence of a potential competitor. It is easier to omit groups in synthetic microcosms than to eliminate them from excised portions of natural communities.

Synthetic microcosms may have two problems as test systems for GEMs. First, synthetic communities may not be as closely packed with species and may be more permissive of invasion by novel organisms than a coevolved community. This is a testable hypothesis. Second, many of the organisms used are well-researched "laboratory weeds" whose physiological properties are known and easily expressed mathematically. There is always the danger that the properties of these laboratory weeds will be intensively studied because they are easy to grow in the laboratory and that these organisms may lack important ecological properties. For example, most researchers choose to work with organisms that grow rapidly and have simple nutritional requirements, whereas many ecologically important organisms grow slowly and seem to have important survival mechanisms that allow them to survive long periods without resources and complex nutritional requirements or abilities to switch to alternative resources. Still the use of organisms that allow rapid experiments has provided ecological insights into growth-substrate, competitive, and predator-prey relationships. With caution, these organisms and models may be used to gain understanding of how GEMs may survive and function in natural environments.

Both types of microcosms lack large organisms and large-scale processes. The full-scale ecosphere is the ultimate experimental unit, but we have only one and we do not want to try our high-risk experiments on it until we have satisfied ourselves on smaller, expendable test systems that the GEMs to be released will be harmless.

Note added in proof: See also, Landis, W. G., N. A. Chester, and M. V. Haley. In press. Utility of the standardized aquatic microcosm in the evaluation of degradative bacteria to reduce impacts to aquatic ecosystems. In *Procedings of the 1991 ASTM Aquatic Toxicology and Risk Assessment Symposium.* The standardized aquatic microcosm is also being used to study *Pseudomonas acidovorans* survival and degradation by Beak Consultants Ltd. for Environment Canada SCC file No. 066SS. Ke144-0-6238 (*A Development of Methods to Assess Monitoring Techniques for Environmental Introduction of Microorganisms*).

Acknowledgments

The author wishes to thank Buzz Hoffman, John Matheson III, William van der Schalie, Wayne Landis, and Morris Levin for the research opportunities displayed in this work. This publication is a School of Fisheries Contribution No. 836.

References

1. Awong, J., G. Bitton, and G. R. Chaudhry. 1990. Microcosm for assessing survival of genetically engineered microorganisms in aquatic environments. *Appl. Environ. Microbiol.* 56:977–983.
2. Anderson, R. L., and L. Shannon. 1992. Use of the mixed flask culture microcosm protocol to estimate the survival and effects of microorganisms added to freshwater ecosystems. In M. A. Levin, R. J. Seidler, and M. Rogul (eds.), *Microbial Ecology.* McGraw-Hill, New York, pp. 625–641.
3. Aronson, A. I., W. Beckman, and P. Dunn. 1986. *Bacillus thuringiensis* and related insect pathogens. *Microbiol. Rev.,* 50:1–24.
4. Bej, A. K., M. H. Perlin, and R. M. Atlas. 1988. Model suicide vector for containment of genetically engineered microorganisms. *Appl. Environ. Microbiol.* 54:2472–2477.
5. Blazka, P., T. Backiel, and F. B. Taub. 1980. Trophic relationships and efficiencies. In E. D. Le Cren and R. H. Lowe-McConnell (eds.), *The Functioning of Freshwater Ecosystems.* International Biological Programme 22, Cambridge University Press, pp. 393–410.
6. Burns, L. A. 1985. Models for predicting the fate of synthetic chemicals in aquatic ecosystems. In T. P. Boyle (ed.), *Validation and Predictability of Laboratory Methods for Assessing the Fate and Effects of Contaminants in Aquatic Ecosystems.* ASTM STP 865, ASTM, Philadelphia, Pa., pp. 176–190.
7. Cairns, J., Jr. (ed.) 1986. *Community Toxicity Testing.* ASTM STP 920. ASTM, Philadelphia.
8. Connell, J. H. 1978. Diversity in tropical rain forests and coral reefs. *Science.* 199:1302–1310.
9. Conquest, L. L. 1983. Assessing the statistical effectiveness of ecological experiments: Utility of the coefficient of variation. *Intern. J. Environ. Stud.* 20:209–221.
10. Conquest, L. L., and F. B. Taub. 1989. Repeatability and reproducibility of the standardized aquatic microcosm: Statistical properties. In U. M. Cowgill and L. R. Williams (eds.), *Aquatic Toxicology and Hazard Assessment,* vol. 12. ASTM STP 1027. ASTM, Philadelphia, Pa., pp. 159–177.
11. Cooke, G. D. 1977. Experimental aquatic laboratory ecosystems and communities. In J. Cairns, Jr. (ed.), *Aquatic Microbial Communities.* Garland Publishing, New York, pp. 61–103.
12. Cripe, C. R., and P. H. Pritchard. Site-specific aquatic microcosms as test systems for fate and transport of microorganisms. In M. A. Levin, R. J. Seidler, and M. Rogul (eds.), *Microbial Ecology.* McGraw-Hill, New York, pp. 467–493.
13. Dean-Ross, D. 1986. Release of genetically engineered organisms: Hazard assessment. *ASM News* 52:572–575.
14. Dolfing, J., and J. M. Tiedje. 1987. Growth yield increase linked to reductive

dechlorination in a defined 3-chlorobenzoate/degrading methanogenic coculture. *Arch. Microbiol.* 149:102–105.

15. Dolfing, J., and J. M. Tiedje. 1986. Hydrogen cycling in a three-tiered food web grown on the methanogenic conversion of 3-chlorobenzoate. *FEMS Microbiol. Ecol.* 38:293–298.

16. Dougherty, E. 1959. Axenic culture of invertebrate Metazoa: A goal. *Ann. New York Acad. Sci.* 77(2):25–406.

17. Drake, J. F., and H. M. Tsuchiya. 1976. Predation on *Escherichia coli* by *Colpoda steinii. Appl. Environ. Microbiol.* 31(6):870–874.

18. Dwyer, D. F., K. N. Timmis, and F. Rojo. 1988. Fate and behaviour in an activated sludge microcosm of a genetically-engineered microorganism designed to degrade substituted aromatic compounds. In M. Sussman, C. H. Collins, F. A. Skinner, and D. E. Stewart-Tull (eds.), *The Release of Genetically-Engineered Micro-Organisms.* Academic Press, New York, pp. 77–88.

19. Franco, P. J., J. M. Giddings, S. E. Herbes, L. A. Hook, J. D. Newbold, W. K. Roy, G. R. Southworth, and A. J. Stewart. 1984. Effects of chronic exposure to coal-derived oil on freshwater ecosystems. I. Microcosms. *Environ. Toxicol. Chem.* 3:447–463.

20. Gealt, M. A., M. D. Chai, K. B. Alpert, and J. C. Boyer. 1985. Transfer of plasmids pBR322 and pBR325 in wastewater from laboratory strains of *Escherichia coli* to bacteria indigenous to the waste disposal system. *Appl. Environ. Microbiol.* 49(4): 836–841.

21. Giddings, J. M. 1983. Microcosms for assessment of chemical effects on the properties of aquatic ecosystems. *Hazard Assessment of Chemicals, Current Develop.* 2:45–94.

22. Giddings, J. M. 1986. A microcosm procedure for determining safe levels of chemical exposure in shallow-water communities. In J. Cairns, Jr. (ed.), *Community Toxicity Testing.* ASTM STP 920. ASTM, Philadelphia, pp. 121–134.

23. Giesy, J. P., Jr. 1980. *Microcosms in Ecological Research.* CONF-781101. U.S. Department of Energy, National Technical Information Service.

24. Haley, M. V., D. W. Johnson, W. T. Muse, Jr., and W. G. Landis. 1987. The aquatic toxicity and fate of brass dust. In W. Adams, G. Chapman, and W. G. Landis (eds.), *Aquatic Toxicology and Hazard Assessment,* vol. 10. ASTM STP 971. ASTM, Philadelphia, pp. 468–479.

25. Hamilton, R. D., and J. E. Preslan. 1970. Observations on the continuous culture of a planktonic phagotrophic protozoan. *J. Exp. Mar. Biol. Ecol.* 5:94–104.

26. Harrass, M. C., A. C. Kindig, and F. B. Taub. 1985. Responses of blue-green and green algae to streptomycin in unialgal and paired culture. *Aquat. Toxicol.* 6:1–11.

27. Harrass, M. C., and F. B. Taub. 1985. Comparison of laboratory microcosms and field responses to copper. In T. P. Boyle (ed.), *Validation and Predictability of Laboratory Methods for Assessing the Fate and Effects of Contaminants in Aquatic Ecosystems.* ASTM STP 865. ASTM, Philadelphia, pp. 57–74.

28. Harrass, M. C., and P. G. Sayre. 1989. Use of microcosm data for regulatory decisions. In U. M. Cowgill and L. R. Williams (eds.), *Aquatic Toxicology and Hazard Assessment,* vol. 12. ASTM STP 1027. ASTM, Philadelphia, pp. 204–223.

29. Herbert, D. 1964. Multi-stage continuous culture. In I. Malek, K. Beran, and J. Hospodka (eds.), *Continuous Cultivation of Microorganisms.* Proceedings of the 2nd symposium held in Prague, June 18–23, 1962. Publishing House of the Czechoslovak Academy of Science, Prague, pp. 23–44.

30. Hook, L. A., P. J. Franco, and J. M. Giddings. 1986. Zooplankton community responses to synthetic oil exposure. In J. Cairns, Jr. (ed.), *Community Toxicity Testing.* ASTM STP 920. ASTM, Philadelphia, pp. 291–321.

31. Jain, R. K., G. S. Sayler, J. T. Wilson, L. Houston, and D. Pacia. 1987. Maintenance and stability of introduced genotypes in groundwater aquifer material. *Appl. Environ. Microbiol.* 53:996–1002.

32. Jost, J. L., J. F. Drake, A. G. Fredrickson, and H. M. Tsuchiya. 1973. Interactions of *Tetrahymena pyriformis, Escherichia coli, Azotobacter vinelandii* and glucose in a minimal medium. *J. Bacteriol.* 113(2):834–840.

33. Kubitschek, H. E. 1970. *Introduction to Research with Continuous Cultures.* Prentice-Hall, Englewood Cliffs, N.J.
34. Landis, W. G. 1987. Resource competition modeling of the impacts of xenobiotics on biological communities. In T. M. Poston and R. Purdy (eds.), *Aquatic Toxicology and Environmental Fate,* vol. 9. ASTM STP 921. ASTM, Philadelphia, pp. 55–72.
35. Landis, W. G., N. A. Chester, M. V. Haley, D. W. Johnson, W. T. Muse, Jr., and R. M. Tauber. 1989. Utility of the standardized aquatic microcosm as a standard method for ecotoxicological evaluation. In G. W. Suter II and M. A. Lewis (eds.), *Aquatic Toxicology and Fate,* vol. 11, ASTM STP 1007. ASTM, Philadelphia, pp. 353–367.
36. Larsen, D. P., F. DeNoyelles, Jr., and F. Stay. 1986. Comparisons of single-species, microcosm and experimental pond responses to atrazine exposure. *Environ. Toxicol. Chem.* 5:179–190.
37. Lenski, R. E., and J. E. Bouma. 1987. Effects of segregation and selection on instability of plasmid pACYC184 in *Escherichia coli* B. *J. Bacteriol.* 169(11):5314–5316.
38. Lincoln, R. J., G. A. Boxshall, and P. F. Clark. 1982. *A Dictionary of Ecology, Evolution and Systematics.* Cambridge University Press, New York, p. 298.
39. Mancini, P., S. Fertels, D. Nave, and M. A. Gealt. 1987. Mobilization of plasmid pHSV106 from *Escherichia coli* HB101 in a laboratory-scale waste treatment facility. *Appl. Environ. Microbiol.* 53(4):665–671.
40. May, R. M. 1973. *Stability and Complexity Model in Ecosystems.* Princeton University Press, Princeton, N.J.
41. May, R. M. 1976. Stability and complexity. In R. M. May (ed.), *Theoretical Ecology Principles and Application.* Blackwell Scientific, pp. 158–162.
42. McPherson, P., and M. A. Gealt. 1986. Isolation of indigenous wastewater bacterial strains capable of mobilizing plasmid pBR325. *Appl. Environ. Microbiol.* 51(5):904–909.
43. Nixon, S. W. 1969. A synthetic microcosm. *Limnol. Oceanogr.* 14(1):142–145.
44. Ogunseitan, O. A., G. S. Sayler, and R. V. Miller. 1990. Dynamic interactions of *Pseudomonas aeruginosa* and bacteriophages in lake water. *Microb. Ecol.* 19:171–185.
45. Pace, N. R., D. A. Stahl, D. J. Lane, and G. J. Olsen. 1985. Analyzing natural microbial populations by rRNA sequences. *ASM News* 51:4–12.
46. Rose, K. A., G. L. Swartzman, A. C. Kindig, and F. B. Taub. 1988. Stepwise iterative calibration of a multi-species phytoplankton-zooplankton simulation model using laboratory data. *Ecol. Model.* 42:1–32.
47. Sambanis, A., and A. G. Fredrickson. 1988. Persistence of bacteria in the presence of viable, nonencysting, bacterivorous ciliates. *Microb. Ecol.* 16:197–211.
48. Saye, D. J., O. Ogunseitan, G. S. Sayler, and R. V. Miller. 1987. Potential for transduction of plasmids in a natural freshwater environment: Effect of plasmid donor concentration and a natural microbial community on transduction in *Pseudomonas aeruginosa. Appl. Environ. Microbiol.* 53:987–995.
49. Shannon, L. J., M. C. Harrass, J. D. Yount, and C. T. Walbridge. 1986. A comparison of mixed flask culture and standardized laboratory model ecosystems for toxicity testing. In J. Cairns, Jr. (ed.), *Community Toxicity Testing.* ASTM 920. ASTM, Philadelphia, pp. 135–137.
50. Shannon, L. J., T. E. Flum, R. L. Anderson, and J. D. Yount. 1989. Adaptation of mixed flask culture microcosm for testing the survival and effects of introduced microorganisms. In U. M. Cowgill and L. R. Williams (eds.), *Aquatic Toxicology and Hazard Assessment,* vol. 12. ASTM STP 1027. ASTM, Philadelphia, pp. 224–239.
51. Sheehan, P. J., R. P. Axler, and R. C. Newhook. 1986. Evaluation of simple generic aquatic ecosystem tests to screen the ecological impacts of pesticides. In J. Cairns, Jr. (ed.), *Community Toxicity Testing.* ASTM STP 920. ASTM, Philadelphia, pp. 158–179.
52. Sheehan, P. J. 1989. Statistical and nonstatistical considerations in quantifying pollutant-induced changes in microcosms. In U. M. Cowgill and L. R. Williams

(eds.), *Aquatic Toxicology and Hazard Assessment*, vol. 12. ASTM STP 1027. ASTM, Philadelphia, pp. 178–188.

53. Stay, F. S., T. E. Flum, L. J. Shannon, and J. D. Yount. 1989. An assessment of the precision and accuracy of SAM and MFC microcosms exposed to toxicants. In U. M. Cowgill and L. R. Williams (eds.), *Aquatic Toxicology and Hazard Assessment*, vol. 12. ASTM STP 1027. ASTM, Philadelphia, pp. 189–203.

54. Stay, F. S., D. P. Larsen, A. Katko, and C. M. Rohm. 1985. Effects of atrazine on community level responses in Taub microcosms. In T. P. Boyle (ed.), *Validation and Predictability of Laboratory Methods for Assessing the Fate and Effects of Contaminants in Aquatic Ecosystems*. ASTM STP 865. ASTM, Philadelphia, pp. 75–90.

55. Stotzky, G., and H. Babich. 1984. Fate of genetically-engineered microbes in natural environments. *Recomb. DNA Tech. Bull.* 7(4):163–188.

56. Swartzman, G. L., and S. P. Kaluzny. 1987. *Ecological Simulation Primer.* Macmillan, New York.

57. Swartzman, G. L., and K. A. Rose. 1984. Simulating the biological effects of toxicants in aquatic microcosm systems. *Ecol. Model.* 22:123–134.

58. Swartzman, G. L., K. Rose, A. Kindig, and F. Taub. 1989. Modeling the direct and indirect effects of streptomycin in aquatic microcosms. *Aquat. Toxic.* 14:109–130.

59. Taub, F. B. 1969. A continuous gnotobiotic (species defined) ecosystem. In J. Cairns, Jr. (ed.), *Structure and Function of Freshwater Microbial Communities*. Res. Monogr. 3. Virginia Polytech. Inst. and State Univ., Blacksburg, Va., pp. 101–120.

60. Taub, F. B. 1969. Gnotobiotic models of freshwater communities. *Verh. Internat. Verein. Limnol.* 17:485–496.

61. Taub, F. B. 1969. A biological model of a freshwater community: A gnotobiotic ecosystem. *Limnol. Oceangr.* 14(1):136–142.

62. Taub, F. B. 1977. A continuous gnotobiotic (species defined) ecosystem. In J. Cairns, Jr. (ed.), *Aquatic Microbial Communities*. Garland Reference Library of Sci. Technol., vol. 15. Garland, New York, pp. 105–138.

63. Taub, F. B. 1984. Synthetic microcosms as biological models of algal communities. In L. E. Shubert (ed.), *Algae as Ecological Indicators*. Academic Press, New York, pp. 363–394.

64. Taub, F. B. 1985. Toward interlaboratory (round-robin) testing of a standardized aquatic microcosm. In J. Cairns, Jr. (ed.), *Multispecies Toxicity Testing*. Pergamon Press, United Kingdom, pp. 165–186.

65. Taub, F. B. 1989. Standardized aquatic microcosm development and testing. In A. Boudou and F. Ribeyre (eds.), *Aquatic Ecotoxicology*, vol. II, pp. 47–92. CRC Press, Boca Raton, Fla.

66. Taub, F. B., and D. H. McKenzie. 1973. Continuous cultures of an alga and its grazer. *Bull. Ecol. Res. Comm. (Stockholm)* 17:371–377.

67. Taub, F. B., and M. E. Crow. 1980. Synthesizing aquatic microcosms. In J. P. Giesy, Jr. (ed.), *Microcosms in Ecological Research*. CONF-781101, DOE Symposium Ser. 52, pp. 69–104.

68. Taub, F. B., M. E. Crow, and H. J. Hartmann. 1980. Responses of aquatic microcosms to acute mortality. In J. P. Giesy, Jr. (ed.), *Microcosms in Ecological Research*. CONF-781101, DOE Symposium Ser. 52, pp. 513–535.

69. Taub, F. B., M. C. Harrass, H. J. Hartmann, A. C. Kindig, and P. L. Read. 1981. Effects of initial algal density on community development in aquatic microcosms. *Verh. Internat. Verein. Limnol.* 21:197–204.

70. Taub, F. B., P. L. Read, A. C. Kindig, M. C. Harrass, H. J. Hartmann, L. L. Conquest, F. J. Hardy, and P. T. Munro. 1983. Demonstration of the ecological effects of streptomycin and malathion on synthetic aquatic microcosms. In W. E. Bishop, R. D. Cardwell, and B. B. Heidolph (eds.), *Aquatic Toxicology and Hazard Assessment.* Sixth Symposium. ASTM STP 802. ASTM, Philadelphia, pp. 5–25.

71. Taub, F. B., and A. C. Kindig, and L. L. Conquest. 1986. Preliminary results of interlaboratory testing of a standardized aquatic microcosm. In J. Cairns, Jr. (ed.), *Community Toxicity Testing*. ASTM STP 920. ASTM, Philadelphia, pp. 93–120.

72. Taub, F. B., A. C. Kindig, and L. L. Conquest. 1988. Interlaboratory testing of a standardized aquatic microcosm. In W. J. Adams, G. A. Chapman, and W. G. Landis

(eds.), *Aquatic Toxicology and Hazard Assessment,* vol. 10. ASTM STP 971, ASTM, Philadelphia, pp. 384–405.

73. Taub, F. B., A. C. Kindig, L. L. Conquest, and J. P. Meador. 1988. Results of interlaboratory testing of the standardized aquatic microcosm protocol. In G. W. Suter II and M. A. Lewis (eds.), *Aquatic Toxicology and Environmental Fate,* vol. 11. ASTM STP 1007, ASTM, Philadelphia, pp. 368–394.

74. Thomas, W. H., O. Holm-Hansen, D. L. R. Siebert, F. Azam, R. Hodson, and M. Takahashi. 1977. Effects of copper on phytoplankton standing crop and productivity: Controlled ecosystem pollution experiment. *Bull. Mar. Sci.* 27:34–43.

75. Tiedje, J. M., and T. O. Stevens. 1988. The ecology of an anaerobic dechlorinating consortium. In G. S. Omenn (ed.), *Environmental Biotechnology.* Plenum Press, pp. 3–14.

76. Trevors, J. T., J. D. van Elsas, L. S. van Overbeek, and M. Starodub. 1990. Transport of a genetically engineered *Pseudomonas fluorescens* strain through a soil microcosm. *Appl. Environ. Microbiol.* 56:401–408.

77. Wellington, E. M. H., N. Cresswell, and V. A. Saunders. 1990. Growth and survival of streptomycete inoculants and extent of plasmid transfer in sterile and nonsterile soil. *Appl. Environ. Microbiol.* 56:1413–1419.

78. Whiteley, H. R., and E. Schnepf. 1986. The molecular biology of parasporal crystal body formation in *Bacillus thuringiensis. Ann. Rev. Microbiol.* 40: 549–576.

Chapter 33

Distribution of Metal-Resistant Microorganisms in the Environment: Significance and Methodology

Tamar Barkay

Cynthia Liebert

Introduction

This chapter examines and describes methods to study how microbes affect and are affected by metals in their environment. Metal-resistant microorganisms determine the potential of metal-exposed environments to sustain their integrity, via two microbially mediated mechanisms: (1) Resistant microorganisms maintain the geochemical cycling necessary for equilibrium in the biosphere in the presence of metals (Barkay and Pritchard, 1988); (2) metals are transformed by the activity of microorganisms (both resistant and sensitive) to less toxic chemical forms. Some of these are detoxification processes because inhibiting effects are alleviated (Barkay, 1987; Erardi et al., 1987; Silver and Misra, 1988). These processes have practical implications for the use of microorganisms (natural or genetically engineered) in bioremediation of metal-polluted environments and in processes employed by the mining industry (Halvorson et al., 1985). Successful bioremediation depends upon understanding factors and processes that control distribution of metal-resistant microbes in the environment. Such knowledge will assist not only in the construction of microorganisms that colonize, survive, and function in the site of application, but will ensure safe release. Factors that permit containment (described in Part 6) of released microorganisms and their activities could be included in the design of field applications.

Distribution studies explain how metal-resistant microorganisms affect the response of the environment to metal insult. The ecological significance of specific populations depends on their taxonomy, catabolic capabilities, and place in the microbial community of the studied environment (Tate, 1986). In addition, microbial processing depends on interactions with physicochemical factors and with other biota in the ecosystem. These factors are revealed by studies on microbial distribution and its dependence on environmental factors.

The distribution of metal-resistant microorganisms has been investigated extensively in a large variety of polluted and unpolluted environments (Duxbury, 1986). A few examples are summarized in Table 33.1. Because growth of colonies on media was the method employed in the cited studies, the data are pertinent to aerobic culturable heterotrophic bacteria, but not to all microorganisms in the community. Comparisons of data obtained by the different studies are difficult because of a variability in methods, growth media, and other test conditions. For example, seven $HgCl_2$ concentrations were used by nine different studies cited in Table 33.1. However, some trends emerge from such studies. First, metal-resistant bacteria are found in all environments regardless of previous exposure to metals. This is not surprising, as metals are natural elements that were present in the biosphere at the time that microorganisms evolved. Second, abundance of resistant bacteria is increased among microbial communities of metal-impacted ecosystems as compared to control samples from unpolluted environments due to selection of resistant microorganisms. This phenomenon has been utilized to indicate metal pollution in the environment (Olson and Thornton, 1982; Duxbury and Bicknell, 1983). As such, the distribution of metal-resistant microbes could become a practical tool in environmental management. Although the interpretation of bacterial distribution studies is difficult, approaches and methods to acquire data and their successful analyses have been developed. The simplest and most commonly used approach is to calculate the fraction of resistant microorganisms among all microorganisms in a given sample. Since enumerations of resistant and total bacteria are performed under the same conditions, with the supplemented metal being the only variable in the procedure, a basis for comparison among samples treated similarly exists. The percentage resistance values presented in Table 33.1 were obtained by this approach.

TABLE 33.1 Occurrence and Distribution of Metal-Resistant Aerobic Heterotrophic Bacteria in Aquatic Environments*

Environment	Test metal concentration	% resistance† in communities of		Reference
		Contaminated samples	Control samples	
Salt marsh sediment	$HgCl_2$—23 μM	37.5	8.5	Hamlett (1986)
Salt marsh water	$HgCl_2$—10 $\mu g/mL$	87	9	Barkay (1987)
Estuarine water	$HgCl_2$—6 $\mu g/mL$	8.3	0.08	Nelson and Colwell (1975)
	PMA‡—0.3 $\mu g/mL$	19.6	0.14	Nelson and Colwell (1975)
Estuarine water	$CoSO_4 \cdot 7H_2O$—100 $\mu g/mL$	43.0	1	Mills and Colwell (1977)
Estuarine sediment	$HgCl_2$—6 $\mu g/mL$	6.0	0.31	Nelson and Colwell (1975)
Estuarine sediment	$SnCl \cdot 5H_2O$—75 $\mu g/mL$	38.0	6.6	Hallas and Cooney (1981)
Freshwater	$HgCl_2$—10 $\mu g/mL$	72	3	Barkay (1987)
Freshwater sediment	$HgCl_2$—50 $\mu g/mL$	0.55	0.007	Barkay and Olson (1986)
River sediment	$Cr(VI)$—25 $\mu g/mL$	88	3.5	Luli et al. (1983)
Mosses grown on river bank	$CdSO_4$—180 μM	34	16	Houba and Remacle (1980)
Coastal marine water	$HgCl_2$—10 $\mu g/mL$	124	23	Barkay (1987)
Coastal marine sediment	$HgCl_2$—20 $\mu g/mL$	23.2	0.2	Timoney et al. (1979)
Coastal marine sediment	$ZnSO_4$—50 $\mu g/mL$	47.0	14.1	Pickaver and Lyes (1981)
Coastal marine sediment	$HgCl_2$—40 $\mu g/mL$	35.2	0.3	Nakamura et al. (1986)

*All cited studies employed solid media to enumerate bacterial colonies.
†Percent resistance = (No. of resistant CFUs/No. of total CFUs) × 100.
‡PMA = phenylmercuric-acetate.

Since our experience is primarily with mercury-resistant microorganisms, examples of work with this metal will be described. An introduction on general properties of mercury is warranted.

Mercury is an earth crust element, where it exists as cinnabar (HgS) at an average concentration of 0.5 μg/g (Rose et al., 1979). Naturally occurring mercury deposits are not evenly distributed; some locations have higher cinnabar contents. In addition, although the release of mercury into the environment was banned more than 20 years ago in the United States, old sources still contaminate nearby ecosystems (R. Turner, Oak Ridge National Laboratory, personal communication), and mercury of natural and anthropogenic sources may accumulate in environments where physicochemical conditions allow reactions that immobilize this element (Rada and Winfrey, 1987).

Toxicity of mercury to living organisms is due to four unique chemical properties: (1) exceptional affinity to thiol groups, (2) strong capacity to associate in covalent bonding, (3) high stability of the Hg-C bond, and (4) a linear two-coordinate stereochemistry with proteins and nucleic acids. The process by which such interactions bring about observed toxic effects of mercury is under investigation (Delnomdedieu et al., 1989).

The response of bacteria to mercury is of a particular interest because mercury is the only metal for which biotransformation is the resistance mechanism. Other bacterially mediated transformations are known but are unrelated to the mechanism of resistance [for example, the reduction of chromate Cr(VI) to chromite Cr(III) (Bopp and Ehrlich, 1988)]. The enzymatic reduction of Hg^{2+} and of several organomercurials to $Hg°$ is the resistance mechanism in most bacteria studied to date (Silver and Misra, 1988). Mercury resistance genes provide a unique opportunity for studies on the interaction between microbes and their environment, because they code for the survival of the organism in the polluted environment as well as for the detoxification process.

A complete review of microbial metal resistance is beyond the scope of this chapter. The reader is referred to review articles on the ecology of metal resistance (Duxbury, 1986), the molecular mechanisms of metal resistance and biotransformation (Silver and Misra, 1988), the effect of environmental factors on metal-microbe interactions (Babich and Stotzky, 1986), and adaptation to metal stress (Klerks and Weis, 1987). This chapter will describe methods used in distribution studies of metal-resistant microbes, their advantages, weaknesses, possible improvements, and value to microbial ecology.

Indirect Approaches for Detection and Enumeration of Metal-Resistant Bacteria

Procedures

Indirect methods are based on enumeration of metal-resistant culturable bacteria by growth in the laboratory. The spread-plate or pour-plate techniques (Koch, 1981) are most commonly used for this purpose. Diluted environmental samples are inoculated to growth medium supplemented with the test metal. Following incubation, colonies that grew from cells present in the original sample appear on the surface of the solid medium. These colonies are counted, and their number multiplied by the dilution factor of the plated sample to determine how many culturable bacteria resistant to the metal are present in the original sample. This number is compared to the number of colonies grown on unsupplemented medium to quantitate the resistant fraction of the studied community. The most-probable-number (MPN) procedure, sometimes employed for the enumeration of functional groups of bacteria, is not used often in metal-resistance studies. Although the enumeration procedure is straightforward and does not require sophistication in equipment or technical skill, a few criteria are to be considered before conclusions pertinent to the studied community can be ascertained.

Growth media. The growth medium and incubation conditions must be selected according to the studied environment. For example, an estuarine sample collected at a geographical location with temperate climate is tested by growth on medium containing an elevated salt concentration and the plates are incubated at 25°C. Most commonly, growth media designed to select heterotrophic bacteria are composed of rich undefined mixtures of proteins and carbohydrates, such as yeast extract and peptone. Media have been composed carefully for enrichment of bacteria from specific environments: for marine organisms, Medium 2216 (Zobell, 1941) is recommended; estuarine organisms have been enumerated on several commercially available media, such as protease peptone (Nelson and Colwell, 1975) and plate-count agar (Barkay, 1987); media based on soil extracts supplemented with low concentrations of yeast extract and peptone are used to enumerate soil organisms (Martin, 1975).

One problem commonly encountered is that the concentration of organic nutrients in many of these media may inhibit rather than stimulate growth of organisms taken from environments low in nutrients. As a result, 10- or even 50-fold dilution of commercial media may support growth of more organisms compared to the undiluted media. If the experimental goal is to isolate the highest number of bacteria from a given sample, the selection procedure should be optimized by testing several media. However, as described below and in Part 1, no one medium can support growth of all bacteria. Such an approach in the study of distribution of metal-resistant organisms introduces a bias toward a certain part of the community.

Metal supplements.   The major question in preparing metal containing media is, what concentration of metal should be used? The answer is complicated by several factors. (1) Bacteria vary in their tolerance to metals. For example, Babich and Stotzky (1977) showed cadmium toxicity as follows: gram-negative bacteria > gram-positive bacteria > actinomycetes. (2) Some metals are more toxic than others, toxicity being related to their affinity to chelating agents since metals act by interacting with constituents (proteins, lipids, nucleic acids) of the living cell (Sterritt and Lester, 1980). (3) Binding to ingredients of the media results in an unknown fraction of the metals that is biologically unavailable (and thus nontoxic). Ramamoorthy and Kushner (1975) used ion-specific electrodes to determine the proportion of free ions in solution in the presence of several commonly used media components. When 20 ppm of Hg^{2+}, Pb^{2+}, or Cu^{2+} were added, only 80 ppb (0.4 percent) remained in solution.

Metal-supplemented growth media are prepared by addition of filter-sterilized metal stock solutions to autoclaved media cooled to 45 to 55°C, because some metals (such as Hg^{2+}) are volatile and others may precipitate out at elevated temperatures. The pH of the medium must be checked because addition of metals may alter it dramatically (B. Olson, University of California, Irvine, personal communication). The pH can be adjusted to neutrality either by the addition of acid or base, or by buffer systems suited to maintain biological activities (for example, phosphate buffer). Since a multitude of factors (see above) interact to determine the growth of bacteria in the presence of metals, results are usually unique to the specific combination of environmental sample investigated, the medium selected, and the test metal. The investigator has to determine empirically the metal concentrations that distinguish the resistant segment of the bacterial community under study. Olson and Thornton (1982) correlated resistance level of bacterial communities with the indigenous concentration of the tested metal in the sampled soil. Resistance was determined by plating on media supplemented with three different concentrations of metal, and the concentration at which positive correlations were obtained was used to select "real" resistant microorganisms. This concentration often selects a small part (<10 percent) of the culturable community. For example, in a study of the response of sediment communities to mercury pollution (Barkay and Olson, 1986), resistance to Hg^{2+} detected by plating on three metal concentrations was correlated with mercury contamination of sediment samples (Fig. 33.1). As Hg^{2+} concentrations in the media increased, a smaller portion of the bacterial community was resistant, but the correlation with mercury levels in sediments was stronger.

An elaboration of this approach has been proposed by Duxbury and Bicknell (1983) and Duxbury (1986). When the line described by plotting log 10 of the number of colony-forming units (CFUs) against metal concentration in the growth medium was subjected to curve-fitting analysis, an exponential function emerged:

$$Y = ae^{-bx}$$

where Y is the number of organisms growing in the presence of x mM of the test metal, a is the number of organisms growing in the absence of metal, and b is a constant expressing the intrinsic resistance of the soil bacterial community. Resistance is inversely related to b. The b constant was also used to demonstrate the high resistance of communities that evolved in metal-contaminated soil (Duxbury and Bicknell, 1983). It has been suggested that soil communities can be divided into two subgroups according to their metal resistance. The first subgroup is resistant to low concentrations of the test metal, and resistance did not correlate strongly with metal contamination in the soil. This resistance level represents intrinsic tolerance, owing to such factors as less permeable cell walls (for example, in gram-positive bacteria). The second subgroup, with high resistance levels that correlate strongly with soil contamination, could be mediated by genes carried on plasmids (Duxbury, 1986).

The selection of test concentrations cannot be based on metal levels that are commonly found in polluted environments. These concentrations are usually too low because of interactions of metals with media constituents (Ramamoorthy and Kushner, 1975). Testing of media supplemented with low concentrations of metals may lead to the erroneous conclusion that "all organisms were resistant" (Olson et al., 1979; Colwell et al., 1986). Selection of metal concentrations could be based on studies of resistance mechanisms in bacterial strains and their sensitive mutants. For example, it is well established that growth in rich media supplemented with 10 μg Hg^{2+} per milliliter (50 μM) distinguishes between mercury-resistant and sensitive bacteria (Foster et al., 1979), and this level of Hg^{2+} in the growth medium

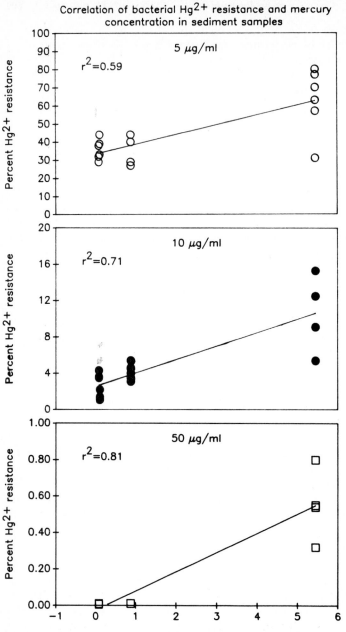

Figure 33.1 Effect of Hg^{2+} concentration in test medium on the correlation between bacterial resistance and mercury contamination of sediment samples. Diluted sediment samples were plated on Plate Count Agar amended with 5 μg Hg^{2+}/mL (○); 10 μg Hg^{2+}/mL (●); 50 μg Hg^{2+}/mL (□). Plates were incubated for 4 days before colonies were counted and percentage resistance calculated as

$$\frac{\text{CFUs on Hg}^{2+}\text{-supplemented medium}}{\text{CFUs on medium without Hg}^{2+}} \times 100$$

Mercury levels in sediment samples were determined by plasma emission spectrometry. *(Kozuchowski, 1978.)*

yielded a strong correlation between mercury in sediment samples and the resistance of the indigenous community (Fig. 33.1). According to the criteria discussed above, this portion probably represented "real" resistance. Concentrations of metals in growth media required to distinguish between resistant and sensitive strains are found in the scientific literature (summarized by Silver and Misra, 1988).

Comments

The advantage of the indirect method to study distribution of metal-resistant bacteria lies in the ease and rapidity by which test results are obtained. In addition, the method is cost-effective and does not require a great deal of training or expertise. As such, there is no replacement for measurements based upon plate count procedures. However, their two major limitations must be kept in mind when interpreting data: (1) The results do not represent microbial responses to metals in the sampled environment; (2) at best, the data represent that part of the community that can grow on the media and under the conditions that were selected (Mills and Bell, 1986). How the results relate to in situ response of organisms to metals remains totally unknown. The resistance level of the organisms to a given metal, as determined by viable counts, is a laboratory artifact (see above), with no relevance to metal tolerance in situ. Environmental samples treated identically can be compared but results do not mean that the indigenous microbial community of the sample that yielded the greatest number of resistant colonies is less affected (that is, more tolerant of metal insults). If the major concern in studying the ecology of microbial metal resistance is understanding the capacity of the environment to maintain its function in presence of pollution, indirect enumeration procedures are of a limited utility. However, these procedures may be useful in practice to indicate presence of metals in the environment in which the tested bacteria evolved (Olson and Thornton, 1982; Olson and Barkay, 1986).

Direct Approaches for the Detection of Metal-Resistant Microorganisms

Procedure

Direct approaches are based on quantification of resistant microorganisms without removal from environmental samples. The methodology to directly enumerate metal-resistant bacteria in aquatic samples is available (Zelibor et al., 1987b; Liebert and Barkay, 1988). The metal-resistant direct viable counts (MRDVC) procedure is based on the direct viable counts (DVC) method of Kogure et al. (1979, 1980, 1984). This procedure enumerates actively metabolizing metal-resistant microorganisms after a period of incubation with growth substrates, nalidixic acid (NA), and inhibitory concentrations of test metals. The method is based on the fact that nalidixic acid specifically inhibits DNA synthesis but not RNA or protein synthesis (Goss, 1964). Therefore, actively metabolizing cells continue to grow but not to divide. As a result, elongated cells are formed that are visualized by epifluorescence microscopy after acridine orange staining (Hobbie et al., 1977). Good correlations between results of DVC and measurements of microbial activity (i.e., electron transport system activity and active substrate uptake) have been demonstrated (Orndorff and Colwell, 1980; Maki and Remsen, 1981; Tabor and Neihof, 1984; Kogure et al., 1987; Roszak and Colwell, 1987; Liebert and Barkay, 1988).

Samples are incubated in the dark with yeast extract (YE), NA, HEPES buffer (pH 7.2), and varying concentrations of the test metal at the in situ temperature. Final concentrations of the added ingredients are determined for each water sample as described below. Following incubation, samples are fixed by the addition of formalin (final concentration, 3.7% formaldehyde). They may be stored at 4°C in the dark for up to two weeks before staining and counting. Samples are stained with a solution of 0.01% acridine orange (in 0.02 M Tris, pH 8.0, and 2.0% formaldehyde, final concentration), filtered through prewet 0.2-μm, 25-mm-diameter, black polycarbonate membranes (Nucleopore Corp., Pleasanton, California), and membranes are placed between microscope slides and coverslips with a drop of low-fluorescent immersion oil (Cargille Type A, R. P. Cargille Laboratories, Cedar Grove, New Jersey). Preparations are examined under epifluorescence illumination (Daley and Hobbie, 1975) at 1000 × magnification (oil immersion), using a high-pressure mercury or halogen lamp fitted with an excitation filter allowing spectral transmission in the 420- to 485-nm wavelength range. Consultation with manufacturers regarding selection of epifluorescent attachments is recommended. All enlarged substrate-responsive bacterial cells fluorescing orange or green are counted (at least 200 cells or 40 randomly selected fields). Two independent subsamples are counted to provide statistically significant results, as suggested by Kirchman et al. (1982). All solutions used are filtered (0.2 μm) to reduce background fluorescence and false-positive counts that may be created by particles.

The number of active cells per ml is calculated as follows:

$$\text{Cells/mL} = \frac{\text{No. of elongated cells/field} \times \text{No. of fields/filter}}{\text{sample volume (ml)} \times \text{dilution}}$$

Results are converted to percentage resistance by dividing values observed for metal-treated samples by values for corresponding metal-free samples and multiplying by 100. The metal concentration that inhibits 50% of cell elongation (inhibitory concentration 50 = IC_{50}) is calculated by regression analysis of percentage resistance versus metal concentration data.

The IC_{50} value is the parameter used to define the metal tolerance level of a community. Thus, communities from polluted waters may have a higher IC_{50} value than those from unimpacted control waters. For example, the MRDVC procedure was used to study the effect of mercury pollution on freshwater microbial communities (Liebert et al., 1991). Samples were collected at a polluted site (1750.0 ± 70.7 ppt total mercury) and at a control site (4.5 ± 0.7 ppt), and the number of substrate-responsive cells in subsamples treated with a gradient of Hg^{2+} concentrations was determined (Fig. 33.2). Regression analysis of data indicated that the community that evolved in polluted water was 47.3 times more tolerant to Hg^{2+} (with an IC_{50} = 3464.6 ± 311.9 µg Hg^{2+}/liter) than the control community (IC_{50} = 73.3 ± 17.9 µg Hg^{2+}/liter). Similarly, Zelibor et al. (1987a) showed that communities from arsenic-polluted well waters had a higher tolerance to arsenate than communities of control pristine well waters.

The efficiency of the DVC procedure depends to a great extent on the concentrations of the growth substrate and NA. Unfortunately, optimal conditions may vary with samples originating in different environments, so that preliminary experiments are needed for successful application to newly investigated microbial communities (Liebert, unpublished observation). Criteria and experiments to determine optimal test conditions are outlined in the following sections.

Buffers supplements. Zelibor et al. (1987b) introduced the use of HEPES buffer (25 mM, pH 7.2) to ensure stable pH. Stable pH is es-

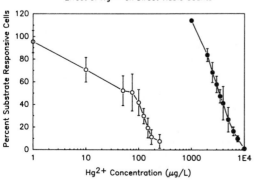

Figure 33.2 MRDVC for the determination of Hg^{2+} resistance of freshwater microbial communities. Percentage resistance at increasing concentrations of Hg^{2+} for communities of mercury-polluted water (●) and for communities of control unpolluted water (○). Results are presented as the means of duplicate samples ±1 SD (error bars).

sential because some metals may drastically affect the pH of the test media. Chelating buffer systems have been proposed to control the concentration of free metal ions in natural waters (Sanders et al., 1983). Thus, they overcome one major problem that complicates studies on biological responses to metals in aquatic systems: unknown levels of bioavailable metals. Assuming that free metal ions are the bioavailable forms of metals, one can calculate their concentrations by theoretical considerations (Sunda et al., 1978). The use of these buffers in the MRDVC procedures has not been attempted.

Nalidixic acid supplement. The concentration of NA commonly used for DVC is 0.002 percent, as was selected by Kogure et al. (1979), because it suppressed exponential growth of indigenous bacteria during 6 h of incubation. However, depending on the nature of samples, higher concentrations may be needed to surpress cell division. For example, Tabor and Neihof (1984) observed that efficacy of NA inhibition was diminished by dissolved compounds associated with a high density of phytoplankton. High densities of phytoplankton rendered NA inactive in samples otherwise affected by 0.002 percent of the drug. Concentrations of NA had to be increased to 0.01 percent to obtain effective inhibition of cell division and the maximum number of substrate responsive bacteria. Substrate-responsive NA-resistant microorganisms that continue to divide under the test conditions may result in false-negative counts.

The optimal concentration of NA is determined by gradually increasing the concentration of the drug and selecting the concentration that gives the highest number of substrate-responsive cells without an appreciable increase in total counts. An increase of total counts indicates that the NA concentration is too low and that cell division has occurred. For example, Liebert and Barkay (1988) supplemented estuarine samples with increasing concentrations of NA from 0.002 to 0.02 percent (w/v). Cell division was prevented by 0.01% NA, a concentration that resulted in an optimal number of elongated cells and was subsequently used in all experiments.

The MRDVC procedure as outlined here is useful with microbial communities comprised largely of gram-negative bacteria because of the specificity of NA. NA does not inhibit DNA replication of gram-positive bacteria (Davis et al., 1980). DNA replication inhibitors of gram-positive bacteria, such as pirmidic acid, are available. At present, the DVC procedure is mostly applied to aquatic communities because direct counts are performed with difficulty in samples that contain particulate matter, such as sediments or soils. Since the majority of aquatic bacteria are gram-negative (Rheinheimer, 1985), the investigation of alternative DNA replication inhibitors has been limited to date (Kogure et al., 1984).

Growth substrate supplements. YE is used commonly as a growth substrate in the DVC procedure. YE and tryptone were the best substrates for marine bacteria as compared with several combinations of amino acids, carbohydrates, and vitamins (Peele and Colwell, 1981; Kogure et al., 1987). However, it should be kept in mind that no one substrate can support growth of all metabolically active microorganisms. False-negative results (i.e., active cells that cannot use the provided substrate) can be minimized by selection of appropriate substrate, but cannot be eliminated completely. An optimal YE concentration is determined as the concentration that results in the highest number of enlarged cells. For example, YE at 0.02 percent (w/v) was too diluted when added to estuarine samples because increasing the concentration to 0.03 percent produced a 64 percent increase in the number of substrate-responsive cells (Liebert and Barkay, 1988).

Background fluorescence (due to acridine orange–YE complexes) presents a problem if a large volume of samples is filtered to obtain statistically significant counts. Removal of YE before staining with acridine orange prevents this problem. The filtered sample is stained in the funnel of the filtration apparatus (Liebert and Barkay, 1988).

Incubation time. The incubation period may be critical if the activity of organisms in the sample results in reduced metal toxicity, alleviating growth inhibition of sensitive cells. These cells will then enlarge and produce false-positive counts. Microorganisms in natural samples may remove metals, as with the reduction of mercurial compounds to the volatile Hg° (Barkay, 1987; Silver and Misra, 1988). In addition, divalent metal cations are rendered less toxic by precipitation with microbially formed H_2S (Erardi et al., 1987), chelation with organic ligands, and adsorption to particulates (Babich and Stotzky, 1985). This potential of microbial activities to reduce toxicity of metals should be considered when the MRDVC procedure is applied for metal tolerance measurements. Comparisons may be achieved by contrasting the increase in numbers of enlarged cells in metal-treated samples with metal-free samples. Normally, the number of enlarged cells increases with time in an exponential mode before it levels off (Kogure et al., 1979). Metal-treated samples respond similarly, albeit with lower counts. If a large increase in number of enlarged cells is observed after a period of incubation in the metal-treated sample, but not in the metal-free control, alleviation of metal inhibition has occurred (Liebert and Barkay, 1988).

Comments

The MRDVC procedure is a tool for estimating metal-tolerant microbial biomass in aquatic environments. This procedure could be used to evaluate in situ effects of metals and possible ecosystem perturbations resulting from metal contamination, because IC_{50} values are similar to concentrations of metals that inhibit microbial processes. For example, a Hg^{2+}-exposed estuarine community had IC_{50} values of 9528 ± 300 µg Hg^{2+} per liter and 2360 ± 239 µg Hg^{2+} per liter determined by the DVC and the ^{3}H-methyl-thymidine incorporation methods (Jonas et al., 1984), respectively. The latter is a measure of microbial activity adapted by Jonas et al. (1984) for toxicity measurements. The unexposed communities had corresponding values of 1.8 ± 1.6 µg Hg^{2+} per liter obtained by the MRDVC procedure, and 0.48 ± 0.02 µg Hg^{2+} per liter obtained by the activity measurement test (Liebert and Barkay, 1988). Orndorff and Colwell (1980), who studied the toxic effect of Kepone on microbial function and biomass in the Chesapeake Bay, reported similar sensitivity levels obtained by the DVC approach and as measured by the effect of Kepone on the utilization of ^{14}C-L-amino acids. Thus, the strength of the MRDVC procedure is its relevance to in situ effects of metals.

At present, the direct method does not provide information on specific resistant populations within the community. In addition, the MRDVC procedure is time-consuming and tedious, and allows processing of only a limited number of samples. Thus, its inclusion in environmental screening approaches is not feasible; its utility is limited to studies on effects of metals in selected environments.

Bacteria in natural waters may assume small-cell morphology. As a result, bacteria may respond to YE and NA but still be too small to be considered enlarged. It is recommended that a sample fixed immediately upon collection be observed microscopically to familiarize the researcher with the size and appearance of indigenous cells. In addition, this sample can be used to obtain total direct counts and background counts of indigenous long cells (that are subtracted from treated samples). Total direct counts of the sample are used as a control to ensure the efficacy of NA. These counts should not increase in samples incubated with growth substrate if NA inhibition is effective.

The approach described here for estimation of metal-tolerant microbial biomass may have a much broader application to study of the distribution of specific functional traits in natural communities. It could be used to study responses to other toxicants, such as organic compounds, as was suggested for Kepone (Orndorff and Colwell, 1980). In addition, it could be modified to estimate microbial biomass that can utilize specific substrates. Microorganisms have evolved metabolic pathways that enable acquisition of nutrients and energy from chemically complex and highly toxic environmental pollutants (Ghosal et al., 1985). Biotechnologists consider these microorganisms and their genes to be a potential tool in bioremediation (Halvorson et al., 1985).

Although degradation of such compounds has been demonstrated and pure cultures with the desired metabolic capabilities are readily isolated (Spain et al., 1980), studies on the ecology of these microorganisms are limited because methods for reliable estimations of active biomass are scarce (Somerville et al., 1985). The DVC procedure could be used to quantify substrate-responsive populations after incubation of natural samples with NA and the specific growth substrates (i.e., replace YE with the studied compound). In this case, only cells that can obtain their nutritional needs by utilization of the available substrate would enlarge. The relative abundance of active cells in the community could be used to predict their potential to degrade the pollutant.

New Approaches to Study Distribution of Metal-Resistant Organisms in the Environment

Ideally, specific populations of microorganisms and their metal resistance should be studied in situ to learn how unique groups contribute to the ability of the community to withstand metal stress. Populations of metal-resistant organisms are currently investigated by the isolation of pure cultures, which results in a biased sample that does not represent the entire community and that is not pertinent to in situ microbial responses to metals. Approaches and methods have been developed to study specific populations of microorganisms without removal from their environment. Cell-specific antibodies and nucleic acid probes are used to label unique cells in samples of natural waters and soils (Bohlool and Schmidt, 1980; Giovannoni et al., 1988; DeLong et al., 1989; Amann et al., 1990). These techniques, in combination with DVC and detection of cells performing specific metabolic activities by autoradiography (Tabor and Neihof, 1983), allow distinction of specific populations in mixed microbial assemblies (Mills and Bell, 1986). For example, they have been used to follow population dynamics of *Thiobacillus ferrooxidans* in acid mine leachate and on coal particles (Muyzer et al., 1987).

Detection by the immunofluorescence technique is based on reaction with cell-specific antibodies labeled by conjugating a fluorescent tag to the antibody molecule. Cell-specific antibodies are produced by immunizing a rabbit with cell suspensions of the bacterium to be detected, or if a unique cell wall component of this organism can be isolated in a purified form, monoclonal antibodies can be obtained. The latter approach has the advantage of higher specificity because cross reactions with organisms that share cell wall components are prevented (Garvey et al., 1977). Antiserum produced by injecting a rabbit with a bacterial suspension should be tested for cross reactivity with species that inhibit the studied ecosystem (Mills and Bell, 1986).

Oligonucleotide probes are short (18 to 20 nucleotides in length) DNA fragments that complement a unique sequence found in the genome of the organism to be detected. The probes are synthesized in vitro and are end-labeled with either α-^{32}P-dATP (Giovanonni et al., 1988) or fluorescent tags (D. Stahl, personal communication). End-labeled oligonucleotide probes are superior to the nick-translated probes. They easily cross the cell wall due to their small size and are more sensitive because of a higher specific activity (that is, more label per probe molecule) and greater specificity because the small number of necleotides ensures exclusive hybridization with sequences unique to the target organism. Probes can be directed to sequences that are unique to specific strains, such as variable parts of 16s rRNA molecules, thus detecting specific populations of microorganism (Stahl et al., 1988). Alternatively, they can be directed toward sequences that code for enzymes involved in specific reactions, such as nitrogen fixation, to detect microorganisms that have the potential to carry out essential ecological functions (F. Genthner, personal communication; Ogram et al., 1987).

The combination of techniques for direct detection of organisms, their genes and activities, and application to communities in natural samples could start a new era in the study of the ecology of microbial resistance to metals. For example, the active sites of the mercuric reductase enzymes of five bacterial strains (three gram-negative and two gram-positive), whose mercury-resistance operons have been sequenced to date, have homologous amino acid sequences (S. Silver, University of Illinois, Chicago, personal communication). Conceivably, a mercuric reductase gene-directed oligonucleotide probe could be designed to hybridize to mercury-resistant microorganisms. Specific taxonomic groups could be detected by immunofluorescence of 16s rRNA-directed probes, and function-specific probes could be employed to detect populations with specific metabolic capabilities. A combination of these techniques applied to a mercury-exposed natural sample could reveal (1) which populations of microorganisms are resistant to mercury, (2) how mercury affects a specific microbial function, and (3) how resistance may alleviate toxic effects.

Conclusions

Two approaches, a direct and an indirect, are currently used to study distribution of metal-tolerant microorganisms in the environment. Each has its advantages and disadvantages, but they complement each other. The indirect culturing method provides pure cultures that serve as a source of information on microbial processes that enable metal tolerance. Yet this approach lacks applicability to in situ microbial processes and thus information on the ecology of impacted communities is limited. Ecological information is provided by the direct approach (MRDVC), enabling evaluation of metal impacts on microbial processes that are essential for maintaining the geochemical cycling of major elements. However, at present, the MRDVC procedure is not refined enough to provide information at the population level. Future developments, using novel technologies, will combine the advantages of both approaches; probes designed to detect specific taxonomic and functional groups will be integrated with DVC procedures to investigate the role of unique groups of microorganisms in the response of natural communities to metal stress.

References

Amann, R. I., L. Krumholz, and D. A. Stahl. 1990. Fluorescent-oligonucleotide probing of whole cells for determinative, phylogenetic, and environmental studies in microbiology. *J. Bacteriol.* 172:762–770.

Babich, H., and G. Stotzky, 1986. Environmental factors affecting the utility of microbial assays for the toxicity and mutagenicity of chemical pollutants. In B. J. Dutka and G. Bitton (eds.), *Toxicity Testing Using Microorganisms*, vol. 2. CRC Press, Boca Raton, Fla., pp. 9–42.

Babich, H., and G. Stotzky. 1985. Heavy metal toxicity to microbe mediated ecologic processes: A review and potential application to regulatory policies. *Environ. Res.* 36: 111–137.

Babich, H., and G. Stotzky. 1977. Sensitivity of various bacteria, including Actynomycetes and fungi, to cadmium and the influence of pH on sensitivity. *Appl. Environ. Microbiol.* 33:681–695.

Barkay, T. 1987. Adaptation of aquatic microbial communities to Hg^{2+} stress. *Appl. Environ. Microbiol.* 49:686–692.

Barkay, T., and B. H. Olson. 1986. Phenotypic and genotypic adaptation of aerobic heterotrophic sediment bacterial communities to mercury stress. *Appl. Environ. Microbiol.* 52:403–406.

Barkay, T., and P. Pritchard, 1988. Adaptation of aquatic microbial communities to pollutant stress. *Microbiol. Sci.* 5:165–169.

Bohlool, B. B., and E. L. Schmidt. 1980. The immunofluorescence approach in microbial ecology. *Adv. Microb. Ecol.* 4:203–241.

Bopp, L. H., and H. L. Ehrlich. 1988. Chromate resistance and reduction in *Pseudomonas fluorescens* strain LB300. *Arch. Microbiol.* 150:426–431.

Colwell, R. R., D. Allen-Austin, T. Barkay, J. Barja, and J. D. Nelson, Jr. 1986. Antibiotic resistance associated with heavy metal mineralization. In W. B. Carlisle, J. Watterson, and I. Kaplan (eds.), *Mineral Exploration, Biological Systems and Organic Matter*, vol. 5. Prentice-Hall, Englewood Cliffs, N.J., pp. 282–300.

Daley, R. J., and J. E. Hobbie. 1975. Direct counts of aquatic bacteria by modified epifluorescent technique. *Limnol. Ocean.* 20:875–882.

Davis, B. D., R. Dulbecoo, H. N. Eisen, and H. S. Ginsberg. 1980. *Microbiology*, 3d ed. Harper & Row, Philadelphia.

Delnomdedieu M., A. Boudou, J. P. Desmazés, and D. Georgescauld. 1989. Interaction of mercury chloride with the primary amine group of model membranes containing phosphotidyserine and phosphotidylethanolamine. *Biochim. Biophys. Acta* 986:191–199.

DeLong, E. F., G. S. Wickham, and N. R. Pace. 1989. Phylogenetic stains: Ribosomal RNA-based probes for the identification of single cells. *Science* 243:1360–1363.

Duxbury, T. 1981. Toxicity of heavy metals to soil bacteria. *FEMS Microbiol. Lett.* 11: 217–220.

Duxbury, T. 1986. Ecological aspects of heavy metal responses in microorganisms. *Adv. Microb. Ecol.* 8:185–235.

Duxbury, T., and B. Bicknell. 1983. Metal-tolerant bacterial populations from natural and metal-polluted soils. *Soil Biol. Biochem.* 15:243–250.

Erardi, F. X., M. L. Failla, and J. O. Falkinham, III. 1987. Plasmid-encoded copper resistance and precipitation by *Mycobacterium scrofulaceum.* *Appl. Environ. Microbiol.* 53:1951–1954.

Foster, T. J., H. Nakahara, A. A. Weiss, and S. Silver. 1979. Transposon A-generated mutations in the mercuric resistance gene of plasmid R100-1. *J. Bacteriol.* 140:167–181.

Garvey, J. S., N. E. Cramer, and D. H. Sussdorf. 1977. *Methods in Immunology: A Laboratory Tool for Instruction in Research*, 3d ed. Benjamin, Reading, Mass. 545 p.

Ghosal, D., I. S. You, D. K. Chatterjee, and A. M. Chakrabarty. 1985. Microbial degradation of halogenated compounds. *Science* 228:135–142.

Giovannoni, S. J., E. F. DeLong, G. J. Olsen, and N. R. Pace. 1988. Phylogenetic group-specific oligonucleotide probes for identification of single microbial cells. *J. Bacteriol.* 170:720–726.

Goss, W. A., W. H. Dietz, and T. M. Cook. 1964. Mechanism of action of nalidixic acid on *Escherichia coli. J. Bacteriol.* 88:1112–1118.

Hallas, L. E., and J. J. Cooney. 1981. Tin and tin-resistant microorganisms in Chesapeake Bay. *Appl. Environ. Microbiol.* 241:446–471.

Halvorson, H. O., D. Pramer, and M. Rogul. 1985. *Engineered Organisms in the Environment: Scientific Issues.* Proceedings of a cross-disciplinary symposium held in Philadelphia, Pennsylvania, 10–13 June 1985. American Society for Microbiology, Washington, D.C.

Hamlett, N. V. 1986. Alternation of a salt marsh bacterial community by fertilization with sewage sludge. *Appl. Environ. Microbiol.* 52:915–923.

Hobbie, J. E., R. J. Daley, and S. Jasper. 1977. Use of nucleopore filters for counting bacteria by fluorescence microscopy. *Appl. Environ. Microbiol.* 33:1225–1228.

Houba, C., and J. Remacle. 1980. Composition of the saprophilic bacterial communities in freshwater systems contaminated by heavy metals. *Microb. Ecol.* 6:55–69.

Jonas, R. B., C. C. Gilmour, D. L. Stoner, M. M. Weir, and J. H. Tuttle. 1984. Comparison of methods to measure acute metal and organometal toxicity of natural aquatic microbial communities. *Appl. Environ. Microbiol.* 47:1005–1011.

Kirchman, D., J. Sigda, R. Kapuscinski, and R. Mitchell. 1982. Statistical analysis of the direct count method for enumerating bacteria. *Appl. Environ. Microbiol.* 44:376–382.

Klerks, P. L., and J. S. Weis. 1987. Genetic adaptation to heavy metals in aquatic organisms: A review. *Environ. Pollut.* 45:173–205.

Koch, A. L. 1981. Growth measurement. In P. Gerhardt (ed.), *Manual of Methods for General Bacteriology.* American Society for Microbiology, Washington, D.C., pp. 179–207.

Kogure, K., U. Simidu, and N. Taga. 1979. A tentative direct microscopic method for counting living marine bacteria. *Can. J. Microbiol.* 25:415–420.

Kogure, K., U. Simidu, and N. Taga. 1984. An improved direct viable count method for aquatic bacteria. *Arch. Hydrobiol.* 102:117–122.

Kogure, K., U. Simidu, and N. Taga. 1980. Distribution of viable marine bacteria in neritic seawater around Japan. *Can. J. Microbiol.* 26:318–323.

Kogure, K., U. Simidu, N. Taga, and R. R. Colwell. 1987. Correlation of direct viable counts with heterotrophic activity for marine bacteria. *Appl. Environ. Microbiol.* 53: 2332–2337.

Kozuchowski, J. 1978. Determination of total mercury in sediments by furnace combustion and plasma emission spectrometry. *Analyt. Chemica Acta* 99:293–297.

Liebert, C. A., and T. Barkay. 1988. A direct viable counting method for measuring tolerance of aquatic microbial communities to Hg^{2+}. *Can. J. Microbiol.* 34:1090–1095.

Liebert, C. A., T. Barkay, and R. R. Turner, 1991. Acclimation of aquatic communities to Hg(II) and CH₃Hg⁺ in polluted freshwater ponds. *Microb. Ecol.* 21:139–149.

Luli, G. W., J. W. Talnagir, W. R. Strohl, and R. M. Pfister. 1983. Hexavalent chromium-resistant bacteria isolated from river sediment. *Appl. Environ. Microbiol.* 46:846–854.

Maki, J. S., and C. C. Remsen. 1981. Comparison of two direct-count methods for determining metabolizing bacteria in freshwater. *Appl. Environ. Microbiol.* 41: 1132–1138.

Martin, J. K. 1975. Comparison of agar media for counts of viable soil bacteria. *Soil Biol. Biochem.* 7:401–402.

Mills, A. L., and P. E. Bell. 1986. Determination of individual organisms and their activities in situ. In R. L. Tate III (ed.): *Microbial Autecology. A Method for Environmental Studies.* Wiley, New York, pp. 27–60.

Mills, A. L., and R. R. Colwell. 1977. Microbiological effects of metal ions in Chesapeake Bay water and sediment. *Bull. Environ. Contam. Tox.* 18:99–103.

Muyzer, G., A. C. deBruyn, D. J. M. Schmedding, P. Bos, P. Westbroek, and G. J. Kuenen. 1987. A combined immunofluorescence–DNA-fluorescence staining technique for enumeration of *Thiobacillus ferrooxidans* in a population of acidophilic bacteria. *Appl. Environ. Microbiol.* 53:660–664.

Nakamura, K., T. Fujisaki, and H. Tamashiro. 1986. Characteristics of Hg-resistant bacteria isolated from Minamata Bay sediment. *Environ. Res.* 40:58–67.

Nelson, J. D., and R. R. Colwell. 1975. The ecology of mercury-resistant bacteria in the Chesapeake Bay. *Microb. Ecol.* 1:191–218.

Ogram, A., G. S. Sayler, and T. Barkay, 1987. The extraction and purification of microbial DNA from sediments. *J. Microb. Methods* 7:57–66.

Olson, B. H., and I. Thornton. 1982. The resistance patterns to metals of bacterial populations in contaminated land. *J. Soil Sci.* 33:271–277.

Olson, B. H., and T. Barkay. 1986. Feasibility of using bacterial resistance to metals in mineral exploration. In D. Carlisle, W. Berry, I. Kaplan, and J. Watterson (eds.): *Mineral Exploration: Biological Systems and Organic Matter,* Prentice-Hall, Englewood Cliffs, N.J., pp. 311–327.

Olson, B. H., T. Barkay, and R. R. Colwell. 1979. Role of plasmids in mercury transformation by bacteria isolated from aquatic environment. *Appl. Environ. Microbiol.* 38: 478–485.

Orndorff, S. A., and R. R. Colwell. 1980. Effect of kepone on estuarine microbial activity. *Microb. Ecol.* 6:357–368.

Peele, E. R., and R. R. Colwell. 1981. Application of a direct microscopic method for enumeration of substrate-responsive marine bacteria. *Can. J. Microbiol.* 27:1071–1075.

Pickaver, A. H., and M. C. Lyes. 1981. Aerobic microbial activity in surface sediments containing high or low concentrations of zinc taken from Dubin Bay, Ireland. *Est. Coast. Shelf Sci.* 12:13–22.

Rada, R. G., and M. R. Winfrey. Abstracts of the Society of Environmental Toxicologists and Chemists 8th Annual Meeting, Pensacola, Fla., November 9–12, 1987.

Ramamoorthy, S., and D. J. Kushner. 1975. Binding of mercuric and other heavy metal ions by microbial growth media. *Microb. Ecol.* 2:162–176.

Rheinheimer, G. 1985. *Aquatic Microbiology,* 3d ed. Wiley, Chichester, England.

Rose, A., H. E. Hawkes, and J. S. Webb. 1979. *Introduction to Geochemistry in Mineral Exploration.* Academic Press, New York.

Roszak, D. B., and R. R. Colwell. 1987. Metabolic activity of bacterial cells enumerated by direct viable counts. *Appl. Environ. Microbiol.* 53:2889–2983.

Sanders, B. M., K. D. Jenkins, W. G. Sunda, and J. D. Costlow. 1983. Free cupric ion activity in seawater: Effects on metallothione and growth in crab larvae. *Science* 222: 53–55.

Silver, S., and T. K. Misra. 1988. Plasmid-mediated heavy metal resistance. *Annu. Rev. Microbiol.* 42:717–743.

Somerville, C. C., C. A. Monti, and J. C. Spain. 1985. Modification of the ¹⁴C-most-probable number method for use with nonpolar and volatile substrates. *Appl. Environ. Microbiol.* 49:711–713.

Spain, J. C., P. H. Pritchard, and A. W. Bourquin, 1980. Effects of adaptation on biodegradation rates in sediment/water cores from estuarine and freshwater environments. *Appl. Environ. Microbiol.* 40:726–734.

Stahl, D. A., B. Flesher, H. R. Mansfield, and L. Montgomery. 1988. A molecular assessment of microbial ecosystem perturbation. *Appl. Environ. Microbiol.* 54:1079–1084.

Sterritt, R. M., and J. N. Lester. 1980. Interactions of heavy metals with bacteria. *Sci. Tot. Environ.* 14:5–17.

Sunda, W. G., D. W. Engel, and R. M. Thuotte. 1978. Effect of chemical speciation on toxicity of cadmium to grass shrimp, *Palaemonetes pugio:* Importance of free cadmium ion. *Environ. Sci. Technol.* 12:409–413.

Tabor, P. S., and R. A. Neihof. 1983. Improved method for determination of respiring individual microorganisms in natural waters. *Appl. Environ. Microbiol.* 43:1249–1255.

Tabor, P. S., and R. A. Neihof. 1984. Direct determination of activities for microorganisms of Chesapeake Bay Population. *Appl. Environ. Microbiol.* 48:1012–1019.

Tate, R. L., III. 1986. Importance of autecology in microbial ecology. In R. L. Tate III (ed.), *Microbial Autecology: A Method for Environmental Studies.* Wiley, New York, pp. 1–26.

Timoney, J. F., J. Port, J. Giles, and J. Spanier. 1979. Heavy-metal and antibiotic resistance in the bacterial flora of sediments of New York Bight. *Appl. Environ. Microbiol.* 36:465–472.

Zelibor, J. L., M. W. Doughton, D. J. Grimes, and R. R. Colwell. 1987a. Testing for bacterial resistance to arsenic in monitoring well water by the direct viable counting method. *Appl. Environ. Microbiol.* 53:2929–2934.

Zelibor, J. L., M. Tamplin, and R. R. Colwell. 1987b. A method for measuring bacterial resistance to metals employing epifluorescent microscopy. *J. Microbiol. Methods* 7:143–155.

Zobell, C. E. 1941. Studies on marine bacteria, I. The cultural requirements of heterotrophic aerobes. *J. Mar. Res.* 4:42–75.

Part 5

Effects on Individual Organisms (Nontarget)

34

Overview:
The Effects of Microbial
Pest Control Agents
on Nontarget Organisms

John D. Briggs
David C. Sands

Background

Microbial pest control agents (MPCAs) historically have been subject to governmental regulation. The best example involves the development and release of *Bacillus thuringiensis,* a bacterium for control of larval Lepidoptera on agricultural crops and forests. In 1959, an application for residue exemptions was initiated for viable spores of *B. thuringiensis* on several crops. The U.S. Food and Drug Administration (FDA) requested evidence from the first manufacturer to support claims that the products were safe for field use. What followed was the development of a microbial insecticide and the coevolution and development of regulatory programs covering both conventional chemical control agents and microbiological agents.

Since no guidelines for granting residue tolerance exceptions for microbiological agents were available in 1959, the applicants in the United States selected the most readily available vulnerable mammal for safety demonstrations of *B. thuringiensis.* Utilizing unspecified 20-g white mice, a "mouse safety test" was developed that supported the residue tolerance petition. Results were published in *Agricultural and Food Chemistry* in 1959. The "mouse safety test" involved subjecting a representative mammal to large doses for serial inoculation of the microorganism in question. Inhalation, allergenicity, and ingestion tests were also performed. In addition, human volunteers were tested for ingestion effects and guinea pigs for effects following intraperitoneal injections. The initiative, taken by the producer, and the residue tolerance exemption, granted by the FDA in 1960, provided the opportunity for additional industrial groups to initiate field studies, to develop data and information on the efficacy of *B. thuringiensis* formulations, and to market products.

In order to dispel inherent fears of biological pesticides, early marketing efforts stressed and illustrated the safety aspects of *B. thuringiensis* utilization. This included demonstrations of the host specificity expressed by the microorganism. For example, in California in 1960, a manufacturer initiated successful demonstrations illustrating the safety of the microorganisms for golden trout, pheasants, and quail. Public reception of the results was enthusiastic.

At the time, the initiation of marketing and the field release of products containing *B. thuringiensis* were made under the authority of two federal agencies, the FDA and the U.S. Department of Agriculture (USDA), acting independently yet in collaboration and insisting on agreement in decisions on label registration. Subsequently, the safety and residue tolerance requirements of the FDA merged with the efficacy requirements of the USDA within the new U.S. Environmental Protection Agency (EPA). The simple and direct approach for support of the apparent safety of microbial control agents was then expanded upon and structured for codification in the Federal Insecticide, Fungicide, Rodenticide Act (FIFRA).

Description

Several fundamental differences between chemical and microbial pesticides exist. An inherent characteristic of chemicals is their purification by concentration, in contrast to microorganisms that are purified by dilution. Chemicals can be diluted to harmless concentrations, while microorganisms can grow from a single cell to a large population. Lethal doses in chemical pesticide toxicity are expressed in fixed numbers; conversely, live microorganisms can infect a host organism beginning with a few cells, given appropriate inoculation, predisposition, or susceptibility of the host. In the case of microorganisms, a genetic system is released, capable of response to environmental changes.

Any concerns expressed by society are resolved when there is appreciation by the public of the differences in behavior and effect of a "living agent" in contrast to a "nonliving agent." The concept that a microbial pesticide is alive is an essential precept of biological control. Living insects parasitic on other insects and predacious on other anthropods, and microorganisms that cause diseases of arthropods or weeds, are active in nature with and without manipulation. The appreciation by the public of the maintenance of a "balance of nature" facilitates the acceptance of the use of biological control agents.

Differences in the consequences of the use of living microorganisms in contrast to chemical pesticides can be summarized briefly: (1) chemical use often results in residues that are generally diluted or degraded by environmental factors or that may be selectively concentrated by living organisms; whereas (2) microbial pesticides may succumb to physical factors or multiply and disperse indefinitely in the environment, generally in association with the target pest organisms. This dichotomy does not address the potential of a microbial control agent producing a toxin that could be used as a chemical pesticide in the absence of the microorganism that produces the pesticidal toxin.

Despite existing fundamental differences, the effects of microbial pest control agents (MPCAs) on individual organisms are, out of tradition, investigated using the procedures and techniques that have been accepted for several decades to investigate the effects of conventional chemical pesticides on plants and animals. Some of the parameters that are observed for chemical pesticide screening are valid for MPCAs, but many are irrelevant. Addressing safety considerations for the use of MPCAs with respect to their effect on particular groups of organisms has required that contributors to Part 5 recognize challenges expressed for conventional chemical pesticides plus the unique qualities of a living microbial agent. Additional consideration must be given to the potential of MPCAs for permanent establishment in a new habitat and exposure of growth and development stages of nontarget organisms to different developmental stages of the MPCA.

Perspective

Contributions to this part represent our efforts to evaluate and investigate the impact of MPCAs on plants and animals. Extrapolations may be made to include organisms not specifically treated here. The need to establish procedures for developing data related to target vs. nontarget species is addressed.

Additional arthropods, invertebrates, and associated vertebrates in aquatic communities are discussed regarding the dispersion of MPCAs from terrestrial to aquatic habitats. The two chapters on crustacea, fish, and birds are provided to meet environmental concerns on the impact of introduced agents on desirable species.

The primary use of MPCAs will likely occur in terrestrial habitats on agricultural plants. Because plants play a primary role in food chains, rigorous safety reviews of plants and their products, particularly roots and seeds, exposed to MPCAs need to be conducted. In Part 5 the authors present both safety and efficacy concerns in evaluating MPCAs and their effects on particular organisms.

Conclusions

We believe that the scientific community, in general, continues to assemble valid and useful data on the safety of MPCAs with respect to nontarget organisms. However, due to the lack of repeated use over many years of particular MPCAs, weaknesses do exist in the accumulated data base. These are discussed in each of the chapters in this part. Nevertheless, two exceptions provide excellent research data for the United States. These are the extensive use of two bacilli: *Bacillus popillae* and *Bacillus thuringiensis*. The "wild" isolate of *B. thuringiensis* used in 1960 for initial MPCA products has been superseded by many more effective and safer strains, some derived from the original isolates. *Bacillus thuringiensis* is, in fact, a genetic engineering "workhorse" for MPCA development. A distinction must be made between products containing the initial isolates or derivatives of these isolates and products that may be developed by genetic manipulation of original isolates.

Statements in the news media suggest that society has adequate information to dispute the safety of conventional chemical agents. We are often reassured that these concerns are not valid if proper use is practiced. Yet revelations of undesirable side effects of conventional chemical agents continue to be evident. Haste in releasing microbial pesticide agents may lead to a similar situation.

Chapter 35

Testing the Effects of Microbial Agents on Plants

C. Lee Campbell
David C. Sands

Introduction

Plants are regularly and routinely evaluated for their sensitivity to natural and manufactured chemicals and for their susceptibility to plant pathogenic microorganisms. Sophisticated protocols have been developed, particularly with regard to pesticides (33) for evaluating the efficacy of such compounds in achieving an intended purpose, e.g., disease control or weed kill; for establishing the potential phytotoxicity of chemical substances on crops of economic, environmental, and ecological importance; and for determining the residual levels of such chemicals in living plant materials and in plant by-products. The methodology for establishing phytotoxicity and residual levels of applied chemical substances is well documented (21, 55). Criteria for judging detrimental effects of applied chemicals often include general plant health, growth rate, and quantity and quality of yield.

Plant diseases result from the interaction of a pathogen and host under appropriate environmental conditions over time. The literature on plant pathogenic microorganisms and the diseases they cause is extensive. The basic principles of the development of plant diseases are presented in several recent texts (1, 41, 50), and the diseases of specific plants or groups of plants are documented in the Disease Compendia Series published by the American Phytopathological Society (St. Paul, Minnesota) and in several disease indices (5, 28, 72). Methods ranging in scope from the molecular, biochemical, and physiological to epidemiological techniques are widely available to determine effects of microorganisms, particularly pathogens, on plants (12, 22, 33, 34, 39, 46, 48, 49, 68).

In comparison with chemical pesticides, relatively few microbial pest control agents (MPCAs) have been tested on plants. Some of those tested have been intended to control insect pests and have received only limited plant effects evaluation since no pathogenicity was known for the MPCA. For example, minimal testing of *Bacillus thuringensis* on plants was necessary since members of the genus *Bacillus* are not plant pathogens. Alternatively, other microbes used for biological control of plant pathogens are members of genera known to cause plant diseases and have required more extensive testing (6, 19, 73).

The basic philosophy for testing the effects of MPCAs on plants differs in several important aspects from that for evaluating the effects of chemical pesticides on plants. Chemical pesticides can persist in the plant or environment in an unaltered or altered form, usually with decreasing residual concentrations over time (9, 55). MPCAs, however, can reproduce, remain an active component of the ecosystem, and actually exhibit increases in population numbers over time. Chemical pesticides are mobile in the environment only to the extent that they are moved as contaminants in water, soil, or plant materials; spread may rarely occur over long distances but is usually localized. MPCAs can reproduce and be dispersed over relatively large distances either by physical factors in the environment (24) or by biological vectors (14, 75) including humans. Finally, the dose-response relationship may differ between chemical pesticides and MPCAs. With a chemical pesticide, the dosage applied is fixed and will either remain constant or decrease over time due to degradation, decomposition, etc. (9, 32). Because an MPCA is subject to host or environmental conditions that can affect reproduction and survival, the dose-response relationship may change over time.

As a final introductory note, it must be acknowledged that a fundamental difference exists between determining whether a specific microorganism is a pathogen on one host species or cultivar and determining that there is no effect of an MPCA on any plant. In the first case, the null hypothesis is that microorganism X is not a pathogen on host A or cultivar Z of host A. The desired result is actually the rejection of the null hypothesis, and Koch's postulates can then be completed for the microorganisms in question. In the second case, that of testing the effects of an MPCA on plants, the null hypothesis is much broader, i.e., there is no effect of MPCA X on plants. This second null hypothesis poses an understandable problem in designing experiments to achieve generality. Also, what is actually desired is acceptance of the null hypothesis. Such an acceptance is quite different from failing to reject the null hypothesis. The latter is scientifically feasible, while the former is problematic. The goal in testing the effects of MPCAs on plants must be, then, to obtain a sufficient amount of data on an acceptable spectrum of plants under appropriate environmental conditions to be able to have sufficient confidence in the conclusions made.

At present there is not a standard methodology or set of accepted protocols for performing tests to determine the effects of MPCAs on plants. The Environmental Protection Agency (EPA) recommends a tiered approach similar to the hazard assessment protocol used in the registration of pesticides (7). The test procedures for plants are essentially adapted from chemical (e.g., pesticide) bioassays, because most procedures that exist to examine the effects of microorganisms are designed to establish phytopathogenicity rather than to establish a lack of phytotoxicity or phytopathogenicity. The following discussion represents a proposal for testing the effects of MPCAs on plants and attempts to highlight areas in which additional knowledge is required. The tiered system recommended by the EPA has been utilized in this discussion. The use of the term *tier* is intended to be similar to that used in the EPA Subdivision M guidelines.

General Considerations

Many plant pathogens have been identified and characterized. The host range of some plant pathogens is extensive, while for others it is very limited. Indeed, the concept of *formae speciales* for fungi and pathovar for bacteria relates to the high degree of host specificity in certain microorganisms. Alternatively, many microorganisms have never been associated with plant diseases.

Because of the extensive knowledge of the pathogens of most important plants, the accurate taxonomic identification of an MPCA is essential and will allow determination of its similarity to known plant pathogens. MPCAs that do not resemble any known plant pathogen may require extensive plant testing to establish the absence of effects that are detrimental to plants. MPCAs that are similar to plant pathogens with very narrow host ranges, such as certain *Puccinia* spp. (rust diseases), *Erysiphe* spp. (powdery mildews), and *Pseudomonas* spp. (bacterial blights and wilts), may only require testing for adverse effects on plants similar to the known hosts. Alternatively, MPCAs that are similar to pathogens which have a wide host range, such as *Rhizoctonia solani* (damping-off and root rots), *Alternaria solani* (leaf spots), *Pythium* spp. (damping-off and root rots), and *Meloidogyne* spp. (root knot), may require extensive testing to identify the complete host range. Microbial herbicides, which by definition are designed to be pathogenic or toxic to target plants, may require special scrutiny to ensure that nontarget plants are not affected adversely.

The intended pattern of use, fecundity of the MPCA, potential for dissemination, and possibility of persistence or survival of the MPCA in the environment will also be determining factors in the extent of testing required on nontarget plants. For example, MPCAs that will not be disseminated to or that will not survive in soil or aquatic environments should not need to be tested as potential pathogens of roots or aquatic plants. MPCAs endemic to an area will not require testing that is as extensive on nontarget plants as will be needed for exotic microorganisms.

Species of plants that represent commercial agricultural and forestry concerns in the area of intended use and in areas where the MPCA could potentially be disseminated should be given priority in evaluating potential pathogenicity. For plants on which the candidate MPCA or closely related microorganisms are not known to cause disease, cultivars selected should be representative of those grown in the potential dissemination area. For MPCAs that are known to belong to species pathogenic on nontarget plants, susceptible cultivars of those plants and closely related plants should be selected for testing. In all cases, the growth stage or stages of the nontarget plants that are most likely to be susceptible to the MPCA should be used, even if plants in this growth stage will not be present at the intended time of use of the MPCA.

Environmental conditions selected for evaluations should be conducive for potential disease development. Unrealistic environmental extremes that will unduly stress plants should be avoided. Conditions should, however, be representative of those that might be expected to occur in the area of potential dissemination of the MPCA and over the seasons during which plants will be exposed to the MPCA.

Procedures

Tier 1 tests: Acute toxicity and pathogenicity

General approach.   Initial tests with nontarget plants are designed to detect the acute phytotoxicity or pathogenicity of the MPCA under "worst-case" conditions of high inoculum density and direct plant contact. The dosage of the MPCA should represent the maximum possible intended label rate to be used for pest control to which a plant could be exposed under conditions of proper application. The technical grade of the MPCA can be used in the initial portion of tier 1; however, eventually, the proposed commercial formulation should be utilized so that microbial contaminants, by-products, or toxic substances in the formulation can be evaluated. Method of application of the MPCA to test plants should be similar to that intended under actual use conditions.

Test species. The test species selected should be representative of the area and habitat of intended use and, in the case of an MPCA being related to a known plant pathogen, representative of the host range of the pathogen. The major agronomic crops in the United States for 1985 are listed in Table 35.1 (8). A general rule could be that plants selected should include six species of *Dicotyledoneae* from at least four families and four species of *Monocotyledonae* from at least two families. This is the general rule that is currently being applied to pesti-

TABLE 35.1   Value of Important Crops in the United States in 1985

Crop	Value, thousands of $
Field Crops	
Corn	21,327,048
Soybeans	10,823,330
Hay (all)	9,706,161
(alfalfa)	5,850,963
Wheat	7,652,444
Cotton	3,530,532
Tobacco	2,489,369
Sorghum	2,407,580
Barley	1,185,421
Rice	1,074,732
Peanuts	947,693
Sugar beets	750,162
Oats	649,834
Vegetables	
Potatoes	1,563,359
Tomatoes (fresh and processing)	1,195,554
Lettuce	673,107
Sweet corn (fresh and processing)	368,069
Onions	347,247
Broccoli	239,108
Carrots	208,325
Celery	188,995
Cauliflower	169,266
Sweet potato	142,936
Green peas	138,489
Cucumber	123,639
Snap beans	118,920
Fruit	
Oranges	1,598,663
Grapes	960,646
Apples	908,794
Strawberries	460,383
Grapefruit	321,786
Peaches	308,532
Pears	200,633

SOURCE: *Agriculture Statistics 1986,* USDA.

cide testing (7). It may be equally valid in the case of MPCAs, possibly with the exception of their use as mycoherbicides where more inclusive host-range studies would be needed.

The species selected should be from plants of high commercial value. In most cases the plants for initial tests should be chosen from those listed in Table 35.1. Subsequent tests may be needed on noncommercial plants that are representative of the intended release area. In the case of nursery or greenhouse crops or forest trees, the test plants should include the most important horticultural or forest species as well as a reduced number of plants from the agronomic crops. When the MPCA is intended to control plant growth and development or is closely related to a known plant pathogen, the plants selected for testing should include susceptible cultivars of known hosts of economic importance, plants beneficial to the maintenance of the ecosystem that could reasonably be expected to serve as hosts, as well as representatives of the *Dicotyledonae* and *Monocotyledonae*.

Test environment. Various factors influence the establishment of plant diseases. On plant foliage, physical factors such as temperature, relative humidity, surface wetness, and light and biological factors such as phyllosphere microflora can have stimulatory, inhibitory, or neutral effects on the development of disease. For diseases of roots, environmental factors affect the pathogen-host interaction, but interactions with soil microbes may also play a very significant role in disease establishment. The intensity of the environmental effects identified will vary from area to area and, especially, from soil to soil. Therefore, for tests to establish the possible effects of MPCAs on plants, the environmental conditions selected should eliminate as many potential interacting factors as possible. The only potential interaction that should persist in the system is that between the MPCA and plant.

Plants selected should be tested under environmental conditions that are optimum for the growth and reproduction of the plant and the MPCA. Production of host plants must be standardized as a part of the total quality management procedures for tests. Extremes in environmental conditions should be avoided because stresses due to moisture or temperature can alter susceptibility of plants to pathogens (30). Ideally, plants should be tested under controlled environment conditions to reduce variation in variables such as temperature, light, and moisture. Temperature-controlled tanks (74) would be the site of choice for testing MPCAs with intended application to soils. Variability within environmental chambers (23, 31, 47) should also be considered as a design factor in evaluations, and data quality objectives should be established accordingly. The numbers of plants to be tested will, however, often preclude the completion of MPCA evaluations in environmental chambers, and the greenhouse would be the next test location of choice for tier 1 studies. It should be recognized, however, that while controlled conditions may be created in greenhouses, due to the size and construction of such facilities, the environmental conditions are less uniform and control is less accurate (67).

Plant age at inoculation can influence the expression of symptoms and of resistance or susceptibility in certain plants (15, 18, 58, 61, 65, 69, 79). If the MPCA is related to a known pathogen, application to test plants should occur when plants are at the most susceptible stage. If the MPCA is not closely related to any known plant pathogens, test inoculations should be made to both seedling and adult plants.

Application of MPCAs to plants. The point of application of the MPCA to test plants should be the site of application to be specified on the product label and in the most probable, natural environment for that microbe. The rationale for the first application site is clear. The second application site will be important only if a naturally occurring soil-inhabiting microbe is used as a pest control agent on foliar plant parts or vice versa. In this latter case, even though the MPCA is intended for application in one particular plant environment, e.g., phyllosphere or rhizosphere, test environments should include all environments in which the MPCA could reasonably be expected to occur.

Application should be made (1) with the propagative unit of the MPCA intended for use in the commercial preparation, (2) with all methods of application to be specified on the label, and (3) in amounts that could be reasonably expected to occur under maximum intended label rates. Inoculum should be quantified (22, 43, 77). For comparison among potential host plants, the quantity of inoculum applied must be standardized and the same quantity applied to each plant (4, 10, 16).

Incubation of plants after application of the MPCA is an extremely important factor influencing potential infection and subsequent symptom development. Incubation conditions should be optimum for propagule germination and penetration into plants. For example, if penetration of related known pathogens occurs through stomata, then stomata should be open during the period following application (22). With the exception of the fungi that cause powdery mildew diseases, germination (fungi) or reproduction (bacteria) of most plant pathogens is dependent on a film of water in the phyllosphere or adequate moisture in the rhizosphere (40). Although most foliar pathogens penetrate plant tissues within 24 h when placed in a moisture-saturated atmosphere at 20 to 25°C, for practical purposes, a 48- to 96-h period will ensure the maximum likelihood that any infections which are possible will occur (22). Methods of creating free leaf moisture are discussed by Rotem (67).

Statistical design.   In tier 1 screening, the application of a single concentration of the MPCA results in an effect–no effect endpoint compared to controls. The effect may be any visually discernible symptom varying from a single lesion, mild chlorosis, or slight stunting to extensive necrosis, wilting, or plant death. Alternatively, the effect may be the accumulation of a metabolic by-product or end product of pathogenesis, such as an aflatoxin. The detection of such metabolites may require sophisticated assay procedures. The methods of determination and the designation of a positive effect should be in accordance with recognized physiological effects of known pathogens similar to the MPCA. Care should be taken when assessing the plant reaction. A rapid necrosis with loss of cell turgor, i.e., hypersensitive reaction, may occur as the result of an incompatible reaction between a pathogen and a nonhomologous host. This defense reaction should not be confused with a pathogenic reaction which usually occurs over a longer period of time and often results in an expanding area of infection.

In tests of this type where a positive result, i.e., disease development, is of such importance, both negative and positive controls should be included in the experimental design. Negative controls should be pest-free. If the candidate MPCA is aerially or splash disseminated, it may be necessary to maintain negative controls and treated plants in separate environmental units under identical conditions so that reliable negative controls can be achieved. An alternative would be to apply nonphytotoxic chemical pesticides known to provide effective control of the MPCA to negative control plants.

Positive controls are included to ensure that environmental conditions and inoculation procedures are appropriate to allow for disease development in a susceptible host. These positive controls are particularly important when the MPCA is closely related to a known plant pathogen or is intended for use as a microbial herbicide. If the MPCA is not intended for use as a microbial herbicide, the microorganisms selected for the positive control should resemble the subject MPCA in terms of taxonomy and optimal conditions for disease development and the plant should be susceptible to that pathogen. If the MPCA is intended for use as a microbial herbicide, the logical positive control would be the MPCA and the target weed.

The experimental design and data analysis for the establishment of an effect or no effect of the MPCA on test plants can be simple and straightforward. Experimental designs can be similar to those for testing chemical pesticides (56, 57). Plants should be observed weekly or more frequently until normal harvest time or plant death. Perennial plants should be observed for at least two growing seasons. Because no dose-response information is gained from tier 1 studies, a simple statistical test for determining significant differences between treatments is sufficient. Multiple plants of a particular species and cultivar should be included in each replication of the test if host variability is encountered in preliminary studies. If host variability is not a problem, one plant per species-cultivar combination can be used in each replication with three to four replications per treatment.

Tier 2 tests: Persistence, survival, and reproduction in the environment

General approach. Microbial pest control agents that show any adverse effects with regard to the symptomatic or asymptomatic, but detectable development of infection and disease in tier 1 tests require further testing. Information should be obtained concerning the nature of the response and probability of occurrence of detrimental effects if such MPCAs are released into the environment. Population dynamics are examined in tier 2 testing to quantify the capacity for reproduction and survival of the MPCA in the environment. By extension, the potential for dissemination of the MPCA is also examined in tier 2 testing. The general goal of the tier 2 tests discussed here is to determine the expression of effects due to an MPCA on nontarget plants in the terrestrial environment. In the opinion of the authors, if no adverse effects are demonstrated in tier 2 data, no further testing may be required.

Evaluation procedures. The tests to evaluate the potential for the MPCA to survive and reproduce in the plant environment will vary with the type of MPCA under consideration (fungal, bacterial, etc.) and the intended application environment (phyllosphere or rhizosphere). A typical end-use formulation or technical grade of the MPCA should be used in all tests at the maximum dosage likely to be utilized for field application. Most survival and reproduction tests should be conducted in contained facilities such as growth chambers or sealed greenhouses. The variation likely to be encountered in environmental conditions in the area of eventual intended application should be taken into account in the experimental design. Perhaps the best expression of survival and reproductive ability in the environments included in tier 2 studies is a population growth or decline curve when the MPCA is applied to one or more test species that exhibited adverse effects in tier 1 studies.

The experimental design should consist of soil and vegetation representative of the proposed application site. In general, the use of pasteurized or otherwise disinfested soil should be discouraged in favor of natural soil characteristic of the intended application site. Plants should be grown in containers that are as large as possible, and the planting density should be realistic for the use area. Environmental conditions, including temperature, relative humidity, amount of water applied and frequency of application, pH, light and nutrient status, should be varied to realistically mimic the range of expected environmental conditions at the intended use site. Chlorinated tap water and water with a low pH should be avoided. Extremes in any environmental variable should be avoided.

When an MPCA is applied to plant leaves or stems and if discrete lesions are produced, then characteristics such as latent period (i.e., the time from inoculation to reproduction), reproduction rate (i.e., the number of reproductive units produced per unit of host area per day), and infectious period (i.e., the time from the end of the latent period to the cessation of reproduction) can be monitored. Latent period can be expressed as time to the onset of sporulation (63) or as the time to the sporulation of 50 percent of those lesions that will sporulate (70). Reproductive rate can be expressed as the mean number of units produced per day or as cumulative reproduction.

If monitoring of individual lesions is impractical for determining rate of reproduction, impaction or suction collecting devices may be used. A wide range of devices for sampling microorganisms in the atmosphere is available (13, 22, 24, 29, 66). If the propagules of the MPCA are easily identifiable, e.g., fungal spores, the Hirst spore trap (44, 45) may be appropriate. If a culture medium is needed to identify the propagules, either because the MPCA is a bacterium or the selectivity of a particular culture medium is needed to separate the MPCA from other airborne microorganisms, the Andersen sampler (3, 52) can be utilized. The method of choice will depend on the type of microorganisms to be quantified, the relative cost of quantification, and the need for continuous or intermittent sampling.

For an MPCA that will be applied to the soil surface, incorporated into the soil, or eventually reside in the soil after having been washed off plant leaves or shoots, several methods for quantifying the survival and reproduction in soil are available. The methods of estimating the number of viable propagules in soil include (1) direct enumeration coupled with subsequent viability testing, (2) selective culture media, (3) substrate colonization, and (4) plant bioassay. Each group of methods has certain advantages and disadvantages when related to specific soilborne microorganisms, and careful consideration should be given to the appropriateness of a particular method before it is adopted.

Direct enumeration is appropriate for microorganisms with relatively large reproductive or survival structures such as the sclerotia of *Sclerotinia* spp. or *Sclerotium rolfsii* (53, 62) or most nematodes (12).

Selective culture media are available for a large number of known soilborne plant pathogens (22, 33, 39, 68, 76) and may be appropriate for a given MPCA. These culture media can be utilized in conjunction with various soil-washing or soil dilution techniques to increase ease and reliability of the assay. Substrate colonization and plant bioassay rely on the ability of the target microorganisms to colonize or infect plant tissue (12, 20, 54). A population index can be developed in relation to the number of substrate or plant units colonized, or a most-probable number technique (51, 60, 71) can be utilized to estimate populations.

The duration of the survival or reproduction tests should be at least until two population half-life determinations have been made or until it is clear that the population of the MPCA has the ability to persist in the plant environment at or above the levels present when the test was initiated. Three to four replicate plants or plant environments should be included in each test and each test should be repeated. The number of samples necessary to determine the population level of an MPCA at any given time will depend on (1) the variability in the population level within the habitat being sampled and (2) the variability inherent in the assay method being used. Both sources of variation should be examined in preliminary experiments, and the variation encountered should be used in designing the sampling program (43, 46).

Tier 3 tests: Dose-response relationships

General approach.   If the results of the tier 2 studies indicate that the MPCA is able to persist in the plant environment such that susceptible nontarget plants (as established in tier 1) are likely to be exposed, then tier 3 testing is needed. A range of dose levels of the MPCA should be utilized to determine if the population levels observed in tier 2 studies are able to produce adverse effects on nontarget plants. Another goal of tier 3 testing is to determine if there is a minimum infective dose for any of the adverse effects determined on nontarget plants in tier 1 tests.

Experimental methods.   The tests should be performed in contained environments, such as growth chambers or greenhouses, and should, if possible, include a range of nontarget plants from the most sensitive to the least sensitive, but still adversely affected. The use of a range of sensitive plants will provide a more complete evaluation of the dose-response relationship for the MPCA than utilizing only one adversely affected species. Environmental conditions should represent the range of conditions anticipated in the projected MPCA application area; extremes should be avoided.

Initially, the dosage levels of the MPCA should range in graded steps from the maximum dosage expected to be applied under commercial conditions to a level one-half that of the lowest survival level encountered in tier 2 testing. If adverse effects are observed at the lowest dosage used, another decreasing series of doses should be utilized in a second test. If the MPCA is easily dispersed among experimental units, care should be taken to prevent the contamination of lower-dosage-level treatments from higher-dosage-level treatments. Observations of plants should be made regularly with particular emphasis on the time of symptom occurrence observed in tier 1 studies. Observations should continue past the tier 1 symptom development point, because at lower dosage levels, development of obvious symptoms may be delayed.

Tier 4 tests: Hazard testing under field conditions

Field applications of MPCAs that could potentially affect nontarget plants have been limited. Several possible examples include biocontrol agents for plant diseases such as *Agrobacterium radiobacter* (42), *Trichoderma harzianum* (78), *Peniophora gigantea* (64), and hypovirulent strains of *Cryphonectria parasitica* (2). Further information on biocontrol microorganisms for biocontrol of plant diseases can be found in Cook and Baker (19).

Field testing MPCAs for the potential hazard to nontarget plants should be done with the utmost caution and, if at all possible, in an area isolated from agronomic, horticultural, or forest plants that may be affected adversely by the MPCA. Isolation of test plots may be accomplished geographically, e.g., in an area where the MPCA is known not to survive and the adversely effected nontarget plants do not normally occur, by border crops designed to prevent dispersal of the MPCA, or temporally, e.g., by performing tests in a season of the year when the nontarget hosts and the MPCA would not be expected to occur.

Field tests for evaluating MPCAs should be designed with a full knowledge of field plot techniques used in studying epidemics of plant diseases (11, 80) and with an appreciation of the potential problems associated with such studies. The effects of plot shape and size (17, 25–27, 37) and implications of interplot interference (35, 36, 38, 59) should be evaluated before initiating field investigations on the effects of any MPCA. Failure to account for such factors can result in inappropriate or invalid test results and may pose unnecessary environmental risks.

The ability to detect the adverse effects of an MPCA on nontarget plants under field conditions will depend on the population level of the MPCA, the virulence of the MPCA, the conduciveness of the environmental conditions for plant growth and disease development, and the number of plants examined. If population levels are low or environmental conditions are suboptimal for infection by the MPCA, large numbers of plants will need to be exposed and examined.

Conclusions

The question of the occurrence (and observation) of adverse effects caused by MPCAs on nontarget plants is not one to be resolved only by tests of statistical significance, but by tests of biological significance. Statistical tests will be important to provide indications of the likelihood of occurrence of events such as possible stunting of plants or differences in yield between treated plants and positive and negative controls. Statistical tests may also be of value in some tier 3 tests to establish the linearity or curvilinearity of a given dose-response relationship. Many of the results obtained in tests in tiers 1, 2, and 4, however, will be largely qualitative in nature and their significance will be more reasonably judged biologically than statistically. The presence of only a few lesions, for example, on a nontarget plant may not result in a statistically significant reduction in plant biomass production or plant "health." However, if such a nontarget plant served as an inoculum reservoir for the MPCA, this occurrence could have biological-ecological significance.

The testing of microbial pest control agents to establish the absence of adverse effects on nontarget plants is relatively new. Such testing requires a conceptually different approach than evaluations to confirm the pathogenicity of a particular microorganism on a particular plant. With the MPCA the goal of the testing program is to "confirm" the null hypothesis that the candidate microorganism has no effect on the nontarget plant. While this is not actually possible scientifically, it must be made possible from a practical standpoint.

The need, then, is to test a sufficiently large number of representative nontarget plants in replicated trials under biologically and environmentally realistic conditions in order to have an acceptable level of confidence in the conclusions derived concerning the candidate MPCA. The design and initiation of such tests should be done with a full knowledge of the similarity or difference of the MPCA to known plant pathogens and of the pathosystems of those pathogens which are similar. Consultation with plant pathologists and statisticians will be beneficial in successful implementation and interpretation of tests to determine the effects of microbial agents on plants.

References

1. Agrios, G. N. 1988. *Plant Pathology,* 3d ed. Academic Press, New York.
2. Anagnostakis, S. L. 1982. Biological control of chestnut blight. *Science* 215:466–471.
3. Andersen, A. A. 1958. A new sampler for the collection, sizing, and enumeration of viable airborne particles. *J. Bacteriol.* 76:471–484.
4. Andres, M. W., and R. D. Wilcoxson. 1984. A device for uniform deposition of liquid-suspended urediospores on seedling and adult cereal plants. *Phytopathology* 74:550–552.
5. Anonymous. 1960. *Index of Plant Diseases in the United States,* USDA Handbook No. 160. Government Printing Office, Washington, D.C.
6. Anonymous. 1978. *Biological Agents for Pest Control: Status and Prospects,* USDA Special Report. Washington, D.C.
7. Anonymous. 1982. *Pesticide Assessment Guidelines, Subdivision J, Hazard Evaluation: Nontarget Plants,* PB83-153940, EPA 540/9-82-020. U.S. Environmental Protection Agency, Washington, D.C.
8. Anonymous. 1986. *Agricultural Statistics 1986.* U.S. Department of Agriculture, Washington, D.C.
9. Audus, L. J. 1976. *Herbicides: Physiology, Biochemistry, Ecology,* 2d ed. Academic Press, New York.
10. Aust, H. J., and J. Kranz. 1974. Eine automatisches Sporenfalle fur den Gebrauch in Klimaschranken. *Angew. Bot.* 48:267–272.
11. Aust, J., and J. Kranz. 1988. Experiments and procedures in epidemiological field studies. In J. Kranz and J. Rotem (eds.), *Experimental Techniques in Plant Disease Epidemiology.* Springer-Verlag, Berlin, pp. 7–17.
12. Barker, K. R., C. C. Carter, and J. N. Sasser. 1985. *An Advanced Treatise on Meloidogyne.* Vol II. *Methodology.* North Carolina State University Graphics, Raleigh.
13. Bartlett, J. T., and A. Bainbridge. 1978. Volumetric sampling of micro-organisms in the atmosphere. In P. R. Scott and A. Bainbridge (eds.), *Plant Disease Epidemiology.* Blackwell Scientific, Oxford, pp. 23–30.
14. Boag, B. 1986. Detection, survival and dispersal of soil vectors. In G. D. McLean, R. D. Garrett, and W. G. Ruesink (eds.), *Plant Virus Epidemics: Monitoring, Modelling and Predicting Outbreaks.* Academic Press, Sydney, pp. 119–145.
15. Brokenshire, T., and B. M. Cooke. 1978. The effect of inoculation with *Selenophora donacii* at different growth stages on spring barley. *Ann. Appl. Biol.* 89:211–217.
16. Brown, J. F., and J. F. Fittier. 1981. A quantitative method of inoculating plants with uniform densities of fungal spores. *Aust. Plant Pathol.* 10:51–53.
17. Burleigh, J. R., and M. J. Loubane. 1984. Plot size effects on disease progress and yield of wheat infected by *Mycosphaerella graminicola* and barley infected by *Pyrenophora teres. Phytopathology* 74:545–549.
18. Chi, C. C., and E. W. Hanson. 1962. Interrelated effects of environment and age of alfalfa and red clover seedlings on susceptibility to *Pythium debaryanum. Phytopathology* 52:985–989.
19. Cook, R. J., and K. F. Baker. 1983. *The nature and practice of biological control of plant pathogens.* American Phytopathological Society, St. Paul, Minn.
20. Dance, D., F. J. Newhook, and J. S. Cole. 1975. Bioassay of *Phytophthora* spp. from soil. *Plant Dis. Rep.* 59:523–527.
21. Das, K. G. (ed.). 1981. *Pesticide Analysis.* M. Dekker, New York.
22. Dhingra, O. D., and J. B. Sinclair. 1985. *Basic Plant Pathology Methods.* CRC Press, Boca Raton, Fla.
23. Downs, R. J., and H. Hellman. 1975. *Environment and the Experimental Control of Plant Growth.* Academic Press, London.
24. Fitt, B. D. L., and H. A. McCartney. 1986. Spore dispersal in relation to epidemic models. In K. J. Leonard and W. E. Fry (eds.), *Plant Disease Epidemiology: Population Dynamics and Management.* Macmillan, New York, pp. 311–345.

25. Fleming, R. A., L. M. Marsh, and H. C. Tuckwell. 1982. Effect of field geometry on the spread of crop disease. *Prot. Ecol.* 4:81–108.
26. Gilligan, C. A. 1980. Size and shape of sampling units for estimating incidence of stem canker on oil-seed rape stubble in field plots after swathing. *J. Agric. Sci. Camb.* 94:493–496.
27. Gilligan, C. A. 1982. Size and shape of sampling units for estimating incidence of sharp eyespot, *Rhizoctonia cerealis,* in plots of wheat. *J. Agric. Sci. Camb.* 99:461–464.
28. Grand, L. F. (ed.). 1985. *North Carolina Plant Disease Index, Technical Bulletin* 240 (rev.). North Carolina Agricultural Research Service, Raleigh.
29. Gregory, P. H. 1973. *Microbiology of the Atmosphere,* 2d ed. Wiley, New York.
30. Hale, M. G., and D. M. Orcutt. 1987. *The Physiology of Plants under Stress.* Wiley Interscience, New York.
31. Hammer, P. A., and R. W. Langham. 1972. Experimental design consideration for growth chamber studies. *Hort. Sci.* 7:481–483.
32. Hassall, K. A. 1982. *The Chemistry of Pesticides: Their Metabolism, Mode of Action and Use in Crop Protection.* Macmillan, New York.
33. Hickey, K. D. (ed.). 1986. *Methods for Evaluating Pesticides for Control of Plant Pathogens.* APS Press, St. Paul, Minn.
34. Hill, S. A. 1984. *Methods in Plant Virology.* Blackwell Scientific, Oxford.
35. James, W. C., C. S. Shih, W. A. Hodgson, and L. C. Callbeck. 1976. Representational errors due to interference in field experiments with late blight of potato. *Phytopathology* 66:695–700.
36. Jenkyn, J. F. 1977. Interference between plots as a source of error in experiments with *Erysiphe graminis. Pest. Sci.* 8:428–429.
37. Jenkyn, J. F., and A. Bainbridge. 1974. Disease gradients and small plot experiments on barley mildew. *Ann. Appl. Biol.* 269–279.
38. Jenkyn, J. F., A. Bainbridge, G. V. Dyke, and A. D. Todd. 1979. An investigation into inter-plot interactions, in experiments with mildew on barley, using balanced designs. *Ann. Appl. Biol.* 92:11–28.
39. Johnson, L. F., and E. A. Curl. 1972. *Methods for Research on the Ecology of Soilborne Plant Pathogens.* Burgess, Minneapolis, Minn.
40. Jones, A. L. 1986. The role of wet periods in predicting foliar diseases. In K. J. Leonard and W. E. Fry (eds.), *Plant Disease Epidemiology: Population Dynamics and Management.* Macmillan, New York, pp. 87–100.
41. Jones, D. G. 1987. *Plant Pathology: Principles and Practice.* Prentice-Hall, Englewood Cliffs, N.J.
42. Kerr, A. 1974. Soil microbiological studies on *Agrobacterium radiobacter* and biological control of crown gall. *Soil Sci.* 118:168–172.
43. Kiraly, Z., A. Klement, F. Solmosy, and J. Voros. 1970. *Methods in Plant Pathology with Special Reference to Breeding for Disease Resistance.* Academiae Kiado, Budapest.
44. Kramer, C. L., M. G. Eversmeyer, and T. I. Collins. 1976. A new 7-day spore sampler. *Phytopathology* 66:60–61.
45. Kramer, C. L., and S. M. Pady. 1966. A new-24 hour spore sampler. *Phytopathology* 56:517–520.
46. Kranz, J., and J. Rotem (eds.). 1988. *Experimental Techniques in Plant Disease Epidemiology.* Springer-Verlag, Berlin.
47. Lee, C. S., and J. O. Rawlings. 1982. Design of experiments in growth chambers: Uniformity studies in the NCSU Phytotron. *Crop Sci.* 22:551–558.
48. Lelliot, R. A., and D. E. Stead. 1987. *Methods for Diagnosis of Bacterial Diseases of Plants.* Blackwell Scientific, Oxford.
49. Leonard, K. J., and W. E. Fry (eds.). 1986. *Plant Disease Epidemiology: Population Dynamics and Management.* Macmillan, New York.
50. Lucas, G. B., C. L. Campbell, and L. T. Lucas. 1985. *Introduction to Plant Diseases: Identification and Management.* Van Nostrand Reinhold, New York.
51. Maloy, O. C., and M. Alexander. 1958. The "most probable number" method for es-

timating populations of plant pathogenic organisms in the soil. *Phytopathology* 48: 126–128.

52. May, K. R. 1964. Calibration of a modified Andersen bacterial aerosol sampler. *Appl. Microbiol.* 12:37–43.

53. Menzies, J. D. 1963. The direct assay of plant pathogen populations in soil. *Annu. Rev. Phytopathol.* 1:127–142.

54. Mitchell, D. J., M. E. Kannwischer-Mitchell, and G. A. Zentmeyer. 1986. Isolation, identifying and producing inoculum of *Phytophthora* spp. In K. D. Hickey (ed.), *Methods for Evaluating Pesticides for Control of Plant Pathogens.* APS Press, St. Paul, Minn, pp. 63-66.

55. Moye, H. A. (ed.). 1981. *Analysis of Pesticide Residues.* Wiley-Interscience, New York.

56. Nelson, L. A. 1978. Use of statistics in planning, data analysis and interpretation of fungicide and nematicide tests. In *American Phytopathological Society Methods for Evaluating Plant Fungicides, Nematicides and Bactericides.* American Phytopathological Society, St. Paul, Minn., pp. 2–14.

57. Nelson, L. A. 1986. Use of statistics in planning, data analysis and interpretation of fungicide and nematicide tests. In K. Hickey (ed.), *Methods for Evaluating Pesticides for Control of Plant Pathogens.* APS Press, St. Paul, Minn., pp. 11–23.

58. Ostazeski, S. A., D. K. Barnes, and C. H. Hanson. 1969. Laboratory selection of alfalfa for resistance to anthracnose, *Colletotrichum trifolii. Crop Sci.* 9:351–354.

59. Paysour, R. E., and W. E. Fry. 1983. Interplot interference: A model for planning field experiments with aerially disseminated pathogens. *Phytopathology* 73:1014–1020.

60. Pfender, W. F., D. I. Rouse, and D. J. Hagedorn. 1981. A "most probable number" method for estimating inoculum density of *Aphanomyces euteiches* in naturally infested soil. *Phytopathology* 71:1169–1172.

61. Populer, C. 1978. Changes in host susceptibility with time. In J. G. Horsfall and E. B. Cowling (eds.), *Plant Disease: An Advanced Treatise,* vol. 1. Academic Press, New York, pp. 239–262.

62. Punja, Z. K., V. L. Smith, C. L. Campbell, and S. F. Jenkins. 1985. Sampling and extraction procedures to estimate numbers, spatial pattern, and temporal distribution of *Sclerotium rolfsii. Plant Dis.* 69:469–474.

63. Ricker, M. D., M. K. Beute, and C. L. Campbell. 1985. Components of resistance in peanut to *Cercospora arachidicola. Plant Dis.* 69:1059–1064.

64. Rishbeth, J. 1963. Stump protection against *Fomes annosus.* III. Inoculation with *Peniophora gigantea. Ann. Appl. Biol.* 52:63–77.

65. Roncadori, R. W., and S. M. McCarter. 1972. Effect of soil treatment, soil temperature and plant age on Pythium root rot of cotton. *Phytopathology* 62:373–376.

66. Rotem, J. 1988. Quantitative assessment of inoculum production, dispersal, survival and infectiousness in airborne diseases. In J. Kranz and J. Rotem (eds.), *Experimental Techniques in Plant Disease Epidemiology.* Springer-Verlag, Berlin, pp. 69–83.

67. Rotem, J. 1988. Techniques of controlled-condition experiment. In J. Kranz and J. Rotem (eds.), *Experimental Techniques in Plant Disease Epidemiology.* Springer-Verlag, Berlin, pp. 19–31.

68. Schaad, N. W. (ed.). 1988. *Laboratory Guide for Identification of Plant Pathogenic Bacteria,* 2d ed. APS Press, St. Paul, Minn.

69. Seem, R. C. 1988. The measurement and analysis of the effects of crop development on epidemics. In J. Kranz and J. Rotem (eds.), *Experimental Techniques in Plant Disease Epidemiology.* Springer-Verlag, Berlin, pp. 51–68.

70. Shaner, G. 1980. Probits for analyzing latent period data in studies of slow rusting resistance. *Phytopathology* 70:1179–1182.

71. Sivasithamparam, K., C. A. Parker, and C. S. Edwards. 1979. Bacterial antagonists of the take-all fungus and fluorescent *Pseudomonas* in the rhizosphere of wheat. *Soil Biol. Biochem.* 11:161–165.

72. Smith, I. M., J. Dunez, R. A. Lelliot, D. H. Phillips, and S. A. Archer. 1988. *European Handbook of Plant Diseases*. Blackwell Scientific, Oxford, UK.
73. Spurr, H. W., Jr. 1985. Bioassays—critical to biocontrol of plant disease. *J. Agric. Entomol.* 2:117–122.
74. Steele, A. E. 1967. A constant temperature bath for pot-grown plants. *Plant. Dis. Rep.* 51:171–173.
75. Taylor, L. R. 1986. The distribution of virus disease and the migrant vector aphid. In G. D. McLean, R. G. Garrett, and W. G. Ruesink (eds.), *Plant Virus Epidemics: Monitoring, Modelling and Predicting Outbreaks*. Academic Press, Sydney, pp. 35–57.
76. Tsao, P. H. 1970. Selective media for isolation of pathogenic fungi. *Annu. Rev. Phytopathol.* 8:157–168.
77. Tuite, J. F. 1969. *Plant Pathological Methods: Fungi and Bacteria*. Burgess, Minneapolis, Minn.
78. Wells, H. D., and D. K. Bell. 1979. Variable antagonistic reaction in vitro of *Trichoderma harzianum* against several pathogens. *Phytopathology* 69:1048–1049.
79. Yarwood, C. E. 1959. Predisposition. In J. G. Horsfall and A. E. Dimond (eds.), *Plant Pathology: An Advanced Treatise*. Academic Press, New York, pp. 521–562.
80. Zadoks, J. C. 1978. Methodology of epidemiological research. In J. G. Horsfall and E. B. Cowling (eds.), *Plant Disease: An Advanced Treatise*, vol. 1. Academic Press, New York, pp. 64–96.

Chapter 36

Testing the Effects of Microbial Agents on Fish and Crustaceans

Anne Spacie

Introduction

Fish and crustaceans have been used extensively to test the toxicity of naturally occurring and manufactured chemicals in water. Test methods developed over the last 25 years are available for measuring acute lethality, sublethal effects on reproduction and development, behavioral changes, disposition of residues in tissues, physiological and cellular changes, and effects on multispecies model aquatic ecosystems. The principles and practice of aquatic toxicology are amply described and discussed in several general references (59, 66, 75). Specific protocols for acute and long-term tests with fish and other aquatic organisms are also available (2–5, 20, 27, 62).

In comparison to chemicals, few microbial pest control agents (MPCAs) or other introduced microorganisms have ever been tested for effects on either fish or crustaceans. This lack of experience with MPCAs reflects the relatively recent history of their commercial development and environmental release. Compared to chemical agents, microbial agents have much greater selectivity for target organisms. Nontarget organisms, especially those in families outside the host range, are not usually susceptible. The MPCA used most extensively in aquatic habitats, *Bacillus thuringiensis* servar. *israelensis,* has not shown effects on fish or crustaceans at application rates used to control mosquitoes and blackflies (40). Consequently there has been little impetus to conduct broad testing programs on the effects of MPCAs on aquatic taxa. The need for testing, however, is expected to increase as new MPCAs are discovered and developed through genetic engineering.

Aquatic microbial tests differ from chemical toxicity tests in several important respects. While chemical bioassays are performed with closely controlled exposure concentrations, microbial tests allow for the possibility of MPCA growth and replication in the test chamber or host. Chemical tests are designed to produce a range of responses (such as 0 to 100 percent mortality) that is proportional to exposure level. Pathogenic agents can act in a variety of ways that do not follow classic dose-response relationships. For example, a latency period may be followed by rapid infectivity. The severity of infection may depend on crowding of the test organisms—a factor not encountered in standard chemical bioassays. Finally, most MPCA screening tests, which are essentially host-range challenges, produce no response. Such an accumulation of negative results makes it difficult to develop the type of quantitative structure-activity relationships that are valuable predictive tools in chemical toxicology. Without more extensive experience testing MPCAs in aquatic systems, the importance of such factors is difficult to judge.

At present there are no standard methods for conducting MPCA tests with fish or crustaceans. The Environmental Protection Agency (26) recommends a tiered approach similar to the hazard assessment scheme used to register new chemical pesticides (10). The test protocols for fish and crustaceans are essentially adapted from standard chemical bioassays, since little background information is available on specific microbial tests. The following discussion reflects the current state of MPCA testing with aquatic organisms and highlights the aspects of testing that require additional research.

Procedures

Tier 1: Acute toxicity and pathogenicity tests

General approach. Initial screening tests with nontarget organisms including fish and crustaceans are designed to detect any possible acute toxicity, infectivity, or pathogenicity of the MPCA under worst-case conditions of high dosage and direct contact. The dosage does not represent an expected field concentration, but rather a generous overestimate of the amount used for pest control. The technical grade of the MPCA can be tested as part of the initial toxicity screening. Eventually, however, the commercial formulation should be administered so that microbial contaminants, by-products, or toxic carrier substances in the formulation can be evaluated.

Choice of test species.   Because of the extensive use of aquatic organisms in ecotoxicology, there is now a wide choice of suitable fish and crustacean species. Table 36.1 lists North American species for which culture conditions and laboratory handling methods are well documented. Most species have been chosen for a combination of traits such as sensitivity to chemicals, ease of culture, rapid growth, commercial availability, or importance in a particular ecosystem. Small size is also a consideration for fishes held in the laboratory.

TABLE 36.1   Recommended Species and Test Temperatures for Tests with Microbial Agents

Recommended species	Recommended test temperature, °C
Freshwater fish:	
Coho salmon, *Oncorhynchus kisutch*	12
Rainbow trout, *Oncorhynchus mykiss*	12
Brook trout, *Salvelinus fontinalis*	12
Fathead minnow, *Pimephales promelas*	20
Channel catfish, *Ictalurus punctatus*	20
Bluegill, *Lepomis macrochirus*	20
Mosquitofish, *Gambusia affinis*	20
Freshwater crustaceans:	
Daphnids, *Daphnia* spp.,	20
Ceriodaphnia spp. (55)	20
Amphipods, *Gammarus* spp. (7)	20
Hyalella azteca (73)	20
Crayfish, *Orconectes* spp., *Cambarus* spp.,	20
Procambarus spp.	20
Marine and estuarine fish:	
Sheepshead minnow, *Cyprinodon variegatus*	20
Killifish, *Fundulus heteroclitus*	20
Fundulus similis	20
Silverside, *Menidia* spp.	20
Threespine stickleback, *Gasterosteus aculeatus*	20
Pinfish, *Lagodon rhomboides*	20
Spot, *Leiostomus xanthurus*	20
Flounder, *Paralichthys dentatus, P. lethostigma*	20
Sanddab, *Citharichthys stigmaeus*	12
Winter flounder, *Pseudopleuronectes americanus*	12
English sole, *Parophrys vetulus*	12
Marine and estuarine crustaceans:	
Penaeid shrimp, *Penaeus setiferus, P. aztecus,*	20
P. duorarum	20
Grass shrimp, *Palaemonetes* spp.	20
Sand shrimp, *Crangon* spp.	20
Mysid shrimp, *Mysidopsis bahia* (60),	20
Neomysis spp.	20
Oceanic shrimp, *Pandalus jordani*	12
Copepod, *Acartia tonsa* (82)	20
Blue crab, *Callinectes sapidus*	20
Dungeness crab, *Cancer magister*	12

SOURCE: From Refs. 20 and 62, and as noted.

Since most MPCAs are intended for release in specific habitats, the choice of test organism should be related to the application site. For example, the rice field crayfish (*Procambarus clarkii*) and the mosquitofish are logical choices for testing pest control agents of rice. Similarly, the benthic mysid shrimp *Pontoporeia hoyi* would be appropriate for testing substances entering the Great Lakes because of its importance to that particular ecosystem.

Throughout the development of aquatic toxicology, there has been an effort to select test species that are sensitive to chemicals and to avoid species that are especially hardy such as carp, goldfish, or eel. This has been a reasonable approach because the few bioassay species must represent many other species that cannot be tested (25). The surrogate species concept has been used to establish water quality criteria. It has also led to useful toxicity correlations among test species (49). Pathogenic microbial products may fall outside this general pattern, however, since sensitivity to chemicals is no sure predictor of susceptibility to microorganisms. Pathogenic agents are often selective, and their host ranges generally fall within taxonomic groups. Thus the choice of surrogate species may need to be modified according to the type of MPCA tested.

Another principle of aquatic testing is that early life stages are generally more susceptible than adults. Young animals typically lack the detoxifying mechanisms that function in the adult. The generalization may well hold for microbial agents that act as pathogens since early life stages may lack immune defenses. Parasites are often more damaging to young aquatic organisms because of the proportionately smaller difference in body size between parasite and host. While these are reasonable expectations, little evidence from tests with MPCAs is available to support the idea of increased susceptibility of early life stages.

Regardless of age or size, bioassay organisms should be disease-free at the time of testing. Species cultured in the laboratory can be kept free of most diseases. However, if test animals are collected in the field, a determination of their health status should be made before testing.

Test temperature. Test temperature is an additional factor to consider when selecting a test species. Fish and crustaceans should be held and tested at a temperature suitable for their long-term survival. Trout, for example, are cold-water forms adapted to temperatures somewhat below room temperature. While this preference causes no particular problems in conventional toxicity tests, it might discourage the use of trout in tests with live microorganisms that are dormant or inactive in cold water. Ideally, both the microbial agent and the test aquatic species should be adapted to a similar temperature range.

Route of administration. The MPCA could be given to fish or crustaceans by any of four routes: (1) through exposure to an aqueous solution or suspension, (2) through dietary exposure to treated food, (3) by gastric intubation, or (4) by injection. A combination of the aqueous and dietary routes is appropriate for mosquito control agents since these could enter both the habitat and food chain of aquatic organisms directly. Aqueous exposure can be used with aquatic organisms of all sizes and life stages. The exposure concentration can be adjusted easily by dilution, and the toxic effect can be expressed in units of concentration (see the discussion of LC50 and EC50 below). Whereas fish and crustaceans can acquire most dissolved chemicals through direct uptake at the gill, other routes, particularly dietary, may be necessary for effective exposure to certain MPCAs. Therefore, if aqueous exposure is the only route tested, steps should be taken to verify that the MPCA in fact enters the organism, as was done in the study of Brazner and Anderson (14). Furthermore, the actual dose would need to be estimated in order to calculate a rate of increase of the pathogen in host tissue. Such dose calculations based on aqueous exposure are rarely done in aquatic chemical testing.

Administration of the agent in treated food may be most effective, depending on the test species chosen. Many fish including trout, sunfish, and mosquitofish are insectivorous and can be fed infected larvae or pelleted food treated with the MPCA. Larger crustaceans (crabs, crayfish, adult penaeid shrimp) can be treated similarly. If the test organisms are isolated and observed carefully during feeding, the exposure dose through food can be calculated directly. Daphnids and other filter-feeding microcrustaceans consume bacterial, algal, or yeast suspensions directly, making the water and food routes indistinguishable for such organisms.

The gastric intubation and direct injection routes are not realistic in environmental terms, but may be useful for preliminary screening studies with fish. Test materials that are difficult to handle in the laboratory or are available only in small quantities (rare strains, radiolabeled products) could be screened first by injection into a few test organisms. Unlike feeding or gastric intubation, injection cannot be used to define lethal exposure concentrations for aquatic organisms. Injection of nonoccluded viruses into the coelomic fluid of arthropods is possible, but not recommended for bioassays because of the labor required and high probability of injury or secondary infection in the host (37).

Exposure system.   The static or static-renewal test is the simplest and most adaptable for general screening purposes. In such a test, each group of organisms is held in treated water, with or without additional feeding of MPCA. No filtration or other treatment of the water is necessary during the test other than aeration and, if needed, heating or cooling to maintain the desired temperature. The MPCA suspension is maintained in the aquarium by exchanging the water periodically—typically every 2 to 4 days. The size of the tank and volume of test water are determined by the "loading" of test organisms, calculated as the total grams of organism per liter of test water. Recommended loading rates for static tests are usually 1 g/liter or less, depending on the temperature and dissolved oxygen level (20). Test organisms should not be held for more than a few days in completely static aquaria because of the build-up of toxic waste products produced by the organisms themselves.

Ideally, the concentration and viability of the microbial agent should be verified daily in each treated test container. If direct enumeration is impractical, it may be possible to use a bioassay with susceptible target aquatic organisms such as mosquito larvae.

Test water quality.   Fish and crustaceans should be tested in water that is similar to the culture water of their stock tanks or to the water from their field collection site. Synthetic hard or soft freshwater and seawater of an appropriate salinity can be reconstituted from deionized water with added salts, according to published formulations (20, 62). Distilled, deionized, or chlorinated tap water is not suitable for the culture or testing of fish or crustaceans.

Exposure duration.   Acute chemical tests are normally conducted for 96 h with fish and large crustaceans, and 48 h with daphnids. However, no theory or appropriate studies have been reported that would help to establish the appropriate exposure length for MPCAs. The few published studies with fish and crustaceans generally have involved short exposures of one to several days. It would appear from experiments with highly virulent entomopathogens that a few hours are sufficient to infect susceptible hosts. Exposure times might need to be longer for toxin-producing microorganisms since several steps are involved in the process of toxin production. Still other MPCAs such as slow-acting viruses or fungi with several reproductive stages might produce adverse effects only after long latent periods within the host or other carrier organisms. Clearly, any general recommendations about exposure duration would be premature at this time. Until additional research is done, MPCA tests should be designed on a case-by-case basis using information on the life cycle of the microorganism, its mode of expression, and expected viability in the aquatic environment.

Statistical design. A tier 1 screening test with a single exposure concentration gives an effect–no effect endpoint when compared to controls. The endpoint may be mortality or immobilization (which are quantal responses), or any type of graded or ranked response, such as degree of pathology or infection. A simple statistical test for significant difference between treatments suffices. The design may be factorial, with aqueous and dietary routes of administration or exposure and observation times treated as factors. No dose-response information can be gained from a single exposure level. Consequently a tier 1 test uses fewer organisms and replicate tanks than a dose-response test, which requires a range of exposure concentrations, as described for tier 3 tests below. While there is no rule regarding the number of organisms in a screening test, a common choice is a minimum of perhaps 20 individuals in each of 2 to 4 replicate treatments, plus controls.

Evaluation of pathogenesis. Standard examination techniques for necropsy, histology, serology, and pathology are available in references on fish pathology (8, 16, 50, 64, 68) and to a lesser extent on crustacean pathology (72). Histologic studies of fish and invertebrates are also reviewed in relation to chemical toxicity (52). Whenever possible, MPCA accumulation in tissue should be quantified in addition to tissue damage. Techniques that apply to pathogenic MPCAs include microscopic identification and enumeration of microbial spores with or without staining. Verification of spore viability may be possible by reisolation from tissue followed by bioassay with a susceptible host.

Tiers 2 and 3: Dose-response, chronic, and multispecies tests

Microbial agents showing adverse effects on nontarget aquatic species in tier 1 tests require additional information on the nature of the response and the probability of environmental exposure. Estimations of the expected environmental exposure level, persistence, and geographic distribution (environmental expression characteristics) are made as part of a tier 2 step in the hazard evaluation. Further tests on the adverse effects to nontarget organisms are considered in a tier 3 evaluation. The types of tests chosen depend on the type of tier 1 effects observed. Typically, they may include dose-response studies of toxicity and pathogenicity, quantification of infectivity rates, chronic or subchronic exposures to various life stages, additional host range tests, and/or multispecies microcosm tests.

Dose-response tests.   A microorganism that acts through the production of a toxin may be characterized by a dose-response curve analogous to chemical toxicity. A dose-response toxicity test uses a geometric series of exposure concentrations spanning the expected median lethal concentration (LC50), median effective concentration (EC50), or median lethal dose (LD50) for the population of test organisms (61). Immobilization is one type of EC50 response frequently used as the endpoint for small crustaceans such as *Daphnia* since lethality may be difficult to determine by direct observation. Statistical determination of the median response and its confidence limits is made using a standard probit or logit model (30, 74). The slope of the dose-response curve, indicating whether the toxicity occurs over a broad or narrow range of concentrations, may be useful for planning additional tests.

Long-term tests.   Chronic or subchronic exposures would be indicated if the tier 1 tests are positive and if the exposure assessment suggests persistence or repeated input into the environment. Partial chronic or embryo-larval tests typically last 30 days for fish and 21 days for daphnids. They are expected to show effects not seen in acute adult exposures because early life stages are probably more susceptible than adults to microbial agents. However, embryo-larval, chronic, or partial-chronic tests have rarely if ever been done with aquatic organisms exposed to MPCAs.

Standard long-term chemical exposure techniques using fish and crustaceans, as described in several references (2–4, 9, 51, 63), could be adapted for use with MPCAs if the unique aspects of microbials are recognized. First, the choice of exposure route (aqueous, dietary, or combination) must be made, as discussed for tier 1 tests. If the dietary route is chosen, dosing procedures need to be adapted to the changing food requirements and preferences of the various life stages tested. Second, the exposure should be long enough to allow for the development of any latent pathogenesis. Finally, the potential transfer of microbial pathogen from host parent to progeny should be evaluated in addition to typical sublethal effects such as reduction in growth rate, fecundity or hatchability, and abnormal development.

Microcosm and other multispecies tests.   Multispecies tests to measure aquatic ecosystem disruption have been recommended for certain tier 3 evaluations of MPCAs (26). Since a wide variety of model aquatic ecosystem tests have been designed for various purposes (17, 21), the specific objective of such a test needs to be clearly defined before an appropriate choice can be made. Model ecosystem tests are most useful for hazard assessment when designed to test a specific ecosystem process or function, as discussed by Gillett (32). After considerable review of state-of-the-art microcosms, Geisy (31) concluded that multispecies tests are not necessarily more accurate or sensitive than single-species tests. He emphasized the limitations of such systems and argued strongly against adopting a standard protocol for new product registration.

Many of the current microcosm designs are excellent for studying the fate of introduced chemicals or changes in microbial community structure and function caused by such introductions. However, most are unsuitable for exposure-response testing with fish and larger crustaceans. Fish are not readily incorporated into aquarium-sized microcosms because of limitations of scale (31, 71). Small fish added to 65-liter microcosms by Harrass and Taub (34) caused unacceptably high rates of community exploitation unless they were restricted spatially or temporally. The presence of fish in this system did not interfere with observations on direct algicidal effects to the phytoplankton community, but did reduce the detectability of the secondary effects on cladocerans and ostracods caused by the algicide.

Small crustaceans such as *Daphnia* are more amenable than fish to bench-scale systems. The standard aquatic microcosm of Taub and associates (77), for example, contains cladocerans (*Daphnia magna*), amphipods (*Hyalella* sp.), and ostracods (*Cyprinotus incongruens*) as well as several protozoan and algal species in a chemically defined medium. A microcosm of this type is particularly good for studying the effects of broad-spectrum chemicals on system properties including primary production, carbon cycling, and species diversity. Copper, for example, is toxic in varying degrees to both phytoplankton and zooplankton, so that the population shifts and food chain alterations in a Taub system exposed to copper are particularly interesting (35).

Microbial pest control agents, in contrast to copper, are quite species selective. If an MPCA kills or inhibits a single species within the microcosm, the effect on overall system function will probably be indistinguishable from the loss of that species through any other factor, such as predation. Aquatic microcosms have been shown to be resilient to moderate exploitation and to acute mortality of a single species (78).

Aquatic microcosms can be effective tools for studying the bioavailability of chemical pollutants to aquatic organisms—something that cannot generally be simulated in simple aquarium bioassays. The combination of water, sediment, and biotic phases produces a realistic distribution of chemical in dissolved, adsorbed, and complexed fractions that ultimately determines availability and uptake. Microbial agents may prove to be less complicated than chemical agents in this regard. The uptake and food chain transfer to fish or crustaceans can probably be more efficiently tested in simple feeding studies.

Microcosm design and interpretation of results from multispecies tests are both continuing issues of debate among aquatic toxicologists. With the introduction of new MPCAs, the need for evaluation of test designs and for field validation of microcosm predictions will only increase.

Tier 4: Field testing

Field applications of MPCAs thus far have been limited to a few selective microorganisms such as *B. thuringiensis* that show little effect on nontarget aquatic organisms. Experimental releases in the future are also expected to be restricted to well-studied agents under suitable safeguard conditions. Lake or marine enclosures (44), artificial streams, or small natural ponds or embayments would be possible mesoscale systems for testing caged or free-swimming aquatic organisms. An example of a field application of the chemical insecticide fenthion to caged crustaceans and fish in near-shore estuarine habitats was described by Clark et al. (18, 19). Differences in mortality among the nontarget species (pink shrimp, mysids, and sheepshead minnow) were correlated to observed fenthion concentrations in the water after application by spraying.

In a factorial test plot design, Mather and Lake (48) used an artificial "micromarsh" system consisting of a series of shallow pools (57 liters each) to test the lethality of an experimental insect growth regulator to killifish and grass shrimp. Caging was unnecessary in the small enclosures because water levels could be lowered during daily observations of the animals.

Regardless of the field plot design used, the ability to detect adverse effects on nontarget aquatic organisms will depend on two factors: the virulence of the MPCA agent and the number of nontarget organisms sampled. If the MPCA has a low (but finite) rate of infectivity to nontarget organisms, then a rather large number of organisms will need to be exposed and examined to detect the effect statistically. Examples of numbers of fish to be examined at various rates of infection are given by McDaniel (50).

Test Methods Applied to Specific Microbial Agents

Application to viral agents

The discovery of naturally occurring *Baculovirus* (a nuclear polyhedrosis virus, or NPV) in penaeid shrimp (23) raised the concern that NPV insecticides could affect crustaceans as well. Couch and Martin (22) developed a practical exposure system for testing the infectivity and pathogenicity of such NPV agents to the estuarine grass shrimp, *Palaemonetes vulgaris,* by the dietary route. Shrimp were held in submersed holding cups (2 per cup) in a recirculating static aquarium. The NPV agent isolated from alfalfa leaf hopper, *Autographa californica,* was incorporated at a known rate into prepared shrimp pellets, which were then fed twice weekly to the shrimp for 4 weeks. The viability of the virus in the feed was checked with a bioassay using cabbage looper larvae. At 30 days, the shrimp were killed and processed for histologic, serologic, and microscopic examination. No infection or pathology was caused by the entomopathogen in this study. The authors recommended dietary exposure for viral tests with nontarget species such as grass shrimp because of the method's simplicity and similarity to an actual environmental exposure.

An earlier test of the effect of *A. californica* virus on penaeid shrimp, *Penaeus aztecus* and *P. setiferus,* used injection as the method of exposure (43). After intramuscular injection of virus suspension, shrimp were sampled at 5, 10, 20, and 30 days and processed for histologic examination. Additional shrimp were exposed to the virus through the diet for 10 days. Again, no pathologic effects of the NPV were detected in the shrimp by either route.

A third method of administration was tried in a test of the spruce budworm NPV on two species of fish (70). Rainbow trout and white sucker (*Catostomus commersoni*) were given viral preparations by gastric intubation. The viral material consisted of freeze-dried powered infected insect larvae, dried purified polyhedra, or virions in saline solution. Controls receiving uninfected larvae were included in the analysis. The fish were killed after 28 days and various tissue samples were scored for pathological changes. A significant effect was found for the white suckers treated with virions or purified polyhedra, as compared with controls, but the results were not considered conclusive.

The Baculoviridae are not known to cause lethal infections in vertebrates (53), so that extensive screening with fish may prove unnecessary. Vertebrate hosts of other viral agents could be identified by using fish cell lines as a complement to whole-animal studies. Cell lines are widely used to study viral fish pathogens and have been applied to tests of chemical pollutants as well (12, 41, 65). Cell lines currently available for North American species include those of rainbow trout, fathead minnow, chinook salmon, northern pike, mudminnow, channel catfish, brown bullhead, and bluegill sunfish (13, 41, 85). Such an approach might be a cost-effective method for initial screening of several fish species simultaneously. Eventually, though, in vitro exposure would have to be verified with whole-animal tests. Little is known about the response to MPCAs of cell lines in comparison to complete organisms. Furthermore, continuous cell lines have not been developed or tested for the crustaceans, which are more likely than fish to respond to viral entomopathogens.

Application to bacterial agents

No group of MPCAs has been tested more thoroughly on nontarget aquatic organisms than the spore-forming bacterial insecticides of the family Bacillaceae. To date, only *B. thuringiensis* serovar. *israelensis* (Bti) has been developed commercially for mosquito and blackfly control (45). Its use on a large scale in several countries has shown little or no documented effect on nontarget fish and crustaceans in field situations (40, 53). Among the crustaceans monitored for effects of Bti in actual or simulated field trials were *Ceriodaphnia* sp., *Moina* sp., Chydorinae, *Cyclops* sp., *Cyprois* sp., and *Hyalella azteca* (54); *Daphnia, Ceriodaphnia, Cyclops,* and *Pontogammarus* (69); and ostracods (56). More limited research with *B. sphaericus* applications also showed no effects on similar organisms (47, 57, 58).

Several laboratory studies have established dose-response relationships for Bti with fish or crustaceans. Holck and Meek (36) tested both *B. sphaericus* and Bti (trade name Bactimos) toxicity to the crayfish *Procambarus clarkii* as well as to three mosquito species. The test design was that of a standard 96-h static bioassay, with four replicates at each of five test concentrations plus control. Ten immature crayfish were exposed to the *Bacillus* per 5-liter aquarium and the mortality results were analyzed by the probit method. The acute toxicity of *B. sphaericus* (LC50 = 75.19 ppm) to crayfish proved to be greater than for Bti (LC50 = 103.24 ppm), but both strains were several thousand fold more toxic to mosquitoes. Thus, although the *Bacillus* agents are somewhat toxic to crayfish, the safety margin for the effective mosquito control rate is quite large. Reish et al. (67) reported that the estuarine amphipod *Elasmopus bampo* is much more susceptible to Bti (Bactimos), with a 96-h LC50 of only 12.8 ppm.

The Teknar formulation of Bti used in blackfly control was found to be surprisingly toxic to brook trout fry (29). Exposures of 2 h to 3000 mg/liter or higher produced signs of toxicity, while 6000 mg/liter caused 86.4 percent mortality in just 45 min. Chemical analysis of the formulation and additional experimentation with freeze-dried Teknar showed that the toxicity was apparently caused by xylene, a chemical used as a preservative in the formulation. The difference between the toxicity results of the fresh and freeze-dried Bti underscores the importance of testing commercial formulations as well as reagent-grade material.

The most detailed research on the behavior of Bti in nontarget aquatic organisms concerned a feeding study with the freshwater amphipod *Gammarus lacustris* reported by Brazner and Anderson (14). Amphipods were exposed to Bti spores in static aqueous exposures lasting 1 or 24 h. The spore content of fecal pellets was subsequently analyzed by microscopy and insect bioassay to estimate the total dose ingested by the animals during the test. Spores were found in the feces during the 30 days following exposure, demonstrating that Bti was effectively taken up from water and ingested by the amphipods. The dosage was highly dependent on both exposure concentration and test duration. Fecal pellet analysis indicated that spores attached to the amphipod's carapace, where they were removed and ingested during preening activity. Transfer of spores from parent to progeny also occurred.

The work of Brazner and Anderson (14) illustrates several of the unique features that distinguish MPCA tests from traditional chemical bioassays. The intake of microbial agents from water does not occur by the same route as for dissolved chemicals, which typically diffuse across gills and other membranes. The uptake of Bti, as an example, involves attachment of spores to outer surfaces and ingestion through feeding activities. The amount of feeding that occurs during and after microbial exposure can affect the dosage received and any subsequent toxic effects. For arthropods, molting may either serve to rid the animal of attached spores, or it may provide a more susceptible period of exposure.

Application to fungal agents

The only fungal agent tested to any extent for effects on nontarget crustaceans and fish is the mosquito larvicide *Lagenidium giganteum*. It was originally isolated from a planktonic cladoceran (*Daphnia* sp.)

but apparently has no pathogenic effect on it (24). Safety tests on other microcrustaceans (ostracods and copepods), amphipods (*Gammarus lacustris*), and rice-field crayfish (*Procambarus clarkii*) produced no detectable toxicity or pathogenicity at levels of 50 to 200 times the maximum standard application dose used for field conditions (38). The microcrustaceans in these tests were exposed to mycelium and its growth medium added every fourth day to static containers of water from the collection site. The total exposure lasted 14 days. *Gammarus* were held in distilled water (100 organisms per 500 mL), fed green algae, and exposed similarly. Fifty adult crayfish were held for 30 days in individual containers of distilled water (2 liters each). The water was exchanged every fourth day, at which time new mycelium was added to the test water. Each crayfish was fed prepared fish food daily plus five infected mosquito larvae on every fourth day. The presence of zoospores, the infective asexual stage of *L. giganteum*, was verified visually in the various treatments. Controls without mycelium and with uninfected mosquitoes as food were carried out as well. The absence of fungal infection in the test organisms was determined by examining individuals removed from the tests at 4-day intervals and at the conclusion of each exposure. Both visual examination and staining with methylene blue were used to check for mycelial proliferation. Essentially no mortality (<4 percent) was observed during the crustacean tests.

Mosquitofish, green sunfish (*Lepomis cyanellus*), and rainbow trout were also exposed to the fungal agent in 21-day static renewal tests as part of the same safety evaluation. The mosquitofish, in aerated tap water at 25°C, and the trout, at 10°C, were fed infected mosquito larvae (late instar *Culex tarsalis*) every sixth day. Mycelium was also renewed in each tank on a 6-day schedule. The sunfish were held at 21 to 25°C in distilled water in which mycelium was renewed every 4 days. Sunfish were also fed infected mosquito larvae every sixth day. No significant mortality or signs of fungal infection were found among the three fish species tested. Viable *L. giganteum* was recovered from the mosquitofish and sunfish tanks. However, the fungus deteriorated in the colder trout tanks because it becomes dormant below 18°C (38). This loss of viability illustrates the importance of selecting a test species within the temperature range of the microbial agent.

Applications of *L. giganteum* to rice fields produced no observable effects on cladocerans (*Daphnia pulex*), copepods (*Cyclops* sp.), or crayfish (*Procambarus clarkii*) at concentrations that control mosquitoes (28). Such field trials have also shown that temperature and water quality (oxygen content, pH, and organic content) can affect the viability of the fungal agent (33). Thus the composition of the test water used for exposing fish or other aquatic organisms can have an effect on the viability of the MPCA during the test.

While it appears unlikely that *L. giganteum* would ever prove pathogenic to fish and has produced no effects on crustaceans tested to date, other members of the genus *Lagenidium* are known to be parasitic on important commercial shellfish species. Blue crab (*Callinectes sapidus*) (11), Dungeness crab (*Cancer magister*) (6), and white and brown shrimp (*Penaeus setiferus* and *P. aztecus,* respectively) (42) are all susceptible to mycosis produced by other species of *Lagenidium.* The most susceptible life stage for these crustaceans appears to be the second zoeal stage of the larvae, prior to the mysis stage. This suggests that tests of related fungal agents on such crustaceans should include an exposure of the early life stages of the animal.

The appropriate length of pathogenicity tests with fungal agents has not been established. If zoospores are present in the inoculum, infection should occur relatively rapidly, if at all. In contrast, a test conducted with the oospore stage of a fungal agent may require a lag time of a week or more to allow for germination and production of zoospores before infection can occur (40). The rate of infection is temperature dependent and may also be concentration dependent as well.

A good example of the importance of such factors is shown in the study by Alderman et al. (1) of the commercially important fungal crayfish plague. The zoospore stage of the pathogen *Aphanomyces astaci* was given in water to crayfish (*Astacus leptodactylus* and *A. pallipes*) at several temperatures and concentrations. At a high dosage (13 zoospores per milliliter at 20°C), a lag phase of several days was followed by rapid mortality. The time to 50 percent mortality of infected crayfish was approximately 9 days. At one-tenth that concentration (1.3/mL at 20°C), a longer lag phase was observed, followed by a more gradual rate of mortality. The time to 50 percent mortality was about 25 days. Both exposures eventually led to complete mortality. Low test temperatures also prolonged the lag phase and time to death. In addition to the importance of dosage and water temperature, the authors noted that removal of dead animals during the test probably prolongs the course of the infection in the population by removing infected hosts that are the source of additional zoospores.

The *Coelomomyces* fungi, which are obligate parasites of mosquitoes, have received attention as possible pest control agents. The group is unusual in that the parasite requires a crustacean as intermediate host (83). To date, five *Coelomomyces* species requiring particular copepod hosts and two requiring ostracods have been identified (84). Margalit and Evenchik (46) experimentally infected copepods (*Acanthocyclop vernalis*) with *C. psorophorae* zoospores in water. The rate of infection, made visible by methylene blue staining, increased to about 25 percent of the copepods after 1 week. A second copepod species, *A. viridis,* was less susceptible, with only about 10 percent infected after 1 week. The Israeli copepod *Mesocyclop leuckarti* most often associated with mosquito larvae was not susceptible to fungal infection. In a similar series of trials, five species of copepods from Fiji were used to experimentally infect *Aedes* mosquito larvae with *Coelomomyces* (79). Only one of the five copepods proved to be an effective intermediate host. Such research illustrates not only that fungal MPCAs are highly species selective, but that the choice of test species is extremely important.

One additional fungal MPCA, *Culicinomyces clavisporus,* has been tested for effects on freshwater shrimp (Atyidae) and mosquitofish using both water and infected food as routes of exposure (76). No toxicity or pathogenicity was observed for these species after an exposure of unspecified length.

Application to protozoan agents

Since parasitic protozoans for mosquito control have not yet been developed to the point of commercial application, few tests with nontarget fish or crustaceans have been reported. Spores of the microsporidian *Nosema algerae* were administered to eight crayfish (unspecified species) by injection into the hemocoel (80). The crayfish developed heavy infection in muscle and gill tissue, causing 100 percent mortality about 14 days after injection. Viability of the spores recovered from injected arthropods was shown by feeding the spores back to mosquito larvae. In another study of *N. algerae,* immature crayfish (*Procambarus* sp.) and mosquitofish held in individual jars were fed infected fourth-instar mosquito larvae (81). Water was exchanged weekly, and each test organism was observed daily to determine the number of larvae consumed. The total number of spores ingested per animal was calculated based on the average spore concentration in the mosquito larvae. Exposure of the crayfish and fish by feeding lasted about 3 weeks. No evidence of infection was present in either species at the end of the trials. Thus it appears that administration by injection is a more severe exposure route than administration through food.

The protozoan entomopathogen *Mattesia trogodermae* was tested with several nontarget species including brine shrimp (*Artemia salina*), *Daphnia magna,* and the minnow *Notropis atherinoides* (J. G. Pounds, Ph.D. thesis, University of Wisconsin—Madison, 1977). No toxicity or infectivity of the animals was detected. However, viable spores were concentrated in the guts of both the daphnid and brine shrimp, suggesting a possible hazard to susceptible consumer organisms higher in the food chain.

Research Needs

Because the practice of aquatic testing with microbial agents is relatively new and untried, a myriad of questions arise concerning the choice of test protocols and interpretation of results. The following issues are identified as important topics needing additional research or validation before extensive use of standardized test protocols is warranted.

1. *Choice of representative test species:* Fish and crustacean species commonly used for chemical testing (Table 36.1) have been selected for a combination of traits including chemical sensitivity, ease of culture, and importance in particular aquatic ecosystems. Is the current surrogate species approach adequate for the hazard assessment of microbial agents? Is chemical sensitivity a good predictor of MPCA susceptibility, or is taxonomic group the most important factor? To what extent can pathology results for one species be extrapolated to others in the same genus or family? How can the number of required test species be minimized without overlooking important nontarget effects?

2. *Design of exposure systems:* Pathogenic viruses, bacteria, fungi, and protozoans vary greatly in their ability to survive in the aquatic environment. Their virulence and modes of pathogenesis also vary widely. Can all types of microbial agents be tested adequately by using an aqueous exposure route, or is the dietary route more efficient and/or realistic for some agents? What is the optimum length of an acute microbial exposure, and what length of exposure is necessary to produce latent effects of slow-acting microbial agents?

3. *Importance of dose vs. exposure concentration:* Is it necessary to calculate toxicity and pathogenicity of MPCAs to aquatic animals on the basis of ingested dose (LD50), or is the more conventional method of using aqueous exposure concentration (LC50) sufficient for routine screening purposes?

4. *Role of in vitro tests:* There is a need to compare protocols using fish cell lines to the results of in vivo bioassays for microbial agents. If tests with fish cell lines prove to be accurate and efficient screening tools, additional efforts to develop in vitro techniques for crustaceans would be warranted.

5. *Applicability of microcosms for fish and crustaceans:* At present there are no standard multispecies tests designed to detect aquatic ecosystem disruption caused by MPCAs. The utility of such systems in

the case of extremely host specific pathogens has not been established. Can effects on fish and crustaceans be adequately observed in microcosms designed to test community structure and function? Can trophic transfer of microbial agents be tested equally well in simpler dietary exposures?

6. *Design of aquatic field trials:*   Aquatic field trials will need to be designed on a case-by-case basis until considerably more experience with MPCAs is accumulated. Mesocosms such as manufactured pools, artificial channels, and lake enclosures may prove useful for preliminary trials. Field studies of microbial agents with low toxicity or pathogenicity to nontarget organisms will require especially large sample sizes (numbers of nontarget organisms collected and examined) in order to establish effect levels that can be interpreted statistically.

7. *Test development using a model microorganism:*   Many of the questions and concerns discussed above have arisen because of a lack of experience with positive tests—that is, with microbial agents that produce significant toxicity or pathogenicity to nontarget aquatic organisms. Such agents would usually be rejected early in the commercial development process. An agent of this type would be useful nevertheless as a positive control for the purpose of test method development. Therefore, a final research recommendation is the consideration of one or more representative microbial agents capable of producing effects across several taxa. Naturally occurring disease vectors might serve as suitable models for this purpose. In any case, the positive endpoints for each type of bioassay need to be defined.

Many of the issues raised here will be clarified as new MPCAs are developed and tested. Until additional experience is gained with a wider variety of microbial agents, the adoption of a single aquatic testing protocol is premature. In the meantime, studies designed to compare the performance of various bioassay designs will be worthwhile.

Acknowledgments

The author wishes to thank John Briggs for helpful discussions and Linda Baril for assistance with literature references. Preparation of this chapter was supported by the Purdue School of Agriculture, Hatch Project No. IND-59042.

References

1. Alderman, D. J., A. L. Polglase, and M. Frayling. 1987. *Aphanomyces astaci* pathogenicity under laboratory and field conditions. *J. Fish Dis.* 10:385–393.

2. American Society for Testing and Materials. 1990. Standard guide for conducting renewal life-cycle toxicity tests with *Daphnia magna*, E 1193. ASTM, Philadelphia.

3. American Society for Testing and Materials. 1990. Standard guide for conducting life-cycle toxicity tests with saltwater mysids, E 1191. ASTM, Philadelphia.

4. American Society for Testing and Materials. 1990. Standard guide for conducting early life-stage toxicity tests with fishes, E 1241. ASTM, Philadelphia.

5. American Society for Testing and Materials. 1990. Standard guide for conducting acute toxicity tests with fishes, macroinvertebrates, and amphibians, E 729. ASTM, Philadelphia.

6. Armstrong, D. A., D. V. Buchanen, and R. S. Caldwell. 1976. A mycosis caused by *Lagenidium* sp. in laboratory-reared larvae of the dungeness crab, *Cancer magister,* and possible chemical treatments. *J. Invert. Pathol.* 28:329–336.

7. Arthur, J. W. 1980. Review of freshwater bioassay procedures for selected amphipods. In A. L. Buikema, Jr., and J. Cairns, Jr. (eds.), *Aquatic Invertebrate Bioassays.* American Society for Testing and Materials, Philadelphia, pp. 98–108.

8. Austin, B., and D. A. Austin. 1987. *Bacterial Fish Pathogens.* Wiley, New York.

9. Benoit, D. A., F. A. Puglisi, and D. L. Olson. 1982. A fathead minnow *Pimephales promelas* early life stage toxicity test method evaluation and exposure to four organic chemicals. *Environ. Pollut.* (Series A) 28:189–197.

10. Betz, F. S. 1986. Registration of baculoviruses as pesticides. In R. R. Granados and B. A. Federici (eds.), *The Biology of Baculoviruses,* vol. 2. CRC Press, Boca Raton, Fla., pp. 203–222.

11. Bland, C. E., and H. V. Amerson. 1973. Observations on *Lagenidium callinectes:* Isolation and sporangial development. *Mycologia* 65:310–320.

12. Bols, N. C., S. A. Boliska, D. G. Dixon, P. V. Hodson, and K. L. E. Kaiser. 1985. The use of fish cell cultures as an indication of contaminant toxicity to fish. *Aquat. Toxicol.* 6:147–155.

13. Bowser, P. R., and J. A. Plumb. 1980. Channel catfish virus: Comparative replication and sensitivity of cell lines from channel catfish ovary and the brown bullhead. *J. Wildl. Dis.* 16:451–454.

14. Brazner, J. C., and R. L. Anderson. 1986. Ingestion and adsorption of *Bacillus thuringiensis* subsp. *israelensis* by *Gammarus lacustris* in the laboratory. *Appl. Environ. Microbiol.* 52:1386–1390.

15. Buikema, A. L., Jr., and J. Cairns, Jr. (eds.). 1980. *Aquatic Invertebrate Bioassays.* ASTM STP 715, Am. Soc. Testing Materials, Philadelphia.

16. Bullock, A. M. 1978. Laboratory methods. In R. J. Roberts (ed.), *Fish Pathology.* Bailliere Tindall, London, pp. 235–267.

17. Cairns, J., Jr. (ed.). 1985. Multispecies toxicity testing. Pergamon, New York.

18. Clark, J. R., P. W. Borthwick, L. R. Goodman, J. M. Patrick, Jr., E. M. Lores, and J. C. Moore. 1987. Comparison of laboratory toxicity test results with responses of estuarine animals exposed to fenthion in the field. *Environ. Toxicol. Chem.* 6:151–160.

19. Clark, J. R., P. W. Borthwick, L. R. Goodman, J. M. Patrick, Jr., E. M. Lores, and J. C. Moore. 1987. Effects of aerial thermal fog applications of fenthion on caged pink shrimp, mysids, and sheepshead minnows. *J. Am. Mosq. Control Assoc.* 3:466–472.

20. Committee on methods for acute toxicity tests with aquatic organisms. 1975. Methods for acute toxicity tests with fish, macroinvertebrates and amphibians. EPA-660/3-75-009. U.S. Environmental Protection Agency, Washington, D.C.

21. Conway, R. A. 1982. *Model Ecosystems in Environmental Risk Analysis for Chemicals.* Van Nostrand-Reinhold, New York.

22. Couch, J. A., and S. M. Martin. 1984. A simple system for the preliminary evaluation of infectivity and pathogenesis of insect virus in a nontarget estuarine shrimp. *J. Invert. Pathol.* 43:351–357.

23. Couch, J. A. 1974. An enzootic nuclear polyhedrosis virus of pink shrimp: Ultrastructure, prevalence, and enhancement. *J. Invert. Pathol.* 24:311–331.

24. Couch, J. N. 1935. A new saprophytic species of *Lagenidium giganteum. Mycologia* 27:376–387.

25. Dean-Ross, D. 1986. Release of genetically engineered organisms: Hazard assessment. *Am. Soc. Microbiol. News* 52:572–575.

26. Environmental Protection Agency. 1989. *Pesticide Assessment Guidelines, Subdivision M: Biorational Pesticides*, NTIS No. PB 89-211676. Washington, D.C.
27. Ewell, W. S., J. W. Gorsuch, R. O. Kringle, K. A. Robillard, and R. C. Spiegel. 1986. Simultaneous evaluation of the acute effects of chemicals on seven aquatic species. *Environ. Toxicol. Chem.* 5:831–840.
28. Fetter-Lasko, J. L., and R. K. Washino. 1983. In situ studies on seasonality and recycling pattern in California of *Lagenidium giganteum* Couch, an aquatic fungal pathogen of mosquitoes. *Environ. Ent.* 12:635–640.
29. Fortin, C., D. Lapointe, and G. Charpentier. 1986. Susceptibility of brook trout (*Salvelinus fontinalis*) fry to a liquid formulation of *Bacillus thuringiensis* serovar. *israelensis* (Teknar) used for blackfly control. *Can. J. Fish. Aquat. Sci.* 43:1667–1670.
30. Gelber, R. D., P. T. Lavin, C. R. Mehta, and D. A. Schoenfeld. 1985. Statistical analysis. In G. M. Rand and S. R. Petrocelli (eds.), *Fundamentals of Aquatic Toxicology*. Hemisphere, New York, pp. 110–123.
31. Giesy, J. P., Jr. 1985. Multispecies tests: Research needs to assess the effects of chemicals on aquatic life. In R. C. Bahner and D. J. Hansen (eds.), *Aquatic Toxicology and Hazard Assessment*, 8th Symposium, ASTM STP 891, ASTM, Philadelphia, pp. 67–77.
32. Gillett, J. W. 1986. Risk assessment methodologies for biotechnology impact assessment. *Environ. Manag.* 10:515–532.
33. Guzman, D. R., and R. C. Axtell. 1987. Temperature and water quality effects in simulated woodland pools on the infection of *Culex* mosquito larvae by *Lagenidium giganteum* (Oomycetes: Lagenidiales) in North Carolina. *J. Am. Mosq. Control Assoc.* 3:211–218.
34. Harrass, M. C., and F. B. Taub. 1985. Effects of small fish predation on microcosm community bioassays. In R. D. Cardwell, R. Purdy, and R. C. Bahner (eds.), *Aquatic Toxicology and Hazard Assessment*, 7th Symposium, ASTM STP 854, ASTM, Philadelphia, pp. 117–133.
35. Harrass, M. C., and F. B. Taub. 1985. Comparison of laboratory microcosms and field responses to copper. In T. P. Boyle (ed.), *Validation and Predictability of Laboratory Methods for Assessing the Fate and Effects of Contaminants in Aquatic Ecosystems*. ASTM STP 865, ASTM, Philadelphia, pp. 57–74.
36. Holck, A. R., and C. L. Meek. 1987. Dose-mortality responses of crawfish and mosquitoes to selected pesticides. *J. Am. Mosq. Control Assoc.* 3:407–411.
37. Hughes, P. R., and H. A. Wood. 1986. *In vivo* and *in vitro* bioassay methods for baculoviruses. In R. R. Granados and B. A. Federici (eds.), *The Biology of Baculoviruses*, vol. 2. CRC Press, Boca Raton, Fla., pp. 1–30.
38. Jaronski, S., and R. C. Axtell. 1983. Effects of temperature on infection, growth and zoosporogenesis of *Lagenidium giganteum*, a fungal pathogen of mosquito larvae. *Mosq. News.* 43:42–45.
39. Kerwin, J. L., D. A. Dritz, and R. K. Washino. 1988. Nonmammalian safety tests for *Lagenidium giganteum* (Oomycetes: Lagenidiales). *J. Econ. Entomol.* 81:158–171.
40. Lacey, L. A., and A. H. Undeen. 1986. Microbial control of black flies and mosquitoes. *Annu. Rev. Entomol.* 31:265–296.
41. Landolt, M. L., and R. M. Kocan. 1983. Fish cell cytogenetics: A measure of the genotoxic effects of environmental pollutants. In J. O. Nriagu (ed.), *Aquatic Toxicology*. Wiley, New York, pp. 335–353.
42. Lightner, D. V., and C. T. Fontaine. 1973. A new fungus disease of the white shrimp *Penaeus setiferus*. *J. Invert. Pathol.* 22:94–99.
43. Lightner, D. V., R. R. Procter, A. K. Sparks, J. R. Adams, and A. M. Heimpel. 1973. Testing penaeid shrimp for susceptibility to an insect nuclear polyhedrosis virus. *Environ. Entomol.* 2:611–613.
44. Lundgren, A. 1985. Model ecosystems as a tool in freshwater and marine research. *Arch. Hydrobiol.* (Suppl. 70)2:157–196.
45. Margalit, J., and D. Dean. 1985. The story of *Bacillus thuringienses* var. *israelensis* (*B.t.i.*). *J. Am. Mosq. Control Assoc.* 1:1–7.

46. Margalit, J., and Z. Evenchik. 1983. Mosquito and copepod host range tests with *Coelomomyces psorophorae* (Blastocladiales, Chytridiomycetes). *Insect Sci. Appl.* 4: 383–385.
47. Mathavan, S., and A. Velpandi. 1984. Toxicity of *Bacillus sphaericus* strains to selected target and non-target aquatic organisms. *Indian J. Med. Res.* 80:653–657.
48. Mather, T. N., and R. W. Lake. 1982. Small plot evaluations of the toxicity of an experimental IGR to salt marsh mosquitoes and non-target organisms. *Mosq. News* 42:190–195.
49. Mayer, F. L., Jr., and M. R. Ellersieck. 1986. Manual of acute toxicity: Interpretation and data base for 410 chemicals and 66 species of freshwater animals. U.S. Fish and Wildlife Service Resource Publ. 160, Washington, D.C.
50. McDaniel, D. (ed.). 1979. Procedures for the detection and identification of certain fish pathogens. Fish Health Section, American Fisheries Society, Washington D.C.
51. McKim, J. M. 1985. Early life stage toxicity tests. In G. M. Rand and S. R. Petrocelli (eds.), *Fundamentals of Aquatic Toxicology*. Hemisphere, New York, pp. 58–95.
52. Meyers, T. R., and J. D. Hendricks. 1985. Histopathology. In G. M. Rand and S. R. Petrocelli (eds.), *Fundamentals of Aquatic Toxicology*. Hemisphere, New York, pp. 283–331.
53. Miller, L. K., A. J. Lingg, and L. A. Bulla, Jr. 1983. Bacterial, viral, and fungal insecticides. *Science* 219:715–721.
54. Miura, T., R. M. Takahashi, and F. S. Mulligan, III. 1980. Effects of the bacterial mosquito larvicide, *Bacillus thuringiensis* serotype H-14 on selected aquatic organisms. *Mosq. News* 40:619–622.
55. Mount, D. I., and T. J. Norberg. 1984. A seven-day life cycle cladocern test. *Environ. Toxicol. Chem.* 3:425–434.
56. Mulla, M. S., B. A. Federici, and H. A. Darwazeh. 1982. Larvicidal efficacy of *Bacillus thuringiensis* serotype H-14 against stagnant water mosquitoes and its effects on nontarget organisms. *Environ. Entomol.* 11:788–795.
57. Mulla, M. S., H. A. Darwazeh, E. W. Davidson, H. T. Dulmage, and S. Singer. 1984. Larvicidal activity and field efficacy of *Bacillus sphaericus* strains against mosquito larvae and their safety to nontarget organisms. *Mosq. News* 44:336–342.
58. Mulligan, F. S., III, C. H. Schaefer, and T. Miura. 1978. Laboratory and field evaluation of *Bacillus sphaericus* as a mosquito control agent. *J. Econ. Entomol.* 71:774–777.
59. Murty, A. S. 1986. *Toxicity of Pesticides to Fish*. CRC Press, Boca Raton, Fla.
60. Nimmo, D. R., L. H. Bahner, R. A. Rigby, J. M. Sheppard, and A. J. Wilson, Jr. 1977. *Mysidopsis bahia:* An estuarine species suitable for life-cycle toxicity tests to determine the effects of a pollutant. In F. L. Mayer and J. L. Hamelink (eds.), *Aquatic Toxicity and Hazard Evaluation*. ASTM STP 634, Am. Soc. Testing Materials, Philadelphia, pp. 109–116.
61. Parrish, P. R. 1985. Acute toxicity tests. In G. M. Rand and S. R. Petrocelli (eds.), *Fundamentals of Aquatic Toxicology*. Hemisphere, New York, pp. 31–57.
62. Peltier, W. H., and C. I. Weber (eds.). 1985. *Methods for Measuring the Acute Toxicity of Effluents to Freshwater and Marine Organisms*, 3d ed. EPA/600/4-85/013. U.S. Environmental Protection Agency. Cincinnati, Ohio.
63. Petrocelli, S. R. 1985. Chronic toxicity tests. In G. M. Rand and S. R. Petrocelli (eds.), *Fundamentals of Aquatic Toxicology*. Hemisphere, New York, pp. 96–109.
64. Post, G. W. 1983. *Textbook of Fish Health*. T. F. H. Publications, Neptune City, N.J.
65. Rachlin, J. W., and A. Perlmutter. 1968. Fish cells in culture for study of aquatic toxicants. *Water Res.* 2:409–414.
66. Rand, G. M. 1985. Introduction. In G. M. Rand and S. R. Petrocelli (eds.), *Fundamentals of Aquatic Toxicology*. Hemisphere, New York, pp. 1–29.
67. Reish, D. J., J. A. Lemay, and S. L. Asato. 1985. The effect of *Bacillus thuringiensis* var. *israelensis* (H-14) and methoprene on two species of marine invertebrates from southern California estuaries. *Bull. Soc. Vector Ecol.* 10:20–22.

68. Roberts, R. J. (ed.) 1982. *Microbial Diseases of Fish*. Academic Press, for Society for General Microbiology, New York.
69. Rogatin, A. B., and M. Baizhanov. 1984. Laboratory study of the effect of an experimental series of bacterial preparation of *Bacillus thuringiensis* (serotype 14) on various groups of hydrobionts. *Izv. Akad. Nauk. Kaz. Ssr. Ser. Biol.* 0:22–25 (English abstract).
70. Savan, M., J. Budd, P. W. Reno, and S. Darley. 1979. A study of two species of fish inoculated with spruce budworm nuclear polyhedrosis virus. *J. Wildl. Dis.* 15:331–334.
71. Spacie, A., and J. L. Hamelink. 1985. Bioaccumulation. In G. M. Rand and S. R. Petrocelli (eds.), *Fundamentals of Aquatic Toxicology*. Hemisphere, New York, pp. 495–525.
72. Sparks, A. K. 1985. *Synopsis of Invertebrate Pathology: Exclusive of Insects*. Elsevier, New York.
73. Spehar, R. L., D. K. Tanner, and J. H. Gibson. 1982. Effects of kelthane and pydrin on early life stages of fathead minnow (*Pimephales promelas*) and amphipods (*Hyalella azteca*). In J. G. Pearson, R. B. Foster, and W. E. Bishop (eds.), *Aquatic Toxicology and Hazard Assessment*, 5th Conference. ASTM STP 766, ASTM, Philadelphia, pp. 234–244.
74. Stephan, C. E. 1977. Methods for calculating an LC50. In F. L. Mayer and J. L. Hamelink (eds.), *Aquatic Toxicity and Hazard Evaluation*. ASTM STP 634. ASTM, Philadelphia, pp. 65–84.
75. Stokes, P. M. 1981. *Ecotoxicology and the Aquatic Environment*. Pergamon, New York.
76. Sweeney, A. W. 1975. The insect pathogenic fungus *Culicinomyces* in mosquitoes and other hosts. *Aust. J. Zool.* 23:59–64.
77. Taub, F. B., and M. C. Crow. 1980. Synthesizing aquatic microcosms. In J. P. Giesy, Jr. (ed.), *Microcosms in Ecological Research*. Dept. of Energy Symposium Series, Conf. 781101, National Tech. Info. Service, Springfield, Va., pp. 69–104.
78. Taub, F. B., M. E. Crow, and H. J. Hartmann. 1980. Responses of aquatic microcosms to acute mortality. In J. P. Giesy, Jr. (ed.), *Microcosms in Ecological Research*. Dept. of Energy Symposium Series, Conf. 781101, National Tech. Info. Service, Springfield, Va., pp. 513–535.
79. Toohey, M. K., G. Prakash, and M. S. Goettel. 1982. *Elaphoidella taroi:* The intermediate copepod host in Fiji for the mosquito pathogenic fungus *Coelomomyces*. *J. Invert. Pathol.* 40:378–382.
80. Undeen, A. H., and J. V. Maddox. 1973. The infection of nonmosquito hosts by injection with spores of the microsporidan *Nosema algerae*. *J. Invert. Pathol.* 22:258–265.
81. Van Essen, F. W., and D. W. Anthony. 1976. Susceptibility of nontarget organisms to *Nosema algerae* (Microsporida: Nosematidae), a parasite of mosquitoes. *J. Invert. Pathol.* 28:77–85.
82. Ward, T. J., E. D. Rider, and D. A. Drozdowski. 1979. A chronic toxicity test with the marine copepod *Acartia tonsa*. In L. L. Marking and R. A. Kimerle (eds.), *Aquatic Toxicology*. ASTM STP 667. ASTM, Philadelphia, pp. 148–158.
83. Whisler, H. C., S. L. Zebold, and J. A. Shemanchuk. 1974. Alternate host for mosquito parasite *Coelomomyces*. *Nature* (London) 251:715–716.
84. Whisler, H. C. 1985. Life history of species of *Coelomomyces*. In J. N. Couch and C. E. Bland (eds.), *The Genus Coelomomyces*. Academic Press, New York, pp. 9–22.
85. Wolf, K., and J. A. Mann. 1980. Poikilothermic vertebrate cell lines and viruses: A current listing for fishes. *In Vitro* 16:168–179.

Chapter 37

Testing the Effects of Microorganisms on Birds

James L. Kerwin

Introduction

Registration of chemical agents for control of insect and plant pests has required evaluation of their potential effects on birds since the passage of the Federal Insecticide, Fungicide, and Rodenticide Act (FIFRA) in 1947. Chemical, biochemical, and microbiological pest control agents are subjected to a variety of similar tests to assess their effects on bird populations prior to registration by the federal government. Although these tests are primarily designed to satisfy requirements outlined by FIFRA as amended by the Pesticide Control Act of 1972, the FIFRA amendments of 1975, and the Federal Pesticide Act of 1978, regulation of potentially toxic agents is covered by at least eight major acts of federal legislation. These guidelines include legislation which directly affects birds, e.g., the Marine Protection Research and Sanctuaries Act of 1972, the Toxic Substances Control Act of 1976, and the Comprehensive Environmental Response, Compensation, and Liability Act of 1980 (Levin et al., 1984). Discussion that follows primarily focuses on testing outlined by FIFRA and related policy statements and documents such as U.S. Environmental Protection Agency (USEPA) hazard evaluation guidelines.

Environmental effects tests for chemicals have included avian toxicology evaluations; therefore, many precedents have been set regarding species of birds to be used and general protocols to be followed for microbial testing based on years of experience from chemical toxicology (Kenaga, 1979; Lamb and Kenaga, 1981). Until the last months of 1987, the majority of microbial safety tests using birds have been based on general guidelines promulgated by USEPA (1982). Two distinct tests involving avian oral toxicity and intravenous (for bacteria, viruses, and protozoa) or intraperitoneal (for fungi) injection of concentrated solutions of microorganisms, while slightly modified to accommodate unique characteristics of these organisms, are essentially those used for assessing the safety of chemicals for the last several decades. Recent proposed revisions of these guidelines attempt to provide more viable techniques for assessing potential pathogenicity of microorganisms to nontarget species. For birds the oral toxicity tests are unaltered except for several minor changes in suggested experimental protocols. Also outlined more specifically are reporting requirements which should mitigate confusion arising from the previous guidelines. In-house reports generated by USEPA personnel provide excellent guidance for the majority of avian safety evaluations which will have to be performed to support registration of MPCAs (Knittel, 1988a, b; Knittel and Fairbrother, 1988; Fairbrother et al., 1988).

Intraperitoneal or intravenous injection tests have been dropped as a requirement for most organisms on the basis that this is an unlikely route of introduction of a microbial agent. Intratracheal installation of all microorganisms, designated the avian respiratory tract instillation test, is proposed in the revisions as an alternative, since injection is considered to be an unlikely means of exposure under natural conditions. Implications of these proposed revisions are discussed below.

Limitations of toxicological tests for chemical and biological agents for making regulatory decisions are universally acknowledged (Levin et al., 1984). As data are accumulated from many scientific disciplines, impetus is provided for regulatory guidelines and protocols, e.g. the recent proposed revisions in guidelines for microbial safety tests. The major problems with using toxicological testing as the sole or primary basis for registration of a biological control agent are briefly listed here and will form the basis for subsequent detailed discussion. Aspects relating specifically to avian testing are emphasized.

First, selected species are chosen as being representative of birds in general; unfortunately, there are numerous examples of interspecific and, indeed, intraspecific differences in susceptibility of birds to parasitism. Choice of species for evaluation is more often based on ease of handling in the laboratory, availability, and economic importance than on biological characteristics unique to the specific parasite being tested. Second, the highly controlled laboratory conditions commonly used are designed for precision rather than accuracy and do not represent environmental conditions encountered by natural populations. The host-parasite relationship can be greatly altered in differing environments, making extrapolation of laboratory data to the field suspect. Third, effects of massive introduction of a microbial control agent into the environment can affect the ecological balance in ways that have no relation to the presence or absence of toxicity to a small number of species. Fourth, subtle effects on bird behavior that may be absent or ignored during laboratory tests may have profound implications for survival and reproductive capabilities in wild populations. Finally, for those microorganisms producing toxins, there may be little relationship between chronic and acute effects on animal reproduction or survival, as has been demonstrated for a variety of chemicals (Kenaga, 1979a, b). Implications of these limitations for avian testing of microbial agents are detailed in subsequent sections.

Statistical considerations

Design of MPCA safety tests, although conforming generally to procedures outlined in subdivision M guidelines and supporting documents, should be guided in part by preliminary observations on pathogenicity to nontarget organisms. Especially if microbial toxins are responsible for the efficacy of the MPCA, sublethal or even lethal effects could be expected following challenge of nontarget organisms, including birds. In extreme cases, generation of LC_{50}s similar to those for chemical pesticides will be required; however, if a microbial toxin has obvious adverse effects on nontarget organisms, it is unlikely that it will be developed for operational control. Research and development, registration, and production costs in conjunction with restricted host range of an MPCA with the disadvantages of a classical chemical control agent would probably be prohibitive.

For most MPCAs seriously considered for commercial development, no adverse effects will be documented during avian or other pathogenicity tests. Exposure of test animals to very high concentrations of an MPCA as outlined in subdivision M will, however, occasionally result in mortality or sublethal effects. In many instances interpretation of adverse effects will be straightforward. For instance, intratracheal instillation of dried oospore preparations of the mosquito pathogenic fungus *Lagenidium giganteum* can result in asphyxiation of or anaphylactic shock to birds because of the large size of the fungus. Increase in the number of animals treated with varying concentrations of an MPCA in, for example, avian intratracheal instillation evaluations, will often be sufficient to ferret out effects due to anaphylactic shock following introduction of large quantities of foreign protein.

Although safety tests are designed to be performed under highly controlled laboratory conditions, there are circumstances such as those outlined in the previous section under which tests may have to be completed in a more natural environment. In these rare cases the effects of the MPCA on the test organisms will have to be separated from environmental and other unanticipated or unknown components which could influence the outcome of the safety evaluation. Techniques commonly used by epidemiologists would provide the best guidance for design, monitoring, and implementation of the tests (Lilienfield and Lilienfield, 1980; Schlesselman, 1982).

It is known that parasites from many phyla have evolved mechanisms to avoid host defenses and to survive in a dormant (and difficult to detect) state for extended periods of time (Ogilvie and Wilson, 1977; Parkhouse, 1984; Behnke, 1987). Even in highly controlled laboratory tests, then, unexpected mortality or slowly developing sublethal effects may be manifested. A well-designed experimental protocol coupled with detailed understanding of the developmental plasticity of the MPCA being tested are necessary to accurately assess effects on nontarget organisms.

Those not familiar with basic concepts in epidemiology dealing with research strategies and methods of analysis, estimation of various sources of bias in experimental systems, criteria for causation, and methods for assessing the relative contributions of two or more variables to an observed effect are referred to the epidemiology texts cited above and to those dealing with ecological and statistical methods of analysis (Sokal and Rohlf, 1981; Davies and Goldsmith, 1984; Rollinson and Anderson, 1985).

Even the simplest host-parasite systems are best described by nonlinear, dynamic mathematical models (Anderson and May, 1979, 1981; Dobson and Hudson, 1986). Recently it has been recognized that apparent random noise or irregular fluctuations in these systems may in fact be due to chaotic behavior (Gleick, 1987). General characteristics of chaotic behavior include the lack of random inputs, complex dynamics, fractal (noninteger) dimension, and sensitive dependence on initial conditions (Schaeffer and Kot, 1986). If seemingly random fluctuations occur during the course of pathogenicity tests, the investigator should be familiar with the concept of chaos, the techniques of verifying this nonlinear behavior, and the interpretation of the manipulated data.

Rationale for Avian Toxicity Tests

Avian wildlife toxicology has been included in the assessment of risk in pest control since the inception of legislatively mandated guidelines. The deleterious effects of, for example, commonly used organochlorine compounds either directly or indirectly via concentration in food sources is well documented (Heath et al., 1972; Hill et al., 1975; Brown, 1978; Lamb and Kenaga, 1981).

Whether avian testing of microorganisms targeted for operational use in pest control is necessary is a legitimate issue. A primary consideration is the probability of exposure of birds to microbial pest control agents in sufficient quantities to initiate infection and the possible susceptibility to infection by these organisms to justify the effort of toxicological tests. Since the targets of biological control are usually pest insects and plants, it is obvious that many species of birds adjacent to treated areas will encounter at least trace quantities of these agents. As an indication of the extent of exposure, it is estimated that of 640 species of North American birds, about 44 percent of them forage in or near forests and include insects as 10 percent or more of their diet at least during one season (Jackson, 1979). This has implications for any microorganisms applied for control of, for example, spruce bud worm, *Choristoneura fumiferana,* or the Engelmann spruce beetle, *Dendroctonus engelmanni.* This general observation can be refined for forest ecosystems by examination of the literature for species of birds known to prey upon specific insects. For instance, woodpeckers (*Piciformes*) are the only species which have demonstrated effects on subsurface bark beetles (Jackson, 1979); therefore, this order of birds would include the best species for safety tests involving a microorganism targeted for control of this group of coleopterans.

The majority of birds feed upon invertebrates, seeds, or a combination of these two sources. The remaining major groups feed upon rodents or fish, which are notorious for concentrating chemicals in their tissues (Poston and Purdy, 1986). Any microbial agent applied with the aim of controlling an organism serving as a food source for bird species must be considered as a potential pathogen. Birds are susceptible to infection by all groups of microorganisms (Wobeser, 1981; Edwards and McDonnell, 1982), and the biological similarities of host and potential pathogen or parasite to known pathologies should be an initial consideration in the design of microbial safety tests.

An obvious approach to toxicology and infectivity studies involves the identification of specific species or higher-order taxonomical or ecological groupings most likely to be affected by large-scale application of a given microbial agent. Use of these species for supplementary testing serves the dual purpose of limiting the number of tests while optimizing the utility and reliability of the data for regulatory decisions. This approach has limitations similar to those of the mandated USEPA tests as discussed below.

Historical Perspectives of Avian Microbial Safety Tests

Except for the group of microbial insecticides produced by various serotypes of the bacterium *Bacillus thuringiensis, Bacillus popillae,* and a limited number of baculoviruses, interest in the operational use of microbial agents is a very new phenomenon, and this is reflected in a paucity of references, especially in the public domain, to safety and toxicity evaluations for microorganisms using birds (Ignoffo, 1973; Groner, 1986; Burges, 1981). Available literature is selectively reviewed here to provide examples of approaches taken for specific groups of microorganisms. The relation of these tests to current USEPA requirements are summarized, their strengths and weaknesses will be discussed, and alternative approaches suggested.

Viruses

By far the most investigated group of microorganisms for avian toxicity and infectivity are the viruses, primarily the baculoviruses (Ignoffo, 1973; Groner, 1986), which include the nuclear polyhedrosis (NPV), nonoccluded (NOV), and granulosis (GV) viruses (Bilimoria, 1986). A major characteristic of this group supporting its safety to nontarget organisms is the limitation of the approximately 300 isolates to arthropods (Longworth, 1983; Bilimoria, 1986). This optimism must be tempered by the well-documented ability of birds to ingest and transmit infective virus (Gitay and Polson, 1971; Lautenschlager et al., 1980; Evans, 1986). Ingestion of detectable quantities of virus and failure to inactivate it indicates host-specific pathogenicity but also suggests the possibility of occult, covert, or latent infections. This should be carefully considered in the case of baculoviruses, since detection of polyhedra in tissue using light microscopy is often used as the criterion for infection. Failure to form observable polyhedra in nontarget organisms is not sufficient to prove the absence of virions.

A related complication could possibly arise by the activation by noninfectious NPVs of occult virus present in bird hosts. This phenomenon is well documented among insect baculoviruses (Longworth and Cunningham, 1968; Jurkovicova, 1979; McKinley et al., 1981).

A second argument supporting the use of baculoviruses is the absence of toxin production by this group (Ignoffo, 1973; Groner, 1986); however, this dogma must be considered in light of the recent isolation of a protein toxin from the armyworm, *Pseudaletia unipuncta,* when it is infected with the Hawaiian strain of granulosis virus, which adversely affects the development of an internal parasitiod, *Glyptapanteles militaris* (Hotchkin and Kaya, 1985).

There is also less direct evidence for a viral toxin in armyworms infected with the hypertrophy strain of nuclear polyhedrosis virus (Kaya and Tanada, 1972).

A further complication with viruses arises when considering the bacterium *Clostridium botulinum,* which kills millions of waterfowl each year (Smith, 1976). There are seven types of *C. botulinum,* designated A to G, one of which, type C, is most harmful to birds (Hariharan and Mitchell, 1977). There are two subtypes of this strain, each with different proportions of C1, C2, and D components of the neurotoxin produced by the bacterium. Production of these components is mediated by a bacteriophage; strains lacking the bacteriophage do not synthesize the toxin (Eklund et al., 1971; Inoue and Iida, 1971) and are not lethal to birds. This example of viral-mediated toxicity is not likely to be imitated at random or with high frequency, but serves to indicate the potential for viral toxigenicity to nontarget organisms by indirect and unexpected means.

Ignoffo (1973, 1975), Burges (1981), and more recently Groner (1986) summarize bird toxicity studies using NPVs and GVs; the majority of published studies are based on the standard oral toxicity and intravenous injection tests. Related studies on the passage of NPVs through birds following ingestion of infected insects have demonstrated lack of toxicity or infectivity although the virus remained viable in the feces.

Bacteria

The first insect pathogen registered, circa 1950, was *Bacillus popillae* for control of scarabeid beetles (Burges, 1981). Spores ingested by chickens and starlings remained viable during passage through the gut but did not adversely affect the birds. Subsequent tests have primarily involved various strains of *Bacillus thuringiensis,* which produce a variety of proteinaceous crystalline toxins with varying degrees of host-specific action. Due to the production of these toxins, both infectivity tests and protocols essentially the same as those used for chemicals were used to support registration (Heimpel, 1971; Ignoffo, 1973), i.e., intraperitoneal injection and subacute feeding tests. Since *B. thuringiensis* grows well at 37°C and is related to several *Bacillus* spp. toxic to mammals, initial tests for registration were comprehensive and included attempts to induce mutations leading to mammalian toxicity (Fisher and Rosner, 1959). Subsequent studies revealed the existence of the beta-exotoxin in addition to the delta-endotoxin in some strains of these bacteria. The latter proteins play a major role in insecticidal activity of these isolates, while the former are toxic to mammals parenterally and to birds perorally (Burges, 1981). Strains producing the beta exotoxin are banned from use in the United States.

More recently a strain specific to mosquito and black fly larvae, *B. thuringiensis* var. *israelensis* (Bti), was registered based on standard protocols for microbial agents, including avian oral and injection tests. Subsequent tests have demonstrated broad spectrum mammalian toxicity of the Bti delta-endotoxin (Thomas and Ellar, 1983; Cheung et al., 1985, 1987). Although not tested, similar toxicity to birds is possible. Most Bt toxins apparently exhibit a high degree of host specificity, but the documented mammalian toxicity of the beta-exotoxin and the Bti proteins suggest that the full spectrum of safety evaluations should be undertaken for new strains of this bacterium.

Fungi

The avian toxicity of a limited number of entomopathogenic fungi has been investigated, usually following the USEPA guidelines for acute oral toxicity or monitoring the effects of ingestion of fungal suspensions from drinking water (Ignoffo, 1973; Hartmann et al., 1980; Hartmann and Wasti, 1980; Wasti et al., 1980). Tests have usually involved the toxin-producing deuteromycetes *Metarhizium anisopliae* and *Beauveria bassiana* or other imperfect fungi such as *Nomurea rileyi* and *Paecilomyces fumoso-roseus* which have been considered likely candidates for operational control efforts. More recently tests for an oomycetous fungal pathogen of mosquitoes, *Lagenidium giganteum*, has followed established protocols for avian injection evaluations and modified tests for oral toxicity (Kerwin et al., 1988). Since the infective stage of this fungus is a motile biflagellate zoospore which can only be induced by dilution of the vegetative stage in water, an initial oral dose of the vegetative plus sexual stages administered by gavage was followed by prolonged exposure of the mallard ducks to zoospore suspensions as the sole water source. No adverse effects were noted in treated animals for either series of tests.

Most of the fungi currently under investigation for operational control of insect or plant pests are either deuteromycetes or members of the entmophthorales (Hall, 1981; Ignoffo, 1981; Carruthers and soper, 1987). Many of the former produce toxins and several species of the latter are known mammalian pathogens (Bras et al., 1965; Andrade et al., 1967; Chauhan et al., 1973). Many fungi also produce large numbers of powdery spores which are readily taken into animal lungs; even if the spores do not germinate in lung tissue, the potential for nonspecific allergic response cannot be ignored.

Protozoa

The phylum Protozoa has recently been divided into seven phyla, four of which contain species that parasitize insects (Maddox, 1987). By far the most important of these groups for potential development for biological control are the Microspora. Microsporidia have complex life cycles, often involving obligate alternation of hosts (Sweeney et al., 1985), and have been described from most orders of insects. Little is known beyond descriptive morphology for most species, and there are known mammalian pathogens (Heimpel, 1971). There are also reports of microsporidian infections of birds (Kemp and Kluge, 1975; Novilla and Kwapien, 1978; Randall et al., 1986). The absence of any extensive safety or toxicity data for these unicellular eukaryotes, the lack of understanding of their basic biology, and their parasitism of birds suggests that comprehensive evaluation should be undertaken for any species being considered for operational control.

Specific Recommendations

With the inclusion of an avian respiratory or nasal installation test in the proposed revisions of the USEPA guidelines, the standard protocols for avian testing are greatly improved. The respiratory passages are a common mode of entry for many microbial organisms and provide a more definitive evaluation of potential and likely toxicity to birds than the intraperitoneal and intravenous injection tests previously required. In combination with a logically designed and carefully performed acute oral test and careful attention to any unique biological characteristics of the organisms, these tests may be sufficient to assess possible adverse effects on birds. The following sections discuss possible approaches to optimize the utility of the standard USEPA tests and suggest supplementary tests when adverse effects on birds are encountered or suspected.

Background literature

Familiarity with the types of organisms infecting birds, their modes of transmission, pathology, and etiology allows for rational design and implementation of avian safety tests. The taxonomic and biological similarity of the test organism to any known bird pathogens should be investigated. Background in this area is provided by compendia covering diseases of poultry (Hofstad et al., 1984) and of cage and aviary birds (Petrak, 1982). These comprehensive books cover a broad range of diagnostic techniques and parasite descriptions and can be supplemented by manuals on the identification and isolation of common avian pathogens (Hitchner et al., 1980; Randall, 1985). Reviews on diseases of wild birds (Davis et al., 1971; Wobeser, 1981) provide a broader perspective on the diversity of bird diseases which are likely to be encountered. Finally, an update on new wild bird diseases is provided by the Wildlife Disease Review, a monthly database which searches more than 6000 scientific journals for articles on wildlife pathology. Citations are conveniently separated into major taxonomic groupings of hosts, including birds.

Choice of test species

USEPA protocols for avian tests require the use of either mallard ducks, *Anas platyrhynchos,* or bobwhite quail, *Colinus virginianus,* primarily due to their availability, ease of handling, and the extensive use of these species for chemical toxicity testing. Some type of standard must be established, and for many microbial agents, tailoring the standard protocols to optimize the chance for expression of toxicity or pathogenicity may be sufficient; however, there are numerous examples of differing susceptibility of sympatric species of birds or even subspecies to viruses (Palmer and Trainer, 1971; Karstad, 1971; Wobeser, 1981), bacteria (Burkardt and Page, 1971), and hematozoa (van Riper et al., 1986). Changes required by the adaptation of zoopathogenic fungi to their hosts (Dei Cas and Vernes, 1985) suggest that native hosts will also vary greatly in their response to fungal invasion.

If preliminary tests suggest that a given microbial agent may have adverse effects on avian hosts, evaluation of pathogenicity to species likely to encounter the organism following application should be considered. This will require careful selection of target species based on dietary preferences and behavioral characteristics that will optimize encounters between the pest control agent and different species of birds. These efforts will often require the use of wild birds or their laboratory-reared progeny. Given the high incidence of hematozoa in wild populations (Jones, 1985; Rock, 1984; van Riper et al., 1986) and the possible presence of all major phyla of parasites, this approach requires careful examination of treated and control groups to minimize the effect of cryptic infections on the interpretation of test results.

Criteria for choosing specific species for a given microbial agent should include considerations of economic importance; the possibility of encounters involving endangered species; birds which may be predisposed to infection due to taxonomic affiliation with known hosts or are under stress due to unrelated environmental conditions; species known to dominate sites which are likely to be exposed to the microbial control agent; the availability of sufficient numbers of relatively parasite-free birds; and some consideration of the feasibility of capture, maintenance, and subsequent laboratory handling necessary for reliable safety testing.

Close adherence to recent guidelines covering the use of wild birds in research (Oring et al., 1988) will minimize the number of animals required for this type of investigation. These guidelines include protocols for trapping, marking, transporting, housing, and experimental manipulation of wild birds and provide useful information even for testing involving the more commonly used bird species. Tests using wild species most likely to be affected by a microbial pest control agent provide perhaps the best supplement to the standard protocols.

Administration of the agent

The primary modes of entry of most avian parasites are via the alimentary or respiratory tract (Davis et al., 1971). Ingestion of insects or seeds infected with microbial pest control agents is a likely route of exposure for birds. Infection can be initiated by several major avian viral parasites such as duck plaque virus (Spieker, 1978) and avian influenza virus (Easterday and Tumova, 1972; Webster and Schild, 1978) by ingestion. These two types of virus can also be transmitted by respiratory secretions. Other types of virus that infect birds, such as the well-known arthropod-borne members of the Togoviridae and Bunyaviridae, require an insect vector, and avian poxviruses are transmitted either mechanically by arthropods or skin abrasion (Wobeser, 1981). Most bacteria, fungi, and probably the microsporidia initiate infection of birds following ingestion or following nasal or respiratory exposure. Probable modes of transmission should be considered when designing avian safety tests.

USEPA protocols under the proposed revisions cover the two major routes of entry. The major concern in performing avian safety tests should be careful selection of the stage(s) of the microbial agent to be used and the form in which it is administered. Specifics of any series of tests have to consider any unique properties of the microbial agent which require modification of the standard protocols. For instance, fungi for which zoospores are the infective stage often require extreme dilution in water to induce zoosporogenesis (Kerwin et al., 1988). If a test organism is to be exposed to large quantities of zoospores, the usual methods of administration are precluded due to the large volumes of water needed for induction. For the mosquito parasitic fungus *Lagenidium giganteum,* this problem was obviated by supplying zoospore suspensions to the treated animals as the sole source of water for several weeks (Kerwin et al., 1988).

An obvious alternative or supplementary approach to the acute oral toxicity test is to provide birds with insects or seeds infected with the microbial agent. This usually encourages ready ingestion of the control agent while mimicking a likely mode of field exposure. Careful controls are required as outlined in the previous discussion of cryptic infections in wild birds. Properties of a microbial agent may be altered following infection of target pests and this type of safety evaluation, when combined with a discriminate choice of the bird species to be tested, provides an added dimension to avian tests.

Microbial toxins

A major problem when evaluating microbial agents which produce known or suspected toxins is the lack of any consistent relationship between acute and chronic effects; failure of acute tests to accurately predict higher-level effects has been demonstrated for many chemicals (Kenaga, 1979a, b). This failure can be devastating on wild bird populations for which it has been stated that covert effects that reduce reproductive efficiency by as little as 10 percent could have a much greater (though completely unnoticed) effect relative to sporadic medium-scale epizootic disease outbreaks (Wobeser, 1981).

Circumvention of this problem is difficult and perhaps not practical. Long-term toxicity studies followed through completion of even a single generation, especially when dealing with wild bird species which are more likely to encounter a given microbial agent in the field, are prohibitively expensive, require a great deal of expertise in basic avian sciences, and simply cannot be performed for many bird species. Environmental modification of toxins also occurs, which can either mitigate or amplify adverse effects. Site-specific design of toxicity tests, although a very new area of inquiry, offers a rational but expensive approach to this problem (Levin et al., 1984). Whether this methodology can be incorporated into future avian microbial safety tests awaits further development of this fledgling area of science.

Conclusions

Previous sections of this chapter have presented a discussion of basic avian pathology in relation to the major groups of microbial pest control agents. A general outline of previous avian safety tests using microbial agents included strengths and weaknesses of current and proposed revisions of USEPA standard protocols for avian testing. Rather than attempt to present very specific proposals for all major phyla of microbial agents, discussion was limited to presentation of general concepts integral to the development and implementation of avian safety tests. It is hoped that this information will lead to avian testing procedures which ensure minimal impact of large-scale operational application of all groups of microbial pest control agents on bird populations.

Acknowledgment

Support for this chapter was provided in part by NIH grant 5 RO1 AI22993-02 and 2 RO1 AI22993-04A3.

References

Anderson, R. M., and R. M. May. 1979. Population biology of infectious disease. *Nature* (London) 280:361–367, 455–461.

Anderson, R. M., and R. M. May. 1981. The population dynamics of microparasites and their invertebrate hosts. *Phil. Trans. R. Soc. Lond.* B291:451–524.

Anonymous. 1982. Pesticide assessment guidelines, Subdivision E, Hazard evaluation: Wildlife and aquatic organisms. USEPA publ. no. PB83-153908. National Technical Information Service, Springfield, Va.

Andrade, A. S., L. A. Prula, I. A. Sherlock, and A. W. Cheever. 1967. Nasal glaucoma caused by *Entomophthora coronata*. *Am. J. Trop. Med. Hyg.* 16:31–33.

Behnke, J. M. 1987. Evasion of immunity by nematode parasites causing chronic infections. *Adv. Parasitol.* 26:1–71.

Bilimoria, S. C. 1986. Taxonomy and identification of baculoviruses. In R. R. Granados and B. A. Federici (eds.), *The Biology of Baculoviruses, Vol. 2, Biological Properties and Molecular Biology*. CRC Press, Boca Raton, Fla., pp. 37–59.

Bras, C., C. O. Gordon, C. M. Emmons, K. M. Pendergast, and M. Sugar. 1965. A case of phymocosis observed in Jamaica: Infection with *Entomophthora coronata*. *Am. J. Trop. Med. Hyg.* 14:144–145.

Brown, A. W. A. 1978. *Ecology of Pesticides*. Wiley, New York, pp. 218–269.

Burges, H. D. 1981. Safety, safety testing and quality control of microbial pesticides. In H. D. Burges (ed.), *Microbial Control of Pests and Plant Diseases 1970–1980*. Academic Press, New York, pp. 218–269.

Burkhart, R. L., and L. A. Page. 1971. Chlamydiosis (ornithosis-psittacosis). In J. W. Davis, R. C. Anderson, L. Karstad, and D. O. Trainer (eds.), *Infectious and Parasitic Diseases of Wild Birds*. Iowa State University Press, Ames, Iowa, pp. 118–140.

Carruthers, R. I., and R. S. Soper. 1987. Fungal diseases. In J. R. Fuxa and Y. Tanada (eds.), *Epizootiology of Insect Diseases*. Wiley, New York, pp. 357–416.

Chauhan, H. N., G. L. Sharma, D. S. Kaira, F. C. Malhotra, and M. P. Kapur. 1973. A fatal cutaneous granuloma in a mare. *Vet. Rec.* 92:425–427.

Cheung, P. Y. K., R. M. Roe, B. D. Hammock, C. L. Judson, and M. A. Montague. 1985. The apparent in vivo neuromuscular effects of the delta-endotoxin of *Bacillus thuringiensis* var. *israelensis* in mice and insects of four orders. *Pestic. Biochem. Physiol.* 23:85–94.

Cheung, P. Y. K., D. Buster, B. D. Hammock, R. M. Roe, and A. R. Alford. 1987. *Bacillus thuringiensis* var. *israelensis* delta-endotoxin: Evidence of neurotoxic action. *Pestic. Biochem. Physiol.* 27:42–49.

Davies, D. L., and P. L. Goldsmith (eds.). 1984. *Statistical Methods in Research and Production*, 4th ed. Longman Group, London.

Davis, J. W., R. C. Anderson, L. Karstad, and D. O. Trainer (eds.), 1971. *Infectious and Parasitic Diseases of Wild Birds*. Iowa State University Press, Ames, Iowa.

Dei Cas, E., and A. Vernes. 1985. Parasitic adaptation of pathogenic fungi to mammalian hosts. *CRC Crit. Rev. Microbiol.* 13:173–217.

Dodson, A. P., and P. J. Hudson. 1986. Parasites, disease and the structure of ecological communities. *Tr. Ecol. Evol.* 1:58–63.

Easterday, B. C., and B. Tumova. 1972. Avian influenza. In M. S. Hofstad, B. W. Calnek, C. F. Helmboldt, W. M. Reid, and H. W. Yoder, Jr. (eds.), *Diseases of Poultry*,

6th ed. Iowa State University Press, Ames, Iowa, pp. 449–482.

Edwards, M. A., and U. McDonnell (eds.). 1982. *Animal Disease in Relation to Animal Conservation.* Symposium of the Zoological Society of London No. 50. Academic Press, New York.

Eklund, M. W., F. T. Poysky, S. M. Reed, and G. A. Smith. 1971. Bacteriophage and the toxicity of *Clostridium botulinum* type C. *Science* 172:480–482.

Evans, H. F. 1986. Ecology and epizootiology of baculoviruses. In R. R. Granados and B. A. Federici (eds.), *The Biology of Baculoviruses, Vol. 2, Practical Application for Insect Control.* CRC Press, Boca Raton, Fla., pp. 89–132.

Fairbrother, A., P. Buchholz, and M. D. Knittel. 1988. Validated protocols for pathogenicity testing of viruses in quail. EPA/ERL/Corvallis, Oregon in-house report.

Fisher, R., and L. Rosner. 1959. Toxicology of the microbial insecticide, Thuricide. *Agric. Food Chem.* 7:686–688.

Gitay, H., and A. Polson. 1971. Isolation of a granulosis virus from *Heliothis armigera* and its persistence in avian faeces. *J. Invertebr. Pathol.* 17:288–290.

Gleick, J. 1987. *Chaos: Making a New Science.* Viking, New York.

Groner, A. 1986. Specificity and safety of baculoviruses. In R. R. Granados and B. A. Federici (eds.), *The Biology of baculoviruses, Vol. 1, Biological Properties and Molecular Biology.* CRC Press, Boca Raton, Fla., pp. 177–202.

Hall, R. A. 1981. The fungus *Verticillium lecanii* as a microbial insecticide against aphids and scales. In H. D. Burges (ed.), *Microbial Control of Pests and Plant Diseases 1970–1980.* Academic Press, New York, pp. 483–498.

Harihaten, H., and W. R. Mitchell. 1977. Type C botulism: The agent, host spectrum and environment. *Vet. Bull.* 47:95–103.

Hartmann, G. C., S. S. Wasti, and D. L. Hendrickson. 1979. Murine safety of two species of entomogenous fungi, *Coryceps militaris* (Fries) Link and *Paecilomyces fumosoroseus* (Wize) Brown and Smith. *Appl. Ent. Zoo.* 14:217–220.

Hartmann, G. C., and S. S. Wasti. 1980. Avian safety of three species of entomogenous fungi. *Comp. Physiol. Ecol.* 5:242–245.

Heath, R. G., J. W. Spann, E. F. Hill, and J. F. Kreitzer. 1972. Comparative dietary toxicities of pesticides to birds. Special Scientific Report—Wildlife, No. 152, U.S. Fish and Wildlife Service, Washington, D.C.

Heimpel, A. M. 1971. Safety of insect pathogens for man and invertebrates. In H. D. Burges and N. W. Hussey (eds.), *Microbial Control of Insects and Mites.* Academic Press, New York, pp. 469–489.

Hill, E. F., R. G. Heath, J. W. Spann, and J. D. Williams. 1975. Lethal dietary toxicities of environmental pollutants to birds. Special Scientific Report—Wildlife, No. 191, U. S. Fish and Wildlife Service, Washington, D.C.

Hitchner, S. B., C. H. Domermuth, H. G. Purchase, and J. E. Williams (eds.). 1980. *Isolation and Identification of Avian Pathogens.* American Association of Avian Pathologists, College Station, Tex.

Hofstad, M. S., H. J. Barnes, B. W. Calnek, W. M. Reid, and H. W. Yoder, Jr. (eds.). 1984. *Diseases of Poultry,* 8th ed. Iowa State University Press, Ames, Iowa.

Hotchkin, P. G., and H. K. Kaya. 1985. Isolation of an agent affecting the development of an internal parasitoid. *Arch. Biochem. Physiol.* 2:375–384.

Ignoffo, C. M. 1973. Effects of entomopathogens on vertebrates. Ann. N.Y. Acad. Sci. 217:141–164.

Ignoffo, C. M. 1975. Evaluation of in vivo specificity of insect viruses. In M. Summers, R. Engler, L. A. Falcon, and P. V. Vail (eds.), *Baculoviruses for Insect Pest Control: Safety Considerations.* American Society for Microbiology, Washington, D.C., pp. 52–ff.

Inoue, K., and H. Iida. 1971. Phage conversion of toxigenicity in *Clostridium botulinum* types C and D. *Japan J. Med. Sci. Biol.* 24:53–56.

Jackson, J. A. 1979. Insectivorous birds and North American forest ecosystems. In J. G. Dickson, R. N. Conner, R. R. Fleet, J. C. Knoll, and J. A. Jackson (eds.), *The role of Insectivorous Birds in Forest Ecosystems.* Academic Press, New York, pp. 1–7.

Jones, H. I. 1985. Hematozoa from montane forest birds in Papua New Guinea. *J. Wildlife Dis.* 21:7–10.

Jurkovicova, M. 1979. Activation of latent virus infections in larvae of *Adoxophyes orana* (Lepidoptera: Tortricidae) and *Barathra brassicae* (Lepidoptera: Noctuidae) by foreign polyhedra. *J. Invertebr. Pathol.* 34:213–223.

Karstad, L. 1971. Arboviruses. In J. W. Davis, R. C. Anderson, L. Karstad, and D. O. Trainer (eds.), *Infectious and Parasitic Diseases of Wild Birds.* Iowa State University Press, Ames, Iowa, pp. 17–21.

Kaya, H. K., and Y. Tanada. 1972. Response of *Apanteles militaris* to a toxin produced in a granulosis-virus-infected host. *J. Invertebr. Pathol.* 19:1–17.

Kemp, R. L., and J. P. Kluge. 1975. *Encephalitozoon* sp. in the blue-masked lovebird, *Agapornis personata* (Reichenow): First confirmed report of the microsporidian infection in birds. *J. Protozool.* 22:489–491.

Kenaga, E. E. (ed.). 1979a. Avian and mammalian wildlife toxicology. ASTM Special Technical Publication No. 693. ASTM, Philadelphia, Pa.

Kenaga, E. E. 1979b. Acute test organisms and methods useful for assessment of chronic toxicity of chemicals. In K. L. Dickson, A. W. Maki, and J. Cairns, Jr. (eds.), *Analyzing the Hazard Evaluation Process.* American Fisheries Society, Washington, D.C., pp. 101–111.

Kenaga, E. E. 1979c. Acute and chronic toxicity of 75 pesticides to various animal species. *Down Earth* 35:25–32.

Kerwin, J. L., D. A. Dritz, and R. K. Washino. 1988. Nonmammalian safety tests for *Lagenidium giganteum* (Oomycetes: Lagenidiales). *J. Econ. Entomol.* 81:158–171.

Knittel, M. D. 1988a. Interim protocol for acute oral exposure of avian species to microbial pest control agents. EPA/ERL/Corvallis, Oreg., in-house report.

Knittel, M. D. 1988b. Interim protocol for intravenous exposure of avian species to microbial pest control agents. EPA/ERL/Corvallis, Oreg., in-house report.

Knittel, M. D., and A. Fairbrother. 1988. Interim protocol for nasal/respiratory exposure of avian species to microbial pest control agents. EPA/ERL/Corvallis, Oreg., in-house report.

Lamb, D. W., and E. E. Kenaga. (eds.). 1981. Avian and mammalian wildlife toxicology: Second conference. ASTM Special Technical Publication No. 754. ASTM, Philadelphia, Pa.

Lautenschlager, R. A., J. D. Podgwaite, and D. E. Watson. 1980. Natural occurrence of the nucleopolyhedrosis virus of the gypsy moth, *Lymantria dispar* (Lep.: Lymantriidae) in wild birds and mammals. *Entomophaga* 25:261–267.

Levin, S. A., K. D. Kimball, W. H. McDowell, and S. F. Kimball. 1984. New perspectives in ecotoxicology. *Environ. Manag.* 8:375–422.

Lilienfield, A. M., and D. E. Lilienfield. 1980. *Foundations of Epidemiology.* Oxford University Press, New York.

Longworth, J. L. 1983. Current problems in insect virus taxonomy. In R. E. F. Matthews (ed.), *A Critical Appraisal of Virus Taxonomy.* CRC Press, Boca Raton, Fla., pp. 123–138.

Longworth, J. F., and J. C. Cunningham. 1968. The activation of occult nuclear polyhedrosis viruses by foreign nuclear polyhedra. *J. Invertebr. Pathol.* 10:361–367.

Maddox, J. V. 1987. Protozoan diseases. In J. R. Fuxa and Y. Tanada (eds.), *Epizootiology of Insect Diseases.* Wiley, New York.

May, R. M. 1986. When two and two do not make four: Nonlinear phenomena in ecology. *Proc. R. Soc. London* B228:241–266.

McKinley, D. L., D. A. Brown, C. C. Payne, and K. A. Harrap. 1981. Cross-infectivity and activation studies with four baculoviruses. *Entomophaga* 26:79–84.

Novilla, M. N., and R. P. Kwapien. 1978. Microsporidian infection in the pied peach-faced lovebird (*Agapornis roseicollis*) *Avian Dis.* 22:198–204.

Ogilvie, B. M., and R. J. M. Wilson. 1977. Evasion of the immune response by parasites. *Br. Med. J.* 32:177–181.

Oring, L. W., K. P. Able, D. W. Anderson, L. F. Baptista, J. C. Barlow, A. S. Gaunt, F. B. Gill, and J. C. Wingfield. 1988. Guidelines for use of wild birds in research. *Auk* 105: In press.

Palmer, S. F., and D. O. Trainer. 1971. Newcastle disease. In J. W. Davis, R. C. Anderson, L. Karstad, and D. O. Trainer (eds.), *Infectious and Parasitic Diseases of Wild Birds.* Iowa State University Press, Ames, Iowa, pp. 3–16.

Parkhouse, R. M. E. (ed.). 1984. Parasite evasion of the immune response (Symposia of the British Society for Parasitology, 21). *Parasitology* 88:571–682.

Petrak, M. L. 1982. *Diseases of Cage and Aviary Birds*, 2d ed. Lea & Febiger, Philadelphia.

Poston, T. M., and R. Purdy (eds.). 1986. Aquatic toxicology and environmental fate: Ninth volume. ASTM Technical Publication 921. ASTM, Philadelphia.

Randall, C. J. 1985. *Diseases of the Domestic Fowl and Turkey*. Iowa State University Press, Ames, Iowa.

Randall, C. J., S. Lees, R. J. Higgins, and N. H. Harcourt-Brown. 1986. Microsporidian infection in lovebirds (*Agapornis* spp). *Avian Path.* 15:223–231.

Rock, M. L. 1984. A preliminary survey of avian haematozoa at the Wildlife Rehabilitation Service in Fairfield, California. *Wildl. Rehab. Council J.* 7:13–15.

Rollinson, D., and R. M. Anderson. 1985. *Ecology and Genetics of Host-Parasite Interactions.* Academic Press, New York.

Schaeffer, W. M., and M. Kot. 1986. Chaos in ecological systems: The coals that Newcastle forgot. *Tr. Ecol. Evol.* 1:58–63.

Schlesselman, J. J. 1982. *Case-Control Studies—Design, Conduct, Analysis.* Oxford University Press, New York.

Smith, G. R. 1982. Botulism in waterfowl. In M. A. Edwards and U. McDonnell (eds.), *Animal Disease in Relation to Animal Conservation.* Symposium of the Zoological Society of London, Vol. 50. Academic Press, New York.

Sokal, R. R., and F. J. Rohlf. 1981. *Biometry—The Principles and Practice of Statistics in Biological Research,* 2d ed. W. H. Freeman, New York.

Speiker, J. O. 1978. Virulence assay and other studies of six North American strains of duck plague virus tested in wild and domestic waterfowl. University of Wisconsin-Madison, Madison, Wis. Dissertation.

Sweeney, A. W., E. I. Hazard, and M. F. Graham. 1985. Intermediate host for an *Amblyospora* sp. (Microspora) infecting the mosquito *Culex annulirostris. J. Invertebr. Pathol.* 46:98–102.

Thomas, W. E., and D. J. Ellar. 1983. *Bacillus thuringiensis* var. *israelensis* crystal delta-endotoxin: Effects on insect and mammalian cells in vitro and in vivo. *J. Cell Sci.* 60:181–197.

van Riper III, C., S. G. van Riper, M. L. Goff, and M. Laird. 1986. The epizootiology and ecological significance of malaria in Hawaiian land birds. *Ecol. Monogr.* 56:327–344.

Wasti, S. S., G. C. Hartmann, and A. J. Rousseau. 1980. Gypsy moth mycoses by two species of entomogenous fungi and an assessment of their avian toxicity. *Parasitology* 80:419–424.

Webster, R. G., and G. C. Schild. 1978. Summary of the international workshop on the ecology of influenza viruses. *J. Infect. Dis.* 138:110–113.

Wobeser, G. A. 1981. *Diseases of Wild Waterfowl.* Plenum, New York.

Chapter 38

Testing the Effects of Microbial Pest Control Agents on Mammals

Joel P. Siegel
John A. Shadduck

Introduction

Historical overview

The development and use of entomopathogens as insecticides has increased, as chemical insecticides have become less popular due to problems associated with the resistance of target insect species, nontarget organism toxicity, and environmental persistence. Entomopathogens, often referred to as microbial pest control agents (MPCAs), share areas of mammalian safety concern with their chemical counterparts, such as acute toxicity, irritancy, and allergenicity, but are unique because they can multiply in suitable hosts. Historically, MPCAs were subjected to chemical safety tests as well as infectivity studies, but the testing scheme was revised when it became apparent that certain chemical safety tests such as carcinogenicity studies were not applicable (Ignoffo, 1973; Shadduck, 1983). A need was recognized for tests that dealt with the unique safety challenges posed by living organisms. This chapter will first discuss the general issues surrounding infection, acute toxicity of MPCAs, and animal procurement and housing, then address specific mammalian safety tests. This chapter will not discuss methods to determine the toxicity of exotoxins, as these are readily handled by standard chemical toxicology guidelines.

Evaluating the safety of MPCAs differs from chemical toxicants, because while chemical safety tests assume that a measurable biological effect can be achieved if the dose of the chemical administered is high enough, it is often impossible to produce mortality in mammals using MPCAs. In order to produce death by conventional routes of exposure such as oral, dermal and aerosol administration, massive quantities of material are needed and death results from suffocation or blockage of the gastrointestinal tract. This problem of producing measurable effects by MPCAs without using massive doses of test material has been met by incorporating invasive administration procedures such as intravenous, intraperitoneal, and intracerebral injection into the mammalian safety protocols, so that the dose of test material can be lowered.

Conduct of tests

Prior to testing, a literature search is conducted to find records of isolation of the MPCA or species closely related to it from mammals, and the tissues that the microorganism was recovered from are then included in the testing scheme. An extensive literature review for mammalian isolates of a candidate organism may also give insight into the animal species to include in these tests. A literature search on microsporidia would reveal that the microsporidium *Encephalitozoon cuniculi* is a pathogen of rabbits and primates and causes brain lesions. This information suggests that the rabbit may be a good test animal for the MPCA, and that the brain would be an excellent organ to target in the testing scheme, in order to maximize the opportunity to observe deleterious effects caused by the MPCA.

In practice, current mammalian safety testing protocols for MPCAs combine elements of conventional toxicology such as oral administration with more invasive routes of exposure such as intravenous and intraperitoneal injection. The emphasis in tier 1 of the U.S. Pesticide Assessment Guidelines, Subdivision M, is on short-term (4 weeks or less) acute tests (single administration), and the prolonged exposure tests of tiers 2 and 3 in Subdivision M are reserved for difficulties that arise during tier 1 testing (Anonymous, 1988). Immunocompromised animals have been used in tier 1 tests as well, and their inclusion is compatible with the philosophy of seeking the most sensitive system possible to evaluate an MPCA. The pros and cons concerning the use of immunocompromised animals will be described in more detail later in this chapter.

General Procedures

Infectivity studies

The question of mammalian infection by microbial insecticides has been sidestepped in the Environmental Protection Agency (EPA) guidelines, which fail to provide an operational definition for this term. Historically, there have been two schools of thought concerning the meaning of infection, with each school's definition having a different implication for the relationship between microorganism and host. One school defines infection as the simple presence of microorganisms within living tissue and notes that their presence does not inevitably lead to damage. In this view, all animals are colonized shortly after birth by myriad microorganisms, which are restricted by host defenses to areas where they can be tolerated, such as the gastrointestinal tract or the upper respiratory tract (Davis et al., 1973; Kissane, 1977; Weinstein and Swartz, 1974). The equilibrium between the host and its flora is maintained by a variety of tissue structures and defense mechanisms. Infectious disease results when microorganisms penetrate these defenses and disrupt the status quo. The term *pathogenicity* denotes the intrinsic capability of a microorganism to penetrate host defenses, and the term *virulence* refers to the speed by which this is accomplished (Davis et al., 1973).

The second school of thought inextricably links infection to disease by defining infection as occurring when living agents "enter an animal body and set up a disturbance" (Bruner and Gillespie, 1973) either through multiplication in tissue, production of toxins, or both (Apperly, 1951; Boyd, 1977). From the viewpoint of safety testing, the second definition of infection, which links the presence of a microorganism to tissue damage, is the more useful one. Safety tests by design introduce MPCAs into the host through a variety of routes. Thus, transient disturbances in the normal flora should be expected, and recovery of some portion of the inoculum from host tissue may occur over a variable length of time, depending on the route of administration and the magnitude of the dose. Under these circumstances, concluding that an MPCA is infectious merely because it is recovered is not particularly enlightening.

In this chapter, infection in mammalian safety studies is considered established when there is evidence of multiplication of the MPCA in mammalian tissue coupled with tissue damage. Evidence of multiplication includes a measurable increase in the total amount of MPCA recovered above the amount administered, recovery of vegetative stages of an MPCA when the inoculum consisted solely of spores or conidia, and/or failure of the inoculum to clear over time. Infection cannot be determined solely on the basis of lesions, since the injection of foreign material can elicit an inflammatory response (Siegel et al., 1987; Siegel and Shadduck, 1987). For example, MPCAs produced in insects may contain insect fragments as contaminants in the test preparation, which can then produce inflammation independent of the MPCA.

There is an additional need for a term to describe the ability of an MPCA to remain viable in mammalian tissue without multiplying, and in this chapter the term *persistence* will be used for this situation. The concept of persistence is important because even under ideal circumstances, clearance of foreign objects from mammals is not instantaneous. Adlersberg et al. (1969) found that radioiodinated latex particles injected intravenously in mice could be recovered as long as 6 months after administration, and that the particles were redistributed during this period between the spleen, lungs, and liver. Unlike latex, microorganisms are biodegradable so that their residence time in host tissue may be shorter. However, one would also expect that environmentally resistant life stages of microorganisms such as spores may not lose their viability in mammalian tissue and may be recovered for some period of time. This has been demonstrated in laboratory studies, where *Bacillus thuringiensis* spp. *israelensis* was recovered from the spleens of mice 7 weeks after injection and *Nosema locustae* spores were observed in rabbit tissue 42 days after injection, although their viability was not determined (Siegel et al., 1987; Siegel and Shadduck, 1990). It may be difficult to distinguish between persistence and multiplication, when recovery methods for MPCAs, such as plating organ homogenates, serve as the sole basis of recovery. These simple plating methods cannot determine the identity of the life stages recovered. Hadley et al. (1987) used heat treatment at 65°C combined with plating to differentiate between spores and vegetative stages of *B. thuringiensis,* and this approach may be applied to other MPCAs as well. Additionally, histological techniques may be used to determine if infection occurred, through examination of tissue for the presence of vegetative stages of a MPCA.

The use of immunocompromised animals may be a valuable component of infectivity studies for the following reasons. Currently, outpatient treatment of humans undergoing immunosuppressive cancer therapies is increasing in western Europe and in North America, and these individuals may be at risk for infection by MPCAs as these agents become increasingly used in urban settings and recreational areas. Additionally, the prevalence of acquired immunodeficiency syndrome (AIDS) in regions of the world where MPCAs serve as important components of vector control programs may place local populations at risk. Successful clearance of MPCAs from immunocompromised animals is convincing evidence of a candidate organism's lack of infectivity to this population, and an understanding of the clearance dynamics of an MPCA from immunocompromised animals may allow one to predict future hazard of related MPCAs to humans.

These tests are opposed by researchers who argue that both immunocompromised animals and humans will succumb to opportunistic pathogens of vertebrates before MPCAs could cause infection (Burges, 1981). In addition, the various methods of chemical immune suppression, as well as the variety of laboratory animal strains with genetic defects, makes comparison between studies difficult. For example, among its many effects in mice, the immunosuppressive drug cyclophosphamide causes a reduction in both T lymphocytes and B lymphocytes, but affects B lymphocytes to a much greater degree, while glucocorticosteroids deplete T lymphocytes and B lymphocytes but have a greater effect on T lymphocytes (Dumont, 1974; Parillo and Fauci, 1979; Pierson et al., 1976). The use of chemical immunosuppressants such as cyclophosphamide may complicate safety studies because they cause general bone marrow suppression, resulting in lowered resistance to opportunistic infections, and this makes interpretation of mortality difficult. This is illustrated by the work of Pierson et al. (1976), who found that euthymic mice injected intraperitoneally with cyclophosphamide (300 mg/kg) had a tenfold decrease in resistance to *Pseudomonas aeruginosa*. While the arguments against incorporating immunocompromised animals into safety studies are valid, the authors believe that the major argument in favor of the use of immunocompromised animals, which is the increased sensitivity of the testing scheme, justifies their inclusion in safety protocols.

In summary, the issue of infectivity is complicated by the need to distinguish between true infection and persistence, because each category has different safety implications. Infectivity is grounds for outright rejection of a candidate. On the other hand, some persistence for any MPCA can be expected, because clearance is not instantaneous, but the significance of prolonged persistence is best evaluated on a case-by-case basis.

Acute toxicity studies

Unlike infectivity studies, where infection can serve as grounds for immediate rejection of an MPCA, toxicity does not immediately disqualify an organism, depending on the dose at which it occurs and the route of exposure. Historically, toxicity that occurred below 10^6 colony forming units (CFU) per test animal was considered worrisome, and MPCAs that only were lethal to rodents exposed to concentrations greater than 10^6 CFU have been used safely (Ignoffo, 1973). Fisher and Rosner (1959) reported that *B. thuringiensis* was toxic to mice at a level of 3×10^8 CFU when injected intraperitoneally, and that it was toxic to guinea pigs as well at an unspecified concentration. Lamanna and Jones (1963) reported that vegetative stages of *B. thuringiensis* were more toxic than spores, but that the toxicity did not occur below the concentration of 10^7 CFU. To place this toxicity in perspective, Lamanna and Jones also evaluated the toxicity of the virulent mammalian pathogen *B. anthracis* (anthrax), and found an LD50 of three spores per mouse, so that there was a millionfold difference in the toxicity of bacterial MPCAs compared to mammalian pathogens. This vast difference in virulence has been further demonstrated in laboratory studies of *B. thuringiensis* and *B. thuringiensis* spp. *israelensis,* in which intracerebral injection of both bacteria did not kill rats at concentrations below 10^6 CFU (Siegel et al., 1987).

Determining the mammalian dose of an MPCA may be problematic. Ignoffo (1973) suggested testing an MPCA at 10 to 100 times the average field dose per acre, with a conversion ratio of the weight of the test animal to the weight of a human, yet Burges (1981) suggested an upper limit of 10^6 CFU per mouse and 10^7 CFU per rat. Ideally, the principles of safety testing call for the highest practical dose, but it is difficult to establish a limit as individual species of MPCAs differ in their physical characteristics. For example, oospore preparations of *Lagenidium giganteum,* a fungal mosquito pathogen, caused mortality in rats when instilled intratracheally at a concentration of 10^5 oospores in 0.5 mL. The deaths occurred within 3 min and were due to mechanical blockage of the airways, necessitating a 75 percent reduction in the dose to eliminate this effect (Siegel and Shadduck, 1987). This experiment illustrates the fact that whatever the a priori test dose chosen for an MPCA, there may be circumstances that justify reducing the inoculum for a particular route of exposure.

Currently, acute mammalian toxicity is a matter of concern for bacterial and fungal MPCAs, as both groups kill insects with toxins, and these toxins may have vertebrate activity as well. There appears to be a physiological barrier to entomopathogenic bacterial endotoxin activation in mammals, because entomopathogenic bacterial endotoxins require alkaline conditions (pH 9) for activation in the insect gut and mammalian conditions have not been suitable to date. However, endotoxin that is activated in the laboratory and then injected can have mammalian activity. The solubilized delta endotoxin of *B.*

thuringiensis spp. *israelensis* is toxic to mice and is cytolytic to human erythrocytes, mouse fibroblasts, and primary lymphocyte cultures in vitro (Armstrong et al., 1985; Gill et al., 1987; Thomas and Ellar, 1983). Other alkaline solubilized endotoxins such as that of *B. sphaericus* have shown no mammalian activity, so it is difficult to generalize concerning the risk posed to mammals by these endotoxins (Davidson, 1986). The mammalian safety tests currently in use expose mammals to MPCAs by a variety of routes, so that there is ample opportunity to determine if an endotoxin is activated in mammals.

Animal procurement and housing

Animals should be procured from reputable suppliers in order to ensure their overall quality and freedom from disease. Infection by noncandidate microorganisms complicates the interpretation of test results, and stressed animals may respond abnormally when administered the MPCA. Animals must be housed in such a way as to minimize stress and in accordance with National Institutes of Health guidelines (Anonymous, 1985). Special precautions must be taken with immunocompromised animals in order to reduce their risk of infection by microorganisms that are not of interest.

Special consideration must be given to housing rodents. As a general rule, female mice are less likely to fight than males and may be housed without incident, while males not raised as littermates will fight to establish their position in the social hierarchy. Consequently, male mice must be carefully monitored to remove less dominant individuals if they become injured, and one should be aware that since these animals are more stressed than their dominant cage mates, they may react differently to an MPCA. If mice are used, careful monitoring of the animals is necessary after treatment because dead mice may be eaten by their cage mates, which can destroy tissue of interest or otherwise interfere with the experiment.

Consideration must also be given to housing control animals, in order to prevent the controls becoming exposed to the MPCA by coprophagous activity or communal grooming. Coprophagous activity must be taken into consideration for treated animals as well, because if the MPCA is shed in the feces, the ingestion of fecal material can result in renewed exposure to the MPCA. The use of separate cages for control animals is a simple method to minimize their exposure to the MPCA.

In summary, many problems can be avoided by using animals from reputable suppliers. Mammalian species differ in their tolerance to crowding and other stress, and this must be considered when housing animals. The animals' social structure may impose additional stress, and the animals' eating and grooming activities must be considered as well.

Tier 1 Testing

This section will briefly address key issues associated with some of the mammalian safety tests of subdivision M—Tier 1, as well as some drawbacks to the safety protocols as they are currently written. For each major mammalian safety test, the Code of Federal Regulations Heading is given in bold type, and in certain instances, suggestions will be given concerning sample size or additional data that may be generated from these experiments. Where possible, references for the procedures mentioned are provided.

Acute oral toxicity-pathogenicity study (152A-10). The purpose of this test is to determine if the MPCA causes mortality after a single dose received by gavage. This is a standard toxicology protocol and the guidelines state that mice or rats are the preferred animals for use in this test. At least five animals of each group must be exposed to the MPCA, not counting the control group, and control animals dosed with inactivated MPCA may prove useful. Incorporation of these controls is necessary to help determine if any lesions observed in rodent tissue result from the presence of foreign protein rather than infection.

An MPCA that causes mortality in this test is an unlikely candidate for further testing. The emphasis of this test is on toxicity, although infectivity is also evaluated in this test by examination of rodent tissues for lesions or other evidence that the MPCA multiplied. It is unlikely that the MPCA will be found in tissue outside the gastrointestinal tract, although lesions in the gastrointestinal tract or passive transport through the gut may allow the MPCA entry into the bloodstream, where the MPCA may be taken up by phagocytic organs such as the spleen and liver as well as circulate in the peripheral blood.

If there is reason to suspect that the MPCA will disseminate to other organs, it is advisable to periodically kill animals and examine their organs to determine if the MPCA has cleared. The recommended sample size of five animals per group is inadequate to determine the rate of clearance, so that if this determination is of interest, a greater number of animals must be exposed to the MPCA. Clearance of many entomopathogens in mammals is greatest during the first 2 weeks (personal observation), so it is important to concentrate the majority of the sampling in this period in order to accurately determine clearance rate. Sacrifice of three treated animals each at days 1, 4, 7, 17, and 21 after injection should give a meaningful sample size for regression analysis (Zar, 1974). All the tissues stated in the protocol should be examined for the MPCA, and its recovery quantified per unit tissue. Further information on oral administration techniques can be found in Paget and Thomson (1979).

Acute dermal toxicity study (152A-11). The purpose of this test is to determine if a single dermal exposure to the MPCA will result in mortality. This test is a standard toxicology protocol, but its applicability to MPCAs is doubtful, because it is unlikely that an MPCA will prove toxic when applied to unabraded skin, and a candidate MPCA that can do so may be unsuitable for registration. This test can also be used to evaluate the dermal irritancy of the MPCA as well as toxicity, and if this test is conducted, dermal irritancy should also be evaluated in order to maximize the information obtained from this test. It is advisable that the initial treatment groups contain more than five rabbits, so that animals whose skin is irritated or cut when shaved may be dropped from the study before administration of the MPCA. A control group of rabbits receiving inactivated MPCA is recommended, in order to determine if dermal irritancy is associated with viable MPCAs. Further information on evaluating skin irritancy may be found in McCreesh and Steinberg (1983).

Acute pulmonary toxicity-pathogenicity study (152A-12). The purpose of this test is to determine if a single exposure to the MPCA results in acute toxicity or infection. Previously, aerosol exposure was used, but due to uncertainties regarding the amount of material delivered to the test animal, intranasal or intratracheal administration of the MPCA is now recommended. The physical characteristics of the inoculum may cause mortality in this study, due to mechanical obstruction of the airways or foreign body pneumonia. Therefore, a trial study should be undertaken before the main study, in order to determine if the dose and/or method of administration will cause excessive mortality. Treatment groups should include extra animals because of the possibility that administration procedures will cause mortality. If rats are used for intratracheal instillation, personal experience indicates that rats weighing at least 200 g are easier to dose than smaller rats, and additionally, for many strains of rats, females rarely exceed 250 g; this sex-related difference in adult weight should be considered when evaluating weight gain.

As previously stated, it has been the authors' experience when evaluating the clearance of bacterial MPCAs, that most clearance occurred during the first 2 weeks after exposure. It may be beneficial to perform the majority of the sampling for the MPCA during this period in order to accurately determine the clearance rate. One possible sampling scheme that minimizes the number of animals used, involves the sacrifice of at least two animals per group immediately after exposure, as stipulated in the regulations, at 2-day intervals for the following 6-day period, and at 7-day intervals thereafter for the 21-day observation period. Regression analysis can then be used to determine clearance rate. Particular attention should be focused on recovery of the MPCA from the lungs, spleen, and liver, although all organs specified in the guidelines should be tested.

The use of a control group comprising animals exposed to inactivated MPCAs is particularly important in this test, because inflammation may occur in the lungs due to the presence of foreign protein or normal flora from the pharynx introduced into the lungs by the administration process. It is important to determine if the lesions are caused by viable MPCAs or result from the simple presence of foreign protein. Further information on intratracheal instillation can be found in Nicholson and Kinkead (1982).

Acute intravenous toxicity-pathogenicity study (152A-13). The purpose of this test is to determine if a single intravenous injection of the MPCA will result in acute toxicity or infection. Intravenous injection has been used to evaluate chemical toxicants and this test is the most invasive test in tier 1. The physical nature of the inoculum may cause difficulties during administration of the MPCAs, as large particles may clog the hypodermic needle or cause emboli if high pressure is needed for injection. The guidelines state that in some cases, intravenous injection may be impractical and that exemption may be granted, provided that it can be justified. If intravenous injection is unsuitable, intraperitoneal injection may be an acceptable alternative. This test evaluates toxicity as well as infectivity, but a dose of 10^7 MPCAs is unlikely to cause mortality (Shadduck et al., 1980; Shadduck et al., 1982; Siegel et al., 1987; Siegel and Shadduck, 1987). If mortality does result in this test, it may be beneficial to quantify it by conducting an LD50 experiment. There is some ambiguity in the guidelines regarding toxicity, because the suggested test dose is not given on a cfu per kilogram test animal basis. Rats may weigh at least 10 times more than mice and may not be killed by an MPCA dose that is toxic to mice, simply because rats receive a lesser dose unless the weight difference is compensated for. If both rats and mice are tested, rats should receive a 1 log greater dose than mice in order to ensure that they receive comparable exposure.

The EPA guidelines suggest that the observation period last a minimum of 21 days. If the goal of this experiment is to follow the clearance of an MPCA until it can no longer be detected, an observation period of 42 days is a more realistic time frame to evaluate clearance.

As previously stated, the majority of clearance occurs during the first 2 weeks, so most samples should be concentrated in this period for a precise estimate of clearance rate. For example, in a model experiment, the treatment group would contain 24 rodents, with 3 rodents sacrificed at 3-day intervals beginning 1 day after injection for the first 16 days, and on days 22 and 28 thereafter. Emphasis should be placed on quantifying the MPCAs present in the spleen and liver, although other organs specified in the guidelines must be assayed. Quantification is important in order to determine if the numbers of the MPCA recovered are increasing, because a simple system of MPCA "present. absent" $(+, -)$ will not provide this information. The rate of clearance for the MPCA would then be determined by regression analysis (Zar, 1974).

Immunocompromised animals can be incorporated in this test, provided that the inoculum is not contaminated with other microorganisms, and use of these animals can provide insight into the components of the immune system necessary for MPCA clearance. It is critical to determine which life stages of the MPCA are present in the inoculum prior to injection, because the presence of vegetative stages may complicate the determination of infection. Ordinarily, the presence of vegetative stages implies multiplication, but if they were present in the end-use product, then concluding that the MPCA multiplied may be wrong. Caution should be taken in handling tissue samples, because nonvegetative stages of MPCAs such as spores may germinate when an organ they are trapped in such as the liver or spleen is removed from the test animal. This can be prevented by chilling tissue samples before the MPCAs are recovered.

When conducting these tests, one must be aware that mammalian tissue may react to the presence of large quantities of foreign protein, regardless of whether infection occurred. A control group of animals injected with inactivated MPCAs should be included in this experiment, in order to determine if any inflammation or lesions were truly caused by viable MPCAs. These control animals do not have to be killed at intervals and may be sacrificed at the end of the experiment.

The following references may be helpful when assessing the clearance of MPCAs from mammalian tissues: Adlersberg et al., 1969; Pierce et al., 1952.

Primary eye irritation-infection study (152A-14).   The purpose of this test is to determine if a single dose of the MPCA results in ocular irritation or ocular infection. The testing scheme used is a modification of the Draize ocular irritation test and is a standard toxicology procedure. It is important to remember that the physical characteristics of the inoculum may cause irritation in this study (personal observation). Incorporation of inactivated MPCAs into this test may prove invaluable in assessing whether irritation is dependent on MPCA viability. The guidelines mention that a local anesthetic may be used to alleviate pain, but since these anesthetics can reduce the blink response of rabbits, resulting in dry eyes, any irritation caused by the candidate MPCA may be exacerbated. If an anesthetic is used, consideration should be given to moistening the rabbits' eyes after treatment with an ocular lubricant.

The test guidelines state that untreated control rabbits are unnecessary, since the contralateral eye of each rabbit receiving the MPCA can serve as a control. Test material can be transferred from the treated to the untreated eye by activities such as grooming, and this transfer occurred with *B. sphaericus* and *B. thuringiensis* spp. *israelensis* (personal observations). Consideration should be given to having an additional group of untreated rabbits as controls.

The MPCAs are supposed to be enumerated periodically from the treated eyes and periocular structures, in order to give an estimate of clearance during this study. If there is no multiplication of the MPCA in situ, swabbing or flushing the eye may remove the entire inoculum the first time that the eye is sampled. If the minimum group of six rabbits is used, the authors suggest sampling the eyes of two rabbits on days 1, 4, and 7 after treatment, with no treated rabbit eye sampled twice in order to get a valid estimate of the persistence of the MPCA. Previous persistence studies indicated that both *B. sphaericus* and *B. thuringiensis* spp. *israelensis* were recovered for as long as 8 weeks after ocular instillation, so that if assessing the maximum length of persistence is a test goal, more than six rabbits should be treated with the MPCA and the sample period extended to at least 6 weeks.

It is imperative that persistence is not equated with multiplication, due to the reasons discussed earlier in this chapter, and that evidence of multiplication in tissue is necessary to establish infection. Many factors, such as inflammation of the conjunctiva, can affect recovery of MPCAs, so that careful interpretation of persistence data is necessary. Upon termination of the experiment, at a minimum, the conjunctival cul-de-sac and eyelids should be collected for histology in order to check for infection. Further information on the conduct of ocular irritation tests can be found in McDonald et al., 1983.

Hypersensitivity incidents with microbial pest control agents (152A-15)
There are no tests required for assessing hypersensitivity incidents in tier 1. Rather, data on incidents of hypersensitivity, including immediate and delayed reactions in humans and animals that occur during production and testing, should be reported. If hypersensitivity is an issue, it must be evaluated according to the guidelines stated in tier 2.

Conclusions

It is difficult to make sweeping statements concerning individual safety tests, because each candidate MPCA will be evaluated on a case-by-case basis. Clearly, all of the tests in tier 1 are not applicable for every MPCA without some modification, and in some cases, additional tests may be necessary to satisfy public concerns. However, there are two issues, infection and persistence, that lie at the core of these tests. The operating definition of infection must be decided prior to testing, because the definition used will affect the interpretation of test data. If persistence is regarded as synonymous with infection, long-term clearance studies must be undertaken to quantify the length of recovery, and regression analysis is proposed as the method of choice to establish clearance rate. MPCAs recovered should be quantified where possible on a cfu per milligram tissue basis, as well as life stage recovered, in order to determine if multiplication occurred. Ultimately, the issue of the mammalian safety of a candidate MPCA can only be determined when all the test results are in. Even then, it is wise to consider the admonition of H. D. Burges (1981) that "a no risk situation does not exist, certainly not with chemical pesticides, and even with biological agents, one cannot absolutely prove a negative. Registration of a chemical is essentially a statement of usage in which risks are acceptable, and the same must be applied to biological agents."

Acknowledgment

Preparation of this chapter was supported in part by the World Health Organization Special Program for Research and Training in Tropical Diseases. The opinions of the authors do not reflect the endorsement of the World Health Organization.

References

Adlersberg, L., J. M. Singer, and E. Ende. 1969. Redistribution and elimination of intravenously injected latex particles in mice. *J. Reticuloendothel. Soc.* 6:536–560.

Anonymous. 1988. *Toxicology Guidelines for Microbial Pest Control Agents,* Subdivision M. U.S. Environmental Protection Agency. Office of Pesticide and Toxic Substances.

Anonymous. 1985. Guide for the care and use of laboratory animals. NIH Publication No. 85–23. U.S. Dept. Health and Human Services, Washington.

Apperly, F. L. 1951. *Patterns of Disease on a Basis of Physiologic Pathology.* Lippincott, Philadelphia.

Armstrong, J. L., G. F. Rohrman, and G. S. Beaudreau. 1985. Delta endotoxin of *Bacillus thuringiensis* subsp. *israelensis. J. Bacteriol.* 161:39–46.

Bruner, W. B., and J. H. Gillespie. 1973. *Hagan's Infectious Diseases of Domestic Animals.* 6th ed. Cornell University Press, Ithaca, NY.

Boyd, W., and H. Sheldon. 1977. *An Introduction to the Study of Disease,* 7th ed. Lea & Febiger.

Burges, H. D. 1981. Safety, safety testing, and quality control of microbial pesticides. In H. D. Burges (ed.), *Microbial Control of Pests and Plant Diseases.* Academic Press, New York, pp. 738–769.

Davidson, E. W. 1986. Effects of *Bacillus sphaericus* 1593 and 2362 spore/crystal toxin on cultured mosquito cells. *J. Invertebr. Pathol.* 47:21–32.

Davis, B. D., R. Dulbecco, H. N. Eisen, H. S. Ginsberg, and W. B. Wood, Jr. 1973. *Microbiology,* 2d ed. Harper & Row, New York.

Dumont, F. 1974. Destruction and regeneration of lymphocyte populations in the mouse spleen after cyclophosphamide treatment. *Int. Arch. Allergy* 47:110–123.

Einstein, L., and M. N. Swartz. 1974. Pathogenic properties of invading microorganisms. In W. A. Sodeman, Jr., and W. A. Sodeman (eds.), *Pathologic Physiology— Mechanisms of Disease,* 5th ed. Saunders, Philadelphia, pp. 454–471.

Fisher, R., and L. Rosner. 1959. Toxicology of the microbial insecticide, thuricide. *J. Agr. Food chem.* 7:686–688.

Gill, S. S., G. J. P. Singh, and J. M. Hornung. 1987. Cell membrane interaction of *Bacillus thuringiensis* subsp. *israelensis* cytolytic toxins. *Inf. Immunol.* 55:1300–1308.

Hadley, W. M., S. W. Burchiel, T. D. McDowell, J. P. Thilsted, C. M. Hibbs, J. A. Whorton, P. W. Day, M. B. Friedman, and R. E. Stoll. 1987. Five month oral (diet) toxicity/infectivity study of *Bacillus thuringiensis* insecticides in sheep. *Fund. Appl. Toxicol.* 8:236–242.

Ignoffo, C. M. 1973. Effects of entomopathogens on vertebrates. *Ann. N.Y. Acad. Sci.* 217:141–164.

Kissane, J. M. 1977. Bacterial diseases. In W. A. D. Anderson and J. M. Kissane (eds.), *Pathology,* vol. 1. C.V. Mosby, St. Louis, pp. 369–414.

Lamanna, C., and L. Jones. 1963. Lethality for mice of vegetative and spore forms of *Bacillus cereus* and *Bacillus cereus*-like insect pathogens injected intraperitoneally and subcutaneously. *J. Bacteriol.* 85:532–535.

McCreesh, A. H., and M. Steinberg. 1983. Skin irritation testing in animals. In F. N. Marzulli and H. I. Maibach (eds.), *Dermatotoxicity,* 2d ed. Hemisphere Publishing, New York, pp. 147–166.

McDonald, T. O., V. Seabaugh, J. A. Shadduck, and H. F. Edelhauser. 1983. Eye irritation. In F. N. Marzulli and H. I. Maibach (eds.), *Dermatotoxicity,* 2d ed. Hemisphere Publishing, New York, pp. 555–610.

Nicholson, J. W., and E. R. Kinkead. 1982. A simple device for intratracheal injections in rats. *Lab. Anim. Sci.* 32:509–510.

Paget, G. E., and R. Thomson. 1979. *Standard Operating Procedures in Toxicology.* University Park Press, Baltimore.

Parillo, J. E., and A. S. Fauci. 1979. Mechanisms of glucocorticoid action on immune processes. *Annu. Rev. Pharmacol. Toxicol.* 19:179–201.

Pierce, C. H., R. J. Dubos, and W. B. Schaeffer. 1952. Multiplication and survival of tubercle bacilli in the organs of mice. *J. Exp. Med.* 97:189–205.

Pierson, C. A., G. Johnson, and I. Feller. 1976. Effect of cyclophosphamide on the immune response to *Pseudomonas aeruginosa* in mice. *Inf. Immunol.* 14:168–177.

Shadduck, J. A., S. Singer, and S. Lause. 1980. Lack of mammalian pathogenicity of entomicidal isolates of *Bacillus sphaericus. Environ. Entomol.* 9:403–407.

Shadduck, J. A., D. W. Roberts, and S. Lause. 1982. Mammalian safety of *Metarhizium anisopliae:* Preliminary results. *Environ. Entomol.* 11:189–192.

Shadduck, J. A. 1983. Some considerations on the safety evaluation of nonviral microbial pesticides. *Bull. W.H.O.* 61:117–128.

Siegel, J. P., J. A. Shadduck, and J. Szabo. 1987. Safety of the entomopathogen *Bacillus thuringiensis* var. *israelensis* for mammals. *J. Econ. Entomol.* 80:717–723.

Siegel, J. P., and J. A. Shadduck. 1987. Safety of the entomopathogenic fungus *Lagenidium giganteum* to mammals. *J. Econ. Entomol.* 80:994–997.

Siegel, J. P., and J. A. Shadduck. 1990. Safety of microbial insecticides to vertebrates— humans. In M. Laird, L. A. Lacey, and E. W. Davidson (eds.), *Safety of Microbial Insecticides.* CRC Press, Boca Raton, Fla., pp. 101–114.

Thomas, W. E., and D. Ellar. 1983. *Bacillus thuringiensis* var. *israelensis* crystal delta endotoxin: Effect on insect and mammalian cells in vitro and in vivo. *J. Cell Sci.* 60: 181–197.

Zar, J. H. 1974. *Biostatistical Analysis*. Prentice-Hall, Englewood Cliffs, N.J.

Chapter 39

Testing of Microbial Pest Control Agents In Nontarget Insects and Acari

Susan W. Fisher
John D. Briggs

Introduction

The use of microbial agents as substitutes for chemical insecticides has increased dramatically during the last 20 years. Today particular protozoa, bacilli and their products, fungi, viruses, and nematodes are available for use as biological control agents. Furthermore, the array of microorganisms which might enjoy commercial use continues to expand. The expansion has come with the advent and development of recombinant nucleic acid techniques to modify and specifically extend the susceptible host range of pest-mitigating species of microoganisms. With each wild type or genetically modified microorganism which is proposed for release there is the responsibility of ensuring that the release of the microorganisms into the environment will not result in destructive effects to any type of nontarget organisms. Since many but not all of the microorganisms proposed for commercial development are to be used for insect control, it is appropriate to consider the hazards, if any, which may exist for species closely related to the target as well as unrelated nontarget insects and mites. The purpose of this chapter is to review the considerations which must be respected when proposing and planning tests for nontarget insects principally. Further, this chapter is intended to provide recommendations for testing of specific nontarget organisms and the adoption of particular protocols.

Issues in Toxicity Testing of Microbial Pest
Control Agents (MPCA) for Insects

Agent characteristics

Developmental stages of inoculum. Protocols for measuring effects of pathogenic microorganisms on insects are based generally on experience with chemical insecticides. It is an error to assume that an array of isolates of a specific MPCA is a homogeneous series of elements without variation in pathogenicity. Further, pathogens of insects are living organisms with developmental stages. Any one of the stages could be uniquely pathogenic in pests or nontarget organisms. For example, viruses could be used which are occluded within nuclear polyhedral inclusion bodies (PIBs), or as infectious DNA. The spores of fungi (conidia) and protozoa (whether sexually or asexually formed), as well as proliferative (vegetative) forms of fungi and protozoa are all candidates for testing. Among entomopathic bacteria, consideration must be given to the effects of spores, proliferative vegetative cells, and both endotoxins and exotoxins. As a general rule, it is essential to test the form of the pathogen which is toxic to target insects against nontarget organisms. There appears to be a large range of interspecific and intraspecific susceptibility of insects to the different life stages of particular pathogens. This can be understood as we consider the mode of action and pathogenesis of microbial pathogens and the nutritional preference of insect developmental life-forms and the changing susceptibility within and between host life-forms (e.g., larvae vs. adults). For example, Hitchcock et al. (1979) found that either spores or the vegetative cells of *Bacillus pulvifaciens* were pathogenic to 1- to 2-day-old larvae of the honey bee *Apis mellifera*. However, older larvae were susceptible only to spores. In contrast, Lightner et al. (1973) tested shrimp using an entomopathic nuclear polyhedrosis virus (NPV) using two different inocula: nonoccluded virus rods and polyhedra with occluded virus rods. Neither form caused infection of shrimp. The variability in such measurements led Kerwin et al. (1988) to test all developmental stages of the fungus *Lagenidium giganteum* in nontarget insects whenever possible. Clearly, this is the ideal solution. However, with practical limitations of time, labor, availability of inoculum, and hosts, frequently conditions are less than ideal. Testing of the stage which is infectious to the target host must be the principal testing procedure.

Toxins. The analysis of tests using a particular bacteria MPCA can be complicated by the presence of more than a single insecticidal element. Using the entomopathic bacillus *Bacillus thuringiensis,* Burges and Thompson (1971) noted that the purity of preparations of *B. thuringiensis* spore and toxic crystal delta-endotoxin affected activity. In addition, the insecticidal activity can also vary depending on the kind(s) of toxin(s) produced. The amount of toxin elaborated by toxigenic strains can vary with the chemical nature of the medium (Rogoff et al., 1969). As a matter of principle, the toxin free of spores, in addition to with spores should be tested. Further, the procedures for obtaining and handling inoculum for testing must follow standardized protocols.

An entomopathic bacterium or fungus can secrete toxic substances including enzymes into the medium in which the microorganisms are grown. In cases where the relative contribution of insecticidal principals are in question, nontarget species should be exposed to the growth medium at different stages of microbial growth to estimate toxicity with and without the known active elements. This procedure is particularly important when the growth medium may be recovered in the final formulated product.

Quantification. Quantification of inoculum, regardless of pathogen type or life stage used, is of paramount importance in every protocol. Dulmage et al. (1971) confirmed that for preparations of *B. thuringiensis,* spore count alone is not a valid indication of pathogenicity. The authors wisely proposed the use of International Units for activity. Earlier, Ignoffo (1964) argued that occlusion body counts ("polyhedra" and "granules") do not measure fully the insecticidal activity of virus inocula. Variation in activity was attributed to the presence of nonoccluded virus particles, possible contaminants, the number of virus particles within a polyhedron, and the size and chemical composition of occlusion bodies particularly polyhedra. Methods for standardizing activity include cell or particle counts, serological analyses of preparations, plaque assays, and quantitative bioassays. For all protocols, quantification of microbial agent activity for particular life stages used should be confirmed by more than one method and compared to known and accepted standards (Hoskins and Craig, 1962). The early acceptance of standardization of *B. thuringiensis* products internationally (Dulmage et al., 1971) established an important precedent for all MPCA.

Route of exposure to inoculum. Both target and nontarget insects might be exposed to a pathogen in the laboratory or in the field by different routes, e.g., exclusively by ingestion, topical, through the respiratory system or in combinations of these routes. Burges and Thompson (1971) argued that the route of exposure for nontarget organisms in screening tests must be those that are expected in the field. Thus, the focus for testing for viruses, bacteria, and protozoa should be ingestion. For fungi, topical application of nontarget subjects to integument or by respiratory routes are appropriate and analogous to penetration of target insects by mycelia through the cuticle. On the other hand, Kurstak et al. (1978) considered it inappropriate in safety testing to limit routes of exposure to those which may occur in nature. For example, the toxicity or pathogenicity of an organism which can only occur by injection into a nontarget species could be examined. However, experimental controls should include a saprophytic organism from the habitat of the nontarget species to detect facultative pathogenic action to compare the action of an infectious biological control agent. Because of the labor-intensive nature of injections process and the liability to the organisms involved, it is most appropriate to include for tests using mammals and in those nontarget insect species and forms for which there is some reasonable suspicion of toxicity, e.g., in species phylogenetically related to the target species. In all other cases, routes of exposure to the pathogen can be limited to ingestion and surface contact exposure of the nontarget species in a nutrient support medium or a simulation of the habitat of the nontarget organisms.

When testing is restricted to introduction of an insect pathogen into a nutrient substrate for nontarget organisms, additional issues must be considered. For instance, if the pathogen is incorporated into the diet, examinations must demonstrate that the pathogen remains active after incorporation to detect whether the pathogen multiplies in the substrate (Ignoffo and Gard, 1970). These needs can be met by bioassay of the medium or substrate using target species (Couch and Martin, 1984). In addition, a measurement of the amount of food consumed is essential for an accurate estimate of dosage when the pathogen is added to a nutrient substrate. This is essential for pathogenic agents which paralyze the alimentary system and prevent further feeding. The assay of microbial pathogenicity by incorporation into a diet can be complicated if unconsumed food is not removed. The latter may serve as substrate for microbial growth which may be toxic (Couch and Martin, 1984).

Formulation. The validity of safety testing of insect pathogens for nontarget insects depends on the identity and insecticidal quality of the inoculum. A body of literature demonstrates the variability in results of bioassay with the use of different commercial formulations of entomopathogens. In the case of *B. thuringiensis,* for example, formulations possessing *B. thuringiensis* beta-exotoxin can be more toxic than preparations containing only the delta-endotoxin (Dubois, 1986; Moar et al., 1986; Morris, 1988). These results are due to an additive or synergistic effect of the two toxins in certain insects. In addition, the "inert" ingredients in a formulation can possess toxic properties (WHO, 1972). These observations underscore the need for rigorous quality control for the nature of additives in formulations including screening assays for individual additives. Further, quality control is necessary on the active ingredients in MPCA preparations, both unformulated and formulated with respect to insecticidal action.

Characteristics of test insects

The physiological similarity of individuals in a population of test subjects is of fundamental importance in the measurement of safety of pathogens in nontarget species. Whereas the genetic similarity may be a practical assumption, the state of health of individuals with respect to parasitization can be dissimilar within the population. Characteristics of individuals is a particularly pressing matter in the testing of nontarget insects because many existing protocols call for the use of field-collected insects or other organisms which can vary considerably in the number of available subjects in age categories of the same developmental stage, their individual health, physiological state of susceptibility, and genotype and phenotype. In the following paragraphs, parameters of the host insect which are known to affect the outcome of toxicity tests are discussed as a preface to recommendations for standardized protocols.

Age and life stage of insect.   There is a generalization implied or stated in the invertebrate pathology literature that young individual insects within a life-form (stage) are more susceptible to pathogens than are the older specimens of the life-form of the same species (USEPA, 1975; Canning, 1970; Kurstak et al., 1978; Dulmage et al., 1971). However, there are important exceptions reported to this generalization. For instance, Ignoffo (1964) found that 75 percent of all mortality of larval *Trichoplusia ni* exposed to an NPV was in the third or fourth instar. More advanced larval instars lacked susceptibility. Similarly, Hitchcock et al. (1979) reported that susceptibility of the larvae of *Apis mellifera* to *B. pulvifaciens* decreased dramatically with increasing larval age. In contrast, all postembryonic life stages of mosquito larvae, pupae, and adults were reported to be susceptible to parasitization by the microsporidian *Nosema algerae* (Undeen and Maddox, 1973). These investigations suggest that all life stages of nontarget species should be screened for sensitivity to a pathogen to identify one stage of development as a sentinel form. In cases where different multiple life-forms of the nontarget insect cannot be used, emphasis should be placed on an immature form early in development, and/or on the developmental stage which has the highest probability of exposure to the microbial control agent. This is a particularly important consideration in studies when a biological agent is incorporated into the diet. The nature and quantity of diet ingested and the nature and quantity of inoculum ingested can vary between instars and life-forms in a life cycle (Burges and Thompson, 1971).

Sex of insect

The relative susceptibility of individual insects to chemical insecticides is, in part, a function of the sex of individuals. The levels of certain enzymes can vary between the sexes, thus affecting the rates of detoxification and ultimately the susceptibility of each sex to the insecticide (El Aziz et al., 1969). Other physiological differences between male and female insects can alter susceptibility to pathogens. A potentially important difference is cuticular composition. In those cases where penetration of the pathogen through the integument or fore or hindgut is a necessary step for an insecticidal activity, any differences arising in cuticular composition between the sexes can make a difference in the data from toxicity tests. Such differences have been observed (Hoskins and Craig, 1962). Thus, animals of one sex, probably female, should be used as test subjects whenever possible.

Host habitat

In deciding what insect species should be used to construct safety protocols, it is necessary to perform preliminary screening of a wide variety of species, especially those which are likely to be exposed (Ignoffo, 1973). Virtually no part of a given environment can escape exposure to a released insect pathogenic microorganism. The agents may be subject to air and water movements, on inert particulates, or passed through the alimentary system of predators (Bergold, 1963). Consequently, insects drawn from both terrestrial and aquatic habitats should be included in safety evaluations. In addition, a variety of abiotic environmental factors, e.g., moisture, heat, organic constituents, and other organisms in the community, can affect pathogen stability and availability. Both terrestrial and aquatic tests should be included for those insects which occupy a range of habitats in environments exposed to the MPCA.

Host behavior

Beneficial hosts. Nontarget insects which are considered to be beneficial aesthetically, economically, or ecologically should be tested for susceptibility to pathogens. The choice of beneficial species should be broadened further to include species which are active in different ecological niches and with characteristic behavior. For instance, predatory insects are likely to be exposed to pathogens through ingestion of contaminated prey in addition to substrate contamination. Parasitoid species will be exposed in a contaminated habitat or in an infected host, and could perish when the host dies or as the direct result of the biological agent. The latter is a substantially different route than exposure outside of a host, i.e., by oral and substrate contamination. Both predators and parasitoids must be included in safety protocols if populations are subject to exposure. In addition, there is the possibility of living infected parasitoids, or topically contaminated parasitoids, infecting hosts through oviposition. The possibility of this action of the parasitoid should be evaluated (Andreadis, 1980).

Injurious hosts. The evidence that a pathogen released to mitigate one target also can have activity against another injurious yet nontarget insect may be perceived as a beneficial phenomenon. There can be unexpected consequences. For instance, if a pathogen can multiply within or upon a nontarget susceptible host, whether injurious or not, the level of inoculum available to additional hosts in the ecosystem can increase. Exposure of phytophagous insects to contaminated plant material is an important component of safety testing. It is a conspicuous contrast to consumption of contaminated prey or a diet serving as substitute substrate because the interactions between the waxy cuticle of plants and the pathogen may influence availability of the pathogen. The contamination of nutritional substrate is a route of exposure which is common in the field with microbial pesticide agents.

Characteristics of tests

Nature of experimental controls. The number and the nature of experimental control organisms varies considerably in published protocols. In some cases, the control organisms are not treated, particularly if formulations are used (Morris, 1988). In others, controls received a dose of heat or otherwise attenuated pathogenic agents (Couch et al., 1975; Couch and Martin, 1984). In general, it is preferred to administer a dose of inactive inoculum to controls. This is particularly important in aquatic systems where an increase in the turbidity of the water with higher concentrations of an inoculum can cause mortality for which the pathogenicity of the microbial control agent is not responsible. The effects of the action of the active principal(s) of the agent must be ascertained by using properly designed controls. In addition, when inoculum is incorporated into diet, it is essential to demonstrate that pathogenicity of the active agent is maintained in the diet for the target species (Couch and Martin, 1984).

Number of subjects per dose of inoculum.   The use of MPCA in toxicity assays can be more complicated than the testing of insecticidal chemicals. With the former, in addition to the risk of toxic effects of formulation elements other than the active principal, or contamination by other species, there is the high probability of the presence of additional life stages of a pathogen, with differences in pathogenicity (see Developmental Stages of Inoculum and Toxins in this section). In addition, other factors must be considered, e.g., reduced fecundity and teratologies, which can require new protocols rather than conventional chemical insecticide bioassay techniques. In deciding how many animals and the age of the subjects to be employed in MPCA bioassays, the endpoint which is expected to be achieved must be identified and defined. If mortality is the desired effect, the number of animals used in a protocol is generally less than when another endpoint is expected. The use of a greater number of individuals and more than one sex and life-form can be necessary when developmental or reproductive effects are selected as endpoints (see Endpoints in this section).

Nontarget insects which are considered to be closely related to the target species can be included to further quantitate mortality or other effects. Mortality is the common endpoint in protocols for testing nontarget insects. Dulmage et al. (1971) recommends the use of 50 subjects at each level of inoculum and several replicates of each exposure when recognizing death as an endpoint. Other investigators have recommended using 15 to 30 animals per dose of inoculum (Ignoffo, 1964; Rogoff et al., 1969; Gardner et al., 1982; Kerwin et al., 1988). The number of animals to be used must be judged by the reproducibility of the results. The 95 percent fiducial limits should fall within 30 percent of the LD_{50} with control mortality not exceeding 4 percent (Dulmage et al., 1971).

When parasitism or substantial genetic or physiological variability (e.g., age, enzootic disease) occurs between individuals, as might be expected for field-collected specimens, a larger number of individuals per dose must be used to obtain results within an acceptable range of probability. As a generality, 30 individuals per dose with two to three replications is a reasonable starting point in a screening procedure to estimate the best range for dosages.

If mortality is not the endpoint, or if mortality data appear not to be dose-responsive, additional procedures which require different numbers of animals can be used. When death is not the result in an appropriate time, the MPCA is obviously not toxic to a nontarget insect, or it is of low toxicity. Either can be characteristic of pathogens with extended incubation periods. In such cases, the endpoint may be an effect which can be quantified, for example, reduced fecundity in surviving adults, or no effect. In either instance the results will be expressed as effect (or lack of it) at the highest dose tested. Because a graduated series of doses is not always necessary for the endpoints other than death, the commitment of animals would appear to be smaller. An initial sample size of 50 animals per microbial agent is likely to be adequate in any case. However, projecting and assuring an expected sample size several weeks hence can require a larger initial population to accommodate mortality during the maintenance period. The assay results may not be dose-responsive. Departure from dose response can occur when mortality is dependent upon the chance that the test subject will obtain the active agent associated with food (Hughes et al., 1984). In the cases depending on food encounter, bioassays employing a large number of animals per dose can be conducted. However, analysis of the data must be conducted according to a procedure other than conventional probit analysis, e.g., using the "one-hit" Poisson model.

Endpoints. As discussed above, a variety of endpoints should be tentatively expected, ranging from no effect (Gardner et al., 1982), mortality of individuals, or to population effects. For nontarget insects, mortality is an immediate concern. For screening of several species, particularly those which are phylogenetically related to the target species and in the same location, communities, and ecosystems, mortality may not provide sufficient data to be a reliable indicator of effect (Rogoff et al., 1969). Developmental and reproductive endpoints are particularly important in populations of universally accepted beneficial insects, e.g., the honey bee (*Apis mellifera*). Protocols are available describing the measurement of the effects of MPCA on larval honey bee longevity, adult survival, and foraging behavior (Morton et al., 1975; Davidson et al., 1977; Hitchcock et al., 1979; Kerwin et al., 1988). The benign survival and/or reproduction of an MPCA in the body of nontarget insects with subsequent dispersal of the agents must be considered. All assays, should include histological subculturing of the agents or serological examinations to estimate the presence and longevity of the active agent in nontarget insect species (Ignoffo, 1973; Carey et al. 1978; Gardner et al., 1982; Kerwin et al., 1988). If the agent is present, the consequences can be the contamination of additional trophic levels in host-prey interaction. There is currently no acceptable protocol to determine whether or not the microorganisms found in tissues are part of the original dose which was applied, or whether it represents reproduction of the organisms in the inoculum. Tests to discriminate the latter are not recommended as part of general screening protocols. The organism is either present or absent.

Length of test periods. The length of time for which tests are conducted varies considerably according to reports in the literature. Generally, the length of acute toxicity tests are relatively short, ranging from several hours to 30 days (Ignoffo, 1964; Lightner et al., 1973; Gardner et al., 1982; Kerwin et al., 1988). Tests conducted on host species for which there is not mortality but which may involve developmental or reproductive effects are generally conducted for a minimum of 15 to 30 days and extending to at least one entire life cycle. Many such tests also include an acclimation period of the test subjects in test units before inoculation. This is a consideration which is especially important for field-collected specimens (Couch and Martin, 1984). For safety testing of MPCA for nontarget insects, it is recommended that a test period of at least 30 days be used, or sufficient time to complete the life cycle, to assure that immature forms will complete several stages in their development. Many MPCA invade target insects through the gut. The cuticular lining of the foregut and hind gut is shed during a molt. Nontarget insects which do not appear to be susceptible to pathogen penetration of the fore or hind gut can possibly be invaded during the molt of the cuticle. Further, the half-life of midgut cells can be a few hours to several days, depending on the life stage of development in a species.

Testing for residual activity. The persistence of many microbial agents in the environment can rival that of the most persistent chlorinated organic compounds. For example, viruses occluded in polyhedra remain active for up to 20 years in a variety of physical conditions (Bergold, 1963). When the residual activity of each MPCA is determined by bioassay, the longevity of different microorganisms in particular circumstances must be recognized. These residual data can be used to modify recommendations with respect to the length of time for which tests should be conducted on nontarget insects.

Environmental factors. Various physical environmental factors have been shown to affect the stability and pathogenicity of MPCA (Undeen and Maddox, 1973). Dulmage et al. (1971) reported that the response of target insects to pathogens was temperature dependent. Ganieva (1978) demonstrated that the increased susceptibility of the silkworm (*Bombyx mori*) to an NPV was correlated positively with temperature. Further, *B. thuringiensis* was more toxic to the Bertha armyworm at 25 than 20°C (Morris, 1988). The absolute humidity of assay units and media, measured with or without changes in temperature, has been found to profoundly alter the toxicity or pathogenicity of *Pseudomonas* spp. in *Oncopeltus fasciatus,* the fungus *Hirsutella thompsonii* in the two-spotted spider mite, and *B. thuringiensis* in lepidopterous larvae (Dorn, 1977; Dulmage et al., 1971). These variables must be controlled carefully and specified in standard protocols if the results of tests are to be comparable. In general, it is recommended that conditions favorable to the nontarget host be selected as test parameters. It is a useful alternative to also include in the protocol extreme conditions which can place both the target and the nontarget insects under stress and consequently facilitate invasion of the pathogen.

Biotic factors, particularly target insect host density and saprophytic microorganisms in the test units, can affect microbial populations and their insecticidal activity. It is a valid premise that infectious microorganisms in insect populations act as density-dependent mortality factors. With increasing numbers of susceptible test subjects, the number of foci for developing additional inoculum are increased. In a field situation the age structure of the population of target insects at risk will provide a spectrum of age groups with varying susceptibility to the pathogens, which can provide additional inoculum to infect additional hosts. The effect in the field is that of temporarily extending the activity of the original inoculum. This is true of both invasive microorganisms which subsequently develop additional infectious entities (spores or conidia) or for microorganisms which are toxigenic. A bioassay in the laboratory is an incomplete reflection of field impact if a single age class of experimental subjects is used. However, the single age class of insects susceptible to an agent is a valid response to questions of infectivity, and dosage mortality data.

The addition of any items or materials to the screening and assay units containing the subject insect and infectious agent combinations can serve to introduce contaminating microorganisms to the units. Microorganisms which behave saprophytically can degrade nitrogenous toxins using the toxins as a nitrogen source. A specific case is *Bacillus thuringiensis israelensis* in laboratory tests. Without the presence of contaminating microorganisms from soil, sod, or water, the inoculum can retain the toxic activity of the delta-endotoxin for 48 h. Assays in water to which unsterilized soil has been added were without insecticidal activity within 6 h (Singer, 1976). In contrast to the *B. thuringiensis israelensis* toxin exposed unprotected in the water, the insecticidal toxin of *B. sphaericus* is protected within the peripheral region of the spore. In assays, *B. sphaericus* retains insecticidal activities in water for periods exceeding 72 h.

Additional factors influencing the activity of microorganisms in assay must be anticipated. For example, the density of the target insect population in assay units and the possible association of other organisms, e.g., other insects, fish, or amphibians, can reduce the sedimentation rate of suspended particulates. The particulates may be spores or toxic entities, a formulation of insecticidal microorganisms, or inert sediment with adsorbed microorganisms. The influence of both abiotic and biotic factors in laboratory assays may or may not reflect the impact of the complex communities encountered in the field.

Environmental media. The physical, abiotic components of a nontarget insect bioassay are important if the inoculum is incorporated into the medium or substrate, or if the nutrient substrate supports and sustains the test subject. In bioassays of insecticidal activity for a terrestrial habitat, the medium can be either natural (a soil) or a substitute, e.g., a piece of filter paper, or a prepared diet. Demonstration of pathogen viability and/or toxin stability on or within the medium is essential if the MPCA is to be incorporated into a medium. In addition, information is essential on how the medium affects the availability and persistence of the MPCA. For instance, if an organic soil is used, adsorption of *B. thuringiensis* particulates, spores, and delta-endotoxin crystalline structures can occur (Ohana et al., 1987). The effect of adsorption is the reduction of the availability of toxic elements from the test subjects and also removes toxic elements from the effects of degradative elements in the environment. The consequence is an apparent increase in the longevity of the MPCA and perhaps also its effects. The corollary is that a perturbation of the soil system could cause release of toxins and active entities which would increase any hazard associated with their presence. In contrast to a soil in organic adsorption sites, the result would be different in a sandy medium where adsorptive surfaces would be reduced. A substitute nutritive medium for assays should have a defined composition and the adsorptive activity of a natural substrate should be recognized.

Aquatic systems provide large variations in test condition. If a sediment is employed, its impact must be considered as noted above. A compelling argument can be made for inclusion of a sediment in tests using organisms in water. As a generalization, all aquatic systems possess some suspended or sedimented particulates. The nature of the sediment is important, as is the quality of water for aquatic tests. The water medium can vary in pH, mineral content, and oxygenation, all of which have been shown to affect bioassays of chemical insecticides (McEwen and Stephenson, 1979). If a natural source is used, analyses are important to estimate the effects of water quality on the toxicity and pathogenicity of MPCA (Couch et al., 1975; Benz, 1975). One approach is to employ standard reference water (distilled and deionized water) to which known quantities of minerals, buffers, and specified organic materials are added. The standard reference waters described by USEPA (1975) can be pH adjusted and adapted for freshwater or marine environments. Use of standard reference waters is recommended.

One final consideration in aquatic tests is to select either static or flowthrough (dynamic) systems. Both types have been employed in the testing of MPCA (Ignoffo et al., 1973; Lightner et al., 1973; USEPA, 1975; Van Essen and Anthony, 1976; Couch and Martin, 1984; Kerwin et al., 1988). Static tests, in which the pathogen is added directly to the water and held for the duration of the test period, are the less complicated to execute. However, with increasing turbidity due to the excretion of test subjects or growth of a microbial agent, coincidental oxygen depletion can result in mortality not due directly to the microbial agent. Growth of microorganisms occurs in all systems, particularly static, whether agitated, aerated, or still. In static designs, heat-attenuated or a nontoxic inoculum must be used in treating controls to assess these variables. If a test is conducted for more than 5 days, a static renewal procedure should be employed. The treated animals are removed from the original contaminated water and put into a newly contaminated system. If long-term tests are employed, a flowthrough system should be used which dilutes the impact of variables. Essentially it is necessary to prevent the accumulation of excretory products from test subject and minimum microbial growth which may accentuate the effect of the agent assayed.

Environmental perturbation. As noted in Environmental Media in this section, a disturbance in the environment can result in the sudden, mass release of adsorbed materials. This event can dramatically increase effects on target organisms and increases possible hazards to susceptible species. This has been demonstrated most convincingly with *B. thuringiensis* and the delta-endotoxin (Ohana et al., 1987; Sheeran and Fisher, unpublished data, 1988). The release of adsorbed toxic elements is a phenomenon which can easily be simulated in state aquatic bioassays and should be investigated if there is evidence of acute toxicity to nontarget insects in bioassays.

Dose of inoculum. As a generalization, for nontarget insects, the MPCA will fail to be acutely toxic at the dose which kills the target host. The question then arises as to what should be the upper limit of inoculum for exposure of suspected non-susceptible hosts. Ignoffo (1973) used the concept of "adjusted host equivalents" to determine reasonable dosage levels. Others have elected to exceed the LD_{50} or LD_{90} dose in the target host by factors of 100 times or 1000 times. The assumption is that failure to show toxicity at those levels was an adequate demonstration of safety (Kerwin et al., 1988). Use of the USEPA "maximum hazard estimation" (USEPA Subdivision M, In Press) is considered superior and is recommended for adoption.

Host Species Recommended for Use in Nontarget Insect MPCA Safety Tests

Based on the considerations elaborated in the Introduction, the following list of recommended nontarget species is assembled.

1. Terrestrial Nontarget Insects
   a. Beneficial species—Honey bee (*Apis mellifera*) adults and larvae. Endpoints = mortality, adult longevity, and changes in foraging behavior.
   b. Predator—Ladybird beetle (family Coccinellidae)—adult. Endpoint = mortality.
   c. Parasitoid—*Trichogramma* spp., larvae in hosts. Endpoints = mortality in hosts and adult emergence.
   d. Phylogenetically related hosts
      (1) Sibling species
      (2) Intrageneric or familial species

Species and endpoints selected will vary with the habitats of the target host. A phytophagous insect in addition to the honey bee should be included in the terrestrial category.

2. Aquatic Nontarget Insects
   a. Predator—Hemiptera (Notonecta) adults and dragonflies (Odonata) nymphs. Endpoints = mortality, growth (of nymphs).
   b. Saprophytic—Midges (*Chrionomus* spp.)—fourth instar larvae. Endpoints = mortality, change in behavior.
   c. Phylogenetically related hosts

Recommended Protocols for Testing Nontarget Species

Terrestrial insects

Honey bee, *Apis mellifera* (Hymenoptera), larvae and adults. A protocol for testing the susceptibility of toxicity and pathogenicity to *A. mellifera* larvae is described by Hitchcock et al. (1979). The protocol for larvae is compelling because it (1) incorporates a very large number of test organisms of the same age class, (2) uses lethal as well as behavioral endpoints, (3) confirms a positive effect by examining tissues for evidence of the pathogen, and (4) examines the toxicity of the microbiological agent for more than one life stage or life-form of the insect.

Only minor modifications of the published protocol are recommended. These include using a topical application of inoculum as well as oral administration, and the use of heat-inactivated aquatic suspensions of inoculum for treatment of controls rather than administration of sterile water only.

Adult *A. mellifera* should be tested in the manner described by Kerwin et al. (1988). Advantages of this method include (1) testing of more than one life stage of the microbial agent, (2) demonstration of the presence of the agent in treated animals, (3) careful monitoring of the consumption of the microorganism, (4) avoid bacterial contamination by changing water supply, (5) a relatively long 30-day test period, and (6) an evaluation of adult longevity. The protocol could be further improved by adding a topical treatment and using inactivated inoculum in the experimental control bees.

Ladybird beetles (Coccinellidae/Coleoptera).   The protocols described in the USEPA Interim Protocol are recommended for use (Anonymous, 1988a). Three modes of application (contact, ingestion, and topical) are recommended. Contact toxicity (method 1) is advised because it is simple to execute and does not require expensive supplies or sophisticated equipment. The protocol is excellent in that it uses multiple life stages, heat-killed inoculum, and accurately assesses a variety of endpoints. Sample sizes are realistic. However, the presence of the pathogen must be demonstrated in the animals which exhibit a positive effect.

Trichogramma **(Trichogrammidae/Hymenoptera).**   The general protocol described in the USEPA Interim Protocol is recommended for use (Anonymous, 1988b). However, the number of animals to be used must not be fewer than 120. Demonstration that the pathogen is present in the affected animals is necessary.

Phylogenetically related hosts.   These will vary with the host and cannot be stated a priori.

Aquatic insects

Predator.   Protocols for testing dragonfly nymphs (Odonata) and Notonecta adults (Hemiptera) are described by Kerwin et al. (1988). These are recommended for use without modification.

Immature forms.   The midge, *Chironomus riparius,* is a cosmopolitan species in many habitats. The larvae are important components of food chains. The use of the midge is recommended. The procedure described for testing MPCA against mosquito larvae by Kerwin et al. (1988) provides a basic framework for testing *C. riparius.* In addition, the behavioral assay described by Fisher and Lohner (1986) should be used.

References

Andreadis, T. G. 1980. *Nosema pyrausta* infection in *Microgentrus grandii,* a Braconid parasite of the European corn borer, *Ostrinia nubilalis. J. Invert. Pathol.* 35:229–233.

Anonymous. 1988a. Interim protocol for assessing the impact of microbial pest control agents on predacious coccinellids (family Coccinellidae). USEPA, unpublished.

Anonymous. 1988b. Interim protocol for testing the effects of microbial pathogens on parasitoid insects. USEPA, unpublished.

Benz, G. 1975. Action of *Bacillus thuringiensis* preparation against larch bud moth, *Zeiraphera diniana* (Gn.), enhanced by β-exotoxin and DDT. *Experimentia* 31:1288–1290.

Bergold, G. 1963. The nature of nuclear polyhedrosis viruses. In E. A. Steinhaus (ed.), *Insect Pathology, An Advanced Treatise,* Vol. 1, Academic Press, N.Y., pp. 413–456.

Burges, H. D., and E. M. Thompson. 1971. Standardization and assay of microbial insecticides. In H. D. Burges and N. W. Hussey (eds.), *Microbial Control of Insects and Mites.* Academic Press, New York, pp. 591–621.

Canning, E. U. 1970. Transmission of microsporida. In *Proceedings of the 4th International Colloquium on Insect Pathology.* College Park, Maryland, pp. 415–424.

Carey, D., K. A. Harrap, T. W. Tinsley, and J. S. Robertson. 1978. Safety tests on the nuclear polyhedrosis viruses of *Spodoptera littoralis* and *Spodoptera exempta,* In *Progress in Invertebrate Pathology.* Proc. Int'l. Colloq. Invert. Path. XIth Ann. Mtg. Soc. Invert. Pathol., Prague, pp. 29–33.

Couch, J. A., and S. M. Martin. 1984. A simple system for the preliminary evaluation of infectivity and pathogenesis of insect virus in a nontarget estuarine shrimp. *J. Invert. Pathol.* 43:351–357.

Couch, J. A., M. D. Summers, and L. Courtney. 1975. Environmental significance of baculovirus infections in estuarine and marine shrimp. In L. A. Bulla and T. C. Cheng (eds.), *Pathobiology of Invertebrate Vectors of Disease.* Annals NY Acad. Sci. 266:528–536.

Davidson, E. W., H. L. Morron, J. O. Moffett and S. Singer. 1977. Effect of *Bacillus sphaericus* strain SSII-1 on honey bees, *Apis mellifera. J. Invert. Pathol.* 29:344–346.

Dorn, A. 1977. Studies on the fat body of *Oncopeltus fasciatus* invaded by *Pseudomonas aeruginosa. J. Invert. Pathol.* 29:347–353.

Dubois, N. R. 1986. Synergism between β-exotoxin and *Bacillus thuringiensis* subspecies kurstaki (HD-1) in gypsy moth, *Lymantria dispar* larvae. *J. Invert. Pathol.* 48:146–151.

Dulmage, H. T., O. P. Boening, C. S. Rehnborg and G. D. Hansen. 1971. A proposed standardized bioassay for formulations of *Bacillus thuringiensis* based on the international unit. *J. Invert. Pathol.* 18:240–245.

El-Aziz, S. A., R. L. Metcalf and T. F. Fukuto. 1969. Physiological factors influencing the toxicity of carbamate insecticides to insects. *J. Econ. Ent.* 62:318.

Fisher, S. W., and T. W. Lohner. 1986. Studies on the environmental fate of carbaryl as a function of pH. *Arch. Environ. Contam. Toxicol.* 15:661–667.

Ganieva, M. R. 1978. The role of ecological factors in the nuclear polyhedrosis infections in the silkworm. In *Progress in Invertebrate Pathology.* Proc. Int'l. Colloq. Invert. Pathol. and XIth Ann. Mtg. Soc. Invert. Pathol., Prague, pp. 71–72.

Gardner, W. A., R. D. Oetting, and G. K. Storey. 1982. Susceptibility of the two-spotted spider mite, *Tetranychus urticae* Kich, to the fungal pathogen *Hirsutella thompsonii* Fisher. *Florida Entomol.* 65:458–465.

Hitchcock, J. D., A. Stoner, W. T. Wilson, and D. M. Menapace. 1979. Pathogenicity of *Bacillus pulvifaciens* to honey bee larvae of various ages (Hymenoptera: Apidae). *J. Kansas Ent. Soc.* 52:238–246.

Hoskins, W. M., and R. Craig. 1962. Uses of bioassay in entomology. *Ann. Rev. Entomol.* 7:437–480.

Hughes, P. R., H. A. Wood, J. P. Burand, and R. R. Granados. 1984. Quantification of the dose mortality response of *Trichoplusia ni*, *Heliothis zea*, and *Spodoptera frugiperda* to nuclear polyhedrosis viruses: Applicability of an exponential model. *J. Invert. Pathol.* 43:343–350.

Ignoffo, C. M. 1964. Bioassay technique and pathogenicity of a nuclear polyhedrosis virus of the cabbage looper, *Trichoplusia ni* (Hübner). *J. Insect Pathol.* 6:237–245.

Ignoffo, C. M. 1973. Effects of entomopathogens on vertebrates. In L. A. Bulla, Jr. (ed.), *Regulation of Insect Populations by Microorganisms*. *Ann. N.Y. Acad. Sci.* 217:141–172.

Ignoffo, C. M., K. D. Biever, W. W. Johnson, H. D. Sanders, H. C. Chapman, J. J. Petersen, and D. B. Woodard. 1973. Susceptibility of aquatic vertebrates and invertebrates to the infective stage of the mosquito nematode *Reesimermis nielseni*. *Mosquito News* 33:599–602.

Ignoffo, C. M., and I. Gard. 1970. Use of an agar-base diet and house fly larvae to assay β-exotoxin activity of *Bacillus thuringiensis*. *J. Econ. Entomol.* 63:1287–1289.

Kerwin, J. L., D. A. Dritz, and R. K. Washino. 1988. Non-mammalian safety tests for *Lagenidium giganteum* (Domycetes: Lagenidiates). *J. Econ. Entomol.* 81:158–171.

Kurstak, E., P. Tijssen, and K. Maramorosch. 1978. Safety considerations and development problems make an ecological approach of biocontrol by viral insecticides imperative. In E. Kurstak and K. Maramorosch (eds.), *Viruses and Environment*. Academic Press, New York, pp. 571–592.

Lightner, D. U., R. R. Proctor, A. K. Sparks, J. R. Adams, and A. M. Heimpel. 1973. Testing penaeid shrimp for susceptibility to an insect nuclear polyhedrosis virus. *Environ. Entomol.* 2:611–613.

McEwen, F. L., and J. Stephenson. 1979. *Use and Significance of Pesticides in the Environment*. Wiley-Interscience, New York.

Moar, W. J., W. L. A. Oslerink, and J. T. Tramble. 1986. Potential of *Bacillus thuringiensis* var. Kurstaki with thuringiensin on beet armyworm. *J. Econ. Entomol.* 79:1443–1446.

Morris, O. M. 1988. Comparative toxicity of delta-endotoxin and thuringiensin of Bacillus thuringiensin and mixtures of the two for the bertha armyworm (Lepidoptera: Noctuidae). *J. Econ. Entomol.* 81:135–141.

Morton, H. L., J. O. Moffett, and F. D. Stewart. 1975. Effect of alfalfa looper nuclear polyhedrosis virus on honey bees. *J. Invert. Pathol.* 26:139–140.

Ohana, B., J. Margalit, and Z. Barak. 1987. Fate of *Bacillus thuringiensis* subsp. *israelensis* under simulated field conditions. *App. Environ. Microbiol.* 53:828–831.

Rogoff, M. H., C. M. Ignoffo, S. Singer, I. Gard, and A. P. Prieto. 1969. Insecticidal activity of thirty-one strains of *Bacillus* against five insect species. *J. Invert. Pathol.* 14:122–129.

Undeen, A. H., and J. V. Maddox. 1973. The infection of non-mosquito hosts by injection with spores of the microsporidian *Nosema algerae*. *J. Invert. Pathol.* 22:258–265.

USEPA. 1975. Methods for acute toxicity tests with fish, macroinvertebrates and amphibians. USEPA Ecol. Res. Series. EPA 660/3-75-009.

USEPA. 1982. Revision of subdivision M of FIFRA pesticide assessment guidelines. NTIS No. EPA-540/9-82-028. October, in press.

Van Essen, F. W., and D. W. Anthony. 1976. Susceptibility of nontarget organisms to *Nosema algerae* (Microsporida: Nosematidae), a parasite of mosquitoes. *J. Invert. Pathol.* 28:77–85.

WHO. 1972. The use of viruses for the control of insect pests and disease vectors. Rept. Joint Mtg. FAO/WHO Meeting on Insect Viruses, Wld. Hlth. Org. Tech. Rep. Ser., No. 531, 48 pp.

Part 6

Decontamination and Mitigation

40

Overview: Confinement, Decontamination, and Mitigation

Anne K. Vidaver

Guenther Stotzky

Introduction

The planned introduction into the environment of microorganisms, especially genetically modified microorganisms, needs to be considered in terms of both their effectiveness and their safety. There has been much speculation about "runaway" microorganisms, especially those produced by recombinant DNA techniques. There has been no such incident, although there have been few introductions of "real" genetically modified microorganisms, that is, those with insertions that are expressed in the environment. Similarly, there is no record of runaway parent or nonengineered microorganisms that have been used in the environment, in both research and the marketplace. Nevertheless, there is a need to examine the bases and possibilities for containment (which is really "confinement"—see below), decontamination, and mitigation, in the event of a runaway microorganism. A runaway microorganism is one that produces or induces undesirable effects in or upon the physical or biological environment. Such effects may range from benign to hazardous, and they need a spatial and a temporal component to describe them. The following are recognized at the outset: absolute confinement is not feasible and, based on experience with both beneficial and detrimental microorganisms, not essential for initial, small-scale evaluative purposes. Total decontamination and eradication of harmful organisms are also rare, but acceptable levels of decontamination and eradication (i.e., below economic or damaging threshold levels) have been possible, effective, and economically feasible with some pathogens of humans, other animals, and plants. Mitigation, defined as the alleviation of deleterious effects caused by a runaway microorganism or by the methods used for its decontamination and eradication, is also not absolute. The body of knowledge about the confinement, decontamination, and mitigation of recognized harmful microorganisms provides the basis for this discussion.

The major purposes of this chapter are to consider the decontamination and potential eradication of deleterious microorganisms introduced into the environment and the mitigation of their effects. The concept of confinement is consistent with these aspects. It is more appropriate to use the term "confinement" rather than "containment" for microorganisms released outside of laboratories or other "contained" facilities. Contained facilities retain microorganisms inside them, albeit not always perfectly. Confinement means to keep an organism within bounds and to restrict its movement, dissemination, and spread. Although confinement does not mean that microorganisms cannot spread from the point of application, it implies that such spread can be effectively managed and minimized. Confinement practices have been the basis for both research on and commercial uses of microorganisms, plants, and animals in the environment for over a century.

The effects of naturally occurring microorganisms in the environment, including those causing disease or other deleterious effects, can seldom be predicted with certainty. Nevertheless, there appears to be no evidence for anthropogenic epidemics or deleterious effects that have resulted from either experimental investigations with microorganisms in the environment or their large-scale use. However, the large-scale commercial use of certain plants and animals can sometimes result in outbreaks of disease attributable to a number of factors, such as an increase in the numbers of naturally occurring deleterious microorganisms to a critical concentration (i.e., inoculum potential); inadvertent selection of a plant or animal that is susceptible to a microorganism previously unrecognized as a problem; mutation of a relatively benign microorganism to a pathogenic stage; and changes in the environment, either natural or anthropogenic, that are conducive to the establishment and spread of undesirable microorganisms.

The effects of deleterious microorganisms are better known and understood than those of beneficial organisms in the environment. Although theoretically, "anything" can happen, the displacement of or the addition to indigenous microorganisms in a natural habitat or one managed for a useful purpose has usually been either transient or unsuccessful. It may be many years before any effects are detected, and outbreaks of disease can occur in virtually any managed or natural ecosystem. Nevertheless, even pathogens introduced into the environment for the purpose of assessing resistance of plants to particular microorganisms do not normally persist without the presence of a susceptible plant. Plant breeding relies heavily on such testing. Similarly, introduced microorganisms beneficial to humans seldom persist in effective numbers for long periods of time, e.g., for more than one growing season. Yet, it is precisely these beneficial microorganisms that are of greatest interest. Paradoxically, therefore, existing knowledge of deleterious microorganisms is the primary guide for predicting the effects of introduced beneficial and useful microorganisms.

Confinement of microorganisms in research situations is well documented for wild-type and even for marked or genetically modified microorganisms (e.g., 27 and Chap. 49), both free-living and those associated with plants and animals. The literature of applied and environmental microbiology since the 1900s indicates that when appropriate experimental design is used (1, 8, 9, 13, 17, 28), no deleterious effects outside such confined areas have been reported, even when pathogens have been tested within the areas.

Methods for decontaminating the environment from microorganisms have been studied and analyzed (1, 3, 5, 10, 12, 15, 19, 22). These methods for decontamination are applicable to both research and commercial applications (Table 40.1). The use of any method depends on the need for alleviating a problem, the feasibility, and the cost effectiveness. Decontamination sometimes can be absolute (i.e., eradication); e.g., microorganisms associated with plants and animals can be destroyed by burying and/or incineration. Incineration of diseased animals is a common procedure with wild animals, such as with migratory birds afflicted with fowl cholera, and domestic animals. However, decontamination is usually sufficient when the numbers of microorganisms are below a level that either produce or could produce an undesirable effect. These threshold levels of effects are usually considered in terms of costs and benefits, whether to humans, commerce, or the environment (7, 11). Aesthetic values may also be considered, such as eutrophication of recreational waters by blue-green algae. The majority of microorganisms that require extensive decontamination and mitigation are either pathogens or nuisance ("contaminating") microorganisms.

The effectiveness of confinement, decontamination, and mitigation

TABLE 40.1   Time Frames and Methods for Controlling or Eliminating Unwanted Effects of Free-Living Microorganisms and Microorganisms Associated with Plants and Animals

Microorganism association	Immediate[a]	Short-term[b]	Long-term[c]
Free-living	Fumigation Flooding Chemicals[d]	Fumigation Flooding Chemicals Erosion control Soil amendments	Fumigation Flooding Erosion control Soil amendments

Plants	Burning (eradication)
	Quarantine
	Tillage
	Chemicals
	Biological control
	Irrigation/flooding
	Insect vector control
	Machinery sanitation
	Runoff water control
	Solarization
	    (cover with plastic)
		Breeding for
		    resistance[e]
		Biological control[f]
		Quarantine
		Chemicals
		Crop rotation
		Cultivar rotation
		Irrigation/flooding
		Heat treatment
		Soil solarization
		Induced resistance
		Meristem/tissue
		    culture
		Insect vector control
		Weed control
		Erosion control
			Breeding for
			    resistance
			Biological control
			Crop rotation
			Cultivar rotation
			Soil amendments
			Weed control
			Erosion control

Animals	Incineration
	Quarantine
	Slaughter
	Bird, rodent, insect
	    control
	Runoff water
	    control (insects)
	Biological control
	    (insects)
		Immunization
		Quarantine
		Antibiotics, drugs
		Bird, rodent, insect
		    control
		
		Biological control
		    (insects)
			Immunization
			Breeding for
			    resistance
			Antibiotics, drugs
			Bird, rodent, insect
			    control
			Biological control
			    (insects)

[a]Hours to several days to achieve effectiveness.
[b]0–3 years to achieve effectiveness.
[c]Longer than 3 years.
[d]Choice and availability of chemical for target microorganisms dictate feasibility and approach.
[e]Germ plasm may be adequately identified for rapid development; otherwise process normally takes 5–10 years.
[f]Few biological control agents are yet available for widespread use; several are under investigation and development for some disease-causing and pest microorganisms.

in commercial and natural situations varies with the microorganism and its habitat and whether it is associated with plants or animals or is free-living. In general, microorganisms associated with domesticated animals can be controlled easily as the result of the specificity of their interaction with the animal; e.g., by quarantine or slaughtering, with subsequent burial or incineration (these all control the "niche";

25). However, not considered here is the administration of microorganisms to free-ranging wild animals or the use of beneficial microorganisms in domesticated aquatic species, because of the relative lack of development of such microorganisms. For example, live vaccines for the immunization of aquatic species have apparently not yet been developed.

In the United States, commercial use of microorganisms in the environment is limited. Microorganisms that are used to promote plant health or mitigate against plant pests and pathogens include rhizobia and bradyrhizobia, applied principally to seed for nitrogen-fixation; *Bacillus thuringiensis* in spray or powder formulation as an insecticidal agent; *Agrobacterium radiobacter* for control of certain crown-gall biotypes of *Agrobacterium tumefaciens,* principally in transplants of ornamental bushes and trees; a fungus, *Colletotrichum gloeosporides,* for biological control of a weed (Northern joint vetch); mycorrhizal fungi for tree establishment; and lactobacilli as silage inoculants. All of these naturally occurring microorganisms are being genetically manipulated to enhance their usefulness. In animals, the commercial usefulness of some recombinants, such as a pseudorabies vaccine, is clear, whereas that of ice-minus *Pseudomonas syringae* for alleviation of frost damage to plants remains to be proved. Free-living wild-type microorganisms, both individually and in consortia, are being used for the degradation of biomass and hazardous, petroleum, and other wastes; metal leaching; sewage treatment; and other environmental applications (6). These organisms appear to be essentially confined by their niche, and there appear to be no problems resulting from their use. The use of genetically modified microorganisms, however, particularly those with added traits, may raise concerns, as discussed elsewhere in Part 6.

Specialists were asked to comment on specific methods for confinement (or containment), decontamination, and mitigation of specific groups of microorganisms that were either (1) already of concern to human health or to the environment, or (2) already used or likely to be used in the environment. These presentations follow this chapter.

This part deals with contingencies for the control of microorganisms, especially genetically modified microorganisms, that may have a deleterious effect. Although a runaway microorganism is possible, the concept of runaway implies unlimited spread, dissemination, and harm over a short or protracted period of time. However, if adverse effects occur, they are likely to be detected first in the place of application. The vast literature on wild-type microorganisms, including those that harm humans and the environment, must necessarily provide a foundation for the control of introduced genetically modified microorganisms. However, it can be argued that extrapolating from known deleterious microorganisms to potential beneficial ones is of limited value. Consequently, the history of use of beneficial microorganisms, including those in everyday foods (e.g., yogurt), that have been manipulated by humans for hundreds of years should be equally considered.

The types and magnitude of deleterious effects are difficult to determine and predict. What criteria should be used? What time frames are appropriate to assess such effects (days, months, years), and what actions should be taken (7, 11, 24)? Overt effects, e.g., symptoms on or even the death of macroscopic plants and animals, are much more likely to be observed than transfer of a genetic trait to an uncultivable microorganism. Nevertheless, it should be feasible to recommend a level of concern for each genetically modified microorganism used, based on the nature of the parent organism, the type of modification, the nature of the recipient environment, and the collective experience of microbiologists. There clearly are microorganisms that are generally regarded as compatible with the environment (GRACE). Development and compilation of a GRACE list would be of assistance to microbiologists studying in the environment and to agencies that have oversight and regulatory responsibility.

Microorganisms Associated with Animals

Current and expected uses of microorganisms associated with animals include amelioration of diseases, insect control, and improved productivity, e.g., in the rumen. Such uses are expected to be direct, such as by inoculation of gnotobiotic animals, and indirect, such as by ingestion from feeding. Insect control by microbes is well established in principle and is likely to increase. With the increasing emphasis on domestication and enhanced productivity of aquatic animals, the use of beneficial microorganisms for disease control and increased food efficiency is also expected in this area.

Microorganisms Associated with Plants

Current and proposed uses of microbes associated with plants range from those that are expected to enhance plant growth, nutritive value, or yield to those that are expected to protect plants from other organisms and environmental stresses. Insecticidal activity and nitrogen fixation are the most prominent uses expected. Some undesirable plants (weeds) can now be controlled by microorganisms, and new or modified agents for this purpose are also expected. Production of foodstuffs by algae and control of undesirable algae are also areas of investigation. These uses of microbes do not include nonreplicating vectors that introduce traits into plants or portions of noninfectious genomes incorporated into plants.

Microorganisms in Natural Environments

Microorganisms in natural environments that are and will be subject to modification include those used in metal extraction, enhanced oil recovery, degradation of toxic compounds, enhanced production and degradation of biomass, and nitrogen fixation. Whether such microorganisms can be effectively modified is the subject of intense investigation.

Current Techniques for Confinement, Decontamination, and Mitigation

Confinement

The confinement of microorganisms associated with animals for disease-control purposes is specified in detail by federal practices (1; see also Chap. 43). These practices should serve equally well for the confinement of beneficial microorganisms associated with domestic animals. However, wild and aquatic animals are not currently considered in these practices.

Microorganisms associated with plants can be confined to experimental test areas, provided that appropriate precautions are taken with respect to the microorganism itself (e.g., pretested for deleterious effects), the plant host, conditions of application of the microorganism, slope of the land, etc. (8, 9, 13, 20). No case of unreasonable adverse effects of experimental use in the field, even with pathogenic microorganisms, is known. Paradoxically, because of the specific host and environmental conditions needed for the successful establishment of a particular plant disease, it may be more difficult to confine nonspecific plant-associated microorganisms.

Many free-living organisms, e.g., those used for degradation of toxic compounds and in mining, are effectively confined by their ecological niche. Again, no cases of wild-type microorganisms producing deleterious effects are known. However, with genetically modified organisms, the introduced trait may extend the niche of the microbe; this has to be determined (e.g., see Chaps. 41, 42, and 45). In all the above cases, spatial and temporal isolation can serve to confine the microorganisms to particular sites.

Decontamination and mitigation

Only general statements can be made about decontamination and mitigation of microorganisms associated with domesticated animals (Table 40.1). Such standard practices are generally effective for pathogenic microorganisms (1), and they are probably also applicable to genetically modified microorganisms. Nevertheless, as stated earlier, different considerations may apply to microorganisms associated with aquatic animals and animals in the wild. Microorganisms used for biological control of insects, prominently *B. thuringiensis,* are considered self-limiting, as experience has shown that without the insect, survival of the microorganisms at high enough concentrations to be effective against subsequent introductions of the insects is low (2). However, care will need to be exercised in monitoring genetically modified microorganisms with a broadened insect-host range. For example, if unintended insect hosts are killed or damaged, it may be necessary to use insecticides, trap-feeding plants for subsequent incineration, and/or insect predators that, in turn, may need to be destroyed to minimize spread of the microorganism.

Microorganisms that are associated with plants and that prove to be deleterious can be controlled or, in some cases, eliminated by use of the techniques listed in Table 40.1. These techniques are in current use with varying degrees of success, depending on the microorganism, its life cycle, the plant host, and the environment, as well as on factors that affect their applicability, such as cost, relative effectiveness of the agent, toxicity, availability, and side effects, both as to type and duration. Although expanded use of biological control is promising, such control is still essentially exploratory (4). Covering test plots with cement (25) is not recommended, even though it might reduce further dissemination, as this approach would result in other deleterious effects on the environment, such as the loss of land, erosion around the cemented area, and dispersion below and outside the area.

Control of some known deleterious microorganisms associated with plants can be achieved with chemicals. These include soil fumigants (i.e., nonspecific agents; Table 40.2), fungicides (Table 40.3), and nematicides (Table 40.4). However, several of these compounds may soon be deregistered, as the result of their nontarget toxicity or other problems. Comprehensive details for use of these and other compounds with specific plants are available (14). Paradoxically, although bacteria are the most likely microorganisms to be introduced initially into the environment, there are fewer effective and acceptable chemical controls for bacteria than for other microorganisms.

One of the most effective long-term solutions for countering the deleterious effects of microorganisms on plants and animals is breeding for disease and pest resistance in both animals and plants (Fig. 40.1; 18). This process can take up to 20 years or more by standard methods, and it requires that appropriate genetic material can be identified.

TABLE 40.2 Soil Fumigants*

Common name	Chemical name (some trade names)	Formulation	Vapor pressure, mm Hg	Boiling point, °C	Specificity	Dosage, amt/ha	Toxicities Plant	Toxicities Mammalian	Application considerations
Methyl bromide	Bromomethane (Dowfume MC-2)	98% + 2% chloropicrin	1420	4.6	General biocide	450–900 kg	Toxic	LD_{50}† = 1 mg/kg	Requires gas-proof seal
Chloropicrin	Trichloronit-romethane (Picfume, Larvacide)	100%	20	112	General biocide	300–500 liters	Toxic	LD_{50} = 1 mg/kg	Best activity with gas proof seal
Chlorinated hydrocarbons (1,3-D)(DD)	1,2-Dichloropropane, 1,3-dichloropropene, and other chlorinated hydrocarbons (Telone, Vidden D)	1,3-D alone or with other chlorinated hydrocarbons	19–25	106–111	Nematicidal	100–500 liters	Toxic	LD_{50} = 140 mg/kg	Requires soil seal
Ethylene dibromide (EDB)	1,2-Dibromoethane (Dowfume W-84, Nematox 100)	60–85% in liquid	8	132	Nematicidal	19–94 liters	Toxic	LD_{50} = 150 mg/kg	Requires soil seal
Methyl isothiocyanate	Methyl isothiocyanate is added directly or is the active breakdown product of several unstable compounds	30–40% liquid or solid (85% WP§)	—	—	General biocide	600–1200 liters or 300–400 kg	Toxic	LD_{50} = 280–650 mg/kg	Injected or rotovated in
Dibromochloropropane‡ (DBCP)	1,2-Dibromo-3-chloropropane (Fumazone, Nemagon, etc.)	Liquid	0.6	199	Nematicidal	19–38 liters	Toxic to some plants	LD_{50} = 172 mg/kg	Injected or drenched

*Data are from several sources, including Peachey and Chapman (1966).
†LD_{50} is the dosage lethal to 50 percent of a test (usually rat) population.
‡Because of toxicities, DBCP is no longer used. It is included here for comparisons only.
§WP = Wettable powder.
SOURCE: From Fry (10).

TABLE 40.3   Fungicides Used Commonly as Seed Treatments in the United States[*]

Common name(s)	Compound	Common formulations	Mammalian toxicity (LD$_{50}$)[†]	Fungal specificity	Major crops	Approximate dosages, g (a.i.)/kg seed[‡]
Captan	N-Trichloromethylthio-4-cyclohexene-1,2-dicarboximide	WP,[§] dust	Low, LD$_{50}$[¶] = 9000 mg/kg	Little specificity	Corn, sorghum, soybeans, peanuts, vegetables, cotton	1–3
Carboxin	5,6-Dihydro-2-methyl-1,4-oxathiin-3-carboxanilide	WP, dust	Low, LD$_{50}$ = 3820 mg/kg	Basidiomycetes	Small grains, cotton	2–3
Etridiazol	5-Ethoxy-3-trichloromethyl-1,2,4-thiadiazole	WP, dust	Low, LD$_{50}$ = 2000 mg/kg	Phycomycetes	Cotton, sorghum, soybeans, small grains	Usually in mixture 0.2–0.5
Maneb	Manganese ethylenebisdithio-carbamate	WP, dust	Low, LD$_{50}$ = 6750 mg/kg	Little specificity	Rice, potatoes	1–2
PCNB	Pentachloronitro-benzene	WP, dust	Low, LD$_{50}$ = 12,000 mg/kg	*Rhizoctonia, Plasmodiophora, Streptomyes,*[**] *Sclerotinia* and others; inactive against Oomycetes and some others	Sorghum, soybeans, small grains	1–2
Thiram	Tetramethylthiuram-disulfide	WP, dust	Low, LD$_{50}$ = 780 mg/kg	Little specificity	Corn, soybeans, peanuts, small grains, vegetables	1–4

[*]Data are from several sources, including Rodriguez-Kabana et al. (1977).
[†]Toxicities were obtained from Thomsen (1979).
[‡]a.i. = active ingredient.
[§]WP = Wettable powder.
[¶]LD$_{50}$ is that dosage lethal to 50 percent of a test (usually rat) population.
[**]Some fungicides also have some bactericidal activity.
SOURCE: From Fry (10).

TABLE 40.4 Some Nonfumigant Nematicides

Common name	Chemical name [some trade names]	Formulations	Common dosages, (a.i.)/ha (kg)*	Mammalian toxicity	Application
Aldicarb	2-Methyl-2(methylthio)propionaldehyde oxime [Temik, Ambush]	10–15% granules	2.2–3.3	LD_{50}[†] = 1 mg/kg	At planting
Carbofuran	2,3-Dihydro-2,2-dimethyl-7-benzofuranyl methylcarbamate [Furadan]	2–10%	3.3	LD_{50} = 11 mg/kg	At planting
Ethoprop	0-Ethyl-5,5-dipropyl phosphorodithioate [Mocap]	EC,[‡] granule	5.6	LD_{50} = 61 mg/kg	At planting
Oxamyl	Methyl-N',N'-dimethyl-N-[(methyl carbamoyl)oxy]-1-thicoxamimidate [Vydate]	24% liquid	5.6	LD_{50} = 5 mg/kg	At planting
Phenamiphos	Ethyl-3-methyl-4-(methylthio)phenyl-(1-methylethyl)phosphoramidate [Nemacur]	Granules, liquids	5.6	LD_{50} = 8 mg/kg	At planting

*a.i. = active ingredient.
†LD_{50} is that dosage lethal to 50 percent of a test (usually rat) population.
‡EC = emulsified concentrate.
SOURCE: From Fry (10).

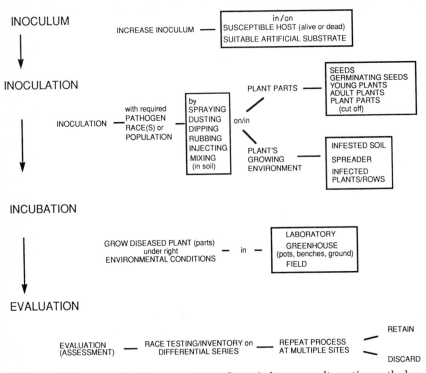

Figure 40.1 Screening for disease resistance. Items in boxes are alternative methods or choices. [*Adapted from Parlevliet (18).*]

For some microbial diseases, no genetic sources of resistance for plants or animals may be known or available.

The survival and dissemination of free-living microorganisms may be controlled by manipulating their niche (Table 40.1). The properties of the parent microorganism (e.g., kind and strain) are more likely to affect the survival of the modified organism than its modification, assuming that the microorganism is not also modified for greater survival in the environment (23).

Prospective Techniques

Except for possible new chemicals, there appear to be no prospective novel means for decontamination and mitigation of microorganisms. New chemicals that may be used for inactivation of biological systems will be limited by their availability, specificity, rate of application, relative effectiveness, and cost. Currently, only chemicals that are used in human and animal medicine and against very damaging plant diseases are probably cost effective for commercialization. This is probably also true for biological control agents. Solar panels may provide energy to heat-kill effectively some microorganisms in small soil plots or bodies of water, but the probability is that desirable macroorganisms, such as fish, would be killed first. Internal controls, e.g., auxotrophy, sporulation-deficiency, host-dependent promoters, suicide genes, that may be introduced along with the desired genes are not yet considered reasonable on a practical scale (24).

Research Needs and Concerns

In addition to the research needs expressed elsewhere in this book, the following information and considerations will be helpful, and even necessary, for safe planned introductions, particularly of genetically modified microorganisms, into the environment.

1. Microorganisms should be classified into categories on the basis of expected risks. For example, most scientists would not group the genera of *Arthrobacter, Rhizobium,* and *Bradyrhizobium* with that of *Xanthomonas* in terms of their risks to plants (see *Federal Register* 1987, 52: 22909). Development of a GRACE list of microorganisms should be considered. Many of such microorganisms are known, and their safety is generally acknowledged. These include the first three bacterial genera mentioned above, as well as numerous others considered by the specialists in the following chapters. For genetically modified microorganisms, the genetic trait to be introduced should also be categorized; for example, the ability to degrade a toxic pesticide more efficiently by alteration of a permease or the introduction of a catabolic enzyme should be differentiated from the ability to synthesize a metabolite toxic to an insect, a plant, or another microbe.

Classification of microorganisms based on risk assessment would affect the type and extent of mitigating steps that would be recommended.

2. Model genetically modified microorganisms may be useful for developing databases that are analogous to those that exist for *Escherichia coli* K12. These models should encompass microorganisms that are most likely to be modified for release to the environment by researchers or companies. At present, primarily strains that will probably be of commercial interest are being tested in the environment. However, such microorganisms may or may not provide the baseline data needed to enable reasonably rapid and relevant evaluation of other genetically modified organisms in the future.

3. Small-scale tests (e.g., the EPA recommendation of 10 land acres and 1 surface-acre in water) appear to be reasonable in terms of confinement, decontamination, and mitigation. However, there is also the need to scale up gradually and safely. Decontamination, mitigation,

and, certainly, confinement of microorganisms used on a large scale, such as on 1 million acres, will be more difficult from a practical basis. Depending on the problem and the microorganism involved, difficulties with decontamination and mitigation may be encountered with even a few hundred acres. Furthermore, it must be emphasized that plant and animal breeders are continually "improving" their respective organisms and that, except for disease resistance, this is normally done independently of microbiological considerations. For example, soybean breeders do not breed for maximizing compatibility of nitrogen-fixing bacteria with their selections. Potential deleterious effects of such improvements on the survival and activity of introduced genetically modified microorganisms may not be known or become obvious for some time after introduction. However, there appear to have been no long-term epidemics or deleterious effects from nongenetically modified microbes from such standard improvement practices in the past.

4. The development of specialty chemicals for specific organisms and limited-target biologicals that would be environmentally compatible and effective for the control of deleterious runaway introductions is desirable. However, the marketplace of today, and undoubtedly of the future, it geared toward the production of broad-spectrum, high-volume biological or chemical agents that maximize cost effectiveness for both the user and the manufacturer. Consequently, agents with small market potential are unlikely to find corporate sponsorship, at least without intervention by the federal government. One mechanism by which to encourage the development of such biological and chemical agents is to foster legislation analogous to the U.S. Orphan Drug Act that provides incentives to pharmaceutical manufacturers to produce drugs for limited markets. Such an act could be legislated for the development of both specialty chemicals and biologicals of high specificity. The availability of such materials would increase the arsenal of potential mitigating agents for microorganisms used in the environment.

5. Predictive forecasting of the behavior of both wild-type and genetically modified microorganisms is needed. The databases for even deleterious organisms are severely limited (23, 25), except for some human- and other animal-associated microorganisms. With increasingly better mathematical tools and computer modeling, more and better predictive systems should be developed. The availability of such systems would be of great potential use.

Concluding Remarks

There is reasonable agreement that present methodology for the confinement, decontamination, and mitigation of wild-type microorganisms used in research and commerce is adequate. Knowledge about the microorganism, whether genetically modified or wild-type, and the environment in which it is to be introduced is crucial for predicting the outcome of the methods used. For microorganisms associated with plants, the most appropriate, feasible, and cost-effective methods are environmental management and the use of inherited host resistance (Table 40.1). For microorganisms associated with higher animals, induced host resistance, i.e., immunization, is extremely effective and economical. Techniques of insect control by microorganisms are highly developed, primarily as the result of the high host specificity of current agents; however, the potential for extended host range may be of concern. Consequently, no significantly new methodologies for the control of "problem" nongenetically modified microorganisms, in general, are foreseen. The genetic material of wild-type microorganisms generally appears to be highly conserved and has very low transferability in the absence of selection pressure (21). Thus, in most cases, the problem with unmodified microorganisms introduced into the environment is not their proliferation but, rather, their rapid death or decreased effectiveness, such as occurs with most biological control agents (2,4). However, this may not be the case with genetically modified microorganisms that are designed to carry out a specific task and to persist and proliferate in the environment.

Because of the lack of data, a more gradual scale-up should be considered with introductions of genetically modified microorganisms into the environment than with wild-type microorganisms. Similarly, attention to the selection of microorganisms that become associated with new cultivars of plants and breeds of animals is desirable and, in fact, may be necessary. Special attention may need to be paid to aquatic species. Transgenic plants and animals (i.e., those with genes incorporated from unrelated microorganisms, plants, or animals) need to be carefully observed for changes in their associated wild-type microbiota. Such changes may presage areas of concern with other wild-type microorganisms and prospective genetically modified microorganisms.

With reasonable precautions, based on knowledge of and experience with the existing use of wild-type microorganisms in the environment, the initiation of the testing phase of genetically modified microorganisms should be able to be expedited. The needs and concerns for safely moving ahead with research in this area have been addressed (16). Although no absolute guarantees can be given, it is apparent from this chapter and the subsequent discussions by specialists on specific groups of microorganisms that means for confinement, decontamination, and mitigation of microorganisms in the environment are available. Such measures vary in effectiveness, but, collectively, they appear to be appropriate and useful for incorporation into recommendations for use with genetically modified microorganisms.

References

1. Anonymous. 1987. Code of federal regulations, 9. Animals and animal products. U.S. Government Printing Office, Washington, D.C.
2. Aronson, A. I., W. Beckman, and P. Dunn. 1986. *Bacillus thuringiensis* and related insect pathogens. *Microbiol. Rev.* 50:1–24.
3. Ajello, L., and R. J. Weeks. 1983. Soil decontamination and other control measures. In A. F. DiSalvo (ed.) *Occupational Mycoses, a Text.* Lea & Febiger, Philadelphia, pp. 229–238.
4. Baker, R., and F. M. Scher. 1987. Enhancing the activity of biological control agents. In I. Chet (ed.) *Innovative Approaches to Plant Disease Control.* Wiley, New York, pp. 1–17.
5. Bruehl, G. W. (ed.). 1975. *Biology and Control of Soil-Borne Plant Pathogens.* The American Phytopathological Society, St. Paul, Minn.
6. Crueger, W., and Crueger, A. 1984. *Biotechnology. A Textbook of Industrial Microbiology.* Science Tech, Madison, Wis.
7. Eversmeyer, M. G., and C. L. Kramer. 1987. Components and techniques of integrated pest management threshold determinations for aerial pathogens. *Plant Dis.* 71:456–459.
8. Ewel, J. J. 1986. Designing agricultural ecosystems for the humid tropics. *Annu. Rev. Ecol. Syst.* 17:245–271.
9. Freeman, G. H. 1981. Methods in plant pathology. Part 3. Design and analysis of field trials in plant pathology. *Rev. Plant Pathol.* 60:439–444.
10. Fry, W. E. 1982. *Principles of Plant Disease Management.* Academic Press, New York.
11. Gholson, L. E. 1987. Adaptation of current threshold techniques for different farm operations. *Plant Dis.* 71:462–465.
12. Horsfall, J. G., and F. B. Cowling (eds.). 1977. *Plant Disease, an Advanced Treatise. Vol. 1. How Disease Is Managed.* Academic Press, New York.
13. LeClerg, E. L., W. N. Leonard, and A. G. Clark. 1962. *Field Plot Techniques,* 2d ed. Burgess Publications, Minneapolis, Minn.
14. McGrath, H., J. Feldmesser, and L. D. Young (eds.). 1986. *Guidelines for the Control of Plant Diseases and Nematodes.* USDA, ARS, Washington, D.C.
15. McDonald, S. A. 1984. Applying pesticides correctly. A guide for private and commercial applicators. USDA/EPA/No. Carolina State Univ., revised. Raleigh, N.C.
16. National Research Council. 1989. *Field Testing Genetically-Modified Organisms: Framework for Decisions.* Committee on Scientific Evaluation of the Introduction of Genetically-Modified Microorganisms and Plants into the Environment. National Academy Press, Washington, D.C.
17. Nelson, L. A. 1987. Experimental design and field plot testing for biological and cultural treatments. In *Biological and Cultural Tests for Control of Plant Diseases,* vol. 2. APS Press, St. Paul, Minn.
18. Parlevliet, J. E. 1981. Disease resistance in plants and its consequences for plant

breeding. In K. J. Frey (ed.), *Plant Breeding II.* Iowa State University Press, Ames, pp. 309–347.

19. Pimentel, D. (ed.) 1990. *Handbook of Pest Management in Agriculture,* 2d ed. vols. 1 to 3. CRC Press, Boca Raton, Fla.

20. Poehlman, J. M. 1979. *Breeding Field Crops.* 2d ed. AVI Publishing, Westport, Conn.

21. Riley, M. 1989. Constancy and change in bacterial genomes. In J. S. Poindexter and E. R. Leadbetter (eds.) *Bacteria in Nature,* vol. 3. Plenum, New York, pp. 359–388.

22. Sharvelle, E. G. 1979. *Plant Disease Control.* AVI Publishing, Westport, Conn.

23. Stotzky, G. 1989. Gene transfer among bacteria in soil. In S. B. Levy and R. V. Miller (eds.) *Gene Transfer in the Environment.* McGraw-Hill, New York, pp. 165–222.

24. Teng, P. S. (ed.) 1987. *Crop Loss Assessment and Pest Management.* APS Press, St. Paul, Minn.

25. Tiedje, J. 1987. Environmental monitoring of microorganisms. In J. W. Gillet (ed.), *Prospects for Physical and Biological Containment of Genetically Engineered Organisms.* Proceedings of the Shackelton Point Workshop on Biotechnology Impact Assessment, Oct. 1–4, 1985. Ecosystems Research Center, Ithaca, N.Y.

26. Vanderplank, J. E. 1963. *Plant Diseases: Epidemics and Control.* Academic Press, New York.

27. Van Elsas, J. D., A. F. Digkstra, J. M. Govaert, and J. A. van Veen. 1986. Survival of *Pseudomonas fluorescens* and *Bacillus subtilis* introduced into two soils of different texture in field microplots. *FEMS Microbiol. Ecol.* 38:151–160.

28. Zehr, E. I. 1978. Methods for evaluating plant fungicides, nematicides, and bactericides. American Phytopathological Society, St. Paul, Minn.

Degradative Bacteria

Peter A. Vandenbergh

Introduction

Each year in the United States, about 14,000 industrial plants generate 265 metric tons of hazardous waste. The U.S. Environmental Protection Agency (EPA) estimates that 80 percent of this waste is landfilled (14). Faced with major disposal problems, there are currently three methods of treatment: (1) place the material in a sealed container and store indefinitely, (2) incineration, or (3) treat in situ.

The use of microorganisms to treat waste material in situ has been known for some time. In 1914, it was observed that the addition of bacteria reduced the organic content of sewage more rapidly. The scientific basis for bacterial degradation of hydrocarbons has been intensely studied. In 1946, ZoBell (22), while studying the action of microorganisms on hydrocarbons, observed that many bacteria in nature have the ability to utilize specific hydrocarbons as sole carbon and energy sources. He also suggested that the metabolism of these hydrocarbons within complex mixtures was highly dependent on environmental factors.

Many investigators continue to study the degradative properties of microorganisms, including aerobic and anaerobic bacteria and fungi. These microbes are either naturally occurring strains or genetically modified microorganisms. Naturally occurring degradative bacterial strains are usually obtained from soils in and around hazardous waste sites. These strains, presumably derived by natural selection, are studied in the laboratory and may be returned to the site for potential use as a biodegradative tool for cleanup. Many investigators believe that by extending the substrate range of a natural strain through the use of genetic engineering, the strain can be made to function more efficiently (16). Rojo et al. (16), through the use of genetic engineering techniques, were able to integrate enzymes from five different catabolic pathways of three distinct soil bacteria into one strain. The new strain was able to utilize, in vitro, mixtures of chloroaromatic compounds that were toxic to the individual bacterial strains that contributed to the integrated metabolism of the engineered strain. To date, however, no applications have been submitted to the EPA for permits to test degradative genetically engineered microbes in the environment (12; F. Batz, EPA, personal communication).

With many research groups working on the development of engineered strains for toxic wastes, these investigators will be faced with many of the same questions encountered by other researchers in the release of genetically modified bacteria. Therefore, prior to the release of genetically modified degradative strains into the environment to enhance degradation of toxic or hazardous waste, extensive research and development into areas concerning confinement, decontamination, and mitigation of these bacteria must be accomplished.

General Considerations

Current methods to determine the environmental effects of the introduction of genetically modified microorganisms for toxic waste degradation focus on answering the following questions.

1. Will the released organism survive in the environment?
2. Can the organism be confined when released into open waste sites?
3. Can it transfer its genetic material to other organisms?
4. Will the original organism or any others that might acquire its genes prove harmful?
5. Could the site of application and surrounding areas be decontaminated if accidental release occurs?
6. Will it be possible to mitigate the adverse effects of the organism in the environment?

Survival in the environment of genetically modified degradative bacteria is not known. These strains must function and survive in environments with selective carbon sources. Some information concerning environmental survival might be obtained in model simulation systems.

Containment methods are either biological or physical. Elimination of an organism on a large scale may not be possible, especially in open environments. With genetically modified microorganisms, the addition of a lethal gene that would be expressed only when the toxic substance is depleted might offer a possible method of biological containment. Physical containment might be easier to achieve through the use of immobilized microbes on carriers placed into columns or through the use of bioreactors. In these situations, the contaminated water would be passed through and the cells would be retained through filtration, and only the end products of the degradation would be released into larger areas.

Information is increasing on the ability of bacteria to transfer genes in the environment. Currently, investigators are examining the transfer of R plasmids coding for antibiotic resistance between soil bacteria (4, 19, 20). Although these studies are useful, they do not address the issue of degradative gene mobilization. Many of the degradative plasmids do not encode for complete utilization of the compound to be degraded, and hence other pathways that are chromosomally encoded are involved in the complete breakdown of the compound. Studies are needed to determine if degradative genes can function in various types of soil bacteria. These laboratory experiments should involve both genetic and metabolic studies to evaluate the risk of degradative transfer. However, experiments to assess the possible transmissibility of a certain trait may be difficult to model in a laboratory that would duplicate the many possible combinations that could occur in nature.

The environmental impact of a genetically modified degradative microbe would be difficult to ascertain in the laboratory. Research has focused on the fate of these organisms and their genetic material after release (17). To date, research has focused on the introduction and recovery of microorganisms (18), bacterial survival (3), and genetic interactions (21). It must be remembered that usually pollutants are observed in nature at the ppm (parts per million, $\mu g/g$) level, and that when they fall below 10 ppb (parts per billion), such bacteria are unable to grow.

From limited laboratory studies with degradative bacteria, it can be predicted that either instability of the recombinant DNA or lack of its expression under differing conditions in the environment may reduce the ecological risks. Instability of the genetic material is the major reason that the original oil-degrading microbe developed by Chakrabarty (15) is not used in the field. Nevertheless, the ecological impact of the degradative strains must also be studied for long-term effects.

If accidental release occurs into either soil or water, the surrounding environment may be decontaminated through the use of hydrogen peroxide if the material could be removed and slurried. Hydrogen peroxide has the advantage of being strongly inhibitory of microbial activity at relatively low concentrations (10). However, it would be difficult to totally eliminate the organism completely from the environment (5). Mitigation of any possible adverse effects on the environment will be difficult to accomplish due to the difficulty in the total elimination of a species from an environment. However, careful construction of genetically modified microorganisms could allow for a lessening of the damage if the strains have poor survivability. Many studies are currently in progress that address these issues, and we must wait until the data are available.

Specific Procedures

The use of bioremediation to clean up hazardous soil and water throughout the United States is under way. Underground storage tank leakage of hydrocarbons into the environment is recognized as a major problem. Naturally occurring degradative bacteria for the dissipation of these toxic hydrocarbons in the surrounding soil and water is now being utilized (9). Several applications of different commercial nongenetically engineered strains have been successful, and wide acceptance of their use is apparent.

The General Electric Company is currently using a strain of *Pseudomonas putida* to degrade polychlorinated biphenyls that were detected in contaminated soil in a greenhouse landfarming study (2). Recent studies have identified bacterial isolates capable of utilizing 4-chlorobiphenyl (1), with the cloning and identification of the genes involved (8). Other investigators are studying the potential use of anaerobes to degrade chloroaromatic and chloroaliphatic hydrocarbons detected in the environment (7). The application of anaerobic bacteria in the field, however, may present technical difficulties. Additionally, some of the anaerobic strains produce toxic compounds, such as vinyl chloride, in their metabolic pathways. Also, the long time necessary for degradation of toxic compounds with anaerobes may be a serious deterrent to their use.

Another trend today in the bioremediation industry is the use of the bioreactor. These reactors are used as an alternative for enhancing degradation in situ. Several companies have developed these containment vessels, and usually the bacteria are fixed to biocarriers or activated carbon (6, 13). The bioreactor is placed at the contamination site, and nutrients, pH, and temperature are maintained. The toxic material being treated, usually a liquid, is passed through the bioreactor and degradative rates are monitored. A wide variety of compounds are degradable in bioreactors including halophenols, halobenzoates, and a number of pesticides (9).

The bioreactor concept would be very useful with genetically modified microorganisms. The final effluent could be filtered or treated with hydrogen peroxide or manipulated by temperature and pH before release into the environment. Enclosed bioreactors could be constructed and engineered to prevent even accidental release. Using this method of containment, the risk associated with biological control agents is reduced because the microbes would be in a closed system. Bioreactors are just being utilized, and engineering improvements in their design, coupled with the development and application of recombinant strains, should be encouraged. The application of the bioreactor concept offers extensive control and containment because the genetically modified microorganisms would not be released directly into the environment. Currently, bacteria have been successfully used to degrade phenol in a variety of reactors (11). These reactors include columnar fluidized-bed bioreactors and batch reactors. In the former, the bacteria are usually maintained as a fixed-film population.

Conclusion

Development of degradative recombinant strains is market-driven. Such strains for normal wastewater sewage treatment are unlikely due to current economic factors in the marketplace. The current wastewater treatment market in the United States with bacterial cultures is approximately $10 million. This market is divided among several companies who also have to absorb part of the cost of environmental risk assessment. The outlook for the development of genetically modified microorganisms for normal industrial wastewater treatment under these conditions is not promising.

However, the use of genetically modified microorganisms in the remediation of hazardous wastes in soil and ground water is currently being investigated by several companies. The remediation of toxic waste sites in the United States is projected to amount to billions of U.S. dollars. Also, stricter requirements regarding decontamination to lower levels likely will result in the application of these strains. Therefore, to successfully decontaminate hazardous waste sites to comply with EPA or state regulations will result in the use of these types of organisms.

The release of degradative genetically modified microorganisms into the environment will require the examination of a multitude of risks, both real and imagined. However, the current economic risks involved in such ventures may well provide the largest hurdle.

References

1. Barton, M. R., and R. L. Crawford. 1988. Novel biotransformations of 4-chlorobiphenyl by *Pseudomonas* sp. *Appl. Environ. Microbiol.* 54:594–595.

2. Bedard, D. L., R. Unterman, L. H. Bopp, M. J. Brennan, M. L. Haberl, and C. Johnson. 1986. Rapid assay for screening and characterizing microorganisms for the ability to degrade polychlorinated biphenyls. *Appl. Environ. Microbiol.* 51:761–768.

3. Compeau, G., B. J. Al-Achi, E. Platsouka, and S. B. Levy. 1988. Survival of rifampin-resistant mutants of *Pseudomonas fluorescens* and *Pseudomonas putida* in soil systems. *Appl. Environ. Microbiol.* 54:2432–2438.

4. Devanas, M. A., and G. Stotzky. 1987. Survival and conjugal transfer of plasmid RP4 from *Escherichia coli* to *Pseudomonas aeruginosa* in soil: Effect of clay minerals and nutrients. *Abstr. Annu. Meet. Am. Soc. Microbiol.*, p. 303.

5. Domsch, K. H., A. H. Driesel, W. Goebel, W. Andersch, W. Lindenmaier, W. Lotz, H. Relse, and F. Schmidt. 1988. Considerations on release of gene-technologically engineered microorganisms into the environment. *FEMS Microbiol. Ecol.* 53:261–272.

6. Ehrhardt, H. M., and H. J. Rehm. 1985. Phenol degradation by microorganisms adsorbed on activated carbon. *Appl. Microbiol. Biotechnol.* 21:32–36.

7. Fathepure, B. Z., J. M. Tiedje, and S. A. Boyd. 1987. Reductive dechlorination of 4-chlororesorcinol by anaerobic microorganisms. *Environ. Toxicol. Chem.* 6:929–934.

8. Khan, A., and S. Walia. 1989. Cloning of bacterial genes specifying degradation of 4-chlorobiphenyl from *Pseudomonas putida* strain OU83. *Appl. Environ. Microbiol.* 55:798–805.

9. Knackmuss, H.-J. 1983. Xenobiotic degradation in industrial sewage: Haloaromatics as target substances. *Biochem. Soc. Symp.* 48:173–190.

10. Lee, M. D., V. W. Jamison, and R. L. Raymond. 1987. Applicability of in-situ bioreclamation as a remedial action altamatic. Proceedings of the NWWA/API Conference on Petroleum Hydrocarbons and Organic Chemicals in Ground Water: Prevention, Detection and Restoration, Nov. 1987, NWWA, Dublin, Ohio.

11. Lee, M. D., and C. H. Ward. 1985. Biological methods for the restoration of contaminated aquifers. *Environ. Toxicol. Chem.* 4:743–750.

12. Marx, J. L. 1987. Assessing the risks of microbial release. *Science* 237:1413–1417.

13. Morgan, P., and R. J. Watkinson. 1989. Hydrocarbon degradation in soils and methods for soil biotreatment. *CRC Crit. Rev. Biotechnol.* 8:305–333.

14. Nicholas, R. B. 1987. Biotechnology in hazardous waste disposal: An unfulfilled promise. *ASM News* 53:138–142.

15. Roberts, L. 1987. Discovering microbes with a taste for PCB's. *Science* 237:975–977.

16. Rojo, F., A. Pieper, K. Engesser, H. Knackmuss, and K. Timmis. 1987. Assemblage of ortho cleavage route and simultaneous degradation of ortho- and methylaromatics. *Science* 238:1395–1398.

17. Torsvik, V. L., and J. Goksoyr. 1978. Determination of bacterial DNA in soil. *Soil Biol. Biochem.* 10:7–12.

18. Trevors, J. T., T. Barkay, and A. W. Bourquin. 1987. Gene transfer among bacteria in soil and aquatic environments: A review. *Can. J. Microbiol.* 33:191–198.

19. Trevors, J. T., and G. Berg. 1989. Conjugal RP4 transfer between pseudomonads in soil and recovery of RP4 plasmid DNA from soil. *Syst. Appl. Microbiol.* 11:223–227.

20. Trevors, J. T., J. D. van Elsas, L. S. van Overbeek, and M. E. Starodub. 1990. Transport of a genetically engineered *Pseudomonas fluorescens* strain through a soil microcosm. *Appl. Environ. Microbiol.* 56:401–408.

21. Van Elsas, J. D., J. T. Trevors, M.-E. Starodub, and L. S. van Overbeek, 1990. Transfer of plasmid RP4 between pseudomonads after introduction into soil; influence of spatial and temporal aspects of inoculation. *FEMS Microbiol. Ecol.* 73:1–12.

22. ZoBell, C. E. 1946. Action of microorganisms on hydrocarbons. *Bacteriol. Rev.* 10:1–49.

42

Metal-Extracting Bacteria

Jessup M. Shively

Introduction

The solubilization (extraction) of metals from ores by leaching dates back to the eighteenth century in the Rio Tinto mines in Spain. However, systematic production has only been in use since the late 1950s. Copper and uranium have been extensively produced by leaching, and the application of the technique to other metals, e.g., gold, nickel, zinc, and lead, appears likely (2, 7, 12, 17).

General Considerations

The most commonly employed bacteria are *Thiobacillus ferrooxidans* and *Thiobacillus thiooxidans* (6). A number of other bacteria could potentially be used for leaching, including members of the genera *Pseudomonas, Achromobacter, Bacillus, Thiobacillus, Thermothrix, Sulfolobus,* and *Leptospirillum* (3, 6). However, until methodologies using these bacteria are developed, it seems premature to discuss the containment, decontamination, and mitigation relating to genetically engineered strains of these genera. Therefore, this report will concern itself with only *T. ferrooxidans* and *T. thiooxidans*. These aerobic, chemolithotrophic, acidophilic bacteria are widespread in nature, and many different strains have been isolated. They obtain their cellular carbon from carbon dioxide and their energy from the oxidation of ferrous ion and/or elemental sulfur or reduced sulfur compounds (3, 10, 13).

The leaching of metals from their ores by these bacteria includes both direct and indirect (bacterial supported) processes (3, 10, 19, 20). The metal sulfide (MS) may be oxidized directly to its metal sulfate, which is soluble in the acidified leach medium

$$MS + 2O_2 \rightarrow MSO_4$$

The metal sulfide may also be oxidized by ferric ion

$$MS + 2Fe^{3+} \rightarrow M^{2+} + 2Fe^{2+} + S^{\circ}$$

The elemental sulfur and ferrous ion are then oxidized by the bacteria

$$2S^{\circ} + 3O_2 + 2H_2O \rightarrow 2H_2SO_4$$

$$4\ Fe^{2+} + O_2 + 4H^+ \rightarrow 4\ Fe^{3+} + 2H_2O$$

The leaching can be accomplished in semienclosed containers (bioreactors), in large dumps (slope-leaching) and heaps, or in situ (3, 7). Acid leach liquor containing ferric ions and bacteria is sprinkled or sprayed on the surface of the ore or pumped into underground passages, and the leach solution is collected at the bottom or lowest point. The solubilized metals are recovered, and the leach liquor recycled.

Our knowledge of the genetics of *Thiobacillus* is meager. Mobilization of DNA into a number of *Thiobacillus* species has been accomplished, but not with *T. ferrooxidans* and *T. thiooxidans* (4, 5, 11, 14, 15, 18, 24). Cryptic plasmids have been isolated from several *T. ferrooxidans* strains and the creation of shuttle vectors seems likely (8, 9, 16). The gene for glutamine synthetase from *T. ferrooxidans* has been cloned and expressed in *Escherichia coli* (1). Studies dealing with gene organization in *T. ferrooxidans* are in progress in a number of laboratories.

The genetic manipulation of these bacteria will generally involve the alteration of existing metabolism, e.g., increasing substrate oxidation and growth rates and resistance to sulfuric acid and metals (21). The DNA will most likely be moved from one strain of *T. ferrooxidans* or *T. thiooxidans* to another or to these species from another *Thiobacillus* species, i.e., intrageneric.

The thiobacilli are essentially ubiquitous entities of our environment and their metabolism usually does not have any deleterious effects. In fact, they are important contributors to the sulfur cycle. However, when their energy substrates become available in relatively large quantities, along with oxygen and water, excessive amounts of sulfuric acid and dissolved metals are produced. This can result in serious environmental problems. Acid drainage from coal mines and coal refuse piles is one of the most serious industrial pollution problems in the United States. Obviously, this acid drainage also results in the mining of pyritic or pyritic-containing metal ores. Significant acid drainage should not take place in appropriately constructed active leaching operations since the leach liquor is recycled. A serious problem could result when leach operations are abandoned.

The creation of a genetically engineered *T. ferrooxidans* or *T. thiooxidans* strain more useful for leaching will require considerable time and effort. When an intrageneric engineered strain is developed, control procedures beyond those required for naturally occurring strains is not expected; a naturally occurring strain possessing many of the characteristics that will be engineered could probably be found if extensively searched for. However, this does not minimize the seriousness of acid mine or acid coal refuse drainage.

Specific Procedures

To date, there are no decontamination procedures available which are specific for the thiobacilli. Furthermore, with the large quantities of ore to be leached, decontamination to prevent the spread of the organism appears difficult if not impossible.

Containment of the engineered *Thiobacillus* would also be difficult with any of the mining methods currently in use with the possible exception of in situ. Even in this case, one should be aware of the possibility of seepage. Containment could probably be achieved with the bioreactor, but there is still the problem of what to do with tons of contaminated spent ore. Containment appears impossible with heap or dump leaching. Air dispersion would result as the leach-liquor containing the bacteria is sprayed or sprinkled on the ore. Storm runoff would also take place.

Mine and refuse drainage can be treated with alkaline materials such as lime, limestone, sodium hydroxide, and/or sodium carbonate. This reduces the acidity and precipitates the metals. The process is expensive, creates large quantities of sludge, and does not reduce the production of the acid drainage. Wetlands and in-pipe aeration can be used to augment the neutralization. Covering the pyritic refuse or spoil with inert materials appears to be of little value; acid drainage is not prevented. Covering the spoil or refuse with an oxygen-consuming layer (soil containing a large, active microbial population and decaying organic matter, and vegetation) is generally a better control strategy. Chemical treatments, including sodium lauryl sulfate, sodium benzoate, and potassium sorbate, may be used to treat certain spoils and refuse. The numbers of thiobacilli are reduced, thus reducing the amount of acid drainage. Improved prediction of acid potential as well as improved mine-planning methodologies are under development. Specific instruction on all of the above control methods can be found in United States Department of Interior, Bureau of Mines Information Circulars (21, 22).

Conclusions

Decontamination and containment of metal-extracting bacteria appear extremely difficult if not impossible. Mitigation procedures for engineered intrageneric strains beyond those necessary for naturally occurring strains will not be necessary. However, it appears that current mitigation methods are not totally adequate and new ones need to be developed.

As our understanding of the physiology, ecology, biochemistry, and genetics of the thiobacilli and other potentially useful microorganisms continues to increase, and methodologies for gene manipulation are developed, the release of genetically engineered strains will need to be reassessed. This could be especially important if intergeneric strains are produced. Specific control methods may be needed before such strains can be utilized.

References

1. Barros, M. E. C., D. E. Rawlings, and D. R. Woods. 1985. Cloning and expression of the *Thiobacillus ferrooxidans* glutamine synthetase gene in *Escherichia coli*. *J. Bacteriol.* 164:1386–1389.
2. Bruynesteyn, A. 1986. Biotechnology: Its potential impact on the mining industry. *Biotech. Bioeng. Symp.* 16:343–350.
3. Crueger, W., and A. Crueger. 1984. Leaching. In T. D. Brock (ed.), *Biotechnology: A Textbook on Industrial Microbiology*. Sinauer Associates, Sunderland, Mass.
4. Davidson, M. S., P. Roy, and A. O. Summers. 1985. Transpositional mutagenesis of *Thiobacillus novellus* and *Thiobacillus versutus*. *Appl. Environ. Microbiol.* 49:1436–1441.
5. Davidson, M. S., and A. O. Summers. 1983. Wide-host-range plasmids function in the genus *Thiobacillus Appl. Environ. Microbiol.* 46:565–572.
6. Ehrlich, H. 1986. What types of microorganisms are effective in bioleaching, bioaccumulation of metals, ore benefication, and desulfurization of fossil fuels. *Biotech. Bioeng. Symp.* 16:227–237.
7. Holmes, D. S. 1988. Biotechnology in the mining and metal processing industries: Challenges and opportunities. *Minerals Metallurg. Process.* 5:49–56.
8. Holmes, D. S., J. H. Lobos, L. H. Bopp, and G. C. Welch. 1984a. Cloning of a *Thiobacillus ferrooxidans* plasmid in *Escherichia coli. J. Bacteriol.* 157:324–326.
9. Holmes, D. S., J. R. Yates, J. H. Lobos, and M. V. Doyle. 1984b. Genetic engineering of biomining organisms. In *The World Biotechnology Report*. Online Publishers, Pinner, United Kingdom, pp. 864–881.
10. Ingledew, W. J. 1986. Ferrous iron oxidation by *Thiobacillus ferrooxidans. Biotech. Bioeng. Symp.* 16:23–33.
11. Kulpa, C. F., M. T. Roskey, and M. T. Travis. 1983. Transfer of plasmid RP1 into chemolithotrophic *Thiobacillus neapolitanus. J. Bacteriol.* 156:434–436.
12. Lakshmanan, V. I. 1986. Industrial views and applications: Advantages and limitations of biotechnology. *Biotech. Bioeng. Symp.* 16:351–361.
13. Lundgren, D. G., M. Valkova-Valchanova, and R. Reed. 1986. Chemical reactions important in bioleaching and bioaccumulation. *Biotech. Bioeng. Symp.* 16:7–22.
14. Plasota, M., E. Piechucka, J. Baj, and M. Wlodarczyk. 1985. Naturally occurring chromosomal gene transfer in *Thiobacillus versutus. Acta Microbiolog. Polonica* 34:81–83.
15. Plasota, M. E. Piechucka, B. Kauc, and M. Wlodarczyk. 1984. R68.45 plasmid mediated conjugation in *Thiobacillus* A2. *Microbios.* 41:81–89.
16. Rawlings, D. E., I.-M. Pretorius, and D. R. Woods. 1986. Expression of *Thiobacillus ferrooxidans* plasmid functions and the development of genetic systems for the thiobacilli. *Biotech. Bioeng. Symp.* 16:281–287.
17. Spisak, J. F. 1986. Biotechnology and the extractive metallurgical industries: Perspectives for success. *Biotech. Bioeng. Symp.* 16:331–341.

18. Summers, A. O., P. Roy, and M. S. Davidson. 1986. Current techniques for the genetic manipulation of bacteria and their application to the study of sulfur-based autotrophy in *Thiobacillus. Biotech. Bioeng. Symp.* 16:267–279.
19. Torma, A. E. 1986. Biohydrometallurgy as an emerging technology. *Biotech. Bioeng. Symp.* 16:49–63.
20. Tuovinen, O. H. 1986. Acid leaching of uranium ore materials with microbial catalysis. *Biotech. Bioeng. Symp.* 16:65–72.
21. U.S. Department of the Interior. 1985. Control of acid mine drainage. Bureau of Mines Information Circulars, IC 9027.
22. U.S. Department of the Interior. 1988. *Mine Drainage and Surface Mine Reclamation, Vol. 1, Mine Water and Mine Waste.* Bureau of Mines Information Circulars, IC 9183.
23. Wichlacz, P. L. 1986. Practical aspects of genetic engineering for the mining and mineral industries. *Biotech. Bioeng. Symp.* 16:319–326.
24. Yankofsky, S. A., R. Gurevich, N. Grimland, and A. A. Stark. 1983. Genetic transformation of obligately chemilithotrophic thiobacilli. *J. Bacteriol.* 153:652–657.

43

Microorganisms Associated with Animals

Timothy J. Miller

Introduction

The development and utilization of microorganisms and fermentation products had played a critical role in the health benefits obtained from research on disease pathogenesis, drug metabolism, and vaccine formulation. However, the use of microorganisms to produce health care products has long been under the scrutiny of regulatory authorities. For microorganisms associated with animals, the regulation of production and marketing of biological agents is under the jurisdiction of the Veterinary Services (VS) Branch of the Animal and Plant Health Inspection Service (APHIS) of the USDA. These regulatory procedures have been effective for mitigation of biological agents produced by conventional approaches and have proved to be effective as guidelines for characterizing the first genetically engineered microorganisms (GEMs) released into the environment.

Most of the first GEMs were constructed using the bacterium, *Escherichia coli,* but now genetic elements are available for other bacteria, viruses, algae, fungi, and protozoa, as well as with plants (39, 47) and large multicellular organisms, particularly transgenic animals (22). In assessing the impact of any organism artificially introduced into the environment, the route or method of introduction becomes important. A GEM introduced into an animal is under different considerations than an organism inoculated into the soil or water supply. The animal becomes a potential carrier that can perhaps amplify the GEM by replication; act as a reservoir for attenuation, modification, or recombination with homologous organisms from which the GEM was constructed; or transmit the GEM to other animals or humans. Although established regulatory testing has adequately provided information on the mitigation of microorganisms, new considerations for risk assessment may be necessary for procedures that are specifically designed to introduce foreign genes into the germline of an animal (transgenic animals). These mitigation procedures for integrated genes should be handled as totally different subjects and will not be the topic of discussion here.

History of the Regulation of Biological Agents Used in Animals

Traditionally, the Virus-Serum-Toxin Act (VSTA) of 1913 enacted by Congress (5) describes the requirements for testing the safety, efficacy, and purity of any biological agent. In 1983, APHIS received the first two applications for licenses to manufacture and market a vaccine derived from recombinant DNA technology for use in animals; the regulatory criteria used to determine the safety of a conventionally produced biological agent was used for these GEMs. These first recombinant proteins were used in combination with inactivated *E. coli* pili vaccines for prevention of swine scours. There was no threat to the environment because the recombinant products were not viable and guidelines for laboratory therapeutics and prophylactics. Category 2 included products consisting of live organisms that had been made avirulent by removing genomic sequence from the organism or adding a foreign gene. The new gene or gene product could not increase or change in any way the known pathology or virulence activity of the parent organism. Category 3 are products using live recombinant vectors containing one or more foreign genes to be used as immunizing or immunomodulating components (38). As described above, two licenses were granted for category 1 recombinant DNA products in 1984. In 1986, the first category 2 license was issued to TechAmerica for a recombinant DNA pseudorabies vaccine (2). In 1988, a second category 2 license was issued to Syntro Vet, Inc., for a similar but more extensively recombinant DNA pseudorabies vaccine (38). APHIS provided an environmental assessment (EA) of both products. In the case of the pseudorabies category 2 GEMs, findings of no significant impact (FONSI) were assessed, and product licenses were issued. The FONSI was supported by standard regulations and guidelines set forth in the Code of Federal Regulations for Animal and Animal Products (5), which is used as a guideline for licensing conventional products. These findings were unprecedented, and for the first time, live genetically engineered viruses (GEMs) were released into the environment through administration of vaccine to animals. As a result of the decision, several considerations, precautions, and procedures were established to evaluate GEMs produced under category 3 (6).

General Issues

There are well-established and proven procedures for physical and chemical means for decontamination and mitigation of microorganisms associated with animals. Descriptions and protocols of these methods can be readily found through ample literature references (1, 3, 8, 9); however, a few concepts for mitigation should be kept in mind when designing a GEM.

Regulatory procedures

The USDA will continue to monitor licensing of GEMs on a case-by-case basis. However, as described above for the live pseudorabies vaccine, an EA will be prepared in accordance with the Natural Environmental Policy Act (NEPA) for 1989 (7), considering possible hazards to humans or the environment. The precedent set by licensing the pseudorabies GEMs established criteria for data that needed to demonstrate safety and efficacy. Materials submitted that do not meet standards for demonstrating efficacy and safety compared with prior data can be eliminated for consideration in field tests or licensing.

Considerations for designing safe genetically engineered organisms

Although GEMs can be designed to express genes encoding any protein sequence (enzymes, hormones, antibiotics, etc.), recombinant DNA as a means to generate novel vaccines has been one of the principal applications of this technology (14, 19, 21, 26, 27, 34, 36, 40, 45, 53) and will be the primary focus in the following sections. In the case of animal health, vaccination is still the preferred method for infectious disease prevention and control because of convenient and inexpensive methods for administering and producing vaccines. When designing a recombinant DNA–derived vaccine, the safety concerns for developing and releasing a GEM can be reduced by selecting potential host organisms that have desirable properties. The following concepts should be considered in selecting a host system that would not only meet therapeutic or prophylactic criteria, but would be much more amenable to mitigation and decontamination procedures under standard practices.

1. The host system should replicate or deliver the gene or gene product to the appropriate tissue without introducing a new or altered

virulence characteristic. The first criteria with which to assess release of a GEM is whether the organism provides a significant health advantage over conventional products. GEMs that do not provide a significant advantage over conventional biological agents previously licensed or are not as effective should not be considered for development. In the case of the pseudorabies GEMs described above (26, 27, 29), gene deletions were used to modify the virus (38).

2. The GEM should be genetically stable and limited in its ability to recombine with tissue or endogenous pathogens of the animal or environment. Established experimental methods are available to ensure that genetic transfer (transformation, conjugation, or transduction) is absent or controllable (16). As described above, the genetically modified pseudorabies virus could not be recovered from any of the target animals (pigs) used in the immunogenicity trials or during the field tests. Reversion of the deleted genes or loss of the reporter gene could also not be demonstrated. Thus, by existing regulatory standards, the virus was not recoverable or present in the animal at the time of slaughter and posed no threat for gene transfer to humans or other nontarget animals (2, 27, 38).

The interaction of the GEM with the immune system of the human or animal is extremely important when considering live vectors for delivery of any health care component (41). It would be advantageous if a GEM could be so designed that it will enter a target animal, introduce the prophylactic or therapeutic component, and then be totally controlled or destroyed by a preplanned or a designed immune response or a chemical, enzymatic, or physiological activity of the animal. A vector or virus, such as pseudorabies, can be eliminated in a short time by the immune response. As a result, the GEM is destroyed, and the animal still acquires an immunity to disease. This is a simple and very cost-efficient concept for mitigation of the GEM in animals. When GEMs are designed to deliver an antibiotic, enzyme, hormone, or other therapeutic drug, a vector could be chosen to allow delivery over a short period of time.

3. The GEM should be restrictive in replication and should not be transmitted easily to nontarget host animals. This feature can be controlled by a number of effective procedures, including the immune response (18, 26, 29, 30, 36, 37), autolysis (24, 39, 49), or auxotrophic mutations (14, 15, 23, 42, 52). When there is no indication of the GEM outside the animal once it is introduced, the remaining consideration is whether the GEM can persist in animal tissue and be transmitted in a consumable food product. The persistence of a GEM in the animal depends on its ability to induce immune tolerance and the resistance that it has to physiological responses. In most situations, samples at

the site of inoculation or at preferred sites of replication of the GEM can be used for isolation of the organism (56, 59). The use of a vector that has a known preferred site of replication would enhance the ability to analyze the activity of the GEM at different times after introduction into the animal.

4. The GEM should be susceptible to commonly used physical or chemical decontamination protocols. A GEM that is used as a vector to introduce a biological product to an animal vector should be chosen because of its favorable characteristics for transmissibility. An organism that is transmitted by aerosol (e.g., respiratory pathogens), infected cells (e.g., sexual transfer, slaughter tissue), or by biting insects (e.g., blood-borne organisms) would not be desirable. An organism that is confined to a nontransmitted tissue or can be controlled by immune surveillance has a greatly reduced adverse impact on the environment.

5. A stable nontransferrable marker or reporter gene(s) should be incorporated to identify easily the GEM from other biological agents. Many procedures are now available for inserting genes into the genome of microorganisms (16) or autonomous replication genetic elements (35).

6. Secondary genes or gene products should also be considered as specific components of the GEM to inactivate genetically, enzymatically, chemically, or immunologically the therapeutic gene, vector, or gene product. In some cases, the reporter or marker gene can also be used as a biological decontamination component. Specific examples of GEMs that have been engineered to allow biological mitigation and/or decontamination are given below.

Physical Procedures for Mitigation and Decontamination

The USDA has long-standing guidelines for limiting the spread of pathogenic organisms. Import and shipping restrictions have been effective in limiting the introduction of potentially biohazardous material into the United States and between states. Quarantine practices for infected herds, flocks, or households have eliminated or reduced transmission of disease during periods when the host animal is highly contagious. Rules mandating disposal of consumables (milk, meat, eggs) that are demonstrated to be carrying transmissible disease are enforced as part of the Food Safety and Inspection Service (FSIS) (8). These practices will continue to be effective as mitigation and decontamination procedures for biological agents in animals and should be considered as the primary directive for containing GEMs. However, in special cases, the means for evaluating virulence, persistence, and transmissibility of GEMs may need to be monitored by improved technology including gene probes, monoclonal antibody tests, or gene amplification.

Chemical Means for Mitigation and Decontamination

A number of organic or inorganic compounds have been identified and routinely used for decontamination of common pathogens of humans and animals (11, 12, 31, 46). Many compounds are inexpensive and can be used to disinfect or decontaminate holding pens, equipment, containment buildings, and even small field lots (3).

Biological Mitigation and Decontamination

The information and techniques that have introduced the age of biotechnology and the potential for novel, more effective health products have brought with them special environmental concerns and restrictions. As new considerations and uses of GEMs are conceived, investigators, as well as regulatory authorities, need to be aware that the technology that can so elegantly produce GEMs can also be used to provide mitigation and decontamination procedures. Biotechnology used for mitigation and decontamination is not a substitute for the standard practices of regulating biological agents but, rather, increases our ability to understand disease and monitor the safety of GEMs.

Host-dependent gene expression systems

The necessity to create bacterial host strains that compete poorly in the environment was recognized early during the formulation of recombinant DNA containment guidelines. As a consequence, a number of laboratory practices were established for handling recombinant DNA. In addition, bacterial host strains were modified to ensure that they would not be hazardous if accidentally released into the environment (43, 44). These bacterial strains have proved to be safe for laboratory procedures, and many have desirable qualities as live delivery systems for expressed gene products of recombinant DNA.

An example of this type of system utilizes an expression plasmid that carries an enzyme required for a nutritional component that is deleted in a *Salmonella typhimurium* vaccine strain (43). Many mutations of *S. typhimurium* exist, and some are known to have reduced virulence in target animals (14, 23, 52) and could be licensed for use as vaccines under current regulatory requirements. However, when

bacterial mutations are induced by using genomic integration protocols and carry an autonomously replicating recombinant expression
plasmid, the host and vector must be assessed for susceptibility to mitigation or decontamination procedures.

One of the best-characterized vaccine strains of *Salmonella* has
mutations in the functions of the adenyl cyclase (*cya*) and cyclic
AMP receptor protein (*crp*) genes, which are involved with regulation of many gene functions used in nutrient utilization and transport, as well as the synthesis of outer membrane proteins. *S.
typhimurium* vaccine strains that are Δ*cya* Δ*crp* have reduced capacity to replicate and are less virulent than wild-type strains in
the environment (14, 15, 42).

To utilize the Δ*cya* Δ*crp* mutants as potential recombinant DNA
vector systems for a foreign gene product, a third mutation was introduced. The gene for aspartate semialdehyde dehydrogenase (*asd*) was
deleted, endowing the mutant with a strict requirement for threonine,
methionine, and diamino pimelic acid (DAP). DAP cannot be supplied
by the mammalian host, therefore, *asd⁻* mutants are restricted for
replication inside the host. Furthermore, expression plasmids carrying the *asd⁺* gene are strongly selected by Δ*asd* mutants. As a result,
Δ*cya* Δ*crp* Δ*asd* mutants carrying a *asd⁺* expression plasmid are stably maintained through many generations and in passage through the
mammalian host. Wild-type microorganisms do not have a selective
pressure to harbor an *asd⁺* plasmid, whereas Δ*asd* *S. typhimurium*
mutants undergo rapid cell death without a complementing *asd⁺*
function (14).

Advantages. When introduced into mice, the Δ*cya* Δ*crp* strains can
populate the gut-associated lymphatic tissue (GALT) and intestines,
but they have as much as a 10,000-fold reduced capacity to populate
the spleen and liver than with other strains. This makes the Δ*cya* Δ*crp*
mutants attractive vaccine candidates, as well as live carriers for recombinant DNA-expressed genes (14, 42). The Δ*asd* mutation causes
rapid cell lysis and death in the absence of DAP which eliminate concerns about the strain surviving without a *asd⁺* expression vector.
Thus, the *asd⁺* gene serves both as a strong selective marker for *asd⁻*
mutants as well as a reporter gene for following the fate of the GEM.
Recombinant strains recovered in the feces cannot only be identified
readily but can be quantitatively enumerated among the microbiota of
the fecal sample. The Δ*cya* Δ*crp* strains also have virtually no chance
of reversion, and transduction would be a rare event because of the
large size of the deletions. The Δ*cya* Δ*crp* could be restored by mating
with a Hfr-type donor; however, the mutations are 11 min apart on
the chromosome, and no such donor has been identified for *E. coli* or *S.
typhimurium* (13).

Disadvantages.   Although the Δ*cya* Δ*crp* mutants do not populate the spleen to the same extent as wild-type *S. typhimurium,* the vaccine strain will populate the intestinal tract and can reside persistently in the GALT (14, 15). As a result, Δ*cya* Δ*crp* Δ*asd* mutants can be shed (although at much lower levels compared to wild type) from the animal and could reinfect a nonvaccinated animal. Contact exposure of nonvaccinated animals may be desirable in controlling some diseases, especially those harbored in wildlife reservoirs. However, the long-term benefit of a GEM vaccine to control disease by continued secretion and contact between vaccinated and nonvaccinated animals would need to be monitored and compared to similar vaccination programs with modified live microorganisms produced by conventional technology. The primary concern of the *S. typhimurium/asd*$^+$ plasmid system would be the assessment of the competitive advantage that the strain or plasmid may have in the environment. Despite the fact that the loss of the *asd*$^+$ plasmid causes death of the Δ*cya* Δ*crp* mutants, the plasmid may be transferred to nontarget organisms. Possible means to limit the *asd*$^+$ plasmid to only the Δ*cya* Δ*crp* mutants could be accomplished by a mutation in a maintenance or replication dependent gene in the plasmid vector. The host cell (*Salmonella*) should provide a complementing function that could not be supplemented in nature (14).

Temperature-induced lysis

Bacteriophages can be used to create lysogenic bacteria that integrate regulative properties into the bacterium for mitigation purposes (24, 49). The bacteriophage lambda λ integrates into the genome of *E. coli* at one unique site. The integrated phage DNA (prophage) can be activated under certain conditions to induce lysis of the bacterium. This acts as an internal control for mitigation of the lysogenic host strain. One such strain, *E. coli* MG3, is a derivative of the cI857 lysogen of *E. coli* strain N99. The λ prophage of MG3 carries the temperature-depressible [λ] repressor gene cI857 (39). Bacteriophage λ cI857 contains two mutations in the cI repressor gene: one makes cI insensitive to protease, and the second causes the cI product to denature and become nonfunctional at 43°C. When grown at 32°C, the cI product is fully functional and represses synthesis of the N-gene product, which controls synthesis of the Q-gene product, which, in turn, controls the expression of the S and R genes. The S- and R-gene products are lysozymes that cause cell lysis. Thus, when the temperature is raised to 43°C, the cI repressor is inactivated, which allows the N protein to be synthesized and ultimately enables the lysis function to be expressed.

In addition to the mutations in the cI gene of the MG3 λ prophage, the O and P genes for phage replication also carry mutations (P3am), 029). Consequently, the MG3 lysogen is double-blocked for replication. Finally, the prophage carries a deletion (generated in vitro with BamHI endonuclease) of the 6.3-kb BamHI fragment from 0.58 to 0.71 of the λ genome. This deletion results in the loss of the λ *int-, xis-, red-, gam-, kil-,* and *cIII*-gene products. The loss of these genes prevents excision of the prophage once it is integrated into the bacterial chromosome. The resulting MG3 lysogen contains functional R, S, N, and Q genes, but it will not replicate or excise its DNA from the bacterial chromosome. Thus, when a culture of MG3 is shifted from 32 to 43°C, no phage is produced, yet cells will lyse, as the S and R genes are functional (39).

Advantages.   The MG3 lysogenic strain contains naturally occurring sequences of the λ prophage and does not introduce a new virulence factor. Upon induction of cell lysis, the bacterial DNA is rapidly degraded, thus reducing the potential for transfer of the bacterial chromosome to other organisms. Because λ can integrate at only one site in the chromosomal DNA, the bacteria are immune to infection by another λ phage (54). The prophage DNA cannot replicate, excise, or integrate with other DNA. As a result, this culture stably maintains the prophage DNA but cannot compete with other microorganisms of phage in the environment. The MG3 strain can be used in combination with any recombinant DNA plasmid carrying an ampicillin-resistance marker. Upon lysis, the recombinant DNA–expressed protein is released from the bacteria. These qualities make the MG3 strain desirable as an oral vaccine delivery system. The high temperature of the mammalian intestinal tract (39°C) would induce lysis, which would reduce survival and secretion from the host animal. This would be a self-limiting modification by which any lysogens surviving the first passage through an orally vaccinated animal would be transmitted to a nonvaccinate. However, on each passage, the strain would be reduced in number.

Disadvantages.   The MG3-λ lysogen strain is ampicillin-sensitive and is commonly used with recombinant plasmids expressing β-lactamase. As a live GEM introduced into the animal, it would provide drug resistance by virtue of the recombinant plasmid. Because the USDA does not allow for new antibiotic-resistant strains to be licensed, this is not a favorable feature of the system. The problem could be remedied by incorporating similar mutations and a nutritional selection system for the plasmid system, as described above for *S. typhimurium.*

Mammalian viruses

Mammalian viruses have been successfully used to express foreign genes in cell culture. The brevity of this discussion does not allow discussion of all systems that are available (48); however, vaccinia and herpes virus systems have been used successfully and extensively as live GEMs through site-directed deletions of the viral thymidine kinase (tk) gene (12, 26, 28, 33, 34). The method exploits the fact that deletion of certain regions of the virus genome strongly attenuates the virus; these regions are nonessential for replication. The tk$^-$ mutants of the pseudorabies virus were considered to be safe, as they were avirulent, and the animal mitigated the virus through an immune response. The tk$^-$ mutants of vaccinia and herpes viruses are dramatically reduced in virulence (13, 26, 36). Moreover, TK$^-$ mutants can readily be identified by growing the viruses in the presence of bromodeoxyuridine BuDR, as the TK$^+$ viruses incorporate BuDR, which is lethal, whereas the tk$^-$ do not utilize the drug. In the mammalian host, tk$^-$ viruses use the thymidine kinase of the cell inefficiently and do not exhibit productive infections (26, 27). For pseudorabies viruses, the sites of pathology as well as of latency occur in the nerve tissue where the cell provides low levels of tk. Viruses that are tk$^+$ can replicate efficiently, whereas tk$^-$ mutants have difficulty in replicating or in being reactivated after latency is established (30).

Recombinant viruses are made by coinfecting a permissive cell line with tk$^+$ virus and a plasmid containing a foreign gene that is activated by an operational transcriptional promoter inserted in the middle of the tk gene of the virus. The tk gene is nonfunctional by virtue of the foreign gene introduced into the coding region. The recombinant plasmid is tk$^-$, but it can recombine into the tk$^+$ loci of the wild-type virus in a coinfection. These tk$^-$ recombinant viruses are avirulent, but these techniques provide a means to introduce a foreign gene into the virus, which can then infect a target cell and deliver an appropriate gene product.

Although tk$^-$ viruses are generally avirulent, new techniques have been developed to attenuate other gene functions of mammalian viruses. One of these techniques uses a reporter gene as a marker to monitor recombination in vaccinia virus. The technique uses the enzyme β-galactosidase (β-gal) which is ligated into selected restriction enzyme fragments such that the β-gal gene is flanked by the restriction fragment sequences. The β-gal gene has a constitutive promoter that continually synthesizes the enzyme which can convert the substrate, 5-bromo-4-chloro-3-indoly-1-β-galactoside (X gal) into a blue chromogen. The recombinant plasmid is then cotransfected with wild-type virus into permissive cells. The restriction fragment sequences flanking β-gal homologously recombine to create a recombinant virus carrying the β-gal in a designated gene fragment. The recombinant virus can be identified as a blue plaque when X gal is present in the plaquing media (20, 55). The production of a blue plaque indicates that the recombination has successfully taken place. In addition, since the β-gal interrupts a vaccinia DNA sequence, the site of integration designates a nonessential region for virus replication. This procedure allows a systematic approval to identify regions for attenuation without compromising replication.

Advantages.   Recombinant vaccinia virus carrying a foreign gene offers several advantages as live GEMs:

1. Viruses, such as vaccinia and herpesvirus, have large genomes and could accommodate large segments of foreign DNA that could express several genes at once (45).

2. Live-virus vaccines usually elicit a much broader spectrum of immunity and can be controlled by the immune system, as described above. Cells infected with tk⁻ pseudorabies are eliminated by the immune system over a period of time; the GEM has a built-in mechanism for the biological decontamination of the animal. Therefore, the virus has a reduced capacity to be shed or to transfer DNA to nontarget organisms. The immune system acts as a natural surveillance system for mitigation of GEMs in the animal.

3. Live-virus vaccines are usually less expensive to manufacture and they can be propagated easily (33, 34, 40).

4. Viruses, such as vaccinia, are stable and can withstand high temperatures and desiccation.

5. Live vaccines are easy to administer under standard practices for vaccination.

6. Viruses, such as vaccinia, have a wide host range and could be useful in the production of products for both human and animal health care.

7. The incorporation of marker or reporter genes offers two advantages. One, the recombinant virus is readily identified by blue-colored plaques when plated with X gal; and two, the reporter gene can be inserted into DNA restriction enzyme fragments of the virus genome that are useful in identifying nonessential gene functions. This technique could potentially identify genes that are involved in important functions of maturation and replication of the virus. For example, the virus could be attenuated to restrict host cell range or

eliminate virus maturation in certain cell types. As a result, selective features of attenuation could be engineered into vaccinia that will allow for more sophisticated methods for biological mitigation or decontamination.

Disadvantages. Despite the usefulness and efficacy of live vaccines, there are several disadvantages to their use.

1. Live viruses can replicate in the target animal and potentially recombine with indigenous virus strains or host DNA.
2. Attenuated live viruses can potentially revert to virulence.
3. For viruses, such as vaccinia, the features of wide host range and stability may be a disadvantage if the health care component is needed in selected tissue for a short period of time. These problems can be alleviated by attenuating specific gene functions to limit the host range specificity or by use of tissue-specific promoters. Recombinant viruses carrying genes under the control of promoter enhancers sequences would be recognized in selected tissue (25, 32, 51, 57).
4. Although risk may be low for individual animals receiving the GEM in the field, containment procedures at the manufacturing level must still be adhered to for handling high concentrations of the vector systems by trained personnel.

The choice of a vector system can be of great significance when considering mitigation or decontamination in the environment. Vaccinia virus is a safe, modified live virus that was used to eradicate smallpox (17, 58), but it has a broad host range (8, 10), which raises concerns about recombinant vaccinia virus affecting nontarget animals. Although other poxviruses (e.g., swinepox and fowlpox) that have limited host range are being developed, they are not as proliferative in vitro, and the immune response is not as well characterized.

Summary and Conclusions

Biotechnology offers a tremendous opportunity to provide improved and novel health care products especially in cases where health care treatment has not been possible by conventional research approaches. However, the criteria for the safe use of any organism as a health care product should be based on highly selective and rational processes. The benefit of the GEM must be balanced against all other technology that could produce similar health care treatment.

The purpose of the technology is to understand and treat diseases induced by microbial infections or by genetic aberrations. The treatment of animals or plants with well-defined and characterized genes or GEMs will better define the disease process rather than to enhance it. Furthermore, the elegant techniques now available for genetic engineering and for biological mitigation and decontamination are not only methods that can ensure safety to the environment but that can also serve as additional reporter tracking agents for epidemiological studies. Methods and guidelines for handling microorganisms that have been successfully used in the past should remain the major effective means for mitigation and decontamination of GEMs. Technology is capable of engineering many features for biological mitigation; however, expense of engineering both safety and health care features must also be cost-effective for manufacturing or development to proceed.

References

1. Anonymous. 1987. Interstate transportation of animals (including poultry) and animal products. Code of Federal Regulations. Animal and Animal Products, Title 9. U.S. Government Printing Office, Washington, D.C., pp. 125–230.
2. Anonymous. 1986. Availability of environmental assessment and finding of no significant impact for the licensing of a recombinant derived pseudorabies virus vaccine. *Federal Register* 51:15657.
3. Anonymous. 1987. Cooperative control and eradication of livestock or poultry diseases. Code of Federal Regulations. Animal and Animal Products, Title 9. U.S. Government Printing Office, Washington, D.C., pp. 102–122.
4. Anonymous. 1984. Proposal for coordination framework of biotechnology. *Federal Register* 51:23303.
5. Anonymous. 1987. Viruses, serums, toxins and analogous products; organisms and vectors. Code of Federal Regulations. Animal and Animal Products, U.S. Government Printing Office, Washington, D.C., pp. 352–533.
6. Anonymous. 1989. Public meeting. Availability of environmental assessment and preliminary finding of no significant impact for field testing a genetically engineered vaccinia vectored rabies vaccine. *Federal Register* 54:9241.
7. Anonymous. 1979. APHIS guidelines implementing National Environmental Policy Act of 1969. *Federal Register* 44:50,381; 51,272.
8. Anonymous. 1988. Food safety and inspection service, meat and poultry inspection, Department of Agriculture. Code of Federal Regulation. Animal and Animal Products, Title 9, U.S. Government Printing Office, Washington, D.C., pp. 89–335.
9. Anonymous. 1985. A concise index of U.S. laws for safety regulation of biological systems and organisms. *Federal Register* 50:47174.
10. Barby, D. 1977. Poxviruses host and reservoirs: Brief review, *Arch. Virol.* 55:169.
11. Bernarde, M. A. 1970. *Disinfection.* Marcel Dekker, New York.
12. Borick, P. M. 1973. *Chemical Sterilization.* Dowden, Hutchinson and Ross, Inc., Stroudsburg, Pa.
13. Buller, R. M. L., G. L. Smith, K. Cremer, A. L. Notkins, and B. Moss. 1985. Decreased virulence of recombinant vaccinia virus expression vectors as associated with a thymidine kinase-negative stereotype. *Nature* 317:813–815.
14. Curtiss, R. III. 1988. Engineering organisms for safety. What is necessary? In *Release of Genetically Engineered Microorganisms.* Academic Press, New York, pp. 8–19.
15. Curtiss, R. III, S. M. Kelly, P. A. Gulig, C. R. Gentry-Weeks, and J. E. Galan. 1988. A virulent *Salmonella* expressing virulence antigens from other pathogens for use as orally administered vaccines. In James A. Roth (ed.), *Virulence Mechanisms of Bacterial Pathogens.* American Society of Microbiology, Washington, D.C., pp. 311–328.

16. Davis, R. W., D. Botstein, and J. R. Roth. 1980. Procedures and appendixes. In *A Manual for Genetic Engineering. Advanced Bacterial Genetics.* Cold Spring Harbor Laboratory, Cold Spring Harbor, N.Y., pp. 70–253.

17. Deria, A., Z. Jazek, K. Markvard, P. Carrasco, and J. Weisfeld. 1980. The world's last endemic case of smallpox. *Bull. WHO* 58:279.

18. Fenner, F. 1985. Poxviruses. In B. Fields (ed.), *Virology,* Raven Press, New York, pp. 661–685.

19. Flexner, C., B. R. Murphy, J. F. Rooney, C. Wohlenberg, V. Yuterov, A. L. Notkins, and B. Moss. 1988. Successful vaccination with polyvalent live vector despite existing immunity to an expressed antigen. *Nature* 335:259–262.

20. Girard, S., D. S. Spehner, R. Drillien, and A. Kirn, 1986. Localization and sequence of a vaccinia virus gene required for multiplication in human cells. *Proc. Natl. Acad. Sci. USA* 83:5573–5577.

21. Hamer, D. H., and P. Leder. 1979. Expression of the chromosomal mouse β majglobin gene cloned in SV-40. *Nature* 281:35–40.

22. Hogan, B., F. Constantini, and E. Lacy. 1986. Introduction of new genetic information into the developing mouse embryo. In *Manipulating the Mouse Embryo.* Cold Spring Harbor Laboratory, Cold Spring Harbor, N.Y., pp. 151–197.

23. Hoiseth, S., and B. A. D. Stocker. 1981. Aromatic-dependent *Salmonella typhimurium* are non-virulent and effective as live vaccines. *Nature* 291:238–239.

24. Johnson, A. D., B. J. Meyer, and M. Ptashne. 1979. Interaction between DNA-bound repressors governs regulation of λ phage repressor. *Proc. Natl. Acad. Sci. USA* 76:5061.

25. Jordano, J., and M. Perucho. 1988. Initial characterization of a potential transcriptional enhancer for the human L-K-ras gene. *Oncogene* 2:359–366.

26. Kit, S., M. Kit, B. Lawhorn, and S. McConnell. 1985. Immunization of pregnant pigs in a quarantined swine herd with a thymidine kinase deletion mutant. In *Virus Vaccine.* American Society for Microbiology, Washington, D.C., pp. 82–102.

27. Kit, S., M. Kit, and E. C. Pirtle. 1985. Attenuated properties of a thymidine kinase-negative deletion mutant of pseudorabies virus. *Am. J. Vet. Res.* 46:1359–1367.

28. Kit, S., M. Sheppard, H. Ichimura, and M. Kit. 1987. Second generation pseudorabies virus vaccine with deletions in thymidine kinase and glycoprotein genes. *Am. J. Vet. Res.* 48:780–793.

29. Kit, S., M. Sheppard, and M. Kit. 1986. Control of Aujesky's disease. *Vet. Rec.* 118:310.

30. Kit, S., H. Qaui, D. R. Dubbs, and H. J. Otsuka. 1983. Attenuated marmoset herpesvirus isolated from recombinants of virulent marmoset herpesvirus and hybrid plasmids. *J. Med. Virol.* 12:25–36.

31. Lelie, P. N., H. W. Reesink, and C. J. Lucas. 1987. Inactivation of 12 viruses by heating steps during manufacture of hepatitis β vaccine—vaccinia virus, encephalomyocarditis virus, sindbis virus, arbovirus, mouse hepatitis virus, influenza virus, vesicular stomatitis virus, SV-40, cytomegolovirus, human immunodeficiency virus, murine leukemia virus, canine parvovirus and phage phiX-174. *J. Med. Virol.* 23:297–302.

32. Lufkin, T., and C. Bancroft. 1987. Identification by cell fusion of gene sequences that interact with positive transacting factors. *Science* 237:283–286.

33. Mackett, M., G. Smith, and B. Moss. 1982. Vaccinia virus, a selectable eucaryotic cloning expression vector. *Proc. Natl. Acad. Sci. USA* 79:7415–7419.

34. Mackett, M., G. Smith, and B. Moss. 1984. General method for production and selection of infectious vaccinia virus recombinants expressing foreign genes. *J. Virol.* 49:857–864.

35. Maniatis, T., E. F. Fritsch, and T. Sambrook. 1982. *Molecular Cloning. A Laboratory Manual.* Cold Spring Harbor Laboratory, Cold Spring Harbor, N.Y.

36. McFarland, M. D., H. I. Hill, and L. B. Tabotabai. 1987. Characterization of virulent and attenuated strains of pseudorabies virus to thymidine kinase activity, virulence and restriction patterns. *Can. J. Vet. Res.* 51:334–339.

37. McFarland, M. D., and H. T. Hill. 1987. Vaccination of mice and swine with a

pseudorabies virus mutant lacking thymidine kinase activity. *Can. J. Vet. Res.* 51: 340–344.

38. Medley, T. L. 1988. Issues in assessing the environmental impact of veterinary biologics produced through biotechnology. *Food Drug Cosmetic Law J.* 43:821–829.

39. Miller, T. J., R. Peetz, A. P. Reed, T. Kost, A. Brown, J. Auerbach, and R. Rosenberg. 1985. High level expression and externalization of the β λ subunit *Escherichia coli* heat labile toxin by a λ lysogen. *Vaccines 85.* Cold Spring Harbor Laboratory, Cold Spring Harbor, N.Y., pp. 95–99.

40. Moss, B., and L. W. Fexner. 1987. Vaccinia virus expression vectors. *Annu. Rev. Immunol.* 5:305–324.

41. Murphy, B. R., and R. M. Chanock. 1985. Immunization against viruses. In B. Fields (ed.), *Virology.* Raven Press, New York, pp. 349–370.

42. Nakayama, K., S. M. Kelly, and R. Curtiss III. 1988. Construction of an asd⁺ expression-cloning vector: Stable maintenance and high level expression of cloned genes in a Salmonella vaccine strain. *Biotechnology* 6:693–697.

43. National Academy of Sciences USA. 1977. *Research with Recombinant DNA: An Academy Forum.* National Academy of Sciences, Washington, D.C.

44. National Institutes of Health. 1976. Recombinant DNA research: Guidelines for recombinant molecule research. *Federal Register* 41:3702–3794.

45. Perkus, M. E., A. Piccini, B. R. Lipinskas, and E. Paoletti. 1985. Recombinant vaccinia virus: Immunization against multiple pathogens. *Science* 229:981–984.

46. Phillips, G. B., and W. S. Miller. 1973. *Industrial Sterilization.* Charles C Thomas, Springfield, Ill.

47. Pouwels, P. H., B. E. Enger Valk, and W. J. Brammer. 1985. *Cloning Vectors.* Elsevier, New York.

48. Ramig, R. F. 1985. Principles of animal virus genetics. In B. Fields (ed.), *Virology.* Raven Press, New York, pp. 101–128.

49. Roberts, J., and R. Devoret. 1983. Lysogenic function in *lambda II.* In W. Hendrix, J. Roberts, F. Stahl, and R. Weisberg (eds.). Cold Spring Harbor Laboratories, Cold Spring Harbor, N.Y., pp. 123–144.

50. Schnieke, A., K. Harbers, and R. Jaenisch. 1983. Embryonic lethal mutation in mice induced into the α(I) collagen gene. *Nature* 304:315–320.

51. Schmidt, E. V., P. K. Pattengale, L. Weir, and P. Leder. 1988. Transgenic mice bearing the human c-myc gene activated by an immunoglobulin enhancer: apre-B-cell lymphoma model. *Proc. Natl. Acad. Sci. USA* 85:6047–6051.

52. Smith, B., M. Reina-Guerra, S. Hoiseth, B. A. D. Stocker, F. Habasha, E. Johnson, and F. Merrit. 1984. Aromatic-dependent *Salmonella typhimurium* as modified live vaccines for calves. *Am. J. Vet. Res.* 45:59–66.

53. Smith, G. L., M. Mackett, and B. Moss. 1983. Infectious vaccinia virus recombinants express hepatitis B virus surface antigens. *Nature* 302:490–495.

54. Smith, G. 1983. General recombination in *lambda II.* In W. Hendrix, J. Roberts, F. Stahl, and R. Weisberg (eds.). Cold Spring Harbor Laboratory, Cold Spring, N.Y., pp. 175–209.

55. Spyropoulos, D. D., B. E. Roberts, D. L. Panicali, and L. K. Cohen. 1989. Delineation of the viral products of recombination in vaccinia virus infected cells. *J. Virol.* 62(3):1046–1054.

56. Valenzuela, R., and S. D. Deodhar. 1986. Tissue immunofluorescence. In N. R. Rose, H. Friedman, and J. L. Fahey (eds.), *Manual of Clinical Laboratory Immunology.* American Society for Microbiology, Washington, D.C., pp. 923–925.

57. Vogt, T. F., R. S. Compton, R. W. Scott, and S. M. Tilghman. 1988. Differential requirements for cellular enhancers in stem and differentiated cells. *Nucleic Acids Res.* 16:487–500.

58. World Health Organization. 1980. The global eradication of smallpox. Final report of the global commission to certification of smallpox eradication. *Geneva Weekly, Epidemiological Record* 55:145–152.

59. Tubbs, R. R. 1986. Lymphocyte markers in solid tissues. In N. R. Rose, H. Friedman, and J. L. Fahey (eds.), *Manual of Clinical Immunology.* American Society for Microbiology, Washington, D.C., pp. 934–937.

44

Bacillus

Jonathan Lamptey

Carol A. Hendrick

Nancy J. Tomes

Susan Brown

Donald H. Dean

History

Description of class

The genus *Bacillus* comprises rod-shaped, gram-positive microorganisms which form endospores as a resting physiological state when exposed to adverse growth conditions (16, 71). Endospores are about 1000 times more resistant to heat, irradiation, and desiccation than vegetative cells. Vegetative cells will grow aerobically and, in some cases, anaerobically. Most are catalase-positive. Most species are widely distributed in nature and are common saprophytic soil isolates. Bacilli produce a wide range of peptide antibiotics, enzymes, and insect toxins. Some species are pathogenic to animals; *B. anthracis* is the causal agent of anthrax, and certain strains of *B. cereus* can cause food poisoning.

Actual and potential uses

Bacilli are widely used commercially either as purified cell-free extracts of metabolic end products or as whole cells or spores. They are important industrially as sources of refined cell-free enzymes and antibiotics. Kornberg et al. (38) list about 60 different enzymes contained within the spore alone. Starch-hydrolyzing enzymes (alpha-amylase, beta-amylase, and amyloglucosidase) have found the broadest use industrially. Extracellular enzymes such as cellulase and xylanase are potentially useful for the bioconversion of waste lignocellulosic materials (e.g., corn stover, paper pulpmill sludge, wheat straw) into fermentable sugars for the production of various useful products (e.g., single-cell protein, ethanol, chemicals) (2, 49, 50).

Bacilli produce peptide antibiotics which are effective against gram-positive bacteria, although some are active against gram-negative bacteria, molds, and yeasts. These purified compounds have been used widely in medicine and as feed additives. Live cells or spores have been studied as biocontrol agents for bacterial and fungal pathogens on seeds, plants, vegetables, and fruits (33, 43, 67). Other strains have been advocated as biofertilizers and are thought to produce plant-growth-promoting substances (19, 39).

B. thuringiensis, B. popilliae, B. sphaericus, and *B. lentimorbus* produce insecticidal crystal protein toxins and are utilized commercially for insect control (5, 15, 17). *B. thuringiensis* varieties, for example, are a group of microbial pesticides that have been widely released to control agricultural and forest pests, as well as the human disease vectors, mosquito, and black fly. About 5 million pounds of *B. thuringiensis* var. *kurstaki* HD1 were produced for insect control in 1986 (64). These biocontrol insecticides are used either as whole broth sporulated cultures or as semipurified toxin.

Bacillus species comprise the major biocontrol agents approved and registered by the U.S. Environmental Protection Agency (12). In the future, safer, inexpensive, and more effective microbial pesticides based on genetically engineered microorganisms (GEMs) can be expected. The GEMs will be designed to have a more potent biocontrol activity with a broader host range and survive better in the environment.

Persistence in the environment

The introduction of a microorganism into soil does not always lead to its establishment. For example, bacteria and fungi not indigenous to a soil type rapidly die out on release (3). Results from extensive studies on the survival of *B. thuringiensis* in the environment show that it does not adapt, grow, or survive beyond a day to a few weeks (59, 66, 76, 81, 82). Based on a cell number to dry weight ratio of 1.96×10^{11} colony-forming units per gram of cells (dry weight), 4.5×10^{20} *B. thuringiensis* cells are released into the environment annually. Assuming that mutation rates are in the range of 1×10^{-6} to 1×10^{-10} per gene per replication (13), approximately 4.5×10^{10} to 10^{14} mutants per gene of *B. thuringiensis* have been released. It is evident that such large-scale release has provided significant potential for adaptation to occur by natural processes, yet variants with increased ability to survive in the environment have not been detected because conditions for prolonged growth or survival are not favorable.

There are several reasons that may account for the failure of introduced *B. thuringiensis* to adapt in the environment. *B. thuringiensis* has been shown to be especially sensitive to ultraviolet light in the sunlight spectrum (18, 32, 34, 35, 56, 76), weather (26), cold temperatures (35, 68), and plant extracts (52, 69, 70). One obvious exception is the fact that *B. thuringiensis* subsp. *israelensis* survives and grows in dead mosquito larvae (4, 65, 83) in the environment.

B. sphaericus 1593, which is also used as a microbial pesticide against mosquitoes, is generally considered to survive better than *B. thuringiensis* in aquatic environments (65). However, the survival of *B. sphaericus* 1593 depends on the presence of *Culex quinquefasciatus* larvae. It is unable to survive in other mosquito larvae, e.g., *Aedes aegypti,* or in the absence of mosquito larvae.

Known Methods of Confinement and Decontamination

General comments

Confinement and decontamination (or sterilization) of bacilli require special attention since the spores formed are very small and many times more resistant to chemical and physical agents than are vegetative cells. For this reason, more severe conditions are needed for decontamination of spores than vegetative cells. For liquids, factors such as the viscosity of the medium, solids loading, and substrate concentration all influence the strategy used for decontamination. For confinement of dried powders or spores, extremely fine filtration systems (<0.2 μm) are required for most microorganisms.

There is a wealth of knowledge in the scientific literature on physical and chemical modes of confinement and decontamination. For this reason, only an outline of each specific technique will be given.

Sterilization kinetics are stochastic, and the success of a sterilization attempt is only a probability. Hence, one can only speak of a probability that sterility will be achieved (7). Sanitization is the reduction of a microbial population to a level which is determined to be safe by public health requirements. Various chemicals and detergents are usually used to achieve this end. It should be noted that sanitary systems are not necessarily sterile.

Laboratory and industrial methods

Heat sterilization

Wet heat sterilization. Heat sterilization is the most useful method for the destruction of both spores and vegetative cells. Since wet heat treatment is the most reliable and, on a large scale, the easiest to control, it is the method of choice throughout the fermentation industry. The technique is effective for both naturally occurring and genetically engineered cells and spores of bacilli. However, successful sterilization must be determined for each particular organism and application.

It is well documented that dramatic differences in sterilization kinetics usually result from differences in the heat resistance between vegetative cells and spores, and between wet heat and dry heat sterilization conditions (1, 7). Spores obtained from different species, strains within species, grown on different media, or under different environmental conditions may also show differences in sterilization kinetics (11, 40, 47, 61).

Death rate–temperature (D) relationships for a given strain are not always linear. The D value (in minutes) is defined as the time required at a given temperature to reduce the microbial population by 90 percent or one order of magnitude. Two major types of deviations from normal linear survival curves have been identified (7), depending on populations of spores with different thermal resistances, spore clumping, and presence of solids. However, once D values are determined for the specific environmental conditions used, retention or sterilization times to kill the population can be determined. In some instances, a lower temperature for sterilization may be desirable. However, death rate constants will decrease and, therefore, longer times will be required.

In practice, satisfactory destruction of spores of bacilli can be achieved by using temperatures of at least 121°C. Under normal operating conditions, the following sterilization conditions (7) are sufficient to kill all organisms including heat-resistant spores: holding temperatures of 121°C (15 psig), 126°C (20 psig), 134°C (30 psig), or 140°C (38 psig) for 15 min, 10 min, 3 min, and 0.67 min, respectively. However, deviations from normal sterilization kinetic behavior due to differences in medium, strain, and age should be expected. For these reasons, heat-sterilization protocols for bacilli have to be designed carefully. Separate protocols for each strain may be necessary.

Dry heat sterilization. For satisfactory sterilization of bacilli by dry heat, significantly higher temperatures and holding times than for wet heat are required. For this reason, dry heat sterilization is used to sterilize materials which are relatively insensitive to high temperatures. The technique is commonly used to sterilize surgical instruments and laboratory glassware. Dry heat is not recommended for glass syringes with metal tips, paper, cotton, plastics, and wood.

Pasteurization. Pasteurization is used for the destruction of commonly occurring pathogenic organisms. The technique does not achieve sterilization. It involves heating a material to 63°C for 30 min or for high-temperature, short-time pasteurization, to 80°C for 15 s. In some fermentation facilities, pasteurization is used in a two-stage sterilization process in which pasteurization is followed by a milder wet heat treatment. Pasteurization is not effective against spores of bacilli.

Filtration. The air supplied to fermenters during aerobic metabolism must be sterilized. The number of particles and contaminants in air varies widely, depending on location of the facility, air movement, and previous treatment of the air. On the average, outdoor air has 10 to 10,000 particles per cubic meter and 5 to 2000 microorganisms per cubic meter. Of these, about 50 percent are fungal spores, and approximately 40 percent are gram-negative bacteria (21). The effluent gases from the fermenter must also be filter-sterilized, especially with GEMs, before being discharged to the environment.

The methods available for sterilizing gases include filtration, gas injection (ozone), gas scrubbing, radiation (UV), and heat. Of these, only filtration and heat treatment are employed at the industrial level. In the past, the method of choice was heat treatment in which the gas, usually air, was sterilized by passing it over electrically heated elements (22). However, as a result of the high cost of electricity, filtration is currently the most widely used technique. In older fermentation plants, depth-type filter devices, such as granulated carbon beds and packed towers containing glass wool, were widely used. These devices have often proved unsatisfactory for several reasons: notably, shrinkage and solidification during steam sterilization. Glass fiber filter cartridges, which do not have these problems, have replaced glass wool filters. However, these are still not very satisfactory.

New cartridge filter systems using pleated membranes are now being widely used. These have the advantage of being significantly smaller than the older units and much easier to operate. Conway (20) has given a detailed description of several characteristics of these state-of-the-art gas filters. Constructed of cellulose ester, polysulfone, polytetrafluoroethylene, or nylon, these membrane filters have the same structure as depth filters, but because they have a membranous structure, they have a near-absolute filter effect.

Filtration of spent fermentation broth is rarely undertaken because of cost. As previously discussed, wet heat is the method of choice in this case.

Other physical methods. For miscellaneous physical methods of confinement and decontamination of bacilli, such as grinding and shaking, pressure, high concentrations of oxygen, sonic and supersonic radiation waves, freezing, desiccation and irradiation, the reader is referred to the reviews by Rahn (58) and Sykes (74). For the most part, these techniques are not very useful in an industrial setting. Ethylene oxide is used in hospitals and in industry for sterilizing heat-sensitive materials, such as pipettes, process equipment, and also for disinfecting heart-lung machines, ophthalmoscopes, blankets, and bedding (74).

Chemical methods. There are only a few chemicals which are reliable sterilizing agents. Only chlorine, formaldehyde, 70% alcohol acidified with mineral acid to pH 2, quaternary ammonium compounds and ethylene oxide have proved useful. Of these, only chlorine, usually as dilute hypochlorite solution, is widely used, often as a disinfectant rather than a sterilant.

Halogens. Chlorine and its compounds (e.g., calcium and sodium hypochlorite, chloramine-T, halazone, and the halogenated methylhydantoin and isocyanuric acid derivatives) are the most important agents for disinfecting and sanitizing food-processing equipment and utensils, and for disinfecting water supplies. Their properties as bactericides have been described in detail by Sykes (73). A hypochlorite solution containing 1000 ppm chlorine will kill 99 percent of a suspension of *B. subtilis* spores at pH 11.3 in 70 min at room temperature. The treatment takes less than 15 min at pH 6.5.

Other chemicals. Formaldehyde is the chemical of choice for space disinfection (72). Aqueous solutions of formaldehyde (formalin) containing 37 to 40% formaldehyde can be used to disinfect biological safety cabinets, incubators, laboratory rooms, buildings, etc. Formaldehyde is highly toxic and potentially mutagenic, with a threshold limit value of only 2 ppm. Hence, caution must be exercised in its handling and use. There is some speculation on the sporicidal activity of formaldehyde, but safe and effective treatments can be achieved with a 1% solution at 37°C, or a 5 to 8% solution at room temperatures. Both require up to 24 h of contact. Acid alcohol will sterilize spores within 4 h and ethylene oxide in 1% (w/v) solution is effective in about 2 h.

Quaternary ammonium compounds (e.g., benzalkonium chloride–based compounds, also known as alkyldimethyl-ammonium chloride, a mixture of alkyls from C_8H_{17} to $C_{18}H_{37}$) are excellent bactericides. They are effective in the destruction or inhibition of various types of bacteria and fungi. When applied to surfaces, a quaternary ammonium solution forms a bacteriostatic film that inhibits bacterial growth. Ions such as calcium, magnesium, and iron, if present in sufficient concentration, as in hard water, tend to inhibit the action of quaternary ammonium compounds.

Chemical agents can be employed for sterilization of packaging materials that are heat-sensitive in order to avoid heat damage. In general, chemical sterilizing agents have little value in practice. Their main application is restricted to the treatment of apparatus and equipment where only surfaces are involved. Ethylene oxide has been recommended for sterilizing liquid culture media. Unfortunately, it also impairs their nutrient properties, possibly because of the formation of ethylene chlorohydrin. Chemical methods are not expected to find large-scale application in the confinement and decontamination of bacilli.

Assessment of sterilization effectiveness. Selected resistant strains of microorganisms are usually used as biological indicators in sterilization cycle development, cycle validation, and routine monitoring of industrial sterilization processes. Various types of biological monitors, including spore suspensions, inoculated paper carriers, and self-contained units are available for these applications. Spores of *B. subtilis* (*globigii*) and *B. stearothermophilus* are used routinely in industry.

The performance of biological indicators is markedly influenced by spore propagation and harvesting procedures, inoculation procedures, carrier and packaging materials, as well as subsequent evaluation and use of the indicators, including test methodologies and environmental conditions during recovery of sterilant-exposed spores. Monographs developed by the U.S. Pharmacopeia deal with the performance of biological indicators. These monographs are updated on a regular basis to reflect current technology. The U.S. Food and Drug Administration also has guidelines which deal with specific issues relating to indicator incubation time.

Quesnel (57) has given a detailed review on the assessment of resistance to sterilization and on the choice of, factors affecting, and the function of biological indicators. A number of excellent sources are available for details on the theory of methods of determining heat sterilization effectiveness (1, 9, 25, 46, 58, 74, 80).

Mitigation of environmental release

Small-scale release. Small-scale release refers to contamination of small quantities of plants, soil, or liquid that can be removed and decontaminated in a controlled environment such as a laboratory autoclave. Contaminated plants and soil can be removed to a depth determined by soil type and extent of contamination (3) and placed in autoclavable biohazard bags. The site then should be monitored to assay the extent of removal. The heat sterilization techniques described in Heat Sterilization (above) are effective for mitigation and decontamination of small quantities of contaminated soil. The depth of soil should be minimized for effective heat treatment. The generally preferred method, however, is gamma radiation with a ^{60}Co source at 300 krad; e.g., 100 krad/h for 3 h or 3 krad/h for 96 h (24). Gamma radiation effects the least damage to soil nutrients (44, 45), thereby allowing the soil to be reused. Contaminated liquids may be treated similarly. It should be noted that cobalt irradiation facilities are generally limited, and the technique may be most effective only for very small quantities of contaminated material. This would tend to limit large-scale use of this technique.

Medium-scale release

Steam sterilization. A variety of techniques employ steam to effect soil sterilization in situ. These include buried perforated steel pipes or Hoddesdon pipes (considered obsolete), spiked grids, and the steam plow. In some of these systems, heat distribution within the contaminated site generally tends to be inadequate (55). Steam treatments are labor-intensive and expensive. The main steam-treatment methods used are a network of buried drain pipes and sheet steaming.

Steam drain pipes. A mechanical trenching machine is used to prepare trenches to a depth of 50 to 55 cm. Drain pipes, made of temperature-resistant tile or plastic, 30 cm long with 5 cm internal diameter (ID) are connected to the main steam source by 10 cm (ID) polypropylene pipes. The pipes are covered with coarse peat to prevent silting of the pipes. Steam condensate is drained from the system through 5-cm (ID) pipes to a land drain. A separation distance of 80 cm between pipes is considered the most efficient spacing distance for effective sterilization. A steam generator or boiler capable of producing 1×10^6 kcal/h is required for effective treatment of one circuit of four drain pipe lines in a treatment area that is 3.20 m wide by 40 m in length (55).

Sheet steaming. Sheet steaming is the least labor-intensive and most effective of the steam treatments. It is one of the most widely used techniques. PVC sheeting, 0.25 mm thick and 3.6 m wide, is smoothed out over a portion of the contaminated area and anchored at the edges. Steam is blown in under the edges for 12 h and left to penetrate the soil. Sandbags, soil, ships' chains, or nylon nets can be used to anchor the sheeting in place (55). Temperatures of 54 to 100°C may be reached at depths of 25 to 45 cm to assure 99 percent decontamination. Pasteurization techniques (see Pasteurization in this section) in which heat treatments are separated by cooling periods to allow spores to germinate can be used to kill spores. Sheet steaming can be expected to be most effective for mitigation and decontamination of relatively shallow subsurface materials where the temperature is likely to be the highest. The effectiveness of the technique is affected by soil moisture content and type and cultivation techniques (55).

Chemical methods. The advantages of chemical methods of treatment, especially methyl bromide and chloropicrin, are the relative ease of use and low cost. Unfortunately, their effectiveness against spore-formers is inadequate, and rebound of undesirable microorganisms is rapid. A major disadvantage of these chemicals is their extreme toxicity to all life-forms. Both methyl bromide and chloropicrin are toxic to plants and mammals, and are at least 100 times more toxic than chlorinated hydrocarbons (27).

Large-scale release. For large-scale releases, the methods available for reduction in the microbial population are limited by the size of the contaminated area. Steam sterilization techniques are not practical for large-scale releases. Chemical methods using volatile chemicals are also unrealistic in open-air environments. Other effective chemicals are usually undesirable because of their toxic effects on other life forms.

Dispersal mechanisms for bacteria, including wind and rain, may result in significant apparent reduction in the local population of the released microorganisms over time. This will depend to a large extent on the hydrogeology of the subsurface material. Sorption properties of the various soils in the subsurface can be expected to influence the rate of reduction in microbial population. However, such natural dilution methods are not acceptable. Other novel techniques, including crop rotation and soil amendments with fertilizer and antagonistic microorganisms have met with little success (28). Biological methods have been evaluated for effectiveness in mitigation of large-scale releases (8). Such methods have generally proved unreliable. An apparent antagonism between *Actinomyces* and bacilli in an aquatic environment has been reported (66) but the cause of this interaction was not determined. It is doubtful that effective biological methods will be available for confinement, mitigation, and decontamination of large-scale releases.

Biological confinement

Biological confinement refers to the use of strains of bacteria that are unable to survive outside of the laboratory and are unable to transfer genetic information to other bacteria. It is very likely that gene transfer between bacilli occurs in nature. Intraspecific and interspecific transfer of chromosomal genes (6) and plasmid DNA (6, 10, 29, 37, 60) has been observed in the laboratory between strains of *B. thuringiensis, B. cereus, B. anthracis,* and *B. subtilis.* Intraspecific and interspecific transfer of the conjugative transposon Tn*916* has been observed between strains of *B. thuringiensis* subsp. *israelensis* and *B. subtilis* (53, 54). Transformation of *B. subtilis* has been observed in sterile soil (30, 31) and on sterile sand grains (41), and plasmid transfer in sterile soil has been reported between *B. cereus* and *B. subtilis* (78). It is important to consider the potential for such gene transfer in the design of biological confinement systems.

Although spores of *Bacillus* are resistant to environmental extremes, poor survival and lack of multiplication of bacilli in the environment have been problems for the development of microbial pesticides. As mentioned previously, several studies suggest that introduced bacilli do not compete successfully with indigenous soil organisms, although low numbers of added spores may persist (36, 79, 81). Therefore, a certain level of biological confinement may be inherent in strains of *Bacillus* sp. that are poor competitors in soil. Biological confinement could be enhanced by using sporulation-deficient mutants (23, 42, 51) when sporulation is not essential for the activity or product of interest. Biological confinement also could include the use of auxotrophic, heat-, cold-, or UV-sensitive mutants, as well as certain conditional lethal mutations which cause cell death in nonpermissive conditions (48).

Monitoring Survival and Mitigation

To assess the persistence of *Bacillus* sp. in the environment and to monitor the effectiveness of confinement or decontamination procedures, selective monitoring methods for the strain in question will be required. The methods most likely to monitor accurately the presence of released microorganisms include

1. The use of genetically marked strains (e.g., with multiple antibiotic-resistant mutations) that can be selected unambiguously from other microorganisms in the sample. Care must be taken so that mutants are stable to ensure that instability does not inadvertently result in underestimation of the microbial population to be monitored (14).

2. The use of semiselective plating followed by serotyping or colony hybridization with probes to specific DNA sequences (62) can be used to definitively identify the released microorganisms.

In environmental monitoring experiments, it is important to include appropriate controls to determine the sensitivity and effectiveness of the methods. For example, background levels of the organism can be determined using samples from uncontaminated locations. Reconstruction experiments should be done by adding known levels of the organism to the natural substrate being tested. One also must use proper sampling techniques, involving multiple sample sites and replicates to ensure statistical significance of the results. Methods of strain identification that depend solely on antibiotic resistance markers or plasmid-borne traits are less desirable than approaches that use a combination of techniques. Indigenous antibiotic resistance can occur in *Bacillus* (77), and gene transfer in *Bacillus* which has been demonstrated in sterile soil (30, 31, 78) could occur under field conditions. Combinations of methods such as selective spore germination (75) and detection of specific chromosomal sequences (62, 63) may provide a useful approach.

Conclusions

In general, the perceived relative risk of different organisms increases in direct proportion to our lack of knowledge about the short- and long-term ecological effects, pathogenicity, immunological effects, persistence in the environment, and size of the population. The realistic approach to evaluating processes for safety relies essentially on an assessment of the risk involved. This analysis involves estimating the magnitude of the potential hazard posed by the contaminant to the public and the environment. The analysis relies on the demonstration that the microbial agent in question is harmful, or is potentially harmful, coupled with the likelihood of persistence in the environment. When the risk involved is more certain, for example, when known pathogens are cultured, then confinement and assured destruction are critical. When pathogenicity has not been reported, or cannot be reasonably demonstrated, as in the case of well-known naturally occurring microorganisms (NOMs) (such as nongenetically engineered *B. subtilis*), then the necessity for eradication is not clear. An organism that is part of the normal biota of an environment can be isolated and safely reintroduced without modification back into that environment. Typical examples of the reintroduction of nonpathogenic NOMs include (1) a seed inoculant to enhance plant vigor (19, 33, 39), or (2) a hay inoculant to preserve feed value (N. J. Tomes, S. Soderlund, J. Lamptey, S. Croak-Brossman, and G. Dana, *J. Prod. Agric.,* in press). No adverse ecological consequences of the aforementioned releases have been demonstrated, and it is unlikely that any will occur. Therefore, confinement and sterilization, while important for industrial production, are not major issues in environmental release of nonpathogenic NOMs. If the organism has been genetically altered, then questions arise regarding the introduced or altered property that scientists must address on a case-by-case basis.

References

1. Aiba, S., A. E. Humphrey, and N. F. Millis. 1973. *Biochemical Engineering*, 2d ed., Academic Press, New York, pp. 239–269.
2. Alani, D. I., and M. Moo-Young (eds.). 1986. *Perspectives in Biotechnology and Applied Microbiology*, Elsevier, London, pp. 1–152.
3. Alexander, M. 1977. *Introduction to Soil Microbiology*, 2d ed. Wiley, New York.
4. Aly, C., M. S. Mulla, and B. A. Federici. 1985. Sporulation and toxin production by *Bacillus thuringiensis* var. *israelensis* in cadavers of mosquito larvae (*Diptera culicidae*). *J. Invertebr. Pathol.* 46:251–258.
5. Aronson, A. I., W. Beckman, and P. Dunn. 1986. *Bacillus thuringiensis* and related insect pathogens. *Microbiol. Rev.* 50:1–24.
6. Aronson, A. I., and W. Beckman. 1987. Transfer of chromosomal genes and plasmids in *Bacillus thuringiensis*. *Appl. Environ. Microbiol.* 53:1524–1530.
7. Bader, F. G. 1986. Sterilization: Prevention and contamination. In A. L. Demain and N. A. Solomon (eds.), *Manual of Industrial Microbiology and Biotechnology.* American Society for Microbiology, Washington, D.C., pp. 345–362.
8. Baker, K. F., and R. J. Cook. 1974. *Biological Control of Plant Pathogens*. W. H. Freeman, San Francisco.
9. Bartholomew, W. H., D. E. Engstrom, N. S. Goodman, A. L. O'Toole, J. L. Shelton, and L. P. Tannen. 1974. Reduction of contamination in an industrial fermentation plant. *Biotechnol. Bioeng.* 16:1005–1013.
10. Battisti, L., B. D. Green, and C. B. Thorne. 1985. Mating system for transfer of plasmids among *Bacillus anthracis*, *Bacillus cereus*, and *Bacillus thuringiensis*. *J. Bacteriol.* 162:543–550.
11. Bayliss, C. E., W. M. Waites, and N. R. King. 1981. Resistance and structure of spores of *Bacillus subtilis*. *J. Appl. Bacteriol.* 50:379–390.
12. Betz, F., M. Levin, and M. Rogul. 1983. Safety aspects of genetically engineered microbial pesticides. *Recomb. DNA Tech. Bull.* 6:135–141.
13. Birge, E. A. 1981. *Bacterial and Bacteriophage Genetics: An Introduction.* Springer-Verlag, New York.
14. Brand, T. D., D. E. Pinnock, K. Jackson, and J. E. Milstead. 1975. Methods for assessing field persistence of *Bacillus thuringiensis* spores. *J. Invert. Pathol.* 25:199–208.
15. Bulla, L. A. (Jr.), R. N. Costilow, and E. S. Sharpe. 1978. Biology of *Bacillus popilliae. Adv. Appl. Microbiol.* 23:1–18.
16. Bulla, L. A. (Jr.), and J. A. Hoch. 1985. Biology of bacilli. In A. L. Demain and N. A. Solomon (eds.), *Biology of Industrial Microorganisms.* Benjamin/Cummings, Menlo Park, Calif., pp. 57–78.
17. Burges, H. D. (ed.), 1981. *Microbial Control of Pests and Plant Diseases 1970–1980.* Academic Press, London.
18. Cantwell, G. E., and B. A. Franklin. 1966. Inactivation by irradiation of spores of *Bacillus thuringiensis* var. *thuringiensis. J. Invert. Pathol.* 8:256–258.
19. Chanway, C. P., L. M. Nelson, and F. B. Holl. 1988. Cultivar-specific growth promotion of spring wheat (*Triticum aestivum* L.) by coexistent *Bacillus* species. *Can. J. Microbiol.* 34:925–929.
20. Conway, R. S. 1984. State of the art in fermentation air filtration. *Biotechnol. Bioeng.* 26:844–847.
21. Crueger, W., and A. Crueger. 1984. *Biotechnology: A Textbook of Industrial Microbiology,* Sinauer Associates, Sunderland, Mass., pp. 83–90.
22. Decker, H. M., F. J. Citek, J. B. Harstad, N. H. Gross, and F. J. Piper. 1954. Time-temperature studies of spore penetration through an electric air sterilizer. *Appl. Microbiol.* 2:33–36.
23. Ellis, D. M., and D. H. Dean. 1981. Characterization of a non-reverting asporogenous strain of *Bacillus subtilis* 168 for use as an HV-1 host. *Recomb. DNA Tech. Bull.* 4:1–3.

24. Eno, C. F., and H. Popenoe. 1964. Gamma radiation compared with steam and methyl bromide as a soil sterilizing agent. *Soil Sci. Soc. Proc.* 28:533–535.

25. Frobisher, M., R. D. Hinsdill, K. T. Crabtree, and C. R. Goodheart. 1974. *Fundamentals of Microbiology*, 9th ed. W. B. Saunders, Philadelphia.

26. Fry, R. D., C. G. Scholl, E. W. Scholz, and B. R. Funke. 1973. Effect of weather on a microbial insecticide. *J. Invert. Pathol.* 22:50–54.

27. Fry, W. E. 1982. *Principles of Plant Disease Management.* Academic Press, London.

28. Gindrat, D. 1979. Biological soil disinfestation. In D. Mulder (ed.), *Soil Disinfestation.* Elsevier, Amsterdam, pp. 251–287.

29. Gonzalez, J. M. (Jr.), B. J. Brown, and B. C. Carlton. 1982. Transfer of *Bacillus thuringiensis* plasmids coding for delta-endotoxin among strains of *B. thuringiensis* and *B. cereus. Proc. Natl. Acad. Sci. USA* 79:6951–6955.

30. Graham, J. B., and C. A. Istock. 1978. Genetic exchange in *Bacillus subtilis* in soil. *Mol. Gen. Genet.* 166:287–290.

31. Graham, J. B., and C. A. Istock. 1979. Gene exchange and natural selection cause *Bacillus subtilis* to evolve in soil culture. *Science* 204:637–639.

32. Griego, V. M., and K. Spence. 1978. Inactivation of *Bacillus thuringiensis* spores by ultraviolet and visible light. *Appl. Environ. Microbiol.* 35:906–910.

33. Handlesman, J., S. Raffel, E. H. Mester, L. Wunderlich, and C. R. Grau. 1990. Biological control of damping-off of alfalfa seedlings with *Bacillus cereus* UW-85. *Appl. Environ. Microbiol.* 56:713–718.

34. Ignoffo, C. M., and C. Garcia. 1978. UV-photoinactivation of cells and spores of *Bacillus thuringiensis* and effects of peroxidase on inactivation. *Environ. Entomol.* 1: 270–272.

35. Ishiguro, T., and M. Miyazono. 1982. Fate of viable *Bacillus thuringiensis* spores on leaves. *J. Pest. Sci.* 7:111–116.

36. Knudsen, G. R., and H. W. Spurr, Jr. 1987. Field persistence and efficacy of five bacterial preparations for control of peanut leaf spot. *Plant Dis.* 71:442–445.

37. Koehler, T. M., and C. B. Thorne. 1987. *Bacillus subtilis* (*natto*) plasmid pLS20 mediates interspecies plasmid transfer. *J. Bacteriol.* 169:5271–5278.

38. Kornberg, A., J. A. Spudick, D. L. Nelson, and M. P. Deutscher. 1968. Origin of proteins in sporulation enzymes nucleic-acids *Bacillus*-spp: Review. *Annu. Rev. Biochem.* 37:51–78.

39. Kucey, R. M. N. 1988. Plant growth-altering effects of *Azospirillum brasiliense* and *Bacillus* C-11-25 on two wheat cultivars. *J. Appl. Bacteriol.* 64:187–196.

40. Leaper, S. 1987. A note on the effect of sporulation conditions on the resistance of *Bacillus* spores to heat and chemicals. *Lett. Appl. Microbiol.* 4:55–57.

41. Lorenz, M. G., B. W. Aardema, and W. Wackernagel. 1988. Highly efficient genetic transformation of *Bacillus subtilis* attached to sand grains. *J. Gen. Microbiol.* 134: 107–112.

42. Losick, R., P. Youngman, and P. J. Piggot. 1986. Genetics of endospore formation in *Bacillus subtilis. Ann. Rev. Genet.* 20:625–669.

43. Maplestone, P. A., and R. Campbell. 1989. Colonization of roots of wheat seedlings by bacilli proposed as biocontrol agents against take-all. *Soil Biol. Biochem.* 21: 543–550.

44. McLaren, A. D. 1969. Radiation as a technique in soil biology and biochemistry. *Soil Biol. Biochem.* 1:63–73.

45. McLaren, A. D., R. A. Luse, and J. J. Skujins. 1962. Sterilization of soil by irradiation and some further observations on soil enzyme activity. *Soil Sci. Soc. Proc.* 26: 371–377.

46. Meynell, G. G., and E. Meynell. 1970. *Theory and Practice in Experimental Bacteriology,* 2d ed. Cambridge University Press, London.

47. Milhaud, P., and G. Balassa. 1973. Biochemical genetics of bacterial sporulation IV: Sequential developments of resistances of chemical and physical agents during sporulation of *Bacillus subtilis. Molec. Gen. Genet.* 125:241–250.

48. Molin, S., P. Klemm, L. K. Poulsen, H. Biehl, K. Gerdes, and P. Andersson. 1987. Conditional suicide system for containment of bacteria and plasmids. *Bio/Technology* 5:1315–1318.

49. Moo-Young, M., J. Lamptey, and P. Girard. 1984. Paper pulpmill sludge utilization: Technoeconomic potential for fuel ethanol, methane and SCP production. *Biotech. Adv.* 2:253–272.
50. Moo-Young, M., S. Hasnain, and J. Lamptey (eds.). 1986. *Biotechnology and Renewable Energy*. Elsevier Applied Science Publishers, London, pp. 46–199.
51. Mottice, S. L., G. A. Wilson, and F. E. Young. 1979. Characterization of an asporogenous strain of *Bacillus subtilis* submitted to RAC for certification as the host component of an HV-1 system. *Recomb. DNA Tech. Bull.* 2:100–105.
52. Muksymiuk, B. 1970. Occurrence of antibacterial substances in plants affecting *Bacillus thuringiensis* and other entomogenous bacteria. *J. Invert. Pathol.* 15:356–371.
53. Naglich, J. G., and R. E. Andrews, Jr. 1988. Introduction of the *Streptococcus faecalis* transposon Tn*916* into *Bacillus thuringiensis* subsp. *israelensis*. *Plasmid* 19:84–93.
54. Naglich, J. G., and R. E. Andrews, Jr. 1988. Tn*916*-dependent conjugal transfer of pC194 and pUB110 from *Bacillus subtilis* into *Bacillus thuringiensis* subsp. *israelensis*. *Plasmid* 20:113–126.
55. Nederpel, L. 1979. Soil sterilization and pasteurization. In D. Mulder (ed.), *Soil Disinfestation*. Elsevier, Amsterdam, pp. 29–50.
56. Pozsgay, P., P. Fast, H. Kaplan, and P. R. Carey. 1987. The effect of sunlight on the protein crystals from *Bacillus thuringiensis* var. *kurstaki* HD1 and NRD12: A Raman spectroscopic study. *J. Invert. Pathol.* 50:246–253.
57. Quesnel, L. B. 1984. Biological indicators and sterilization processes. In M. H. E. Andrew and A. D. Russell (eds.), *The Revival of Injured Microbes*. Academic Press, London, pp. 257–292.
58. Rahn, O. 1945. Physical methods of sterilization of microorganisms. *Bacteriol. Rev.* 1:1–47.
59. Raun, E. S., G. G. Sutter, and M. A. Revelo. 1966. Ecological factors affecting the pathogenicity of *Bacillus thuringiensis* var. *thuringiensis* to the European corn borer and Fall army worm. *J. Invert. Pathol.* 8:365–375.
60. Reddy, A., L. Battisti, and C. B. Thorne. 1987. Identification of self-transmissible plasmids in four *Bacillus thuringiensis* subspecies. *J. Bacteriol.* 169:5263–5270.
61. Roberts, R. A., and A. D. Hitchins. 1969. Resistance of spores. In G. W. Gould and A. Hurst (eds.), *The Bacterial Spore*. Academic Press, London, pp. 611–670.
62. Sayler, G. S., C. Harris, C. Pettigrew, D. Pacia, A. Breen, and K. M. Sirotkin. 1987. Evaluating the maintenance and effects of genetically engineered microorganisms. *Develop. Ind. Microbiol.* 27:135–149.
63. Sayler, G. S., M. S. Shields, E. T. Tedford, A. Breen, S. W. Hooper, K. M. Sirotkin, and J. W. Davis, 1985. Application of DNA-DNA colony hybridization to the detection of catabolic genotypes in environmental samples. *Appl. Environ. Microbiol.* 49:1295–1303.
64. Senuta, A. 1987. Are microbials viable? *Agrichem. Age* 33:21.
65. Silapanuntakul, S., S. Pantuwatana, A. Bhumiratana, and K. Charoensiri. 1983. The comparative persistence of toxicity of *Bacillus sphaericus* strain 1593 and *Bacillus thuringiensis* serotype H-14 against mosquito larvae in different kinds of environments. *J. Invert. Pathol.* 42:387–393.
66. Silvey, J. K. G., and J. T. Wyatt. 1977. The interaction between freshwater bacteria, algae, and actinomyces in Southwestern reservoirs. In J. Cairns, Jr. (ed.), *Aquatic Microbial Communities*. Garland, New York, pp. 161–203.
67. Sinclair, J. B. 1989. *Bacillus subtilis* as a biocontrol agent for plant diseases. In V. P. Agnihotri, N. Singh, H. S. Cahube, U. S. Singh, and T. S. Dwivedi (eds.), *Perspectives in Plant Pathology*. Today and Tomorrow's Printers, New Delhi, pp. 367–374.
68. Smirnoff, W. A. 1963. The formation of crystals in *Bacillus thuringiensis* var. *thuringiensis* Berliner before sporulation at low-temperature incubation. *J. Insect Pathol.* 5:242–250.

69. Smirnoff, W. A. 1972. Effects of volatile substances released by foliage of *Abies balsamea. J. Invert. Pathol.* 19:32–35.

70. Smirnoff, W. A., and P. A. Hutchinson. 1965. Bacteriostatic and bactericidal effects of extracts of foliage from various plant species on *Bacillus thuringiensis* var. *thuringiensis* Berliner. *J. Insect Pathol.* 7:273–280.

71. Sneath, P. A. 1986. Endospore-forming gram-positive rods and cocci. In P. A. Smith, N. S. Mair, E. Sharpe, J. G. Holt (eds.), *Bergey's Manual of Systematic Bacteriology.* Williams & Wilkins, Baltimore, pp. 1104–1139.

72. Songer, J. R., D. T. Brayman, and R. G. Mathis. 1972. The practical use of formaldehyde vapor for disinfection. *Health Lab. Sci.* 19:46–55.

73. Sykes, G. 1965. *Disinfection and Sterilization,* 2d ed. Spon, London.

74. Sykes, G. 1969. Methods and equipment for sterilization of laboratory apparatus. *Meth. Microbiol.* 1:77–121.

75. Travers, R. S., P. A. W. Martin, and C. F. Reichelderfer. 1987. Selective process for efficient isolation of soil *Bacillus* spp. *Appl. Environ. Microbiol.* 53:1263–1266.

76. Vago, C., and M. C. Busnel. 1952. Photosensibilite du *Bacillus cereus* var. *alesti* au rayonnement ultraviolet de 2537 angstrom. *Antonie van Leeuwenhoek* 18:125–127.

77. Van Elsas, J. D., and M. Pereira. 1986. Occurrence of antibiotic resistance among bacilli in Brazilian soils and the possible involvement of resistance plasmids. *Plant Soil* 94:213–226.

78. Van Elsas, J. D., J. M. Govaert, and J. A. van Veen. 1987. Transfer of plasmid pFT30 between bacilli in soil as influenced by bacterial population dynamics and soil conditions. *Soil Biol. Biochem.* 19:639–647.

79. Van Elsas, J. D., A. F. Dijkstra, J. M. Govaert, and J. A. van Veen. 1986. Survival of *Pseudomonas fluorescens* and *Bacillus subtilis* introduced into two soils of different texture in field microplots. *FEMS Microbiol. Ecol.* 38:151–160.

80. Wang, D. I. C., J. Scharer, and A. E. Humphrey. 1964. Kinetics of death of bacterial spores at elevated temperatures. *Appl. Microbiol.* 12:451–454.

81. West, A. W., H. D. Burges, T. J. Dixon, and C. H. Wyborn. 1985. Survival of *Bacillus thuringiensis* and *Bacillus cereus* spore inocula in soil: Effects of pH, moisture, nutrient availability and indigenous microorganisms. *Soil Biol. Biochem.* 17:657–665.

82. West, A. W., H. D. Burges, and C. H. Wyborn. 1984. Effect of incubation in natural and autoclaved soil upon the potency and viability of *Bacillus thuringiensis. J. Invert. Pathol.* 44:121–127.

83. Zaritsky, A., and K. Khawaled. 1986. Toxicity in carcasses of *Bacillus thuringiensis* var. *israelensis*-killed *Aedes aegypti* larvae against scavenging larvae: Implications to bioassay. *J. Am. Mosquito Control Assoc.* 2:555–559.

Decontamination and Mitigation of Baculoviruses

Anne Fairbrother

Introduction and History

Properties of baculoviruses

The family Baculoviridae has one genus, *Baculovirus,* that is divided into three subgroups: nuclear polyhedrosis viruses (NPV), granulosis viruses (GV), and nonoccluded viruses (Carter, 1984; Tweeten et al., 1981). Viruses in this family are rod-shaped, enveloped nucleocapsids that are 30 to 100 nm wide by 250 to 360 nm long and contain double-stranded DNA. NPVs and GVs are found as either free nucleocapsids or occlusion bodies, wherein the nucleocapsids are bound together in a crystalline protein matrix. NPVs form polyhedral occlusion bodies, 1 to 15 μm in diameter, that contain several nucleocapsids each. GVs have only one nucleocapsid embedded in a capsule-shaped occlusion body of 0.3 to 0.5 μm in length.

Replication of baculoviruses

NPVs and GVs are present in the environment in their occluded form. When these forms are ingested by the larval form of their insect hosts, they are disassociated into free virus by alkaline proteases in the foregut. The free nucleocapsids then infect gut epithelial cells, replicate within the nucleus (NPVs) or cytoplasm (GVs) of the cells, and are released into the hemocoel by budding off of the cellular membrane. GVs also infect fat bodies and other hemocoelic tissues, wherein they undergo a second round of replication. Nucleocapsids are aggregated into occlusion bodies within the hemocoel and are released to the environment when the host larva disintegrates after death. Death is probably caused by viral disruption of protein synthesis and proper function of infected cells, plus impairment of hemolymph circulation as the result of the physical presence of large numbers of viruses. Just before death, the larvae turn a milky white color, as a result of the enormous number of viruses in the fat body cells and hemolymph. The entire cycle from ingestion to death of the host is generally 4 to 5 days. Adult insects are infected less frequently, and replication is generally restricted to gut epithelial cells, resulting in large amounts of virus being shed in fecal material. However, transovarial transmission has been documented for at least one baculovirus, the virus of the Jackpine sawfly, *Neodiprion americanus banksianae* (Bird, 1955). Further information about ultrastructures, host ranges, and pathologies of the baculoviruses can be found in reviews by David (1975), Harrap and Payne (1979), Maramarosch (1968, 1977), Smith (1976), and Carter (1984).

Host specificity

There are 65 known types of GVs, all of which infect only lepidopteran insects. Of the 284 types of NPVs, 243 infect *Lepidoptera* sp., and the rest infect various species of *Orthoptera, Neuroptera, Trichoptera, Choleoptera, Hymenoptera,* and *Diptera* (Granados, 1980). Viruses in this family infect only invertebrate hosts; no morphologically similar viruses have been detected in vertebrates. Host specificity is probably due to the requirement for the presence of alkaline proteases to dissolve the protein matrix of the occlusion body and liberate free virus, as well as the specificity of binding sites on target cells. This host specificity, plus their high pathogenicity for economically important insects, make these viruses ideal candidates for biocontrol agents. In the United States, four NPVs are currently registered for use as biocontrol agents: those viruses that infect the cotton earworm (*Heliothis zea*), the gypsy moth (*Lymantria dispar*), the pine sawfly (*Neodiprion* sp.), and the douglas fir tussock moth (*Orgria pseudotsugata*). An NPV that infects *Autographa californica* is close to registration, as are the GVs specific for the coddling moth (*Lymantria pomonella*) and the Indian meal moth (*Plodia interpunctella*). Large-scale applications of other GVs have been used in the Soviet Union, Czechoslovakia, Yugoslavia, The People's Republic of China, Australia, Great Britain, Japan, and Canada (Tweeten et al., 1981). A 1973 report by the World Health Organization (WHO) lists 39 agricultural and forest insect pests for whose control baculoviruses have been tested or used (Smith, 1975). A complete catalog of viral diseases of insects, mites, and ticks has been compiled by Martignoni (1981).

General Considerations

Survival of baculoviruses in nature

Baculoviruses have several physical properties that facilitate their large-scale production and use as biocontrol agents. They are very stable in the occluded form and maintain their infectivity for several years in the dark, especially at 4°C, as either an aqueous suspension or a dried powder. The free viruses are also relatively stable, maintaining infectivity for 120 days at 5°C, 60 days at 37°C, and 30 days at 50°C (David and Gardiner, 1967a). Thermal inactivation occurs at 75 to 80°C for *H. zea* NPV and at 82 to 88°C for *Trichoplusia ni* NPV (Jacques, 1975). All baculoviruses can withstand freezing and maintain their infectivity titer for many years when stored at or below −70°C. The occlusion body protein is degraded by weak alkalis but is resistant to strong acids and is not readily degraded by proteolytic microoganisms commonly found in soil. Additionally, the polyhedra adsorb strongly on soil particles and, therefore, are not leached out by rain or irrigation practices (Davis and Gardiner, 1967b). *I. ni* NPV has been shown to retain 15 percent of its initial activity in soil to a depth of 2.5 cm for up to 318 weeks after application (Jacques, 1975). Occlusion bodies also adhere strongly to foliage and are not easily washed off by rain, and they are not affected by changes in relative humidity (David and Gardiner, 1966).

The baculoviruses do have one area of vulnerability: both the occluded and free forms of the viruses are very susceptible to ultraviolet (UV) radiation, especially in the germicidal portion of the spectrum (254-nm wavelength). Because of this, more than 50 percent of the activity of the virus is lost within 2 days after application to foliage, and little activity remains after 10 days (Tweeten et al., 1981). Material from decomposed infected insect larvae provides some protection from UV radiation to the viruses. The addition of proteinaceous material, black dyes, and/or charcoal to the virus formulation can add protection from UV radiation and increase environmental survival time (Jacques, 1972). Commercial preparations include some type of protectant along with appropriate dispersants to aid in survival and distribution to the virus particles.

Specific Control Procedures

All known baculoviruses are highly host specific, and none has been shown to be pathogenic in vertebrate species. Therefore, it is unlikely that heroic measures will be necessary to contain a spill in either the laboratory or field, and high-level containment facilities are probably not required for working with these viruses. Under contained conditions, normal laboratory safety measures should provide all necessary controls. However, baculoviruses would have greater economic potential as insecticides if their host ranges were broadened so that they could be marketed for use against more than one crop pest. Studies of genetic manipulations to achieve this and other desirable modifications have already begun. Hence, there is a remote possibility that a genetically altered baculovirus with undesirable features could be spilled accidentally in the laboratory or greenhouse or sprayed on a field plot before its dangers were known. In these instances, all necessary measures should be taken to minimize the spread of the virus.

Laboratory

Accidental spills and release of baculoviruses within the laboratory can be controlled easily. Germicidal lamps should be used where possible and will inactivate viruses within 24 h (David et al., 1968). Surface disinfection can be accomplished using commercially available viricides, such as Rocal (National Laboratories, Monvale, New Jersey) or Nolvosan (Ft. Dodge Laboratories, Inc., Tigard, Oregon), or a solution of 10% formalin or of concentrated chlorine. (*Note:* Proper ventilation must be used with these last two solutions.) Contaminated disposables and glassware can be sterilized by autoclaving for 20 min at 121°C and 15 lb/in^2 of pressure. Personnel working in the laboratory during this time should wear protective clothing (coveralls and disposable gloves, boots, face masks, and hats), which should be bagged and autoclaved when removed. After cleanup is completed, areas that had been contaminated should be wiped and cultured to determine whether there is any virus remaining.

Greenhouse

Cleanup of spills in the greenhouse follows the same general procedures as in the laboratory. Personnel entering the greenhouse during cleanup time should remove all street clothes and change into coveralls, boots, hats, gloves, and face masks. Contaminated soil, plants, pots, and other small equipment should be placed into biohazard bags that are surface disinfected with a viricide when leaving. These bags should be autoclaved before they are reopened. Soil should be divided into small aliquots, so that all portions are exposed to proper temperature and pressure. When leaving, personnel should remove contaminated clothing and, if possible, shower before putting street clothes back on. Remaining tables and large apparatus, as well as walls, floor, ceiling, sprinklers, and other parts of the greenhouse, should be disinfected by fumigation with formaldehyde or methyl bromide gas. All vents, intakes, coolers, etc., should be closed and doors and other openings properly sealed before fumigation begins. The greenhouse should remain sealed for at least 24 h before exhausting the gas. Swabs should be taken from randomly selected areas and cultured for the presence of virus to verify that the cleanup procedures were successful.

Field

A modified baculovirus released into the field may be difficult to contain and control, especially if UV-blocking agent has been incorporated into the field formulation. It would be a good practice to eliminate the UV-blocking agents from formulations used in initial field tests of new baculovirus control agents. The extent of the control effort should be related to the perceived threat of the virus to the ecosystem. If the threat is limited to insects, only eradication of insects and other arthropods in the area by repeated applications of broad-spectrum chemical pesticides (e.g., pyrethroids or organophosphates) may be necessary. If the area is kept free of insects for a year, the UV-induced mortality of the virus should result in a viral population density that is too low to cause significant future infections of insects.

If the virus is perceived as a threat to vertebrate wildlife, aquatic life, or humans, more stringent control measures should be taken. These types of control are very costly and require coordination of large numbers of personnel. The most immediate concern is to contain the virus within an area as small as possible. Birds and mammals, as well as infected insects, have the potential to leave the contaminated area and carry the virus with them. They may be contaminated on their skin, fur, or feathers by the spray event or within their gastrointestinal tract through ingestion of infected insects or foliage (Lautenschlager and Podgwaite, 1979; Lautenschlager et al., 1980). Unfortunately, it is not feasible to live-trap and contain wildlife for decontamination procedures. Therefore, some form of humane animal control and/or depopulation would need to be implemented. Reentry of animals into the contaminated area should be prevented by hazing until the remaining control measures are achieved. Specific examples of animal control and field decontamination procedures are available from the U.S. Fish and Wildlife Service (Friend, 1987) through their work to contain duck plague virus (a herpesvirus; Wobeser, 1985) and avian cholera (*Pasteurella multocida*) epizootics in waterfowl. Avian cholera epizootics continue to occur in Texas, California, and

Nebraska but have been contained in Chesapeake Bay (Pursglove et al., 1976) and offshore Maine (Gershman et al., 1964) through depopulation and cleanup measures.

The foliage on the entire contaminated area should be burned, preferably after razing the crop if in an agricultural area. This can be done in more forested areas only during spring and early summer when the vegetation is dry enough to burn but not so dry as to burn out of control. In open fields, the soil should then be turned over at least once a week for a month and then once a month for up to three months, to allow UV radiation to penetrate to the 7- to 8-cm depth that the virus generally permeates. The pH of the soil should be lowered at this time by the addition of sulfuric or nitric acid to attain a pH of 4 or less, the level below which the virus is killed rapidly (Knittel and Fairbrother, 1987). In forests or orchards, where plowing is not feasible and the soil is covered by several inches of leaf litter and other organic material, the heat of the fire should be sufficient to decontaminate the ground. In forested areas, where bare soil predominates, the soil should be treated with acid after the fire, using crop-dusting airplanes and equipment. This also will acidify lakes and streams within the area, thereby decontaminating the water.

Early in the cleanup procedure, the contaminated zone must be delineated. Movement of equipment and personnel in and out of this area should be minimized and closely monitored. All personnel and equipment must be disinfected when leaving this area. This can be accomplished by requiring personnel to remove protective clothing (coveralls, boots, and hats) and to wash thoroughly hands, faces, and other exposed skin. Equipment can be surface disinfected with chlorine or a commercial viricide. This includes large equipment, such as trucks and tractors, as well as animal traps and other small paraphernalia. A buffer zone should be maintained around the contaminated area in which the number of people, wildlife, and insects should be controlled and kept as low as possible.

Once initial control operations are completed, a monitoring system should be initiated to ascertain that eradication of the virus has been successful. Soil, vegetation, animals, and insects within the contaminated area and buffer zone should be sampled and assayed for the presence of virus. Air samplers should be set up at the margin of the contaminated area and within the buffer zone to detect aerial spread of the virus. Although the virus is not likely to percolate into the water table, surface runoff and nearby drainages and wells should be sampled for virus to verify that they remain uncontaminated. After decontamination has been confirmed, efforts should begin to return the area to its natural state.

Summary and Conclusions

Baculoviruses are highly host-specific. Due to their sensitivity to UV radiation, they rarely persist outside their insect hosts for longer than 10 days in an infectious form. However, the occluded form, where individual viruses are clumped together and protected by a protein matrix, adhere strongly to foliage and may persist in the environment for many months. These characteristics make the baculoviruses attractive biocontrol agents. It is unlikely that intensive efforts would be necessary to contain an accidental spill. However, in the event that a virulent virus is accidentally released in the laboratory, greenhouse, or field, the control measures discussed in this chapter should provide sufficient containment and decontamination. These procedures, including those discussed for a field cleanup, have been used successfully in the past for other types of viruses. They are costly and labor-intensive, but if the virus is thought to be a threat to nontarget species, it would be a worthwhile expenditure.

References

Bird, F. T. 1955. Virus diseases of sawflies. *Can. Entomol.* 87:124–127.

Carter, J. B. 1984. Viruses as pest-control agents. *Biotech. Genet. Eng. Rev.* 1:375–418.

David, W. A. L. 1975. The status of viruses pathogenic for insects and mites. *Annu. Rev. Entomol.* 20:97–117.

David, W. A. L., and B. O. C. Gardiner. 1966. Persistence of a granulosis virus of *Pieris brassicae* on cabbage leaves. *J. Invert. Pathol.* 8:180–183.

David, W. A. L., and B. O. C. Gardiner. 1967a. The effect of heat, cold, and prolonged storage on a granulosis virus of *Pieris brassicae. J. Invert. Pathol.* 9:555–562.

David, W. A. L., and B. O. C. Gardiner. 1967b. The persistence of a granulosis virus of *Pieris brassicae* in soil and in sand. *J. Invert. Pathol.* 9:342–347.

David, W. A. L., B. O. C. Gardiner, and M. Woolner. 1968. The effects of sunlight on a purified granulosis virus of *Pieris brassicae* applied to cabbage leaves. *J. Invert. Pathol.* 11:496–501.

Friend, M. 1987. *Field Guide to Wildlife Diseases. Vol. I. General Field Procedures and Diseases of Migratory Birds.* U.S. Fish and Wildlife Service Res. Publ. No. 167. Washington, D.C., 224 pp.

Gershman, M., J. F. Witter, H. E. Spencer, and A. Kalvaitis. 1964. Case report: Epizootic of fowl cholera in the common eider duck. *J. Wildl. Mgmt.* 28:587–589.

Granados, R. R. 1980. Infectivity and mode of action of baculoviruses. *Biotech. Bioeng.* 22:1377–1405.

Harrap, K. A., and C. C. Payne. 1979. The structural properties and identification of insect viruses. *Adv. Virus Res.* 25:273–355.

Jaques, R. P. 1972. The inactivation of foliar deposits of viruses of *Trichoplusia ni* (Lepidoptera: Nocutidae) and *Pieris rapae* (Lepidoptera: Pieridae) and tests on protectant additives. *Can. Entomol.* 104:1985–1994.

Jaques, R. P. 1975. Persistence, accumulation and denaturation of nuclear polyhedrosis and granulosis viruses. In M. Summers, R. Engler, L. A. Falcon, and P. V. Vail (eds.), *Baculoviruses for Insect Pest Control: Safety Considerations.* American Society for Microbiology, Washington, D.C., pp. 90–101.

Knittel, M. D., and A. Fairbrother. 1987. Effects of temperature and pH on survival of free nuclear polyhedrosis virus of *Autographa californica. Appl. Environ. Microbiol.* 53:2771–2773.

Lautenschlager, R. A., and J. D. Podgwaite. 1979. Passage of nucleopolyhedrosis virus by avian and mammalian predators of the gypsy moth, *Lymantria dispar. Environ. Entomol.* 8:210–214.

Lautenschlager, R. A., J. D. Podgwaite, and D. E. Watson. 1980. Natural occurrence of the nucleopolyhedrosis virus of the gypsy moth, *Lymantria dispar* [*Lep.: Lymantriidae*] in wild birds and mammals. *Entomophaga* 25:261–267.

Maramarosch, K. 1968. Insect viruses. *Curr. Topics Microbiol. Immunol.* 42:1–192.

Maramarosch, K. (ed.). 1977. *The Atlas of Insect and Plant Viruses.* Academic Press, New York.

Martignoni, M. E. 1981. A catalogue of viral diseases of insects, mites, and ticks. Appendix 2. In H. D. Burgess (ed.), *Microbial Control of Pests and Plant Diseases 1970–1980.* Academic Press, London, pp. 897–911.

Pursglove, S. R., D. F. Holland, F. H. Settle, and D. C. Gnegy. 1976. Control of a fowl cholera outbreak among coots in Virginia. Conf. Southeastern Assoc. Fish Wildl. Agencies 30:602–609.

Schemnitz, S. D. 1980. *Wildlife Management Techniques Manual,* 4th ed. The Wildlife Society. Washington, D.C.

Smith, K. M. 1976. Virus-Insect Relationships. Longman, New York.

Smith, R. F. 1975. Why are baculoviruses necessary for plant production? In M. Summers, R. Engler, L. A. Falcon, and P. V. Vail (eds.), *Baculoviruses for Insect Pest Control: Safety Considerations.* American Society for Microbiology, Washington, D.C., pp. 6–8.

Tweeten, K. A., L. A. Bulla, and R. A. Consigli. 1981. Applied and molecular aspects of insect granulosis viruses. *Microbiol. Rev.* 45:379–408.

Wobeser, G. A. 1985. *Diseases of Wild Waterfowl.* Plenum Press, New York.

46

Containment, Decontamination, and Mitigation of Plant Viruses in the Environment

Myron K. Brakke

Introduction

A long-term goal of research on plant viruses has been the reduction of loss in crop yield. Losses are controlled by use of resistant cultivars and by prevention of the spread of the virus. Complete elimination of a virus is not necessary. Except for a few viruses infecting greenhouse crops, most experiments have not been directed toward elimination of virus, but to the reduction of virus incidence to the point that losses are less than cost of control. No experiments have been conducted on procedures for elimination of viruses from field plots or for absolute prevention of spread. Recommendations for elimination of viruses from test plots and complete prevention of spread must be based on general principles of virus properties and epidemiology. Therefore, this chapter will include a brief description of pertinent properties of plant viruses and their epidemiology. Viruses that persist and/or spread in soil will be emphasized because they are difficult to eliminate.

Little is known about viruses confined to noncrop plants, although viruses of algae have recently been intensively studied by Van Etten and colleagues (93). This chapter will be limited to viruses of flowering plants. Viruses of algae are considered in Chap. 48.

Plant Viruses in Genetic Engineering

Plant viruses do not appear to be promising vectors for introduction of DNA into plant chromosomes because no genomes of plant viruses are known to integrate into host nuclear or organellar DNA (10). However, foreign genes can be introduced into plant cells and expressed there by inserting them into either RNA or DNA plant virus genomes that are then used to infect plants (1, 45). High levels of expression can be obtained because of the efficient promoters of many plant viruses. In fact, the efficient promoters of cauliflower mosaic virus are frequently introduced upstream from coding sequences that are introduced into plant cells by the Ti plasmid or other methods. Plant viruses are also potential sources of genes of enzymes and other proteins, e.g., of posttranslational processing proteases with specific recognition sites (21).

Individual genes of plant viruses have been introduced into plant genomes by the Ti plasmid to study their function or to produce virus-resistant plants (6, 79). Genomes of satellite viruses have been introduced in the same way to produce tolerant or resistant plants (32, 41). Ecological implications of releasing such genetically engineered plants into the environment should be evaluated on a case-by-case basis. There is little probability that the individual virus genes would escape from the host plants, but the satellite viruses might be spread by superinfecting viruses. Such escape would not necessarily be bad because the satellite viruses occur naturally.

Properties of Plant Viruses

Genomes and genome expression

Plant viruses have genomes of positive-sense single-stranded (ss) RNA, negative-sense ssRNA, double-stranded (ds) RNA, ssDNA, or dsDNA (Table 46.1). Most have genomes of positive, or messenger, sense ssRNA. These have four general types of translation strategy, all of which accommodate the preference of the plant cell for translating monocistronic messenger RNAs (34). The four strategies are (1) production of a polyprotein which is cleaved after translation; (2) divided genomes; (3) translation from subgenomic RNAs; and (4) translation through amber, opal, or ochre stop codons. A single virus may use more than one of these strategies.

Viroids are plant pathogenic RNAs, smaller than plant virus RNAs,

TABLE 46.1 **Plant Virus Groups**

Group	Type of virus	Vector	Mechanically transmissible	Found in soil or water
		dsDNA viruses		
Caulimovirus	Cauliflower mosaic virus	Aphids	+	–
Commelina yellow mottle	Commelina yellow mottle virus			–
		ssDNA viruses		
Geminivirus	Maize streak virus	White flies, leafhoppers	Most –, some +	–
		dsRNA viruses		
Phytoreovirus	Wound tumor virus	Leafhoppers	–	–
Fijivirus	Fiji disease virus	Planthoppers	–	–
Cryptic virus group	Beet cryptic virus	None known	–	–
		ssRNA, negative-sense viruses		
Rhabdoviridae	Lettuce necrotic yellows	Aphids or leafhoppers	One +, rest –	–
		ssRNA, positive-sense, monopartite genome, quasispherical		
Tymovirus	Turnip yellow mosaic virus	Beetles	+	–
Luteovirus	Barley yellow dwarf virus		–	–
Tombusvirus	Tomato bushy stunt virus	None known	+ +	+
Sobemovirus	Southern bean mosaic virus	Beetles	+ + +	+
Necrovirus	Tobacco necrosis virus	*Olpidium* fungus	+ +	+
Carmoviruses	Turnip crinkle	Beetles	+ +	–
Ungrouped	Maize chlorotic dwarf virus	Leafhoppers	–	–
	Maize chlorotic mottle virus	Beetles	+ + + +	+ (?)
	Bean mild mosaic virus	Beetles	+ +	+
	Maize white line virus	Unknown	–	+
Carrot mottle group	Carrot mottle virus	Aphids (helper-dependent)	–	–
Marafiviruses	Maize rayado fino virus	Leafhoppers	–	–
		ssRNA, positive-sense, monopartite genome, rod-shaped		
Closterovirus	Sugar beet yellows virus	Aphids	+	–

Information in this table is from References 11, 19, 23, 30, 61, 68, 76, 90, and 91. Information on ungrouped viruses is incomplete.

+ = mechanically transmissible but low-dilution endpoint.

+ + + + = transmissible with high-dilution endpoint, ca. 10^5.

– = not mechanically transmissible.

TABLE 46.1 Plant Virus Groups (*Continued*)

Group	Type of virus	Vector	Mechanically transmissible	Found in soil or water
	ssRNA, positive-sense, monopartite genome, rod-shaped			
Carlavirus	Carnation latent virus	Aphids	+	−
Potyvirus	Potato virus Y	Aphids	+ +	−
Potyvirus-mite subgroup	Wheat streak mosaic virus	Aeriophyid mites	+	−
Barley yellow mosaic group	Barley yellow mosaic virus	*Polymyxa graminis*	+	+
Potexvirus	Potato virus X	None known	+ + + +	+
Tobamovirus	Tobacco mosaic virus	None known	+ + + +	+
Tenuiviruses	Rice stripe mosaic virus	Leafhopper	+/−	−
Capilloviruses	Apple stem grooving	None known	+	−
	ssRNA, positive-sense, bipartite genome, quasispherical			
Nepovirus	Tobacco ringspot virus	Nematodes	+ + +	+
Pea enation mosaic group	Pea enation mosaic virus	Aphids	+ +	−
Comovirus	Cowpea mosaic virus	Beetles	+ +	−
Dianthovirus	Carnation ringspot virus	None known	+ +	+
Fabaviruses	Broad bean wilt	Aphids	+	−
Velvet tobacco mottle group	Velvet tobacco mottle virus	Beetles and Myrids	+	−
	ssRNA, positive-sense, bipartite genome, rod-shaped			
Tobravirus	Tobacco rattle virus	Nematodes	+ +	+
Furovirus	Soil-borne wheat mosaic	*Polymyxa* or *Spongospora* sp. (fungus)	+	+
	ssRNA, positive-sense, tripartite genome, quasispherical			
Bromovirus	Brome mosaic virus	Beetles	+ +	−
Ilarvirus	Tobacco streak virus	None known	+	−
Cucumovirus	Cucumber mosaic virus	Aphids	+	−
Alfalfa mosaic group	Alfalfa mosaic virus	Aphids	+	−
Tomato spotted wilt group	Tomato spotted wilt virus	Thrips	+	−
	ssRNA, positive-sense, tripartite genome, rod-shaped			
Hordeivirus	Barley stripe mosaic virus	None known	+ +	−

Information in this table is from References 11, 19, 23, 30, 61, 68, 76, 90, and 91. Information on ungrouped viruses is incomplete.

+ = mechanically transmissible but low-dilution endpoint.

+ + + + = transmissible with high-dilution endpoint, ca. 10^5.

− = not mechanically transmissible.

which do not code for any proteins and are not encapsidated (24). Satellite viruses are small RNAs that need another virus for replication. They can either ameliorate or intensify symptoms, depending on the accompanying virus and the host plant (67).

The simplest plant viruses code for three or four proteins. The smallest, usually with a Mr of 20,000 to 40,000 daltons, is the capsid protein. The largest one or two are the RNA replicase(s) (34). One of the other proteins is necessary for cell-to-cell movement in the case of tobacco mosaic virus (TMV) (59). The two largest RNAs of alfalfa mosaic virus and brome mosaic virus (BMV), which have three genomic RNAs, can replicate by themselves in plant protoplasts (50, 75), confirming that these two RNAs code for the RNA replicase. Viruses whose genome is translated to give a polyprotein code for more than four proteins, some of which are the proteases needed to cleave the polyprotein (34). Potyviruses, caulimoviruses, and others produce a "helper" protein necessary for aphid transmission (78). Viruses defective in certain properties can sometimes be "rescued" by coinfection of the host with a closely related virus that produces the needed capsid protein, translocation protein (65), or aphid transmission helper protein (78).

Plant virus RNAs are efficient messenger RNAs. The capsid proteins of the virions of the mechanically transmitted viruses are usually major protein constituents of the plant cell and can be readily detected by one dimensional SDS-PAGE of unfractionated plant sap or after simple centrifugal fractionation. Virions of TMV and BMV reach concentration of 5 mg/g fresh weight of leaf tissue, but others have concentrations less than 1 μg/g. Some plant viruses produce inclusion bodies. Proteins of these inclusion bodies may also be major constituents of the cell.

Genetic stability of plant viruses

Plant virus cultures usually contain mutants (69). Theoretically, RNA plant virus genomes should have a very high mutation rate (43, 80), but it has been impossible to calculate exact mutation rates for most plant viruses because of the inefficiency of mechanical inoculation procedures, the lack of synchronous infection, and difficulty in proving that a culture is free of mutants. Some experimental results support a high mutation rate, but others do not (1, 33, 51, 58, 81, 85, 86.).

A plausible hypothesis is that plant viruses have high mutation rates, but that selective pressures and competition between strains (29) prevent accumulation of mutants. If selective pressures are relaxed, mutants accumulate. Thus, mutants not transmissible by vectors arise in virus cultures maintained without periodic vector transmission. Maintenance of a culture under different environmental conditions or in a different host may result in a different dominant mutant.

Proof for genetic recombination in RNA plant viruses was long sought without conclusive success because of the same reasons that prevent calculation of accurate mutation rates. Recently, Ahlquist et al. (1) have used infectious RNA transcribed from cloned cDNA of BMV RNA to show recombination of the RNA genome.

The gamma RNA of the type strain of barley stripe mosaic virus (a virus with three genomic RNAs, alpha, beta, and gamma) has a 366-base direct tandem repeat which is lacking in the ND18 strain of the virus (37). Although not proven, it is possible that this repeat arose by duplication.

In summary, the evidence suggests that plant viruses can mutate by deletions, point mutations, duplications, and recombination. New strains can also arise by reassortment of multicomponent genomes. In some conditions, mutants may accumulate quickly, but in most conditions, selective pressures favor maintenance of the dominant mutant.

Chemical stability of virions of plant viruses

Virions of some plant viruses lose their infectiousness in plant sap in a few hours in vitro at 20 to 30°C. Others are stable for several weeks. Most are stable for several years in dried, infected leaf tissue at 0°C. Their stability to the effects of heat, nucleases, and chemicals varies. Most of this variation is a function of the protein of the virion. Some virions are resistant to enzymatic degradation. For example, potato virus X survived plant tissue rotted by *Erwinia atroseptica* (89), and barley stripe mosaic, cucumber green mottle, tomato bushy stunt, tobacco rattle, alfalfa mosaic, and brome mosaic viruses were reported to pass through the gut of rabbits or mice without completely losing infectivity (54).

Plant viruses can be inactivated by heat, ionizing radiation, ultraviolet radiation, cationic detergents, acid, alkali, oxidizing agents, and formaldehyde (67). They are also inactivated by combinations of chemicals that disrupt nucleoproteins and those that inactivate nucleic acids. The free positive-strand ssRNAs of many plant viruses are infectious, and simply disrupting the virion does not necessarily destroy the infectiousness. Treatment with phenol or anionic detergents, such as sodium dodecyl sulfate, will give infectious nucleic acid. However, the free nucleic acid can be more readily inactivated by nucleases, alkali, heavy metals, diethylpyrocarbonate, acid, and other agents than when in the protective capsid (67). Virions of some plant viruses swell above pH 7, particularly in the presence of chelating agents, such as ethylenediamine tetraacetic acid, and become permeable to ribonuclease, which hydrolyzes the RNA. Viruses whose free nucleic acid is not infectious (plant reoviruses, rhabdoviruses, and most geminiviruses) should be inactivated by treatments that disrupt the virion.

Many plant virus virions that do not have lipids are stable to chloroform, carbon tetrachloride, and nonionic detergents, all of which are used in purification of virions. Chloropicrin (2500 ppm), metham sodium (5000 ppm), dichloropropene-dichloropropane (DD, 2500 ppm), and methylisothiocyanate (2500 ppm) did not inactivate TMV in plant sap. These chemicals and methyl bromide and formaldehyde did not inactivate TMV when they were used to fumigate soil (15).

Transmission and Spread of Plant Viruses

Plant viruses do not have a mechanism to penetrate plant cell walls nor are they taken into plant cells by pinocytosis. They either penetrate the cell wall through wounds or are placed in cells by vectors such as insects, mites, nematodes, and fungi. Plant viruses are not spread in aerosols, but they can be airborne in vectors, either arthropods or fungal spores.

Mechanical transmission of plant viruses

For experimental purposes, many plant viruses are transmitted "mechanically" by rubbing leaves with a solution of virus containing suspended abrasive. With few exceptions, plant reoviruses, rhabdoviruses, and geminiviruses cannot be transmitted mechanically (68). Most of the viruses with positive-sense ssRNA genomes can be transmitted mechanically, with the exception of viruses confined to the phloem, such as the luteoviruses.

Leaf rubbing is a relatively inefficient method that requires approximately 10^6 virus particles or more per milliliter of solution. However, given this high concentration, infection can be obtained with small volumes containing only 10^3 particles (26, 38, 94). Viruses that cannot be transmitted by leaf rubbing can sometimes be transmitted by injecting them into insect vectors or feeding the vectors with virus through membranes (66, 73).

Plants also become infected after mechanical inoculation with purified genomic, positive-sense ssRNA of many viruses.

TMV, potato virus X, turnip yellow mosaic virus, turnip crinkle mosaic virus, barley stripe mosaic virus, carnation ringspot virus, and potato spindle tuber viroid are transmitted from plant to plant when wind-blown leaves of adjacent plants rub (17, 87, 91) and by root contact, e.g., carnation ringspot virus (91) and potato virus X (17).

Broadbent (13) found that TMV spread to more plants when birds were caged with a mixture of infected and uninfected plants than in their absence. Todd (89) reported that potato virus X could be recovered from fur of rabbits and dogs after they had run among infected plants. However, there is no evidence, nor even anecdotal stories, that birds, rabbits, dogs, or other such animals are responsible for the spread of plant virus in the field.

Vector transmission of plant viruses

The vectors that carry viruses from plant to plant in the field are insects, mites, nematodes, and fungi (Table 46.1). No vectors are known for tobamoviruses, tombusviruses, hordeiviruses, dianthoviruses, ilarviruses, cryptic viruses, potexviruses, and viroids.

The most common insect vectors are aphids, leafhoppers, planthoppers, and beetles (67). Thrips, white flies, and mealy bugs are less common vectors, but some of the viruses they transmit are important economically. Transmission is specific. For example, viruses transmitted by aphids are not transmitted by leafhoppers, planthoppers, beetles, or fungi, but they may be transmitted by several related species of aphids with varying degrees of efficiency. Exceptions have been reported, but only a few of the exceptions have been confirmed.

The pattern of transmission with respect to the feeding time needed for the insect to acquire the virus from an infected plant, the incubation period before the insect can transmit the virus, the length of feeding period needed to transmit the virus, and the length of time the insect is able to transmit varies greatly from one virus to another. Each virus has a characteristic transmission pattern (27, 39, 67).

A few plant viruses multiply in leafhopper and aphid vectors (9). In addition, the virus passes through the egg of some leafhoppers to persist from generation to generation (52, 71).

Fungi of three genera, *Olpidium*, *Polymyxa*, and *Spongospora*, have been reported to transmit viruses. *Olpidium* and *Polymyxa* are obligate parasites of roots (40, 42). *Spongospora* is a soil-inhabiting fungus that infects potato tubers to cause scab (48). Viruses transmitted by these fungi are referred to as "soil-borne," although it is really their vectors that are soil-borne. The relation of these viruses to their fungal vectors is not as well understood as is the relation of other viruses to their insect vectors. Two types of transmission patterns, persistent and nonpersistent, are observed (42). *Olpidium* transmission of necroviruses typifies the nonpersistent pattern. The zoospores can acquire these viruses in vitro, i.e., when purified virus is mixed with zoospores, and, conversely, can lose the virus when washed or treated with antiserum to the virus. *Polymyxa* always, and *Olpidium* sometimes, transmits virus in a persistent fashion, which means that the zoospores can neither acquire virus in vitro nor lose it by washing or by antiserum treatment. Zoospores can acquire any of the viruses they transmit when produced by zoosporangia growing in virus-infected plants. Viruses that cannot be acquired in vitro are apparently carried internally in the fungal spores. It is not known if the viruses multiply in the fungus.

Ectoparasitic nematodes of two families, Trichodoridae and Longidoridae, transmit viruses belonging to two groups, tobraviruses and nepoviruses, respectively (42). Nematodes acquire viruses by feeding on infected plants, and transmit them by feeding on other plants. Nematodes may retain viruses for several months, especially at low temperatures, but not through a molt.

The ability to be transmitted by a vector is a genetic property of the virus, and nontransmissible mutants are common (5, 51, 63, 72, 81, 86). Viral proteins appear to mediate vector specificity. Transmission of potyviruses and caulimoviruses depends on virus-coded accessory proteins that are not part of the virion. The protein is specific and facilitates transmission of closely related viruses only (78). The capsid protein appears to determine vector specificity in the luteoviruses (83). Vector transmission of a plant virus only when a second is present is called "dependent transmission" and has been reviewed by Rochow (83).

Seed and pollen transmission

Transmission of plant viruses by seed and pollen is common, although transmission by seed is more common than by pollen. Matthews (67) reported that 70 of 200 well-studied plant viruses are transmitted by seed, with the rate of transmission varying from 100 to 0.01 percent, the least that can be reliably detected. The same virus may be transmitted by seed in one host and not in another.

Among crop plants, legumes have many seed transmitted viruses including potyviruses and comoviruses (17). Transmission by seed is common among the nepoviruses and tobraviruses that are also transmitted by nematodes (40). Ilarviruses are transmitted by seed produced from infected female or male parent. They also may be transmitted by pollen to the mother plant (28). TMV is seed-transmitted, but as a contaminant of the seed coat rather than as an infection of the embryo. Young tomato plants grown from seed with TMV-contaminated seed coat do not become infected unless injured (14). Cryptic viruses and hordeiviruses are also seed transmitted (4, 76).

People as unintentional plant virus transmitters

People transmit plant viruses unintentionally during the course of agricultural and horticultural operations. If a virus-infected plant is selected for vegetative propagation, all progeny plants will have virus, unless special techniques are used, such as meristem propagation. Viroids are transmitted by pruning knives (24). "Soil-borne" viruses are carried from field to field on machinery and in soil adhering to roots of nursery stock and seedlings used as transplants. Viruses present in high concentration and easily transmitted mechanically may be spread when plants are pruned, transplanted, etc. For example, TMV is transmitted when workers handle tobacco and tomato plants, especially if the workers smoke or otherwise use tobacco products because the virus survives the curing process (15). Potato virus X was transmitted by workers alternately walking in infected and then in uninfected fields of potato (89). BMV is common around farmyards where the grass is frequently mowed with mechanical mowers.

Viruses that may spread by mechanical transmission from soils or debris

It was originally thought that all soil-borne plant viruses infected roots through wounds, i.e., mechanically. Subsequently, nematode and fungal vectors were discovered for most, but not all, of these viruses. The remainder appear to be transmitted mechanically, albeit inefficiently. Only tobamoviruses cause important diseases of the crop plants that they infect by mechanical transmission from soil.

The spread of TMV in soil has been studied more than that of any other virus because of the economic importance of the tomato strain of TMV in tomatoes. In both field and greenhouse, a few plants become infected by mechanical inoculation from contaminated seed, soil, plant debris, or human hands. Roberts (82) showed that tomato plants could be infected through their roots, in the absence of vectors, with TMV. Allen (2) showed that TMV in the soil can be splashed onto leaves where infections are initiated if the leaves are wounded. After a few plants have been infected, the virus is spread rapidly by hands or tools (15).

Tombusviruses, bean mild mosaic (ungrouped), and a dianthovirus (carnation ringspot virus), when present in the soil, can infect roots mechanically (7, 8, 31, 42, 44, 54, 62). Tombusviruses are easily acquired by plants from soil in the absence of vectors, and slightly better in the presence of *Olpidium* (62). Necroviruses also infect through roots in both the presence and absence of *Olpidium* zoospores, but the increase in efficiency of infection in the presence of zoospores is often much greater for the necroviruses than for the tombusviruses (42).

Evidence of soil transmission of potexviruses is equivocal. Potato virus X and other potexviruses in cultivated crops are spread by vegetative propagation of infected hosts and by hand, but these methods seem inadequate to explain the spread in noncrop plants. Transmission of potato virus X by a fungus, *Synchytrium endobioticum,* was reported (77) but could not be confirmed (56). Some possible members of the potexviruses reportedly may be transmitted by aphids in the presence of helper viruses (55). Roberts (82) reported mechanical transmission from soil to roots, but such transmission has not been shown to be important in the field.

Some sobemoviruses can be experimentally transmitted by beetles and mechanically from soil to roots (42), but there appears to be no evidence that mechanical transmission from soil is important in the field. On the other hand, maize chlorotic mottle virus is experimentally transmitted by beetles, but spread in the field suggests soil transmission (47).

Finally, there are some viruses that are soil-borne, but the mechanism of transmission, whether by vector or not, is unknown. Maize white line is an example (23).

Of the viruses discussed in this chapter, TMV, potato virus X, and southern bean mosaic are the best studied and the most apt to be used for genetic engineering experiments. Clearly, soil transmission must be considered in any use or release of genetically modified TMV. The evidence for importance of soil transmission for the other two is contradictory and it is not clear that extreme precautions would be needed to prevent soil transmission of them.

Survival of Plant Viruses in the Environment

To date, the survival of plant viruses in the environment has been determined by infectivity assay. Survival in vectors has been determined by vector transmission. Survival in soil, on clothing, etc., has usually been determined by mechanical inoculation and sometimes by vector transmission. Survival of many viruses has not been tested.

Annual plants

Most plant viruses survive in plant hosts or in vectors. Annual plants, except for winter annuals, are poor hosts for survival from season to season. In some cases, viruses survive in overlapping winter and spring annual crops. Wheat streak mosaic virus, for example, can survive by alternately infecting adjacent winter wheat and summer corn crops (11).

Perennial plants

Most plant viruses probably survive as infections in perennial plants (17). Perennial plants that are propagated vegetatively are almost always virus-infected unless virus-free material is selected for propagation. There is less information on uncultivated perennial plants, but many undoubtedly serve as reservoirs for viruses that infect crop plants. For example, Johnson grass, a wild perennial sorghum, harbors maize dwarf mosaic and other viruses that infect corn and sorghum crops. Weed control is an important control procedure for these viruses.

Seeds

Seed-transmitted viruses are often more stable to storage and to heat in the seed than in other tissue or media (67). They may survive in the seed as long as the seed remains viable and sometimes longer. Some weed seeds remain viable for years, and, conceivably, could provide long-term persistence for seed-borne viruses (17). As pollen remains viable for only a limited time, it probably is not important in virus survival.

Tomato plants grown in sewage sludge became infected with tomato bushy stunt virus, tomato strain of TMV, and cucumber mosaic virus, apparently from viruses present in seed that had passed through the human gut (54).

Arthropod vectors

Some plant viruses multiply in insect vectors and pass through the egg from one generation to the next. These viruses can survive in the insect vectors from season to season, e.g., rice dwarf virus (71) and rice stripe virus (52). Viruses that are transmitted in a persistent fashion (i.e., retained and transmitted for a long time) by arthropod vectors may survive in the vector for the life of the vector, usually a matter of weeks. This includes most viruses transmitted by leafhoppers, planthoppers, and mites and some transmitted by aphids, white flies, thrips, and beetles (27, 39, 67).

Fungal vectors

Viruses transmitted by fungal vectors may persist for years in resting spores. Soil-borne wheat mosaic virus has persisted for 19 years in

Polymyxa spores in a refrigerator (W. G. Langenberg, personal communication). Resting spores of *Polymyxa* germinate sporadically instead of synchronously (12), and, therefore, although some germinate to give zoospores that infect host plants, there are reserve resting spores to germinate at a later time. Fungal spores and the viruses they transmit survive drying of soil at 25°C (40).

The survival of viruses transmitted in a persistent fashion by *Olpidium* is greatly increased by association with resting spores. Tobacco stunt virus is stable for only a few hours in plant sap but for 20 years in *Olpidium* spores (42). Smith et al. (88) obtained good infection after storage for 130 days at 23°C of moist soil containing roots infected with both tobacco necrosis virus and *Olpidium*. The *Olpidium* would have survived as resting spores, and the virus within the *Olpidium* spores. They obtained poor infection after storing moist soil containing purified virus for only 87 days, and then could detect the virus only by adding *Olpidium* zoospores to aid in transmission of the virus.

Nematodes

Nematodes are not known to have a resting stage that survives drying. Soil containing nematode-transmitted viruses becomes noninfectious after drying. Drying soil is a commonly used test to distinguish between fungal- and nematode-transmitted viruses. However, nematodes in the field migrate to greater depths to survive when top layers of soil become dry. They also survive fumigation by migrating to deep soil (40, 42).

Nematodes can remain viruliferous for months, but not through molts. Viruses may survive in nematodes for several months, especially at low temperatures. Tomato black ring virus persisted, though rarely, for 3 months in *Longidorus elongatus* at 10°C and tobacco ringspot virus persisted for almost a year in *Xiphinema americanum* at 10°C (42). Tomato ringspot virus survived 2, but not 3, years in the vector nematode in soil stored in a plastic bag at 1 to 3°C (8).

Viruses that survive in soil without association with known vectors

Several types of evidence show that plant viruses can survive in soil while not associated with a vector. Some evidence suggests viruses may survive better if adsorbed on clay than otherwise (54), but other evidence suggests that adsorption to clay decreases survival (3, 88). Adsorption to clay may be less important for plant viruses than for animal and human viruses (60).

Tobamoviruses, potexviruses, necroviruses, tombusviruses, and cucumber mosaic virus have been recovered from stream and pond waters, which suggests their presence and survival in soil from which the water drained (54). This water is a potential source of disease in crops. Contaminated surface water used for irrigation was shown to be a source of infection of cucumber green mottle virus, a tobamovirus, in greenhouse-grown cucumbers in the Netherlands (92).

Potato virus X, tobacco rattle, southern bean mosaic, and sowbane mosaic viruses can be mechanically transmitted from soil, but there is no evidence that these viruses survive free in soil in the field (42, 55, 82).

It is thought that some viruses with no known vectors may survive in the soil, and experiments have shown that they do survive for short periods. Red clover necrotic mosaic virus survives for at least short periods in soil and can infect plants mechanically when present in soil (31). Drainage water from pots containing roots of infected plants was infectious for only 12 days after the plant tops had been cut off.

Tombusviruses survive in perennial plants, particularly in roots, and probably also in soil (7, 44, 62). There is no evidence that adsorption to clay enhances survival of the virus. Cymbidium mosaic virus survived 16 weeks at 20°C in dry, but not in wet, soil (44). Tomato bushy stunt virus did not survive 6.5 months in moist compost in two tests, but it survived in one of three pots at a depth of 20 cm in a third test (62). Petunia asteroid mosaic virus could not be recovered after 3 days in dry soil (88).

None of the above viruses causes important diseases in crop plants, or, if they do, survival in the soil does not appear to be important in their epidemiology. Therefore, their survival in soil has not been extensively studied. The situation is different with TMV.

TMV survives free in soil, in plant debris, and on door knobs, greenhouse benches, clothing, and hands. Broadbent and Fletcher (16) reported that TMV survived on clothing stored in the dark for 201 weeks, on greenhouse structures for 2 months, and on wires used to support tomato plants for 4 months. These were the longest times tested. TMV readily infects young plants growing in soil that previously supported infected plants. TMV survived for at least 22 months in debris in undisturbed soil below 30 cm and for 2 years in debris in undisturbed soil beneath plastic. These were the longest times tested. Conversely, TMV and potato virus X were recovered after 10 weeks, but not after 15 weeks, from infected roots buried in moist soil in a 25-cm-diameter pot and kept at 15°C (18). Allen (3) found that 90 percent of the infectivity was lost in a day in dry soil. However, a trace of virus survived several days. Smith et al. (88) reported that purified TMV, southern bean mosaic virus, petunia asteroid mosaic virus, cucumber necrosis virus, and tobacco necrosis virus could not be recovered after 3 days of storage in dry soil at 23°C, but they could be recovered after 25 days of storage in moist soil.

The above tests were all done in the laboratory or the greenhouse. At least two reports have appeared on survival of TMV in fields where infected tomatoes had grown. Lanter et al. (57) reported that TMV persisted from fall to spring in soil in a tomato field in Arkansas when the soil was not tilled or was disked only once, but it did not survive if the soil was disked six times or if winter wheat was planted in the field. Green et al. (36) reported that TMV survived 5 or 6 months, but not 7 months, in soil in fields in Taiwan after TMV-infected tomatoes were harvested. The tomatoes were harvested in March, after which the fields were tilled and planted to another crop. Recovery attempts were made during the following months. It should be noted that average air temperatures in Taiwan in May, June, and July are close to 30°C.

Based on experimental data in different reports, TMV appears to survive longer in moist than in dry soil, longer at low temperatures than at high, longer in undisturbed than in tilled soil, and longer if present in infected plant debris than when added as purified virus. Freezing, however, is detrimental (3). The mechanisms by which TMV is inactivated in soil are unknown, and predictions of survival times in untested conditions cannot be made.

Prevention of Spread

With the exception of tobamoviruses, tombusviruses, and a few others, elimination and control of plant viruses are a matter of controlling or eliminating plants and vectors. If a virus is to be eliminated, the first requirement is to prevent it from spreading.

Transmission by vectors involves probability factors. It is experimentally difficult to obtain transmission of many viruses with single insects. Efficient spread and epiphytotics require massive concentrations of vectors. The probability of spread of the virus is less than the probability of occurrence of the vector. Control measures are usually designed to reduce the level of the disease, not to completely eliminate it.

Insects and mites

For an insect or mite to transmit a virus from plant to plant, the insect or mite must land on the infected plant, feed long enough to acquire the virus, and then fly or be blown to another plant. Wingless insects and mites cannot fly, but they can crawl to adjacent plants and be blown to distant plants. Winged or not, vectors are more likely to travel to adjacent than to distant plants. Therefore, it is helpful to put border trap rows of plants that attract the vectors around an infected area. The border rows and plants in the infected area should be treated with insecticides or miticides (97). Experimental plots should not be placed near crops that are infested with large numbers of vector insects.

Insect repellents also control the spread of viruses. Oil sprays and aluminum foil spread on the ground deter insects from landing and feeding. Sticky yellow sheets trap some insects (97).

Soil-transmitted viruses with vectors

If virus in a test plot is to be contained, the test plot should be located where the vector does not occur. If this is not possible, the soil should be fumigated, heated, or dried to destroy the vector or reduce its population. Any transfer of soil from the plot by machinery, on feet of people or animals, and by wind and water erosion should be prevented or limited. A fence to keep out animals and windbreaks to keep soil from blowing would be desirable. A plant cover and mulches are effective in preventing wind and water erosion of soil. Therefore, a border providing dense foliage and mulches around infected plants may be desirable. Methods should be provided to prevent erosion by heavy rains or excessive irrigation.

Because some soil-borne viruses, particularly nematode-transmitted viruses, are frequently transmitted by seed, use of clean seed to prevent introduction of other viruses is desirable.

Soil-borne viruses without vectors

Controlling the spread of these viruses from an infected area depends on prevention of transfer of soil and plant tissue from the area. Water and wind erosion should be prevented. Access for animals and people should be prevented or limited. Machinery and tools should be washed or disinfected after use.

Control of human access and activity

Mechanical inoculation is inefficient. Prevention of spread by contaminated hands and tools does not require extreme measures. Thorough washing of hands with soap is sufficient to prevent spread of most plant viruses. A wash in an alkaline solution, such as trisodium phosphate, is needed, in addition to soap, to remove TMV completely from the hands (15). To prevent spread of mechanically transmitted viruses, it is essential that people not handle infected plants and then handle other plants. Workers must be trained to keep hands off infected plants and to wash hands if they must touch them. Use of disposable gloves should help prevent virus spread, provided that the gloves can be removed and disposed without contaminating the hands. Tools and machinery should be washed or autoclaved after use.

Biological containment

The probability of viruses spreading could be reduced by the use of defective strains or mutants. Potyviruses that do not produce helper protein, and, therefore, are not aphid-transmitted, would be less likely to spread than a wild-type virus. TMV that was unstable because of a defective coat protein would be less apt to be spread by people handling infected plants and less apt to survive in soil than wild-type virus. Defective viruses could be obtained either by selection of naturally occurring strains or constructed by recombinant DNA techniques. Another type of biological containment would be the use of a virus in a geographical area where the vector for that virus does not occur.

Superinfection by a related wild-type virus might "rescue" the defective strain by producing the missing or defective protein, by reassortment (for multicomponent genome viruses), or recombination. Hence, steps should be taken to prevent superinfection by related viruses.

Elimination of Viruses

With the exception of TMV and other viruses that can survive free in the soil, elimination of viruses means elimination of host plants, plant parts, and vectors. There are no viricidal chemicals that will eliminate viruses within plants. Heat treatment, sometimes coupled with viricidal chemicals, will free a certain percentage of plant meristems from virus. Heat treatment is used to obtain virus-free plants for vegetative propagation, but it is not a practical method to eliminate virus in plants (67).

Elimination of host plants

Infected plants can be killed by herbicides, fire, solarization, tillage, drought, freezing, and maturity. Most plant viruses do not survive in plant tissue decomposed by microbes under natural conditions. Tobamoviruses and others that can survive free in soil are exceptions. These are considered below.

Fire, some herbicides, some forms of tillage, and freezing temperatures may kill only the tops of plants and not the roots. Perennial plants will grow back from the roots, as will annual plants under certain conditions. The roots must be killed if the virus is to be eliminated.

Seed-borne viruses can survive extended periods in seeds, especially in long-lived weed seeds. TMV can be eliminated from tomato seeds by heating for 3 h at 78°C (36). As it is easier to keep seed from setting and maturing than to eliminate infected seed from soil, plants should not be allowed to set and mature seed if a seed-borne virus is present. Elimination of seeds from the soil will be considered below.

Elimination of aerial vectors

Insects and mites can be eliminated by use of insecticides and miticides (64). Systemic miticides and insecticides are more effective than nonsystemic ones for insects and mites that live inside curled leaves or in leaf whorls, where they are protected from contact insecticides. Insects and mites also are killed if they cannot migrate when the host plants are killed. If host plants are killed by cultivation, several days may be required for all the mites to die (11). Aeriophyid mites survive on moist wheat leaves for one to two weeks after tillage. Destruction of leafhopper and planthopper eggs is important for those few viruses that are transmitted through the egg.

Competent specialists should be consulted for information on the effectiveness of currently available insecticides and miticides. Any recommendations at the time this is written would soon be outdated.

Elimination of viruses from the soil

Viruses in soil may be free, in seeds, roots, insects, nematodes, fungal spores, or infected leaves and stems incorporated by tillage. The volume of soil infested with viruses from decomposing roots, or in fungi or nematodes that have fed on the roots, is large. Roots of some grasses and crop plants reached 3 to 8 ft deep, but those of one prairie forb, blazing star, reached 16 ft with a lateral spread of 4.5 ft (95). However, most roots, seeds, fungi, and nematodes are within 1 ft of the surface (20, 95).

A few aphid-borne viruses are transmitted by root-feeding aphids (46). These presumably can be killed by systemic insecticides.

Viruses are easily eliminated from small volumes of soil, such as used in the greenhouse. Virus-transmitting nematodes are readily eliminated from small volumes of soil simply by drying (40). Nematodes, seeds, roots, and fungal spores are killed in small volumes of soil by fumigation, e.g., with methyl bromide or chloropicrin (70, 84).

Greenhouse soil is commonly heated with a steam-air mixture to about 85°C for one or two 30-min periods, which kills nematodes, fungal spores, seeds, and roots (3, 15).

Elimination of TMV requires high temperatures or long storage. Allen (3) found that heating at 85°C did not completely inactivate TMV in soil, but two treatments at 121°C for 30 min each did. Reports differ on the length of time needed to inactivate TMV in dry soil at room temperature (3, 88). Attempted inactivation should be followed by assays. Inactivation with chemicals requires thorough treatment. Broadbent (15) found that formaldehyde treatment did not consistently inactivate TMV in greenhouse benches and soil, possibly because it did not penetrate the infected plant roots and other debris.

There is insufficient information to make recommendations to eliminate viruses other than TMV that occur free in the soil.

The situation is not quite so simple in the field. The effectiveness of chemical soil fumigation depends on soil moisture, porosity, adsorptive capacities, temperature, and other factors (35, 40, 74, 96, 67). Nematodes and nematode-transmitted virus levels are greatly reduced by soil fumigation and subsequently may even be absent for a year or two. However, the nematodes and viruses invariably return with time. It is thought that some nematodes survive deep in the soil or in clumps that are not penetrated by the fumigant. Reinfection from adjacent fields may also contribute to the reappearance of the disease. The incidence of soil-borne wheat mosaic virus, and presumably the numbers of resting spores of *P. graminis,* was eliminated in small volumes of soil (about 10 kg) by chemical fumigation, e.g., with methyl bromide, formaldehyde, chloropicrin, or DD (70, 84). Fumigation in the field eliminated measurable loss, but it did not completely eliminate the virus (25). Crop rotation has not eliminated soil-borne wheat mosaic virus (53). Infection of wheat with wheat spindle streak mosaic virus was delayed, but not prevented, by fumigation of soil with methyl bromide (22).

To assure complete elimination of nematodes, seeds, and fungal spores from soil to a certain depth, it might be necessary to spread the soil in a thin layer, so that complete drying, fumigation, or heating occurs. Covering soil with transparent plastic (solarization; 49) is a simple method to heat a thin layer of soil and at the same time prevent it from blowing away. Solarization is nonhazardous to the environment and personnel.

TMV could presumably be eliminated from field soils by heat, provided that the heating was thorough. Less drastic treatments have worked. Repeated tillage (six diskings) or a crop of winter wheat prevented reoccurrence of TMV in tomatoes planted in a field that had

TMV-infected tomatoes the previous year (36, 57). If large root pieces were eliminated, the remaining TMV in the soil should not survive extended storage (perhaps months) in dry soil (3, 88).

Allowing a plot of soil to lie barren for a few years might eliminate all viruses except those in fungal resting spores and weed seeds. Soil and wind erosion would have to be prevented and animals excluded. Nematodes (e.g., *Xiphinema* sp.) have survived for up to 49 months at 10°C in the absence of host plants (20). However, if the nematodes did not feed on infected plants, virus persisted in *Longidorus* spp. for 2 to 3 months and in *Xiphinema* spp. for 1 year (20, 42). Some fungal resting spores survive in field soil for several years in absence of known host plants and in the laboratory for up to 20 years (*Olpidium*; 42).

Conclusion

Containment and elimination of plant viruses are primarily a matter of containing and eliminating plant hosts and animal and fungal vectors. Elimination of aerial vectors and aerial portions of plants is relatively simple. Elimination of roots, seeds, and vectors in soil is more difficult. A few viruses can survive free in soil and infect plants mechanically, but most of these do not cause serious and widespread diseases except for TMV. Viruses that persist the longest in the environment are those that persist in seeds and fungal spores in the soil. With seed-borne viruses, it may be easier to prevent infected plants from shedding seeds than to eliminate seeds from soil. With soil-borne viruses that are transmitted persistently by fungi, it may be necessary to limit the amount of soil that the roots can penetrate in order to limit the amount of soil that has to be treated.

Perhaps the most important step in containment or elimination of a genetically engineered plant virus introduced into the environment is selection of the virus, host plant, and location. Selection of the right system will minimize the probability of spread of the virus and maximize the ease of its elimination. Among the well-studied viruses, genetically engineered tobamoviruses as well as fungal and nematode-transmitted viruses appear to be poor choices for release into the environment. The virus selected will, of course, depend on the experiment, but it should be one with well-understood epidemiology so that protocols for containment, elimination, and monitoring can be devised.

References

1. Ahlquist, P., R. French, and J. J. Bujarski. 1987. Molecular studies of brome mosaic virus using infectious transcripts from cloned cDNA. *Adv. Virus Res.* 32:215–242.
2. Allen, W. R. 1981. Dissemination of tobacco mosaic virus from soil to plant leaves under glasshouse conditions. *Can. J. Plant Pathol.* 3:163–168.
3. Allen, W. R. 1984. Mode of inactivation of TMV in soil under dehydrating conditions. *Can. J. Plant Pathol.* 6:9–16.
4. Atabekov, J. G., and V. K. Novikov. 1971. Barley stripe mosaic virus. CMI/AAB Descriptions of Plant Viruses, No. 68.
5. Badami, R. S. 1958. Changes in transmissibility by aphids of a strain of cucumber mosaic virus. *Ann. Appl. Biol.* 46:554–562.
6. Baughman, G. A., J. D. Jacobs, and S. H. Howell. 1988. Cauliflower mosaic virus gene VI produces a symptomatic phenotype in transgenic tobacco plants. *Proc. Natl. Acad. Sci. USA* 85:733–737.
7. Behncken, G. M., R. I. B. Francki, and A. J. Gibbs. 1982. Galinsoga mosaic virus. CMI/AAB Descriptions of Plant Viruses, No. 252.
8. Bitterlin, M. W., and D. Gonsalvez, 1987. Spatial distribution of *Xiphinema rivesi* and persistence of tomato ringspot virus and vector in soil. *Plant Dis.* 71:408–411.
9. Black, L. M. 1984. The controversy regarding multiplication of some plant viruses in their insect vectors. In K. F. Harris, (ed.), *Current Topics in Vector Research,* vol. 2. Praeger, New York, pp. 1–30.
10. Brakke, M. K. 1984. Mutations, the aberrant ratio phenomenon, and virus infection of maize. *Ann. Rev. Phytopathol.* 22:77–94.
11. Brakke, M. K. 1987. Virus diseases of wheat. In E. G. Heyne (ed.), *Wheat and Wheat Improvement,* 2d ed. American Society of Agronomy, Madison, Wis., pp. 585–624.
12. Brakke, M. K., and A. P. Estes. 1967. Some factors affecting vector transmission of soil-borne wheat mosaic virus from root washings and soil debris. *Phytopathology* 57:905–910.
13. Broadbent, L. 1965. The epidemiology of tomato mosaic virus. IX. Transmission of tomato mosaic virus by birds. *Ann. Appl. Biol.* 55:67–69.
14. Broadbent, L. 1965. The epidemiology of tomato mosaic. XI. Seed transmission of TMV. *Ann. Appl. Biol.* 56:177–205.
15. Broadbent, L. 1976. Epidemiology and control of tomato mosaic virus. *Annu. Rev. Phytopathol.* 14:755–796.
16. Broadbent, L., and J. T. Fletcher. 1963. The epidemiology of tomato mosaic. IV. Persistence of virus on clothing and glasshouse structures. *Ann. Appl. Biol.* 52:233–241.
17. Broadbent, L., and C. Martini. 1959. The spread of plant viruses. *Adv. Virus Res.* 6:93–135.
18. Broadbent, L., W. H. Read, and F. T. Last. 1965. The epidemiology of tomato mosaic. X. Persistence of TMV-infected debris in soil, and the effects of soil partial sterilization. *Ann. Appl. Biol.* 55:471–483.
19. Brown, F. 1986. The classification and nomenclature of viruses: Summary of results of meetings of the International Committee on the Taxonomy of Viruses in Sendai, September 1984. *Intervirology* 25:141–143.
20. Bruehl, G. 1987. Soilborne viruses. In G. Bruehl (ed.), *Soil-Borne Plant Pathogens.* Macmillan, New York, pp. 254–276.
21. Carrington, J. C., and W. G. Dougherty. 1987. Small nuclear inclusion protein encoded by a plant potyvirus genome is a protease. *Proc. Natl. Acad. Sci. USA* 61: 2540–2548.
22. Cunfer, B. M., J. W. Demski, and D. C. Bays. 1988. Reduction in plant development, yield, and grain quality associated with wheat spindle streak mosaic virus. *Phytopathology* 78:198–204.

23. DeZoeten, G. A., and B. B. Reddick. 1984. Maize white line virus. CMI/AAB Descriptions of Plant Viruses, No. 283.
24. Diener, T. O. 1979. *Viroids and Viroid Diseases.* Wiley, New York.
25. Eversmeyer, M. G., W. G. Willis, and C. L. Kramer. 1983. Effect of soil fumigation on occurrence and damage caused by soil-borne wheat mosaic. *Plant Dis.* 67:1000–1002.
26. Fraser, L., and R. E. F. Matthews. 1979. Efficient mechanical inoculation of turnip yellow mosaic virus using small amounts of inoculum. *J. Gen. Virol.* 44:565–568.
27. Fulton, J. P., R. C. Gergerich, and H. A. Scott. 1987. Beetle transmission of plant viruses. *Annu. Rev. Plant Pathol.* 25:111–123.
28. Fulton, R. W. 1983. Ilarvirus group. CMI/AAB Descriptions of Plant Viruses, No. 275.
29. Fulton, R. W. 1986. Practices and precautions in the use of cross protection for plant virus disease control. *Ann. Rev. Plant Pathol.* 24:67–81.
30. Gamez, R., and F. Saavedra. 1986. Maize rayado fino: A model of a leafhopper-borne virus disease in the neotropics. In G. D. McLean, R. G. Garrett, and W. G. Ruesink (eds.), *Plant Virus Epidemics.* Academic Press, Orlando, Fla., pp. 315–326.
31. Gerhardson, B., and V. Insunza. 1979. Soil transmission of red clover mosaic virus. *Phytopathol. Zeit.* 94:67–71.
32. Gerlach, W. L., D. Llewellyn, and J. Haseloff. 1987. Construction of a plant disease resistance gene from the satellite RNA of tobacco ringspot virus. *Nature* (London) 328:802–805.
33. Goelet, P., G. P. Lomonossoff, P. J. G. Butler, M. E. Akam, M. J. Gait, and J. Karn. 1982. Nucleotide sequence of tobacco mosaic virus RNA. *Proc. Natl. Acad. Sci. USA* 79:5818–5822.
34. Goldbach, R. W. 1986. Molecular evolution of plant RNA viruses. *Annu. Rev. Plant Pathol.* 24:289–310.
35. Gording, C. A. I. 1967. Physical action of the soil in relation to the action of soil fungicides. *Ann. Rev. Phytopathol.* 5:285–318.
36. Green, S. K., L. L. Hwany, and Y. J. Kuo. 1987. Epidemiology of tomato mosaic virus in Taiwan and identification of strains. *Zeit. Pflanzenkrankheiten Pflanzenschutz.* 94:386–397.
37. Gustafson, G., B. Hunter, R. Hanau, S. L. Armour, and A. O. Jackson. 1987. Nucleotide sequence and genetic organization of barley stripe mosaic virus RNA gamma. *Virology* 158:394–406.
38. Halliwell, R. S., and W. S. Gazaway. 1975. Quantity of microinjected tobacco mosaic virus required for infection of single cultured tobacco cells. *Virology* 65:583–587.
39. Harris, K. F. 1981. Arthropod and nematode vectors of plant viruses. *Annu. Rev. Plant Pathol.* 19:391–426.
40. Harrison, B. D. 1977. Ecology and control of viruses with soil-inhabiting vectors. *Ann. Rev. Phytopathol.* 15:331–360.
41. Harrison, B. D., M. A. Mayo, and D. C. Baulcombe. 1987. Virus resistance in transgenic plants that express cucumber mosaic virus satellite RNA. *Nature* (London) 328:799–802.
42. Hiruki, C., and D. S. Teakle. 1987. Soil-borne viruses of plants. In K. F. Harris (ed.), *Current Topics in Vector Research.* Springer-Verlag, New York, pp. 178–215.
43. Holland, J., K. Spindler, F. Horodyski, E. Graban, S. Nichol, and S. Vande Pol. 1982. Rapid evolution of RNA genomes. *Science* 215:1577–1585.
44. Hollings, M., O. M. Stone, and R. J. Barton. 1977. Pathology, soil transmission and characterization of cymbidium ringspot, a virus from cymbidium orchids and white clover (*Trifolium repens*). *Ann. Appl. Biol.* 85:233–248.
45. Hull, R., and J. W. Davies. 1983. Genetic engineering with plant viruses, and their potential as vectors. *Adv. Virus Res.* 28:1–34.
46. Jedlinski, H. 1981. Rice root aphid. *Rhopalosiphum rufiabdominalis,* a vector of barley yellow dwarf virus in Illinois and the disease complex. *Plant Dis.* 65:975–978.

47. Jensen, S. G. 1985. Laboratory transmission of maize chlorotic mottle virus by three species of corn rootworms. *Plant. Dis.* 69:864–868.

48. Jones, R. A. C., and B. D. Harrison. 1969. The behaviour of potato mop-top virus in soil, and evidence for its transmission by *Spongospora subterranea* (Wallr.) Lagerh. *Ann. Appl. Biol.* 63:1–17.

49. Kalan, J. 1981. Solar heating (solarization) of soil for control of soilborne pests. *Ann. Rev. Phytopathol.* 19:211–236.

50. Kibertis, P. A., L. S. Loesch-Fries, and T. C. Hall. 1981. Viral protein synthesis in barley protoplasts inoculated with native and fractionated brome mosaic virus RNA. *Virology* 112:804–808.

51. Kimura, I. 1976. Loss of vector-transmissibility in an isolate of rice dwarf virus. *Ann Phytopath. Soc. Japan* 42:322–324.

52. Kisimoto, R., and Y. Yoshihiro. 1986. A planthopper-rice virus epidemiology model: rice stripe and small brown planthopper, *Laodelphax striatellus* Fallen. In G. D. McLean, R. G. Garrett, and W. G. Ruesink (eds.), *Plant Virus Epidemics*. Academic Press, Orlando, Fla., pp. 327–344.

53. Koehler, B., W. M. Bever, and O. T. Bonnet. 1952. Soil-borne wheat mosaic. *Illinois Agric. Exp. Sta. Bull.* 556.

54. Koenig, R. 1986. Plant viruses in rivers and lakes. *Adv. Virus Res.* 31:321–333.

55. Koenig, R., and D.-E. Leseman. 1978. Potexgroup. CMI/AAB Descriptions of Plant Viruses, No. 200. 5 pages.

56. Lange, L. 1978. *Synchytrium endobioticum* and potato virus X. *Phytopathol. Zeit.* 92:132–142.

57. Lanter, J. M., J. M. McGuire, and M. J. Goode. 1982. Persistence of tomato mosaic virus in tomato debris and soil under field conditions. *Plant Dis.* 66:552–555.

58. Lemaire, O., D. Merdinoglu, P. Valentin, C. Putz, V. Ziegler-Graff, H. Guilley, G. Jonard, and K. Richards. 1988. Effect of beet necrotic yellow vein virus RNA composition on transmission by *Polymyxa betae*. *Virology* 162:232–235.

59. Leonard, D. A., and M. Zaitlin. 1982. A temperature-sensitive strain of tobacco mosaic virus defective in cell-to-cell movement generates an altered viral protein. *Virology* 117:416–424.

60. Lipson, S. M., and Stotsky, G. 1988. Interactions between clay minerals and viruses. In V. C. Rao and J. L. Melnick (eds.), *Human Viruses in Sediments, Sludges, and Soils*. CRC Press, Boca Raton, Fla., pp. 197–230.

61. Lockhart, B. E. L., and N. Khaless. 1988. Commelina yellow mottle virus—a non-enveloped bacilliform virus containing double-stranded DNA. Abstract No. 288, 1988 Annual Meeting of American Phytopathological Society.

62. Lovisolo, O., O. Bode, and J. Volk. 1965. Preliminary studies on the soil transmission of petunia asteroid mosaic virus (= "Petunia" strain of tomato bushy stunt virus). *Phytopathol. Zeit.* 53:323–342.

63. Lung, M. C. Y., and T. P. Pirone. 1973. Studies on the reason for the differential transmission of cauliflower mosaic virus by aphids. *Phytopathology* 63:910–914.

64. Maelzer, D. A. 1986. Integrated control of insect vectors of plant virus diseases. In G. D. McLean, R. G. Garrett, and W. G. Ruesink (eds.), *Plant Virus Epidemics*. Academic Press, Orlando, Fla., pp. 483–512.

65. Malyshenko, S. I., L. G. Lapchic, O. A. Kondakova, L. L. Kunetzova, M. E. Taliansky, and J. G. Atabekov. 1988. Red clover mottle virus B-RNA spreads between cells in Tobamovirus-infected cells. *J. Gen. Virol.* 69:407–412.

66. Maramorosch, K., M. K. Brakke, and L. M. Black. 1949. Mechanical transmission of a plant tumor virus to an insect vector. *Science* 110:162–163.

67. Matthews, R. E. F. 1981. *Plant Virology*, 2d ed. Academic Press, New York.

68. Matthews, R. E. F. 1982. Classification and nomenclature of viruses. Fourth report of the International Committee on Taxonomy of Viruses. *Intervirology* 17:4–199.

69. McKinney, H. H. 1935. Evidence of virus mutation in common mosaic of tobacco. *J. Agric. Res.* 51:951–981.

70. McKinney, H. H., W. R. Paden, and B. Koehler. 1957. Studies on chemical control and overseasoning of, and natural inoculation with, the soil-borne viruses of wheat and oats. *Plant Dis. Rep.* 41:256–266.

71. Miyai, S., K. Kiritani, and F. Nakasuji. 1986. Models of epidemics of rice dwarf. In G. D. McLean, R. G. Garrett, and W. G. Ruesink (eds.), *Plant Virus Epidemics.* Academic Press, Orlando, Fla., pp. 459–480.

72. Mossop, D. W., and R. I. B. Francki. 1977. Association of RNA 3 with aphid transmission of cucumber mosaic virus. *Virology* 81:177–181.

73. Mueller, W. C., and W. F. Rochow. 1961. An aphid-injection method for the transmission of barley yellow dwarf virus. *Virology* 14:253–258.

74. Munnecke, D. E. 1972. Factors affecting the efficiency of fungicides in the soil. *Annu. Rev. Phytopathol.* 10:375–398.

75. Nassuth, A., F. Alblas, and J. F. Bol. 1981. Localization of genetic information involved in the replication of alfalfa mosaic virus. *J. Gen. Virol.* 53:207–214.

76. Natsuaki, T., K. T. Natsuaki, S. Okuda, M. Tiranaka, R. G. Milne, G. Boccardi, and E. Luisoni. 1986. Relationships between cryptic and temperate viruses of alfalfa, beet and white clover. *Intervirology* 25:69–75.

77. Nienhaus, F., and B. Stille. 1966. Ubertragung des Kartoffel-X-Virus durch Zoospore von *Synchytrium endobioticum. Phytopathol. Zeit.* 54:335–337.

78. Pirone, T. P., and K. F. Harris. 1977. Nonpersistent transmission of plant viruses by aphids. *Annu. Rev. Plant Pathol.* 15:55–73.

79. Powell Abel, P., R. S. Nelson, B. De, N. Hoffman, S. G. Rogers, R. T. Fraley, and R. N. Beachy. 1986. Delay of disease development in transgenic plants that express the tobacco mosaic virus coat protein gene. *Science* 232:738–743.

80. Reanney, D. C. 1982. The evolution of RNA viruses. *Annu. Rev. Microbiol.* 36:47–73.

81. Reddy, D. V. R., and L. M. Black. 1974. Deletion mutations of the genome segments of wound tumor virus. *Virology* 61:458–473.

82. Roberts, F. M. 1950. The infection of plants by viruses through roots. *Ann. Appl. Biol.* 37:385–396.

83. Rochow, W. F. 1972. The role of mixed infections in the transmission of plant viruses by aphids. *Annu. Rev. Phytopathol.* 10:101–124.

84. Saito, Y., K. Takanashi, Y. Iwata, and H. Okamoto. 1964. Studies on the soil-borne virus diseases of wheat and barley. III. Influence of chemicals on the infected soils and the viruses. *Bull. Natl. Inst. Agric. Sci. Japan, Ser. C* 17:41–59.

85. Shirako, Y., and M. K. Brakke. 1984. Spontaneous deletion mutation of soil-borne wheat mosaic virus RNA II. *J. Gen. Virol.* 65:855–858.

86. Simons, J. N. 1976. Aphid transmission of a nonaphid-transmissible strain of tobacco etch virus. *Phytopathol.* 66:652–654.

87. Slack, S. A., R. J. Shepherd, and D. H. Hall. 1975. Spread of seed-borne barley stripe mosaic virus and effects of the virus on barley in California. *Phytopathology* 65:1218–1223.

88. Smith, P. R., R. N. Campbell, and P. R. Fry. 1969. Root discharge and soil survival of viruses. *Phytopathology* 59:1678–1687.

89. Todd, J. M. 1958. Spread of potato virus X over a distance. In F. Quak, J. Dijkstra, A. B. R. Beemster, and J. P. H. Van der Want (eds.), *Proceedings of the 3rd Conference on Potato Virus Diseases.* H. Veenman & Zonen, Wageningen, pp. 132–143.

90. Toriyama. S. 1983. Rice stripe virus. CMI/AAB Descriptions of Plant Viruses, No. 269.

91. Tremaine, J. H., and J. A. Dodds. 1985. Carnation ringspot virus. CMI/AAB Descriptions of Plant Viruses, No. 308.

92. Van Dorst, H. J. M. 1988. Surface waters as source in the spread of cucumber green mottle mosaic virus. *Neth. J. Agric. Sci.* 36:291–299.

93. Van Etten, J. L., A. M. Schuster, and R. H. Meints. 1988. Viruses of eukaryotic Chlorella-like algae. In Y. Koltin and M. J. Leibowitz (eds.), *Viruses of Fungi and Simple Eukaryotes*. Marcel Dekker, New York, pp. 411–427.
94. Walker, H. L., and T. P. Pirone. 1972. Particle numbers associated with mechanical and aphid transmission of some plant viruses. *Phytopathology* 62:1283–1288.
95. Weaver, J. E. 1965. *Native Vegetation of Nebraska*. Univ. of Nebraska Press, Lincoln.
96. Wilhelm, S. 1966. Chemical treatments and inoculum potential of soil. *Annu. Rev. Phytopathol.* 4:53–78.
97. Zitter, T. A., and J. N. Simons. 1980. Management of viruses by alteration of vector efficiencies and by cultural practices. *Annu. Rev. Plant Pathol.* 18:289–310.

Environmental Control of Cyanobacteria

J. Skujiņš

Introduction

The prokaryotic, photosynthetic cyanobacteria (cyanophytes, "blue-green algae") are ubiquitous in nature, occurring in waters with a wide range of salinity and on soils having a great range of physicochemical characteristics. Most species are well adapted to survive environmental extremes, such as high temperatures and desiccation. Their adaptability to extreme environments is exemplified by their presence and growth in alkaline hot springs and in the rocks of Antarctica. All heterocystous and a number of nonheterocystous cyanobacteria exhibit nitrogenase activity. As primary producers (CO_2-fixers) and N_2-fixers, they commonly are the first major colonizers of the parent materials of soil. A number of cyanobacteria form symbiotic relations with other organisms: they are the N_2-fixing phycobionts of lichens (i.e., in association with fungi) and of some liverworts, water ferns, angiosperms, and protozoans ("cyanelles").

In terrestrial systems, cyanobacteria are considered beneficial organisms, especially under adverse environmental conditions, such as in arid soils, in human-impacted denuded soils, and in soils of generally poor agricultural value. As primary producers and N_2-fixers, they are instrumental in improving the organic matter content and nitrogen status of soil. In some developing countries, some cyanobacterial species (such as *Spirulina* sp.) are cultivated for food production (6). For thousands of years, rice yields in southeast Asia and other rice-producing countries have been maintained by the nitrogen input of cyanobacteria into rice paddies, especially in symbiotic association with *Azolla*.

Cyanobacteria have contributed significantly to the eutrophication of natural (marine and lacustrine) and anthropogenic water systems and to undesirable contamination of sewage treatment plants and swimming pools, among others. Their growth responds particularly well to the presence of phosphate in the ambient environment. Various methods have been tried and developed to limit, to a certain extent, the proliferation of cyanobacteria in these situations. It is notable that the presence of cyanobacteria decreases considerably in aquatic and terrestrial environments, including soil, at pH <6, and they are absent at pH <5.

The adaptability of cyanobacteria to environmental extremes generally presents difficulties in their containment. The control of the proliferation of cyanobacteria in natural environments depends on the characteristics of individual species or strains.

Genetics

Studies on cyanobacterial genomes, plasmids, mutations, and gene transfer have been reviewed by Herdman (8). The genetic engineering of cyanobacteria has been addressed by a number of laboratories (e.g., 7, 18, 19). Most of the recent work has been directed toward deciphering the *nif,* photosystem, and other gene-coding sequences and characteristics (e.g., 11, 16).

Control Measures: Decontamination and Mitigation

Aside from some environmental control measures, such as limiting the phosphate influx into water systems, three types of approaches can be used for directly controlling cyanobacterial proliferation: (1) application of chemicals; (2) use of biological agents; and (3) physical methods.

Chemical agents

Metals. The most commonly used method for the elimination of cyanobacteria has been the application of Cu^{2+} compounds, usually in the form of dissolved $CuSO_4$. In aquatic systems, the nominal rates for control of cyanobacteria range from 0.1 to 0.5 ppm $CuSO_4$ (0.03 to 0.15 ppm Cu^{2+}) (13). In comparison, about 10-fold higher concentrations are required for the control of eukaryotic algae. Currently, a number of algicides, formulated from Cu-organic complexes, are available for the control of cyanobacteria and eukaryotic algae (Table 47.1). Some of these algicides, such as Ricertrine, have been formulated for the control of algal growth in rice fields.

The cyanobacteria appear to be highly sensitive to some transition and heavy metals, but these are not used as control agents because of obvious environmental hazards. It is interesting that silver (Ag) compounds were used for eliminating cyanobacteria from swimming pools in Washington, D.C., during the 1930s. Conversely, cyanobacteria are comparatively insensitive to some other metals, such as mercury (Hg), that are highly toxic to other bacteria (6, 13).

Other chemical agents. The use of oxidizing materials (e.g., ozone, chlorine) in enclosed systems such as swimming pools has been reasonably successful, but these agents also have a wide spectrum of biocidal properties, and chlorination may produce carcinogenic organic compounds. Nominally, the presence of 1 ppm chlorine is the accepted value for an effective elimination of aquatic plankton, includ-

TABLE 47.1   Currently Available Algicides*

Chemical character and brand names	Primary manufacturers
Cu-ethanolamine complexes	
Aquatrine	Applied Biochemists, Inc.
Cutrine-Plus	Applied Biochemists, Inc.
Stocktrine II	Applied Biochemists, Inc.
Cu-triethanolamine complexes	
Algae-Rhap Cu-7	CP Chemicals
A & V-70	A & V, Inc.
K-Tea	Kocide Chemical Co.
Ricertrine (California registration for rice-field use)	Applied Biochemists, Inc.
Cu-ethylenediamine complexes	
Komeen	Kocide Chemical Co.
Salts of fatty acids (for control of cryptogams on solid surfaces)	
DeMoss Cryptocidal Soaps	Safer Agro-Chem, Inc.
Dyes, soluble (plankton inhibitors)	
Aquashade	Aquashade, Inc.
Organic herbicides, suitable for plankton control	
Endothall	Pennwalt Corp.
Weedtrine-Plus (mixture of Endothall and Cu-ethanolamine complex, California registration only)	Applied Biochemists, Inc.

*Most act as photosynthetic inhibitors on cyanobacteria. Application rates vary according to manufacturer's directions.

ing cyanobacteria. Chlorine is usually delivered to water as Cl_2, ClO_2, or hypochlorite (ClO^-). In contact with water, HClO is formed, which partially dissociates to $H^+ + ClO^-$. The dissociation is pH-dependent: 4 percent is dissociated at pH 6 and 97 percent at pH 9. The toxic principle is thought to be the nondissociated HClO. Thus, the actual effective rates of application are variable and dependent on the pH and other water characteristics (10, 12). Ozone has been considered to be a valuable agent for the removal of cyanobacteria and other planktonic components, as it does not require residue removal. A dosage of 0.1 mg (or 0.0467 cm^3 STP) O_3 per liter is considered sufficient for disinfection. Ozone is commonly generated in place by an electrical discharge or electrolysis before its release into water systems (12).

Permanganate has been tested and used on several occasions. Although Mn per se may be considered as having a low toxicity, the amounts of permanganate needed to control cyanobacteria also affect the rest of the biota, resulting, for example, in fish kills.

Pesticides. The cyanobacteria, being prokaryotes, are rather insensitive to a wide range of pesticides. Although it is evident that cyanobacteria are sensitive to certain photosynthesis inhibitors (herbicides), the possibility of their use for control of specific cyanobacteria has been insufficiently addressed. It appears that the herbicide, Endothall, is the only one that has commercial applications for the control of cryptogams (Table 47.1). Additionally, soap formulations have been used for the removal of cryptogams from surfaces (DeMoss Cryptocidal Soaps, Table 47.1). There are indications of the sensitivity of cyanobacteria to Propanil, 3,4-dichloroaniline, and 2,3-dichloro-1,4-naphthoquinone, among other similar compounds (6, 13). The latter compound appears to be a specific inhibitor of cyanobacteria, useful at 1 ppm; it is not toxic to fish (5). Also notable is the high sensitivity of cyanobacteria to ethionine (<10 ppm), which, however, is a strong carcinogen (1).

Biological agents

Antibiotics. As antibiotics are useful for obtaining eubacteria-free cyanobacterial cultures, numerous cyanobacteria have been tested for their sensitivities to various antibiotics. It appears that the highest sensitivity of various cyanobacterial species is to the broad-range antibiotics, such as penicillin, streptomycin, and polymyxin B, usually in the below 1- to 10-ppm range. Considerable generic differences in sensitivities to specific antibiotics, however, have been observed (6).

Predators.   Cyanobacteria have comparatively well-developed defense mechanisms against various predators. Certain cyanobacterial types may be consumed, however, by specific protozoans, nematodes, mites, and earthworms. However, few systematic studies have been directed to the use of predators for the control of cyanobacteria. For example, Atlavinyte and Pociene (2) reported on the elimination of cyanobacteria (and of eukaryotic algae) from organic matter–enriched soils by earthworms. Taylor and Duerring (14) indicated that anuran larvae (tadpoles) are efficient consumers of cyanobacteria and may eliminate them from aquatic environments.

Phages.   Another approach to control cyanobacteria in the environment might be the use of cyanophages (3, 4). A variety of cyanophages with varying degrees of host specificity have been isolated from natural environments (e.g., 9). As judged from the available literature, however, no field trials with cyanophages have been conducted. There appear to be numerous difficulties with the potential use of phages for biocontrol (17).

Physical methods

Ultraviolet irradiation.   A technique for special uses is a mobile UV unit developed for the control of algal and cyanobacterial growth on historical monuments (15). This method, however, provides only incomplete decontamination, as the radiation would not affect organisms in shaded microcrevices. However, the radiation-generated O_3 may enhance the sterilizing effect.

Conclusions

There is no specific environmentally or economically suitable methodology available for the efficient elimination of cyanobacteria from the environment. Numerous diverse, sometimes exotic, methods have been tested, most with variable success, which is a result of the adaptability of cyanobacteria and their "recalcitrance" in the environment. In critical situations (e.g., the release of unwanted cyanobacteria to the environment), more powerful methods, such as the use of elevated concentrations of Cu compounds or chlorination, would be required in water systems and in soil. There are research needs for finding and developing a specific cyanobacterial inhibitor(s), which preferably would not affect other prokaryotes, eukaryotic algae, and other organisms. This area of inquiry has been neglected because (a) the control (but not total elimination) of undesirable growth of cyanobacteria has been sufficiently successful by using practices applied for the control of eukaryotic algae in water basins, and (b) the proliferation of cyanobacteria on soils has been considered a beneficial phenomenon.

References

1. Aaronson, S., and G. Ardois. 1971. Selective inhibition of blue-green algal growth by ethionine and other amino acid analogs. *J. Phycol.* 7:18–20.
2. Atlavinyte, O., and C. Pociene. 1973. The effect of earthworms and their activity on the amount of algae in the soil. *Pedobiologia* 13:445–455.
3. Barnet, Y. M., M. J. Daft, and W. D. P. Stewart. 1981. Cyanobacteria-cyanophage interactions in continuous culture. *J. Appl. Bacteriol.* 51:541–552.
4. Desjardins, P. R., and G. B. Olson. 1982. Viral control of nuisance blue-green algae. II. Cyanophage strains, stability studies of phages and hosts, and effects of environmental factors on phage-host interactions. Final Report No. W83-02232 to OWRI, Washington, D.C. Water Resources Center, University of California, Davis.
5. Fitzgerald, G. P., and F. Skoog. 1954. Control of blue-green algae blooms with 2,3-dichloronaphthoquinone. *Sewage Ind. Wastes* 26:1136–1140.
6. Fogg, G. E., W. D. P. Stewart, P. Fay, and A. E. Walsby. 1973. *The Blue-Green Algae.* Academic Press, London.
7. Golden, S. S., and L. A. Sherman. 1984. Optimal conditions for genetic transformation of the cyanobacterium *Anacystis nidulans* R2. *J. Bacteriol.* 158:36–42.
8. Herdman, M. 1982. Evolution and genetic properties of cyanobacterial genomes. In N. G. Carr and B. A. Whitton (eds.), *The Biology of Cyanobacteria.* University of California Press, Berkeley, pp. 263–305.
9. Hu, N.-T., T. Thiel, T. H. Giddings, Jr., and C. P. Volk. 1981. New *Anabaena* and *Nostoc* cyanophages from sewage settling ponds. *Virology* 114:236–246.
10. Jolley, R. J., R. J. Bull, W. P. Davis, S. Katz, M. H. Roberts, and V. A. Jacobs (eds.). 1985. *Water Chlorination: Chemistry, Environmental Impact and Health Effects.* Lewis Publishers, Chelsea, Mich.
11. Kallas, T., T. Coursin, and R. Rippka. 1985. Different organization of *nif* genes in nonheterocystous and heterocystous cyanobacteria. *Plant Mol. Biol.* 5:321–329.
12. Katz, J. (ed.). 1980. *Ozone and Chlorine Dioxide Technology for Disinfection of Drinking Water.* Noyes Data Corp., Park Ridge, N.J.
13. Salvato, J. A. 1972. *Environmental Engineering and Sanitation,* 2d ed. Wiley-Interscience, New York.
14. Taylor, J. T., and C. Duerring. 1983. Effects of anuran larvae on aquatic processes associated with water quality. Res. Report RR-42 to Office of Water Research and Technology. Water Resources Research Center, University of New Hampshire, Durham.
15. Van der Mollen, J. M., J. Garty, G. W. Aardema, and W. E. Krumbein. 1978. Growth control of algae and cyanobacteria on historical monuments by a mobile UV unit (MUVU). *Stud. Conserv.* 25:71–77.
16. Vermaas, W. F. J., J. G. K. Williams, and C. J. Arntzen. 1987. Sequencing and modification of *psbB*, the gene encoding CP-47 protein of Photosystem II, in the cyanobacterium *Synechocystis* 6803. *Plant Mol. Biol.* 8:317–326.
17. Vidaver, A. K. 1976. Prospects for control of phytopathogenic bacteria by bacteriophages and bacteriocins. *Ann. Rev. Phytopath.* 14:451–465.
18. Williams, J. G. K., and A. A. Szalay. 1983. Stable integration of foreign DNA into the chromosome of the cyanobacterium *Synechocystis* R2. *Gene* 24:37–51.
19. Wolk, C. P., A. Vonshak, P. Kehoe, and J. Elhai. 1984. Construction of shuttle vectors capable of conjugative transfer from *Escherichia coli* to nitrogen-fixing filamentous cyanobacteria. *Proc. Nat. Acad. Sci. USA* 81:1561–1565.

48

Eukaryotic Algae

Russel H. Meints
James L. Van Etten

Overview

Eukaryotic algae are ubiquitous in nature and comprise an extremely diverse group of organisms. They range in size from small single-celled organisms (microalgae) to large multicellular organisms, such as seaweeds (macroalgae). Most eukaryotic algae are free-living organisms, and members of the algal community can grow and survive in a variety of soils and aquatic environments. A few algae, however, are only found in symbiotic relationships with other organisms such as coelenterates, protozoa, and lichens. Algae play a significant role in aquatic environments, both as primary producers in the food chain and, conversely, as pollutants when growth becomes uncontrolled. Because most algae only require simple salts and a nitrogen source for rapid growth, it is likely that certain algae will eventually be domesticated to produce proteins, carbohydrates, and fats as foodstuffs for humans and animals. In addition, algae are and will continue to be used to produce beneficial products for humans (1, 16, 18).

Eukaryotic algae have also been used by scientists as model organisms with which to study many biochemical and physiological processes. For example, members of the genus *Chlorella* were used in the original studies to determine the biochemical pathways involved in photosynthesis (6). *Chlamydomonas* sp. have been and are presently used as model organisms for many genetic studies—especially studies on the genetics of organelles (8).

Prospective Uses

Algae convert sunlight to organic carbon more efficiently than higher plants, at least on a per-area basis. Therefore, it is likely that algae will be grown in large outdoor culture ponds for biomass production. This is important, as some scientists believe that the destruction of the ability to produce biomass by land plants is a greater problem to humankind than the lack of food (9). Algal biomass, the production of which has already begun in Israel, Taiwan, and Japan, can be used for many purposes, including food for human consumption, animal feed, or as a solid or liquid energy source. Several third world countries anticipate burning algae directly to produce heat (9). In the United States, there is a modest program to identify algae that contain high lipid concentrations (Solar Energy Research Institute, Golden, Colorado). The intention is to use these lipids for liquid fuel. Algae may also be used, alone or in combination with other microorganisms, in the treatment of commercial wastewater and raw sewage. Thus, depending on their intended use scientists will genetically engineer algae to perform different functions.

In addition to these examples, algae are used to produce other valuable commodities, such as beta carotene, alginic acid, and agar (4). Some algae are also sold in health food stores.

Introducing Genetic Variability in Algae

Traditionally, plant breeding was used to improve higher plants. However, the use of conventional genetics to improve algae is precluded. The reasons include (1) the difficulty of working with the sexual stage of many algae, (2) the general lack of knowledge about the sexual stage of many algae, and (3) many algae only replicate asexually (e.g., *Chlorella* sp.). Consequently, alternate methods need to be developed to transfer genes into algae. Somatic hybridization is one method for obtaining genetic recombination between organisms that reproduce asexually or for organisms that are sexually incompatible. Somatic hybridization requires the fusion of two different protoplasts and the successful regeneration of the somatic cells into a stable, reproducing organism.

Other methods, including the use of recombinant DNA techniques, must be developed for transferring desired genetic traits into algae. In order to transform algae successfully with recombinant DNA, a number of procedures will have to be developed, including (1) methods for removing algal cell walls to produce viable protoplasts, (2) methods for the regeneration of cell walls of algal protoplasts and for the resumption of normal cell growth, (3) methods to identify and isolate desirable genes to transform into algae, and (4) the development of appropriate vectors with all the necessary properties, including autonomous replicating sequences and/or regions allowing DNA integration into host chromosomal DNA, promoters, and selectable markers.

Several procedures are used to introduce foreign genes into higher plants, including direct DNA uptake, microinjection of purified DNA, use of viruses and *Agrobacterium tumefaciens* as vectors (10, 17, 20), and more recently, the forcing of transformation competency upon cells by high-voltage pulses over short time periods, i.e., by electroporation (7, 14) or microprojectiles (12). Whether any of these procedures will be directly adaptable to nuclear transformation in eukaryotic algae is unknown. Since it is known that procedures that work for one plant do not necessarily work for another, different procedures and strategies will probably have to be developed for different algae.

It is exciting, however, that a high-frequency nuclear transformation method (ca. 10^3 transformants per microgram of DNA) was recently described for the unicellular green alga, *Chlamydomonas reinhardtii* (11). The method is surprisingly simple. A cell-wall-deficient *Chlamydomonas* was agitated with glass beads, DNA, and polyethylene glycol. The nitrate reductase gene from wild-type *Chlamydomonas* was used to complement a mutation in a corresponding gene in the cell-wall-deficient alga lacking a functional nitrate reductase gene. Transformants were selected by growth with nitrate as the sole source of nitrogen. When cells lacking the nitrate reductase gene were agitated in the presence of two plasmids, one with the gene for nitrate reductase and the second with an unselected gene, the unselected gene was present in 10 to 50 percent of the cells containing the nitrate reductase gene. Because of the high frequency of cotransformation, one should be able to introduce any cloned gene into *Chlamydomonas*. The nuclear transformation system in combination with a previously described system for transforming *Chlamydomonas* chloroplasts (2) should make *Chlamydomonas* an ideal organism for studying chloroplast-nuclear interactions.

Attempts have also been made to transform the macroalga *Acetabularia mediterranea* by microinjecting nuclei with SV40 DNA. Expression of the T-antigen, as measured by immunofluorescence, was detected several days later (13). Based on the timing of expression, it was concluded that the algal cell transcription machinery recognized the SV40 enhancer sequence. The results suggest that SV40, a mammalian virus, has broad-spectrum capability as a vector system and that it might be used to introduce foreign genes into at least some algae.

In summary, with the exception of *Chlamydomonas,* there is not much information on transformation of algae. However, the development of an efficient transformation system for *Chlamydomonas* implies that transformation systems eventually will be developed for other algae.

Detection of Released Algae

Procedures will need to be developed to detect released algae containing recombinant DNA. In some cases morphological identification will be reasonably simple, particularly with the large macroalgae. That is, the organism will look out of place such as a kelp growing in a cornfield. In the case of microalgae, the problems will be the same as those that occur in identifying fungi or bacteria that have recombinant DNA. Thus the recombinant algae will need to contain some easily identifiable feature, trait, or marker DNA. For example, antibiotic resistance gene(s) or a bioluminescence gene(s) could be used as cotransforming markers. Confirmation that the organism is a transformed variant could be accomplished by hybridizing its DNA with appropriate DNA probes.

Lytic viruses that are specific for certain *Chlorella* sp. have recently been discovered (see Control), and it is possible that these or similar viruses could be used to identify certain genetically engineered algae.

Containment

Growing genetically engineered algae in outdoor culture ponds will require physical containment. It is doubtful that these cultivated algae will present a hazard to the environment, because it is unlikely that they will compete successfully with indigenous saprophytic microorganisms in a field environment. However, experiments will need to be conducted to test this assumption.

Control

Copper, applied in the form of $CuSO_4$ at 0.2 to 2 mg/liter, is the most common compound used to control algal growth. Algae are also sensitive to other metals, such as mercury- and silver-containing compounds. The use of heavy metals has severe environmental limitations, however, and only impounded waters, such as swimming pools, can be treated safely.

Oxidizing compounds, such as hypochlorite or permanganate, that are broadly biocidal can also be used to kill algae. Again, only enclosed systems could be treated with these compounds, as the concentrations of these agents required for control in large areas would be costly, dangerous to the environment, and ultimately interact with other materials in the environment to form dangerous secondary compounds, such as carcinogens.

Inasmuch as algae are photosynthetic eukaryotes, they are sensitive to a variety of herbicides that act on the photosynthetic pathway. For example, the photosystem II inhibitor, 3-(3,4-dichlorophenyl)-1,1-dimethyl urea (DCMU), inhibits *Chlorella* strain N1a. This *Chlorella* strain is also inhibited by the potent inhibitors of microtubule formation, benomyl and oryzalin, and the cell-wall synthesis inhibitor, dichlorobenzonitrile (R. Meints, unpublished data).

There is very little information on the use of antibiotics to control algae. They are not likely to be used because of nontarget effects on other organisms. We have tested the effect of tetracycline, kanamycin, the kanamycin analog G-418, hygromycin, and methotrexate for their effect on the growth of *Chlorella* N1a. All of these compounds inhibited growth in the laboratory at about 5 to 50 μg/ml (R. Meints, unpublished data).

Some algae are parasitized by *Bdellovibrio* and other unidentified bacteria (3). Whether these organisms could serve as effective biological control agents is not known. There are clear indications that within natural populations of algae, growth of one alga is at the expense of others, e.g., algal blooms. Although toxins, among other factors, have been suggested to have a role in this selective growth, the reasons why algal blooms occur and rapidly disappear are still largely unknown. However, there is no reason to believe that genetically altered algae will offer significant problems in this regard over natural populations.

The recent discovery of large dsDNA-containing lytic viruses that infect certain isolates of exsymbiotic *Chlorella*-like green algae raises the possibility that algal viruses exist in nature and that they might be used to control certain algal blooms. One advantage to using viruses is that they are very specific for their hosts. In fact, the viruses found to date infect only certain *Chlorella* isolates as they do not attach to other *Chlorella* sp., i.e., cell walls of non-host algae lack the virus receptor(s) (15, 21). It is not known whether lytic viruses of other eukaryotic algae are present in nature. However, ultrastructural studies have detected viruslike particles in several other eukaryotic algae (5, 19).

Conclusions

In all likelihood, eukaryotic algae containing recombinant DNA will be important to the future of humanity. Further, genetically engineered algae will probably be grown for a variety of reasons in large outdoor culture ponds. At present, there is no reason to believe that the growth of these organisms under these conditions will present any special hazard to the environment.

References

1. Borowitzka, M. A., and L. J. Borowitzka (eds.). 1988. *Micro-Algal Biotechnology.* Cambridge University Press, Cambridge, England.
2. Boynton, J. E., N. W. Gillham, E. H. Harris, J. P. Hosler, A. M. Johnson, A. R. Jones, B. L. Randolph-Anderson, D. Robertson, T. M. Klein, K. B. Shark, and J. C. Sanford. 1988. Chloroplast transformation in *Chlamydomonas* with high velocity microprojectiles. *Science* 240:1534–1538.
3. Coder, D. M., and M. P. Starr. 1978. Antagonistic association of the Chlorellavorus bacterium (*"Bdellovibrio" chlorellavorus*) with *Chlorella vulgaris. Curr. Microbiol.* 1:59–64.
4. Cohen, Z. 1986. Products from microalgae. In A. Richmond (ed.), *Handbook of Microalgae Mass Culture.* CRC Press, Boca Raton, Fla., pp. 421–454.
5. Dodds, J. A. 1983. New viruses of eukaryotic algae and protozoa. In R. E. F. Matthews (ed.), *A Critical Appraisal of Viral Taxonomy.* CRC Press, Boca Raton, Fla., pp. 177–188.
6. Emerson, R., and W. Arnold. 1932. A separation of the reactions in photosynthesis by means of intermittent light. *J. Gen. Physiol.* 15:391–420.
7. Fromm, M., L. P. Taylor, and V. Walbot. 1985. Expression of genes transferred into monocot and dicot plant cells by electroporation. *Proc. Natl. Acad. Sci. USA* 82: 5824–5828.
8. Gillham, N. W. 1978. *Organelle Heredity.* Raven Press, New York.
9. Hall, D. O. 1986. The production of biomass: A challenge to our society. In A. Richmond (ed.), *Handbook of Microalgae Mass Culture.* CRC Press, Boca Raton, Fla., pp. 1–24.
10. Howell, S. H. 1982. Plant molecular vehicles: Potential vectors for introducing foreign DNA into plants. *Annu. Rev. Plant Physiol.* 33:609–650.
11. Kindle, K. L. 1990. High-frequency nuclear transformation of *Chlamydomonas reinhardtii. Proc. Natl. Acad. Sci. USA* 87:1228–1232.
12. Klein, T. M., E. D. Wolf, R. Wu, and J. C. Sanford. 1987. High velocity microprojectiles for delivering nucleic acids into living cells. *Nature* 327:70-73.
13. Neuhaus, G., G. Neuhaus-Uri, P. Gruss, and H. Schweiger. 1984. Enhancer-controlled expression of the simian virus 40 T-antigen in the green alga *Acetabularia. EMBO J.* 3:2169–2172.
14. Neumann, E., N. Schaefer-Ridder, Y. Wang, and P. H. Hofschneider. 1982. Gene transfer into mouse lyoma cells by electroporation in high electric fields. *EMBO J.* 1:841–845.
15. Reisser, W., D. E. Burbank, S. M. Meints, R. H. Meints, B. Becker, and J. L. Van Etten. 1988. A comparison of viruses infecting two different *Chlorella*-like green algae. *Virology* 167:143–149.
16. Richmond, A. (ed.). 1986. *Handbook of Microalgal Mass Culture.* CRC Press, Boca Raton, Fla.
17. Schell, J., M. Van Montagu, M. Holsters, A. Depicker, P. Zambryski, P. Dhaese, J. P. Hernalsteens, J. Leemans, H. De Greve, L. Willmitzer, L. Otten, J. Schroder, and G. Schroder. 1982. Plant cell transformations and genetic engineering. In I. K. Vasil, W. R. Scowcroft, and K. J. Frey (eds.), *Plant Improvement and Somatic Cell Genetics.* Academic Press, New York, pp. 255–276.

18. Shelef, G., and C. J. Soeder (eds.). 1980. *Algae Biomass: Production and Use.* Elsevier/North-Holland Biomedical Press, Amsterdam.
19. Sherman, L. A., and R. M. Brown. 1978. Cyanophage and viruses of eukaryotic algae. In H. Frankel Conrat and R. R. Wagner (eds.), *Comprehensive Virology,* Vol. 12. Plenum, New York, pp. 145–253.
20. Steinbiss, H. H., and W. J. Broughton. 1983. Methods and mechanisms of gene uptake in protoplasts. In *International Review of Cytology,* Suppl. 16. Academic Press, New York, pp. 191–208.
21. Van Etten, J. L., A. M. Schuster, and R. H. Meints. 1988. Viruses of eukaryotic *Chlorella*-like algae. In Y. Koltin and M. J. Leibowitz (eds.), *Viruses of Fungi and Simple Eukaryotes.* Marcel Dekker, New York, pp. 411–428.

Containment, Decontamination, and Mitigation of Recombinant *Pseudomonas* Species in Environmental Settings

S. E. Lindow

J. Lindemann

D. Haefele

Introductory Information on the Genus *Pseudomonas*

Pseudomonas is a genus with very diverse members, which are found in many different habitats, such as animals, including humans; wounds; foods; water; air; on plant surfaces and roots; and in the bulk soil (9, 27, 37, 67, 73). "One of the delightful absurdities to emerge from this involuted system [i.e., the current system of bacterial taxonomy] is the taxon *Pseudomonas*, perhaps the best known, most studied, and most pedagogically utilized 'representative genus,' which actually is a collection of at least five separate groups of bacteria, whose name derives from the Greek *pseudes* and *monas,* i.e., false unit!" (90). Woese's admonition should be kept in mind while reading this chapter on the pseudomonads. Many *Pseudomonas* strains are solely saprophytic, while other strains may exhibit toxic effects on plants or animals. Those strains that are toxic or pathogenic to plants or animals are better characterized than most saprophytic members (67, 73). Examples of the former include *P. aeruginosa,* which can cause a toxemia in animals, including nosocomial infections in immuno-compromised human patients, and is commonly associated with infections of burn wounds in humans. Some strains, classified as *P. syringae, P. solanacearum, P. viridiflava, P. cepacia, P. cichorii, P. fluorescens* biovar 2 (*P. marginalis*), and *P. tolasii*, have been isolated from diseased plants (73). Strains of *P. fluorescens, P. putida, P. stutzeri, P. alcaligenes, P. chlororaphis*, and *P. aureofaciens*, among other species, are seldom isolated from diseased plants or animals.

Members of the genus *Pseudomonas* are characterized by a wide range of catabolic capabilities, particularly among nonpathogenic members (13, 67, 69, 71, 73). *Pseudomonas* species are gram-negative, straight or slightly curved rods with a diameter of 0.5 to 1.0 μm and 1.5 to 5.0 μm in length. No resting stages are known. Cells are motile by one or several polar flagella. Most species are strict aerobes, having respiratory metabolism with oxygen as the terminal electron acceptor; however, a few species can grow anaerobically using nitrate as an alternate electron acceptor. Most species fail to grow under acidic conditions (pH <4.5) and do not require organic growth factors. Some species are facultative chemolithoautotrophs, able to use H_2 or CO as energy sources and CO_2 as a carbon source. The type species for this genus is *P. aeruginosa*. A large number of subdivisions within the genus is possible, based on these and other biochemical characterizations. For example, the G + C content varies from 57 to 70 percent. However, strains grouped on the basis of biochemical characteristics may not share ecological similarities. This complicates attempts to generalize ecological features from biochemical characteristics. Little taxonomic consideration has been given to nonpathogenic *Pseudomonas* species, including most *P. fluorescens* strains, because of their unimportance in clinical settings. Generalizations of ecological attributes of nonpathogenic members are difficult to discern from studies on strains important in human and animal health.

Actual and Potential Uses

Considerable literature exists on the proposed uses of pseudomonads for a variety of purposes. For example, there are numerous experimental reports of pseudomonads active in biological control of plant pathogens (8, 10, 11, 14, 16, 17, 26, 27, 30, 32, 39, 40, 49, 74, 78, 83, 89). At least two *Pseudomonas* strains are used commercially for biological control. The biological control potential of many of these strains has been suggested to result from competitive exclusion of deleterious bacteria including other pseudomonads as well as from direct inhibition of pathogens by the production of one or more antimicrobial agents (5, 11, 14, 25, 39, 74). Biological control of both bacterial and fungal pathogens have been reported, and plant pathogenic species of *Pseudomonas* may be utilized in the management of susceptible weed plant species.

Three different species of *Pseudomonas* are known to predispose plants to frost damage (3, 44, 48, 56–58, 68), and only the ice-nucleation active (Ice$^+$) strains appear to have this property. Naturally occurring non-Ice$^+$ strains and chemically induced or in vitro constructed Ice$^-$ derivatives of these species are capable of preventing freezing injury to plants by various mechanisms, including antibiosis and the competitive exclusion of Ice$^+$ strains of these and other bacterial species (44, 46, 47, 49–52, 54). Ice nuclei produced by various Ice$^+$ strains of *Pseudomonas* have many potential uses, including frozen-food processing and storage, and other freezing-related processes (2, 53). An Ice$^+$ *P. syringae* strain is sold commercially for artificial snow production (2).

Pseudomonads can also beneficially or detrimentally alter plant growth and development, independent of inciting disease symptoms. Some *Pseudomonas* species affect plant growth by influencing root development (11, 27, 74): some species enhance the development of roots and, thereby, increase plant growth, whereas others are inhibitory to root and plant growth (62, 74, 81). Various mechanisms have been proposed for these effects on plant growth, including the inhibition of deleterious organisms by the elaboration of antimicrobial agents, as well as direct effects on root elongation, resulting from the production of 3-indoleacetic acid and other plant growth regulators (25, 37, 39, 62, 74, 81).

Due to their unique anabolic capabilities, certain *Pseudomonas* strains have been utilized in the large-scale manufacture of various specialty chemicals (5, 36, 80, 82). Because of their catabolic and anabolic capabilities and relative ease of genetic manipulation (13, 43, 70, 84), pseudomonads will probably be used more frequently in the future in the production of unique biochemicals and in other fermentation processes. Inasmuch as naturally occurring pseudomonads degrade toxic chemicals into less toxic or less persistent analogues (4, 7, 12, 18, 20–24, 38, 91), increasing use of genetically modified strains can be anticipated for hazardous waste management. Since certain strains colonize specialized ecological habitats such as root or leaf surfaces, they may be utilized to produce pesticidal compounds or conduct processes, such as degradation of toxic pesticides, that may be of importance in agricultural uses.

Influence of Dispersal Modes on the Choice
of Containment Methods

Several methods exist for the dispersal of *Pseudomonas* species from one habitat to another: e.g., aerosols, mass movement of water and other materials, transfer on mobile vectors; but the magnitude or efficiency of such transfer processes varies greatly (34, 35, 43, 45). Each individual cell of a *Pseudomonas* strain has a low, but finite, probability of either moving from one habitat to another or surviving such movement. Therefore, the efficiency and magnitude of different modes of transport should be differentiated on their basis of successful dispersal.

While *Pseudomonas* species can be dispersed by soil, water, or plant material (habitat dispersal), they have few adaptations for efficient or long-distance transport. For example, members of this genus do not produce heat- or desiccation-tolerant spores, which would be efficiently transported via aerosols. Unlike many eukaryotic species, including many fungal genera, no passive structures that facilitate dispersal into the atmosphere or other transfer media are elaborated nor are active processes utilized, such as electrostatic repulsion, etc., which facilitate dispersal between habitats (73). Although many *Pseudomonas* species are motile via their production of flagella, the significance of this phenomenon to long-distance transport, especially in nonaqueous environments, is unclear and unlikely. Containment strategies for *Pseudomonas* species should, therefore, be similar to those for other gram-negative bacteria and other passively dispersed microorganisms.

Different methods of physical containment will be addressed, considering both the efficiency of containment and the survival of dispersed microorganisms during and after transport. Three common types of experimental sites, with differing patterns and controls of the movement of organisms and with different efficiencies of survival of *Pseudomonas* species, will be addressed.

Natural Constraints on the Colonization of
New Habitats

Many biological constraints exist to minimize the survival of, or successful colonization of new habitats by, organisms dispersed from different experimental sites. *Pseudomonas* species exhibit different niche specialization patterns. For example, some species (e.g., *P. fluorescens, P. putida*) are competent colonizers of rhizosphere and/or bulk soil (61). Some of these strains, however, do not persist well in aerial habitats, such as on plant surfaces, in aerosols, or on various vectors (73). In contrast, other species (e.g., *P. syringae*) are effective colonizers of aerial plant surfaces but few other habitats (37, 50). It is unlikely that modest genetic modifications anticipated in the near future will significantly affect the habitat limitations characteristic of a given strain. For strains restricted to certain habitats, physicochemical, biological, and environmental constraints exist. There is considerable evidence that biological interactions such as interactions with the organism or with a plant serving as a host are important in determining population size of a given strain in a habitat (67). Both environmentally competent and nonpersistent *Pseudomonas* strains may be utilized experimentally and possibly commercially.

It is anticipated that most agricultural uses will involve environmentally competent *Pseudomonas* species, because of the need to interact with spatially or temporally dynamic targets, such as plants or pests of various types at least for a short time. In contrast, for certain uses, such as in the degradation of toxic chemicals, the growth rate of a microorganism or its longevity in a habitat, may be less important. In such cases, microorganisms selected for unique catabolic capabilities rather than environmental competence will likely be utilized.

Many forms of biological debilitation (e.g., sensitivity to UV light, osmosensitivity, dependence on antibiotics, or unusual nutrients or toxicants) that can be introduced into strains not required to persist in the environment have been proposed to restrict the longevity or the habitats that engineered microorganisms can utilize. Such debilitated strains, however, would probably not have the desired efficacy for use in dynamic agricultural systems and in other habitats where environmental competence is desirable. Even environmentally competent strains, however, have restricted population sizes as the result of natural processes, such as competition, predation, parasitism, and availability of limiting environmental resources, which should be considered in the design and assessment of their use in contained and uncontained experiments. Examples of such biological processes effective in minimizing the spread of natural and recombinant strains under natural settings will be addressed. Such processes are often quantitative in nature and will be most effective in population management rather than in the eradication of such organisms. Therefore, this discussion will focus on the mitigation of *Pseudomonas* species. Mitigation in this context will most commonly be achievable by a reduction in the population size of the organism, which in turn reduces the probability of deleterious effects.

Specific Procedures for Containment, Decontamination, and Mitigation

Mitigation and decontamination (eradication) of *Pseudomonas* species will be addressed in three distinct environmental settings that are differentiated primarily on the level of physical containment (Table

49.1). Differentiation will also be made between microorganisms with different habitat restrictions, such as rhizosphere and epiphytic organisms. Completely enclosed experimental settings, such as microcosms and other habitat simulators for which dispersal and decontamination (outside of normal laboratory practices) present no special problems, will not be considered as part of this chapter. Although such systems may be valuable in obtaining important information on physical, chemical, and biological constraints of organisms to be tested subsequently in more complex environments, they pose no special containment or eradication problems not already adequately addressed by standard laboratory practice.

Semienclosed structures

Biological experimentation often involves studies in structures with some but incomplete physical containment before studies are conducted in uncontained field locations (Table 49.1). These structures are most often greenhouses, or other semienclosed buildings, or bioreactors with appropriate controls over the physical and biological environment.

Containment

Normal modes of transport of microorganisms, such as passively by wind and water and through biological vectors, are virtually eliminated in such facilities. For example, aerosolization of *Pseudomonas* species from plant surfaces or soil usually requires sufficient wind speed to dislodge such organisms (43, 45). Such conditions are normally precluded within greenhouse or other semienclosed structures, and therefore aerosolization is unlikely within this setting. Insect and rodent control procedures, such as screening of air inlets, will normally minimize the presence of such vectors in such buildings. Thus, vectors do not usually gain access to *Pseudomonas* species and, therefore, could not effectively spread them from such a building in significant numbers. Although considerable numbers of *Pseudomonas* species can become aerosolized within greenhouses during commonly used spray inoculation procedures, such aerosols are efficiently physically contained by conducting inoculations in appropriate chambers or filtered hoods or biologically by minimizing air movement away from areas of application, by inoculating plants during seasons unfavorable for the bacteria, or minimizing plant growth around greenhouses. Application of inoculum to soil or other materials in greenhouse settings normally would not produce aerosols.

TABLE 49.1   Comparison of Probable Dispersal Modes and Practical Containment, Confinement, and Decontamination Methods Available for Three Classes of Experimentation with Genetically Engineered *Pseudomonas* Outside of a Contained Laboratory

Type of experimental site	Probable vector or mode of dispersal	Containment methods	Decontamination target and methods
Semienclosed structures (greenhouse)	Aerosol (spray application)	Inoculation chamber or filtered hood	Chamber walls: chemical
	Plant materials and soil	Decontaminate prior to disposal of refuse	Plant materials and soil: heat sterilization
	Tools	Decontaminate prior to removal from greenhouse	Tools: chemical sterilization
	Humans	Limit access, use of laboratory coats, shoe coverings, etc.	Garments: chemical or heat sterilization
	Wastewater	Drain after decontamination	Water: chemical or heat sterilization; filtration
	Insects, rodents		Insecticides, rodenticides
Uncontained site-limited dispersal potential	Insects	Insect netting, choice of site, or timing of experiment	Off-site plants: biocides, removal and sterilization of plants
	Soil aerosols	Keep soil surface wet or use tarps	N/A
	Vertical movement in soil profile	Choice of site, soil type, proximity of water table	Soil: soil sterilants and fumigants, followed by reincorporation of native microbes to fill biological vacuum
	Horizontal spread via plant-to-plant contact	Plant-free zone surrounding plot	N/A
	Mass movement of water	Choice of site, slope of site, use of soil tarps	
	Mechanical movement of soil	Restrict access to authorized personnel, decontaminate equipment	Equipment and boots: chemical or heat sterilization
	Bacterial aerosols from above-ground plant surfaces	Wind break, surround plot with plant-free zone, biological restraints on multiplication	Off-site plants: use of biocides on potential epiphytic host plants outside of plot, removal and sterilization of plants
Confined site-unrestricted dispersal	As above	Biological restraints on multiplication	As above for uncontained experimental site; not practical or advised for off-site locations

Thus, the abundance and dispersal of aerosols from greenhouses is probably inconsequential when proper procedures are utilized.

Decontamination

Large quantities of water are often used in greenhouse settings to maintain proper plant health, and *Pseudomonas* species could become entrained in the wastewater. Decontamination of such wastewater is necessary to ensure containment of test organisms to greenhouse settings. Such decontamination can be accomplished easily with the use of chemical disinfectants, such as alcohol, formaldehyde, phenol, quaternary ammonium compounds, chlorine, and ionophores, or by heat treatment before discarding the water (15, 28, 29, 33, 42, 72, 75, 77, 79). The highest populations of *Pseudomonas* species likely to be encountered in greenhouse studies will be associated with the plant, soil, water, or other experimental habitats under study. It is, therefore, important to restrict the transport of such materials before decontamination to within the experimental area. Thus, care should be taken in the use of tools or other physical contact with habitats containing large numbers of *Pseudomonas* species under experimentation.

Eradication of *Pseudomonas* species from within contained experimental systems is readily accomplished by both chemical and physical methods. *Pseudomonas* species are readily killed by heat sterilization (42), and simple autoclaving of all test materials will eradicate the test organism. Other contaminated surfaces within the enclosure can be effectively disinfected with Mupirocin (81) or the chemical agents suggested above for use in decontamination of water. In addition, gaseous fumigants, such as methyl bromide, chloropicrin, and other sterilizing agents, have been utilized to disinfect entire buildings. Because of the relative numbers of test organisms involved, emphasis should be placed on decontamination of test materials rather than elaborate precautions against dispersal via unlikely methods such as aerosols. Simple procedures, such as limiting access to experimental sites to authorized personnel and the use of appropriate clothing, e.g., laboratory coats, which are left in the experimental area and decontaminated prior to subsequent use, are effective. The vast majority of cells will be contained simply by restricting the transport of experimental materials from the semienclosed structure until after decontamination.

Outdoor (Uncontained) Experimental Sites with Limited Potential for Dispersal

When experimentation with *Pseudomonas* species proceeds from semienclosed structures to field sites, many methods may be used to contain the vast majority of cells, although eradication is much less likely. Experimental designs may differ, however, in the ease with which eradication can be accomplished. Experimental settings in which dispersal to sites where eradication is impractical will be differentiated from those where such eradication is more likely (Table 49.1).

Containment

The modes of dissemination of *Pseudomonas* species discussed above become more likely in field settings than within semicontained structures. The magnitude of transport of *Pseudomonas* species by at least some of these processes can be affected by the experimental design. For example, if vectoring by insects is known to be important in the movement of a given strain, insect netting of an appropriate mesh could be utilized to restrict this mode of transport, leaving other modes unrestricted. Aerosolization of inoculum from aerial plant sources and as windblown soil particles, for example, could continue under such netting, as could movement of inoculum in bulk water through or over soil. The movement of bacterial strains whose habitat is primarily restricted to the rhizosphere or soil could be significantly reduced by preventing surface water runoff (using appropriate physical barriers, such as soil tarps), by minimizing suspension of soil particles in the air [by wetting of the soil surface (31, 43)], by restriction of mechanical soil movement in the experimental area, and by appropriate selection of experimental sites with respect to slope, nearby water sources, and other factors expected to affect passive movement. When used in conjunction with one another, such procedures would be expected to reduce the transport of test organisms so that eradication by procedures to be addressed later might be reasonably contemplated. One example of such a situation might involve the placement of entire potted plants on water-impermeable tarps to facilitate the complete removal of test materials at the completion of the experiment.

The above procedures are designed to restrict the magnitude of dispersal of test organisms beyond the test site. Dispersal can be decreased, although not eliminated, by such practices. Because biological constraints to growth are frequently density-dependent, simply reducing the magnitude of dispersal can have a significant impact on the probability of survival. In some cases, knowledge of the habitat limitation exhibited by a given *Pseudomonas* strain may allow the design of experimental sites for which the survival of inoculum after dispersal is minimized. An example of such an organism would be one for which a living plant (leaf surface or root surface) is required for maintenance of population size (50, 60). For such organisms, the elimination of plants suitable for colonization within a zone dictated by the anticipated range of dispersion from the test site should result in effective confinement. Alternatively, plants can be used as barriers for aerosol transport, vector immigration, and against high wind velocities. Such practices have analogies in existing agricultural practices, such as the plant-free zones used in the production of certified seed and trap plants used in the control of nematodes.

Dissemination of *Pseudomonas* species away from soil habitats can be restricted by easily employed techniques. Continuous wetting of the soil surface can minimize transport of soil particles via the atmosphere. Proper slope selection to minimize surface water runoff will minimize the lateral surface transport via water from unanticipated and uncontrolled precipitation events. Vertical transport in soil profiles is dependent on soil type and porosity (34, 35) and soil types not conducive to vertical movement of bacteria (e.g., clay soils) are readily identifiable. In small experimental sites, the abundance of potential insect vectors and other vertebrate and invertebrate vectors will be greatly influenced by vegetation and other habitats both inside and outside of the experimental area. Suitable experimental sites can be selected by the experimenter.

If a particular vector may be of consequence in the transport of a test organism, the experiment can be conducted only at a time of year or location where such vectors are unanticipated. Analogous with greenhouse settings, the largest numbers of test organisms can be expected where they are directly inoculated. Movement of such organisms outside of the test area can be minimized by restricting access to the experimental site by proper physical barriers and minimizing transport of experimental materials outside of the test area by sanitization of equipment, infested clothing, etc., by the procedures detailed above (60). In arid regions, physical and biological conditions suitable for the growth and/or survival of the test organisms could be maintained only within the experimental site, such as by the application of sufficient water to permit plant growth or microbial activity. As an additional preventive measure to minimize the survival of any organisms dispersed beyond experimental sites, periodic application of biocides could be made to potentially inhabitable sites, such as vegetation (41). Such preventive measures may not be necessary if habitable sites are sufficiently removed from the test area, as a result of the inefficiency of survival of the test organism and minimization of its transport by design of the experimental site.

Decontamination and mitigation

At the termination of some experiments, it may be desirable to significantly reduce population sizes of test organisms from within the experimental site. Procedures were outlined above to minimize dispersal and/or survival of test organisms outside of the area under direct experimentation so that mitigation may be feasible. The population size of *Pseudomonas* species can be greatly reduced, at least in the upper soil profiles, by the application of appropriate soil sterilants, such as methyl bromide and chloropicrin. Considerable technology exists for the incorporation of such sterilants into soils for the reduction of fungal and nematode populations (64, 65). Such materials are generally biocidal and can be expected to kill *Pseudomonas* species at appropriate dosages. Detailed measurements of concentrations of soil fumigants as a function of distance away from the point of incorporation into soil have been made (65). The toxicity of soil fumigants to soil organisms is a function of the product of the concentration of such toxicants and the time of exposure. It is, therefore, necessary to achieve sufficient concentrations of soil fumigants in all areas of the soil for sufficiently long periods for lethal effects. Practical problems exist in the total sterilization of soils, especially at depths greater than 1 m. Soil temperature greatly affects the activity of soil fumigants (64, 65), and since soil temperatures would largely be outside of the control of the experimenter, soil sterilization by the use of fumigation may be restricted to times of year in which sufficiently high soil temperatures exist. In practical usage, soil fumigation has been ineffective in eradicating pathogenic soil fungi and nematodes, especially at depths greater than 2 m (65). Because of the restricted mobility of most *Pseudomonas* species through the matrices similar to soil (34), most cells should be expected to be in the upper 2 m of soil. The population size of test *Pseudomonas* species can, therefore, be greatly reduced in this zone by the use of fumigants. However, all microorganisms would be similarly reduced in this zone, leading to a "biological vacuum." Such treated soils can be rapidly invaded by microbes. Environmentally competent and nondeleterious soil inhabitants could be easily reintroduced into such soil by incorporation of native uncontaminated soils following fumigation. Test *Pseudomonas* species not eradicated by fumigant application should be restricted in the population size that they could achieve subsequent to fumigation, as a result of competition with indigenous organisms reincorporated into the soil.

Complete eradication of *Pseudomonas* species is probably unattainable in most uncontained experimental sites, as a result of the limitations in eradicative procedures and the potential for dispersal of organisms outside of the immediate test site. However, the dispersal of organisms from test sites can be greatly restricted by the several procedures noted above. Although eradication of the remaining test organisms at the experimental site may not be feasible, untoward environmental effects of these test organisms could be precluded subsequent to termination of the experiment by the isolation of the plot, either physically or biologically, from sensitive habitats. For example, because of the difficulty of eradicating microorganisms from soil, a persistent but low population size of test organisms may remain, possibly indefinitely, in experimental sites. Thus, it may be necessary to remove such experimental sites from uses that would sustain or disseminate cells of the residual test strains. If the persistence of plant species maintained the population sizes of test strains, it might be necessary to eliminate such plants for lengthy periods. Physical barriers to prevent dissemination by human activities or other processes, as discussed above, could also be maintained.

The population size of test *Pseudomonas* species can be greatly reduced or eliminated on vegetation within the experimental site by several different procedures. Annual plants can be removed from the test site at the termination of the experiment, either by physical means or by incineration (42, 75). If test *Pseudomonas* strains do not survive in soil, plant material could be incorporated into the soil to permit the elimination of the test strains (42, 60, 75).

If it is deemed important that experimental plants in the test site or surrounding vegetation not be killed in the process of eradication, many fewer options for the use of biocidal chemicals exist. Several chemicals have been shown to reduce the population size of epiphytic bacteria on plants following foliar application under field conditions. Such materials include 70% ethanol solutions in water (77), dilute copper sulfate solutions (91), application of copper bactericides (including copper hydroxide and basic copper sulfate formulations) (47, 49), dilute solutions of sodium hypochlorite, and certain antibiotics, such as streptomycin and oxytetracycline (47, 49). Multiple applications of such materials may be necessary to reduce the numbers of the test strain on infested plant materials, as a result of the nonpersistent nature of many materials and the location of organisms in protected sites on plant tissue. Several of these materials may be phytotoxic to plants, and this must be considered before their use. Population sizes of epiphytic bacteria have been reduced up to 100-fold or more on treated plants by one or more of these materials (47, 49). However, complete eradication of epiphytic bacteria from aerial plant surfaces is unexpected by any of these materials. Population sizes of test *Pseudomonas* strains reduced by eradicant applications of the above materials may be maintained at such low levels by subsequent treatment with indigenous epiphytic bacteria adapted to the test plants (47). As in the soil environment, competition for limited environmental resources on plants between these introduced indigenous organisms and residual test strains should limit their population sizes (44, 46, 51, 54).

It is unlikely that *Pseudomonas* strains could be eradicated from bodies of open water or from aquifers without introducing large quantities of toxicants that could migrate to adjacent sources of potable water. Eradication of *Pseudomonas* species is conceivable only in isolated water bodies of restricted size. Many *Pseudomonas* species are resistant to many commonly used antibiotics to which enteric bacteria are sensitive (50). Such antibiotics include carbenicillin, streptomycin, tetracycline, penicillin, ampicillin, and cephalosporins. Such antibiotics are unlikely candidates for use in the eradication of most *Pseudomonas* species due to cost, intrinsic or acquired resistance, and other factors. Although the use of bacteriophage specific to test *Pseudomonas* strains may cause a transient reduction of population sizes of such strains (86, 88), it is unexpected that populations could be eliminated by their use or that long-term reductions in population size would be sustained due to selection for nonsusceptible mutant strains.

Outdoor (Uncontained) Experiments with Unrestricted Dispersal

Many experiments with *Pseudomonas* species whose survival and environmental behavior are well-characterized may warrant experimental designs for which some survival outside of the site is acceptable and mitigation of the test organism is not mandated (Table 49.1). For example, many environmentally competent organisms may closely mimic natural counterparts in their fitness at such sites. Studies of their natural counterparts will often identify biological processes, such as competition, that limit their population size in habitats similar to that of the test site (44, 50, 60). Studies of the relative fitness, growth rates, habitat preferences, and other ecological attributes of test strains in relation to indigenous counterparts can be used to verify assumptions that processes restricting population size of indigenous organisms operate similarly for the test strain (50, 60). For example, Ice$^-$ mutant *P. syringae* strains were shown to be indistinguishable from Ice$^+$ parental strains, in comparison tests done under controlled conditions, in growth rate, nutrient utilization spectra, epiphytic population sizes on different plant species, and numerous other attributes of ecological significance (44, 50, 60). Because of their similarity in behavior patterns in laboratory measurements, such Ice⁻ mutants were expected to mimic their natural counterparts after dispersal in field trials (50, 60). Data from field studies, using experimental designs that limited dispersal of the released mutants, have verified this expectation (1, 55, 60, 76).

Pseudomonas species may differ greatly in their potential for affecting natural habitats. Organisms with such potential may often be identified by the nature of the native or introduced traits they possess. Many organisms, however, because of their similarity to indigenous organisms, may not pose significant risk to the habitats into which they are applied. The need for eradicative measures against a given test strain will be based on the analysis of hazards of the occurrence or persistence of such strains in natural habitats. One potential hazard of introduced organisms would be their ability to multiply or persist at the expense of other native organisms in a given habitat.

For many organisms which mimic indigenous counterparts in laboratory experiments, such hazards are unanticipated, thereby justifying their study in situations in which their eradication is unnecessary or their dispersal unrestricted. Unrestricted experimentation might simply involve field studies similar to those discussed above but without the conditions employed to minimize dispersal of organisms or their colonization of new sites after dispersal. For example, if soil-inhabiting *Pseudomonas* species were utilized in an agricultural setting, it should not be necessary to restrict the occurrence of habitable plants nearby the plot, minimize soil movement by wetting of the upper soil regions, or to be as rigorous in containment procedures, such as sterilization of experimental equipment, etc., if strains were shown to be similar to indigenous organisms at the test site. Similarly, mitigation of such organisms in or near the experimental site is unlikely to be justified. The population size of such test strains could be expected to be restricted by competition, predation, and parasitism by indigenous organisms (44, 50, 60). Dispersal of such test strains could be expected (at least to modest distances), but their effect on other habitats, at least as measured by their increase in population size relative to indigenous strains, would not be expected. The numbers of bacteria generally decrease greatly with increasing distance away from a source (59, 60). Such dispersal gradients are usually steep both for organisms dispersed within soil profiles and aerially from terrestrial sources. Population size–dependent processes, such as competition, would be expected to preclude great increases in population sizes of organisms dispersed to increasing distances away from experimental sites (50, 60). For example, competition with indigenous strains could be expected to be quantitatively more important in reducing population sizes of those organisms dispersed in the smallest numbers to the greatest distances away from experimental sites. This principle was used successfully in the containment of Ice⁻ *P. syringae* and *P. fluorescens* strains inoculated onto plant surfaces in field sites (1, 55, 60). Vegetation-free buffer zones were effectively utilized to prevent the dissemination of such Ice⁻ *Pseudomonas* strains to suitable plant habitats surrounding the experimental site (60).

References

1. Advanced Genetic Sciences, Inc. 1988. Report on small-scale field efficacy test with INA⁻ strains of *Pseudomonas* for biological control of frost injury to strawberry blossoms. Report submitted to the U.S. Environmental Protection Agency.
2. Anonymous. Snowbugs. *Compressed Air* 91:24.
3. Arny, D. C., S. E. Lindow, and C. D. Upper. 1976. Frost sensitivity of *Zea mays* increased by application of *Pseudomonas syringae*. *Nature* 262:282–284.
4. Baggi, G., P. Barbieri, E. Galli, and S. Tollari. 1987. Isolation of a *Pseudomonas stutzeri* strain that degrades o-xylene. *Appl. Environ. Microbiol.* 53:2129–2132.
5. Baker, R. 1986. Biological control: an overview. *Can. J. Plant Pathol.* 8:218–221.
6. Baldry, M. G., and A. C. Dean. 1981. Environmental change and copper uptake by *Bacillus subtilis* subsp. *niger* and by *Pseudomonas fluorescens*. *Biotech. Lett.* 3:137.
7. Barles, R. W., C. G. Daughton, and D. P. H. Hsieh. 1979. Accelerated parathion degradation in soil inoculated with acclimated bacteria under field conditions. *Arch. Environ. Contam. Toxicol.* 8:647–660.
8. Blakeman, J. P. 1982. Phylloplane interactions. In M. S. Mount and G. H. Lacy (eds.), *Phytopathogenic Prokaryotes*, Vol. I. Academic Press, New York, pp. 307–333.
9. Blakeman, J. P. (ed.). 1981. *Microbial Ecology of the Phylloplane*. Academic Press, London.
10. Blakeman, J. P., and N. J. Fokkema. 1982. Potential for biological control of plant diseases on the phylloplane. *Ann. Rev. Phytopathol.* 20:167–192.
11. Burr, W. J., and A. Caesar. 1984. Beneficial plant bacteria. *CRC Rev. Plant Sci.* 2:1–20.
12. Chatterjee, D. K., J. J. Kilbane, and A. M. Chakrabarty. 1982. Biodegradation of 2,4,5-trichlorophenoxyacetic acid in soil by a pure culture of *Pseudomonas cepacia*. *Appl. Environ. Microbiol.* 44:514–516.
13. Clark, P. H., and M. H. Richmond (eds.). 1975. *Genetics and Biochemistry of Pseudomonas*. Wiley-Interscience, London.
14. Cook, R. J., and K. F. Baker. 1983. *The Nature and Practice of Biological Control of Plant Pathogens*. The American Phytopathology Society. St. Paul, Minn.
15. Cotter, J. L., R. C. Fader, C. Lilley, and D. N. Herndon. 1985. Chemical parameters, antimicrobial activities, and tissue toxicity of 0.1 and 0.5 percent sodium hypochlorite solutions. *Antimicrob. Agents Chemother.* 28:118–122.
16. Crosse, J. E. 1965. Bacterial canker of stone fruits. IV. Inhibition of leaf scar infection of cherry by a saprophytic bacterium from the leaf surface. *Ann. Appl. Biol.* 56:149–160.
17. Crosse, J. E. 1971. Interactions between saprophytic and pathogenic bacteria in plant disease. In T. F. Preece and C. H. Dickinson (eds.), *Ecology of Leaf Surface Microorganisms*. Academic Press, London, pp. 283–290.
18. Dagley, S. 1986. Biochemistry of aromatic hydrocarbon degradation in pseudomonads. In Sokatch, J. R. (ed.), *The Bacteria, Vol. X. The Biology of Pseudomonas*. Academic Press, Orlando, Fla., pp. 527–555.
19. Dye, D. W., J. F. Bradbury, M. Goto, A. C. Hayward, R. A. Lelliott, and M. N. Schroth. 1980. International standards for naming pathovars of phytopathogenic bacteria and a list of pathovar names and type strains. *Ann. Rev. Plant Pathol.* 59:153–168.
20. Focht, D. D., and D. Shelton. 1987. Growth kinetics of Pseudomonas alcaligenes

C-O relative to inoculation and 3-chlorobenzoate metabolism in soil. *Appl. Environ. Microbiol.* 53:1846–1849.

21. Franklin, F. C. H., M. Bagdasarian, M. M. Bagdasarian, and K. N. Timmis. 1981. Molecular and functional analysis of the TOL plasmid pWWO from *Pseudomonas putida* and cloning of the entire regulated aromatic ring meta-cleavage pathway. *Proc. Natl. Acad. Sci. USA* 78:7458–7462.

22. Frantz, B., and A. M. Chakrabarty. 1986. Degradative plasmids in *Pseudomonas*. In J. R. Sokatch (ed.), *The Bacteria*, Vol. X. Academic Press, Orlando, Fla., pp. 295–323.

23. Furukawa, K., and N. Arimura. 1987. Purification and properties of 2,3-dihydroxybiphenyl dioxygenase from polychlorinated biphenyl-degrading *Pseudomonas pseudoalcaligenes* and *Pseudomonas aeruginosa* carrying the cloned *bphC* gene. *J. Bacteriol.* 169:924–927.

24. Furukawa, K., J. R. Simon, and A. M. Chakrabarty. 1983. Common induction and regulation of biphenyl, xylene/toluene, and salicylate catabolism in *Pseudomonas paucimobilis*. *J. Bacteriol.* 154:1356–1362.

25. Gardner, J. M., J. L. Chandler, and A. W. Feldman. 1984. Growth promotion and inhibition by antibiotic producing fluorescent pseudomonads on citrus roots. *Plant Soil* 77:103–114.

26. Garrett, C. M. E., and J. E. Crosse. 1975. Interaction between *Pseudomonas morsprunorum* and other pseudomonads in leaf-scar infection of cherry. *Physiol. Plant Pathol.* 5:89–94.

27. Gaskins, M. H., S. L. Albrecht, and D. H. Hubbell. 1985. Rhizosphere bacteria and their use to increase plant productivity: A review. *Agric. Ecosyst. Environ.* 12:99–116.

28. Gerhardt, P. (ed.). 1981. *Manual of Methods for General Bacteriology*. American Society for Microbiology, Washington, D.C.

29. Gevaudan, M. J., C. Gulian, M. N. Mallet, and P. DeMicco. 1984. Antibacterial activity of chlorhexidine solution in isopropyl alcohol for hand disinfection. *In vitro* and *in vivo* study. *Med. Mal. Infect.* (France) 14/3:94–101.

30. Gibbins, L. N. 1972. Relationship between pathogenic and nonpathogenic bacterial inhabitants of aerial plant surfaces. In H. P. Maas Geesteranus (ed.), *Proceedings of the 3rd International Conference on Plant Pathogenic Bacteria*. Wageningen, The Netherlands, pp. 15–24.

31. Gillette, D. A., I. H. Blifford, Jr., and C. R. Fenster. 1972. Measurements of aerosol size distribution and vertical fluxes of aerosols on land subject to wind erosion. *J. Appl. Microbiol.* 11:977–987.

32. Goodman, R. N. 1967. Protection of apple stem tissue against *Erwinia amylovora* infection by avirulent strains and three other bacterial species. *Phytopathology* 57:22–24.

33. Gosden, P. E., and P. Norman. 1985. Pseudobacteraemia associated with contaminated skin cleaning agent. *Lancet* (England) 2/8456:671–672.

34. Griffin, D. M., and G. Quail. 1971. Movement of bacteria in moist particulate systems. *Austr. J. Biol. Sci.* 21:579–582.

35. Hamdi, Y. A. 1971. Soil water tension and the movement of *Rhizobia*. *Soil Biol. Biochem.* 3:121–126.

36. Higham, D. P., P. J. Sadler, and M. D. Scawen. 1986. Cadmium-binding proteins in *Pseudomonas putida*: pseudothiones. *Environ. Health Perspect.* 65:5.

37. Hirano, S. S., and C. D. Upper. 1983. Ecology and epidemiology of foliar bacterial plant pathogens. *Ann. Rev. Phytopathol.* 21:243–269.

38. Inouye, S., A. Nakazawa, and T. Nakazawa. 1987. Overproduction of the *xylS* gene product and activation of the *xyl*DLEGF operon on the TOL plasmid. *J. Bacteriol.* 169:3587–3592.

39. Kloepper, J. W., and Schroth, M. N. 1981. Relationship of *in vitro* antibiosis by plant growth-promoting rhizobacteria to plant growth promotion and displacement of root microflora. *Phytopathology* 71:1020–1023.

40. Leben, C. 1964. Influence of bacteria isolated from healthy cucumber leaves on two leaf diseases of cucumber. *Phytopathology* 54:405–408.

41. Leben, C. 1981. How plant-pathogenic bacteria survive. *Plant Dis.* 65:633–637.

42. Leben, C., and J. P. Sleesman. 1981. Bacterial pathogens: reducing seed and *in vitro* survival by physical treatments. *Plant Dis.* 65:876–878.

43. Lindemann, J., H. A. Constantinidou, W. R. Barchet, and C. D. Upper. 1982. Plants as sources of airborne bacteria, including ice nucleation active bacteria. *Appl. Environ. Microbiol.* 44:1059–1063.

44. Lindemann, J., and T. V. Suslow. 1987. Competition between ice nucleation-active wild type and ice nucleation-deficient deletion mutant strains of *Pseudomonas syringae* and *P. fluorescens* biovar I and biological control of frost injury on strawberry blossoms. *Phytopathology* 77:882–886.

45. Lindemann, J., and C. D. Upper. 1985. Aerial dispersal of epiphytic bacteria over bean plants. *Appl. Environ. Microbiol.* 50:1229–1232.

46. Lindow, S. E. 1982. Population dynamics of epiphytic ice nucleation active bacteria on frost sensitive plants and frost control by means of antagonistic bacteria. In P. H. Li and A. Sakai (eds.), *Plant Cold Hardiness*. Academic Press, New York, pp. 395–416.

47. Lindow, S. E. 1983. Methods of preventing frost injury through control of epiphytic ice nucleation active bacteria. *Plant Dis.* 67:327–333.

48. Lindow, S. E. 1983. The role of bacterial ice nucleation in frost injury to plants. *Annu. Rev. Phytopathol.* 21:363–384.

49. Lindow, S. E. 1985. Integrated control and role of antibiosis in biological control of fireblight and frost injury. In C. Windels and S. E. Lindow (eds.), *Biological Control on the Phylloplane*. American Phytopathological Society Press, Minneapolis, pp. 83–115.

50. Lindow, S. E. 1985. Ecology of *Pseudomonas syringae* relevant to the field use of Ice⁻ deletion mutants constructed *in vitro* for plant frost control. In "Engineered Organisms in the Environment: Scientific Issues," American Society for Microbiology, Washington, D.C., pp. 23–25.

51. Lindow, S. E. 1986. Strategies and practice of biological control of ice nucleation active bacteria on plants. In N. Fokkema (ed.), *Microbiology of the Phyllosphere*. Cambridge University Press, pp. 293–311.

52. Lindow, S. E. 1987. Competitive exclusion of epiphytic bacteria by Ice⁻ mutants of *Pseudomonas syringae*. *Appl. Environ. Microbiol.* 53:2520–2527.

53. Lindow, S. E. 1989. Practical application of ice nucleation active bacteria. In D. Sands, Z. Klement, and K. Rudolf (eds.), *Methods in Phytobacteriology*. In press.

54. Lindow, S. E. 1986. Construction of isogenic Ice⁻ strains of *Pseudomonas syringae* for evaluation of specificity of competition on leaf surfaces. In F. Megusar and M. Gantz (eds.), *Microbial Ecology*. Slovene Society for Microbiology, Ljubljana, pp. 509–518.

55. Lindow, S. E. 1988. 1987 Final Report—Field testing of two different Ice⁻ deletion mutant strains of *Pseudomonas syringae* for biological control of frost injury to potato. Report submitted to the U.S. Environmental Protection Agency-Experimental Use Permit Progress Report.

56. Lindow, S. E., D. C. Arny, W. R. Barchet, and C. D. Upper. 1978. The role of bacterial ice nuclei in frost injury to sensitive plants. In P. H. Li and A. Sakai (eds.), *Plant Cold Hardiness and Freezing Stress*. Academic Press, New York, pp. 249–261.

57. Lindow, S. E., D. C. Arny, and C. D. Upper. 1978. Distribution of ice nucleation-active bacteria on plants in nature. *Appl. Environ. Microbiol.* 36:831–838.

58. Lindow, S. E., D. C. Arny, and C. D. Upper. 1982. Bacterial ice nucleation: A factor in frost injury to plants. *Plant Physiol.* 70:1084–1089.

59. Lindow, S. E., G. R. Knudsen, R. J. Seidler, M. V. Walter, V. W. Lambou, P. S. Amy, D. Schmedding, V. Prince, and S. Hern. 1988. Aerial dispersal and epiphytic survival of *Pseudomonas syringae* during a pre-test for the release of genetically engineered strains into the environment. *Appl. Environ. Microbiol.* 54:1557–1563.

60. Lindow, S. E., and N. J. Panopoulos. 1988. Design and results of field tests of recombinant Ice⁻ strains of *Pseudomonas syringae* for biological frost control to potato. In M. Sussman, C. H. Collins, and F. A. Skinner (eds.), *The Release of Genetically Engineered Microorganisms*. Academic Press, New York, pp. 121–138.

61. Loper, J. E., C. Haack, and M. N. Schroth. 1985. Population dynamics of soil

pseudomonads in the rhizosphere of potato (*Solanum tuberosum* L.). *Appl. Environ. Microbiol.* 49:416–422.

62. Loper, J. E., and M. N. Schroth. 1986. Influence of bacterial sources of indole-3-acetic acid on root elongation of sugar beet. *Phytopathology* 76:386–389.

63. Miller, T. D., and M. N. Schroth. 1972. Monitoring the epiphytic population of *Erwinia amylovora* on pear using a selective medium. *Phytopathology* 62:1175–1182.

64. Munnecke, D. E. 1972. Factors affecting the efficacy of fungicides in soil. *Annu. Rev. Phytopathol.* 10:375–398.

65. Munnecke, D. E., and S. D. Van Gundy. 1979. Movement of fumigants in soil, dosage responses, and differential effects. *Ann. Rev. Phytopathol.* 17:405–429.

66. O'Brien, R. D., and S. E. Lindow. 1989. Effect of plant species and environmental conditions on epiphytic population size of *Pseudomonas syringae* and other bacteria. *Phytopathology* 79:619–627.

67. Palleroni, N. J. 1984. Pseudomonadaceae. In *Bergey's Manual of Systematic Bacteriology*, Vol. I. Williams & Wilkins, Baltimore.

68. Paulin, J. P., and J. Luisetti. 1978. Ice nucleation activity among phytopathogenic bacteria. In *Proceedings of the 4th International Conference on Plant Pathogenic Bacteria*, Vol. II. Angers, France, pp. 725–733.

69. Phillips, A. T. 1986. Biosynthetic and catabolic features of amino acid metabolism in *Pseudomonas*. In J. R. Sokatch (ed.), *The Bacteria, Volume X. The Biology of Pseudomonas*. Academic Press, Orlando, Fla., pp. 385–437.

70. Ramos, J. L., A. Wasserfallen, K. Rose, and K. N. Timmis. 1987. Redesigning metabolic routes: manipulation of TOL plasmid pathway for catabolism of alkylbenzoates. *Science* 235:593–596.

71. Sands, D. C., M. N. Schroth, and D. C. Hildebrand. 1970. Taxonomy of phytopathogenic pseudomonads. *J. Bacteriol.* 101:9–23.

72. Schoenen, D., B. Stoeck, S. Hienzsch, and B. Emmel. 1986. Decontamination of water taps colonized with *Pseudomonas aeruginosa*. *Zentralbl. Bakteriol. Mikrobiol. Hyg. Ser. B* 182:551–557.

73. Schroth, M. N., D. C. Hildebrand, and M. P. Starr. 1981. Phytopathogenic members of the genus *Pseudomonas*. In M. P. Starr et al. (eds.), *The Prokaryotes; A Handbook on Habitats, Isolation, and the Identification of Bacteria*. Springer-Verlag, Berlin-Heidelberg.

74. Schroth, M. N., J. E. Loper, and D. C. Hildebrand. 1984. Bacteria as biocontrol agents of plant disease. In M. J. Klug and C. A. Reddy (eds.), *Current Perspectives in Microbial Ecology*. American Society for Microbiology, Washington, D.C., pp. 362–369.

75. Schuster, M. L., and D. P. Coyne. 1974. Survival mechanisms of phytopathogenic bacteria. *Ann. Rev. Phytopathol.* 12:199–221.

76. Seidler, R. March, 1988. EPA Special Report: Release of ice-minus recombinant bacteria at California test sites. Environmental Research Laboratory, Corvallis, Oreg.

77. Spurr, H. W., Jr. 1979. Ethanol treatment—a valuable technique for foliar biocontrol studies of plant disease. *Phytopathology* 69:773–776.

78. Spurr, H. W., Jr. 1981. Introduction of microbial antagonists for the control of foliar plant pathogens. In G. C. Papavizas (ed.), *Biological Control in Crop Production*. Beltsville Symp. Agric. Res. Vol. 5. Allanheld, Osmum & Co., Totowa, N.J., pp. 323–332.

79. Stone, L. S., and E. A. Zottola. 1985. Effects of cleaning and sanitizing on the attachment of *Pseudomonas fragi* to stainless steel. *J. Food Sci.* 50:951.

80. Strandberg, G. W., S. E. Shumate, and J. R. Parrott. 1981. Microbial cells as biosorbents for heavy metals: accumulation of uranium by *Saccharomyces cerevisiae* and *Pseudomonas aeruginosa*. *Appl. and Environ. Microbiol.* 41:237.

81. Suslow, T. V., and M. N. Schroth. 1982. Role of deleterious rhizobacteria as minor pathogens in reducing crop growth. *Phytopathology* 72:111–115.

82. Talbot, H. W., L. Johnson, S. Barik, and D. Williams. 1982. Properties of a *Pseudomonas* sp.-derived parathion hydrolase immobilized to porous glass and activated alumina. *Biotech. Lett.* 4:209.

83. Thomson, S. V., M. N. Schroth, W. J. Moller and W. O. Reil. 1976. Efficacy of bactericides and saprophytic bacteria in reducing colonization and infection of pear flowers by *Erwinia amylovora. Phytopathology* 66:1457–1459.

84. Torrey, S. (ed.). 1983. *Microbiological Syntheses. Recent Advances.* Noyes Data Corporation, Park Ridge, N.J.

85. Venette, J. R., and B. W. Kennedy. 1975. Naturally produced aerosols of *Pseudomonas glycinea. Phytopathology* 65:737–738.

86. Vidaver, A. K. 1976. Prospects for control of phytopathogenic bacteria by bacteriophages and bacteriocins. *Annu. Rev. Phytopathol.* 14:451–465.

87. Vidaver, A. K. 1985. Plant-associated agricultural applications of genetically engineered microorganisms: projections and constraints. *Recomb. DNA Tech. Bull.* 8: 97–102.

88. Wiggins, B. A., and M. Alexander. 1985. Minimum bacterial density for bacteriophage replication: Implications for significance of bacteriophages in natural ecosystems. *Appl. Environ. Microbiol.* 49:19–23.

89. Windels, C., and S. E. Lindow. 1985. *Biological Control on the Phylloplane.* American Phytopathological Society, Minneapolis.

90. Woese, C. R. 1987. Bacterial evolution. *Microbiol. Rev.* 51:221–271.

91. Young, J. M. 1978. Survival of bacteria on *Prunus* leaves. Proceedings of the 4th International Conference on Plant Pathogenic Bacteria, Vol. II. Angers, France, pp. 779–786.

Chapter 50

Biological Containment of Genetically Engineered Microorganisms

Stephen M. Cuskey

Introduction and History

The use of molecular methods for genetic rearrangements and the creation of hybrid microorganisms have raised concerns over potential risks to humans and the environment in both the scientific community and the general public. These speculative risks arise from possible unintended ecological consequences resulting from the release of modified organisms to the environment, either intentionally or not. In laboratory experiments, such microorganisms are relatively easily contained from accidental release, mainly through elaborate physical facilities designed to prevent their escape. In early experiments, biological containment mechanisms were also included to ensure that accidental release from a physical enclosure would not cause unintended environmental harm. One strategy involved the construction of "safer" cloning vectors, which would not easily survive or transfer to indigenous microbial populations outside of the laboratory. For example, several of the Charon cosmid vectors, based on derivatives of the bacteriophage, lambda, carry mutations that decrease the host range of recombinant molecules or reduce the possibility that nonlytic recombinant phage will be produced (3).

Plasmid-based cloning vectors have also been constructed with features important for biological containment in the event of an accidental release. In general, these plasmids are small portions of larger molecules and contain only an origin of replication, one or more plasmid genes required for replication, and one or more genes with selectable phenotypes. The plasmids lack transfer functions, and most are not mobilizable in triparental matings (1, 9).

Other proposed containment strategies in vector design include the use of temperature-sensitive replicons and the replacement of antibiotic resistance genes with markers complementing auxotrophic mutations for selection of transformed cells (9). Other versions of "safe" vector strategies have been used in genetically engineered microorganisms designed for release to the environment. Even though most plasmid cloning vectors in common use today are not mobilizable in triparental matings and are, thus, less likely to be transferred to a member of the indigenous bacterial community, an additional safety feature to prevent transfer is to place the engineered traits in the chromosome of the desired host strain. For example, researchers at Monsanto, Inc., have developed an impaired transposon delivery system that will insert genetically engineered DNA into the chromosome, and the DNA is then unable to undergo further transposition (14, 15).

The most widely used biological containment strategy in early experiments involved debilitated host bacteria, which presumably would not survive outside of a laboratory environment (4, 8). One of the most widely used strains for this purpose, *Escherichia coli* χ^{1776}, was developed by Curtiss and colleagues (4). This strain obligately requires exogenous diaminopimelic acid and thymine (or thymidine) and is very sensitive to bile (thereby precluding intestinal passage or colonization), detergents, antibiotics, and DNA-damaging agents (4, 10). Accidental releases of genetically engineered DNA in this and similar microorganisms would be contained because of the extreme lability of the bacterial host.

General Considerations for Control of Planned Release of Microorganisms

Containment strategies that utilize debilitated hosts are obviously not practical for planned environmental introductions ("releases"). Introduced organisms must be able to compete successfully with the indigenous community in order to survive and perform their desired function. Biological containment systems for environmental releases must, therefore, be "conditionally lethal" to the host organism. With these types of controls, the organisms would be viable only under permissive growth conditions and would, theoretically, retain the ability of the parental strain to compete in the natural environment. The survival of the introduced organisms would be controlled by selecting the conditions under which the lethal gene(s) would be activated.

E. coli χ^{1776} and similar strains contain mutations that render them conditionally lethal, but these are designed to ensure death upon release to the environment. There are other examples where conditionally lethal phenotypes are employed to maintain desired populations. For example, in enclosed fermentations using genetically engineered organisms, all of the cells must carry the recombinant plasmid for maximum product formation, and several systems have been developed in which the bacterial cells die upon plasmid loss (12, 16).

Use of Lethal Bacterial Genes for Biological Containment

There has been little research, however, on the construction of conditionally lethal systems designed for bacteria introduced into the environment. One potentially useful system has been developed by Molin and coworkers (11), who make use of a gene from plasmid R1 of *E. coli* that is involved in plasmid maintenance. The protein product of the plasmid-borne *hok* gene destroys the cell membrane in cells without the protecting *sok* gene (7), which encodes an antisense mRNA that prevents *hok* translation, and differential mRNA stabilities ensure that plasmidless segregants are killed through *hok* action (7).

In one system, the *hok* gene has been placed under control of the *E. coli trp* promoter, and cells harboring this construct require tryptophan in the growth medium to repress *hok* activation. This and similar constructs would be useful in containment of accidentally released genetically engineered bacteria, as free tryptophan is not present in the environment in sufficient quantity to ensure cell survival.

In a second construct, the *hok* gene is under the control of the revertible *E. coli* promoter, *fimA,* involved in pilus formation, which randomly flip-flops and is active in only one orientation. Genetically engineered bacteria containing this system would, thus, eventually die as the result of *hok* activation. This system contains several attractive features: the *hok* gene is located on a small DNA fragment, the introduction of which should not present a great metabolic burden to the host cell; the action of the containment gene, *hok,* is understood; and the gene is lethal to a broad spectrum of unprotected Gram-positive and Gram-negative bacterial cells.

The use of the *fimA* promoter, however, may pose practical problems to introduced bacteria, because populations of cells carrying the construct appear to have a slower growth rate than isogenic cells without the construct. However, the apparent lower growth rate probably reflects the activation of the *hok* gene through the flip-flop of the *fimA* promoter in a portion of the population and not a lower growth rate of cells carrying the containment construct. In addition, Atlas and coworkers (2) have placed the *hok* gene downstream from the *E. coli lac* promoter, thereby rendering the cells susceptible to additions of the gratuitous *lac* inducer, isopropyl-thio-β-D-galactoside (IPTG). Tests in soil microcosms showed the utility of this type of containment strategy, as most of the added genetically altered bacteria did not survive the addition of IPTG (2).

Several conditionally lethal systems for control of released genetically engineered bacteria have also been designed, constructed, and tested in this laboratory. One system that is relatively easy to envision involves temperature sensitivity. This type of control might be especially appropriate for agricultural products needed only during the warm weather of a growing season. Mutants with cold sensitivity to DNA-damaging agents were isolated. These are not conditionally lethal mutations per se, but the mutants are more susceptible to DNA damage in the environment and are theoretically less likely to survive than the wild type. The mutant strains, ACG3 and ACG4, are derivatives of *Pseudomonas aeruginosa* strain PAO1. They grow at wild-type rates at 37°C in the presence of normally subinhibitory concentrations of the DNA-damaging agent, methylmethane sulfonate, but not at 25°C. However, this type of control system has several disadvantages. Many temperature-sensitive mutations are point mutations that are easily revertible to the wild-type phenotype. Revertants of mutants ACG3 and ACG4 are readily obtained upon exposure to methylmethane sulfonate at 25°C. These mutations are also not easily transferrable to other strains, and new mutations must be made and tested for all released strains, which is a time-consuming and expensive process.

A second class of conditionally lethal constructs involves the introduction of a partial metabolism pathway that results in the accumulation of a toxic compound. The control of strains carrying this type of mechanism would be dependent on the addition of an innocuous chemical to the environment at a certain time. Examples include strains engineered to be supersensitive to subinhibitory levels of mercury. The *mer* operon of transposon Tn*21* contains genes required for the reductive volatilization of Hg^{2+}. Supersensitivity is conferred by subcloning the positive regulatory gene, *merR,* and *merT,* a gene required for mercury transport, but not the *merA* gene, which codes for mercuric reductase. Thus, the cells accumulate mercury with no mechanism for its detoxification (6, 13).

Plasmid pEPA19 contains the *merR* and *merT* genes subcloned into the unique *Eco*R1 site of the plasmid cloning vector, pRO1769 (5). Tests on lethality and stability were conducted using *P. aeruginosa* PAO1 as the host bacterium and filter-sterilized water from the Escambia River in northwestern Florida as the medium. Transformed cell populations of 10^3 per milliliter were completely killed upon addition of 0.5 to 1.0 μg/liter Hg^{2+} (as mercuric chloride) to the growth medium. The numbers of nontransformed cells were not affected by these additions. Higher concentrations (10 μg/liter) of Hg^{2+} killed both pEPA19-transformed and nontransformed cells. There were no survivors of transformed cells 24 h and 21 days after the addition of Hg^{2+}.

This system was useful in demonstrating that a population could be contained under defined growth conditions. However, the model contains several drawbacks. Higher cell concentrations could not be used, because supersensitivity to mercury was dependent on cell density. Test systems containing 10^7 transformed cells per milliliter were not affected by the addition of up to 10 mg/liter Hg^{2+}. Furthermore, mercury is an extremely toxic compound, and although mercuric chloride is not being considered for control of released genetically engineered bacteria, the concentration ranges that can be used before toxicity is seen in control indigenous bacterial populations are quite low.

The third type of system being developed in this laboratory (as well as by Molin, Atlas, and their coworkers) is based on natural genes that have a "suicide" effect on unprotected cells and is probably more useful for containment of released organisms (2, 7, 11). Examples of natural suicide gene–protecting gene tandems include restriction endonuclease-methylase genes and the suicide genes carried by many large, self-transmissible plasmids. Examples of the latter suicide gene–protecting gene tandems include the *hok/sok* genes from plasmid R1, mentioned above, and the *kil/kor* genes from plasmid RK2 (8). The function of these suicide genes is not known, but they are thought to ensure plasmid stability by killing bacteria that do not acquire the plasmid during cell division. The mechanisms of action of the *kil/kor* genes are probably different from those of the *hok/sok* genes, but they are not as well understood.

There are numerous opportunities for genetic manipulation of these natual suicide systems for the control of released genetically engineered microorganisms. For example, in organisms designed to degrade a toxic waste, the lethal portion of the tandem could be manipulated to be constitutively expressed (probably at a low level), whereas the protecting gene could be controlled by a promoter that is activated by the presence of the toxic waste. In this situation, the organism would be protected while it degraded the toxic waste, but it would die when the compound was no longer present. Constructs based on this principle and using DNA from the TOL plasmid, pWWO (which allows plasmid-bearing cells to utilize toluene, xylenes, and related aromatic hydrocarbons as carbon sources), and the *kilA/korA* genes from plasmid RK2 are currently being tested in this laboratory.

The OP2 promoter from plasmid pWWO is activated by benzoate and related aromatic acids in the presence of the *xylS* regulatory gene. This promoter has been used in the construction of two different types of containment plasmids. Plasmid pEPA88 contains the promoter upstream from the *kilA* gene, rendering the cells sensitive to the addition of benzoate to the growth medium. Alternatively, plasmid pEPA86 contains the OP2 promoter upstream from the *korA* protecting gene, and cells with this plasmid and a functional *kilA* gene require benzoate for survival. Strains bearing these plasmids are currently being tested.

Design Concerns

Several general aspects must be considered in the design and construction of suicide systems. These include the stability of the construct, the extent of lethality to the population, and the effects of the added DNA on the growth rate and survival of the engineered strain. Of these concerns, stability may be the most troublesome, and it will be necessary to determine the rates of spontaneous mutation that inactivate the suicide gene(s) in the environment. Typical spontaneous reversion rates of auxotrophic mutations are 10^{-7} to 10^{-8}, but spontaneous mutations causing inactivation of suicide genes may be selected at much higher frequencies. It may be possible to overcome high rates of spontaneous mutations to a nonconditionally lethal phenotype, and their subsequent selection, by inclusion of two or more conditionally lethal systems in the released organisms. In this case, the rate of reversion to a nonconditionally lethal phenotype would be the product of the separate reversion frequencies and, thus, may be more suitable for containment of the released strain. This may also be a solution to incomplete killing of the released population, but it may increase adverse effects on competitiveness and host cell survival and cause death before it is desired.

If the suicide genes are plasmid-borne, an additional source of instability can occur as the result of plasmid loss from individual cells through improper segregation. Therefore, it may be necessary to place biological containment genes on the chromosome of strains engineered for environmental release.

Summary and Conclusions

Biological containment of genetically engineered microorganisms that are intentionally introduced into the environment may be achieved through the inclusion of conditionally lethal genes in the released strain. Examples include temperature sensitivity, inclusion of a partial metabolic pathway leading to the accumulation of a toxic metabolic intermediate upon addition of a normally innocuous compound to the growth medium, and use of lethal or protecting bacterial gene tandems whose expression has been engineered to respond to an environmental signal. Examples of the latter gene tandems include restriction endonuclease/methylase genes and plasmid-borne genes, such as *hok/sok* and *kil/kor* genes, the latter of which are presumed to be involved in plasmid maintenance.

Practical considerations suggest that the gene tandem-containment strategy may have the widest potential utility. Examples are given where expression of the lethal genes are controlled randomly (e.g., through the *fimA* promoter) or through the presence or absence of specific compounds (e.g., through the *E. coli trp* or *lac* promoters or the promoter of the TOL plasmid, OP2).

However, using this concept, survival of a released bacterium can theoretically be controlled by being coupled to any environmental signal for which regulatory DNA sequences can be isolated. In designing a biological containment strategy, stability of the regulatory DNA, extent of lethality of the total released population upon onset of nonpermissive conditions, and effects of the regulatory DNA on host survival and competitiveness during permissive conditions must be considered.

References

1. Armstrong, K. A., V. Hershfield, and D. R. Helinski. 1977. Gene cloning and containment properties of plasmid ColE1 and its derivatives. *Science* 196:172–174.
2. Bej, A. K., M. H. Perlin, and R. M. Atlas. 1988. Model suicide vector for containment of genetically engineered microorganisms. *Appl. Environ. Microbiol.* 54:2472–2477.
3. Blattner, F. R., B. G. Williams, A. E. Blechl, K. Dennistron-Thompson, H. E. Baber, L. A. Furlong, D. J. Grunwald, D. O. Kiefer, D. D. Moore, J. W. Schumm, E. T. Sheldon, and O. Smithies. 1977. Charon phages: safer derivatives of bacteriophage lambda for DNA cloning. *Science* 196:161–169.
4. Curtiss III, R., M. Inoue, D. Pereira, J. C. Hsu, L. Alexander, and L. Rock. 1977. Construction and use of safer bacterial host strains for recombinant DNA research. In W. A. Scott and R. Werner (eds.), *Molecular Cloning of Recombinant DNA.* Academic Press, New York, pp. 99–111.
5. Cuskey, S. M., J. A. Wolff, P. V. Phibbs, Jr., and R. H. Olsen. 1985. Cloning of genes specifying carbohydrate catabolism in *Pseudomonas aeruginosa* and *Pseudomonas putida. J. Bacteriol.* 162:865–871.
6. Foster, T. J., and H. Nakahara. 1979. Deletions in the r-determinant *mer* region of plasmid R100-1 selected for loss of mercury hypersensitivity. *J. Bacteriol.* 140:301–305.
7. Gerdes, K., F. W. Bech, S. T. Jorgensen, A. Lobner-Olesen, P. B. Basmussen, T. Atlung, O. Karlstrom, S. Molin, and K. von Meyerburg. 1986. Mechanism of postsegregational killing by the *hok* gene product of the *parB* system of plasmid R1 and its homology with the *relF* gene product of the *E. coli relB* operon. *EMBO J.* 5:2023–2029.
8. Leder, P., D. Tiemeier, and L. Enquist. 1977. RK2 derivatives of bacteriophage

lambda useful in the cloning of DNA from higher organisms: The gt*WES* system. *Science* 196:175–177.

9. Levine, M. M., J. B. Kaper, H. Lockman, R. E. Black, M. L. Clements, and S. Falkow. 1983. Recombinant DNA risk assessment studies in man: efficacy of poorly mobilizable plasmids in biologic containment. *Recomb. DNA Tech. Bull.* 6:89–97.

10. Maturin Sr., L., and R. Curtiss III. 1977. Degradation of DNA by nucleases in intestinal tracts of rats. *Science* 196:216–218.

11. Molin, S., P. Klemm, L. K. Poulsen, H. Biehl, K. Gerdes, and P. Anderson. 1987. Conditional suicide system for containment of bacteria and plasmids. *Biotechnology* 5:1315–1318.

12. Miwa, K., S. Nakamori, K. Sano, and H. Momose. 1984. Novel host-vector system for selection and maintenance of plasmid-bearing, streptomycin dependent *Escherichia coli* cells in antibiotic free media. *Gene* 31:275–277.

13. Nakahara, H., S. Silver, T. Miki, and R. H. Rownd. 1979. Hypersensitivity to Hg^{2+} and hyperbinding activity associated with cloned fragments of the mercurial resistance operon of plasmid NR1. *J. Bacteriol.* 140:161–166.

14. Obukowicz, M. G., F. L. Perlak, K. Kusano-Kretzmer, D. J. Mayer, S. L. Bolten, and L. S. Watrud. 1986. Tn*5* mediated integration of the delta-endotoxin gene from *Bacillus thuringiensis* into the chromosome of root-colonizing pseudomonads. *J. Bacteriol.* 168:982–989.

15. Obukowicz, M. G., F. J. Perlak, K. Kusano-Kretzmer, E. J. Mayer, and L. S. Watrud. 1986. Integration of the delta-endotoxin gene of *Bacillus thuringiensis* into the chromosome of root-colonizing strains of pseudomonads using Tn*5*. *Gene* 45: 327–331.

16. Rostek, Jr., P. R., and C. L. Hersberger. 1983. Selective retention of recombinant plasmids coding for human insulin. *Gene* 25:29–38.

Index

ABOUT THE EDITORS

MORRIS A. LEVIN is a senior research scientist at the University of Maryland's Maryland Biotechnology Institute. He was coordinator for biotechnology risk assessment with the U.S. Environmental Protection Agency, has served as science advisor to the U.S. House of Representatives' Science and Technology Committee, and currently teaches microbial ecology at the National Institutes of Health's Foundation for the Advancement of Education in the Sciences.

RAMON J. SEIDLER is a research microbiologist for the U.S. Environmental Protection Agency and holds a courtesy appointment as professor of microbiology at Oregon State University. He has served as the team leader of the Agency's terrestrial biotechnology risk assessment program.

MARVIN ROGUL is the former director of the Maryland Biotechnology Institute's Center for Public Issues in Biotechnology and is currently director of the Institute's Office of External Affairs. He has held administrative positions at the Walter Reed Army Institute of Research and served as deputy director for the Office of Strategic Assessments and Special Studies at the U.S. Environmental Protection Agency.